Lecture Notes in Artificial Intelligence 3020

Edited by J. G. Carbonell and J. Siekmann

Subseries of Lecture Notes in Computer Science

Daniel Polani Brett Browning
Andrea Bonarini Kazuo Yoshida (Eds.)

RoboCup 2003: Robot Soccer World Cup VII

Springer

Volume Editors

Daniel Polani
University of Hertfordshire
School of Computer Science
Hatfield, Herts AL10 9AB, United Kingdom
E-mail: d.polani@herts.ac.uk

Brett Browning
Carnegie Mellon University
Newell Simon Hall 4000B
5000 Forbes Avenue, Pittsburgh, PA 15213, USA
E-mail: brettb@cs.cmu.edu

Andrea Bonarini
Politecnico di Milano
Department of Electronics and Information
Milan, Italy
E-mail: bonarini@elet.polimi.it

Kazuo Yoshida
Keio University
Department of System Design Engineering
3-14-1, Hiyoshi, Kohoku-ku, Yokohama, 223-8522, Japan
E-mail: yoshida@sd.keio.ac.jp

The picture on the cover: ©The RoboCup Federation

Library of Congress Control Number: Applied for

CR Subject Classification (1998): I.2, C.2.4, D.2.7, H.5, I.5.4, J.4

ISSN 0302-9743
ISBN 3-540-22443-2 Springer-Verlag Berlin Heidelberg New York

Springer-Verlag is a part of Springer Science+Business Media

springeronline.com

Printed in Germany

Typesetting: Camera-ready by author, data conversion by Olgun Computergrafik
Printed on acid-free paper SPIN: 11017448 06/3142 5 4 3 2 1 0

Preface

RoboCup 2003, the 7th Robot World Cup Soccer and Rescue Competitions and Conferences, was held at PadovaFiere, in Padua, Italy during July 2–11, 2003. Following the trend established in recent years, the competition continued to grow, with 244 teams from 30 countries making up the 1244 participants. These teams were distributed across different leagues, where each league conducted one or more competitions. The league reports contained in this book summarize the scientific advancements made in each league as well as the results of the competition. Additionally, the supplemental CD coupled with this book contains the Team Description Papers for each team competing in RoboCup. The leagues, in alphabetical order, were:

- RoboCup Humanoid League
- RoboCup Junior League soccer, rescue, and dance competition
- RoboCup Legged League
- RoboCup Middle-Size League
- RoboCup Rescue Real Robot League
- RoboCup Rescue Simulation League
- RoboCup Simulation League soccer, coach, and visualization competition
- RoboCup Small-Size League

This book begins with an overview over the RoboCup competition together with a vision statement for the future development of RoboCup until 2050 and three invited papers by internationally leading researchers of the robotics field. The core part of the book contains papers accepted for oral or poster presentation at the International RoboCup Symposium 2003, which was held directly after the RoboCup competitions. The RoboCup team descriptions which, traditionally, have been part of the proceedings are now provided on a supplementary CD. This enabled us to allocate significantly more space for the fast-growing number of participating teams, thus rendering the team descriptions more informative and thus valuable.

Of the 125 symposium paper submissions received, an increase of 64% over RoboCup 2002, 39 papers were accepted for oral presentations and 35 papers were accepted for poster presentations. The International Program Committee, which contained both RoboCup researchers from around the world as well as researchers from outside the community, selected two papers for the jointly awarded RoboCup Engineering Challenge, and one paper for the RoboCup Scientific Challenge Award. The award winners were:

- Scientific Challenge Award awarded to Andrea Miene, Ubbo Visser and Otthein Herzog (University of Bremen, Germany) for *Recognition and Prediction of Motion Situations Based on a Qualitative Motion Description.*

– Engineering Challenge Award awarded to Daniel Cameron and Nick Barnes (University of Melbourne, Australia) for *Knowledge-Based Autonomous Dynamic Color Calibration*, and to Michael J. Quinlan, Craig L. Murch, Richard H. Middleton, and Stephan K. Chalup for *Traction Monitoring for Collision Detection with Legged Robot.*

For the keynote speakers, five internationally renowned researchers accepted our invitation to present special talks at the RoboCup Symposium. The speakers were:

– Manuela Veloso, Carnegie Mellon University, USA
– Masahiro Fujita, SONY ID Lab, Japan
– Ulrich Nehmzow, University of Essex, UK
– Paolo Dario, University of Pisa, Italy
– Maja Mataric, University of Southern California, USA

As a final note, the editors of this book are grateful to Enrico Pagello, PadovaFiere, and the RoboCup Federation for making the RoboCup Symposium and RoboCup 2003 as a whole, possible. The next international RoboCup event will be held in Lisbon, Portugal in 2004, followed by RoboCup 2005 in Osaka, Japan. All details regarding RoboCup 2004 can be found at *http://www.robocup2004.pt* or at the main RoboCup website *http://www.robocup.org*.

July 2003

Daniel Polani
Brett Browning
Andrea Bonarini
Kazuo Yoshida

RoboCup International Federation

The RoboCup Federation is an international non-profit organization formed to coordinate international effort to promote science and technology using soccer games played by robots and software agents.

RoboCup 2003 Organization

General Organization

General Chair
Enrico Pagello (Italy)
Associate Co-chairs
Giovanni Adorni (Italy)
Daniele Nardi (Italy)

International RoboCup Symposium

Symposium Chairs
Daniel Polani (UK)
Brett Browning (USA)
Andrea Bonarini (Italy)
Kazuo Yoshida (Japan)
Electronic Proceedings and CD
Emanuele Menegatti (Italy)
Enrico Pagello (Italy)

Competitions and Presentations

RoboCupSoccer Simulation League
Martin Riedmiller (Germany) – Chair
Patrick Riley (USA)
Junji Nishino (Japan)
Oliver Obst (Germany)
Maria Simi (Italy) – Local Chair
RoboCupSoccer Small-Size League
Paulo Costa (Portugal) – Chair
Ryad Benosman (France)
Beng Kiat Ng (Singapore)
RoboCupSoccer Middle-Size League
Ansgar Bredenfeld (Germany) – Chair
Yasutake Takahashi (Japan)
Fernando Ribeiro (Portugal)
Emanuele Menegatti (Italy) – Local Chair
RoboCupSoccer Four-Legged Robot League
Alessandro Saffiotti (Sweden) – Chair
Claude Sammut (Australia)
Scott Lenser (USA)
Luca Iocchi (Italy) – Local Chair

RoboCupSoccer Humanoid League
Thomas Christaller (Germany) – Chair
Hajime Asama (Japan)
Lars Asplund (Sweden)
Giuseppe Oriolo (Italy) – Local Chair

RoboCupRescue Real Robot League
Adam Jacoff (USA) – Chair
Satoshi Tadokoro (Japan)
Andreas Birk (Germany)
Paolo Fiorini (Italy) – Local Chair

RoboCupRescue Simulation League
Tomoichi Takahashi (Japan) – Chair
Lin Padgham (Australia)
H. Levent Akin (Turkey)
Ranjit Nair (USA)
Rosario Sorbello (Italy) – Local Chair

RoboCupJunior: Soccer, Rescue, and Dance
Jeffrey Jonhson (UK) – Chair
Gerhard K. Kraetzschmar (Germany)
Elizabeth Sklar (USA)
Luigi Pagliarini (Denmark/Italy) – Local Chair

International Symposium Program Committee

T. Arai
M. Asada
T. Balch
J. Baltes
W. Banzhaf
M. Bowling
L. Canamero
H. Christensen
B. Clement
J. Cooperstock
R. D'Andrea
K. Dautenhahn
J. Dias
D. Fox
C. Goldman
S. Gutmann
J. Hespanha
A. Howard
H. Hu
M. Jamzad
P. Jonker
H. Jung
G. Kaminka
B. Kiat
G. Kraetzschmar
J. Lallinger
S. Lenser
M. Long
I. Matthews
I. Noda
M. Ohta
S. Onn
D. Pynadath
M. Riedmiller
A. Saffiotti
P. Scerri
S. Sen
E. Sklar
B. Smart
L. Sonenberg
O. Sporns
P. Stone
J.-P. Tarel
A. Tews
W. Uther
R. Vaughan
J. Wendler
M. Wiering
L. Winer
G. Wyeth
F. Wörgötter

Table of Contents

Overview and Roadmap

Invited Papers

Technical Papers

Posters

Overview of RoboCup 2003 Competition and Conferences

Enrico Pagello, Emanuele Menegatti*, Ansgar Bredenfeld, Paulo Costa, Thomas Christaller, Adam Jacoff, Jeffrey Johnson, Martin Riedmiller, Alessandro Saffiotti, and Takashi Tomoichi

Intelligent Autonomous Systems Laboratory (IAS-Lab)
Department of Information Engineering, Faculty of Engineering
University of Padua, Padova, Italy
emg@dei.unipd.it

1 Introduction

RoboCup 2003, the seventh RoboCup Competition and Conference, took place between July the 2nd and July the 11th 2003 in Padua (Italy). The teams had three full days to setup their robots. The competitions were held in the new pavilion n°7 of the Fair of Padua (Fig. 1). Several scientific events in the field of Robotics and Artificial Intelligence were held in parallel to the competitions. The RoboCup Symposium was held in the last two days. The opening talks took place in the historical Main Hall of the University of Padua and the three parallel Symposium sections in the conference rooms of the Fair of Padua.

Fig. 1. The entrance of the RoboCup-2003 Competition Site.

* Corresponding author.

D. Polani et al. (Eds.): RoboCup 2003, LNAI 3020, pp. 1–14, 2004.

Fig. 2. The ancien Main Hall of The University of Padua where the opening talks of the RoboCup-2003 Symposium took place.

RoboCup was born with the goal of *"building by 2050, a team of fully autonomous humanoid robot to beat the human winning team of the FIFA Soccer World Cup"*. This is a long term goal that someone saw as utopian or with a limited scientific appeal, but in the mind of the promoters of RoboCup, this is a mean to foster Robotics and AI research among the scientists, the students, and the general public. RoboCup already produced the result of disseminating interest and knowledge about Robotics and Artificial Intelligence. This is witnessed by the ever increasing number of people and institutions that get involved in RoboCup and by the offspring of new initiatives within the RoboCup community. RoboCup is no longer only the Soccer World Cup for autonomous robots, but it is a container for different robotics event: Soccer Robotics, Rescue Robotics, Educational Robotics and a Scientific Symposium on Robotics.

RoboCup 2003 was a new record milestone in the history of robotic events. We had 243 teams for a total of 1244 registered participants coming from more than 30 countries from four of the five Continents, the only missing was Africa. Fig. 3 gives a clear understanding of the tremendous growth of the number of participants along the years. During RoboCup 2003, a small industrial exhibit took place, where some international companies showed their commercial and research products. Among the others, we had the presence of COMAU one of the few "total" suppliers for the automotive industry of automation systems. COMAU has a long tradition in the robotics and automation industry. Another important exhibitor was "Polo Robotico di Genova" a research and technological Consortium of Genoa (Italy).

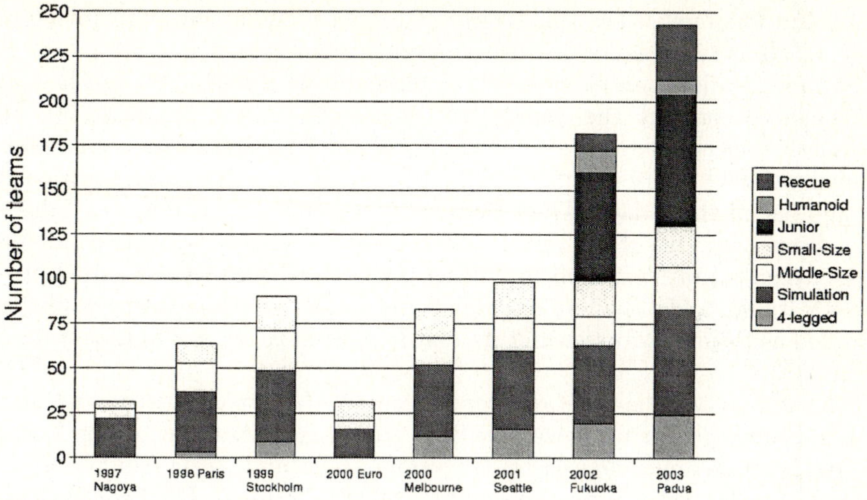

Fig. 3. The trend in the number of participating teams in the RoboCup Competitions.

2 RoboCup Symposium and Correlated Scientific Events

The RoboCup 2003 Symposium opened up in the ancient Main Hall of The University of Padua. This is the hall where Galileo Galilei tought and Manuela Veloso gave her invited talk on the evolution and achievements of seven years of RoboCup, under the family crests of the ancient students of the University of Padua. The organisers wanted a female researcher to open the Symposium to commemorate Elena Lucrezia Cornaro Piscopia, the first graduated woman in the world, that graduated in Philosophy at the University of Padua in 1678.

The second invited talk was from Masahiro Fujita who gave an overview of humanoid robots developed in Japan with an impressive demonstration of the new prototype of companion humanoid robot of Sony.

The other two invited talk where held in the Conference Center of the Fair of Padua and were given by Ulrich Nehmzow, on the use of chaos theory in the study of the interactions between the robot and its environment, and by Paolo Dario, a President of the Robotics and Automation Society of IEEE, on the use of robotics in medicine and other application fields. The invited talks were completed by Maja Mataric talking about multi-robot cooperation, unfortunately this was just a video contribution, because she could not attend the Symposium.

For the first time in the history of RoboCup, the number of papers selected for oral presentation and the number of Symposium attendants were so high, the organisers decided to split the Symposium on three parallel tracks.

The Symposium was closed by the RoboCup roadmap discussion. The roadmap discussion is aimed to identify the intermediate milestone to be reached in the next five, ten or twenty years in order to achieve the final goal of 2050. The roadmap discussion was only started in Padua and it continued by e-mail after

Padua and was finalised in Blaubeuren (Germany) in October 2003 (as detailed in the Roadmap Discussion contribution in this book).

Several parallel scientific events took place during RoboCup2003. The mostly articulated event was the Japan-Italy bilateral seminar organised by JSPS (Japanese Society for the Promotion of Science) and CNR (National Research Council of Italy). This seminar was chaired by Minoru Asada and Enrico Pagello and lasted three days. The schedule of the seminar was dense of talks and panel discussion. One of the results of the bilateral seminar was the participants pinpointed a set of "*hot*" and promising topics in which to start joined project between Italian and Japanese research centers. The most promising topics were identified as Rescue Robotics and standardised Simulation Environment for Mobile Robots.

We had also two one-day Conferences: one on *Multi-robot systems: trends and industrial applications* organized by SIRI (the Italian Association for Robotics and Automation) and chaired by Giuseppina Gini and Rezia Molfino and another one on *Synthetic Simulation and Robotics to Mitigate Earthquake Disaster* chaired by Daniele Nardi.

3 Results of Competitions

As we said in the introduction, nowadays the RoboCup competitions are divided in three main branches: RoboCup Soccer, RoboCup Rescue and RoboCup Junior.

3.1 RoboCup Soccer

Soccer Simulation: This year, the games of Soccer Simulation league showed a big advance in the performances of the teams. For the first time in the history of the Simulation League, all games were started automatically. This resulted in the possibility to have a very tight time schedule with the possibility to play more games during the tournament. The automatic start of the game forced the developers to provide more autonomy to their teams (e.g. by effectively using the 'coach'). From 56 teams that were qualified, 46 teams participated in the tournament. In the first round, 8 groups of 5 to 6 teams participated, from which the first 3 teams of each group advanced. All participating teams showed a good level of individual skills. The teams that advanced to the second round additionally showed a good level on team play abilities. In the second round, 4 groups of 6 teams played, from which again the first 3 advanced. For the first time, a 3rd round was also played in groups. The level of play of the last 12 teams this year was very mature and close to each other. Unlike in previous years, games often were not decided until the end. Most of the games where decided just by one or two goals. Exciting games happened among the teams. In Fig. 4 are reported the results of the final stage of the tournament. The winning teams of the soccer tournament were: first place: UVA Trilearn (Netherlands), second place: Tsinghuaelous (China), third place: Brainstormers (Germany). The winners of the online coach competition were: UT Austin Villa (USA) first place, and FC Portugal (Portugal), second place. The winner of game presentation and analysis competition was the team Caspian (IRAN).

In the word of the organising chair of the Soccer Simulation league "the top teams showed mature capabilities in team play, in stamina management, in active vision, in the use of heterogeneous players and communication. The main reason for the successfulness of the winning teams is a highly elaborated software design that considers all of the above issues"[1].

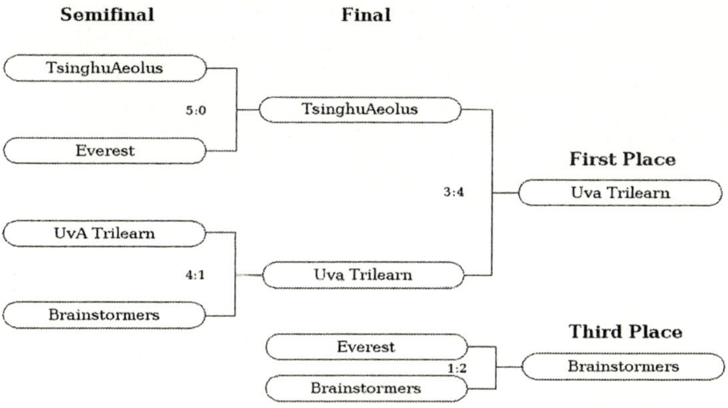

Fig. 4. The results of the games of the final phase of the **Soccer Simulation League** tournament.

Four-Legged League: The RoboCup four-legged league began in 1998, and it was managed by Sony until 2002. In 2003, the management of the league was taken over by the RoboCup Federation for the first time. The transition went rather smoothly, thanks to the kind help received from Sony. In RoboCup 2003, 24 teams from 15 countries participated in the Four Legged League. Teams were evenly distributed across continents, except Africa: 8 coming from Europe, 7 from the USA, 5 from Asia, and 4 from Australia. The teams were divided into 4 pools of 6 teams each. The games were organized in a preliminary round robin phase, followed by a single elimination championship tournament. The winning team was rUNSWift (Australia). This team was already champion in 2000 and 2001, and 2nd place in 2002. UPennalizers (USA) placed second, and NUbots (Australia) third. This year the Sony Prize was awarded to rUNSWift (Australia). The winners of the technical challenge competition are first place German Team (Germany), second place rUNSWift (Australia), third place Araibo (Japan).

In the four-legged league, two different philosophies of robot programming are measuring them-self, i.e. learned behaviours and controls vs. hand-coded

[1] E. Pagello, E. Menegatti, D. Polani, A. Bredenfel, P. Costa, T. Christaller, A. Jacoff, M. Riedmiller, A. Saffiotti, E. Sklar, and T. Tomoichi. Robocup 2003: New scientific and technical advancements. *AI Magazine*, (to appear), 2004.

Fig. 5. A phase of a game of the **Four-Legged League** tournament.

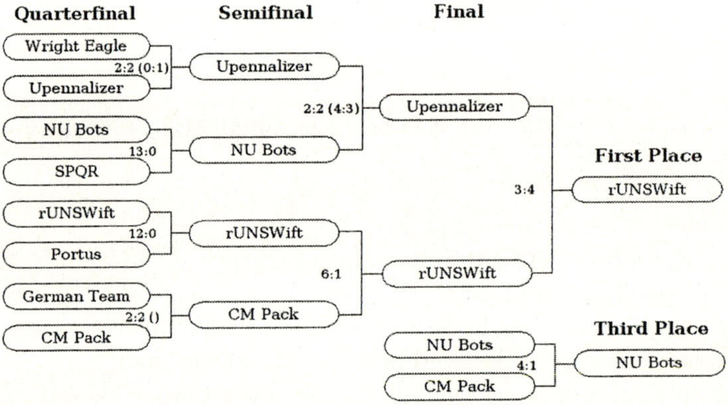

Fig. 6. The results of the games of the final phase of the **Four-Legged League** tournament.

robot programs. In the Four-Legged League the research focus is shifting from lower-level functionalities to higher level skills like planning, coordination, and adaptation. Most teams in 2003 used some form of multi-robot cooperation, including dynamic role assignment and information sharing. In fact, most teams showed fast and stable walking, accurate ball control, reliable ball perception, and good self-localization. This is derived also from the code sharing policy adopted by the league. "A drawback of this policy is a potential reduction in diversity, since many teams prefer to improve on existing successful techniques rather than try to invent radically new ones" as written by the organising chair of the Four-Legged League[2]. Code sharing is possible because, all teams uses a common platform: the Sony AIBO robot. This year there was three technical challenges: 1) to score with a black and white soccer ball in an empty field; 2) to visit 5 points defined by their (x, y) coordinates with the colored landmarks removed; 3) to traverse the field while avoiding collisions with 7 static robots. The result of the first challenge was rather deceiving. Only 9 teams out of the 20 who tried the challenge managed to perceive the ball and to make contact with it. The second challenge showed that localization without colored landmarks can be achieved, and several teams managed to get around the target points. The third challenge was much more successful, showing that the league is ready to get more serious about collision avoidance. Of the 20 teams who tried this challenge, none collided with more than 3 obstacles.

Small-Size League: This year competition saw 20 teams from all over the world. The results of the final stage of the tournament is reported in Fig. 8. This year there was no quarter finals because there was a second round robin with four groups of three teams. The winners of each group progressed to the semifinals. This was to maximize the games for each team while minimizing field changes. In Small Size League field changes are hard because the teams have to unmount, mount and recalibrate their cameras.

Fig. 7. A phase of a game of the **Small Size League** tournament.

The winners were: first place, BigRed'03 from Cornell University U.S.A., second place RoboRoos from The University of Queensland, Australia and third place FU Fighters Freie Universität Berlin, Germany.

[2] ibidem.

This year a "referee box" was introduced, i.e. all commands which the referee can communicate to the teams, were sent directly to the software controlling each robot team from a laptop operated by the assistant referee. The result was that there was no human intervention during the game and the game flow was greatly improved. All teams used one or two cameras, placed 3m above the field to extract the position of the ball and of the robots.

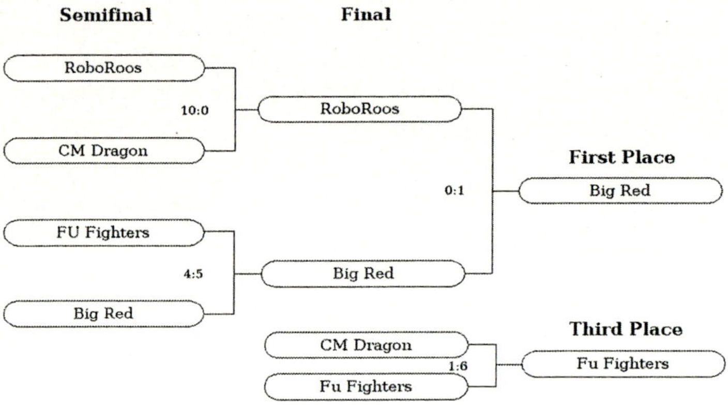

Fig. 8. The results of the games of the final phase of the **Small Size League** tournament.

This year there was a certain convergence on the robot design as most teams adopted an optimized solution. Almost all teams used omnidirectional wheels with three or four of those wheels per robot. The additional maneuverability of this solution made the two-wheel configuration almost obsolete on this league. Most top teams focused on having an efficient dribbler and kicker. A dribbler is a set of rotating rubber cylinders that transmits a backspin to the ball keeping it almost glued to robot even when it is travelling on the field. It was a general concern that this feature was overused and some kind of limitation should be imposed for next years' competitions.

Middle-Size League: The Middle-Size tournaments saw 24 teams from 11 countries participating in 2003. Although 32 teams qualified for the games, finally 8 teams could unfortunately not take part. The main reason was lack of financial resources or not-finished robots. Only a few teams did not take part due to missing student resources. The tournament was played on four fields, thus opening the opportunity to play four games in parallel using one hour time slots. The organizing committee decided to play two round robins in order to maximize the number of games for each team. The results of the final section of the tournament are reported in Fig. 10

As pre-condition for qualification, each team had to submit a team description paper. These papers concentrate on the research focus of the team. All

Fig. 9. A picture of the match between the AIS-Musashi team and the AllemaniACs team in the **Middle-Size League**.

hardware and software details of the robots - which had been included in the team description paper in the last years - were collected systematically by a newly introduced Team Questionnaire. The intention of the Questionnaire is to support information exchange between existing teams and to lower the entrance barrier for new teams that want to join the Middle-Size-League. In addition, the material collected in the Questionnaire provides a concise overview of the methods and technologies used by the teams. The questionnaires of all participating teams are contained on the CD-ROM of this book.

A larger field of play (i.e. $10 \times 8m$) and removed poles around the field were the major rule changes for 2003. Nearly all participating teams did not have problems with this changed field set-up. The security bar around the field turned out to be suitable and sufficient to prevent robots from leaving the field.

The challenge competition consisted of two challenges. Challenge 1 was performed as described in the Middle-Size-League rules of 2003. The team leaders decided during the tournament to perform the second part of the challenge competition as free challenge. The free challenge was a five minutes oral presentation and a short demonstration of innovative results each team wanted to demonstrate. A jury consisting of all team leaders voted on the performance of this challenge and awarded points from 0 to 6 to each presentation. Some teams demonstrated challenges like proposed in the rules, i.e., co-operative behavior or the ability to play with a standard FIFA ball. Other teams gave an insight into ongoing research, new robot developments or special behaviour capabilities of their robots. This includes for example studies on new ball stopping mechanisms, robots playing continuously passes or soccer playing behaviors that had been evolved in a physical robot simulator. The winner was the team Attempto! Tübingen from Germany.

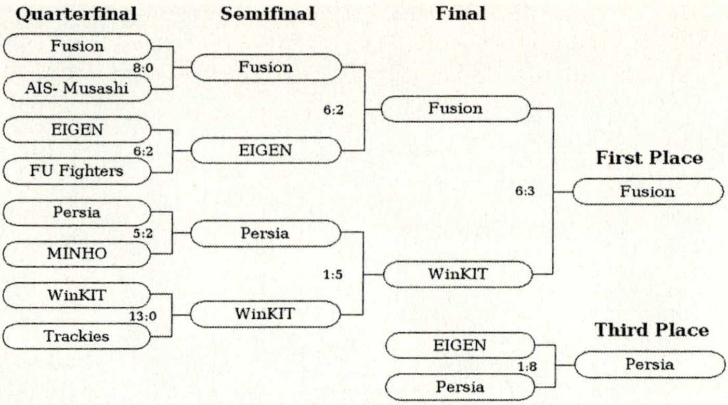

Fig. 10. The results of the games of the final phase of the **Middle Size League** tournament.

Playing the challenge competition at the end of the tournament turned out to be problematic, since not all teams were able to participate. The main reason was broken robots. In future, it should be considered to have the challenges before the start of the round robin and to use their results at least as an additional criterion for the assignment of teams to groups for the first round robin.

Humanoid League: Started in the previous year, the Humanoid League is still rapidly developing. The Humanoid league has different research issue to face with respect to the other leagues. The main difference is that the dynamic stability of robots needs to be well maintained while the robots are walking, running, kicking and performing other tasks. Furthermore, the humanoid soccer robot will have to coordinate perceptions and biped locomotion, and be robust enough to deal with challenges from other players. Test-games could be performed. However, the competition consisted of four non-game disciplines, namely standing on one leg, walking, penalty kick and free style. A number of excellent robots were presented in the competition.

After a good competition with tight results the team HITS-Dream of the Honda International Technical School's received the Best Humanoid Award. In the Walk Competition HITS-Dream (Japan) won the first place, Senchans (Japan) the second place, and Foot-Prints (Japan) the third place. In the Penalty-Kick Competition, Foot-Prints (Japan) ranked first in the class of the robot shorter than 40cm and Senchans (Japan)ranked first in the class of the robot under the 80cm. In the Free Performance Competition the winner was Robo Erectus (Singapore), the second place was of Isaac (Italy) and the third place of Tao Pie Pie (Canada).

Humanoid soccer robots are complex machines, which should have advanced abilities from very different fields of technology, namely materials, locomotion, manipulation, power, communication, perception and intelligence.

Fig. 11. A picture of the five minutes talk allowed to the teams for explaining the free performance their are about to demonstrate in the **Middle-Size League**.

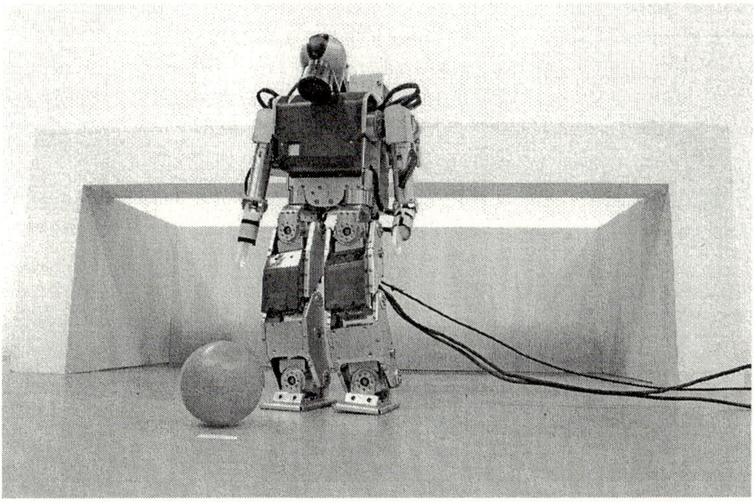

Fig. 12. A picture of a game in **Humanoid League** Soccer Competition.

3.2 RoboCup Junior

RoboCup Junior 2003 involved 74 teams (258 participants) from 16 countries world-wide. In Padua, teams could enter four different challenges: one-on-one soccer, two-on-two soccer, dance and rescue. Three different age groups were

Fig. 13. A picture of a game in **RoboCup Junior Soccer Competition**.

represented: primary (up to age 12), secondary (age 12-18, or end of high school) and undergraduates. The biggest changes in the event from 2002 were the introduction of a newly designed rescue challenge and the development of a new entry-level soccer league for undergraduates, called the *ULeague*. Note that some teams entered more than one challenge within their age group.

At RoboCup Junior 2003, soccer remained the most popular challenge, engaging 67% of teams overall. Some of the secondary students took advantage of state-of-the-art technological improvements and used, for example, magnetic sensors for direction and ultrasonics for collision avoidance. LEGO Mindstorms continues to be the most popular medium for robot construction but many teams, particularly in Asia, use the Elekit SoccerRobo. More advanced teams, most notably from Australia and Germany, even constructed their hardware completely from scratch.

RoboCup Junior has seen strong growth in the number of female participants, particularly in the dance challenge, which provides a unique outlet for creativity. While RCJ attracts in total an average of 15% female students overall (increased from 10% in 2000), the dance challenge at RCJ-2003 had 31% female participation.

3.3 RoboCup Rescue

Real Robot League: RoboCup 2003 hosted the third Rescue Robot League competition, which included 12 teams from 8 countries. The winning teams were: first place ROBRNO team from the Czech Republic, second place CEDRA from Iran, and third place was IUT-MICROROBOT from Iran. Only one team demonstrated autonomous mapping during the competition, but did not

Fig. 14. The blimp used by the UVA -Zeppelin team of the University of Amsterdam in the **Real Robot Rescue League** tournament.

contribute quite enough points to earn a place award. There were other interesting approaches: fully autonomous robots, a robot almost directly from the Middle-Size League, and even a blimp. Although two teams demonstrated fully autonomous robots capable of navigating parts of the yellow arena, they didn't produce maps showing victim identifications so did not score well. Meanwhile, the remotely tele-operated teams showed very few autonomous behaviors to assist their efforts, although several teams were working toward such capabilities.

To evaluate the performances of the teams, the metric of Fig. 15 was used. This takes into account the quality of the output map, the quality of the robot sensing and the motion skill of the robot.

$$\text{ARENA WEIGHTING}\left(\frac{\text{MAP QUALITY} + \text{VICTIM LOCATION} + \text{VICTIM TAG} + \text{VICTIM SITUATION} + \text{VICTIM STATE} - \text{ARENA BUMPING} - \text{VICTIM BUMPING}}{\left[1 + \text{NUMBER OF OPERATORS}\right]^{2}}\right)$$

Fig. 15. The metric used to calculate the performances in the **Real Robot Rescue League** tournament.

Simulated League: In the RoboCup-2003 Rescue Simulation League tournament, 17 teams participated. Many teams were here competing for the first time. In fact, after RoboCup-2002, useful tools like Java based agent developing kits, JGISEdit, and a Multi-platform map editor with the map of the city of Foligno (Italy) were provided and this helped new comers to join rescue community.

This year the map of Foligno was adopted as an official map at competition. This map was chosen in order to easily convey the importance of RoboCup Rescue to the general audience, especially the Italian audience. In fact, Foligno is an Italian city that was seriously damaged by an earthquake. The Foligno map is bigger twice than the two traditionally used maps, Kobe and Virtual City. The adoption of the Foligno map was a challenge for the teams competing in RoboCup-2003. In the preliminary games, all team did rescue operations at two disaster situations per three different maps. The winners of Simulated Rescue competition were: first place ARIAN team, Sharif University of Technology, IRAN, second place YOWAI, University of Electro-Communications, JAPAN, and third S.O.S, University of Technology, IRAN.

With respect to the games played in RoboCup-2002, the teams showed increased abilities both in the single autonomous agents (fire fighter, police agent, and ambulance) and in the cooperation abilities among the agents. In order to improve the capability of their agents the teams used on-line learning methods for rescue formation, clustering methods or agents group formation mechanism.

4 Special Content of the CD-ROM

Due to the always increasing number of participating teams in the competitions, it is no longer possible to include the Team Description Paper in the RoboCup book. Nevertheless, the teams are the engine that move the RoboCup event and the innovations introduced by the teams are the real thrust that moves forward research. In order not to disperse the knowledge and the innovations proposed by the teams during the RoboCup 2003 competition, we proposed to include with this book a CD-ROM containing the Team Description Papers of the teams that participated in Padua. The Team Description Papers have been edited and revised after the competitions in Padua. We expressly asked the teams to critically analyse the performances of their robots during the competition by discussing which solutions and techniques proved to be effective (or proved not to be effective at all).

Acknowledgments

We wish to thank all people who contributed in the writing of this paper, in particular Elizabeth Sklar, Brett Browning, Changjiu Zhou, Norbert M. Mayer. We also wish to say "Thank You!" to all people who contributed to make RoboCup-2003 a successful event: starting from the several co-chairs and co-organizers to the simple participants. We wish to thank PadovaFiere S.p.A. and in particular Franca Ingrassi and Andrew Abberley .

RoboCup: Yesterday, Today, and Tomorrow Workshop of the Executive Committee in Blaubeuren, October 2003

Hans-Dieter Burkhard[1], Minoru Asada[2,3], Andrea Bonarini[4], Adam Jacoff[5], Daniele Nardi[6], Martin Riedmiller[7], Claude Sammut[8], Elizabeth Sklar[9], and Manuela Veloso[10]

[1] Humboldt University Berlin, D-10099 Berlin, Germany
hdb@informatik.hu-berlin.de
[2] Dept. of Adaptive Machine Systems
[3] HANDAI Frontier Research Center
Graduate School of Engineering, Osaka University
Suita,Osaka 565-0871, Japan
asada@ams.eng.osaka-u.ac.jp
[4] Department of Electronics and Information, Politecnico di Milano, Italy
bonarini@elet.polimi.it
[5] National Institute of Standards and Technology, USA
adam.jacoff@nist.gov
[6] Dip. Informatica e Sistemistica, Univ. "La Sapienza" Roma, Italy
nardi@dis.uniroma1.it
[7] Universtaet Osnabrueck, Institute for Cognitive Science
Albrechtstr. 28, D-49069 Germany
martin.riedmiller@udo.edu
[8] School of Computer Science and Engineering, University of New South Wales, Sydney Australia 2052
claude@cse.unsw.edu.au
[9] Dept. of Computer Science, Columbia University
1214 Amsterdam Avenue, MC 0401, New York, NY 10027-7003, USA
sklar@cs.columbia.edu
[10] Computer Science Department, Carnegie Mellon University
Pittsburgh PA 15213-3890
veloso@cs.cmu.edu

1 Introduction (by Manuela Veloso)

RoboCup has been known for the goal of "creating robots capable of beating the world cup in 2050." Clearly, we stated this goal back in 1996 not as an exact scientific goal, but as an audacious challenge to pursue. We aimed at creating a far away target for RoboCup researchers, as we were well aware that the development of fully autonomous soccer robots capable of competing against human world champion soccer players was a rather long term research project.

Although this goal exists, the real RoboCup characteristic has been the research pursuit of its participants to advance the scientific state of the art of the

D. Polani et al. (Eds.): RoboCup 2003, LNAI 3020, pp. 15–34, 2004.

fields of artificial intelligence and Robotics. And from its beginning, we have organized RoboCup along a set of different leagues that provide different research challenges for multi-robot systems in complex environments. Initially, we created robot soccer leagues. In our research pursuit within robot soccer, we soon realized that our techniques could be applicable to other complex environments, such as search and rescue. Furthermore, we understood that our long term goals required the education of children in robotics. So we started Junior leagues for young children.

The RoboCup competitions include now three major areas, namely RoboCup soccer, RoboCup Rescue, and RoboCup Junior. We present and discuss research contributions in the RoboCup Symposium. Furthermore to better share research interests among different leagues, we created RoboCup Camps for students and Special Interest Groups (SIGs).

In 2003, and about six years after the first official RoboCup competitions at IJCAI'97 in Nagoya, we had a special two-day meeting of the Executive Committee in October 2003. We specifically discussed the immediate directions and roadmap for each league towards our 2050 goal. This article summarizes the results of this meeting. The article is organized along the different leagues, research, and education.

2 **RoboCup Roadmap** (by Hans-Dieter Burkhard)

The workshop (Oct. 4th- 5th, 2003) can be understood as a continuation of the panel discussions in Fukuoka 2002 and Padova 2003. The 2050 goal, to win with a team of fully humanoid robots against the human world champion serves as the long-term vision. It sets up high challenges, which need to be solved step by step, and in corporation with other sciences. We are not done with a perfect kicker for let's say Midsize League - instead we will change the conditions of our leagues and our players year by year, according to ambitious scientific problems.

A questionnaire was sent to the participants before the workshop. The questions concerned the problems to be solved until 2050, the meaning about RoboCup benefits, and the development of our championships and conferences. Here are some results concerning the necessary steps in RoboCup:

- Perception appears as the mainly mentioned and discussed challenge in the questionnaire: Recent shortcomings and future requirements (outdoor field, unpredictable lighting) are next steps to be solved. Expectations range from 10 - 20 years for their solutions.
- Robust (humanoid) hardware that can move on outdoor grounds and can handle the ball like humans do - this will still last 15, 20 or even 30 years.
- Safe interaction with humans is one of the most important problems, again it will require a long time. It is an open questions which restrictions are necessary for fairness (e.g. are robots allowed to use wireless communication - or the other way round: will humans be allowed to use additional technical equipment?).

- Adaptation and learning to play as an individual and as a team against an opponent with unknown (but partially observable) intentions and skills: 10-20 years were expected for really useful solutions.
- Body control will become a hard problem as soon as we are able to build humanoids which can run, jump etc.: We will have to control a large number of parameters under complex dynamics.

The answers in this category of the questionaire were mainly focussed on Robotics from an AI and Informatics centered viewpoint - like perception, learning, cooperation. Topics like new materials and energy are subsumed under the general hardware problems, but are not considered in detail. To play a major role, RoboCup will have to enlarge efforts in these directions. It will need cooperation with other sciences.

Another category was called "Benefits of RoboCup". It was the aim to collect answers which are useful for discussions and for funding. Many answers were related to the solutions of the problems from above. The benefits for education and studies in Robotics were pointed out. RoboCup Junior has an important role for promotion of RoboCup. RoboCup Rescue is a convincing application, and it needs further solutions behind the scope of soccer playing robots.

Commercial applications may concern service robotics and robust solutions for many other problems. It was discussed, how far scientific institutions can and should invest their power in the development of industrial applications. The conditions (and needs) are different in different parts of the world. In any case, it would be good for the image of RoboCup to know about applications with origins in RoboCup. Again, the cooperation with other scientists is necessary to prevent from "reinventing the wheel" - and to promote our results.

A next category was related to the development of RoboCup Championships and Conferences. There was a remarkable tendency for concentration: Not more leagues, but merging of existing ones. The cooperation between the leagues should increase (e.g. by comparable challenges, exchange of solutions etc.). The Symposium could be devoted to special topics of common interest, and people from the related communities should be invited. Most important: RoboCup should become a first-rate, prestigious conference.

During the Workshop, each league gave a report about their development and their plans. These reports together with the related further discussions are part of this article (see below). Additional reports were given about the SIGs.

RoboCup has become a good visibility and reputation in many other communities, and RoboCup is often used for examples and illustration. Our progress is observed from outside, especially for educational efforts in AI and Robotics. Most of the reported comments are positive, but still we have to promote our scientific goals. The best way are presentations of our results at various conferences and journals. The participation of RoboCup researchers in other robotic research projects is another important way.

Besides the more technical viewpoints, the Robotics projects are important for the Human Sciences, too. Soccer is a good scenario to study natural (human) skills, the Robotics perspective opens new insights. More about CDR - the Cognitive Development Robotics - can be found in M. Asada's section below.

As our leagues are pursuing new challenges, it becomes harder for beginners to start in RoboCup. Especially for education, where RoboCup scenarios are used to develop the skills of students in a restricted time, a simple setting is necessary. An "Entry"/"Educational"/"Easy" League (ELeague or Uleague="Undergraduate" League, cf. the section of B.Sklar on the Junior League below), may be on the level of recent Small Size League, could be useful for such purposes. A concrete proposal will be developed.

Small Size League in its recent format seems to be at a final point: New challenges for this league point to the direction of Midsize League. Cooperation in Midsize League is an important future milestone. It can be best realized with more players on the field. The conclusion of these discussions is the merging of Small Size League and Midsize League to one League in the next years until 2006. Concrete proposals are to be discussed between the leagues.

As a next step for better perceptional skills, all real robot leagues will reduce the efforts for special lightings and special field designs step by step in the next years. It is the hope, that there will be an exchange of e.g. successful vision systems between the leagues. Common challenges should be defined by the SIGs.

There were several more important topics discussed in Blaubeuren, e.g. the design and maintenance of a common webpage for tutorials, exchange of useful solutions, discussions etc. It works fine on the level of the leagues, but it would be good to have a common page. Like the pages of the leagues, the project should be realized by volunteering.

3 RoboCup: Yesterday, Today, and Tomorrow – Humanoid Science Approach – (by Minoru Asada)

Abstract. The sectionr presents the summary of talk in the RoboCup Workshop at Blaubeuren, Germany, Oct. 4-5th after the RoboCup-2003 Symposium. The talk starts from the early beginning of RoboCup and raised the future issues towards the final goal, that is, roadmap discussion and its related issue. Finally, new activities for RoboCity CoRE was introduced.

3.1 Introduction

The very early days of RoboCup starting from 1993 to 1995 was introduced and the story about the rejection of the authors' first RoboCup conference paper in 1994, which as a result activated the promotions of RoboCup in small workshops is mentioned as RoboCup yesterday. The RoboCup today was just shown as the number of participating teams at RoboCup 2003, Padova.

Review of future issues in RoboCup from different viewpoints are given. They are research, education, industrialization, and connection to the general society. The rest of the section gives the summary of these activities and issues.

3.2 Research: A Humanoid Science Approach

Two aspects of the research issue are pointed out: funding strategy (how naming "RoboCup" explicitly does help us?), and research topic (what's our unique but not too specific topic?). For the former, data of the past and on-going projects funded from governments, public organizations, and industries will be collected for future funding based on RoboCup. For the latter, a humanoid science approach is proposed as one of the scientific topics towards the final goal.

"Humanoid Science" under which a variety of researchers from robotics, AI, brain science, cognitive science, psychology and so on are seeking for new understanding of ourselves by constructivist approaches, that is expected to produce many applications. The humanoid science turns the research topics in RobopCup as follows:

- mechanical design for individual robots → design of humanoid platforms,
- robust sensing, especially, vision (object discrimination and tracking) → attention mechanism,
- self-localization and map building → body representation (body scheme or body image) and spatial perception,
- control architecture → freezing and releasing DOFS, NLPCA, SOM,
- communication → symbol generation and language emergence,
- multiagent systems in general → social interactions,
- combining reactive and modelling approaches → embedded structure (nature) and interaction with environment (nurture),
- sensor fusion → cross modal association for body representation (body scheme or body image), and
- cognitive modelling in general → theory of mind.

Cognitive Development Robotics (hereafter, CDR) as one approach to humanoid science [1] consists of the design of self-developing structures inside the robot's brain, and the environmental design: how to set up the environment so that the robots embedded therein can gradually adapt themselves to more complex tasks in more dynamic situations.

Brief explanations on developmental learning for Joint attention [2], and vowel imitation [3] are given as typical examples of CDR.

3.3 Design of Humanoid Platforms

The current participating teams are using commercially available or provided platforms such as Honda Firstep and Fujitsu HOAP focusing on behavior generation based on these platforms, or completely home-designed humanoids focusing on the design theory and implementation (mechanical structure, sensor, actuator, controller, and so on). Definitely, the latter is very challenging and very hard. Therefore, collaboration with industry is indispensable.

A small company (Vstone) developed a small humanoid platform: omni head that originally designed for Robo-One with a reasonable price. The height and the weight are 290mm and 1.9kg, the number of DOFs is 22. This sort of cheap platforms is not simply useful for research but also for education as a kit product for high school or college students who are currently not involved in RoboCup activities.

Collaboration with industry including ventures seems necessary for the development of standard parts such as sensors, actuators, controllers, and so on. One can make competition like AIBO league, say, Qrio (SDR-4XII) league or Vstone league will be possible. Robot Technologies (hereafter, RT) incubation center is needed for joint development of robot standard parts and new RT products since RT is an amalgam of various kinds of artifacts.

3.4 RoboCity CoRE: An Inner City RT Base

A basic concept of RoboCup are an international joint research, a landmark project: sharing the dream, and open to different disciplines, open to public. Currently, the competition and conference is once a year, and a natural extension of RoboCup concept is to have a permanent place to deploy our activities.

RoboCity CoRE (Center of RT Experiments) is an inner city labs for symbiotic experiments with robots, new partners of our future life. CoRE aims at only one RT base around the world where simultaneous progresses of research, industrialization, and education carry on simultaneously.

Open to public means that researchers, artists, companies, citizens interchange with each other to emerge new ideas, that leads the development of science, technology, and culture. CoRE will be a new cultural symbol of the future high-technological, ecological city.

(a) (b)

Fig. 1. A humanoid platform:Omnihead (left) and open filed in RoboCity CoRE (right)

References

1. Minoru Asada, Karl F. MacDorman, Hiroshi Ishiguro, and Yasuo Kuniyoshi. Cognitive developmental robotics as a new paradigm for the design of humanoid robots. *Robotics and Autonomous System*, 37:185–193, 2001.

2. Y. Nagai, M. Asada, and K. Hosoda. A developmental approach accelerates learning of joint attention. In *The 2nd International Conference on Development and Learning (ICDL'02)*, pages 277–282, 2002.
3. Yuichiro Yoshikawa, Junpei Koga, Koh Hosoda, and Minoru Asada. Primary vowel imitation between agents with different articulation parameters by parrotry like teaching. In *Proc. of IEEE/RSJ International Conference on Intelligent Robots and Systems 2003 (IROS '03)*, 2003.

4 RoboCupJunior (by Elizabeth Sklar)

RoboCupJunior (RCJ) has been growing and changing since its inception, this year reaching 75 teams from 16 countries for a total of 258 participants (see figure 2 for details). In 2003, there were five challenges: 1x1 SOCCER, 2x2 SOCCER, RESCUE, DANCE, and ULEAGUE. The overall participation rate of female students was 15%, while within the DANCE category, this was concentrated at 31%. As the league continues to mature, its internal structure and its role within RoboCup must adapt.

	2000	2001	2002	2003
number of teams	25	25	65	75
number of countries	3	4	12	16
number of participants	100	118	236	258

Fig. 2. RoboCupJunior participation, 2000-2003

4.1 Organizational Issues

The organizational action items for RoboCupJunior from 2002 were:

- to define a role for undergraduates;
- to continue to close the "gender gap" (i.e., increase participation of female students); and
- to establish national committees in participating countries.

These have been partially achieved. The proposed role for undergraduates is a new challenge, dubbed the ULEAGUE. This is discussed at length below (see section 4.3). The gender gap is still too great, but particularly due to the DANCE challenge, RoboCupJunior is succeeding in attracting and sustaining participation by female students. National committees have been established in several countries. Those most active are Australia, Japan and Germany. National representatives from several countries were chosen in 2003 to serve for 2004.

Two new organizational action items came up in 2003. The first is in regard to age groups and the second concerns record-keeping. Initially, the plan for RoboCupJunior was to provide an introduction to RoboCup for young students — primary through high school age. In 2000, there were three age groups: primary, up to age 12; middle, ages 12–15; and secondary; ages 15–18. For 2001 and 2002, there were two age groups: primary, up to age 12; and secondary, ages 12–18. For 2003, an experimental challenge targeted at the undergraduate age group was created (aka, the ULEAGUE); thus there were three age categories: primary, up to age 12; secondary, ages 12–18; tertiary, ages 18-22.

For 2004, the RoboCupJunior Technical Committee has made the decision to raise the boundary between primary and secondary age groups from 12 to 14 years of age. The reason for this is that at international events, there is very little participation in children below age 12. This is primarily because international travel is expensive and complicated for young children due to issues of language and chaperones. However, at these international events, two groups of students have emerged: one group centering around age 13 and another group centering around age 16–17. Thus for 2004, the age groupings will be divided into three categories: primary, up to age 14; secondary, ages 14–18; tertiary, ages 18-22. Note that individual countries, on the national level, may choose to re-align these boundaries, as is appropriate for their regional events. However, it must be emphasized to participants that the international rules will follow these boundaries and so students must be prepared, at the international level, to adhere to this structure.

The second new issue that arose in 2003 is that record-keeping has become quite difficult. Most of the time, there is very poor correlation between those students registered by a team at pre-registration time and those students who actually come to the event. As a result, it is extremely difficult to provide accurate statistics, and it is very important, particularly for Junior, to be able to produce these figures to the media and potential funders. It is suggested that for 2004, on an international level, registration of all team members be centralized in one database. This registration would pertain not only to students who attend the RoboCupJunior events, but also to all students who participate at home in preparing for RCJ events.

For 2004, the organizational goals include: (1) defining a better structure within the RCJ organization; (2) recognizing and responding to needs of the RCJ audience and participants, who are quite different from their counterparts in the senior RoboCup leagues; (3) developing an on-line forum for teams; and (4) creating a funding mechanism especially for RCJ. The League Chairs for 2004 are Elizabeth Sklar and Jeffrey Johnson. The Organizing Committee Chair for 2004 is Gerhard Kraetzschmar. The Local RoboCupJunior League Chair for 2004 in Lisbon is Carlos Carderia.

4.2 Technical Issues

The primary technical action item for RoboCupJunior from 2002 was to develop stepping stones from RCJ into the senior RoboCup leagues. The ULEAGUE was created in answer to this call and is discussed in detail below in section 4.3.

The RoboCupJunior RESCUE event was resurrected and re-designed. In 2000, there was a line-following SUMO challenge, which provided an intermediate-level task for middle-school age students. The environment is (fundamentally) static[1], however in order to perform line-following accurately, robots must be designed and programmed with skill and precision. In 2001, this event was modified as a RESCUE challenge, following the introduction of the RoboCup Rescue League at the senior level. The original intention for RCJ was that each host country would define a simulated disaster environment relevant to their local region, but always keeping within the structure of a static, line-following event. However, in 2002, this event was dropped, a controversial decision made by the local organizers. In 2003, the RoboCupJunior RESCUE event was re-designed, as a miniature version of the NIST Rescue Arenas used at the senior level. The RCJ RESCUE looks almost like a doll's house and the robots have to follow a line through the house, in and out of rooms, up and down ramps, searching for and identifying "victims". This event was well-accepted and there were numerous participants from several countries. For next year, some standards need to be published for construction and scoring, but overall, the challenge is a success and will undergo much change.

For 2004, other goals include (1) creating an outreach program/incentive for graduate and undergraduate students to mentor RCJ teams; (2) beginning discussion of a RoboCup exchange program; and (3) developing a book on *Educational Robotics through RoboCup*. The use of undergraduate students as mentors for Junior teams has been particularly successful in 2003. One hope for the near future is to create a formal mentoring program in which senior-league student team-members who also participate as mentors for Junior teams receive reduced (or free!) registration fees for the RoboCup event. Another proposal for the near future is to establish a formal exchange program for graduate students from RoboCup labs to visit other RoboCup labs around the world. This is already happening informally, but if a more formal program were established, it might open up the door to further technical exchange, understanding and advancement. Finally, a resource textbook is being developed, focusing on *educational robotics* through RoboCup. The idea is to create a resource for instructors teaching undergraduate courses on topics such as Artificial Intelligence, Programming, Autonomous Agents and Multi Agent Systems, using RoboCup as an example for demonstrating technical topics. The book will be a compilation of experiences, curricula and resources contributed by RoboCup team leaders and participants who have taught these courses. RoboCup and RoboCupJunior are together uniquely posed to be leaders in educational robotics on an international basis; such a book will help achieve this goal.

4.3 Discussion

The most contentious issue which has been raised within RoboCupJunior is the place for undergraduates within RoboCup. Currently, there exist students who have "graduated" from RoboCupJunior and are now embarking on their

[1] In SUMO, both robots perform at the same time on the same field, so technically the environment is dynamic however not on the same scale as on the Soccer field.

undergraduate education. However, there is no obvious place for them within RoboCup, especially if they do not attend a university where there is already an active RoboCup team. This motivated the formation of the experimental league within RoboCupJunior which has been mentioned above — the ULEAGUE. This league takes the existing RoboCupJunior 2x2 soccer game and combines it with the Small-Size League (F180), simplifying issues of vision and communication. It was demonstrated successfully in Padova 2003 with teams from four countries (USA, Canada, Australia and Germany) and is reported in [1].

This ULEAGUE was a topic of much discussion at the Blaubeuren meeting. There was some concern that the name is a misnomer. Many, many senior league RoboCup teams are composed partly, even primarily, of undergraduate students. So calling this new challenge the ULEAGUE and emphasizing that it is for undergraduates is perhaps not consistent with existing practices. There was discussion about merging the existing Small-Size League (Smallsize League) with the ULEAGUE, as it appears that there will be many changes to the Small-Size League setup over the next year or two. Out of the 24 teams that compete in Smallsize League, there are apparently only 8 who consistently perform well, year after year. For these teams, it is appropriate to move to a larger field, begin to move to local vision, to remove special lighting and to add more robots to the field. But the other 16 teams are not ready to meet these challenges.

This opens up the question of the role of RoboCup and RoboCupJunior. If the purpose of RoboCup is to advance the state-of-the-art in Artificial Intelligence and Robotics research, then it is not in the interest of the initiative to hold back advances and "wait" for the masses. On the other hand, if the leagues progress too fast, then they will only be accessible to the elite. If the purpose of RoboCup is to bring together researchers from different fields to work together to achieve a common goal, then it is not in the interest of the initiative to push advances beyond the reach of the "masses".

The goal of RoboCupJunior has been to introduce the RoboCup initiative to young students. As well, once RCJ has succeeded in engaging students, somewhere in RoboCup, there must be a mechanism to keep them engaged, as they grow up beyond high school to undergraduate to graduate school. As the gap between "entry-level" and competitive level in each league widens, somewhere within RoboCup, there must always be a bridge from an entry-level to wherever the senior leagues are.

References

1. Anderson, J. and Baltes, J. and Livingston, D. and Sklar, E. and Tower J., Toward an Undergraduate League for RoboCup. In *Proceedings of RoboCup-2003: Robot Soccer World Cup VII*, 2003.

5 **Education** (by Daniele Nardi)

RoboCup devotes a special effort to Education, not only through the Junior League, but also by organizing specific initiatives, that are targeted to students at University level. The aim of these activities is to support the creation and strengthening of the technical and scientific background and skills, that a student needs to successfully participate in RoboCup.

The design and implementation of a team of autonomous soccer players or rescue agents, certainly relies on basic knowledge and skills that can be acquired during University undergraduate and graduate curricula in several branches of Engineering (e.g. Computer, System, Telecommunication, Electronic, Electric, Mechanical). However, in order to realize RoboCup teams (or better, contribute to the realization of RoboCup teams) technical knowledge on specialized issues, that are often not covered by the University curricula, is required. In addition, RoboCup provides a unique challenge to implement complex systems requiring knowledge on a wide set of disciplines and a large spectrum of technical capabilities. Finally, from an educational standpoint, the RoboCup framework is instrumental to train the students to a scientific development of ideas, which requires the ability to understand the technical knowledge available in the literature and use it as a basis for the development of original and more performant artefacts. The Education effort within RoboCup addresses all of the above aspects.

The main educational activity organized by RoboCup is the RoboCup CAMP. RoboCup CAMPs are directed to newcomers to help them entering RoboCup competitions both from a methodological and from a practical point of view; moreover, RoboCup CAMPs are also addressed to practitioners that need to address some of the technical challenges in a more systematic and solid fashion. The name "RoboCup CAMP" wants to convey the idea that a CAMP is not only a highly specialized school, but it requires the active participation of the students. The goal of RoboCup CAMPs is very ambitious, because the aim is to introduce some background knowledge on specific techniques that are used in the realization of a soccer/rescue team, while trying to fill all the steps to the actual implementation. Usually, the RoboCup CAMPs are targeted to specific leagues, and consequently focus on issues that are more relevant in that context. In addition, at the RoboCup CAMP the students are shown how novel ideas needed to improve on the state of the art techniques for specific problems can be effectively developed. Finally, the RoboCup CAMPs are used as opportunities for successful teams to present in a coherent and systematic way the techniques developed, far from the pressure of the competition.

The RoboCup CAMPs held so far are listed below (more information about them and access to the documentation can be found through the RoboCup web site):

- Padova (Italy), February 2000 - Mid size
- Paris (France), April 2001, Small and 4-Legged
- Paderborn (Germany), April 2002, Mid size
- Bremen (Germany), August 2003, Humanoid and Small size

RoboCup CAMPs in Japanese have also been held in conjunction with the Japan Open.

Another issue concerning RoboCup Education effort that deserves consideration is the use of RoboCup frameworks as a support to the teaching activity in several University courses, Summer Schools and tutorials. There are several examples of courses on Multi Agent Systems, that use the RoboCup soccer simulator as a testbed, as well as courses on robot constructions that focus on the robots of the small-size league. For example, the EURON Summer School on Cooperative Robotics (Lisbon 2002) had significant contributions using RoboCup as scenario. The references are available in the league web pages.

Finally, as a follow-up of the teaching based on RoboCup frameworks, textbooks collecting teaching material as well as teaching experiences based on RoboCup are forthcoming.

6 **Rescue League** (by Adam Jacoff)

The second annual RoboCup - Rescue Robot League competition, which took place in Padova, Italy, showed the gaining strength of our league. Twelve teams participated in a vigorous competition, a 50% increase in teams from the previous year, and we continue to raise awareness of the opportunities for robots in urban search and rescue (USAR) applications. Thus, we expect to maintain an aggressive growth rate for the 2004 event, hosting 16-20 teams in Portugal. In addition, we are actively engaged in efforts to expand the Rescue Robot League into the RoboCup national open competitions throughout the world. This year marked the first such rescue competition at the Japanese open, using the arenas fabricated for last year's Fukuoka competition. And a new year-round arena, hosted by Carnegie Mellon University in the USA, was used for rescue robot demonstrations at the first such American open event. The Italian rescue arenas, fabricated this year in Padova, are being set up at the Instituto Superiori Anticendi in Rome, a fire-rescue training facility, and will be available for year-round robot practice starting this winter. They may even host an Italian rescue event in the near future, either an Italian open competition or maybe a RoboCup camp devoted to Rescue Robot League research issues. Also, we are actively trying to get rescue arenas fabricated in Germany to host a rescue competition at the next German open event, and be available for year-round practice for central European researchers. The current site being considered to host these arenas is the International University in Bremen. So the league is expanding quickly due in large part to the enthusiastic response from researchers looking to test their robot's capabilities in the unstructured world of USAR applications, and work on the cutting edge of human-robot interaction for the betterment of disaster response.

Several changes to the league rules were initiated this year. One change discourages parallel teleoperation, where robot/operator control strategies are replicated within teams simply to inflate scoring. This year, specific starting positions were identified and sequential negotiation of the arenas was enforced, although teams could advance as far as they wanted through all three arenas. The first mission of each round started in the yellow arena, allowing direct comparison of navigation and mapping capabilities across teams. In subsequent missions, the teams were allowed to start directly into more advanced arenas to allow purpose built robots to highlight their specific capabilities. Also, false victim identifications were discouraged for the first time, so teams that mistakenly identified sensor signatures as signs of life suffered point reductions. These changes were generally appreciated by the teams, and produced a balanced competition that promoted the pertinent research issues while discouraging certain teaming strategies. Minor rules modifications proposed for next year may artificially limit the use of radio communications during missions to simulate radio signal dropout and interference that occurs at actual disaster sites. This would also encourage development of more autonomous behaviors and tether management systems, both very beneficial assets in eventual deployment systems. Also regarding radio communications, we may encourage a move toward the 5 GHz frequency range and 802.11a communications protocol to generally improve communications bandwidth and performance in the complex environments of our rescue arenas and avoid conflicts with other leagues at these large competitions (which only hints at the radio spectrum difficulties of a real disaster site).

Also this year, we began systematically capturing each team's human-robot interfaces (HRI) for subsequent analysis. Researchers from the National Institute of Standards and Technology (NIST) performed the data capture which included interviews with the operators, a workload assessment, and continuous video capture of robot performance. These interfaces will be analyzed for effective elements or combinations of elements and overall statistics will be published. This HRI analysis effort will be augmented next year with automatic position tracking of the robots throughout the arenas via a new ultra-wideband tracking system, also provided by NIST. Objective robot tracking data such as this, along with operator interface and workload analysis, will provide researchers with important measures of performance of their robots (and other robots), and help identify "best in class" algorithms, sensors, and mechanisms. Hopefully, this will further encourage collaboration around the most effective components and methods, and quicken the pace of technical advancement in the field.

As our league evolves, we are keenly aware of the urgent need for practical robotic solutions for disaster response. Toward this end, we have appointed the following Technical Committee members with distinguished, diverse backgrounds in robotics and disaster response to help steer our league: Dr. Andreas Birk (International University in Bremen, Germany), Dr. Ali Meghdari (Sharif University of Technology, Iran), Dr. Ted Sumrall (President, Counter Terror International, USA/Japan).

In recent times, it has become ever clearer that robots are needed to support first responders and rescue professionals at disaster sites. Many nations are supporting this endeavor, and Japan is among the leaders. Since their disastrous Hanshin-Awaji earthquake near Kobe (and others), they have aggressively supported research and development of robots for search and rescue applications. In 2002, the Japanese Ministry of Education, Culture, Sports, Science, and Technology (MEXT) started a five year project specifically focused on earthquake disaster mitigation in urban areas aimed at developing advanced robots and information systems for disaster response. The International Rescue System Institute, headed by our league chair Dr. Tadokoro, is one such example in that effort, supporting over forty research projects within Japan.

Efforts such as these will provide the funding required to push the technology and methods forward quickly. Evaluating the progress of these research efforts, and encouraging collaborations between organizations to better leverage advances, is where our league will play it's most important role. Our arenas represent standard, representative problems for the community at large. Our metric provides object evaluation of performance and encouragement toward clearly needed capabilities. And our competitions provide intensive, periodic development efforts and collaboration opportunities as teams react to the representative rescue situation at hand and attempt to follow increasingly realistic operational procedures adopted from at actual disaster sites. Practice sessions such as this, without risk to life or robot, can hardly be over valued. And it can play a pivotal role in increasing the rate of advancement in robotic capabilities.

As robot teams begin demonstrating repeated successes against the obstacles posed in the arenas, the level of difficulty will be increased accordingly so that the arenas continue to provide a stepping-stone from the laboratory to the real world. Meanwhile, the yearly competitions will provide direct comparison of robotic approaches, objective performance evaluation, and a public proving ground for field-able robotic systems that will ultimately be used to save lives.

7 From the Discussions Concerning Smallsize League, Midsize League and Humanoid League (Using Materials from the Slides by Gerhardt Kraetzschmar, Thomas Christaller, and Changjiu Zhou)

Smallsize League discusses changes (individual vision, larger field etc.) which makes this league closer to Midsize League. Smallsize league in its recent form is needed for beginners (cf. B.Sklar's section on Junior League above), while scientific challenges could be better pursued using the rules and settings of Midsize League. Moreover, Midsize robots will be smaller in the future. Planned and/or discussed changes in Midsize League concern:

- Less well-defined lighting.
  No special lighting (maybe even limited natural light influence) in 3-5 years.
- New ways of ball manipulation.
- Increasing players-pre-field ratio (larger fields and smaller robots).
- Game instrumentation: referee box, tools for recording and evaluating game data. The referee box is intended to become a common tool for all leagues. Almost no human interference in 3-5 years.
- Behavioral constraints instead of size rules and shape restrictions.
- Activation of many FIFA rules (corner kicks, throw-ins, goal kicks, free kicks).

As a consequence of the discussions in both leagues, Midsize League and Smallsize League will merge in the next years.

Humanoid League will use Midsize League field in 2004 and perform new challenges (e.g. more complex walking, more complex kicks, passing for soccer and balancing for "Free Style"). True dynamic walk, run and jump are considered for the near future. They may lead to new technical solutions like artificial muscles, flexible joints, endo-sceleton construction The following roadmap is proposed:

2004: Balancing challenge, passing challenge, obstacle walk challenge.
2005: Match 1 vs 1, object following, multiple objects tracking.
2007: Match 2 vs 2, collision avoidance, safety issues.
2010: Cognitive issues, coordination of perception and locomotion.

A lot of different robots of humanoid type are expected for the future. Nevertheless, there should be a limited number of different competitions. One might think of

- Small-size Humanoid League (SHL) with global vision and focus on walking and kicking issues.
- Mid-size Humanoid League (MHL) with local vision and focus on integration.
- Humanoid simulation (cf. M.Riedmillers section on Simulation League below).

A lot of problems are common to all leagues, and many of them are already covered by the SIGs (the below). Common challenges are considered as useful means for common work.

8 Future Directions for the Four-Legged League (by Claude Sammut)

The distinguishing feature of the four-legged league has been that all teams use a common hardware platform, the Sony Aibo. Since the platform is fixed, the teams are freed from hardware design concerns and are able to concentrate on software development. The common platform also means that teams are easily able to share programs. These features have allowed the league to progress rapidly since new teams can quickly become competent by using previous code as examples for their own development and experienced teams are able to understand, in detail, how other competitors have solved similar problems. Code sharing is an essential part of the four-legged league and should remain so for the foreseeable future. Thus, it is important to continue using a common hardware platform and associated operating system.

As in all RoboCup leagues, the intention of the four-legged league is to progressively handle more natural environments and develop cooperation amongst teams of robots. This effort is leading to improved vision and localisation algorithms; a better understanding of information sharing between team members and faster and more stable legged locomotion. An important side effect of this research is that new software tools and methodologies are being investigated, including simulation, higher-level programming environments and machine learning.

The common platform poses its own problems for the organisation of the league. The Sony Aibo is a highly sophisticated robot available to teams at a relatively modest cost. The league has received exceptional support from Sony but inevitably, new models will be produced and older models discontinued. Thus, teams must update their platform every two or three years. To avoid too costly a transition, it may be necessary to tolerate some diversity in the models of robots. However, to preserve the ability to share code, this diversity should minimised.

Considerations for the future are that if a humanoid robot becomes available at a reasonable cost would a new league be formed along the same lines as the four-legged league? What would be the relationship of this league with the present humanoid league and would there be a reason to continue the four-legged league? Considering the difficulties that most leagues have with unpredictable lighting, at what point can we consider outdoor games?

9 Simulation League towards 3D (by Martin Riedmiller)

The major novelty in the soccer simulation league is the development of a new simulator.

The simulator system will be built from scratch and is currently under development. Its main goal is to provide a more realistic simulation of real robots in order to bridge the gap between the real-robot leagues and simulation league. The concept comprises a 3D modelling of the environment, a modular approach that allows to design individual robot actuators and sensors, a more realistic handling of timing issues and collisions. The simulation of the physics will be based on the ODE library.

Due to the modularity and flexibility of the new simulator, it will be possible to eventually simulate even very specialized robots. Therefore, the core simulation system could eventually become the base for simulation of robots through all the leagues - including the humanoid leagues.

A main challenge for the transition phase will be to get the level of abstraction right. The simulator league must still focus on its main goals - the development of scientific approaches for mid- and high-level control (e.g. multi-agent coordination) - without bothering too much on low level (close to hardware) problems. In the ideal case, the new simulator will offer various interfaces that allow to tackle control problems on various levels of abstraction, ranging from a close-to-hardware view to a reasonably abstract view that directly allows to deal with mid-level and high-level issues. Providing such a high-level interface (comparable to the one of the current soccer server interface or even more abstract) will also raise the attractiveness for researchers that are more interested in AI than in robotics or control.

The current schedule is

2004, January : First version of new soccer simulator available, discussions.
2004, June: Tournament at RoboCup 2004, in parallel to a 'classic' (2D) simulator tournament.
2005: Tournament at RoboCup 2005, in parallel to a 'classic' (2D) simulator tournament.
2006: New simulator becomes the standard soccer simulation system

The weighting of the classic (2D) and the 3D simulator competition in 2004 and 2005 will depend on the status of the implementation and the number of teams participating for the respective competition. As an outlook for future directions: With simulated humanoid tournaments, Simulation league will become close to real humanoid robots.

10 Vision SIG (by Andrea Bonarini)

Vision is the primary input for robots, not only in RoboCup, but also in many applications involving mobile and fixed robots. The vision task in real world mobile robot applications should be performed in a short time, so to provide input to the control system, but at the same time should contribute to a reliable and rich world model.

RoboCup offers an important testbed in all the real-robot leagues to test vision algorithms and sensors.

The RoboCup community is working on vision aspects as part of the whole activity of implementing playing robots. The primary aim of this SIG is to support the research activities concerning perception by vision systems in RoboCup. In all the leagues working with real robots, people has to implement effective, real-time, vision systems, facing many different problems which are also of great interest for applications outside RoboCup.

Among the faced topics are:

- color vision,
- real-time algorithms for image analysis, object recognition, localization, and panoramic (omnidirectional) vision,
- stereo (multicamera) vision,
- multi-sensor fusion.

To promote research in vision within and outside RoboCup, the SIG has started the following activities.

- Maintaining a mailing list. A mailing list for this SIG has been created as vision@RoboCup.biglist.com. You may connect to http://www.RoboCup.biglist.com/vision/ and follow the instructions to be added to the list. You may also find there archives of past messages.
- Maintaining a web-based repository of data and tools, available as http://RoboCup.elet.polimi.it/SIG-Vision/.
- Promoting and maintaining a forum to discuss and support the development of vision related topics.
- Organizing vision workshops and special sessions, together with researchers from outside of RoboCup, at major conferences like CVPR, ECCV, ICPR.
- Organizing special events at RoboCup workshops/games. The SIG will work with organizing committees to organize annual events that emphasize research in vision.

- Encouraging community-based development of general, re-usable, code and standards, to facilitate comparative evaluation and to accelerate research, mainly for newcomers on this topic. See the repository and the forum on the web site.
- Proposing rule modifications to steer research within RoboCup, for instance reducing the role of color, or improving the role of other features.

At the last meeting of the SIG in Padova, people present have agreed to share code and experience in developing vision systems used in RoboCup. The web site has been updated accordingly. In particular, we have decided to host also a repository of open questions and problems, so that newcomers or people not wishing to be too much involved in vision could find help from the SIG.

We also have decided to focus on some problems, stating a sort of forum to brainstorm about their possible solutions and to share experiences. The problems currently on the table are mentioned here below.

- Adaptive color classification. Probably, from 2004 many leagues will play on fields with uneven light coming from different sources. Adaptive color calibration may play a key role to face this issue, and its real-time solution is still an important open problem in the whole vision community, with important impact on industrial applications
- Spot light and shadows. On the way to play with natural light, coming from a directional source, the sun, it may be interesting to work with a single spot light, providing, for instance, shadows, which may be considered an important element instead of noise.
- Knowledge and interface between sensors and world model. How to interface the vision system to behaviors? Which kind of information is passed? Would it be useful to include knowledge in the vision system or this should only provide raw data and leave the conceptualization to a world modeller? Is a conceptualization useful at all?
- Vision-based self localization. Is self-localization needed in RoboCup? Why? How do you self-localize your robots? Do you fuse data from different sensors? Which kind of algorithms do you use? Do you merge information from different robots?

Finally, we have decided to promote the implementation of benchmarks on the above and maybe other topics, so to focus and base research with a scientific approach, which is often forgot in the competition activities.

A suggestion of the SIG to the Executive Committee has been the implementation of specific competitions, or scientific challenges common to different leagues, so to provide a stimulus to focus research activities on specific, relevant issues.

11 Multiagent Learning SIG (by Use of Communication with Peter Stone)

Participants from all the soccer leagues and at least rescue simulation are participating. Some benchmark tasks for learning were proposed and discussed with the goal of making them usable by people outside of RoboCup. Potential challenges were proposed that could be applicable across the leagues.

- The keepaway task is already used for learning in simulation, and may be immediately possible in the Smallsize League. Hopefully it can be incorporated into other leagues as well.
- A multiagent goal-scoring task has been proposed and implemented in a Midsize League simulator (by Alex Kleiner).
- A further proposal could concern a challenge task for vision learning that would require teams to be able to automatically calibrate their vision. They would need to send in their code for the challenge task BEFORE seeing the lighting conditions at RoboCup.

A big challenge is defining tasks that not only CAN be learned, but that REQUIRE learning for success. There is a common understanding that a RoboCup team with learning has advantages over one without. But the best way to create a competent team quickly is still to hand-code it. That's true for most subtasks that we can think of. By 2010 we hope to have a well-defined and popular suite of challenge tasks both for RoboCuppers and for the general ML community who are not RoboCuppers.

12 Other SIGs (by Use of Notes from Thomas Christaller)

SIG Configurable and Modular Robotics:
A web page using wiki-web software was installed to make it easy for everybody to contribute without any administrative overhead. It serves for e.g.

- Setting up shopping infos in a data base.
- Mailing list.
- Steering committee.

SIG Simulation Tools for Real Robots:
Many teams make use of simulation tools. Current simulation league simulator is not useful to robot teams (but cf. M. Riedmiller's section on Simulation League above). Research challenges are e.g.

- realistic dynamics,
- configurability, extensibility,
- vision as a primary sensor.

The major goal is an open source simulator with ODE as technical basis. It is intended to be useful outside RoboCup, too.

13 Final Remarks (by Hans-Dieter Burkhard)

The workshop took place in the beautiful little village Blaubeuren, near Ulm in the southern part of Germany. Blaubeuren has a well-preserved medieval architecture, and it is especially known by the circular Blautopf ("Blue Pot"), the legendary underground source of the river Blau. This karst spring is among Germany's most mysterious nature prodigies, and it is the setting for the "Legend of the Beautiful Lau". Thanks goes to Gerhard Kraetzschmar for proposing and organizing the meeting in Blaubeuren.

Two days if intensive work helped to clarify recent developments and to outline future challenges and requirements. The main goal of RoboCup is scientific research, hopefully with useful applications. In several countries, funding is directly connected with impacts for such applications. In fact, the aim of our community is not to build special purpose soccer machines, but to come up with new results and new solutions for a broad range of problems. The competitions serve for evaluation and for demonstration of successful developments. The vision of the 2050 year goal serves for the identification of problems which are important for Robotics and AI. Therewith, RoboCup stands for a community with a longterm project.

Thomas Christaller has compiled a list of general capabilities that a humanoid soccer robot has to met for the 2050 year goal:

- Playing over two times 45 minutes plus possible extensions of 2 times 15 minutes.
- Running more then 20 km during one game.
- Playing under severe weather and ground conditions sunshine, rain, snow, slippery, uneven ground, different and changing ground conditions.
- Controlling the ball with nearly all parts of the body (excluding arms and hands).
- Jumping, falling down, touching, body check.
- Artistic movements lay persons are uncapable to do.
- Size, weight, and force similar to an adult man (170cm, 65kg, 100m/12sec, 20m/sec ball speed).
- Forecasting and recognizing intention of movements before it is "manifest".
- Knowing team members individually.
- Knowing members of opponent teams.
- Knowing a lot of past/historical games.

Nobody knows today, if these problems can be solved at all. Most difficult problems may be

- Body construction and energy.
- Body control towards artistic/professional movements.
- Forecasting intended behaviour/movements of other players. (opponent as well as team members)

RoboCup research and competitions will help to clarify the related problems step by step. That forces us to define new goals year by year. The roadmap will be further discussed on the symposium in Lisbon 2004, and the next workshop of the Executive Committee is planned for autumn 2004.

Challenges in Robust Situation Recognition through Information Fusion for Mission Critical Multi-agent Systems*

Hiroaki Kitano[1,2,3]

[1] ERATO Kitano Symbiotic Systems Project, Japan Science and Technology Corporation, Suite 6A, 6-31-15 jingumae, Shibuya Tokyo 150-0001 Japan
kitano@symbio.jst.go.jp
[2] The Systems Biology Institute
Suite 6A, 6-31-15 jingumae, Shibuya Tokyo 150-0001 Japan
[3] Sony Computer Science Laboratories, Inc.
3-14-13 Higashi-gotanda, Shinagawa, Tokyo 141-0022 Japan

Abstract. The goal of this paper is to highlights one of emergent scientific issues in RoboCup task domains that has broader applications even outside of the RoboCup task domains. This paper particularly focuses on robust recognition through information fusions issue among numbers of others issues that are equally important. The robust recognition through information fusion is selected because it is one of the most universal issues in AI and robotics, and particularly interesting for domains such as soccer and rescue that has high degree of dynamics and uncertainty, as well as being resource bounded. The author wish to provide a conceptual framework on robust perception from single agent to multi-agent teams.

1 Robustness and Information Fusion

The RoboCup project is an interesting research platform because it has multiple domains that have both difference and similarities in basic domain features. Soccer is a dynamic game with real-time collaboration within teammate, but adversarial against opponents. Major uncertainties are generated by (1) opponent strategies, (2) uncertain and stochastic nature of physics on ball movements and other factors that affect course of each action, and (3) uncertainty of real-time behaviors of each player. On the contrary, rescue domain is a dynamic and mission critical domain with high degree of uncertainty and partial information under hostile terrain that are totally different in each case. Major uncertainties are generated by (1) unknown terrain and victim information, (2) uncertain and stochastic nature of incoming information, success of each operations, numbers of external perturbations, and other social and political factors, and (3) limited information on individual victims and their situations.

* This work was partially supported by the Air Force Office of Scientific Research/ Asian Office of Aerospace Research & Development, under the contract number AOARD-03-4035.

D. Polani et al. (Eds.): RoboCup 2003, LNAI 3020, pp. 35–42, 2004.

It is interesting to note that despite differences in the task domain, there are substantial commonalities in structures of uncertainty. There are uncertainty and limited information at the macroscopic level at the level of entire terrain or theater of operation and at the microscopic level which is the scale of individual players or victims. In addition, there are issues of unknown and unpredictable perturbations throughout the operation.

In order to best accomplish the task, a team (or teams) of agents, either robotics or informational agents, need to be robust in perceptions and actions, as well as their team behaviors. It should be well coordinated to reconstruct a model of the terrain and other agents in the scene against various noise, uncertainty and perturbations, so that effective actions can be taken. A set of actions need to be robust so that failure of one or more of such actions do not leads to catastrophic failure of the overall mission.

Robustness of the system is generally exhibited as capability of the system to (1) adapt to environmental fluctuation, (2) insensitivity against fluctuations in system's internal parameters, and (3) graceful degradation of performance, as opposed to catastrophic failure of the system.

This applies from sensory level to higher multi-agent team level. A brief example at the team level shall make clear what does robustness means. For the soccer team, this means that the team should be able to cope with differences in strategy and physical performance of opponent team, not being seriously affected by changes in fatigue and other factors of players, and removal of one or more players does not results in complete loss of its team performance. For the rescue team, it means that the team can cope with various different disaster scenario and dynamical changes in the situation, ability of cope with unexpected fatigue, damage, and resource shortage, and capability to carry out missions even if some of its rescue teams have to be withdrawn from the scene. For the rescue team that has to cope with extremely hostile environment, robustness is one of the essential features of the system.

Robustness of the system is usually attained by (1) the use of proper control, such as negative feedback, (2) redundancy, or overlapping functions of multiple subsystems, (3) modular design, and (4) structural stability.

Extensive research has been made on properties of robustness in biological systems, and these traits were shown to be universal using numbers of examples, including bacteria chemotaxis, robustness of gene transcription against point mutations and noises, stable body segment formation, cell cycle and circadian period, etc [4, 5].

Bacteria, for example, swim toward chemoattractants by sensing graduent of concentration. This capability is maintained regardless of concentration level, steepness of the gradient, and keep track of graduent changes consistently. Integral feedback has been identified as a key intrinsic mechanism in which bacterial behaviors are controlled by activation of receptor complex, but deactivated by a negative feedback loop with integral components. This feedback control enables behavior of bacteria dependent on the level of concentration changes took place, but independent of absolute concentration level of chemical in environment

[1, 10]. Similar mechanisms are observed widely amoung different speceies. Feedback control is only one of several mechanisms behind biological robustness.

On the contrary, artificial systems tend to be less robust, and reply on rigid build-in design that may easily fail under fluctuations. How to build robust systems from sensory-level to strategy-level in a consistent manner is one of the major issues in RoboCup research.

In the rest of the paper, possible research directions for robust systems particularly focusing on information fusion aspects are discussed. Information fusion is raised here because it is relatively universal in different domains, and critical for strategic actions to follow. Issues of robust strategic decisions and executions will be the other robustness issues in multi-agent teams, but will not be discussed due to space limitations.

2 Dimensions of Information Fusion for Robust Systems

Information fusion for robust systems has to be considered for multiple aspects:

Abstraction: An interactive processing of different abstraction levels, such as interactive processing of low-level sensory information and high-level recognition and strategy, enhances robustness by providing interlocking feedback of information thereby hypotheses may converge to more plausible one.

Multiple Sensory Channels: Integration of multiple modal perception channels can contribute to improve robust perception by complementing missing information by perception channels with different characteristics.

Perception-Action Loop: Active involvement of probing actions into perception, that is an integration of perception-action loop to enhance recognition by actively manipulating the environment so that ambiguity can be resolved.

Spatio-Temporal Distribution: Integration of spatio-temporally distributed sensory information, as well as absteacted symbolic information is essential to create overall picture of the situation, thereby robust adaptation to the situation can be done with overall re-evaluation of the situation.

Figure 1 illustrates first three dimensions of information fusion for robust perception.

Information fusion in these aspects contribute robust perception of theater of operation through one or more of four mechanisms of robustness.

3 Interaction of Low-Level and High-Level Perception and Actions

Given the multi-scale nature of the domain that requires identification of situation at both macroscopic and microscopic levels, distributed and coordinated information fusion need to be performed that are ultimately combined to create a coherent model. Information fusion at the macroscopic level is called the high level information fusion (HiLIF) and that of the microscopic level is called sensory level information fusion (SLIF).

In the sequential model of AI, a famous Sense Model Plan Act cycle (SMPA cycle) has been used. This paradigm has been criticized as not being able

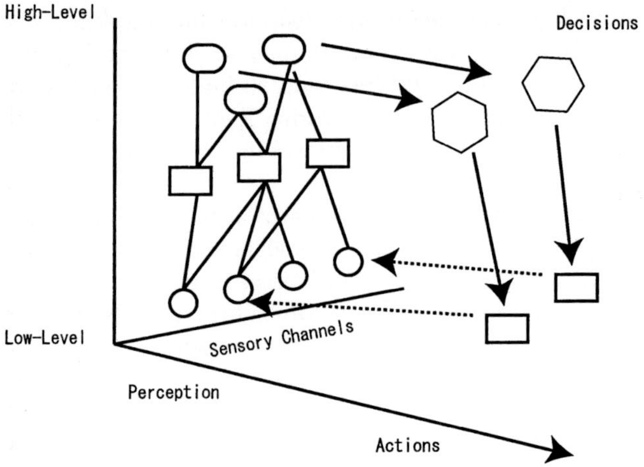

Fig. 1. Integration at Abstraction, Perception channel, and Perception-Action loop

to respond to environment in real-time, and an alternative approach named "behavioral-based approach" has been proposed [3]. While the behavior-based AI demonstrated effectively how simple robotics systems and virtual agents can behave in the real world without creating an internal model, it never scaled to perform complex tasks. In both soccer and disaster rescue, coupling of hierarchy of sensing and actions from low-level sensory units to high-level strategy generators is essential. Reactive control of leg movement to kick a ball must be coordinated with strategic choice of attacking patterns and a pass trajectory to enable such a strategy. Similarly, local actions and decision for the recovery of some critical life-lines has to be carefully coordinated with overall strategy. In this context, low-level perception and action module is not merely behavior-based module. It must be able to recognize local situation that can be aggrigated at a higher level. SLIF should have a certain level of internal model.

While these examples are how overall strategy may constrain local actions, there are cases local actions influence overall strategy. Basically, it is interaction of bottom-up and to-down processes in which the architecture enabling it has been long standing theme in multi-agent systems. While this is well recognized issue, the author would not make further discussions than stating that RoboCup task domains, particularly humanoid league and rescue, are one of ideal platform for seriously tacking this problem. This aspect of integration essentially exploits feedback control of the system for adaptation at certain abstract levels. Low-level perceptions and local decisions are aggrigated to higher-level that feedback differences between desired local situations and actions and actual situation and actions to reduce such descrepancies.

4 Information Fusion of Multiple Sensory Channels

Navigation, localization, and object identification of robotic agents tends to rely on visual information. While vision provides rich information, it is limited in several aspects. First, visual perception can be occluded by obstacles, including dust and smokes. Second, it has to have certain quality of light sources. These features impose serious limitations of visual perception for disaster rescue situation, because disaster scenes are generally highly unstructured, dusty, and may have serious smokes from fire. Lights are often not available in confined environment where victims may be embedded. Auditory perception, on the other hands, has different characteristics. It can transmit over obstacles, do not require light sources. In fact, various noises victims may make is a critically important signals for a victim location identification. Other sensory channels, such as odorant, CO2, and vibrations have different characteristics that complement each other. Information of multiple modalities of perceptions may provide high level of robustness in perceiving the environment. Some of early efforts have been done using integration of auditory and visual perception [6–9]. Vision system often generates false positive identification of objects that is supposed to recognize due to similarity of color and shape in irrelevant objects. When the object is creating certain auditory signals, the use of auditory information to track the sound stream can effectively eliminates false positives. By the same token, objects could be occluded by obstacles so that vision system lost tracking, which can be compensated by keep tracking auditory signals if the object is making some sound streams. Early experiments indicate that well designed integration of multiple modal perception channels based on multiple sensory steam integration is effective for robust perception. Now, the research topic shall be to create basic principle of robust object identification, tracking, and scene understanding by using multiple perception channels, most likely by defining a dynamical defined invariance that corresponds to each object as its signature.This approach is expected to attain robustness by exploring redundancy, or overlapping functions, of perception channels, so that degradation or failure of one perception channel is well compensated by other perception channels with overlapping functions.

5 Integrating Actions to Perception

It is important that the concept of active perception to be integrated into the system, so that ambiguities of information can be resolved by actively directing sensory devices and information gathering agents. Early research of this direction has been proposed as active vision and animated vision [2], but it has to be extended to include not only vision and other perception channels, but also to more high level information collections. Conceptually, this is feedback control to minimize unknown, or ambiguous part of scene to zero.

In the resource constraint situation such as in disaster rescue, the concept of cost of active probing has to be introduced. Any action to obtain information is associated with cost, the use of resources, including time. Decision has to be made on whether actively probe new information or to proceed with ambiguities.

Suppose that a ball that is rolling from right to left is occluded by an opponent player, decision has to be made to use predictions of ball trajectory or actively probe the ball position. While actively proving the ball position may resolve ambiguity of information, it may loose time window to intercept the ball.

By the same token, in the disaster scenario, spreading of fire or cascading collapse of buildings may not be fully monitored by available sensors or information agents. A tactical decision has to be made to dispatch a unit to counter such incidence by predicting possible front of chain reactions or to mobilize information gathering agents to make sure the status of incidents. The cost of information gathering is the use of additional agents and time to wait until the situation to be disambiguated. On the contrary, a quick deployment of counteraction units runs a risk that the prediction is false and deployments are deemed ineffective. Always, there is a trade-off between cost of knowing and risk of not knowing. One of the research topics may be to find out principles of decision on cost of knowing versus risk of not knowing.

6 Spatio-temporal Integration

6.1 Heuristics and Knowledge-Based Estimation

Integration of spatio-temporal information is essential, particularly for disaster rescue operations. Theater of operation is widely distributed, and information is collected only at limited rate from limited locations. Basically it is a problem of making the best estimate of the situation by sparse sampling of complex terrain in 4-D (XYZ+T) space. Certain heuristics, such as continuity of the unfolding events, and a priori on structure of urban infrastructure are expected to be highly useful to constrain possible hypotheses. This applies to both low-level and high-level information fusion, but particularly useful for high-level information fusion where "fog of war" has to be resolved as soon as possible with limited resources. Advantage of this approach is that you can make reasoned estimate of the situation even for the area that cannot be directly measured. The drawback is that it requires substantial knowledge of the urban structures in usable form that are generally not available. At the same time, how to increasing sampling points is the other big issue. One of the best ways is to make sure sensory systems are ubiquitously present in disaster scene. This can be achieve only by creating multi-functional systems that are useful in daily life, but can act as sensory units in emergency. Traffic signals and various monitoring cameras are possible resources that can cover public space. Home security systems, home entertainment robots, and a series of home electronic products are ideal for covering situations in each household. However, there are issues of securing telecommunication with such devices, as well as protection of privacy that are critical, but are outside of AI and robotics issues.

6.2 Adaptive Airborne Spatial Information Fusion

One of possible approach to solve this problem is to develop a small disposable sensory unit and deploy them in large numbers. Figure 2 illustrates one example of such systems which could be deployed airborne. The goal is quickly understand target terrain/city situation for the purpose of disaster rescue.

Phase-I: Large number of small IF (information fusion) devices will be deployed mid-air

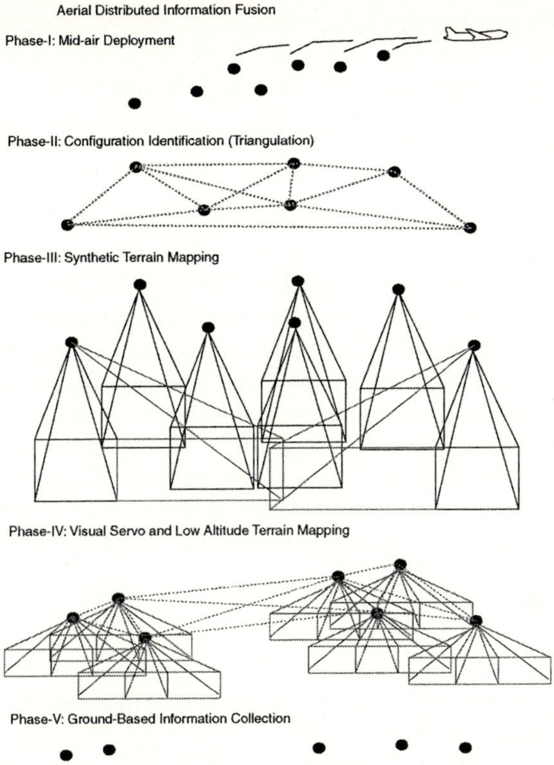

Fig. 2. Aerial Deployment of Small Information Fusion Units for Disaster Area Scanning

Phase-II: Speed break is activated and speed is stabilized. Then each unit has some visual beacon or visual ID so that relation location of units can be determined by triangulation.

Phase-III: First aerial photo/video will be taken using multiple camera unit, but mostly using cameras that are facing terrains below. and send back to airborne server or other servers. Focus of attention is determined and rough model will be created. If processing power is sufficient, this can be done in each unit. Infra-red sensors or other sensors can be used in addition to CMOS imager to identify specific signature on terrain.

Phase-IV: Each unit deploys small and simple flaps to control trajectory, so that visual servo will be made effective to regroup IF units, so that areas that are more significant will be assigned with dense IF units. Photo/video analysis and reconstruction continues. In low altitude, all 360 angle camera as well as microphone and other sensors are activated to capture maximum information on the ground. If possible, frame rate will be increase dramatically.

Phase-V: For those units that safely reached the ground and survived impact, 360 degree vision and microphone systems, as well as other sensors will be activated to identify objects and terrain structure.

This approach integrates self-organization of agents and sensory feedback on-the-fly, and can be enormously useful for rapid deployment at disaster site, as well as being an excellent test of modern AI techniques.

7 Conclusion

This paper addressed several issues in robust systems for RoboCup domains. Several aspects of robustness have been identified, and four aspects of information fusion have been discussed briefly. Both soccer and rescue provides an excellent platform for such research, and several research issues have been raised. Integration of three dimensions (abstraction levels, perception modality, and perception-action loops) of information fusion poses particularly interesting problems that community shall tackle. Spatio-temporal integration applies to both soccer and rescue, but more seriously to rescue scenario. Resolving this issue requires not only improvement of each robotic and AI agents, but also how such systems are deployed before and after the onset of the disaster. A systematic approach for various levels of information fusion is necessary for robust perception of multi-agent teams.

References

1. Alon, U., Surette, M., Barkai, N., and Leibler, S., (1999) "Robustness in bacterial chemotaxis," *Nature,* 397(6715):168-71
2. Ballard, D., (1989) "Reference Frames for Animate Vision," *Proc. of International Joint Conference on Artificial Intelligence* Detroit.
3. Brooks, R., (1986) "A robust layered control system for a mobile robot," *IEEE J. robotics and Automation,* RA-2, 14-23.
4. Kitano, H.,(2002) "Systems biology: a brief overview, *Science,* 295(5560):1662-1664
5. Kitano, H., (2002) "Computational systems biology, *Nature,* 420(6912):206-210
6. Nakagawa, Y., Okuno, H.G., and Kitano, H. (1999) "Using Vision to Improve Sound Source Separation," *Proceedings of the Sixteenth National Conference on Artificial Intelligence (AAAI-1999),* pp.768-775, AAAI, Orlando, Jul. 1999.
7. Nakadai, K., Lourens, T., Okuno, H.G., and Kitano, H. (2000) "Active Audition for Humanoid," *Proceedings of the Seventeenth National Conference on Artificial Intelligence (AAAI-2000),* 832-839, AAAI, Austin, Aug. 2000.
8. Nakadai, K., Hidai, K., Mizoguchi, H., Okuno, H.G., and Kitano, H. (2001) "Real-Time Auditory and Visual Multiple-Object Tracking for Robots," *Proceedings of the Seventeenth International Joint Conference on Artificial Intelligence (IJCAI-01),* IJCAI and AAAI, Vol.2, 1425-1432, Seattle, Aug. 2001.
9. Nakadai, K., Hidai, K., Okuno, H.G., and Kitano, H. (2002) "Real-Time Speaker Localization and Speech Separation by Audio-Visual Integration," *Proceedings of IEEE/RSJ International Conference on Robots and Automation (ICRA-2002),* IEEE, pp.1043-1049, Washington DC., May 2002.
10. Yi, T., Huang, Y., Simon, M., and Doyle, J., (2000) "Robust perfect adaptation in bacterial chemotaxis through integral feedback control," *Proc. Natl Acad Sci USA,* 97(9):4649-4653

On Role Allocation in RoboCup*

Brian P. Gerkey[1] and Maja J. Matarić[2]

[1] Robotics Laboratory
Stanford University, Stanford CA 94305, USA
gerkey@robotics.stanford.edu
http://robotics.stanford.edu/~gerkey
[2] Computer Science Department
University of Southern California, Los Angeles CA 90089, USA
mataric@cs.usc.edu
http://robotics.usc.edu/~maja

Abstract. A common problem in RoboCup is *role allocation*: given a team of players and a set of roles, how should be roles be allocated to players? Drawing on our previous work in multi-robot task allocation, we formalize the problem of role allocation as an iterated form of *optimal assignment*, which is a well-studied problem from operations research. From this perspective, we analyze the allocation mechanisms of a number of RoboCup teams, showing that most of them are greedy, and that many are in fact equivalent, as instances of the canonical Greedy algorithm. We explain how *optimal*, yet *tractable*, assignment algorithms could be used instead, but leave as an open question the actual benefit in terms of team performance of using such algorithms.

1 Introduction

Over the past decade, a significant shift of focus has occurred in the field of mobile robotics as researchers have begun to investigate problems involving multiple, rather than single, robots. From early work on loosely-coupled tasks such as homogeneous foraging (e.g., [1]) to more recent work on team coordination for robot soccer (e.g., [2]), the complexity of the multi-robot systems being studied has increased. This complexity has two primary sources: larger team sizes and greater heterogeneity of robots and tasks. As significant achievements have been made along these axes, it is no longer sufficient to show, for example, a pair of robots observing targets or a large group of robots flocking as examples of coordinated robot behavior. Today we reasonably expect to see increasingly larger robot teams engaged in concurrent and diverse tasks over extended periods of time.

As a result of the growing focus on multi-robot systems, multi-robot coordination has received significant attention. In particular, *multi-robot task allocation*

* This paper is based in part on "A Framework for Studying Multi-Robot Task Allocation" by B.P. Gerkey and M.J Matarić, appearing in Schultz, A., et al., eds.: *Multi-Robot Systems: From Swarms to Intelligent Automata, Volume II*, Kluwer Academic Publishers, the Netherlands (2003) 15–26.

D. Polani et al. (Eds.): RoboCup 2003, LNAI 3020, pp. 43–53, 2004.

(MRTA) has recently risen to prominence and become a key research topic in its own right. As researchers design, build, and use cooperative multi-robot systems, they invariably encounter the fundamental question: "which robot should execute which task?" in order to cooperatively achieve the global goal.

In the RoboCup domain, this problem is usually referred to as *role allocation*, with the concept of a time-extended "role" replacing that of a transient "task". Regardless, the underlying problem remains the same, so we will use the terms task and role interchangeably throughout the paper. As in most other multi-robot systems, the question of how to assign roles in RoboCup is of significant importance for most teams, as it forms the very core of their teamwork strategy[1]. However, the methods demonstrated to date remain primarily *ad hoc* in nature, and relatively little has been written about the general properties of the role allocation problem. The field still lacks a prescription for how to design a role allocation mechanism and there has been little attempt to analytically evaluate or compare the proposed techniques. They have, of course, been extensively evaluated *empirically*.

In this paper we attempt to fill this gap with a formal discussion of role allocation in RoboCup. Our goal is to provide a framework in which to understand the underlying problem, as well as existing solutions. solutions. We show how role allocation can be cast as an iterated form of the long-studied Optimal Assignment Problem (OAP) [4]. In this light, we find that many allocation mechanisms are in fact algorithmically equivalent, as instances of the canonical Greedy algorithm, and thus have known worst-case bounds on solution quality. We also explain how *provably optimal* allocation techniques could be used instead, with a negligible increase in computational overhead. We leave as an open question the actual performance benefit of choosing an optimal allocation algorithm over a greedy one.

The rest of this paper is organized as follows. In the next section, we formalize the problem of role allocation as it is encountered in RoboCup, and discuss greedy and optimal algorithms for solving this problem. In Section 3, we provide a number of examples from the RoboCup literature of teams performing role allocation, and analyze their algorithms in the context of our formalization. We conclude in Section 4 with a discussion of future directions for this kind of analysis in multi-robot coordination problems.

2 The Problem

When studying the problem of multi-robot role allocation we take inspiration from operations research, a field that concerns itself with human organizations. In particular we claim that role allocation can be reduced to an iterated form of the *Optimal Assignment Problem* (OAP) [4], a well-known problem from operations research. A recurring special case of particular interest in several fields of study,

[1] Of course, not all teams use role-based strategies, nor do all role-based teams employ *explicit* role assignment (an example of *intentional* cooperation [3]); we restrict our attention in this paper to those teams that do both.

this problem can be formulated in many ways. Given our application domain, it is fitting to describe the problem in terms of jobs and workers. There are n workers, each looking for one job, and n available jobs, each requiring one worker. The jobs can be of different priorities, meaning that it is more important to fill some jobs than others. Each worker has a nonnegative skill rating estimating his/her performance for each potential job (if a worker is incapable of undertaking a job, then the worker is assigned a rating of zero for that job). The problem is to assign workers to jobs in order to maximize the overall expected performance, taking into account the priorities of the jobs and the skill ratings of the workers. This problem was first formally studied in the context of assigning naval personnel to jobs based on the results of aptitude tests [5].

Our multi-robot role allocation problem can be posed as an assignment problem in the following way: given n robots, n prioritized (i.e., weighted) single-robot role, and estimates of how well each robot can be expected to play each role, assign robots to roles so as maximize overall expected performance. However, because the problem of role allocation is a dynamic decision problem that varies in time with phenomena including environmental changes, we cannot be content with this static assignment problem. Thus we complete our reduction by iteratively re-solving the static assignment problem over time.

Of course, the cost of running the assignment algorithm must be taken into account. At one extreme, a costless algorithm can be executed arbitrarily quickly, ensuring an efficient assignment over time. At the other extreme, an expensive algorithm that can only be executed once will produce a static assignment that is only initially efficient and will degrade over time. Finally there is the question of how many roles are considered for (re)assignment at each iteration. In order to create and maintain an efficient allocation, the assignment algorithm must consider (and potentially reassign) every role in the system. Such an inclusive approach can be computationally expensive and, indeed, some implemented approaches to role allocation use heuristics to determine a subset of roles that will be considered in a particular iteration.

Together, the cost of the static algorithm, the frequency with which it is executed, and the manner in which roles are considered for (re)assignment will determine the overall computational and communication overhead of the system, as well as the solution quality. Before continuing with a formal statement of the role assignment problem as an instance of OAP, we first introduce the vital concept of *utility.*

2.1 Utility

Utility is a unifying, if sometimes implicit concept in economics, game theory, and operations research, as well as multi-robot coordination. The underlying idea in all fields is that each individual can somehow internally estimate the value (or the cost) of executing an action. It is variously called fitness, valuation, and cost. Within multi-robot research, the formulation of utility can vary from sophisticated planner-based methods [6] to simple sensor-based metrics [7]. In the RoboCup domain, it is common to compute utility as the weighted some of several factors, such as distance to a target position, distance from the ball, whether the team is on offense or defense, etc. We posit that utility estimation of this kind is carried out somewhere in every autonomous task or role allocation system.

Regardless of the method used for calculation, the robots' utility estimates will be inexact for a number of reasons, including sensor noise, general uncertainty, and environmental change. These unavoidable characteristics of the multi-robot domain will necessarily limit the efficiency with which coordination can be achieved. We treat this limit as exogenous, on the assumption that lower-level robot control has already been made as reliable, robust, and precise as possible and thus that we are incapable of improving it. We discuss "optimal" allocation solutions in the sense that, under the assumption that all information available to the system (with the concomitant noise, uncertainty, and inaccuracy) is contained in the utility estimates, it is impossible to construct a solution with higher overall utility; this notion of optimality is analogous to optimal scheduling [8].

It is important to note that utility is an extremely flexible measure of fitness, into which arbitrary computation can be placed. The only constraint on utility estimators is that they must each produce a single scalar value such that they can be compared for the propose of ordering candidates for tasks. For example, if the metric for a particular task is distance to a location and the involved robots employ a probabilistic localization mechanism, then one reasonable utility estimator would be to calculate the center of mass of the current probability distribution. Other mechanisms, such as planning and learning, can likewise be incorporated into utility estimation. No matter the domain, it is vital that *all* relevant aspects of the state of the robots and their environment be included in the utility calculation. Signals that are left out of this calculation but are taken into consideration when evaluating overall system performance are what economists refer to as *externalities* [9]; their effects can be detrimental, if not catastrophic.

2.2 Formalism

We are now ready to state our role allocation problem as an instance of the OAP. Formally, we are given:

- the set of n robots, denoted $I_1, \ldots, I_n$
- the set of n prioritized roles, denoted $J_1, \ldots, J_n$ and their relative weights $w_1, \ldots, w_n$
- U_{ij}, the nonnegative utility of robot I_i for role J_j, $1 \leq i, j \leq n$

We assume:

- Each robot I_i is capable of executing at most one role at any given time.
- Each role J_j requires exactly one robot to execute it.

These assumptions, though somewhat restrictive, are necessary in order to reduce role allocation to the classical OAP, which is given in terms of single-worker jobs and single-job workers. It is worth noting that in most existing role allocation work, especially in RoboCup, these same assumptions are made.

The problem is to find an optimal allocation of robots to roles. An allocation is a set of robot-role pairs:

$$(i_1, j_1) \ldots (i_n, j_n)$$

Given our assumptions, for an allocation to be *feasible* the robots $i_1 \ldots i_n$ and the roles $j_1 \ldots j_n$ must be unique. The benefit (i.e., expected performance) of an allocation is the weighted utility sum:

$$U = \sum_{m=1}^{n} U_{i_m j_m} w_{j_m}$$

We can now cast our problem as an integral linear program [4]: find n^2 nonnegative integers α_{ij} that maximize

$$\sum_{i,j} \alpha_{ij} U_{ij} w_j \tag{1}$$

subject to

$$\sum_{i} \alpha_{ij} = 1, \forall j$$
$$\sum_{j} \alpha_{ij} = 1, \forall i \tag{2}$$

The sum (1) is just the overall system utility, while (2) enforces the constraint that we are working with single-robot roles and single-role robots (note that since α_{ij} are integers they must all be either 0 or 1). Given an optimal solution to this problem (i.e., a set of integers α_{ij} that maximizes (1) subject to (2)), we construct an optimal role allocation by assigning robot i to role j only when $\alpha_{ij} = 1$.

By creating a *linear* program, we restrict the space of role allocation problems that we can model in one way: the function to be maximized (1) must be linear. Importantly, there is no such restriction on the manner in which the components of that function are derived. That is, individual utilities can be computed in any arbitrary way, but they must be combined linearly.

2.3 Solutions

If the robots' utilities can be collected at one machine (or distributed to all machines), then a centralized allocation mechanism can be employed. This is often the case in RoboCup, where locally-computed utility values are generally either unicast to a single player/coach or broadcast to all players (so that they each have the same data). In the former case, one machine can execute the allocation algorithm and inform the players of the results; in the latter, all players can execute the allocation algorithm in parallel[2].

Perhaps the most common role allocation technique is the following

1. Assemble the utility values into an $n \times n$ matrix.
2. Find the highest utility u_{ij}, assign robot i to role j, and cross out row i and column j from the utility matrix.
3. Repeat step 2 until all roles have been assigned.

Intuitively, this technique is greedy, in that it always selects the next best choice. In fact, this technique is an instance of the canonical Greedy algorithm, studied in several areas of optimization [10]. Unfortunately, the OAP does not satisfy the *greedy choice property* [11] and so the Greedy algorithm will not necessarily produce an optimal solution[3]. The worst-case performance of the Greedy algorithm on the OAP is well-known: it is 2-competitive [13]. An algorithm is said to be α-competitive if, for any input, it finds a solution that is no worse than $\frac{1}{2}$ of the optimum. Thus for an assignment problem, the Greedy algorithm will in the worst case produce a solution with utility that is $\frac{1}{2}$ of that given by an optimal solution.

In place of this greedy approach, it is possible to employ *optimal* solutions, a great many of which can be found in the literature (for a representative list, see [14]). The best-known approach is the Hungarian method [15], a linear programming algorithm that exploits the special structure of the OAP. This algorithm will find the optimal solution to a role allocation problem in $O(n^3)$ time. We have demonstrated empirically [16] that the constant factor for the Hungarian method is so small that this algorithm could easily be used for real time role allocation in RoboCup teams, where $n \leq 11$.

It is also possible to solve assignment problems in a completely distributed fashion. An optimal distributed algorithm that is relevant to role allocation in the RoboCup domain was derived by viewing the OAP as a "stable marriage" problem [17]: given a group of n boys and a group of n girls, each with preferences (i.e., utilities) over the members of the opposite sex, find a set of pairings such that there exists no boy and girl pair that is not paired together but prefers each other to their current mates. Gale and Shapley developed an intuitive algorithm in which each boy proposes to his favorite available girl and each girl conditionally accepts her favorite proposer. This process is repeated in a series of stages until, when the last girl has received a proposal, the "courtship" is declared to be over and the result is an optimal solution to the problem of assigning boys to girls so as to maximize total utility.

This distributed algorithm obviates the need to communicate the individuals' utilities among the individuals and requires $O(n^2 - 2n + 2)$ stages, which may

---

[2] Due to communication issues, different players may occasionally have different utility information. Because reallocation is performed so frequently, the effects of such inconsistency are usually transient and negligible.

[3] Equivalently, the OAP is a maximization problem over a subset system, but is not a *matroid* [12].

seem better than the $O(n^3)$ running time for centralized solutions. Unfortunately for high-speed, real-world domains like RoboCup, each stage in the proposal algorithm requires a synchronous exchange of messages between some or all of the participants. Given the latency characteristics of real networks, especially wireless networks in congested environments, the time required at each stage to guarantee the receipt of all messages and maintain synchronization makes this algorithm far too slow.

In practice, a team seeking optimal role allocation will be better off periodically broadcasting utility values to all players and having the players run the same centralized assignment algorithm in parallel. Assuming that the robots are linked with a wireless network and that a robot's utility values for all roles can be included in a single packet (this is almost certainly the case, as utilities are scalars), the communication cost of informing n robots of each others' utilities is just n messages. This overhead is negligible, as is the ensuing computational cost of executing even an optimal assignment algorithm.

3 Some Examples

Inspired by the pioneering work of Stone and Veloso [2], many RoboCup teams employ role-based coordination, in which the players can take on different roles within the team. Although it is possible to statically assign roles once at the beginning of the game, this strategy is brittle and unable to exploit unexpected opportunities [18]. Thus most role-based teams employ some kind of dynamic role allocation, in which the current allocation is reevaluated periodically, usually on the order of 10Hz. Thus they are solving the iterated assignment problem, though often without recognizing it as such (a notable exception is RoboLog Koblenz [19], who identify the problem of assigning players to block opponents as one of stable marriage, though it is not clear whether they actually use an optimal algorithm to solve it). Furthermore, most teams use greedy algorithms, despite the fact that tractable optimal algorithms exist.

For example, the assignment algorithm of ART99 consists of ordering the roles in descending priority and then assigning each to the available robot with the highest utility [20]. This algorithm is clearly an instance of the Greedy algorithm, and thus is 2-competitive. As is the case with most teams, utilities in ART99 are computed in a role-specific manner. The same approach, with utility referred to as *function Q*, is used in [21]. Vail and Veloso also employ the Greedy algorithm, with fixed priority roles [22]. Another team [23] describes a more complex approach in which role allocation is carried out differently depending on the game situation, such as who has the ball. We conjecture that their technique can be reduced to the Greedy algorithm by combining the different cases and appropriately modifying the utility functions.

A common problem with dynamic role assignment is that small changes in utility estimates can cause roles to be reassigned very frequently, often in an oscillating fashion. Changing from one role to another is not a costless operation; for example, a player may have to move from its current position to another position on the field. If the team is constantly reassigning roles it will spend less time actually playing and will tend not to perform well. The underlying problem is that the cost of transitioning from one role to another is an externality: it affects system performance but is not included in the utility computation. Thus the solution is to explicitly consider the cost of role transition in the utility computation, which will tend to induce a degree of hysteresis. Some teams, such as RMIT United [24], do this by adding a fixed amount to the utility of a robot retaining its current role

While the teams discussed so far employ centralized assignment algorithms, distributed role-swapping approaches are also used. The winners of the F2000 league at RoboCup 2000, CS Freiburg [25], use a distributed role allocation mechanism in which two players may exchange roles only if both "want" to, in that they will both be moving to higher-utility roles for themselves. This algorithm is similar to the marriage proposal algorithm discussed previously, but is not optimal, a fact that can easily be shown by counterexample. This result is not surprising because an optimal role-swapping algorithm, which considers all feasible assignments, would be too slow in practice. A similar pairwise exchange mechanism is used by FC Portugal [26], who won the simulation league at RoboCup 2000.

Unfortunately, without more detailed information about these role-exchange algorithms (e.g., how often are exchanges attempted? how many are attempted?), it is impossible to derive any performance bounds, other than to say that there exist situations in which they will produce sub-optimal solutions. Furthermore, we conjecture that little is saved in terms of communication or computation by executing such a distributed allocation algorithm. Instead of pursuing pairwise exchanges, the robots could periodically broadcast their utility values for all roles, then each execute an optimal centralized assignment algorithm.

4 Discussion

In this paper we have presented a formal framework for studying the RoboCup role allocation problem. We have shown how this problem can be understood as an iterated Optimal Assignment Problem (OAP), and how many existing team coordination strategies can be seen as assignment algorithms that are used to solve this problem. We have also explained how tractable optimal assignment algorithms developed in operations research can be immediately applied to RoboCup role allocation problems, in place of the existing greedy and/or *ad hoc* solutions.

A natural question arises: what will actually be gained by employing an optimal solution? Consider the analysis showing that the allocation mechanism used by ART99, as an implementation of the canonical Greedy algorithm, is 2-competitive. This kind of competitive factor gives an algorithm's *worst-case* behavior, which may be quite different from its *average-case* behavior. In fact, for many small assignment problems, the Greedy algorithm will find an optimal or near-optimal solution. In this respect, the solution quality bounds established for existing allocation architectures are rather loose. One way to tighten these bounds is to add domain-specific information to the formalism. By capturing and embedding models of how real RoboCup teams behave and evolve over time, it should be possible to make more accurate predictions about algorithmic performance. For example, while the classical theory of the OAP makes no assumptions about the nature of the utility matrices that form the input, role allocation problems are likely to exhibit significant structure in their utility values. Far from randomly generated, utility values generally follow one of a few common models, determined primarily by the kind of sensor data that are used in estimating utility. If only "local" sensor information is used (e.g., can the robot currently see the ball, and if so, how close is it?), then utility estimates tend to be strongly bifurcated (e.g., a robot will have very high utility if it can see the ball, and zero utility otherwise). On the other hand, if "global" sensor information is available (e.g., how close is the robot to a goal position?), then utility estimates tend to be smoother (e.g., utility will fall off smoothly in space away from the goal). A promising avenue for future research would be to characterize this "utility landscape" as it is encountered in RoboCup, and make predictions about, for example, how well a greedy assignment algorithm should be expected to work, as opposed to a more costly optimal assignment algorithm.

Another important issue concerns the way in which the role allocation problem is traditionally framed. Throughout this paper, we have considered the problem of assigning a single robot to a single role. In the context of our previously developed taxonomy of such problems [16], we are studying an iterated instance of ST-SR-IA (single-task robots, single-robot tasks, instantaneous assignment). However, in some situations, such as when two or more defenders are needed to block an attacker, the problem actually involves assigning *multiple* robots to a single role or task. It is generally not the case that the utility of a collection of robots for a given task will be equal to the sum of their individual utilities. Thus the classical assignment formulation is only an approximation of the real problem, which is an instance of ST-MR-IA (same as ST-ST-IA, but with multi-robot tasks). Unfortunately, problems of this kind, which involve *coalition formation*, are rather difficult to solve. In the most general case, ST-MR-IA problems are a form of set partitioning, which is strongly NP-hard [27]. As such, using the single-robot task approximation is a parsimonious choice.

We have also assumed thus far that robots' utilities are *independent*. That is, a robot's utility for a particular role does not depend on the other robots' allocated roles. This assumption is rarely, if ever, true in multi-robot systems. In general, a robot's utility for a role can in fact be affected by the overall allocation of roles to robots. This situation can arise any time that physical interference contributes significantly to task performance [28]. Unfortunately, the state of the art for capturing interrelated utilities of this kind is the Markov Decision Process [29], which remain too difficult to solve quickly enough for domains like RoboCup (especially if partial observability is considered).

Because it is well-defined and requires researchers to empirically compare their proposed techniques, RoboCup is an excellent domain in which to study role allocation. It is our hope that the formal perspective on the problem that we have given in this paper will aid in the understanding of how and why existing techniques work, as well as guide the design of new ones.

References

1. Matarić, M.J.: Designing Emergent Behaviors: From Local Interactions to Collective Intelligence. In Meyer, J.A., Roitblat, H., Wilson, S., eds.: From Animals to Animats 2, Second International Conference on Simulation of Adaptive Behavior (SAB-92), MIT Press (1992) 432–441
2. Stone, P., Veloso, M.: Task Decomposition, Dynamic Role Assignment, and Low-Bandwidth Communication for Real-Time Strategic Teamwork. Artificial Intelligence **110** (1999) 241–273
3. Parker, L.E.: ALLIANCE: An architecture for fault-tolerant multi-robot cooperation. IEEE Transactions on Robotics and Automation **14** (1998) 220–240
4. Gale, D.: The Theory of Linear Economic Models. McGraw-Hill Book Company, Inc., New York (1960)
5. Thorndike, R.L.: The Problem of Classification of Personnel. Psychometrika **15** (1950) 215–235
6. Botelho, S.C., Alami, R.: M+: a scheme for multi-robot cooperation through negotiated task allocation and achievement. In: Proc. of the IEEE Intl. Conf. on Robotics and Automation (ICRA), Detroit, Michigan (1999) 1234–1239
7. Gerkey, B.P., Matarić, M.J.: Sold!: Auction methods for multi-robot coordination. IEEE Transactions on Robotics and Automation **18** (2002) 758–768
8. Dertouzos, M.L., Mok, A.K.: Multiprocessor On-Line Scheduling of Hard-Real-Time Tasks. IEEE Transactions on Software Engineering **15** (1983) 1497–1506
9. Simon, H.A.: The Sciences of the Artificial. 3rd edn. MIT Press, Cambridge, Massachusetts (2001)
10. Ahuja, R.K., Magnanti, T.L., Orlin, J.B.: Network Flows: Theory, Algorithms, and Applications. Prentice Hall, Upper Saddle River, New Jersey (1993)
11. Cormen, T.H., Leiserson, C.E., Rivest, R.L.: Introduction to Algorithms. MIT Press, Cambridge, Massachusetts (1997)
12. Korte, B., Vygen, J.: Combinatorial Optimization: Theory and Algorithms. Springer-Verlag, Berlin (2000)
13. Avis, D.: A Survey of Heuristics for the Weighted Matching Problem. Networks **13** (1983) 475–493
14. Bertsekas, D.P.: The Auction Algorithm for Assignment and Other Network Flow Problems: A Tutorial. Interfaces **20** (1990) 133–149
15. Kuhn, H.W.: The Hungarian Method for the Assignment Problem. Naval Research Logistics Quarterly **2** (1955) 83–97
16. Gerkey, B.P., Matarić, M.J.: A formal framework for the study of task allocation in multi-robot systems. Technical Report CRES-03-013, Center for Robotics and Embedded Systems, School of Engineering, University of Southern California (2003)
17. Gale, D., Shapley, L.S.: College Admissions and the Stability of Marriage. American Mathematical Monthly **69** (1962) 9–15
18. Howard, A.: MuCows. In Stone, P., Balch, T., Kraetzschmar, G., eds.: RoboCup 2000, LNAI 2019. Springer-Verlag, Berlin (2001) 535–538

19. Murray, J., Obst, O., Stolzenburg, F.: RoboLog Koblenz 2001. In Birk, A., Coradeschi, S., Tadokoro, S., eds.: RoboCup 2001, LNAI 2377. Springer-Verlag, Berlin (2002) 526–530
20. Castelpietra, C., Iocchi, L., Nardi, D., Piaggio, M., Scalzo, A., Sgorbissa, A.: Communication and Coordination among heterogeneous Mid-size players: ART99. In Stone, P., Balch, T., Kraetzschmar, G., eds.: RoboCup 2000, LNAI 2019. Springer-Verlag, Berlin (2001) 86–95
21. Ferraresso, M., Ferrari, C., Pagello, E., Polesel, R., Rosati, R., Speranzon, A., Zanette, W.: Collaborative Emergent Actions between Real Soccer Robots. In Stone, P., Balch, T., Kraetzschmar, G., eds.: RoboCup 2000, LNAI 2019. Springer-Verlag, Berlin (2001) 297–302
22. Vail, D., Veloso, M.: Dynamic Multi-Robot Coordination. In Schultz, A., et al., eds.: Multi-Robot Systems: From Swarms to Intelligent Automata, Volume II. Kluwer Academic Publishers, the Netherlands (2003) 87–98
23. Jamzad, M., Chitsaz, H., Foroughnassirai, A., Ghorbani, R., Kazemi, M., Mirrokni, V., Sadjad, B.: Basic Requirements for a Teamwork in Middle Size RoboCup. In Birk, A., Coradeschi, S., Tadokoro, S., eds.: RoboCup 2001, LNAI 2377. Springer-Verlag, Berlin (2002) 621–626
24. Brusey, J., Makies, M., Padgham, L., Woodvine, B., Fantone, K.: RMIT United. In Stone, P., Balch, T., Kraetzschmar, G., eds.: RoboCup 2000, LNAI 2019. Springer-Verlag, Berlin (2001) 563–566
25. Weigel, T., Auerback, W., Dietl, M., Dümler, B., Gutmann, J.S., Marko, K., Müller, K., Nebel, B., Szerbakowski, B., Thiel, M.: CS Freiburg: Doing the Right Thing in a Group. In Stone, P., Balch, T., Kraetzschmar, G., eds.: RoboCup 2000, LNAI 2019. Springer-Verlag, Berlin (2001) 52–63
26. Reis, L.P., Lau, N.: FC Portugal Team Description: RoboCup 2000 Simulation League Competition. In Stone, P., Balch, T., Kraetzschmar, G., eds.: RoboCup 2000, LNAI 2019. Springer-Verlag, Berlin (2001) 29–40
27. Garey, M.R., Johnson, D.S.: "Strong" NP-Completeness Results: Motivation, Examples, and Implications. J. of the ACM **25** (1978) 499–508
28. Goldberg, D., Matarić, M.J.: Interference as a tool for designing and evaluating multi-robot controllers. In: Proc. of the Natl. Conf. on Artificial Intelligence (AAAI), Providence, Rhode Island (1997) 637–642
29. White, D.J.: Markov decision processes. John Wiley & Sons, Chichester, England (1993)

On the Role of Quantitative Descriptions of Behaviour in Mobile Robotics Research

Ulrich Nehmzow

Dept. of Computer Science, University of Essex, Colchester CO4 3SQ, UK

Abstract. This paper – a summary of a keynote address given at the Robocup 2003 symposium – argues i) that mobile robotics research would benefit from a *theoretical* understanding of robot-environment interaction, ii) that independent replication and verification of experimental results should become common practice within robotics research, and iii) that *quantitative* measures of robot behaviour are needed to achieve this.

The paper gives one example of such quantitative measures of behaviour: the reconstruction of the phase space describing a robot's behaviour, and its subsequent analysis using chaos theory.

1 Mobile Robotics Research

1.1 Robot Engineering versus Robot Science

Theory Supports Design. Arguably, there are (at least) two independent objectives of robotics research: on the one hand, to create artefacts that are capable of carrying out useful tasks in the real world – for example industrial, service, transportation or medical robots, to name but a few, and on the other hand to obtain a theoretical understanding of the design issues involved in making those artefacts – for example sensor and actuator modelling, system identification (modelling of entire systems), or sensor, actuator and behaviour analysis. The former can be referred to as 'robot engineering', the latter as 'robot science'. It is robot science that this paper is concerned with.

While robot engineering ultimately produces the 'useful' artefacts, there is a lot that robot science can contribute to this process. Without theoretical understanding, any design process is largely dependent upon trial-and-error experimentation and iterative refinement. In order to design in a principled way, a hypothesis of some kind (a justified expectation) is needed to guide the design process. The hypothesis guides the investigation: results obtained are fed back into the process and brought into alignment with the theory, to lead to the next stage of the experimentation and design. The better the theory underlying the design process, the more effective and goal-oriented the design process will be.

Every process of designing technical artefacts is based on some kind of assumptions (a 'theory'), even if very little is known at all about the object being designed. This is true for current mobile robotics research, too. When asked to design a wall-following robot, the designer will not start with an *arbitrary*

D. Polani et al. (Eds.): RoboCup 2003, LNAI 3020, pp. 54–66, 2004.

program, but with a 'reasonable guess', sensibly speculating on which sensors might be useful to achieve the desired behaviour, which general kind of control program will perform acceptably, *etc.* But, given our current understanding of robotics, he is unable to design the entire behaviour off-line!

Instead, mobile robotics researchers to-date are crucially dependent on trial-and-error procedures. A 'reasonable prototype' has to be tested in the target environment, and refined based on observations and underlying theory ('hunch' is often the more appropriate term for such theories). Here is a practical example: to design the Roomba commercial robot floor cleaner (relying on very simple sensing, and not involving any sophisticated navigation), thirty prototypes had to be built over a period of twelve years [EXN 03]!

The first argument we would make in favour of a better theoretical understanding of the principles underlying a mobile robot's operation in its environment, therefore, is that robot engineering (the process of designing a technical artefact that will perform useful tasks in the real world) will benefit from theory through the resulting more effective, rigorous and goal-oriented development methods.

Science Requires Replication and Verification. Current mobile robotics research practice not only differs from that of established disciplines in its lack of theories supporting design, but also in a second aspect: independent replication and verification of experimental results is uncommon. While in sciences such as biology or physics, for instance, reported results are only taken seriously once they have been verified independently a number of times, in robotics this is not the case. Instead, papers often describe experimental results obtained in specific environment, under specific experimental conditions. These experiments therefore are 'existence proofs' – the demonstration that a particular result can be achieved – but they do not state in general terms under which conditions a particular result can be obtained, nor which principles underlie the result. Existence proofs are useful, they demonstrate that something can be achieved, which is an important aspect of science, but they do not lead towards general principles and theories.

The second argument we make, therefore, is that mobile robotics research is now at a stage where we should move on from existence proofs to a research culture that habitually includes independent replication and verification of experiments.

The Role of Quantitative Descriptions. Theories, experimental replication and experimental verification all depend crucially on *quantitative* descriptions: quantitative descriptions are an essential element of the language of science.

The third argument we make, therefore, is that an essential first step towards a sounder theoretical understanding of robot-environment interaction is to develop and apply quantitative descriptions of robot-environment interaction. The experiments reported in this paper are one example of how to achieve this.

1.2 Theory

Introduction. When referring to 'theory', we mean a coherent body of hypothetical, conceptual and pragmatic generalisations and principles that form the general frame of reference within which mobile robotics research is conducted.

There are two key elements that make a theory of robot-environment interaction useful, and therefore desirable for research:

1. A theory will allow the formulation of hypotheses for testing. This is an essential component in the conduct of 'normal science' [Kuhn 70].
2. A theory will make predictions (for instance regarding the outcome of experiments), and thus serve as a safeguard against unfounded or weakly supported assumptions.

A theory retains, in abstraction and generalisation, the essence of what it is that the triple of robot-task-environment does (see also figure 2). This generalisation is essential: it highlights the important aspects of robot-environment interaction, while suppressing unimportant ones. Finally, the validity of a theory (or otherwise) can then be established by evaluating the predictions made applying the theory.

Benefits of Theory. Two significant advantages of a theory have been given above: generating hypotheses and making testable predictions. But there are practical advantages, too, particularly for a discipline that involves the design of technical artefacts: Theory supports off-line design, i.e. the design of technical artefacts through the use of computer models, simulations and theory-based calculations.

Example: Aircraft Design. Because of a theoretical understanding of aeronautics, it is now possible to design aircraft (such as the Airbus A380, shown in figure 1) almost completely 'off-line'. This incurs considerable advantages.

In a report on 'virtual manufacturing' to the UK Foresight Programme, J. Coyle states that virtual manufacturing and virtual prototyping – validating product design and production processes in a synthetic environment – enabled aerospace manufacturing companies to reduce design to first-unit build time by 33%, using 25% less manpower. In addition to that, the amount of rework is reduced by 90% (Boeing claim), sustained engineering work is reduced by 50%, and production time cycles and planning costs are also reduced. In aircraft design, there is the additional advantage that simulations are reusable for operator work instructions and for maintenance tasks.

These considerable benefits are, obviously, based on sound theoretical knowledge of the governing laws of physics, mechanics, aerodynamics, material science *etc.* underlying the operation of the designed artefact.

In mobile robotics, such theoretical knowledge to construct sufficiently accurate computer models and computer simulations is not yet available, but it is obvious that robot design would benefit significantly if it were.

Fig. 1. Aircraft design is largely based on theory, and computer modelling and considerations based on theoretical understanding have reduced development time and cost considerably.

2 Quantitative Characterisations of Robot-Environment Interaction

A key element of the objectives outlined above are *quantitative* descriptions of the robot's behaviour. They will support replication and independent verification of experimental results, principled experimental design and analysis of the influence of individual experimental factors. In this paper we present a way of using methods from deterministic chaos and dynamical systems theory to achieve this.

2.1 Robot-Environment Interaction Constitutes a Dynamical System

The behaviour of a mobile robot is governed by three main factors: (i) the robot itself: its hardware, physical makeup, inertia, sensor and motor characteristics *etc.*; (ii) the environment the robot is operating in: the colour, texture and structure of walls, floors and ceilings, the temperature and humidity, speed of sound in the environment, noise *etc.*; and (iii) the control program (the 'task') the robot is executing (figure 2).

Therefore, we argue, a mobile robot, interacting with its environment, could be viewed as an *analog computer.*

Similar to an optical lens, which takes light rays as its input and 'computes' the Fourier transformation of the input image as its output (thus acting as an analog computer), or a cylindrical lens taking the visual image of the environment as its input and 'computing' the positions of vertical edges in the image as its output (again acting as an analog computer), the mobile robot, executing some control program in its environment 'computes' behaviour as its output from the three input components shown in figure 2, i.e. robot-specific components, environment-specific components and the task (see figure 3).

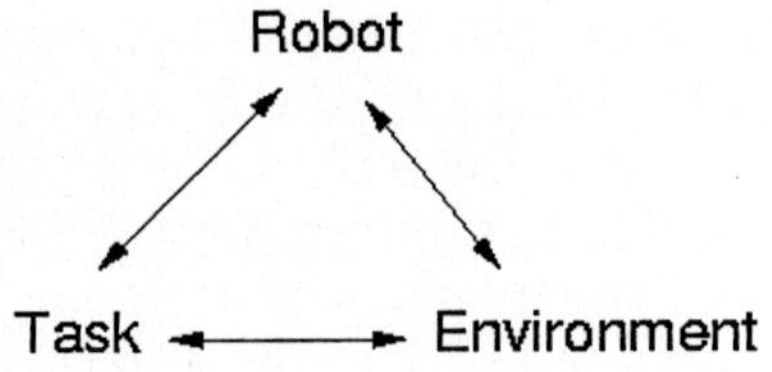

Fig. 2. The fundamental triangle of robot-environment interaction: A robot's behaviour always has to be seen in the context of robot, task and environment.

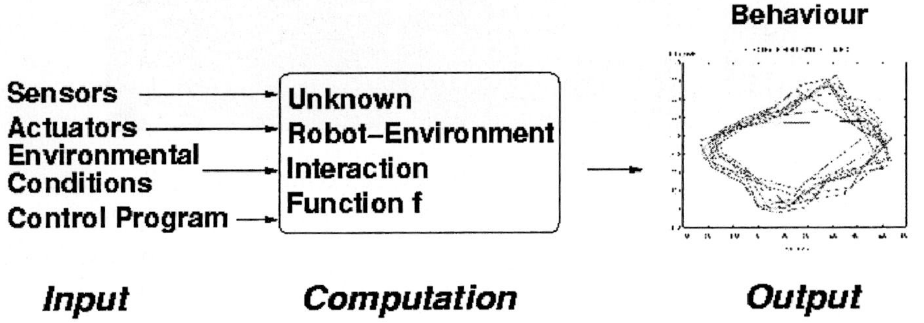

Fig. 3. A mobile robot interacting with its environment can be described as an analog computer, taking environmental, morphological and task-related data as input, and 'computing' behaviour as output (see also figure 2).

We furthermore argue that one of the most fundamental manifestations of the analog computation carried out by a mobile robot interacting with its environment, is the trajectory taken by the robot. Although the trajectory does not encompass *every* aspect of a mobile robot's behaviour, it is the dominant result of the robot's perception, control architecture and actuation. To analyse the trajectory, therefore, means analysing the dominant aspect of the 'output' of the robot's 'computation'. (Note: the methods described in this paper have been applied to the analysis of mobile robot behaviour. However, it is conceivable that other kinds of robot behaviour can be analysed in the same way, as our analysis is a time series analysis, which can be used to analyse *any* time series describing a robot's behaviour, not just trajectories.)

The trajectory of a mobile robot essentially constitutes two time series – one for the robot's x-coordinate, and one for the y-coordinate. In our experiments we analyse these time series for the presence or absence of deterministic chaos, and use the quantitative descriptors of deterministic chaos as quantitative descriptions of the robot's interaction with its environment.

2.2 Chaos Theory to Characterise Robot-Environment Interaction Quantitatively

Introduction to the Approach. Our aim, then, is to obtain a quantitative description of a mobile robot's interaction with its environment. We achieve this by applying methods from dynamical systems theory: We first reconstruct the attractor underlying the robot's behaviour, then analyse the attractor using chaos theory, and describe it quantitatively by computing the Lyapunov exponent (section 2.3). In this way we are able to establish, for instance, that robot environment interaction does exhibit deterministic chaos, as well as the influence of individual experimental parameters (such as objects present in the environment) upon the overall robot behaviour.

Analysis of a Dynamical System in Phase Space. The behaviour of any dynamical physical system is fully described by giving its position $p(t)$ and momentum $m(t)$ at time t, for every degree of freedom this system has. This (p, m) space is the *phase space* of the system – if it is known, the motion of the system is known and analysable[1].

For actual physical systems, the exact nature of the system's phase space cannot be known, due to noise *etc.* It is, however, possible to reconstruct an approximation of the system's phase space very simply from an observed time series of the system's motions, through the so-called time-lag-embedding Y_n given in equation 1 [Peitgen et al 92,Kantz & Schreiber 97,Abarbanel 96].

$$Y_n = (x(t_n - (p-1)h), x(t_n - (p-2)h), \ldots x(t_n - h), x(t_n)), \tag{1}$$

with $x(t)$ being a sequential set of measurements (the time series), p being the embedding dimension and h being the embedding lag (for a full discussion see [Peitgen et al 92,Nehmzow & Walker ny]).

In other words, it is not necessary to have full knowledge of a physical system's phase space, a (sufficiently long) observation of the system's behaviour (for instance by logging a trajectory, using a camera, or by logging other data describing the agent's behaviour) is sufficient to reconstruct its phase space. Once reconstructed, the phase space can be analysed quantitatively (see section 2.3).

In summary: We are interested in analysing some physical system quantitatively. Logging data emanating from the system (such as a robot trajectory) for a reasonably long period of time, we obtain a time series describing the system's behaviour. We know that it is possible to reconstruct the system's phase space from that time series [Takens 81]: instead of analysing the system's motion in physical space, we analyse the system's motion on the 'attractor' in phase space. Finally, attractors (and therefore the system that has generated the attractor) can be described quantitatively, for instance through the Lyapunov exponent or the dimension of the attractor [Peitgen et al 92].

Example from Mobile Robotics. Figure 4 shows the motion of a Pioneer II mobile robot in our laboratory, executing an obstacle-avoidance control program

[1] A very simple illustration of phase space is the ideal pendulum, which has one degree of freedom – the arc along which it swings. The phase space of the ideal pendulum is therefore two-dimensional (speed and position along the arc). The actual trajectory of the pendulum through phase space – the 'attractor' – is a circle.

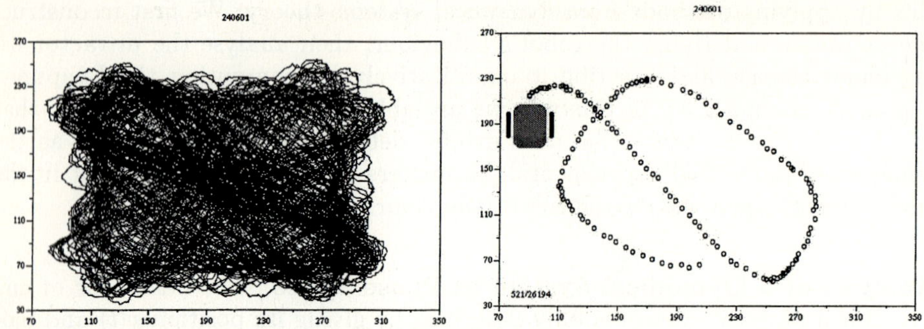

Fig. 4. Quasi 'Billiard Ball' Behaviour in Square Arena – Entire trajectory (left) and 150 data points (right).

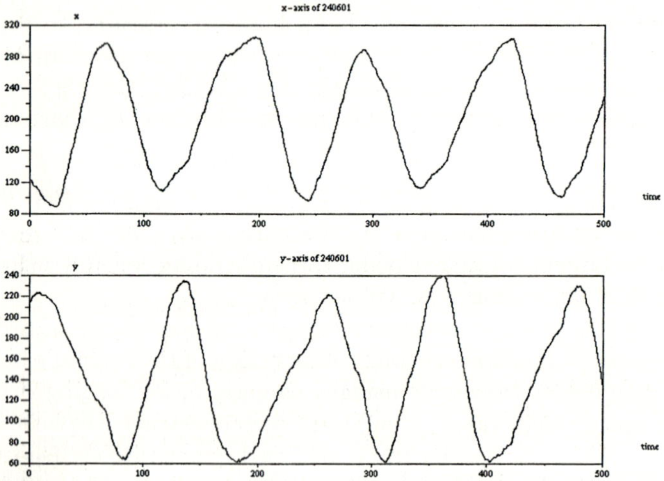

Fig. 5. Part of the x (top) and y (bottom) coordinate of quasi billiard ball behaviour. The attractor describing the robot's behaviour (shown in figure 6) was reconstructed from this data, using time-lag embedding [Peitgen et al 92].

that results in a quasi billiard ball behaviour. The trajectory shown in figure 4 was obtained by logging the robot's motion with an overhead camera for about 2 hours of continuous operation.

Using the time series $x(t)$ or $y(t)$, shown in figure 5, the robot's motion in phase space can be reconstructed through time lag embedding (equation 1). This phase space reconstruction is shown in figure 6.

As the phase space of a physical system describes that system's behaviour fully, we now have a representation of a robot's interaction with its environment that can be analysed quantitatively – one of the key requirements of a theory of robot-environment interaction, as we argued earlier. The following section will discuss how the Lyapunov exponent can be used to achieve this objective.

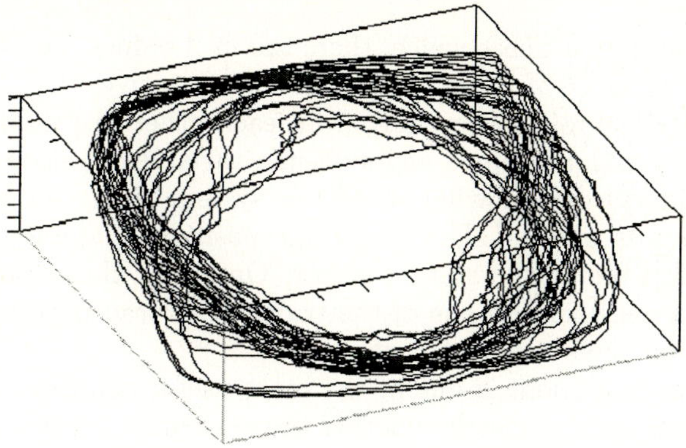

Fig. 6. The phase space reconstruction of the quasi billiard ball behaviour shown in figure 4. The attractor has a fractal dimension of approx. 1.9, it is 'strange'.

2.3 Quantitative Analysis of Phase Space

There are a number of quantitative descriptions of phase space, for instance the dimension of the attractor (correlation dimension), but the description most easily obtained, and which we have used most in our experiments is the Lyapunov exponent.

The Lyapunov Exponent. One of the most distinctive characteristics of a chaotic system is its sensitivity to a variation in the system's variables: two trajectories in phase space that started close to each other will diverge from one another as time progresses, the more chaotic the system, the greater the divergence.

Consider some state S_o of a deterministic dynamical system and its corresponding location in phase space. As time progresses the state of the system follows a deterministic trajectory in phase space. Let another state S_1 of the system lie arbitrarily close to S_o, and follow a different trajectory, again fully deterministic. If d_o is the initial separation of these two states in phase space at time $t = 0$, then their separation d_t after t seconds can be expressed as $d_t = d_o e^{\lambda t}$.

Or, stated differently, consider the average logarithmic growth of an initial error E_0 (the distance $|x_0 - (x_0 + \epsilon)|$, where ϵ is some arbitrarily small value and x_0 a point in phase space) [Peitgen et al 92, p. 709]. If E_k is the error at time step k, and E_{k-1} the error at the previous time step, then the average logarithmic error growth can be expressed by equation 2.

$$\lambda = \lim_{n\to\infty} \lim_{E_0\to 0} \frac{1}{n} \sum_{k=1}^{n} log|\frac{E_k}{E_{k-1}}|. \quad (2)$$

λ (which is measured in Hz or s^{-1}, or sometimes in bits/s) is known as the Lyapunov exponent.

For a m-dimensional phase space, there are m λ values, one for each dimension. If any one or more of those components are positive, then the trajectories of nearby states diverge exponentially from each other in phase space and the system is deemed chaotic. Since any system's variables of state are subject to uncertainty, a knowledge of what state the system is in can quickly become unknown if chaos is present. The larger the positive Lyapunov exponent, the quicker knowledge about the system is lost. One only knows that the state of the system lies somewhere on one of the trajectories traced out in phase space, i.e., somewhere on the strange attractor.

The Lyapunov exponent is one of the most useful quantitative measures of chaos, since it will reflect directly whether the system is indeed chaotic, and will quantify the degree of that chaos. Also, knowledge of the Lyapunov exponents becomes imperative for any analysis on prediction of future states.

One method to determine the Lyapunov of an attractor describing the behaviour of a physical system is to estimate it from an observed time series of the system's motion [Peitgen et al 92]. The methods we have used in our research are the one proposed by Wolf in 1985 [Wolf et al 95], and the one proposed by Abarbanel [Applied Nonlinear Sciences 03]. Software packages implementing either method are readily available.

3 Experiments

3.1 Experimental Setup

In our experiments, we observed autonomous mobile robots (figure 7), performing simple sensor-motor tasks such as obstacle avoidance, in a laboratory environment (see also figure 8).

The robot's trajectory was logged every 250 ms, using an overhead camera. Figures 4 and 5 show examples of the kind of trajectories observed, and the kind of data subsequently analysed.

Results. We carried out a number of different experiments, in which our robots performed different behaviours, in a range of different environments. As this paper is concerned with method, we only give an overview of results here, full results are given in [Nehmzow & Walker 03,Nehmzow & Walker 03b] [Nchmzow & Walker ny].

First, all our experiments showed that the interaction of our mobile robots with their environment exhibited deterministic chaos. Lyapunov exponents were small, but always positive (indicating the presence of chaos), and typically between 0.1 and 0.3 bits/s. All attractors were 'strange', i.e. had a fractal dimension of typically between 1.5 and 2.5.

Second, we were able to investigate the influence of individual experimental parameters quantitatively. For instance, changing the robot's control program

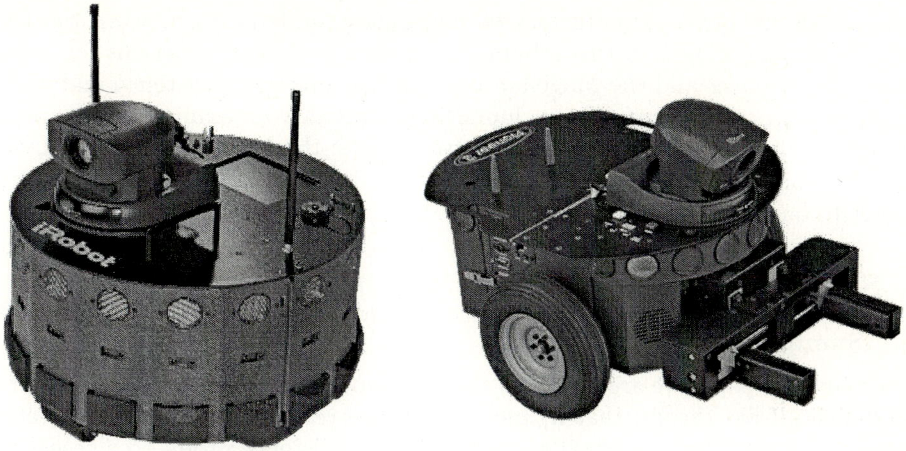

Fig. 7. The Magellan Pro and Pioneer II mobile robots used in the experiments discussed here.

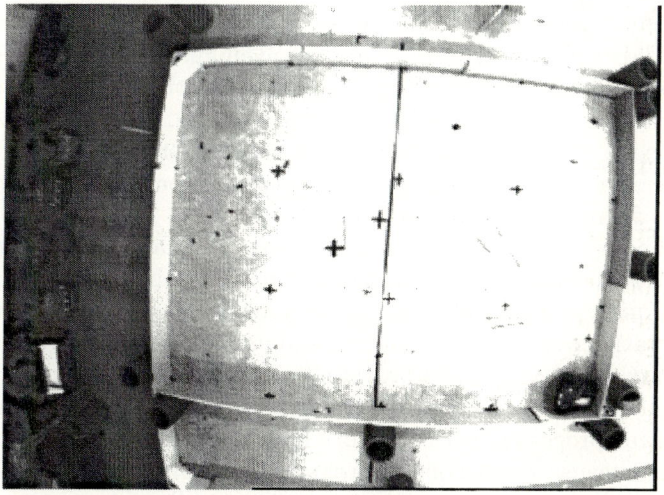

Fig. 8. Bird's eye view of the experimental arena used in our experiments. The robot is visible in the bottom right hand corner of the arena.

from 'wall following' to 'billiard ball behaviour' increased the Lyapunov exponent by a factor of four, whereas changing the environment (by adding a central obstruction to it) whilst leaving the behaviour (billiard ball) unchanged, resulted in no measurable change of the Lyapunov exponent [Nehmzow & Walker ny] [Nehmzow & Walker 03]. In other words: in the experiments we conducted, the control code influenced robot-environment interaction much more noticeably than the nature of the environment!

In a different set of experiments we investigated differences between the Pioneer II and the Magellan Pro robots. Using similar control programs for both robots (the Pioneer and the Magellan use different operating systems, and code can therefore not be *identical*), we found in preliminary experiments that there is a noticeable difference between these two robots: the Magellan attained Lyapunov exponents that were smaller by about a factor of 2. This observation is subject to ongoing research.

4 Summary and Conclusion

4.1 Summary

In this keynote paper we argue that i) the design of technical artefacts (engineering) benefits from theory, that ii) mobile robotics research has reached a stage of maturity that allows the field to move on from existence proofs to a research culture that habitually involves independent replication and verification of experimental results, and that iii) quantitative descriptions of behaviour are the foundation of theory, replication and verification.

An engineering design process that is not guided by a theoretical understanding of the subject matter has to resort to trial-and-error methods and iterative refinement. This is very costly, and does not even guarantee optimal results. On the other hand, design supported by theory can use the hypotheses and predictions afforded by the theory to reduce the design space, resulting in more efficient and faster design cycles. The aircraft industry, for instance, reports substantial gains in efficiency through 'virtual manufacturing', based on a theoretical understanding of aircraft design [Coyle 03].

Mobile robotics research, so far, largely relies on trial-and-error methods and uses iterative refinement techniques to develop task-achieving robot controllers. Existing computer models of robot-environment interaction are such simplified representations of sensors, actuators and environments that their predictive power is of little value for real world applications. Instead of mainly designing robot controllers off-line, control programs have to be developed through iterative refinement processes, which require large amounts of time and are costly. Furthermore, because robot control programs are usually developed in specific target environments, experimental results are typically existence proofs, rather than generally applicable findings. This limits their usefulness.

Finally, mobile robotics research currently does not benefit from a research practice in which experimental results are replicated and independently verified. This, of course, increases the risk that results stemming from accidental singularities go undetected.

We argue that one of the reasons for the current practice of mobile robotics research is that we simply haven't got the means to communicate results in such a way that theoretical descriptions, independent replication and verification are possible – we lack the 'language', i.e. quantitative descriptions of robot behaviour. Statistical descriptions are useful in this regard to some extent, but they still only represent a statistical description of *overall* robot behaviour, rather than a quantitative description of a particular single run.

In this paper, we present one method to obtain a quantitative description of robot behaviour, using dynamical systems theory. We first reconstruct the attractor underlying the robot's behaviour, then analyse the attractor using chaos theory, and describe it by computing the Lyapunov exponent (section 2.3). In this way we were able to establish, for instance, that robot environment interaction does exhibit deterministic chaos, as well as the influence of individual experimental parameters (such as objects present in the environment) upon the overall robot behaviour.

4.2 Conclusion

The experiments presented in this paper and elsewhere [Nehmzow & Walker 03] [Nehmzow & Walker ny,Nehmzow & Walker 03b] demonstrate that dynamical systems theory can be used to obtain quantitative descriptions of robot-environment interaction. Computing the Lyapunov exponent of the phase space underlying the robot's behaviour provides a means of measuring the influence of individual experimental parameters upon the overall robot behaviour. By changing just one of the three components shown in figure 2, for instance, the Lyapunov exponent can be used to describe the robot itself, the environment it is acting in, or the task it is performing. Likewise, it can be used to describe the influence of parameters such as illumination, colour, surface structure *etc*, by modifying the parameter in question in a principled way, and measuring the resulting Lyapunov exponent.

Acknowledgements

Much of the work reported in this paper was conducted in collaboration with my colleague Keith Walker from Point Loma Nazarene University, San Diego, I gratefully acknowledge his contribution.

The research presented in this paper was supported by a grant from the Research Promotion Fund of the University of Essex.

I thank the RoboCup organising committee, which invited me to present a keynote address at the 2003 RoboCup symposium in Padua, subsequently leading to this paper.

Thank You also to Phillip McKerrow and Claudia Nehmzow for detailed and helpful comments on earlier versions of this article.

References

[Abarbanel 96] H. Abarbanel, *Analysis of Observed Chaotic Data*, Springer-Verlag, New York 1996.

[Applied Nonlinear Sciences 03] Applied Nonlinear Sciences, *Tools for Dynamics*, URL=`http://www.zweb.com/apnonlin`, last accessed May 2003.

[EXN 03] EXN.CA discovery channel, 11.3.2003, `http://www.exn.ca/Stories/2003/03/11/57.asp`, last accessed October 2003.

[Coyle 03] J. Coyle, Virtual Manufacturing, UK Foresight Programme Materials and Structures National Advisory Committee, *Aerospace 2020*, AR360 Vol. III, `http://www.iom3.org/foresight/nac/html_docs/AGARD%203.htm`, Last accessed October 2003.

[Kaplan & Glass 95] D. Kaplan and L. Glass, *Understanding Nonlinear Dynamics*, Springer Verlag London, 1995

[Kantz & Schreiber 97] H. Kantz and T. Schreiber, *Nonlinear Time Series Analysis*, Cambridge University Press, 1997.

[Kuhn 70] Thomas Kuhn, *The Structure of Scientific Revolutions*, University of Chicago Press, 1970.

[Nehmzow & Walker ny] Ulrich Nehmzow and Keith Walker, Quantitative Description of Robot-Environment Interaction using Chaos Theory, submitted to *J Robotics and Autonomous Systems.*

[Nehmzow & Walker 03] Ulrich Nehmzow and Keith Walker, The Behaviour of a Mobile Robot Is Chaotic, *AISB Journal*, Vol 1 No 4, 2003.

[Nehmzow & Walker 03b] Ulrich Nehmzow and Keith Walker, Quantitative Description of Robot-Environment Interaction Using Chaos Theory, *Proc. European Conference on Mobile Robotics (ECMR)*, Warsaw 2003.

[Peitgen et al 92] H.-O. Peitgen, H. Jürgens and D. Saupe, *Chaos and Fractals*, Springer Verlag 1992.

[Takens 81] F. Takens, *Detecting Strange Attractors in Turbulence*, Lecture Notes in Mathematics, Vol. 898, Springer Verlag, Berlin 1981.

[Wolf et al 95] A Wolf, J. Swift, H. Swinney and J. Vastano, Determining Lyapunov exponents from a time series, *Physica 16D*, 1995. Code available at `http://www.cooper.edu/~wolf/chaos/chaos.htm`, last accessed November 2003.

Complexity Science and Representation in Robot Soccer

Jeffrey Johnson[1] and Blaine A. Price[2]

[1] Design and Innovation Department and
The Open University, Milton Keynes, MK7 6AA, UK
j.h.johnson@open.ac.uk
[2] Computing Department
The Open University, Milton Keynes, MK7 6AA, UK
b.a.price@open.ac.uk

Abstract. Complexity science is characterised by computational irreducibility, chaotic dynamics, combinatorial explosion, co-evolution, and multilevel lattice hierarchical structure. One of its main predictive tools is computer-generated distributions of possible future system states. This assumes that the system can be represented inside computers. Robot soccer provides an excellent laboratory subject for complexity science, and we seek a lattice hierarchical vocabulary to provide coherent symbolic representations for reasoning about robot soccer systems at appropriate levels. There is a difference between constructs being human-supplied and being abstracted autonomously. The former are implicitly lattice-hierarchically structured. We argue that making the lattice hierarchy explicit is necessary for autonomous systems to abstract their own constructs. The ideas are illustrated using data taken from the RoboCup simulation competition.

1 Introduction

Robot soccer is an excellent laboratory subject for the emerging new science of complexity characterised by computational irreducibility, chaotic dynamics, combinatorial explosion, co-evolution, and multilevel lattice hierarchical structure. The earlier benchmark problem of computer chess also has many of these properties, with the exception of not being chaotic. Start a game of chess in a given position, compute each of a sequence of moves, and the result is always the same.

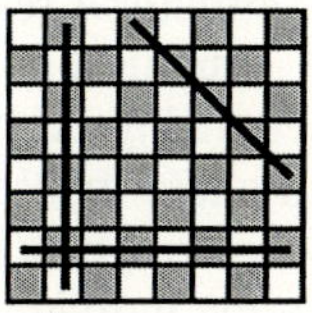

Fig. 1. Structured space in robot chess

Chess players use *structure* in order to play the game. Some of these structures are so fundamental that they have been given names such as the ranks, files, and diago-

D. Polani et al. (Eds.): RoboCup 2003, LNAI 3020, pp. 67–76, 2004.

nals illustrated in Figure 1. These names reflect spatial properties such as contiguity, and the functional properties of the pieces, e.g. rooks move on ranks and files, bishops move on diagonals, while kings and queens do both.

The *space* of the chessboard is divided into micro-units at the level of a player. These micro-units are aggregated into structures (*e.g.* ranks, files, diagonals) reflecting the modes of movement of the players. Thus the space of chessboard is hierarchically structured, with a well-defined set of areas, the squares, at the lowest level. Soccer and robot soccer do not have such an obvious lowest level of aggregation. At higher levels the space is structured by the goal area, the halves of the pitch, the penalty spots, the centre spot, and the centre circle, etc. All of these areas are defined because they play rolls in the rules.

In robot soccer the positions of the players and ball are assumed to be on an *x-y* grid. Even when the positions are represented by floating point numbers, this grid is finite. In this respect, the soccer pitch is like an enormous chessboard. The chessboard has 8 x 8 squares, while in robot soccer there are typically 1680 x 1088 pixels.

In chess, the players use a hierarchical *representation* that includes the ranks, files, diagonals and other more ambiguous areas such the right, left and centre of the board. Even though the number of squares and the number of these constructs is relatively small, chess is characterised by combinatorial explosion as chess players attempt to predict future system states. The 1.8 million pixels of the robot soccer pitch present an even more formidable combinatorial explosion in the way that the pixels can be grouped to form coherent and *relevant* areas of the pitch.

Our research is based on the premise that complex systems have hierarchically structured vocabulary reflecting the relational structure at micro- and macro-levels. By their nature, complex systems have to be investigated using computers, and this means that their representation must be explicit. In the case of robot soccer, this means there must be vocabulary for representing relationships between dynamically forming parts of the pitch, and dynamically forming relationships between players, opponents, and the ball.

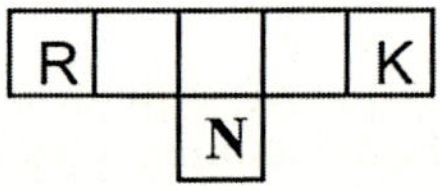

N – kNight, K – King, R – Rook

Fig. 2. The Knight-Fork

Just as in chess there are ‘interesting’ structural relationships with names such as the ‘knight fork’, there are ‘interesting’ structural relationships in robot soccer. In Figure 2 the knight, N, checks the opponents king, *K*, and threatens the more valuable rook, *R*. This structure is so dangerous in chess that it has its own name, the ‘knight fork’. When players reason about chess, the ‘knight fork’ is an entity in its own right, with emergent properties not possessed by the individual pieces.

Figure 2 illustrates a similar well-known structure in soccer. In this case, defending player 3 threatens to take the ball from player-1. If player-1 feigns a pass to player-2, then player-3 must move to intercept that pass. In so-doing, player-2 moves out of position, and player-1 can slip past. This structure has its own vocabulary, *e.g.* player-2 ‘draws out of position’ player-3, allowing player-1 to pass.

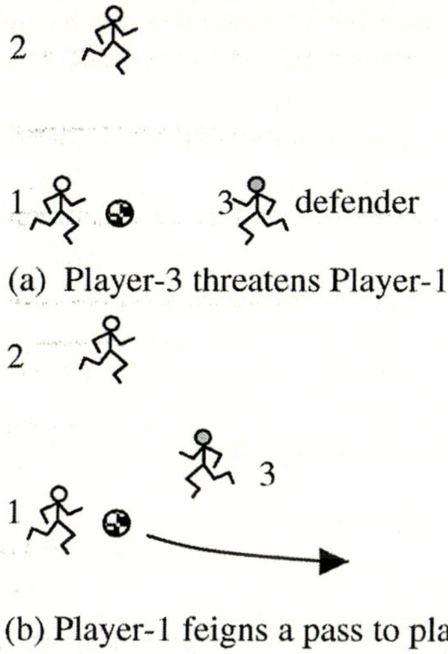

Fig. 3. Player 3 is drawn out of position by the relationship between players 1 and 2

The richer the vocabulary of these structures, the greater will be the advantage possessed by teams using that vocabulary. There are of course astronomic numbers of possible configurations of players and the ball on a soccer pitch. How can the 'interesting' structures be found? One answer to this is analyse many soccer games, and observe which configurations occur at 'interesting' times. These include the scoring of goals, but also include events such as the ball being lost, or even large areas of space opening up.

Early work in the analysis of RoboCup agents concentrated on the offline analysis of statistics gleaned from game logs, such as the work of Takahashi and Naruse [1] who measured statistics for teams at RoboCup 1997 such as number of goals, assists, kicks, own goals, and so on. Takahashi [2] continued this work and found no relationship between scoring and collaboration between agents when looking at the basic statistics as above. He did find, however, that collaborative actions, such as the number of 1-2 passes (player A passes to team-mate B in order to avoid defending player D, then B passes back to A once A has passed D), correlated highly with team ranking.

Tanaka-Ishii and colleagues [3] did a detailed offline statistical analysis of teams from 1997 with teams from 1998 with respect to 32 evaluations features, such as number of pass chains, average distance covered by one play, average pass length, and so on. They also compared the robustness of teams by replaying games with a reduced squad and found that some teams performed better with fewer players. They conclude that teams that perform poorly may not be the worst teams, but merely teams that have been let down badly by one aspect of their play. They argue for a collaborative modular team which can take the best performing parts of each team and also point towards the benefits of the online coach, which was introduced the following year.

Raines et al. [4] developed a system called ISAAC for post-hoc offline analysis of the events leading up to *key events*, such as *shots on goal* in the case of the RoboCup soccer simulation. ISAAC analyses the situations when the defence of the goal succeeds or fails with respect to a number of variables, such as the distance of the closest defender, the angle of the closest defender with respect to the goal, and the angle of the attacker from the centre of the field, the angle of the shot on goal and the force of the kick. The user is able to do a perturbation analysis to determine which changes in a rule will increase the goal success rate (e.g. changing the angle at goal, increasing the force of the kick). This enables analysing teams to seek improvements.

A similar off line approach is described by Wünstel et al. [5] who analysed player movement with respect to the ball to determine what kinds of movements a given player tends to make, although without reference to the context the player is in.

The online coach of Visser et al. [6] compared the formation patterns of opponent players from past games with a set of pre-defined formation patterns in an attempt to predict opponent formations. This allowed the coach to direct its own players to deal with the anticipated behaviour.

The online coach of Riley and Veloso [7] used pre-defined movement models and compared them with the actual movement of the players to predict future behaviour and advise its players accordingly.

Recent work in the analysis of agents in RoboCup has centred on predicting opponent behaviour. Kaminka et al. [8] used a system to identify and learn sequences of coordinated agent behaviour over one or more games for a given team. This was a post-hoc offline method which analysed logs of games after they were played. They ran experiments to show that the system was able to pick sequences that were characteristic of the team rather than arbitrary. Visser and Weland [9] developed a system that works on live games as opposed to a post-hoc analysis. Their system looks at the behaviour of the opponent goalkeeper as it leaves the goal as well as the passing behaviour of opponent players in order to find rules which characterise these agents. It updates these rules every 1000 cycles with the intention of making it available to the on-line coach to take advantage of the data.

Our work differs from that discussed above in our search for a coherent vocabulary through algebraic structures. This reflects our motivation in complexity science, and the desire to discover a methodology for representing complex systems in general, using robot soccer as a well defined, well researched, and replicable laboratory subject.

2 Complexity Science

Complexity science is characterised by computational irreducibility, chaotic dynamics, combinatorial explosion, co-evolution, and multi-level lattice hierarchical structure. Each of these suggests that predicting the future behaviour of complex systems will require relatively high levels of computation:

- computational irreducibility means that the computational load on making predictions is relatively high.
- deterministic chaos means that a high level of computation will be required for making useful predictions. A single point sample in the space of future possibilities has almost no useful information, and many future states have to be computed to gain information on the distributions of possible future system states.
- co-evolving systems tend to be both chaotic and computationally irreducible.

- combinatorial explosion, of its nature, implies high levels of computation for the search techniques used to explore large spaces of possibilities.
- complex systems are usually multi-level with micro- and macro-subsystems. The vocabulary to represent these systems should be coherent with respect to hierarchical aggregation, so that higher and lower level data are consistent.

This means that the representation of the system must be absolutely explicit to support computation. Paradoxically, many complex social systems are currently administered by people, with nearly all the computation and much of the data being in their heads. Such intuitive human processing supports many predictions in business and social administration. It even characterises human soccer. By comparison, the representation for robot soccer is explicit – it has to be because the system is autonomous and implemented on machines. For this reason we are interested in the simulation competition of RoboCup, since we believe that this will give new insights applicable more generally in complexity science and its applications.

3 The Lattice Hierarchy and Representation

Systems are characterised by wholes assembled from parts. This is illustrated in Figure 4, in which a set of three blocks is assembled to form a structure that we will call an *arch*. The arch is clearly more than the sum of its parts since it has emergent features, such as the possibility of walking through it on the path between α and β.

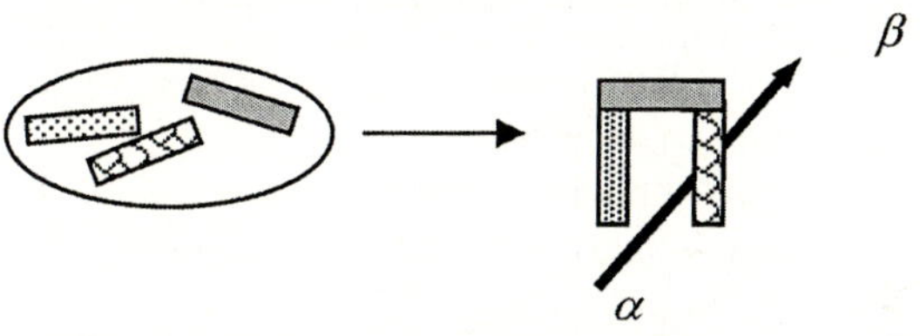

Fig. 4. The arch is a structure built from a set of parts, and has emergent properties not possessed by its parts

In Figure 4, the arrow labelled *R* indicates that the set is mapped to the whole by the relation *R* between the blocks. If the blocks and labelled *a*, *b*, and *c*, then the set is represented in the usual way by the notation { a, b, c }. We will denote the *R*-structured set of blocks as $\langle a, b, c; R \rangle$. Then, { a, b, c } $\neq \langle a, b, c; R \rangle$.

In Figure 5, the set of blocks is represented by an *Euler ellipse*, a variant of the Euler circle used to represent the set properties of intersection, subset, and union. The *hierarchical cone construction* then has the Euler ellipse as base and the name of the structure as its apex.

When analysing any system there is the problem of building a coherent representation between the highest level construct , 'the system', and the lowest level atoms such as the players or the pixels in simulated robot soccer. Generally there is a pre-existing vocabulary in vernacular language made up of terms that are more or less well defined. For example, in soccer the terms 'goal area' and 'Red's half' can be defined precisely, while 'the left wing' and 'the goal mouth' may be less well defined.

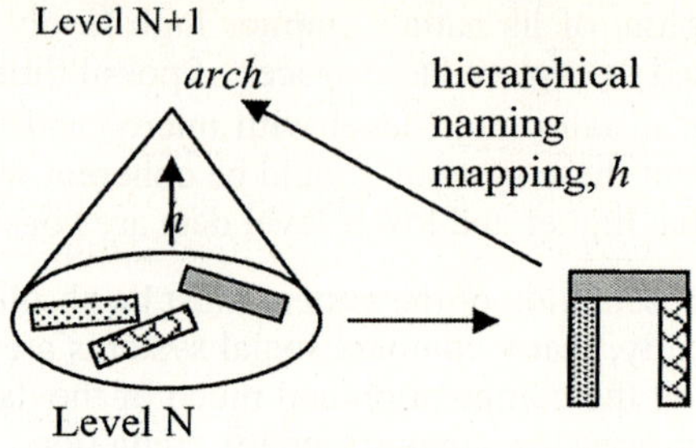

Fig. 5. The hierarchical cone construction

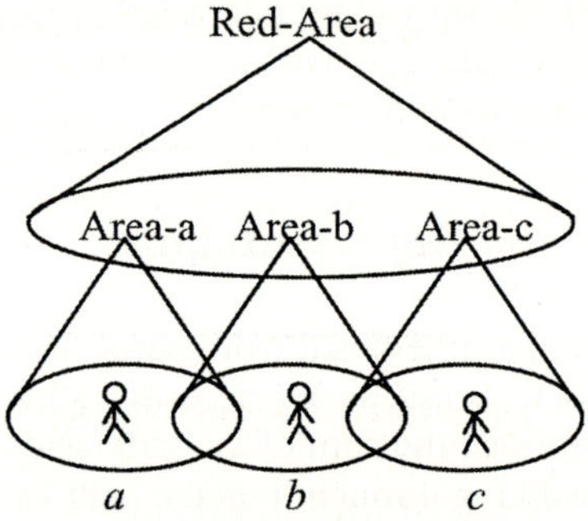

(a) Player areas aggregate into side's area

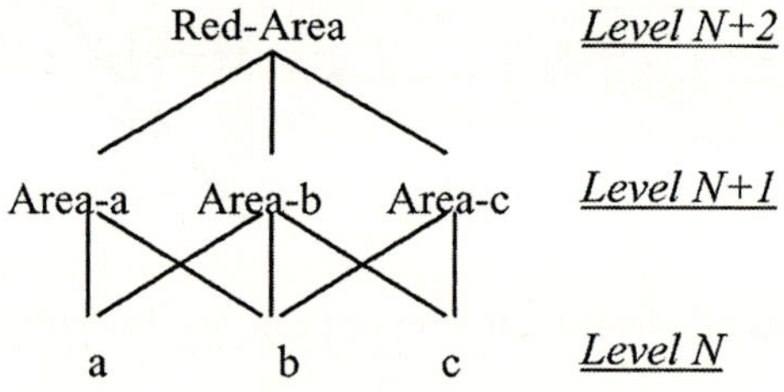

(b) lattice hierarchy of sub-areas

Fig. 6. A lattice hierarchy

Terms in the vocabulary may exist at many levels and the 'set' containing them will be called the *hierarchical soup*. As with the computer analysis of other complex systems, robot soccer has *The Intermediate Word Problem* of lifting a coherent hierarchically structured vocabulary out of the soup.

The term hierarchy is often misunderstood to mean a tree-like structure. More commonly, hierarchies have a 'lattice' structure, since things may aggregate into more than one structure at higher levels. Figure 6 provides a simple example in which three players define a subset of the pitch according to the pixels they are closest to. For example, the pixels closest to player *a* are shown by a Euler ellipse in Figure 6(a). Some pixels will be equidistant to some players, and so belong to both their areas, as shown by the intersecting ellipses for the players *a* and *b*, and the players *b* and *c*.

These areas have been given the names Area-a, Area-b, and Area-c. The union of these areas, together with that for all the other Red team members, makes up the part of the pitch controlled by the Red team, called the Red-Area. This simple hierarchy has three level.

When constructing hierarchical vocabularies, often there is no obvious bottom level. Also it sometimes necessary to define new structures between existing levels, thereby creating new levels. For this reason, levels are usually given the denotation *Level N+k*, emphasising the relative nature of the levels (Fig. 6(b)).

Figure 6(b) shows a graph of the 'part-of' relation implicit in the assembly structure in Figure 6(a), illustrating the notation of lattice. A *lattice hierarchy* is defined to be a class of objects with a *part-of* relation. This is *anti-symmetric*, so that *x part-of y* implies *not y-part-of x*. This is *a quasi-order* on the class, and in graph theory its graph is called a *lattice*. The quasi order is weaker than a partial order, since in a partial order $x \leq y$, or $y \leq x$, or both (equality). For example, there is no line between Area-a and Area-c in Figure 6(b) in the part-of relation between Levels N and N+1.

4 Structure in Robot Soccer

We have experimented with a number of RoboCup games, investigating 'interesting' relational structure.

One kind of structure concerns passes between players. A pass, ⟨player-1, player-2; R_{pass}⟩ is a pair of players and a relation between them. A set of passes between player of the same side is clearly an interesting structure in soccer. Here, a pass is structure between a pairs of players, and the pass-sequence is a structure on the set of passes.

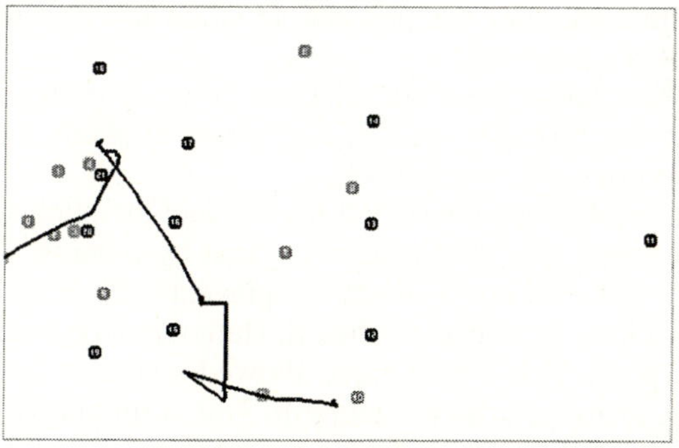

Fig. 7. A set of passes as a structure

Figure 7 shows a particularly long sequence of passes, which results in a goal being scored. The path forms as a consequence of the movements of the players, both on and off the ball, and the relationships that this creates between them.

Figure 8(a) shows a 'nearest opponent' relationship between the players. As can be seen, the graph has five components, reflecting the interactions between the players.

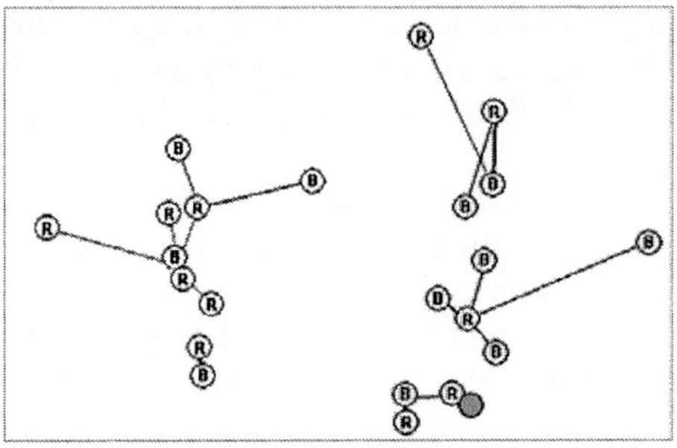

(a) nearest-opponent relation

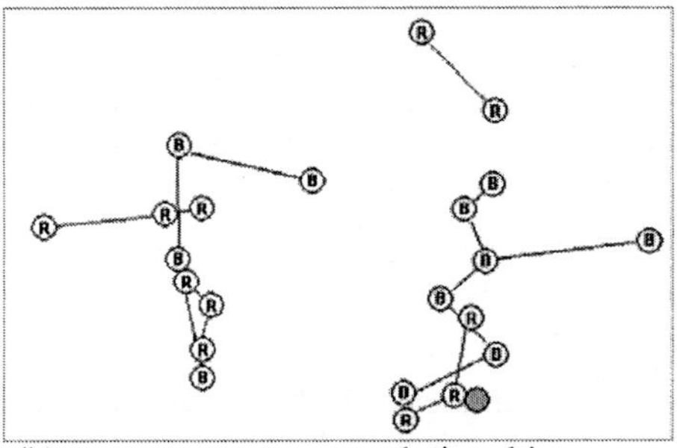

(b) nearest team-mate relationship

Fig. 8. Relational structure in robot soccer

Figure 8(b) shows a 'nearest team-mate' relationship. Inspection of the graph shows that Blue's structure is most highly connected, with two components. By comparison, Red's structure has four components. In some sense, Blue's structure hangs together better that Red's structure.

Related to these structures, there are relations between parts of the pitch and the players. The parts of the pitch are squares, where sets of pixels make up the squares, and the set of squares cover the whole pitch.

Figure 9(a) shows a relation between the players and the pitch squares. Figure 9(b) shows how the blue team, (B), dominates the game by owning almost all the pitch. This resonates with positional chess, when the players are not seeking tactical material advantage, but seeking to control the board. Here it can be seen that the blue team owns almost all the pitch. This whole game showed a similar dynamic pattern, with the blue area rapidly growing after the kick-off. Not surprisingly the blue team won by many goals.

5 Construct Formation

Each of the names or words in a lattice hierarchy vocabulary represents a structure. In most of the programs that people craft, the constructs exist *a priori* in the human mind. One of the goals for intelligent systems is to have them abstract their own constructs from their interaction with their environment, to create their own vocabulary.

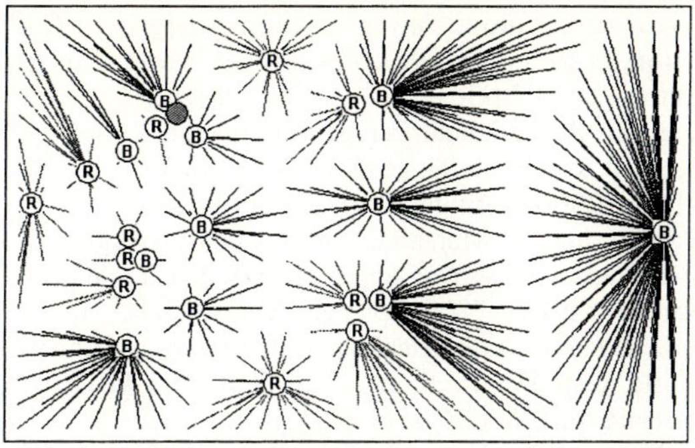

(a) the relation between players and the pitch

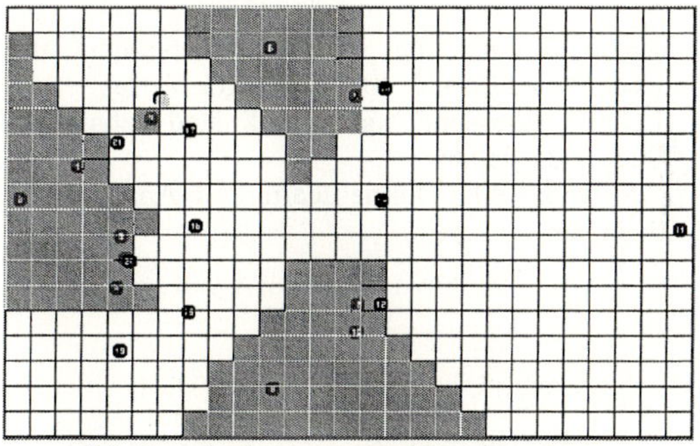

(b) the pitch structured by the teams

Fig. 9. Spatial relational structure

The lattice hierarchical structure is potentially a *meta-representation* for this process. With this architecture, autonomous systems can investigate structured sets and keep information on those that are 'interesting'. We would argue that the relational structure underlying the lattice hierarchy will characterise any vocabulary, and therefore the lattice hierarchy will be fundamental in automatic construct abstraction.

6 Conclusions

In this paper we have defined lattice hierarchies as fundamental structures in complex system. For us, the intermediate word problem is fundamental in robot soccer. We have investigated relational structure in simulated robot soccer games, and shown how it fits into lattice hierarchies. The ultimate goal of the research is to have the lattice hierarchical vocabulary emerge automatically as the robots interact with their environment. This would make a significant contribution to complexity science and its application in other areas.

References

1. Takahashi, T. and T. Naruse, From Play Recognition to Good Plays Detection: Reviewing RoboCup 97 Teams from Logfile, in RoboCup-98: Robot Soccer World Cup II. 1998, Springer-Verlag. p. 187-192.

2. Takahashi, T., LogMonitor: From Player's Action Analysis to Collaboration Analysis and Advice on Formation, in RoboCup-99: Robot Soccer World Cup III, M. Veloso, E. Pagello, and H. Kitano, Editors. 2000, Springer-Verlag. p. 103-113.
3. Tanaka-Ishii, K., et al., A Statistical Perspective on the RoboCup Simulator League: Progress and Prospects, in RoboCup-99: Robot Soccer World Cup III, M. Veloso, E. Pagello, and H. Kitano, Editors. 2000, Springer-Verlag. p. 114-127.
4. Raines, T., M. Tambe, and S. Marsella, Automated Assistants to Aid Humans in Understanding Team Behaviours, in RoboCup-99: Robot Soccer World Cup III, M. Veloso, E. Pagello, and H. Kitano, Editors. 2000, Springer-Verlag. p. 85-102.
5. Wünstel, M., et al., Behavior classification with self-organizing maps, in RoboCup 2000, Robot Soccer World Cup IV, P. Stone, T. Balch, and G. Kraetschmar, Editors. 2001, Springer-Verlag: Berlin. p. 108-118.
6. Visser, U., et al., Recognizing Formations in Opponent Teams, in RoboCup-2000: Robot Soccer World Cup IV, P. Stone, T. Balch, and G. Kraetschmar, Editors. 2001, Springer-Verlag.
7. Riley, P. and M. Veloso, Recognizing probabilistic opponent movement models, in RoboCup-01, Robot Soccer World Cup V. 2002, Springer-Verlag.
8. Kaminka, G., et al. Learning the Sequential Coordinated Behavior of Teams from Observations. in RoboCup 2002. 2002. Japan.
9. Visser, U. and H.-G. Weland. Using Online Learning to Analyze the Opponents Behavior. in RoboCup 2002. 2002. Japan.

Recognition and Prediction of Motion Situations Based on a Qualitative Motion Description

Andrea Miene, Ubbo Visser, and Otthein Herzog

TZI - Center for Computing Technologies
University of Bremen
Universitätsallee 21-23
D-28359 Bremen, Germany
{andrea,visser,herzog}@tzi.de

Abstract. High-level online methods become more and more attractive with the increasing abilities of players and teams in the simulation league. As in real soccer, the recognition and prediction of strategies (e.g. opponent's formation), tactics (e.g. wing play, offside traps), and situations (e.g. passing behavior) is important. In 2001, we proposed an approach where spatio-temporal relations between objects are described and interpreted in order to detect some of the above mentioned situations. In this paper we propose an extension of this approach that enables us to both interpret and predict complex situations. It is based on a qualitative description of motion scenes and additional background knowledge. The method is applicable to a variety of situations. Our experiment consists of numerous offside situations in simulation league games. We discuss the results in detail and conclude that this approach is valuable for future use because it is (a) possible to use the method in *real-time*, (b) we can *predict* situations giving us the option to refine agents actions in a game, and (c) it is *domain independent* in general.

1 Motivation and Related Work

When asking professional coaches in the soccer domain what they do after a game has started they tell us that the analysis of the opponents team is very important. First, the strategic information is considered. This can be the overall formation (e.g. 4-4-2) or whether the team is playing more offensive or defensive. The next step is to gather tactical information. One example is wing play or frequent use of the offside trap. Once this information is obtained the coach decides optional changes with regard to his own team.

If we would like to apply this to the RoboCup scenario, high-level online methods have to be developed. They become also more and more attractive with the increasing abilities of players and teams, preferably in the simulation league. The recognition or even better the prediction of strategies, tactics, and situations is an important feature that will improve a teams' performance.

In 2001, we proposed a method that interprets spatio-temporal relations based on motion direction and speed of single objects and spatial relations between two objects given by direction and distance. The approach assumes that

D. Polani et al. (Eds.): RoboCup 2003, LNAI 3020, pp. 77–88, 2004.

such information can be seen as time series. A threshold-based segmentation method is then used to derive temporal intervals from each time series. In addition, qualitative temporal relations between time intervals such as *before, meets, during* have been used. As a result, simple events such as *approaching, departing* and first complex events such as *player 1 passes ball to player 2* can be interpreted [8].

In this paper we describe a significant extension to this approach. First, a new monotonicity-based segmentation method will be described to derive more appropriate temporal intervals. Second, additional background knowledge about the problem domain is used for a better interpretation of the considered situation in a game.

Our approach is related to the work from Raines and colleagues [9] who describe an approach to automate assistants to aid humans in understanding team behaviors for the simulation league. Their approach ISAAC analyzes a game off-line using a decision tree algorithm to generate rules about the success of individual players. Also, the cooperation within a team is considered with the help of a pattern matching algorithm. ISAAC supports the analysis of so-called 'key events'. Key events are events which directly effect the result of the game. Therefore, single players are analyzed that directly shoot towards the goal. In case of the whole team, kicks of the ball by certain players which lead to a goal are analyzed. ISAAC has to be used off-line, thus the program is not able to support real-time conditions. The rules produced by ISAAC are intended to support the development of the analyzed team. Therefore, they show how successful the team is in certain situations. The approach is designed for the analysis of games to gain new experiences for the next game. The main difference to our approach is that this approach can be used off-line only. Also, key events are limited (e.g. only a single key event is used in the single player scenario).

Riley and Veloso in 2002 [10] use a set of pre-defined movement models and compare these with the actual movement of the players in set play situation. In new set play situations the coach then uses the gathered information to predict the opponent agent's behavior and to generate a plan for his own players. The approach can be used both off-line and on-line. The main difference to our approach described in this paper is that they analyze the movement of all players in set play situations.

Frank and colleagues [3] presented a real time approach which is based on statistical methods. The approach gathers information such as the percentage of ball-holding of a certain player or which player passes the ball to which team mate. The result is a thorough statistical analysis which can then be used to derive information about a game being played. This can help for new future developments of a team. The main difference to our approach is that this approach is designed to gather information that can be used after the game.

A hybrid approach to learn the coordinated sequential behavior of teams was presented by Kaminka and colleagues in 2002 [7]. The idea is to take time-series of continuous multi-variate observations and then parse and transform them into a single-variable categorial time-series. The authors use a set a behavior recognizers that focus only on recognizing simple and basic behaviors or the agents (e.g. pass, dribble). The data are then represented in a trie (a tree-like data structure) to support two statistical methods: (a) frequency counting and (b) statistical dependency detection. Experiments showed that the latter method is more suitable to discover sequential behavior. The main difference to our approach are the data the approach is based on and the fact that this approach is designed for unsupervised learning.

Huang and colleagues [6] recently published an approach for plan recognition and retrieval for multi-agent systems. The approach is based on observations of agents' coordinative behaviors. The basis are players' element behaviors sequences (e.g. pass, dribble, shoot) which are sorted in a temporal order. The field is decomposed into cells where each cell denotes one agent's behavior at a time slice. Interesting and frequent behavior sequences are considered as the team's plans on the assumption that the team's plan is embedded in those sequences. The discovery of significance of sequence patterns are based on statistical evidences. The promising results are plans based on observation. The difference to our approach is the analysis of the sequences. Huang and colleagues use a statistical-based analysis. Also, the interpretation of the results are different. The rules are obtained manually.

The remaining sections are organized as follows: the next section provides information about the qualitative description of motion scenes. Section 3 gives an overview about the background knowledge used and how we can use this knowledge to interpret the scene. The application and results of our approach within the soccer domain are discussed in section 4. Conclusions and future work are pointed out in the last section.

2 Qualitative Description of Motion Scenes

In this section we present our extended approach on a qualitative description of motion scenes that we presented first in [8]. The basic assumption of our approach is that we have a bird view of a motion scene. We further need a set of coordinates describing the positions of the moving objects for each moment (or cycle). Motion causes change not only for a single moving object but also for its spatial relations to other surrounding objects. To take into account both absolute (individual) movement and change in spatial relations (i.e., relative movement) of objects we calculate four types of time series from the raw positional information: the motion direction and speed of each object, and the spatial direction and distance for each pair of objects (see fig. 1).

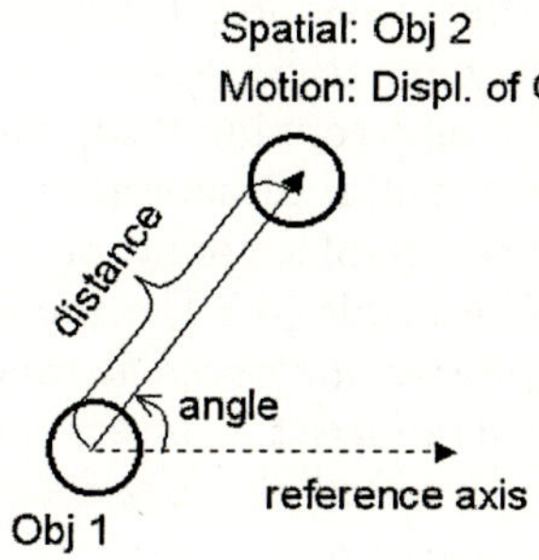

Fig. 1. Motion and spatial relations via direction and length.

These time series describe the motion within a scene on a quantitative level. In order to describe the motion on a qualitative level two steps of abstraction are performed:

- a temporal **segmentation** of the time series into time intervals of homogeneous motion and
- a **mapping** of the attribute values describing the intervals to qualitative classes.

The entire process is carried out online, i.e., at each time cycle one set of positional data is processed. Intervals are either extended or a new interval is started with the actual value if the homogeneity criterion fails.

Segmentation

In order to segment the time series into time intervals two different segmentation methods are used: a threshold-based segmentation method and a monotonicity-based segmentation method, which groups together strictly monotonic increasing intervals, strictly monotonic decreasing intervals and intervals of constant values. Each threshold-based segmented interval is described by a single attribute: the average of its values. A monotonicity-based segmented interval is described by its start value, its end value, and the run direction of values: increasing, decreasing or constant.

Both segmentation methods allow various interpretations of the resulting intervals. The monotonicity-based segmentation is useful to recognize dynamic aspects of motion, e.g., acceleration of a moving object. But due to the fact that the values are measured only at the start and the end of an interval its intermediate values are not known. Therefore, the threshold-based segmentation is more useful to find, e.g. an object that moves with a certain average speed.

Mapping into Classes

The second step of abstraction classifies the attributes of the intervals onto qualitative values for direction, speed or distance, respectively. The mapping functions have to be defined with respect to the domain. For the soccer domain the following mapping functions are used: For the directions (motion direction as well as spatial direction) eight classes as indicated by the dotted lines in fig. 2, i.e., from the viewpoint of object A, object B is in direction 5, object C is in direction 8 and so on. For the distances five classes are valid: *meets, very close, close, medium distance, far* and *very far* as indicated by the dashed circles in fig. 2. There object A meets object B and is very close to C, close to D and so on. For the speed also five classes are distinguished: *no motion, very slow, slow, medium speed, fast* and *very fast.* The speed and distance classes are organized in distance systems [5]. The radius of each distance class is double the size of the radius of the previous one.

For each pair of objects 12 sequences of temporal intervals describe their individual and relative motion: for each of the two objects we obtain one time

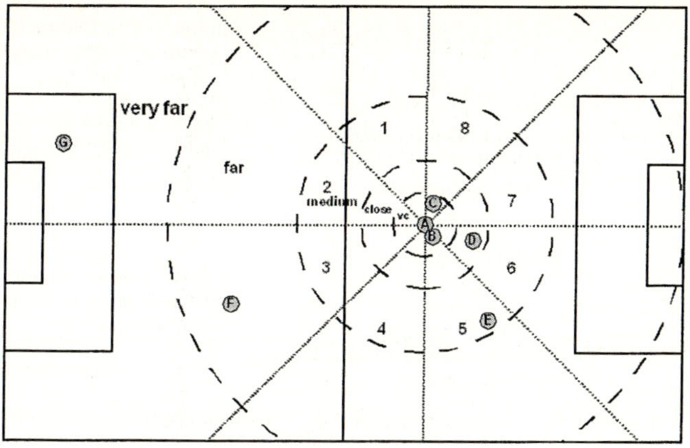

Fig. 2. Spatial relation classes in the soccer domain.

series concerning its motion direction and one concerning its speed (4 time series) and for the two objects one concerning the spatial direction and one concerning the distance (2 time series). Each of these 6 time series is segmented with the two different segmentation methods described in the previous section. Therefore, altogether we obtain 12 interval sequences.

The entire generation of motion descriptions is shown in fig. 3. The example shows the raw positional input data at the left. The time series calculated from the raw data and the results of the monotonicity-based segmentation method are illustrated in the middle (here: a single time series, the distance of two objects). One of the resulting intervals is shown with its attribute values as well as the mapping of values to classes. The single interval already allows a simple interpretation of the movement of the two involved objects: they approach each other and finally meet, which is expressed by the term HOLDS(approach-and-meet(p, q), $\langle t_n, t_{n+k}\rangle$). The predicate HOLDS expresses the coherence between a certain situation (movement or spatial relation), here approach-and-meet and the time interval $\langle t_n, t_{n+k}\rangle$ in which it is taking place or is valid [1].

3 Rule-Based Interpretation of Motion Scenes

These motion description intervals are used to recognize as well as predict motion situations with the help of a logic-based interpretation approach.

Domain knowledge is required for an interpretation of the motion. To know about the function or type of objects involved in a situation leads to more appropriate interpretations. For example, in the soccer domain the interpretation that two objects approach each other and finally meet can than be interpreted that a player gets in contact with the ball. To specialize the interpretation even more, different types of players can be distinguished, e.g., goalkeepers, defenders and offenders.

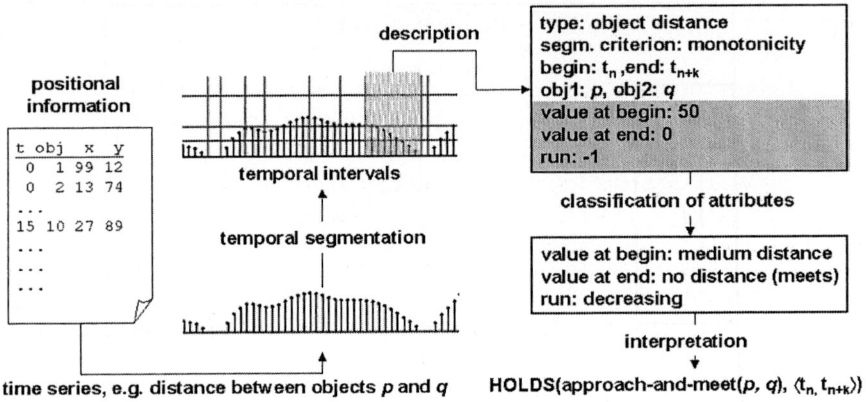

Fig. 3. Overview: Generation of motion description.

In some domains the location is important in which a certain motion situation takes place. For the soccer domain such locations are certain regions on the field of play, e.g. each half of the field, penally area, goal area and so on. E.g. the term HOLDS(region(*player, left-half*), $\langle t_n, t_{n+k} \rangle$) denotes that the object *player* is in the left half of the field of play during the time interval $\langle t_n, t_{n+k} \rangle$.

Currently, we have defined rules to recognize and predict 10 situations from the soccer domain. They include simple situations like a player kicking the ball as well as more complex ones like a one-two situation, a fight for the ball and offside.

For an experiment, we will have a closer look at the offside situation, because it is possible to predict an impending offside situation, that may occur while a team mate is planing to pass the ball. And, as we will show, both aspects of motion information – absolute and relative – are needed to detect and predict offside situations. In addition, further finer interpretations are possible, e.g., if an offside situation occurs it is possible to distinguish an offside trap from a situation that was caused by the offender himself.

Experiment: Offside Position

A player is in an offside position if he is nearer to his opponents' goal line than both the ball and the second last opponent. But he is not in an offside position in his own half of the field of play. For more details on the official offside rule refer to the FIFA rules [2], law 11 and appendix.

In order to recognize, whether a player is in an offside position we have to check if he is in the opponents' half of the field of play. If so, we have to analyze his spatial relation to the ball and the players of the opposite team. In detail we must determine if the ball is *behind* the player and count the amount of opponents that are *in front of* the player. If less than two opponents remain in front of the player, he is in an offside position:

$$
\begin{aligned}
&\textsc{holds}(\mathsf{offsideposition}(player), \langle max(s_i), min(e_i)\rangle) \Leftrightarrow \\
&\exists \langle s_i, e_i\rangle, i \in \{1,2,3\}: \\
&\quad ((\textsc{holds}(\mathsf{region}(player, right\text{-}half), \langle s_1, e_1\rangle) \land \mathsf{team}(player) = 1) \lor \\
&\quad\ (\textsc{holds}(\mathsf{region}(player, left\text{-}half), \langle s_1, e_1\rangle) \land \mathsf{team}(player) = 2)) \land \\
&\quad \textsc{holds}(\mathsf{behind}(ball, player), \langle s_2, e_2\rangle) \land \\
&\quad \textsc{holds}(\mathsf{number\text{-}of\text{-}opponents\text{-}in\text{-}front\text{-}of}(player, n), \langle s_3, e_3\rangle) \land n < 2 \land \\
&\quad \forall i, j \in \{1, \ldots, 3\}: s_i < e_j.
\end{aligned}
\tag{1}
$$

The term $\textsc{holds}(\mathsf{number\text{-}of\text{-}opponents\text{-}in\text{-}front\text{-}of}(p, n), \langle max(s_i), min(e_i)\rangle)$ denotes the number n of opponents located in front of a player p during the time interval $\langle max(s_i), min(e_i)\rangle$, where k is the number of players belonging to the opposite team:

$$
\begin{aligned}
&\textsc{holds}(\mathsf{number\text{-}of\text{-}opponents\text{-}in\text{-}front\text{-}of}(p, n), \langle max(s_i), min(e_i)\rangle) \Leftrightarrow \\
&\exists \langle s_i, e_i\rangle, i \in \{1, \ldots, k\}: \\
&\quad \forall g_i \in \{g_1, \ldots, g_n\}: \textsc{holds}(\mathsf{in\text{-}front\text{-}of}(g_i, p), \langle s_i, e_i\rangle) \land \\
&\quad \forall g_i \in \{g_{(k-n)}, \ldots, g_k\}: \textsc{holds}(\mathsf{behind}(g_i, p), \langle s_i, e_i\rangle) \land \\
&\quad \forall g_i \in \{g_1, \ldots, g_k\}: \mathsf{team}(g_i) \neq \mathsf{team}(p) \land \\
&\quad \forall i, j \in \{1, \ldots, k\}: s_i < e_j.
\end{aligned}
\tag{2}
$$

A complex situation like the above definition of offsideposition($player$) combines several time intervals. The term $\forall i, j \in \{1, \ldots, n\} : s_i < e_j$ postulates that all n intervals involved in the situation are contemporary. $\langle max(s_i), min(e_i)\rangle$ specifies the sub-interval covered by all n time intervals $\langle s_i, e_i\rangle, 1 \leq i \leq n$.

The spatial relations behind and in-front-of are generalizations of the 8 directions shown in fig. 2. Another object is in-front-of a certain player if it is between the player and the opponents' goal and otherwise behind the player. Therefore, the evaluation of the generalization rule depends on the team the player belongs to.

Eq. 3 specifies the spatial relation in-front-of. The spatial relation behind as well as the motion directions forward and backward are specified similarly.

$$
\begin{aligned}
&\textsc{holds}(\mathsf{in\text{-}front\text{-}of}(object, player), \langle s, e\rangle) \Leftrightarrow \\
&\quad \textsc{holds}(\mathsf{spatdir}(player, object, dir), \langle s, e\rangle) \land \\
&\quad ((dir \in \{5, 6, 7, 8\} \land \mathsf{team}(player) = 1) \lor \\
&\quad (dir \in \{1, 2, 3, 4\} \land \mathsf{team}(player) = 2)).
\end{aligned}
\tag{3}
$$

The term $\textsc{holds}(\mathsf{spatdir}(player, object, dir), \langle s, e\rangle)$ denotes that $object$ is located in the direction dir from the viewpoint of $player$ during interval $\langle s, e\rangle$. This information is obtained from the threshold-based segmentation.

In order to predict an offside situation for player p, he has to be located in his own half, actually have the ball behind him and a small remaining number of k opponent defenders (e.g., k=3-4) in front of him. Then it depends on the relative movement of p and q if an offside position is impending or not. Therefore, we have to take into account the actual spatial direction between p and q (spatdir), obtained from the threshold based segmentation, and the development of the spatial direction between p and q (clockwise (change-spatdir-cw) or counterclockwise (change-spatdir-ccw), obtained from the monotonicity-based

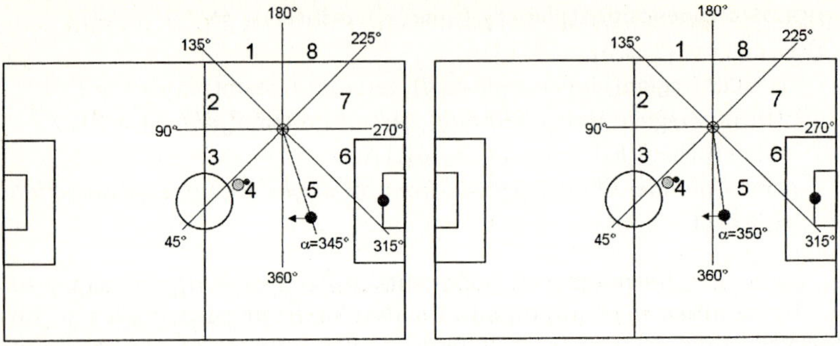

Fig. 4. Development of spatial directions between offender and defender announcing an impending offside position.

segmentation). If the spatial direction is already close to the change between the directions *in front of* and *behind*, and the values are increasing (clockwise change of spatial directions) or decreasing (counterclockwise change of spatial directions) an offside position is impending. For an illustration of this situation refer to fig. 4. The illustration shows the case of an increasing development of values. If the present trend lasts for some further time, an offside situation will occur in the moment the spatial relation changes to the next class (in the illustration from 5 to 4) and at the same point in time from *in front of* to *behind*.

$$\begin{aligned}
&\textsc{holds}(\mathsf{offside\text{-}danger}(p,q), \langle max(s_i), min(e_i)\rangle) \Leftrightarrow \\
&\exists \langle s_i, e_i\rangle, i \in \{1,\ldots,6\}: \\
&\quad ((\textsc{holds}(\mathsf{region}(p, \textit{right-half}), \langle s_1, e_1\rangle) \wedge \mathsf{team}(p) = 1) \vee \\
&\quad\ (\textsc{holds}(\mathsf{region}(p, \textit{left-half}), \langle s_1, e_1\rangle) \wedge \mathsf{team}(p) = 2)) \wedge \\
&\quad \textsc{holds}(\mathsf{behind}(\mathit{ball}, p), \langle s_2, e_2\rangle) \wedge \\
&\quad \textsc{holds}(\mathsf{in\text{-}front\text{-}of}(q,p), \langle s_3, e_3\rangle) \wedge \mathsf{team}(p) \neq \mathsf{team}(q) \wedge \\
&\quad \textsc{holds}(\mathsf{number\text{-}of\text{-}opponents\text{-}in\text{-}front\text{-}of}(p,n), \langle s_4, e_4\rangle) \wedge 2 \leq n < k \wedge \\
&\quad ((\textsc{holds}(\mathsf{change\text{-}spatdir\text{-}cw}(p,q), \langle s_5, e_5\rangle) \wedge \\
&\quad\ \textsc{holds}(\mathsf{spatdir}(p,q, 1 \vee 5), \langle s_6, e_6\rangle)) \vee \\
&\quad\ (\textsc{holds}(\mathsf{change\text{-}spatdir\text{-}ccw}(p,q), \langle s_5, e_5\rangle) \wedge \\
&\quad\ \textsc{holds}(\mathsf{spatdir}(p,q, 4 \vee 8), \langle s_6, e_6\rangle))) \wedge \\
&\quad \forall i,j \in \{1,\ldots,6\} : s_i < e_j.
\end{aligned} \tag{4}$$

Within the prediction phase we distinguish offside traps (see (6)) from offside situations caused solely by the movement of the offender himself. The temporal relation *contemporary* is defined as in [4]:

$$\begin{aligned}
&\textsc{holds}(\mathsf{prediction\text{-}offside\text{-}own\text{-}motion}(p,q), \langle max(s_1,s_2), min(e_1,e_2)\rangle) \Leftrightarrow \\
&\exists \langle s_1, e_1\rangle, \langle s_2, e_2\rangle : \\
&\quad \textsc{holds}(\mathsf{offside\text{-}danger}(p,q), \langle s_1, e_1\rangle) \wedge \textsc{holds}(\mathsf{forward}(p), \langle s_2, e_2\rangle) \wedge \\
&\quad \mathit{contemporary}(\langle s_1, e_1\rangle, \langle s_2, e_2\rangle).
\end{aligned} \tag{5}$$

An offside trap is caused by a forward movement of an opponent q remaining between the goal and the offender p. The offender is brought into an offside position by the movement of his opponents (with or without moving forward by himself).

$$\begin{aligned}&\textsc{holds}(\mathsf{prediction\text{-}offsidetrap}(p,q),\langle max(s_1,s_2),min(e_1,e_2)\rangle) \Leftrightarrow\\&\exists\langle s_1,e_1\rangle,\langle s_2,e_2\rangle :\\&\quad\textsc{holds}(\mathsf{offside\text{-}danger}(p,q),\langle s_1,e_1\rangle) \wedge \textsc{holds}(\mathsf{forward}(q),\langle s_2,e_2\rangle)\wedge\\&\quad contemporary(\langle s_1,e_1\rangle,\langle s_2,e_2\rangle).\end{aligned} \tag{6}$$

A player p is in a punishable offside position, if he is in an offside position in the moment when the ball is kicked by his team mate p_2, and he approaches the ball while the ball is free, i.e. before another player obtains the ball (7).

$$\begin{aligned}&\textsc{holds}(\mathsf{offside\text{-}punishable}(p),\langle max(s_j,s_m),min(e_l,e_m)\rangle) \Leftrightarrow\\&\exists j,k,l,m,p_2 :\\&\quad\textsc{occur}(\mathsf{kick}(p_2),j) \wedge \textsc{holds}(\mathsf{offside\text{-}position}(p),k)\wedge\\&\quad\textsc{holds}(\mathsf{ball\text{-}free},l) \wedge \textsc{holds}(\mathsf{approaching}(p,ball),m)\wedge\\&\quad starts(j,l) \wedge in(j,k) \wedge contemporary(l,m) \wedge \mathsf{team}(p)=\mathsf{team}(p_2).\end{aligned} \tag{7}$$

with $j = \langle s_j, e_j\rangle, k = \langle s_k, e_k\rangle, l = \langle s_l, e_l\rangle$ and $m = \langle s_m, e_m\rangle$.

The predicate $\textsc{occur}(e,t)$ states that an event $\mathsf{kick}p_2$ occurs at the moment j [1]. The temporal relations *starts* and *in* are defined in [1], *contemporary* in [4].

If the player comes to close to the ball this behavior should be penalized by the referee by interrupting the game for a free kick of the opponent team.

4 Results

To evaluate our approach we have chosen three games from the Robocup Worldcup 2002, which contain a reasonable amount of offside situations: FC Portugal vs. Puppets, TsinghuAeolus vs. FC Portugal and VW2002 vs. Cyberoos. Therefore, we have analyzed these three games in order to predict and recognize the occurring offside situations. Table 1 and 2 show the offside situations that occur in the games FC Portugal vs. Puppets and TsinghuAeolus vs. FC Portugal. For lack of space a table showing the table concerning the game VW2002 vs. Cyberoos is not included in this paper.

The first column (*ball contact*) denominates the player who kicked the ball before the penally offside situation occurred together with the time interval, at which he was in contact with the ball. The column *offside* contains the player numbers of his team mates which were already in an offside position before he obtained the ball. The next two columns contain our systems prediction of impending offside positions of further players. The column *mo.* contains the numbers of the players, who are moving in a direction that will possibly bring them into an offside position before the ball will be passed. The fourth column lists players who are possibly running into an *offside trap*. The column *offside/kick* lists the players who are in an offside position at the moment the ball is kicked. The column *penally* contains the players who have been in an offside position at the moment the ball is kicked and are approaching the ball during the following cycles. They are penalized by the referee if they come to close to the ball. The column (*break*) contains the cycle in which the game was interrupted by the referee. The tables contain all situations in which the game was interrupted by the referee due to offside. The last column *s* is marked with an ✓ if the situation was recognized by the system.

Concerning the prediction of offside situations there are also cases of impending offside situations that do not lead to an offside position before the ball is kicked the next time. Also there are situations in which a player, who was in an offside position at the moment the ball was kicked, starts approaching the ball in a penally way but another player gets the ball before the situation becomes critical. To keep them short, these situations are not included in the tables.

The game FC Portugal (FCP) against Puppets (see table 1) was interrupted 9 times by the referee due to offside. In 7 cases our system detected the offside situation. In two situations our systems is not in line with the referee. The first situations occurred from cycle 531 to 560. When player 7 of team Puppets kicks the ball (cycle 534), players 10 and 11 are in an offside position. In the following cycles player 11 approaches the ball, until he is in a very close distance to the ball (cycle 559), which was detected by our system. But in cycle 559 player 3 of the opponent team (FCP) gets in contact with the ball. In this moment our system stops looking for an penally offside for team Puppets, because FCP is already in possession of the ball. Nevertheless, the referee decided on offside and free kick in favor of FCP in cycle 560. According to our operationalization of the FIFA rules the referee should have interrupted the game before cycle 559 or should have let it go on after player 3 of FCP has reached the ball in cycle 559.

A comparable situation can be found from cycle 3844 to 3852. In cycle 3847 the ball is kicked by player 9 of FCP. Player 8 is in an offside position and approaches the ball in cycle 3850. This is detected by our system. But in the same cycle the ball touches player 4 of team Puppets. As before, we stop watching player 8 of FCP. But although the ball has touched a player of the opponent team the referee decides offside and penalizes player 8 of FCP in cycle 3852, which is obviously not in compliance to the FIFA rules.

Table 1. Offside situations FC Portugal (FCP) vs. Puppets (Pup).

Ball contact	offside	mo.	offside trap	offside/kick	punishable	break	s
Pup 7 474-478	10	11	$11_{Pup} \leftrightarrow 2_{FCP}$	10, 11	11	487	✓
Pup 7 531-534	10, 11			10, 11	11	560	
FCP 8 889-891	9, 10, 11			9, 10, 11	11	915	✓
FCP 10 1166-1168	6	7	$7_{FCP} \leftrightarrow 4_{Pup}$	6	6	1172	✓
Pup 5 2417-2422	10, 11			10, 11	11	2429	✓
FCP 9 3091-3094	10			10	10	3102	✓
FCP 9 3385-3386	10	11	$11_{FCP} \leftrightarrow 4_{Pup}$	10	10	3404	✓
FCP 9 3844-3847		8, 10	$8, 10_{FCP} \leftrightarrow 2_{Pup}$	8, 10	–	3852	
FCP 6 5589	10			10	10	5599	✓

Table 2. Offside situations TsinghuAeolus (TsA) vs. FC Portugal (FCP).

Ball contact	offside	mo.	offside trap	offside/kick	punishable	break	s
TsA 4 1279-1289		10	$9, 10, 11_{TsA} \leftrightarrow 2, 3, 5_{FCP}$	11	11	1304	✓
TsA 6 1547-1551		11	$11_{TsA} \leftrightarrow 2, 3_{FCP}$	9, 11	9	1554	✓
TsA 6 1575-1578		9	$9, 10, 11_{TsA} \leftrightarrow 4, 5_{FCP}$	9, 10, 11	9, 10	1584	✓
FCP 4 1650-1657	10, 11			10, 11	11	1673	✓
FCP 7 2944	9, 10, 11			9, 10, 11	9	2950	✓
TsA 8 3300-3302	9, 10		$11_{TsA} \leftrightarrow 5_{FCP}$	9, 10, 11	10, 11	3322	✓
TsA 2 4035-4038			$9, 10, 11_{TsA} \leftrightarrow 4_{FCP}$	10	10	4047	✓
TsA 6 4219-4226	9		$10, 11_{TsA} \leftrightarrow 5_{FCP}$	9, 10, 11	9	4228	✓
TsA 6 5047-5053	9, 10, 11			9, 10, 11	9	5056	✓

The game TsinghuAeolus (TsA) vs. FC Portugal (FCP) (see table 2) was interrupted 9 times by the referee due to offside. In all cases our system detected the offside situation.

The game VW2002 (VW) vs. Cyberoos (Cyb) was interrupted 35 times by the referee due to offside. In 29 cases our system detected the offside situation. In six situations the referee decides offside against a team A although a player of team B has touched the ball before the game was interrupted.

5 Conclusion and Future Directions

Spatio-temporal relations between objects within real-time environments are challenging by nature. We presented an approach for tracking single objects motion in combination with the changes in their pairwise spatial relations over time. The resulting motion description builds the basis for a qualitative interpretation of the dynamic scene.

This approach is domain independent and can therefore be used in various applications. We applied this idea to the soccer domain and argue that an implementation of this method within the online coach could enhance teams abilities. However, tests have been made off-line only at the moment. The additional background knowledge helps to interpret the analyzed motion scenes and significantly improves the results.

The described approach is valuable because it not only analyzes a past situation, it also is able to *predict* the next few steps of the opponents team to a certain extent. This will help the players of the own team to make better decisions at a certain cycle provided they have the information and can act accordingly. Also, when using this approach in an online scenario, the position data of the players have to be considered. For off-line analysis we use the data processed by the soccer server. These data can be quite different than those in the world model of a single player. Future tests have to be made in order to obtain valuable information about this problem. A possible solution to get all the information about the positions of both the opponents and the own team players is the based on the turn_neck-command and the aggregation of positions over a few cycles.

One of the biggest advantages of this approach is the independence from the domain. In the near future, we will also test other domains such as cell tracking in biological systems. Here, the objects are monitored with a camera and the method is able to track the objects over time and describe and store the spatial relations between them as well.

References

1. James F. Allen. Towards a general theory of action and time. *Artificial Intelligence*, pages 123–154, 1984.
2. FIFA. Laws of the game. Féderation Internationale de Football Association, FIFA House, 11 Hitzigweg, 8030 Zurich, Schweiz, July 2002.
3. Ian Frank, Kumiko Tanaka-Ishi, Katsuto Arai, and Hitoshi Matsubara. The statistics proxy server. In Tucker Balch, Peter Stone, and Gerhard Kraetschmar, editors, *4th International Workshop on RoboCup*, pages 199–204, Melbourne, Australia, 2000. Carnegie Mellum University Press.
4. C. Freksa. Temporal reasoning based on semi-intervals. *Artificial Intelligence*, 54(1):199–227, 1992.
5. D. Hernández, E. Clementini, and P. Di Felice. Qualitative distances. In *Proceedings of COSIT'95, LNCS 988*, Semmering, Austria, 1995. Springer.
6. Zhanxiang Huang, Yang Yang, and Xiaoping Chen. An approach to plan recognition and retrieval for multi-agent systems. In Mikhail Prokopenko, editor, *Workshop on Adaptability in Multi-Agent Systems, First RoboCup Australian Open 2003 (AORC-2003)*, Sydney, Australia, 2003. CSIRO.
7. Gal Kaminka, Mehmet Fidanboylu, Allen Chang, and Manuela Veloso. Learning the sequential coordinated behavior of teams from observation. In Gal Kaminka, Pedro Lima, and Raul Rojas, editors, *Proceedings of the RoboCup-2002: Robot Soccer World Cup VI*, pages 104–118, Fukuoka, Japan, 2002.
8. Andrea Miene and Ubbo Visser. Interpretation of spatio-temporal relations in real-time and dynamic environments. In Andreas Birk, Silvia Coradeschi, and Satoshi Tadokoro, editors, *RoboCup 2001: Robot Soccer World Cup V*, volume 2377 of *Lecture Notes in Artificial Intelligence*, pages 441–446. Springer, Seattle, WA, 2002.
9. Taylor Raines, Millind Tambe, and Stacy Marsella. Automated assistants to aid humans in understanding team behaviors. In *Fourth International Conference on Autonomous Agents (Agents 2000)*, Barcelona, Spain, 2000.
10. Patrick Riley and Manuela Veloso. Recognizing probabilistic opponent movement models. In *RoboCup-01, Robot Soccer World Cup V*, Lecture Notes in Computer Science, Seattle, Washington, 2002. Springer-Verlag. in print.

Evaluating Team Performance at the Edge of Chaos

Mikhail Prokopenko and Peter Wang

CSIRO Mathematical and Information Sciences
Locked Bag 17, North Ryde, NSW 1670, Australia
{mikhail.prokopenko,peter.wang}@csiro.au

Abstract. We introduce a concise approach to teamwork evaluation on multiple levels – dealing with agent's *behaviour spread* and multi-agent *coordination potential*, and abstracting away the team decision process. The presented quantitative information-theoretic methods measure *behavioural* and *epistemic* entropy, and detect phase transitions – the *edge of chaos* – in team performance. The techniques clearly identify under-performing states, where a change in tactics may be warranted. This approach is a step towards a unified quantitative framework on behavioural and belief dynamics in complex multi-agent systems.

1 Introduction

The emergence of system-level behaviour out of agent-level interactions is a distinguishing feature of complex multi-agent systems – making them very different from other complicated multi-component systems, where multiple links among the components may achieve efficient interaction and control with fairly predictable and often pre-optimised properties. In robotic soccer, the emergent behaviour is dependent on agents architecture and skills, the employed communication policy, the opponent tactics and strategies, and not least on various unknown factors present in the environment. In short, it appears to be extremely difficult to rigorously investigate and evaluate multi-agent teamwork, coordination, and overall performance. One possible avenue for measuring team performance is to use information-theoretic methods. In particular, we suggest to characterise dynamics of multi-agent teams in terms of generic information-theoretic properties, such as entropy, and correlate it with the overall team performance metrics.

Information-theoretic methods are applied in many areas exhibiting multi-agent interactions. For instance, Cellular Automata (CA) are a well-studied class of discrete dynamical systems, where information-theoretic measures of complexity (such as Shannon entropy of certain frequency distributions) were effectively used to categorise and classify distinct emergent configurations and phase transitions between them [17, 6]. Langton has shown in his seminal work [6] that an increase in the mutual information (defined as a function of individual cell entropies for a particular value of the λ parameter) is an indication of a phase transition from "order" to "chaos". Wuensche [17] has used a similar quantitative metric – variance of input-entropy over time – in classifying rule-space of 1-dimensional CA into ordered, complex and chaotic cases, related to Wolframs's qualitative classes of CA behaviour [16].

It could be argued that the complexity of emergent behaviour increases with a) the complexity of the agents, b) the diversity of the agents, achieved either by origi-

D. Polani et al. (Eds.): RoboCup 2003, LNAI 3020, pp. 89–101, 2004.

nal design or by a learning process, and c) the variety of the communication connections among agents. In the context of RoboCup, the behavioural diversity (the second component of our argument) was extensively analysed by Balch [1], who suggested a new metric – hierarchic social entropy – to characterise the heterogeneity of agents behaviours across a team. Robots are said to be *absolutely behaviorally equivalent* if and only if they select the same behaviour in every perceptual state. Balch introduced the concept of hierarchical clustering as "a means of dividing a society into subsets of behaviorally equivalent agents at a particular taxonomic level" [1], and developed a measure of *behavioral difference*, enabling agent categorisation and subsequent calculation of the hierarchic social entropy.

We will initially focus on the first component – the diversity of a single agent's behaviour in different situations. In other words, we analyse a relation between entropy of an individual agent's behaviour and the team performance. Our conjecture, supported by experimental results, is that each agent is able to express more versatile behaviour when faced with easier opposition. Conversely, when opposing stronger teams, each agent may not be able to realise its behaviour in full – leading to lower *behavioural entropy*. Intuitively, these two extremes resemble the "ordered" and "chaotic" states: when the opponent is too strong then the agent's behaviour is limited to "fixed point" or "limit cycle" attractors, while weak opponents do not put significant constraints allowing the agent to achieve "strange" attractors symptomatic of chaotic behaviour. If this conjecture is true, then "complex" behaviour lies at *the edge of chaos*, and the behavioural entropy would point to a phase transition in this region. Put simply, when playing opponents of similar strength the agents exhibit most interesting "complex" behaviour, but at the same time it becomes much harder to evaluate the performance.

In the second part of the paper we study the third component – the complexity of the inter-agent communications, related to potential of multi-agent coordination. The analysis is focused on entropy of joint beliefs – the *epistemic entropy* – and complements the results reported earlier [10]. The epistemic entropy approach uses the information entropy as a precise measure of the degree of randomness in the agents' joint beliefs. Intuitively, the system with near-zero epistemic entropy (almost no "misunderstanding" in joint beliefs) has a higher multi-agent coordination potential than the system with near-maximal entropy (joint beliefs are almost random). In addition, we identified and considered two coupled levels of dynamic activity (following the Kugler-Turvey model) – showing that self-organisation and the loss of epistemic entropy occur at the macro (agent coordination) level, while the system dynamics on the micro level (within the communication space) generates increasing disorder. The entropy within the communication space is also traced against team performance metrics, showing that phase transitions occur in coordination-communication dynamics as well.

In summary, the developed metrics allow us to evaluate team performance on multiple levels: from individual "behavioural spread" to multi-agent coordination potential.

2 Input-Entropy and the Edge of Chaos

2.1 Mutual Information and Phase Transitions

The information-theoretic analysis of phase transitions is typically based on the notions of *entropy* and *mutual information*. The *entropy* is a precise measure of the amount of freedom of choice in the object – an object with many possible states has high entropy. Formally, the entropy of a probability distribution $P = \{p_1; p_2; \ldots; p_m\}$ is defined by

$$H(P) = \sum_{i=1}^{m} p_i * \log(1/p_i) \tag{1}$$

The ratio of the actual to the maximum entropy is called the *relative entropy* of the source [14]. Langton [6] investigated the mutual information of CA, defined as a function of the individual cell entropies, $H(A)$ and $H(B)$, and the entropy of the two cells considered as a joint process, $H(A,B)$, that is: $I(A;B) = H(A) + H(B) - H(A,B)$, and related it to phase transitions. The average mutual information $I(A;B)$ has a distinct peak at the transition point: "the jump ... clearly indicates the onset of the chaotic regime, and the decaying tail indicates the approach to effectively random dynamics".

Peaks or discontinuities are symptomatic of phase transitions in complex multi-agent systems. For instance, Miramontes [7] analysed artificial ant societies, composed of interacting agents that can generate regular cycles in the activity of the colony, and pointed out that the information capacity of the colony is maximal at certain nest densities – in the neighbourhood of a chaos-order phase transition. In other words, the maximum in the average information capacity of the colony, given by the classical Shannon entropy, corresponds to "the density at which the nest reaches its highest diversity of activity states". When the nest density is increased beyond some critical density and the phase transition has occurred, "the number of ants becomes sufficiently large to facilitate and support the existence of long-range correlated behaviour that manifests itself as coherent collective oscillations in the number active ants" [7].

Another way to identify phase transitions is to use *a variance of input-entropy.* Wuensche [17] characterised rule-spaces of 1-dimensional cellular automata with the Shannon entropy of rules' frequency distribution. More precisely, given a rule-table (the rules that define a CA), the input-entropy at time step t is defined as

$$S^t = -\sum_{i=1}^{m} \frac{Q_i^t}{n} \, log \frac{Q_i^t}{n} \, ,$$

where m is the number of rules, n is the number of cells (system size), and Q_i^t is the look-up frequency of rule i at time t – the number of times this rule was used at t across the CA. The input-entropy settles to fairly *low* values for ordered dynamics, but fluctuates irregularly within a narrow *high* band for chaotic dynamics. For the complex CA, order and chaos may predominate at different times causing the entropy to vary. A measure of the variability of the input-entropy curve is its variance or standard deviation, calculated over time. Wuensche has convincingly demonstrated that only complex dynamics exhibits high variance of input-entropy, leading to automatic classification of the rule-space. Importantly, the peak of input-entropy variance points to a phase transition again, indicating the edge of chaos (complexity).

2.2 Measuring Agent's Behavioural Spread

Having identified the appropriate metrics and the forces shaping the space-time dynamics, we now proceed to the analysis of the heterogeneity of a single agent's behaviour in different situations. As mentioned earlier, we shall explore a relation between the entropy of an agent's behaviour and the team performance. Our intention is to characterise an agent's behaviour without a loss of generality, and thus we would prefer to abstract away a possibly convoluted decision-making mechanism. In other words, we intend to consider only the action rules (condition-action pairs) that the agent triggered at a given time step. In other words, we may use (and did use) the agents designed in accordance with the Deep Behaviour Projection (DBP) agent architecture [10] or another (multi-layered) architecture, but without taking into account the depth of the agent behaviour in terms of multiple decision-making layers. This allows us to apply the developed techniques to any rule-based approach capable of identifying action rules taken by an agent at a given time step. To perform our analysis we employ the input-entropy of a particular frequency distribution B_i^k, where k is a game index, and i is an action rule index: $1 \leq i \leq m$, where m is the number of rules. Analogously to the CA analysis conducted by Wuensche [17], we define the behavioural input-entropy as

$$E^k = -\sum_{i=1}^{m} \frac{B_i^k}{n} \, log \frac{B_i^k}{n} ,$$

where n is the system size (the total number of rule invocations), and B_i^k is the look-up frequency of rule i during the game k. The difference between S^t and E^k is that the former is calculated for each temporal state of the CA in point, while the latter is determined for each game in a multi-game experiment. Both metrics, however, characterise the distribution of rules – either across the CA lattice or during the game.

We intend to show that agents express more diverse behaviour when faced with easier opposition. Formally, the average behavioural input-entropy, calculated for K games against the opponent j: $E_j = \sum_{k=1}^{K} \frac{E^k}{K}$, should in general increase with the average score difference g_j, defined as the average difference between the agent's team score and the opponent team score. Importantly, a standard deviation σ_j of the behavioural entropy E^k calculated across all games against the opponent j, will be shown to be an indicator of a phase transition, reaching a maximum at some g_j close to zero.

3 Experiments: Behavioural Entropy and Phase Transitions

We have carried out our experiments in the RoboCup Simulation League [4], where the platform simulates essentially a pseudo real-time environment, providing each agent with fragmented, localised and imprecise (noisy and latent) information about the environment. Each experiment included 30 games between the test team and a particular opponent, producing a value for the behavioural entropy E^k, $1 \leq k \leq 30$, and a score difference g_j (a negative score difference represents losing the game). Figure 1 shows input-entropy trajectories for 3 experiments, ranging from a much stronger opponent (the average score difference $g = -6.33$), to an opponent of about the same strength as the test team ($g = -0.07$), to a much weaker opponent ($g = +10.17$). It is easy to observe that not only the the behavioural entropy E^k of a test agent (the left mid-fielder of the test team, in this case) decreases on average with the strength of the opponent, but also that E^k fluctuates in a much wider band in the medium case.

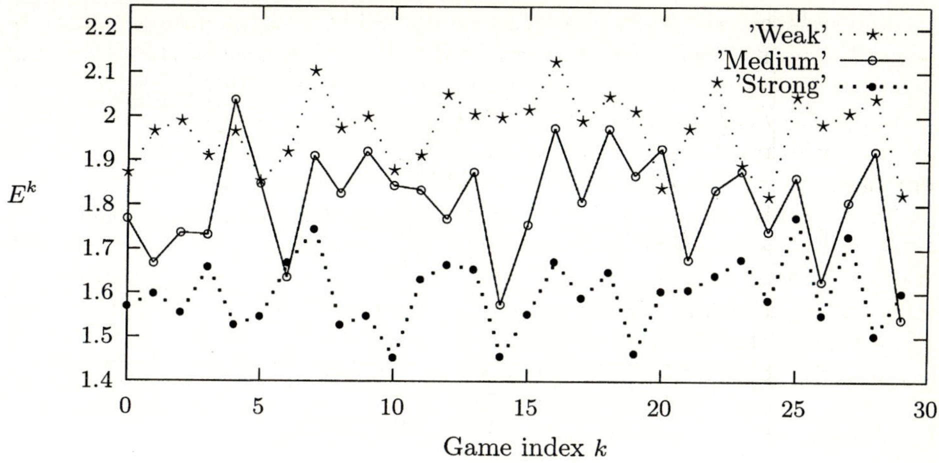

Fig. 1. Behavioural entropy E^k.

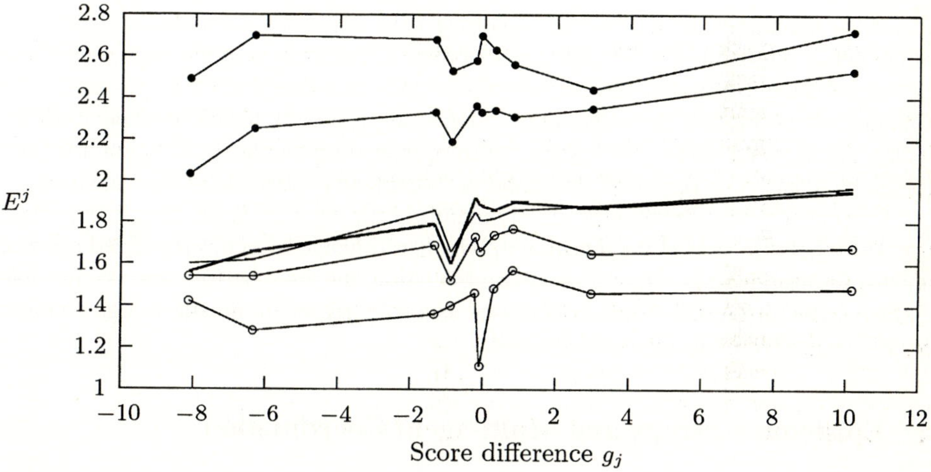

Fig. 2. Average behavioural entropy E_j. Two top plots represent forwards, two middle plots – midfielders, and two bottom plots – defenders.

To support this claim and to verify our conjecture that there is a phase transition, however, we conducted more experiments – against 10 opponents, collecting the statistics for 6 agents (wing- and centre-forwards, wing- and centre-midfielders, and wing- and centre-defenders). Figure 2 shows the average behavioural entropy (after $K = 30$ games), plotted for these 6 agents and for each of the opponents.

The tendency of the behavioural entropy to increase when faced with a weaker opposition is obvious in most field positions, and especially in the midfield. There is also an evident discontinuity exactly in the range we expected – when competing with the opponents of similar strength ($-0.87 \leq g_j \leq 0.83$). This discontinuity is indicative

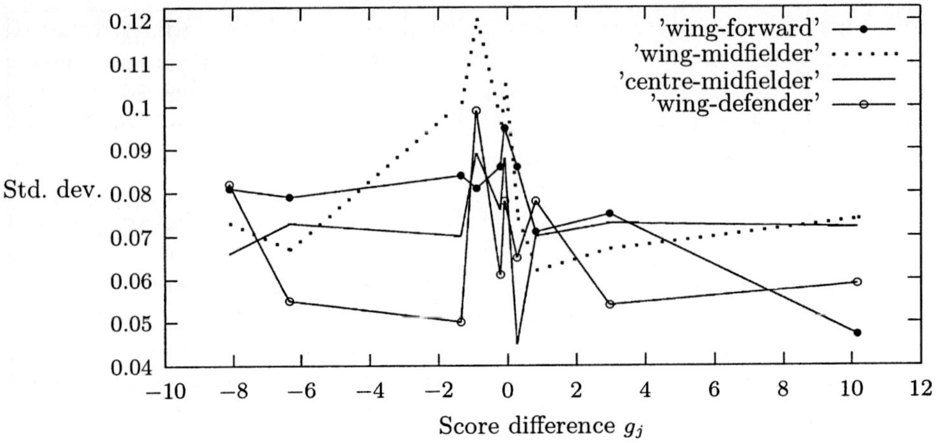

Fig. 3. Standard deviation of behavioural entropy.

of a phase transition. To confirm this, we observe the trajectory of standard deviation σ_j of the behavioural entropy E^k, calculated across all games against the opponent j, and shown in Figure 3. Standard deviation peaks in the expected region for all positions, and conclusively points to a phase transition. Interestingly, precise locations of peaks differ within the narrow range ($-0.87 \leq g_j \leq 0.83$), indicating that the peak is not a feature forced by a particular opponent, but rather a "complexity" attribute of the transition.

This entropy-based technique clearly identifies the *edge of chaos*. This is important because it helps to answer the question of whether a change in tactics is needed in some under-performing cases. During the phase transition, the team performance is unstable and the score difference should not be used as a sole trigger for drastic design changes or on-line interventions by a coach-agent.

4 Epistemic Entropy and Multi-agent Coordination

4.1 Epistemic Entropy on Macro-level

In this section we analyse the complexity of inter-agent communications. As pointed out in the literature [8, 5], emergent self-organisation or *extropy* may seem to contradict the second law of thermodynamics that captures the tendency of systems to disorder. The "paradox" has been gracefully explained in terms of multiple coupled levels of dynamic activity (the Kugler-Turvey model [5]) – self-organisation and the loss of entropy occurs at the macro level, while the system dynamics on the micro level generates increasing disorder. One convincing example is described by Parunak and Brueckner [8] in context of pheromone-based coordination. Their work defines a way to measure entropy at the macro level (agents' behaviours lead to orderly spatiotemporal patterns) and micro level (chaotic diffusion of pheromone molecules). In other words, the micro level serves as an entropy "sink" – it permits the overall system entropy to increase, while allowing self-organisation to emerge and manifest itself as coordinated multi-agent activity on the macro level. The epistemic entropy presented here is analysed in the terms of the Kugler-Turvey model [5] as well. We intend to show that the higher team coordination potential is related to lower entropy of multi-agent joint beliefs (macro level). At the same time, it is explained by increased entropy on a micro level. This micro level is the communication space where the inter-agent messages are exchanged (the process that is similar to diffusion of the pheromone molecules).

For convenience, we reproduce here definitions and two characterisations presented in [10], and follow with extended results. Let us consider a simple protocol $\mathcal{P}_1$ allowing an agent to communicate data about only one agent precisely. In other words, each agent is able to encode either the data about itself or about the other agent. Without loss of generality, we may assume that the protocol $\mathcal{P}_1$ has enough symbols to encode the data about agents (objects) $a_1, \ldots, a_n$ in such a way that they are explicitly distinguishable.

A binary relation $S(a_i, a_j)$ represents that the agent a_i *sends* a message containing the object a_j. A function C maps an agent name (symbol) to another agent name (symbol), and the abbreviation $C(a_i) = a_j$ denotes that the *content* of the message from the agent a_i is the object a_j. We intend that $C(a_i) = a_j$ if and only if $S(a_i, a_j)$.

Definition 1. *A multi-agent agreement $L_1(n)$ is called* selfish *if and only if $S(a_i, a_i)$ for all agents a_i, $1 \leq i \leq n$.*

A multi-agent agreement $L_2(n)$ is called transitively-selfish *if and only if $S^*(a_i, a_i)$ for all agents a_i, $1 \leq i \leq n$.*

Equivalently, $C(a_i) = a_i$ for all agents a_i, $1 \leq i \leq n$, under the selfish multi-agent agreement – each agent symbol is *a fixed-point* of the function $C(a)$. A transitively-selfish agreement suggests that the agents are more cooperative, and may communicate the data about some other agent (when available). Notice, however, that given the transitive closure $S^*(a_i, a_i)$, everyone is in the "loop". By definition, a selfish multi-agent agreement is transitively-selfish. The difference, however, may lie in the presence or absence of fixed-points $C(a_i) = a_i$. The transitively-selfish agreements without fixed-points, where each agent is *cooperative*: $C(a_i) \neq a_i$, will be of special interest. Non transitively-selfish agreements are called *mixed*.

Definition 2. *A multi-agent agreement $L_3(n)$ among $n \geq 2$ agents, where some agents are selfish and the others are cooperative, is called* mixed. *There are (αn) agents such that $S(a_i, a_i)$, and $(1 - \alpha)n$ agents such that $S(a_i, a_j)$ where $i \neq j$.*

The α parameter is called the team composition parameter.

The mixed agreement $L_3^1(n)$ where all cooperative agents provide information about the selfish team-mates is called the mixed agreement of the 1st kind.

The mixed agreement $L_3^2(n)$ where all cooperative agents are transitively communicating among themselves is called the mixed agreement of the 2nd kind.

In order to capture the distinction among selfish, transitively-selfish and mixed agreements in a formal information-theoretic setting we shall analyse the joint "output" of inter-agent communication at the end of each period of team synchronisation [10]. More precisely, we analyse joint beliefs represented by the sequence K_t of individual beliefs at time t: $K(a_i, a_j)$, where $1 \leq i \leq n$ and $1 \leq j \leq n$; the belief-function K is defined for each agent pair. In order to estimate how much information is contained in the whole team after a period of team synchronisation – as the team information progresses from K_t to $K_{t'}$ – we need to answer how much choice would be there if one were to describe $K_{t'}$. To do so we calculate the relative entropy H_r of $K_{t'}$. The following representation results for protocol $\mathcal{P}_1$ were reported in [10].

Theorem 1. *Selfish agreements attain minimal entropy. Transitively-selfish agreements without fixed-points attain maximal entropy asymptotically when $n \rightarrow \infty$.*

The first part of the theorem basically states that whenever agents agree to communicate the data about themselves only, they eventually leave nothing to choice, always maximising their joint beliefs. The intuition behind the second part is that the pair-wise "ignorance" of agents grows faster than the transitively-selfish agreement can cope with.

The next results for protocol $\mathcal{P}_1$ are the extensions produced for mixed agreements.

Theorem 2. *Mixed agreements of the 1st kind attain bounded epistemic entropy.*
Mixed agreements of the 2nd kind attain bounded epistemic entropy, and attain the epistemic entropy of mixed agreements of the 1st kind asymptotically when $n \rightarrow \infty$.

In other words, the lower limit is not 0, meaning that absolute order is never achievable regardless of the team composition or the number of agents, while the upper limit is not 1, so that absolute randomness is avoidable as well. Following [10] and interpreting the extended results, we would like to point out that the relative epistemic entropy of joint beliefs in multi-agent teams serves as a generic indicator of the team coordination potential. In general, the following series is established for the epistemic entropy:

$$H_r(L_1(n)) \leq H_r(L_3^2(n)) \leq H_r(L_3^1(n)) \leq H_r(L_2(n)) \tag{2}$$

while the respective coordination potentials follow the reverse dependency.

4.2 Epistemic Entropy on Micro-level

The epistemic entropy may now be analysed in terms of the Kugler-Turvey model [5]. The higher coordination potential of the team following the selfish agreement with near-zero epistemic entropy can be explained by an increased entropy on the micro level – the communication space where the inter-agent messages are exchanged. Clearly, in the case of the selfish agreement the communication space is quite saturated, and the entropy on the micro level increases dramatically. On the contrary, the transitively-selfish agreement may use the communication channel(s) rather sparingly, resulting in a lesser increase of entropy on the micro level – while attaining near-maximal epistemic entropy on the macro level (joint multi-agent beliefs are almost random).

A characterisation of the micro level (the entropy "sink") can be obtained if one estimates the "regularity" of the communication space. In order to carry out this analysis we consider low-bandwidth domains requiring real-time response – in other words, environments where heavy communication among team agents is impossible. For example, we may consider Periodic Team Synchronization (PTS) domains introduced by Stone and Veloso [15] for pseudo real-time environments. However, our analysis is applicable to more generic domains as well – what is important is that the communication channel is limited. More precisely, a multi-agent domain should contain a parameter h

determining how many messages can be received by each agent in a cycle. In particular, we are interested in capturing situations (*communication clashes*) where communication messages exceed "hear capacity" h in a given cycle, and measuring the average severity, spread and regularity of clashes. Let us introduce the following notation:

$\theta(a)$ is a function returning 1 if a boolean expression a is true, and 0 otherwise.
$\kappa_{i,j}$ is a function returning the number of communication messages received from the team i at cycle j;
$\delta_{i,j}$ is a boolean function returning $true$ if $\kappa_{i,j} > h$, and $false$ otherwise;

The average severity of clashes in the team i is given then by

$$M_i(h) = \frac{\sum_{j=1}^{m} \theta(\delta_{i,j})\, \kappa_{i,j}}{m}$$

where m is the number of cycles, while regularity of the series $\kappa_{i,j}$ can be measured with the auto-correlation function of an integer delay τ:

$$\gamma_i(\tau) = \frac{\sum_{j=\tau+1}^{m} (\kappa_{i,j} - \overline{\kappa_i})\, (\kappa_{i,j-\tau} - \overline{\kappa_i})}{\sum_{j=\tau+1}^{m} (\kappa_{i,j} - \overline{\kappa_i})^2} ,$$

where $\overline{\kappa_i}$ is the series average. The auto-correlation function is equivalent to the power spectrum in terms of identifying regular patterns – a near-zero auto-correlation across a range of delays would indicate high irregularity, while auto-correlation with values close to one indicate very high regularity. Some of this regularity is, however, spurious and is related to the severity of clashes. Therefore, we believe that a better approximation of the entropy on the micro level (communication space) may be given by the ratio

$$\xi_i(\tau, h) = \frac{M_i(h)}{\gamma_i(\tau)}$$

This new statistics attempts to capture how much *regularity* in the series is there *per communication clash*, and invert the measure. Our conjecture is that there is a dependency complementary to the dependency 2 over the range of possible values of τ, given some hear capacity:

$$\xi_{L_2}(\tau) \leq \xi_{L_3^1}(\tau) \leq \xi_{L_3^2}(\tau) \leq \xi_{L_1}(\tau) \tag{3}$$

The higher entropy on the micro level (*communication*) corresponds to the lower epistemic entropy on the macro level (*coordination*), and in turn to the higher coordination potential.

5 Experimental Results: Bounded Epistemic Entropy

Importantly, clear boundaries limiting the team coordination potential are related to particular communication policies. It is, however, not trivial to demonstrate these limits experimentally. First of all, the coordination potential can not be measured directly – it can only be realised in concrete multi-agent interactions. Moreover, the actual multi-agent

Table 1. Results against 2 benchmarks after 100 games for each test.

Team vs "A"	Goals	Points	$\gamma(3)$	$M(1)$	$\xi(3,1)$	Team vs "B"	Goals	Points	$\gamma(3)$	$M(1)$	$\xi(3,1)$
Press	31–91	65	0.553	0.067	0.12	Press	105–70	152	0.533	0.070	0.13
Zonal	18–107	50	0.657	0.224	0.34	Zonal	114–53	180	0.641	0.217	0.34
Mix	30–127	63	0.522	0.068	0.13	Mix	118–65	172	0.495	0.071	0.14

coordination can be comprehensively evaluated only through the overall team performance over some period of time. Secondly, the coordinated activities corresponding to different communication policies would have to sufficiently differ in order to generate a *pronounced* difference in team performance.

In the RoboCup Simulation League, the "hear capacity" h determines how many messages can be heard by each agent in a cycle – for example, one message per cycle ($h = 1$). To measure coordination potential via team performance, we varied communication policies, while leaving all other factors (agents skills and tactics) unchanged. This focussed the experiment on the dependency between communication policies (and therefore, resultant joint beliefs) and the team coordination potential. In addition, we attempted to engineer, by varying the communication policies, very distinct types of coordinated activities, ranging from very local multi-agent coordination to rather global (zonal) one. For each type we also calculated the statistics $M_i(1)$ and $\xi_i(3,1)$ – in order to estimate the corresponding entropy on the micro level.

We investigated three communication policies based on the protocol $\mathcal{P}_1$. The first policy ("Press") modelled the transitively-selfish agreement, with high relative entropy and very local coordination, enabling a pressing aggressive game. The second policy ("Zonal") followed the selfish agreement, with low relative entropy and very global coordination, enabling a passing non-aggressive game. The third policy ("Mix") was aimed at some mixture of local and global coordination, balancing predominantly pressing game with some passing chances – truly a mixed agreement with (anticipated) bounded relative entropy. The results are presented in the Table 1. The "Press" policy showed the best performance against the stronger benchmark "A", while the "Zonal" policy was the worst. The results against the weaker opponent "B", on the contrary, indicate that a team coordinated zonally (across wider spaces) perform better than the aggressive pressing team. In other words, these two communication policies lead to sufficiently different coordinated activities (local vs global) and generate a pronounced difference in team performance. This is important because, as mentioned earlier, we attempt to trace the effect of coordination potential – a capacity that is measurable only via a difference in produced results. The "Mix" policy achieved intermediate results. Importantly, this mixed policy was *within the boundaries* marked by the first two variants (and closer to the first one), as suggested by the relative epistemic entropy. As expected, the entropy $\xi(3,1)$ on the micro-level supported our hypothesis:

$$\xi_{Press}(3) \le \xi_{Mix}(3) \le \xi_{Zonal}(3) ,$$

contrasting with the epistemic entropy inequalities, where the "Zonal" policy is close to the theoretic minimum and the "Press" policy reaches the maximum.

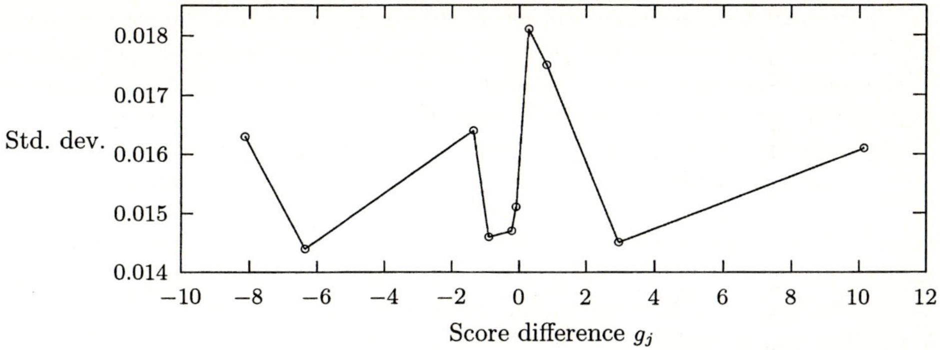

Fig. 4. Standard deviation of communication entropy.

6 Experimental Results: Epistemic Entropy and Phase Transitions

We have also investigated phase transitions in the communication space. This investigation is at preliminary stages, but some promising results were obtained. Again, an experiment included 30 games between the test team (using the "Mix" policy) and an opponent from the pool of the same 10 opponents as in the behavioural entropy experiments. Each experiment produced a value for the communication entropy $\xi_k(3)$, $0 \leq k \leq 30$, and a score difference g_j. The standard deviation σ_j^ξ of the entropy $\xi_k(3)$, calculated across all games against the opponent j, is plotted in Figure 4. As expected, standard deviation σ_j^ξ peaks in the the narrow range ($-0.87 \leq g_j \leq 0.83$), indicating a phase transition in the communication space, and in the coordination potential as well. In other words, epistemic entropy, directly related to the entropy $\xi_k(3)$, also identifies the *edge of chaos*. In summary, the results not only illustrate the dependency between communication policy, the epistemic entropy and the team coordination potential, but also detect a phase transition in the coordination and communication dynamics.

7 Related Work and Conclusions

We presented a set of quantitative techniques for evaluation of team performance on multiple levels: from individual behavioural spread to multi-agent coordination potential. These techniques are based on information-theoretic metrics measuring complexity in multi-agent systems. In particular, we focussed on identifying the "edge of chaos" in team performance – leading to discovery of evident phase transitions. Our conjectures and theoretical results were supported by a number of experiments – over 500 games.

As pointed out by Pynadath and Tambe [11], "despite the significant progress in multiagent teamwork, existing research does not address the optimality of its prescriptions nor the complexity of the teamwork problem". The unified framework suggested by Pynadath and Tambe (COMmunicative Multiagent Team Decision Problem – COM-MTDP model) is general enough to subsume many existing models of multi-agent systems, and provides a breakdown of the computational complexity of constructing optimal teams in terms of observability and communication cost. The COM-MTDP model incorporates the team decision mechanism, and inevitably is rather complex, as almost any unifying framework. In this paper we attempted to introduce a concise approach to teamwork evaluation, dealing with *behaviour spread* and multi-agent *coordination potential*, and excluding the team decision process.

The presented analysis targets our overall goal – development of tools for evaluation of multi-agent adaptability and coordination in comparative terms, rather than methods for designing an "ultimate" intelligent and/or adaptive system. In pursuing this goal, we build up on existing quantitative methods for automated analysis of Simulation League games (eg., the AGL tool – Analysis of Game Logs [2]). We also hope to complement existing teamwork models and techniques impacting the team performance. A pioneering system capable of an automated analysis in the context of RoboCup was the ISAAC system modelling an assistant-coach [12]. ISAAC analyses games off-line and produces structured suggestions to designers, supported by examples. Another autonomous coach agent is recently described by Riley et.al. [13] – it is not only capable of extracting models from past games but may also respond to an ongoing game.

Our quantitative analysis complements this line of research by providing methods and techniques that determine phase transitions in team performance – the *edge of chaos*. These techniques isolate the under-performing cases where a change in tactics is warranted. During the phase transition, the team performance is highly unstable, and the scope for an on-line coach-agent contribution is limited. The quantitative information-theoretic methods presented here incorporate both behavioural and epistemic entropy, and are compatible with the hierarchic social entropy approach developed by Balch [1]. This opens a clear way to a unified quantitative framework on *behavioural and belief dynamics* in multi-agent systems. Another interesting possibility is to explore possible connections between the described techniques and the measures for relevant information investigated by Polani et.al. [9], as well as the conditions reducing "the influence of cognition difference" introduced by Cai et.al. [3] in the context of RoboCup.

Acknowledgements

The authors are grateful to members of the "Entropy and self-organisation in multi-agent systems" discussion group, and in particular to Mark Foreman, Ying Guo, Andrew Lampert, and Philip Valencia for many open and motivating discussions; and to Marc Butler, Thomas Howard and Ryszard Kowalczyk for their exceptionally valuable prior contributions to RoboCup Simulation efforts.

References

1. Tucker Balch. Hierarchic Social Entropy: An Information Theoretic Measure of Robot Team Diversity. *Autonomous Robots*, vol. 8, no. 3, July, 2000.
2. Marc Butler, Mikhail Prokopenko and Thomas Howard. Flexible Synchronisation within RoboCup Environment: a Comparative Analysis. In P. Stone, T. Balch, G. Kraetzschmar (eds.). *RoboCup-2000: Robot Soccer World Cup IV*, LNCS 2019, Springer, 2001.
3. Yunpeng Cai, Jiang Chen, Jinyi Yao, Shi Li. Global Planning from Local Eyeshot: An Implementation of Observation-Based Plan Coordination in RoboCup Simulation Games. In A. Birk, S. Coradeschi, S. Tadokoro (eds.). *RoboCup 2001: Robot Soccer World Cup V*, LNCS 2377, Springer 2002.

4. Hiroaki Kitano, Milind Tambe, Peter Stone, Manuela M. Veloso, Silvia Coradeschi, Eiichi Osawa, Hitoshi Matsubara, Itsuki Noda and Minoru Asada. The RoboCup Synthetic Agent Challenge. In Proceedings of the 15th Int'l Joint Conference on Artificial Intelligence, 1997.
5. Peter N. Kugler and Michael T. Turvey. *Information, Natural Law, and the Self-Assembly of Rhythmic Movement.* Lawrence Erlbaum, 1987.
6. Christopher Langton. Computation at the Edge of Chaos: Phase Transitions and Emergent Computation. In S. Forest (ed.), *Emergent Computation*, MIT, 1991.
7. Octavio Miramontes. Order-Disorder Transitions in the Behavior of Ant societies. *Complexity*, Vol. 1, no. 3, 56-60, 1995.
8. H. Van Dyke Parunak and Sven Brueckner. Entropy and self-organization in multi-agent systems. In Proc. of the 5th Int'l Conference on Autonomous agents, Montreal, 2001.
9. Daniel Polani, Jan T. Kim and Thomas Martinetz. An Information-Theoretic Approach for the Quantification of Relevance. In J. Kelemen and P. Sosik (eds.), *Advances in Artificial Life*, Proceedings of the 6th European Conference on Artificial Life, Prague, Springer 2001.
10. Mikhail Prokopenko and Peter Wang. Relating the Entropy of Joint Beliefs to Multi-Agent Coordination. In Proceedings of the 6th International Symposium on RoboCup, 2002.
11. David V. Pynadath, Milind Tambe. Multiagent teamwork: analyzing the optimality and complexity of key theories and models. In the Proceedings of the First International Joint Conference on Autonomous Agents and Multiagent Systems, AAMAS 2002, Bologna, 2002.
12. Taylor Raines, Milind Tambe, Stacy Marsella. Automated assistants to aid humans in understanding team behaviors. In the Proceedings of the Fourth International Conference on Autonomous Agents, Barcelona, 2000.
13. Patrick Riley, Manuela Veloso, Gal Kaminka. An Empirical Study of Coaching. In H. Asama, T. Arai, T. Fukuda, and T. Hasegawa (eds.), Distributed Autonomous Robotic Systems 5, Springer-Verlag, 2002.
14. Claude E. Shannon and Warren Weaver. *The Mathematical Theory of Communication.* University of Illinois Press, 1949.
15. Peter Stone and Manuela Veloso. Task Decomposition, Dynamic Role Assignment, and Low-Bandwidth Communication for Real-Time Strategic Teamwork. In *Artificial Intelligence*, vol. 110, no. 2, June 1999.
16. Stephen Wolfram. Universality and Complexity in Cellular Automata. *Physica D*, 10 (January 1984), 1–35, 1984.
17. Andrew Wuensche. Classifying Cellular Automata Automatically: Finding gliders, filtering, and relating space-time patterns, attractor basins, and the Z parameter. *Complexity*, Vol. 4, no. 3, 47-66, 1999.

Hidden Markov Modeling of Team-Play Synchronization

Itsuki Noda[1,2]

[1] Cyber Assist Reseach Center
National Institute of Advanced Industrial Science and Technology
[2] PRESTO, Japan Science and Technology Corporation (JST)

Abstract. Imitation Learning is considered both as a method to acquire complex human and agent behaviors, and as a way to provide seeds for further learning. However, it is not clear what is a building block in imitation learning and what is the interface of blocks; therefore, it is difficult to apply imitation learning in a constructive way. This paper addresses agents' intentions as the building block that abstracts local situations of the agent and proposes a hierarchical hidden Markov model (HMM) in order to tackle this issue. The key of the proposed model is introduction of gate probabilities that restrict transition among agents' intentions according to others' intentions. Using these probabilities, the framework can control transitions flexibly among basic behaviors in a cooperative behavior. A learning method for the framework can be derived based on Baum-Welch's algorithm, which enables learning by observation of mentors' demonstration. Imitation learning by the proposed method can generalize behaviors from even one demonstration, because the mentors' behaviors are expressed as a distributed representation of a flow of likelihood in HMM.

1 Introduction

Imitation learning is considered to be a method to acquire complex human and agent behaviors and as a way to provide seeds for further learning [6, 9, 7]. While those studies have focused on imitating behaviors of single agents, few works address imitation for teamwork among multiple agents because the complexity of the world state increases drastically in multi-agent systems. On the other hand, stochastic models like hidden Markov models (HMM) have been studied as tools to model and to represent multi-agent/human interactions [8, 3, 2]. The merit of stochastic models is that we can apply the models in both behavior recognition and generation. However, it is hard to apply these stochastic models to imitate teamworks by observation because of the complexity of the model of multiple agents. This study focuses upon *intentions* of agents as building blocks of an abstract state of the local world for the agent in order to overcome the problem. Using *intention*, I formalize teamwork and propose a hierarchical hidden Markov model for imitation learning of teamwork.

D. Polani et al. (Eds.): RoboCup 2003, LNAI 3020, pp. 102–113, 2004.

2 Teamwork and Imitation

What is *teamwork* in multiagent systems? Consider a case wherein two soccer players pass a ball with each other. When a passer, who is keeping the ball, starts to pass the ball to a receiver, the passer must know that the teammate is ready to receive. Also, the receiver will start free-running when the receiver recognizes that the passer is looking for the receiver for a through-pass. This example illustrates that recognition of the others' intentions is important factor for decision making of player's intention. We enhance usage of intentions to derive teamwork. This section formalizes teamwork from intention and makes a correspondence to imitation learning.

2.1 Intention and Play

We suppose that an *intention* is a short-term idea to achieve a certain condition from another condition. For example, in soccer, the intention 'to guide a ball in a certain direction' is an idea to move to a certain direction with the ball. We assume that an *intention* is an individual idea; therefore, an agent does not pay attention to others' efforts to achieve the intention.

A *play*, as opposed to a *team-play*, is postulated as a sequence of atomic actions to achieve a single *intention*. The *play* is a basic building block of overall behavior of agents. For example, in soccer, a 'dribble' is a *play* to achieve the *intention* 'guide a ball in a certain direction', which consists of atomic actions like 'turn', 'dash', 'kick', and so on. A play for the intention is also an individual behavior without collaboration with other agents because an intention is an individual idea.

We also assume that a play corresponds to just one intention. Therefore, we use the word "play" and "intention" in the same meaning in the remainder of this article.

As shown below, an intention and the corresponding play are used as a main trigger to synchronize team-plays among multiple agents. This means that the intention is treated as a kind of partial condition of the world. For example, the fact that a player has a 'dribble' intention implies the following conditions: the player is keeping the ball; the player is moving toward a certain place; and the player may require teammates for support. In other words, an intention represents abstracted and simplified conditions of the world.

2.2 Team-Play

We suppose that *team-play* is a collection of plays performed by multiple agents to achieve a certain purpose. As mentioned in the previous section, an intention is an individual idea. This means that multiple agents who do not change their intentions can not perform a *team-play* because they have no way to synchronize their plays. Instead, we assume that they can synchronize their plays by changing their intentions according to situations of environments and intentions of other agents. For example, in soccer, when two players (passer and receiver) guide a ball by dribble and pass, players will change their intentions as shown in Fig. 2. In this example, the passer and the receiver initially have intentions 'dribbling' and 'supporting', respectively. Then, the passer changes the intention to 'seek-receiver', followed by the receiver's change to 'free-run', the passer's change to 'pass', and so on. Play synchronization is represented as conditions when agents can change the intention. In the example, the passer changes its intention from 'seek-receiver' to 'pass' when the teammate's intention is 'free-run'. In other words, we can denote the condition as follows:

Intent(passer,seek-receiver) & **Bel**(passer, **Intent**(receiver,free_run))
$\rightarrow$ **Intent**(passer, pass)

2.3 Imitation Learning of Team-Play

Finally, we formalize imitation learning of the team-play.

In general, the imitation learning process is: (1) to **observe** behaviors of a mentor and interpret them into internal representation; (2) to **extract** rules of behaviors from internal representation; and (3) to **generate** a behavior based on rules. In the context of the *team-play* formalized in the previous section, the above process is realized as;

Observation phase: to observe behaviors of mentors and estimate what intention each agent has at each time step.

Extraction phase: to extract conditions prevailing when each agent changes intentions. A condition is represented as a conjunction of others' intentions.

Generation phase: to generate a sequence of intentions according to changes of environment and others' intentions.

In the **Extraction** phase of this process, the *intention* plays an important role: that is, conditions of changes of intentions. As described in Section 2.1, we consider that *intention* can represent world conditions. In addition to it, we use only *intentions* to construct rules for agents to change their *intention.* Although such abstraction reduces performance to express detailed conditions of the world, it provides good features for the machine learning. One important issue in machine learning is how to represent the state of a complex environment. This becomes a serious problem under a multi-agent environment because the number of factors to take into account increases exponentially in such environment. Abstraction of the world state by intentions can reduce the number of factors during the condition significantly. This kind of abstraction is necessary for imitation learning because only a small number of examples are given for imitation learning.

3 Hierarchical Hidden Markov Model for Agents

3.1 Basic Behavior Model

We formalize behaviors of a basic play m performed by a single agent as a Moore-type HMM as follows:

$$\mathbf{HMM^b}_m = \langle \boldsymbol{S}_m, \boldsymbol{V}_m, \boldsymbol{P}_m, \boldsymbol{Q}_m, \boldsymbol{R}_m \rangle ,$$

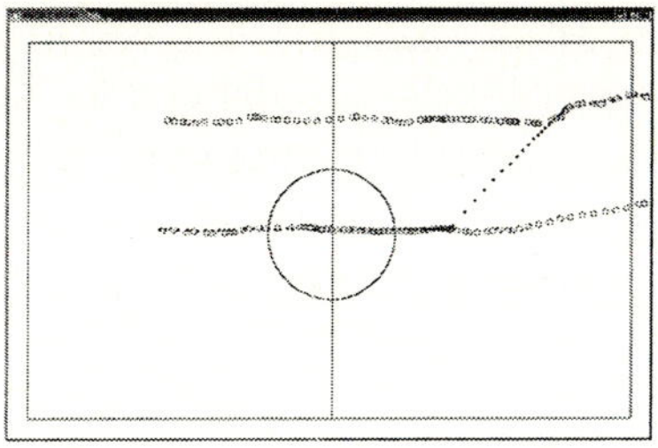

Fig. 1. Dribble and Pass Play

passer	receiver
dribbling	:
↓	supporting
seek-receiver	↓
↓	free-run
pass	↓
:	receive

Fig. 2. Changes of Intentions in Dribble and Pass Play

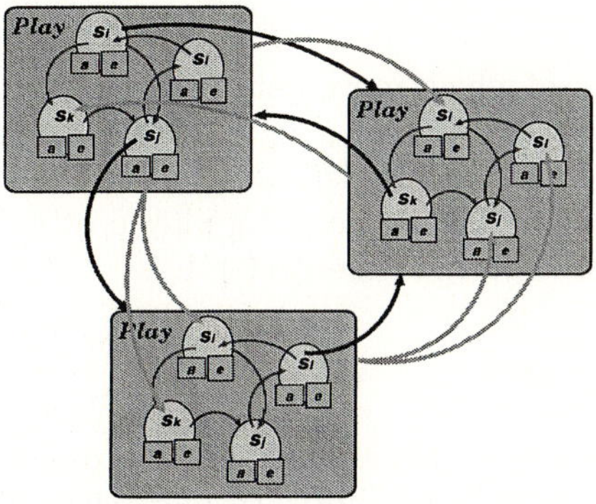

Fig. 3. Complex Behavior Model

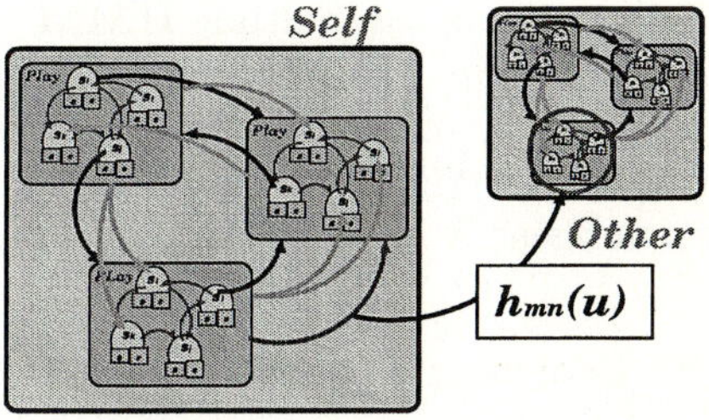

Fig. 4. Joint Behavior Model

where $\boldsymbol{S}_m$ is a set of states for the play m, $\boldsymbol{V}_m$ is a set of a pair of sensor value and action commands which are used as outputs from the state, $\boldsymbol{P}_m = \{p_{mij}\}$, $\boldsymbol{Q}_m = \{q_{mi}(v)\}$, and $\boldsymbol{R}_m = \{r_{mi}\}$ are probability matrixes of state transition, state-output, and initial state, respectively. These probabilities are defined as $p_{mij} = Pr(s_{mj}^{\langle t\rangle}|s_{mi}^{\langle t-1\rangle})$, $q_{mi}(v) = Pr(v^{\langle t\rangle}|s_{mi}^{\langle t\rangle})$, and $r_{mi} = Pr(s_{mi}^{\langle 0\rangle})$, where $s_{xyz}^{\langle t\rangle}$ means $s^{\langle t\rangle} = s_{xyz}$, and $\langle t\rangle$ on the right shoulder of a variable indicate the time t.

3.2 Complex Behavior Model

As discussed in the previous section, we consider that team-play consists of a sequence of intentions of multiple agents. This means that cooperative complex behavior of a single agent in a team of agents is considered as transitions among several basic plays ($\mathbf{HMM^b}$). Therefore, we formalize complex behavior as the following modified Mealy-type HMM (Figure 3),

$$\mathbf{HMM^c} = \langle \boldsymbol{M}, \boldsymbol{U}, \boldsymbol{E}, \boldsymbol{F}, \boldsymbol{G}, \boldsymbol{H} \rangle ,$$

where $\boldsymbol{M} = \{\mathbf{HMM^b}_m\}$ is a set of basic plays and $\boldsymbol{U}$ is a set of output from the model (normally, same as $\boldsymbol{M}$); $\boldsymbol{E} = \{e_m\}$ is a set of initial play probabilities, $\boldsymbol{F} = \{f_{min}\}$ is a set of exiting probabilities from plays, and $\boldsymbol{G} = \{g_{mnj}\}$ is a set of entering probabilities to plays. Also, $\boldsymbol{H} = \{h_{mn}(u)\}$ is a set of gate probabilities between plays. Formally, these probabilities are defined as: $e_m = Pr(\mathbf{HMM^b}^{\langle 0 \rangle}_m)$, $f_{min} = Pr(\mathbf{HMM^b}^{\langle t \rangle}_n | s^{\langle t \rangle}_{mi})$, $g_{mnj} = Pr(s^{\langle t \rangle}_{nj} | \mathbf{HMM^b}^{\langle t-1 \rangle}_m)$, and $h_{mn}(u) = Pr(u^{\langle t-1 \rangle} \in \boldsymbol{U} | \mathbf{HMM^b}^{\langle t-1 \rangle}_m, \mathbf{HMM^b}^{\langle t \rangle}_n)$.

Using these probabilities, an actual probability from state i in play m to state j in play n is calculated as

$$p'_{minj} = Pr(s^{\langle t \rangle}_{nj} | s^{\langle t-1 \rangle}_{mi}) = \begin{cases} f_{mim} p_{mij} & ; m = n \\ f_{min} g_{mnj} & ; m \neq n \end{cases} .$$

3.3 Joint-Behavior Model

Finally, we coupled multiple $\mathbf{HMM^c}$s , each of which represents the behavior of an agent. Coupling is represented by gate probabilities $\boldsymbol{H}$ (Fig. 4). For example, when agent X and agent Y are collaborating with each other, $h_{mn}(u)$ in $\mathbf{HMM^c}$ for agent X indicates the probability that agent Y is performing play u at time t when agent X changes the play from m to n during time $t \rightarrow t+1$.

3.4 Learning Procedure

Using the Baum-Welch algorithm, we can derive an learning procedure to adapt probabilities in $\mathbf{HMM^c}$. The first step is to calculate forward and backward likelihoods for each state and timestep in all plays as follows:

$$\begin{aligned}
\alpha^{\langle 0 \rangle}_{nj} &= e_n r_{nj} q_{nj}(v^{\langle 0 \rangle}) \\
\alpha^{\langle t+1 \rangle}_{nj} &= \left[\sum_i \tilde{f}_{nin} \alpha^{\langle t \rangle}_{ni} p_{nij} + \sum_{m \neq n} \bar{\alpha}^{\langle t \rangle}_{mn} g_{mnj} \right] q_{nj}(v^{\langle t+1 \rangle}) \\
\beta^{\langle T \rangle}_{mi} &= 1 \\
\beta^{\langle t-1 \rangle}_{mi} &= \left[\sum_j \tilde{f}_{mim} p_{mij} q_{mj}(v^{\langle t \rangle}) \beta^{\langle t \rangle}_{mj} + \sum_{m \neq n} \tilde{f}_{min} \bar{\beta}^{\langle t \rangle}_{mn} \right] q_{nj}(v^{\langle t+1 \rangle}),
\end{aligned}$$

where

$$\begin{aligned} \bar{\alpha}^{\langle t \rangle}_{mn} &= \textstyle\sum_i \alpha^{\langle t \rangle}_{mi} \tilde{f}_{min} \\ \bar{\beta}^{\langle t \rangle}_{mn} &= \textstyle\sum_j g_{mnj} q_{nj}(v^{\langle t \rangle}) \beta^{\langle t \rangle}_{nj} \end{aligned} \qquad \tilde{f}_{min} = \begin{cases} \frac{f_{min} + \lambda_{\text{sticky}}}{1 + \lambda_{\text{sticky}}} & ; \text{if} m = n \\ \frac{f_{min}}{1 + \lambda_{\text{sticky}}} & ; \text{otherwise} \end{cases} .$$

Here, λ_{sticky} is a positive value called a *sticky factor*. This factor is introduced because we should consider that an agent retains an intention relatively persistently. If the agent changes its intention repeatedly, it becomes difficult to estimate an agent's intention by observation, rendering complex behavior difficult. The sticky factor λ_{sticky} inhibits such frequent changes of intention during observation and estimation of mentors' behaviors. Note that the sticky factor is used only in the **Estimation** phase, and ignored in the **Generation** phase in the imitation learning.

Using α and β, we can adjust probabilities as follows:

$$e_m \leftarrow \sum_i \gamma_{mi}^{\langle 0 \rangle}, \quad f_{min} \leftarrow \frac{\sum_t \xi_{min}^{\langle t \rangle}}{\sum_t \gamma_{mi}^{\langle t \rangle}}, \quad g_{mnj} \leftarrow \frac{\sum_t \xi_{mnj}^{\langle t \rangle}}{\sum_t \gamma_{mn}^{\langle t \rangle}}, \quad h_{mn}(u) \leftarrow \frac{\sum_{t, u^{\langle t \rangle}=u} \gamma^{\langle t \rangle}}{\sum_t \gamma_{mn}^{\langle t \rangle}},$$

where

$$\begin{aligned} \xi_{min}^{\langle t \rangle} &= \alpha_{Mi}^{\langle t-1 \rangle} f_{min} h_{mn}(u^{\langle t-1 \rangle}) \bar{\beta}_{mn}^{\langle t \rangle} & \gamma_{mi}^{\langle t \rangle} &= \alpha_{mi}^{\langle t \rangle} \beta_{mi}^{\langle t \rangle} \\ \xi_{mnj}^{\langle t \rangle} &= \bar{\alpha}_{mn}^{\langle t-1 \rangle} g_{mnj} q_{nj}(v^{\langle t \rangle}) \beta_{nj}^{\langle t \rangle} & \gamma_{mn}^{\langle t \rangle} &= \bar{\alpha}_{mn}^{\langle t \rangle} h_{mn}(u^{\langle t \rangle}) \bar{\beta}_{mn}^{\langle t+1 \rangle}. \end{aligned}$$

3.5 Imitation Learning Using HMM

Using proposed HMMs, we realize the imitation learning as: (1) to train separately each **HMM$^{\mathbf{b}}$** for each play; (2) to construct **HMM$^{\mathbf{c}}$**s using the trained **HMM$^{\mathbf{b}}$**s ; (3) to observe mentors' behaviors and environmental changes; then estimate likelihoods of each play (and each state in a play) at each time step by calculating forward and backward likelihoods ($\alpha_{nj}^{\langle t \rangle}$ and $\beta_{mi}^{\langle t \rangle}$) as shown in Section 3.4; (4) to adjust initial, exiting and entering probabilities ($\boldsymbol{E}$,$\boldsymbol{F}$ and $\boldsymbol{G}$) according to observation; (5) to repeat steps 3 and 4 until the probabilities converge (**Observation**); (6) to calculate gate probabilities ($\boldsymbol{H}$) using final forward/backward likelihood (**Extraction**); (7) to generate transitions among states and plays according to acquired probabilities (**Generation**). In the last phase, play generation is done as follows: The initial state for each agent is decided according to initial probability $Pr(s_{mi}^{\langle 0 \rangle}) = e_m r_{mi}$. When the play and the state of an agent at time t are m and s_{mi}, respectively and the set of others' plays is $u^{\langle t \rangle}$, then the next play n and the state i of the agent is decided according to the following likelihood L:

$$L(s_{nj}^{\langle t+1 \rangle}) = \begin{cases} f_{mim} p_{mij} q_{mj}(\hat{v}^{\langle t+1 \rangle}) & ; \ m = n \\ f_{min} g_{mnj} h_{mn}(u^{\langle t \rangle}) q_{nj}(\hat{v}^{\langle t+1 \rangle}) & ; \ m \neq n \end{cases} \quad (1)$$

where $\hat{v}^{\langle t+1 \rangle}$ is a partially observed output value in v at time $t+1$.

4 Experiments

4.1 Cyclic Alternating Shift Actions

Several simple experiments were conducted to demonstrate the performance of the proposed model.

In the experiments, two agents change four plays in a certain order by synchronizing them with each other. The four plays are: to move on a round path clockwise (A), to move on a round path counter-clockwise (B), to move in an '∞'-letter-shape path (C) , and to move in an '8'-letter-shape path (D). Actual

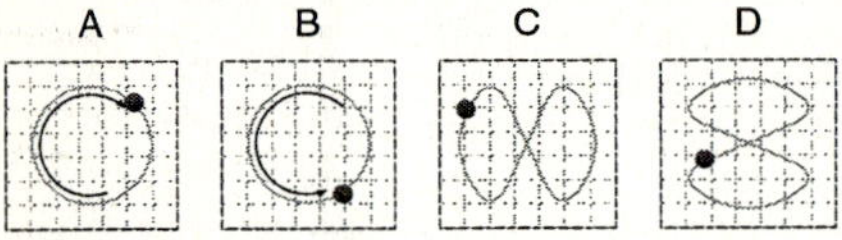

Fig. 5. Basic Plays used in Exp. 1 and Exp. 2

Exp. 1

Agent X	→A→B→C→D→A
Agent Y	D→C→B→A→D→

Exp. 2

Agent X	→A→B→C→D→A→D→C→B→A
Agent Y	A→B→D→A→C→D→B→C→A→

Fig. 6. Change Patterns of Team-plays used in Exp. 1 and Exp. 2

paths are shown in Fig. 5. The agents' actions (movement) are observed as a sequence of positions where the agent is located in each timestep[1].

We suppose that learner agents are already trained for these four plays; that is, the learners' $\mathbf{HMM}^{\mathbf{b}}$s , each of which consists of 20 states, are already trained for these plays. This means that the $\mathbf{HMM}^{\mathbf{b}}$s can generate required movements of corresponding play-A,-B,-C, and -D.

Note that only one example is given to the learner for each experiment in the following experiments, Because of imitation learning, learners should be able to generalize acquired behaviors from the example so that the learners can generate varied behavior according to difference of environments.

Exp. 1: Simple Shift of Plays: In the first experiment, each mentor agent simply changes its plays in a certain order (A →B →C →D for agent X and D →C →B →A for agent Y) alternately with each other as shown in Fig. 6-(a).

Figure 7 shows the relative likelihood of each play state for each agent at each timestep estimated by **Observation** phase. In this figure, there are eight rows of small squares: upper 4 rows correspond 4 plays of the first agent (agent X), and the rest 4 are plays for the second agent (agent Y). Each row corresponds to a play A, B, C or D in Fig. 5 respectively. In each row, a column consists of 20 small squares each of which corresponds a state of $\mathbf{HMM}^{\mathbf{b}}$ for the play A–D at a certain timestep. The ratio of black area in the square indicates the relative likelihood with which the state of the $\mathbf{HMM}^{\mathbf{b}}$ is active at the timestep. Columns are aligned along with time. So, a horizontal line of squares means changes of likelihood of a state of $\mathbf{HMM}^{\mathbf{b}}$. From this figure, we can see that the learner estimate that the agent X behaves according to play-A at the beginning (states for play-A (squares in the most upper row) are active in the left most part of the figure), then adopts play-B, play-C, play-D continuously; it then returns to play-A, followed by the same changes. Similarly, the learner estimates that agent Y behaves play-D first, then changes plays in the reverse order of agent X. In addition to it, the change from play-A to play-B, from play-B to play-C, and from play-C to play-D in the agent X occur while the agent Y is doing play-C,

[1] Actually, the world is quantized into 49 (7×7) blocks when it is observed. Therefore, movements are observed as a sequence of blocks in which the agent is at each timestep.

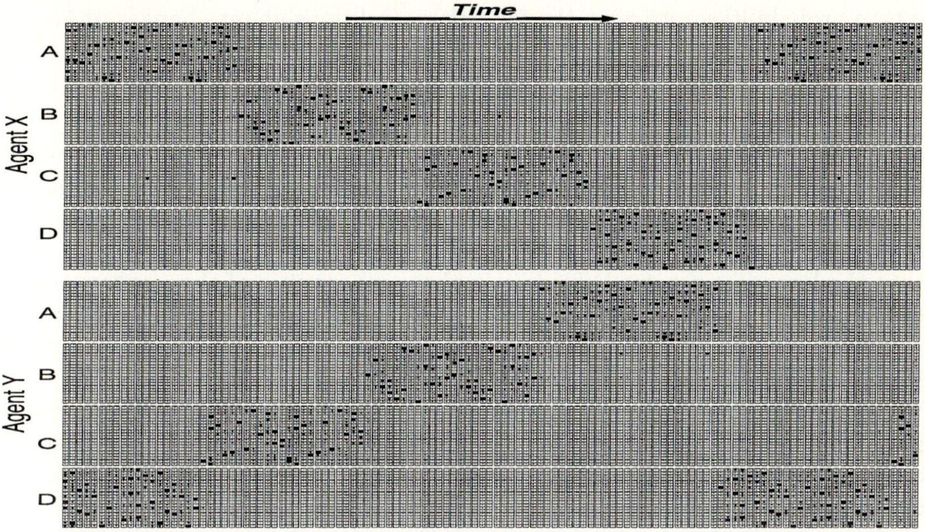

Fig. 7. Exp. 1: Result of Recognition of Mentors' Behaviors

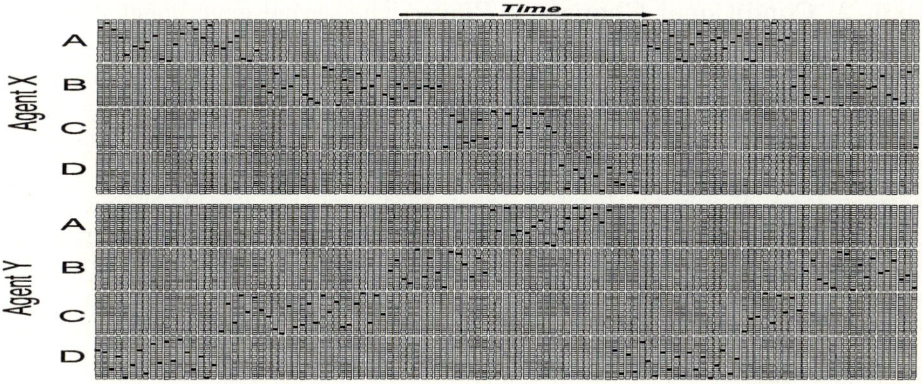

Fig. 8. Exp. 1: State Transitions Generated by Learned HMM

play-B, and play-A, respectively. These conditions are consistent with the change pattern of the mentor shown in Fig. 6.

Using the result of the estimation, the learner acquires probabilities in $\mathbf{HMM}^c$ as conditions of changes of plays. For example, probabilities to transit from play A to play B and from play B to play C in the agent X were acquired in a certain trial as follows[2]:

$$\begin{array}{lll} h_{AB}(C) = 1.00, & f_{A8B} = 0.33, & f_{A14B} = 1.00, \\ h_{BC}(B) = 1.00, & f_{B3C} = 0.71, & f_{B6C} = 0.96. \end{array} \quad (2)$$

[2] Actual values of these probabilities vary according to initial values before learning and random noise added in the mentors' behaviors.

These probabilities represent the following conditions about changes of plays:

Intent(X,A) & **Bel**(X, **Intent**(Y, C)) → **Intent**(X, B)
Intent(X,B) & **Bel**(X, **Intent**(Y, B)) → **Intent**(X, C)

Using these probabilities, the learner can generate similar behaviors to those shown in Fig. 8. This figure is constructed in the same way as Fig. 7, but only one square is filled in a timestep because the learner decides one of the possible states according to the likelihood shown in Eq. 1. From this figure, we can see that the learner generates behaviors in the same manner of the mentor; that is, the order of the generated plays of agent X and Y are 'A →B →C →D' and 'D →C →B →A' respectively. Also, timings of the change of plays are consistent with the mentor's demonstration.

Generated behaviors are not deterministic because the acquired probabilities may take intermediate values as like f_{A8B} and f_{B3C} in Eq. 2. For example, durations of the play-C in agent Y are different in the first cycle and the second cycle in Fig. 8. This means that the learner has the ability to adapt to difference of the environment using methods for HMM such as Vitabi's algorithm.

Exp. 2: Conditional Change of Plays: The second experiment illustrates that the proposed framework can learn conditional transitions of plays using change pattern shown in Fig. 6-(c). The change pattern of Fig. 6-(c) includes conditional branching of the transition of plays. For example, agent X may change its play from A to two possible destination, B or D. The change can be decided encountering agent Y's play. When agent Y is doing play-B, agent X changes its play from A only to B; when agent Y is play-D, agent X changes to D. Figure 10 shows resultant behaviors generated after learning. As shown in this figure, the learner acquires correct conditions of the branching transition. For example, changes from play-A to -B of agent X only occur during agent Y's play-B. Actually, these conditions are represented by gate probabilities: for example, $h_{AB}(B) = 0.97$ and $h_{AD}(B) = 0.00$ for agent X.

4.2 Exp. 3: Dribble and Pass Play in Soccer

Finally, I applied the framework to collaborative play of soccer. The demonstration by mentors is dribble and pass play as shown in Fig. 1: A player starts to dribble from the center of the left half field and brings the ball to the right half. At the same time, another player runs parallel along the upper (or lower) side of the field supporting the dribbling player. Then, the first player slows to look-up the second player; it then passes the ball to that player. Simultaneously, the second player starts to dash to the ball and dribbles after receiving the ball. After the pass, the first player exchanges roles with the teammate so that it becomes a supporting player for the second player.

To imitate this demonstration, I trained six **HMM^b**s to model 'dribble', 'slow-down and look-up', 'pass', 'free-run', 'chase-ball', and 'support'. Each of **HMM^b** has 5 states. The output of these **HMM^b**s consists of local situations

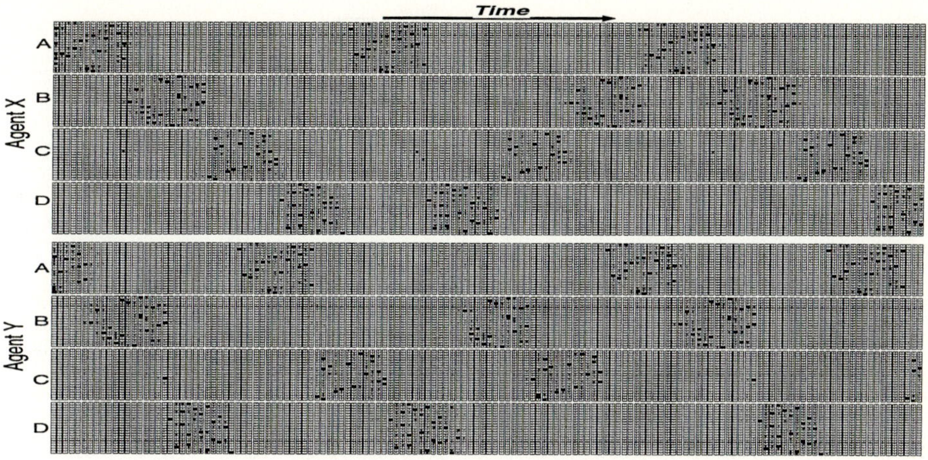

Fig. 9. Exp. 2: Result of Recognition of Mentor's Behaviors

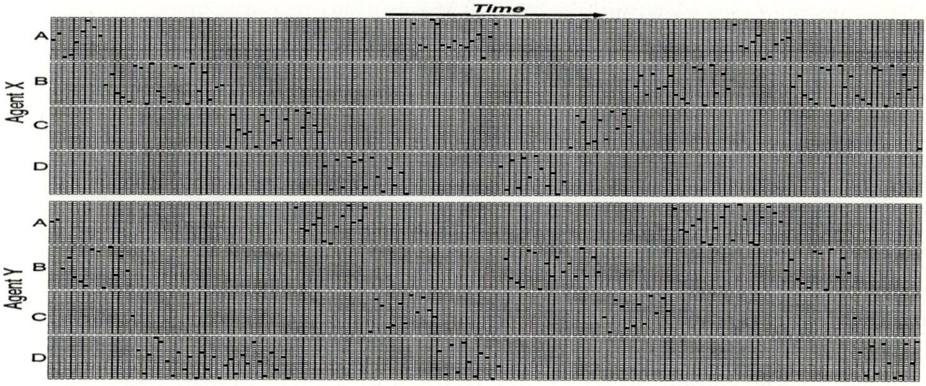

Fig. 10. Exp. 2: State Transitions Generated by Learned HMM

(the relative position and the velocity to the ball) and agent's actions ('turn', 'dash', 'small-kick', 'long-kick', 'trap', and 'look'). Note that there is no information about others' situations for output of $\mathbf{HMM^b}$s . As described in Section 2.2, others' situations are taken into account during the **Extraction** phase in learning.

Two $\mathbf{HMM^c}$s for agent X (the first player) and Y (the second player) are constructed after the training of the $\mathbf{HMM^b}$s . Then, the learner observes behaviors of the mentor and adjusts probabilities of the $\mathbf{HMM^c}$s .

Figure 1 shows result of observation and estimation. There are six $\mathbf{HMM^b}$s in this experiments; therefore, there are six rows (denoted by D, K, P, F, C, and S) for each agent, in which a column consists of five squares. Rows mean 'dribble (D)' , 'slow-down and look-up (K)', 'pass (P)', 'free-run (F)', 'chase-ball (C)', and 'support (S)', respectively. From this figure, we can see that the learner estimates changes of play for agent X and Y are 'dribble' →'slow-down' →'pass'

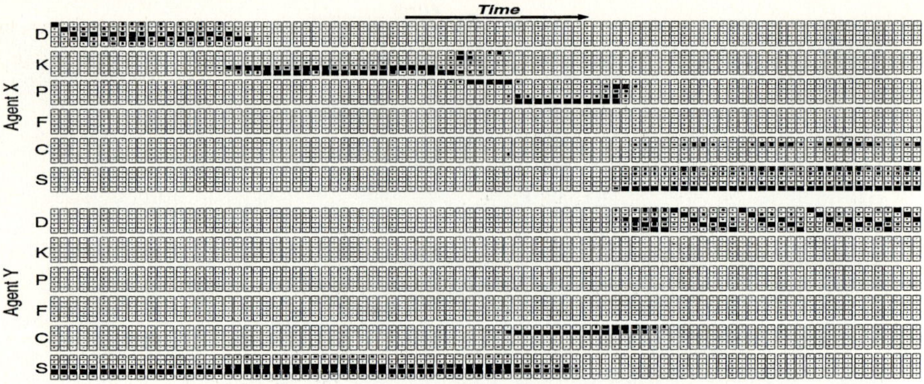

Fig. 11. Exp. 3: Result of Recognition of a Mentor's Behaviors

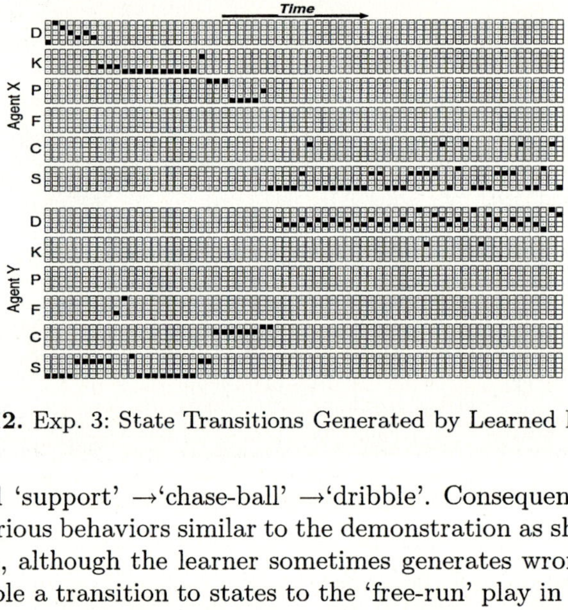

Fig. 12. Exp. 3: State Transitions Generated by Learned HMM

→'support' and 'support' →'chase-ball' →'dribble'. Consequently, the learner can generate various behaviors similar to the demonstration as shown in Fig. 12. In this example, although the learner sometimes generates wrong state transitions, for example a transition to states to the 'free-run' play in agent Y during agent X is doing 'slow-down', it recovers to the suitable transitions and continues to imitate the demonstrator. This shows robustness of the model against accidents. Because the model is coupled loosely with world and other's states by output probabilities of HMM, it can permit variation and misunderstanding of world and others' states.

5 Related Works and Discussion

There are several works on coupling HMMs that can represent combinational probabilistic phenomena like multi-agent collaboration [5, 1, 4]. In these works, probabilistic relation among several HMMs (agents) are represented as state-transition probabilities, such that the amount of memory complexity increases exponentially. This is a serious problem for imitation learning because we assume that the number of examples is small for imitation. In our model, the relation among agents is represented by gate probabilities $\boldsymbol{H}$, in which others' states are treated as outputs instead of as conditions of state transition. Using them, the likelihoods of state-transitions are simplified as products of several probabilities (Eq. 1). In addition, detailed states of other agents are abstracted by play (intention). As a result, the number of parameters is reduced drastically, so that learning requires very small number of examples as shown in above examples. Although such simplification may decrease flexibility of representation as a probabilistic model, experiments show that the proposed model has enough power to represent team-play among agents.

Intention in the model brings another aspect to communication among agents. We assume that there are no mutual communication in the proposed model. However, we can introduce communication as a bypass of observation and estimation of other's intention (play). The proposed model will be able to provide criteria for when an agent should inform their intention to others by comparing agents' actual intentions and estimated intention of the agent itself by simulating its own $\mathbf{HMM^c}$.

One important issue is the design of the intention. In the proposed model, intentions play various important roles like chanking of the actions and conditions of world state. Therefore, we must design intentions carefully so that team-plays can be represented flexibly.

References

1. Zoubin Ghahramani and Michael I. Jordan. Factorial hidden markov models. *Machine Learning*, 29:245–275, 1997.
2. Stephen S. Intille and Aaron F. Bobick. Recognizing planned, multiperson action. *Computer Vision and Image Understanding: CVIU*, 81(3):414–445, 2001.
3. Y. A. Ivanov and A. F. Bobick. Recognition of multi-agent interaction in video surveillance. In *International Conference on Computer Vision (Volume 1)*, pages 169–176. IEEE, Sep. 1999.
4. Michael I. Jordan, Zoubin Ghahramani, Tommi Jaakkola, and Lawrence K. Saul. An introduction to variational methods for graphical models. *Machine Learning*, 37(2):183–233, 1999.
5. Michael I. Jordan, Zoubin Ghahramani, and Lawrence K. Saul. Hidden markov decision trees. In Michael C. Mozer, Michael I. Jordan, and Thomas Petsche, editors, *Advances in Neural Information Processing Systems*, volume 9, page 501. The MIT Press, 1997.
6. Y. Kuniyoshi and H. Inoue. Qualitative recognition of ongoing human action sequences. In *Proc. IJCAI93*, pages 1600–1609, 1993.
7. Hiroyuki Miyamoto and Mitsuo Kawato. A tennis serve and upswing learning robot based on bi-directional theory. *Neural Networks*, 11:1331–1344, 1998.
8. Nuria M. Oliver, Barbara Rosario, and Alex Pentland. A bayesian computer vision system for modeling human interactions. *IEEE Transactions on Pattern Analysis and Machine Intelligence*, 22(8):831–843, 2000.
9. Stefan Schaal. Is imitation learning the route to humanoid robots? *Trends in Cognitive Sciences*, 3(6):233–242, Jun. 1999.

Designing Agent Behavior with the Extensible Agent Behavior Specification Language XABSL*

Martin Lötzsch, Joscha Bach, Hans-Dieter Burkhard, and Matthias Jüngel

Institut für Informatik, LFG Künstliche Intelligenz, Humboldt-Universität zu Berlin
Rudower Chaussee 25, 12489 Berlin, Germany
{loetzsch,bach,hdb,juengel}@informatik.hu-berlin.de

Abstract. Specific behavior description languages prove to be suitable replacements to native programming language like C++ when the number and complexity of behavior patterns of an agent increases. The XML based *Extensible Agent Behavior Specification Language* (*XABSL*) also simplifies the process of specifying complex behaviors and supports the design of both very reactive and long term oriented behaviors. XABSL uses hierarchies of *behavior modules* called *options* that contain state machines for decision making. In this paper we introduce the architecture behind *XABSL*, the formalization of that architecture in XML and the software library *XabslEngine* that runs the formalized behavior on an agent platform. The *GermanTeam* [9] employed *XABSL* in the RoboCup *Sony Four Legged League* competitions in Fukuoka.

1 Introduction

The Sony Four Legged League (as well as the Humanoid League) differs from the "wheel based leagues" in the complexity of physical actions that have to be employed both for interaction and perception. The Sony robots have four legs with 3 DOF each, and a head with 3 DOF. Instead of kicking with a single kicking device like in the middle or small sized league, this allows for a lot of different kicking skills using legs, body, or even head, which often require preparatory movements. Instead of moving on wheels many different styles of walking are used in different situations. With the introduction of Wireless LAN communication in the Sony League in 2002, cooperative strategies became more complex and consequently require adequately formulated high level behavior.

For perception, the Sony robots need a set of perception behaviors, too. Because the field of vision, the image quality and the quality of the other sensors are very limited, information has to be collected over time, the movement of legs and head has to be coordinated with current vision needs and the perception process needs to be supported by methods like active vision. Usually, the related behaviors have to be merged with the other movement skills.

This huge set of abilities results in the need for a complex behavior control architecture that integrates many behavior patterns. It should be modular in

* The Deutsche Forschungsgemeinschaft supports this work through the priority program "Cooperating teams of mobile robots in dynamic environments".

D. Polani et al. (Eds.): RoboCup 2003, LNAI 3020, pp. 114–124, 2004.

the meaning that behavior patterns can be reused in different contexts. It has to support reactive and realtime decision making as well as long term deliberative behaviors. The set of behaviors needs to be easy to extend - adding new behaviors should not have side effects on other ones.

We found C++ not well suited for specifying agent behaviors. Especially extension and maintenance of complex behavior control systems may become a tedious and error prone task. More high level behavior specification languages allow for a separation of the behavior design from the implementation of the agent platform.

1.1 Related Work

In all RoboCup leagues, intentional cooperation and the pursuit of long term strategic behavior remain a challenge. According to the dynamics of soccer, the agents act with only very limited foresight.

Most teams in RoboCup are using **layered architectures**, with comparatively reactive behaviors (basic skills) at the lowest level (cf. e.g. [2, 15, 8]). Ordering different behaviors on layers allows to follow different goals in parallel. Behaviors on higher levels invoke or activate behaviors on lower levels. As long as the architecture has to manage only few basic behaviors, the separation of behaviors in two or three layers may be sufficient. But in our experience, it becomes very difficult to control more than a few basic behaviors without introducting further hierarchies, when their usage depends on a careful analysis of the situation, when they require complex preconditions to be achieved and when their performance needs a considerable amount of time.

There are other attempts to use **behavior languages** in order to simplify the process of behavior development. For example, GOLOG [12] is an logic based robot control language. Funge [11] developed the *cognitive modeling language (CML)* for the domain of computer games. Obst and Stolzenburg [16, 1] employ UML state charts for specifying multiagent systems. They follow a layered state machine approach with a fixed number of layers. They used UML because there exists a rich set of easy-to-adapt editors for editing state charts.

State machines are well suited for behavior modelling (cf. e.g. [16, 1, 7]). The decision which action is executed next depends not only on the environment but also on the last state. That allows to keep behaviors stable and to define hystereses between two behaviors for avoiding oscillations when the sensor readings are noisy.

1.2 Main Contributions

In this paper we present a flexible, open hierarchical behavior control architecture. It consists of state machines which manage the transitions to new behaviors according to the last state and the recent situation. In a flat architecture, the number of transitions between states increases very fast with the number of states. Therefore we use options to encapsulate a limited number of states and transitions according to their abstractness. The options form a rooted directed acyclic graph. In section 2.1 we describe that approach in detail.

Based on that architecture, we introduce *XABSL* as an XML based behavior language that allows to completely specify the behavior of autonomous agents. The development of a robot control includes the design of a hierarchy of options and the implementation of their internal state machines. *XABSL* supports both tasks using the advantages of XML technologies. The *XABSL* framework contains a variety of visualization and debugging tools. The runtime system *XabslEngine* (section 2.3) executes the behaviors written in XABSL.

The *GermanTeam* [9] competes in the Sony Four Legged League and is a national team that consists of separate teams from five German universities, amongst them the Humboldt University in Berlin. Section 3 shows how *XABSL* was employed by that team and which experiences were made in the competitions. *XABSL* could be proven to allow the efficient integration of program parts from different groups. It is possible to develop a new robot control with about 50 different behaviors in only two weeks.

2 Developing Agent Behavior with XABSL

2.1 The Architecture behind XABSL

In the presented architecture an *agent* consists of a number of *behavior modules* called *options*. The options are ordered in a rooted directed acyclic graph, the *option graph* (cf. Fig. 1a), which may be expanded into a tree. The terminal nodes of that graph are called *basic behaviors*. They generate the actions of the agent and are associated with basic skills.

The task of the option graph is to activate and parameterize one of the basic behaviors, which is then executed. Beginning from the root option, each active option has to activate and parameterize another option on a lower level in the graph or a basic behavior. Within options, the activation of behaviors on lower levels is done by state machines (cf. Fig. 1b). Each state has a subsequent option or a subordinated basic behavior. Note that there can be several states that have the same subsequent option or basic behavior.

Each state has a decision tree (cf. Fig. 2) with transitions to other states at the leaves. For the decisions the agent's world state, other sensory information and messages from other agents can be used. As timing is often important, the time how long the state is already active and the time how long the option is already active can be taken into account. Additionally, each state can set special requests, that influence the information processing or determine how and where the robot should point its camera.

2.2 Behavior Specification in XML

In previous RoboCup participations the *GermanTeam* made the experience that implementing such an architecture totally in C++ is error prone and not very comfortable. The source code became very large and it was quite hard to extend

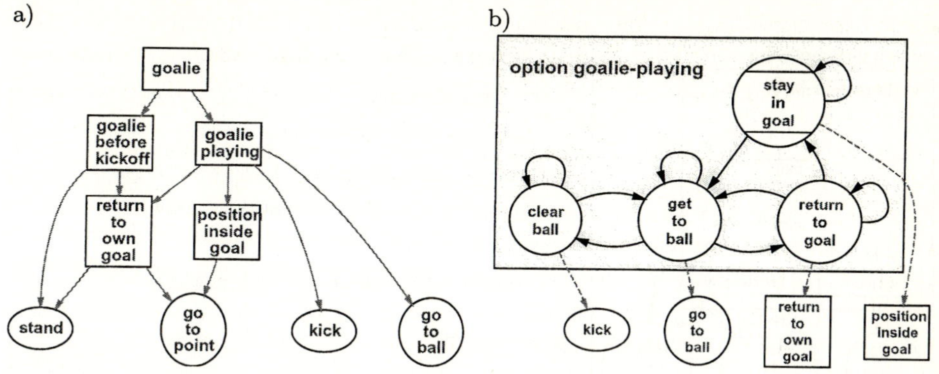

Fig. 1. a) The option graph of a very simplified goalie (this is only a simple example – the option graph developed by the *GermanTeam* for the competitions in Fukuoka contains about 50 options). Boxes denote options, ellipses denote basic behaviors. The edges show which other option or basic behavior can be activated from within an option. b) The internal state machine of the option "goalie-playing". Circles denote states, the circle with the two horizontal lines denotes the initial state. An edge between two states indicates that there is at least one transition from one state to the other. The dashed edges show which other option or basic behavior becomes activated when the corresponding state is active. The charts were generated automatically from the XML source in Fig. 3.

the behaviors. Therefore the *Extensible Agent Behavior Specification Language* (*XABSL*) was developed to simplify the process of specifying behaviors.

XABSL is an *XML 1.0* [4] dialect specified in *XML Schema* [10]. The reasons to use XML technologies instead of defining a new grammar from scratch were the big variety and quality of existing editing, validation and processing tools (many XML Editors are able to check if an *XABSL* document is valid at run-time), the possibility of easy transformation from and to other languages as well as the general flexibility of data represented in XML languages. Behaviors specified in *XABSL* can be easily visualized using XSLT [6] and DotML [13]. Note that the figures 1 and 2 were generated automatically from the XML source in Fig 3.

Agents based on the architecture introduced in the previous section can be completely described in *XABSL*. We have implemented language elements for options, their states, and their decision trees. Boolean logic (||, &&, !, ==, ! =, <, <=, > and >=) and simple arithmetic operators (+, −, ∗,/ and %) can be used for conditional expressions. Custom arithmetic functions (e.g. $distance{-}to(x, y)$) that are not part of the language can be easily defined and used in instance documents.

Symbols are defined in *XABSL* instance documents to formalize the interaction with the software environment. Interaction means access to input functions and variables (e. g. from the world state) and to output functions (e. g. to set requests for other parts of the information processing). For each variable or func-

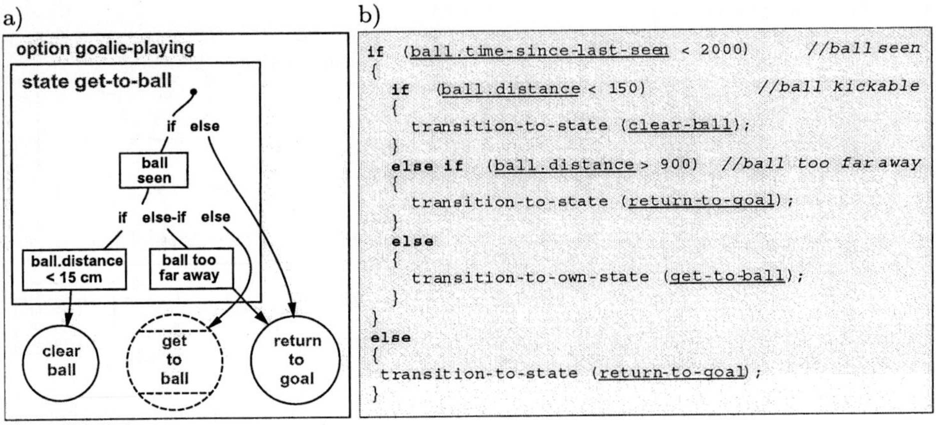

Fig. 2. The decision tree of the state "get-to-ball". a) Graphical notation: The leaves of the tree are transitions to other states. The dashed circle denotes a transition to the own state. b) Pseudo code of the decision tree. Note that both charts were generated automatically from the XML source in Fig. 3.

tion that shall be used for conditions a symbol has to be defined. This makes the *XABSL* framework independent from specific software environments and platforms.

An example:

```
<decimal-input-symbol name="ball.x" measure="mm"
  description="The absolute x position on the field"/>
<decimal-input-symbol name="utility-for-dribbling" measure="0..1"
  description="Utility for dribbling instead of passing/kicking"/>
<boolean-input-symbol name="goalie-should-jump-right"
  description="A ball is right ahead and rolls into to own goal"/>
```

The first symbol *"ball.x"* simply refers to a variable in the world state of the agent, *"utility-for-dribbling"* stands for a member function of an utility analyzer and *"goalie-should-jump-right"* represents a complex predicate function that determines if a fast moving ball is headed to the right portion of the own goal. In options, these symbols then can be referenced.

An example:

```
<if>
  <condition description="behind the ball">
    <less-than>
      <decimal-input-symbol-ref ref="self.x"/>
      <plus>
        <decimal-input-symbol-ref ref="ball.x"/>
        <decimal-value value="200"/>
      <plus>
```

```
    </less-than>
  </condition
  <transition-to-state ref="foo"/>
</if>
<else>
  <if>
    <condition description="clear right">
      <boolean-input-symbol-ref="goalie-should-jump-right"/>
    </condition>
    <transition-to-state ref="clear-right"/>
  </if>
  <else> ... </else>
</else>
```

The developer may decide whether to express complex conditions in *XABSL* by combining different input symbols with boolean and decimal operators or by implementing the condition as an analyzer function in C++ and referencing the function via a single input symbol.

As the *basic behaviors* are written in C++, prototypes and parameter definitions have to be specified in an *XABSL* document so that states can reference them.

An *XABSL* behavior specification can be distributed over many files. The *GermanTeam* uses different XML files for symbol definitions, basic behavior definitions, predefined conditions, agents and options. This helps larger teams of behavior developers to work in parallel. It is easier to keep an overview over the whole agent and a version control system like CVS can be easily used.

We developed tools for generating three different types of documents from an *XABSL* instance document set:

- An *Intermediate Code* which is executed by the *XabslEngine* (*see* sction 2.3). This was done because on many embedded computing platforms (like Sony's AIBO), XML parsers are not available due to resource and portability constraints.
- *Debug Symbols* containing the names for all options, states, basic behaviors and symbols make it possible to implement platform and application dependent debugging tools for monitoring option and state activations as well as input and output symbols.
- An extensive auto-generated *HTML-documentation* containing SVG-charts for each agent, option and state which helps the developers to understand what their behaviors do.

Fig. 3 shows an example for an *XABSL* source file. For more details about the language, the *XABSL* web site [14] contains a complete language reference, the XML schema files and examples.

```
<option name="goalie-playing" initial-state="stay-in-goal"
              description="goalie playing behavior">
  ...
  <state name="get-to-ball">
    <following-basic-behavior ref="go-to-ball"/>
    <set-output-symbol ref="head-control-mode"
                value="search-for-ball"/>
    <decision-tree>
      <if>
        <condition description="ball seen">
          <less-than>
            <decimal-input-symbol-ref
                    ref="ball.time-since-last-seen"/>
            <decimal-value value="2000"/>
          </less-than>
        </condition>
        <if>
          <condition description="ball kickable">
            <less-than>
              <decimal-input-symbol-ref
                        ref="ball.distance"/>
              <decimal-value value="150"/>
            </less-than>
          </condition>
```

```
            <transition-to-state ref="clear-ball"/>
          </if>
          <else-if>
            <condition description="ball too far away">
              <greater-than>
                <decimal-input-symbol-ref
                          ref="ball.distance"/>
                <decimal-value value="900">
              </greater-than>
            </condition>
            <transition-to-state ref="return-to-goal"/>
          </else-if>
          <else>
            <transition-to-state ref="get-to-ball"/>
          </else>
        </if>
        <else>
          <transition-to-state ref="return-to-goal"/>
        </else>
      </decision-tree>
    </state>
  ...
</option>
```

Fig. 3. An example for an *XABSL* XML notation: a source code fragment for the state *get-to-ball* (cf. Fig. 2) of option *goalie-playing* (cf. Fig. 1).

2.3 The Runtime System XabslEngine

For running the compiled behavior on a target agent platform, the runtime environment *XabslEngine* has been developed. The engine is meant to be platform and application independent and can be easily employed on other robotic platforms. This results in a variety of abstract helper classes that have to be adapted to the current software environment.

The *XabslEngine* parses and executes the intermediate code. It links the symbols from the XML specification that were used in the options and states to the variables and functions of the agent platform. Therefore, for each used symbol an entity in the software environment has to be registered to the engine.

The following example connects the C++ variable *worldState.ballPosition.x* to the *XABSL* symbol *"ball.x"*:

```
myEngine.registerDecimalInputSymbol("ball.x",
                                    &worldState.ballPosition.x);
```

While options and their states are represented in XML, basic behaviors are written in C++. They have to be derived from a common base class and registered to the engine.

The engine provides extensive debugging interfaces to be able to monitor the option and state activations, the values of the symbols and the parameters

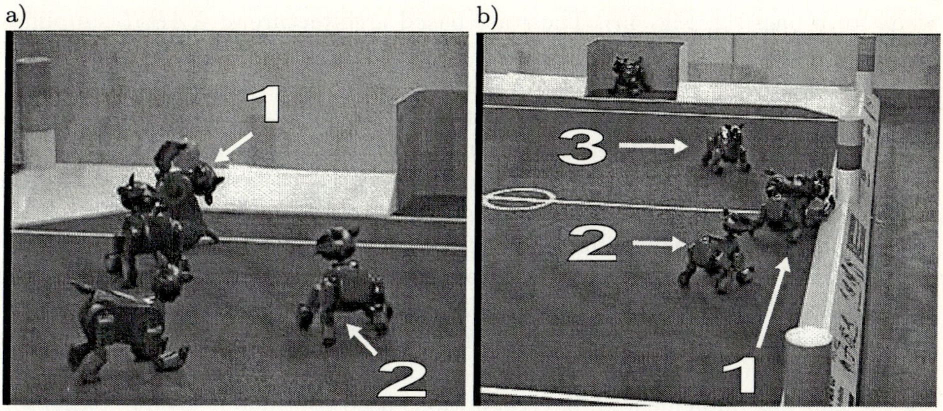

Fig. 4. Scenes from a video of a round robin game against the team *Georgia Tech* (4:1) in Fukuoka: a) *Use of communication.* The first forward (1) performs a bicycle kick directed to the opponent goal. The second forward (2) was told to wait in front of the opponent goal to be able to help if the kick fails. b) *Positioning.* The second forward (1) tries to dribble the ball into the opponent half. The defender (2) stays behind it to support the forward. The first forward (3) waits in the opponent half for a pass.

of options and basic behaviors. Instead of executing the engine from the root option, single options or basic behaviors can be tested separately.

A complete documentation of the engine, along with the code, can be found at the *XABSL* web site [14].

3 Application

XABSL was developed for the participation of the *Aibo Team Humboldt* from the Humboldt-Universität zu Berlin at the *GermanOpen* 2002 in Paderborn. Later on this approach was chosen for the participation of the *GermanTeam* in the RoboCup 2002 in Fukuoka [9]. In the competitions the *GermanTeam* won all its games except against the later finalists from CMU and UNSW.

The strength of the team was based on a big set of different behavior patterns. For instance the players employed over 16 different kicks in different situations. Amongst them the *bicycle kick* is a good method for getting the ball behind the player without previously turning around the ball. (cf. Fig. 4a) All these kicks require different behaviors for approaching the ball. Some work better for bigger ball distances, some require to grab the ball with the both front legs. Varying ball handling behaviors were chosen depending on whether the ball was in the opponent half, in the own half, at the left border, at the right border or in front of the opponent goal. *XABSL* proved to be suitable for implementing and integrating all these different abilities.

On higher levels, a set of team strategies based on communication was implemented. As it is often disadvantageous when two players try to obtain the ball the robots negotiated which of them handles the ball and which stays behind or waits for a pass (cf. Fig. 4a). The state based architecture of *XABSL* simplifies the developing of such strategies. Each robot sends its option and state activations to all other robots so that all players know what the others plan to do. However, since the wireless communication is not always reliable, all strategies have to be able to resort to non-communicative behavior, when necessary.

Complex positioning strategies were also employed. Each player had to care for an area of responsibility which changed depending on the score, the number of own players and the distribution of opponent players on the field (cf. Fig. 4b).

Although *XABSL* is a state based architecture, continuous approaches can easily be integrated into the behaviors. A *potential field* was employed to determine an optimal dribbling direction. This direction was made available to the options by an input symbol. A *Fuzzy Logic* based basic behavior for approaching the ball was implemented. Several options used continuous *utility models* for state transitions.

The hierarchical constitution of *XABSL* allows it to make many both very short-term and reactive decisions and more deliberative and long-term decisions co-instantaneous. The lower behaviors in the option hierarchy that are responsible for ball handling react instantly on changes in the environment. The more high-level behaviors like waiting for a pass, positioning or role changes try to prevent frequent state changes to avoid oscillations.

Altogether the *GermanTeam* implemented over 50 different *options* for the games in Fukuoka. About 10 team members were involved in developing and tuning the behaviors. The modular approach of *XABSL* made it easy to extend or advance the behaviors. New options could easily be added to existing ones without having negative side effects. Better solutions of existing options could be developed in parallel and were easily to compare with the previous ones.

Additionally, to help behavior control developers who want to employ *XABSL* on their own robotic platform, an example agent was implemented for the *Ascii Robot Soccer* environment [3]. In this simple soccer simulation by Tucker Balch the field is displayed in a 78 characters long and 21 lines wide text terminal. A team of four “>” players plays against a team of four “<” players with an “o” as the ball. All agents retrieve the full information about the world and the set of possible actions is very limited. This makes the implemented *XABSL* agent simple and easy to understand. The example implementation containing the *XabslEngine* and the visualization tools can also be downloaded from the *XABSL* web site [14].

4 Conclusion and Outlook

In this paper we present an approach for behavior design for teams of autonomous agents based on hierarchical state machines. The *Extensible Agent Behavior Specification Language (XABSL)* is an XML dialect that allows to conveniently develop behaviors using that architecture. We show how the *GermanTeam* employed that language to develop complex team behaviors for the RoboCup competitions in the Sony Four Legged League. The language and the code library *XabslEngine* are independent from the software platform that the

GermanTeam uses. It is relatively easy to employ *XABSL* on other robotic platforms; the code library is open source and publicly available at our website [14].

Future Work. Current developments of our behavior architecture aim at supporting the pre-deliberation of long-term strategies, which has to take place in parallel with the real-time execution of these strategies. This is done by adopting the *Double Pass architecture* [5], which has been developed for the simulation league team *AT Humboldt* of our work group. The *Double Pass architecture* annotates option hierarchies in its deliberation pass as intended, desirable or inapplicable, and executes the resulting plans in its second pass using a least commitment approach. The accommodation of additional condition types and different run-time requirements ask for extensions of *XABSL* as well as for the *XabslEngine*.

Acknowledgments

The authors would like to thank the members of the *GermanTeam* (especially Uwe Düffert, Jan Hoffmann and Max Risler) for filling the *XABSL* framework with content and Thomas Röfer for technical advice.

References

1. T. Arai and F. Stolzenburg. Multiagent systems specification by UML statecharts aiming at intelligent manufacturing. In *Proceedings of the 1st International Joint Conference on Autonomous Agents & Multi-Agent Systems, C. Castelfranchi, W. Lewis Johnson (Eds.)*, pages 11–18, 2002. Volume 1.
2. R. C. Arkin. *Behavior-Based Robotics.* The MIT Press, 1998.
3. T. Balch. The ascii robot soccer homepage. 1995. http://www-2.cs.cmu.edu/~trb/soccer/.
4. T. Bray, J. Paoli, C. M. Sperberg-McQueen, and E. Maler. W3C recommendation: Extensible markup language (XML) 1.0 (second edition). 2000. http://www.w3.org/TR/REC-xml.
5. H.-D. Burkhard, J. Bach, R. Berger, B. Brunswiek, and M. Gollin. Mental models for robot control. In *M.Beetz et al (Eds.): Advances in Plan-Based Control of Robotic Agents*, Lecture Notes in Artificial Intelligence, pages 71–88, 2002.
6. J. Clark. W3C recommendation: XSL transformations (XSLT) version 1.0. 1999. http://www.w3.org/TR/xslt.
7. Z. Crisman, E. Curre, C. Kwok, L. Meyers, N. Ratliff, L. Tsybert, and D. Fox. Team description: UW huskies-02. In *RoboCup 2002 Robot Soccer World Cup VI, Gal A. Kaminka, Pedro U. Lima, Raul Rojas (Eds.)*, Lecture Notes in Computer Science, 2003. to appear.
8. A. Dahlströhm, F. Heintz, M. Jacobsson, J. Thapper, and M. Öberg. The NOAI team description. In *RoboCup 2000: Robot Soccer World Cup IV, P. Stone, T. Balch, and G. Kraetszchmar (Eds.)*, number 2019 in Lecture Notes in Artificial Intelligence, pages 412–416, 2001.

9. U. Düffert, M. Jüngel, T. Laue, M. Lötzsch, M. Risler, and T. Röfer. GermanTeam 2002. In *RoboCup 2002 Robot Soccer World Cup VI, G. Kaminka, P. Lima, R. Rojas (Eds.)*, Lecture Notes in Computer Science, 2003. to appear. more detailed in http://www.tzi.de/kogrob/papers/GermanTeam2002.pdf.
10. D. C. Fallside. W3C recommendation: XML schema part 0: Primer. 2001. http://www.w3.org/TR/xmlschema-0/.
11. J. Funge, X. Tu, and D. Terzopoulos. Cognitive modeling: Knowledge, reasoning and planning for intelligent characters. In *Siggraph 1999, Computer Graphics Proceedings, Alyn Rockwood (Editor)*, pages 29–38. Addison Wesley Longman, Los Angeles, 1999.
12. H. J. Levesque, R. Reiter, Y. Lesperance, F. Lin, and R. B. Scherl. GOLOG: A logic programming language for dynamic domains. *Journal of Logic Programming*, 31(1-3):59–83, 1997.
13. M. Lötzsch. DotML Documentation. 2003. http://www.martin-loetzsch.de/DOTML.
14. M. Lötzsch. XABSL web site. 2003. http://www.ki.informatik.hu-berlin.de/XABSL.
15. R. R. Murphy. *An Introduction to AI Robotics*. The MIT Press, 2000.
16. O. Obst. Specifying rational agents with statecharts and utility functions. In *RoboCup 2001 Robot Soccer World Cup V, A. Birk, S. Coradeschi, S. Tadokoro (Eds.)*, number 2377 in Lecture Notes in Artificial Intelligence, pages 173–182, 2002.

Feature-Based Declarative Opponent-Modelling

Timo Steffens

Institute of Cognitive Science, Osnabrueck, Germany
timosteffens@gmx.de
http://www.cogsci.uos.de/~tsteffen

Abstract. In the growing area of multi-agent-systems (MAS) also the diversity of the types of agents within these systems grows. Agent designers can no longer hard-code all possible interaction situations into their software, because there are many types of agents to be encountered. Thus, agents have to adapt their behavior online depending on the encountered agents. This paper proposes that agent behavior can be classified by distinct and stable tactical moves, called features, on different levels of granularity. The classification is used to select appropriate counter-strategies. While the overall framework is aimed to be applicable in a wide range of domains, the feature-representation in the case-base and the counter-strategies is done in a domain-specific language. In the RoboCup domain the standard coach-language is used. The approach has been successfully evaluated in a number of experiments.

1 Introduction

An important factor effecting the behavior of agents in MAS is the knowledge which they have about each other [3]. Yet, in open domains like RoboCup agents encounter a variety of opponents which are not known to them beforehand. Traditional methods for inferring the plans or actions are not always applicable. Plan recognition [4] [5] focusses on deliberative agent architectures, so its application in dynamic domains in which (partly) reactive agents are to be encountered is not successful. Tambe attacked the problem of rapidly changing plans, but relies on several simplifying assumptions, e. g. that modelling and modelled agent share the same operator hierarchy and that the exact internal state of the modelled agent is known [10].

A promising approach which this work builds upon is to compare the tactics of new opponents to previously encountered ones and then select a strategy which has been successful back then [7]. It has been successfully applied in setplay-situations [8].

This work extends this case-based-reasoning approach by applying it to other game situations as well, and more importantly, by introducing a more flexible and more compact way of representation.

The remainder of this paper is organized as follows. Section two defines the representation formalism, feature-based opponent models. Section three describes how this framework can be applied in RoboCup and shows first experimental results. Finally, section four concludes.

D. Polani et al. (Eds.): RoboCup 2003, LNAI 3020, pp. 125–136, 2004.

2 Feature-Based Models

This section defines features and feature-based models in a domain-independent way. Also an architecture for agents employing this approach is proposed.

2.1 Features

The general idea is that humans use only the most typical properties of the opponent's behavior to retrieve a similar known opponent strategy. The assumption is that a limited number of opponent models can describe a wide range of opponents [8]. While Riley stores every observation in the opponent-models [7], feature-based models contain only a small number of distinct and stable features. An example of such a feature for the RoboCup domain:

`The opponent often does long passes along the left wing to the forwards.`

And for the MODSAF domain [2], a military air-combat simulator:

`If the aircraft comes into radar range, it turns onto collision course.`

We propose that rules which map actions to situations are the proper means to express such features. It is obvious that the features depend on domain-specific concepts like passing or radar-range. So a domain-specific language is necessary which can formalize situation and action descriptions. In RoboCup this can be accomplished by the standard coach language (CLang) [6].

Definition 1. *Let S be the set of all situation descriptions. Let H be the set of all possible actions. Let $H*$ be the powerset of H. A feature $< s, A >, s \in S, A \in H*$ is a mapping from a situation description s to a set of actions A.*

On this abstract level, situation descriptions may range from raw sensor data to preprocessed conceptualizations with domain-specific concepts. The situation descriptions in S may be incomplete, thereby focussing on only some part of the world. The actions in H are assumed to carry information on which of the agents executes it. As an example, in RoboCup a pass from player 6 to player 11 is different from a pass from player 10 to 11.

As suggested by the above examples, features are of a probabilistic nature. Teams may only tend to execute certain moves to a certain degree of probability. So features may even contradict each other, thereby reflecting the team's ability to decide between alternative options from their repertoire. This is treated in the feature-based declarative opponent-model (FBDOM):

Definition 2. *A feature-based declarative opponent model ω is a set of 2-tupels (F_i, p_i), where the F_i are features and the p_i are probabilities of the features to occur in the strategy which is specified by the model.*
Also the constraint $(s_i = s_j \wedge A_i = A_j) \rightarrow p_i = p_j$ has to be satisfied.

Two aspects of features need to be satisfied before a human deems them typical for a strategy. These are distinctness and stableness which originate from image recognition [12]. Distinctness means that a feature appears only seldom, but if it does the probability of a certain class is high. A stable feature on the other hand appears with a high probability in the class. Not identical notions, but close approximations to them, are used for building the opponent models. In this work, distinctness of a feature f is assumed if $p(M|f) > \alpha$, where M is an opponent model and α is a manually set threshold. Stableness is checked by $p(f|M) > \beta$, where β is another manually set threshold.

2.2 Architecture for FBDOM

This section outlines an architecture for an agent applying FBDOM. As shown in figure 1, the basis are the opponent models containing tactical descriptions on different levels of granularity and specificity. E. g., some might specify the complete strategy of a certain team, others might only specify the marking-assignment of the left-wing defender. In order to detect the features in the opponent models, the observations which come in as raw sensor data have to be processed by the action- and situation-detectors. They try to match the observations into the features of the models. The information if and how the observations match will be passed on to the model selector. It handles for example cases which match only partially (see below for a discussion of the different matching methods), and implements one of the possible selecting methods, e. g. a Bayesian classifier, or Tversky's contrast model for similarity [11]. The opponent model with the best value will be chosen and then a knowledge base will decide which counter-strategy is applicable. Just like the opponent models, the counter-strategies can vary in size. They can either contain full team strategies or just partial specifications, e. g. how a forward should shoot.

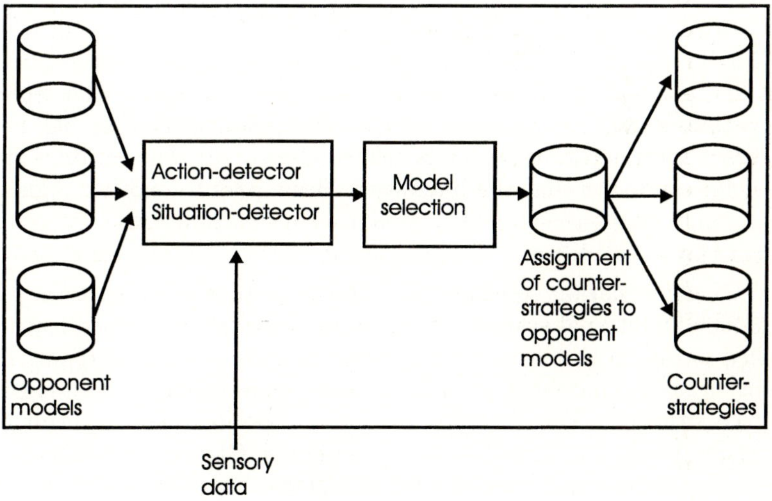

Fig. 1. Architecture of a FBDOM system.

This architecture allows usage as an online- or offline-method. It depends on the domain though, if the method needs only so much data as is provided during the encounter with other agents, or if the histories or logfiles of previous encounters have to be analyzed in order to select the counter-strategy beforehand.

3 Application in RoboCup

This section provides a proof of concept for FBDOM by implementing it in the RoboCup simulation league domain. The advantages and shortcomings of this approach are highlighted, and a series of experiments is described.

3.1 Representing Features for RoboCup

The performance of FBDOM is expected to be highly dependent of the used representation language. It has to be expressive enough to cover a variety of situations and actions, and should provide different levels of granularity (e. g. ranging from general descriptions of pass regions of a team to more specific passing behaviors of a particular player).

For the experiments, the standard coach language (CLang) [6] was chosen, because it fulfills these demands. Initially, CLang was designed to let the online coach inform and advice its field players. But it is also suited to represent strategies, since its messages are basically production rules, mapping situations to actions. The situations are crisp (possibly incomplete) descriptions of the world state like player and ball positions, play modes, score, and time, combined by logical operators. The actions are highlevel-actions which are more abstract than the server primitives and include concepts such as passing-to-regions, passing-to-players, positioning, and marking. As an example, the first example for a feature given before would be expressed in CLang as follows:

```
(definerule rule1 direc
  (and (bowner opp {X}) (bpos BACK_LEFT_WING))
  (do opp {X} (pass FRONT_LEFT_WING)))
```

The first line is due to the coach protocol and irrelevant here. The second line is the situation description and denotes that the opponent has the ball and that the ball is in a specific region (which is defined elsewhere). The action is specified in the last line and says that the ballowner does a pass to another region.

3.2 Situation-Matching

Since the CLang situation concepts are externally observable (i. e. they do not rely on hidden or internal states), the situation descriptions of the features can be matched easily to the actual worldstate. In fact, coachable teams have to implement situation-matching in order to determine when coach advice is applicable to a situation. The situation-matching code for these experiments was slightly adapted from the Dirty Dozen [1]. Although CLang conditions are made up of logical operators and freely definable regions, situation-matching is decidable. It can be seen as checking the includedness of points (concrete observed situations) into regions (general conditions) in the state space.

3.3 Action Detection

While the situation-matching is straight-forward, detection of actions is highly ambiguous. E. g. for an external observer it is hard, if not impossible, to decide if a kicked ball was passed to a player on the ball's trajectory or if it was shot to a certain point on that line. Thus, observed actions are often not singular points in the state-space, but regions. To match actions, we chose an expectation-driven approach. That is, observed actions are matched successfully to a feature's action, if any interpretation of the observed one is not mutually exclusive to the feature's action. To illustrate this:

Consider a ball that was kicked and travels along a certain line. Although the intention of the shooter is not clear, its action would match any feature's action that describes a shot or pass to any point or player on that line. That is, any overlap in situations is considered a match. For a more detailed discussion about operationalizations of CLang actions see [9].

3.4 Building the Models

Having accurate models for different opponent classes is crucial. The models of FBDOM are more complicated to build than those of Riley's approach. While Riley's models can be built by just counting positional observations of the modelled teams [8], the models here need distinct and stable features. That is, building such a model is basically about finding features that reliably describe the behavior of opponents that belong into the class.

Up to now the models are defined by a domain expert. By definition, features are typical moves. For the experiments opponent models for the offensive behavior of five teams were created manually. Surprisingly, it was sufficient to watch one or two games in order to build a model for a given team, because the typical, characteristic behaviors of a team have high saliency. The number of features in the models ranged from two to fifteen. The associated probabilities were then acquired automatically by determining the frequencies in the modelled team and in other teams. These frequencies were also used to ensure distinctness and stableness. If the frequency of a feature f in a model ω of team t was not significantly greater than the frequency of f in all other teams, f was removed from ω, because it was not distinct enough for t. f was also removed if its frequency was not beyond a certain threshold, in order to ensure stableness. In fact, only three features had to be removed (two due to indistinctness and one due to instableness), owing to the expertise of the domain expert. Still, future work aims at acquiring features and models by clustering or rule learning.

3.5 Determining the Best Matching Opponent Model

Feature-based models do not explain every observation, but just a subset. This has to be taken into account when determining the best matching model. Several methods are possible and need to be evaluated.

Matching Types: In MAS pairs of actions can be mutually exclusive (mutex). This depends on the domain, e. g. in RoboCup an agent cannot turn and kick at the same time. While in simple one-agent-domains all pairs of action may be mutex, in MAS most of the action pairs involving two agents are not mutex.

Definition 3. *Two actions a and a' are* `mutually exclusive` *or short* `mutex` *to each other, if they cannot be executed simultaneously. For two mutex actions a and a' we write $a \div a'$.*

Definition 4. *An action a is* `mutex` *to a set of actions A iff $\exists a' : a' \in A \wedge a \div a'$. We write $a \div A$.*

Definition 5. *A situation s is* `subsumed` *by situation s', iff s is true whenever s' is true. We write $s \subset s'$.*

Definition 6. *Let S be the set of all situation descriptions. Let H be the set of all actions. Let $H*$ be the powerset of H. An* `observation` *is a tupel $< s, A >$, $s \in S$, $A \in H*$, where s is the description of a complete observed situation and A is the set of observed actions in that situation.*

Since observations may contain several typical moves at once (say a defender stays in a certain region while the forward passes along a typical line), not observations, but features have to be counted. A feature $< s', A' >$ can be evaluated into the following primitive results wrt. an observation $< s, A >$. In the following list of match-types the probability parameter of the features in the model is abandoned, having no influence on matching. Because two actions are either mutex or not, this list is complete:

- `No-match`: $\neg\ (s' \subset s)$
  The situation of the feature in the model does not match the observation.
- `Partial Match`: $s' \subset s\ \wedge\ \exists a : (a \in A'\ \wedge\ a \in A)$
  The situation matches and there is at least one shared action in the feature action set and the observation action set.
- `Full Match`: $s' \subset s\ \wedge\ \forall a :\ (a \in A'\ \rightarrow a \in A)$
  The situation matches and all actions of the feature are in the observation action set.
- `Partial Mismatch`: $s' \subset s\ \wedge\ \exists a :\ (a \in A'\ \wedge\ a \div A)$
  The situation matches and there is at least one action in the feature action set which is mutex to the observation action set.
- `Full Mismatch`: $s' \subset s\ \wedge\ \forall a :\ (a \in A' \rightarrow\ a \div A)$
  The situation matches and all actions in the feature action set are mutex to the observation action set.

Partial match and partial mismatch can apply together within a given feature. In the selection process, partial matches or mismatches are ignored. On the other hand, a given observation can result in any type for different features, e. g. in a full match in feature 1 and a full mismatch in feature 2. This has to be taken into account when comparing opponent models in terms of the number of matches.

Selection Parameters: The above considerations lead to several parameters that can be combined and need to be evaluated:

- The first parameter (Once vs. All) determines how many matches will be counted for each situation. In case of "Once", the matching process aborts after the first successful full match. In the other case several features may be matched in the same situation. "Once" also means that a match overrules a mismatch of another feature.
- The second parameter (Most vs. Ratio) triggers whether the model that has the highest number of matches or the one that has the best match-mismatch ratio will be selected (cf. Tversky's ratio model [11]).
- The third parameter (Increasing vs. Normalized) specifies if the number of full matches and full mismatches will be divided by the number of features in the model. This way a normalization is done to overcome the variability in the number of features.

Combining these parameters results in eight different selection methods which were evaluated and compared to the Bayesian classifier, which is one of the most common methods in feature-based approaches [12]:

$$P(M_i|\alpha) = \frac{P(\alpha|M_i)P(M_i)}{P(\alpha)}$$

where M_i are the models and α is the observation.

3.6 Benefitting from the Classification

Of course a classification of the opponent alone does not improve the team's performance. This knowledge about the opponent's behavior must be exploited. So for each model a counter-strategy has been created manually. The counter-strategies were built depending on the characteristics of the model. To illustrate this, some examples are listed:

- If the model specified a fixed formation, a counter-formation was used. I. e. in defense players pool around the opponent's forwards and offensive midfielders, and in offense the forwards are located in the free spaces of the opponent's defenders.
- If the positions of the forwards were variable, but the forwards kept their role throughout the game, then the defenders were assigned marking assignments.
- If a model used fixed setplays (positions and/or pass chains), the counter-strategy incorporates marking assignments or positions for kick-offs based on the opponent's positions etc.

For the further experiments it was important to test the counter-strategies' appropriateness against the modelled teams. This was tested by feeding the counter-strategies into the Dirty Dozen (DD) team whose behavior is specified by an extension of CLang. So the counter-strategies are directly executed [1].

It turned out that the counter-strategies indeed achieved significantly higher scores against the modelled team than the baseline strategy which was used by DD during RoboCup 2001 (see [9] for the statistical values). So the strategies could be used in the following experiments.

3.7 Experiments

The experiments were designed to test if feature-based models are able to represent opponent behaviors, if they generalize to previously unseen teams, and what effect the observation length (i. e. the amount of classification data) has. Although FBDOM can be done decentralized, the classification was done by a centralized coach agent in these experiments.

Experiment 1. Considering that the models were built from only one or two games per team, the first experiment had to verify that the models are able to code the behavior of the team they were built for in new games and against other opponents. Additionally, the experiment was used to determine the best parameter settings (cf. section 3.5).

There were five opponent models $OM_1 - OM_5$ which had been built for teams $T_1 - T_5$. Now each team played several times against all other teams, including itself. 20 games thus resulted in 40 test instances for the nine counting methods. A classification was counted as correct, if OM_i was selected for team T_i. To fare better than random, more than 20% accuracy had to be achieved. Interestingly all normalized methods performed better than the increasing ones (see table 1). All normalized methods and the best increasing method were significantly better than random ($p < 0.01$), showing that the models were able to generalize to new games against new opponents. Especially one parameter setting achieved an accuracy of 82.5 %, which is a promising result for the following experiments.

Table 1. Parameters for counting methods and their identification success.

Normalized	**Once**	**All**
Most	82.5 %	80 %
Ratio	75 %	65 %

Increasing	**Once**	**All**
Most	25 %	35 %
Ratio	32.5 %	47.5 %

Bayes
26 %

Surprisingly, the Bayesian classifier performed on random niveau. A possible explanation for this is that the inter-dependence of the features violated the independence demand of Bayes. For example, a feature saying that a forward shoots from the left wing on the goal is highly dependent on a feature that specifies that the midfielders pass the ball on the left wing to the forward. These feature dependencies are likely to render Bayesian classification unsuccessful.

Experiment 2. In order to test if the team can benefit from the models even if it plays against a totally new opponent, experiment 2 had to test if classifying a new opponent and then playing with the appropriate counter strategy yields better goal-differences than playing with the baseline strategy, which was the behavior specification of DD as it competed at RoboCup 2001.

There were six new teams $N_1 - N_6$, the baseline strategy DD, the opponent models $OM_1 - OM_5$, and the counter strategies $CS_1 - CS_5$. For each new team N_j a baseline goal-difference was found by running several games against DD. In order to use the logfiles effectively, these games were also used to classify the new teams. Subsequently, according to each single classification OM_i, the new team played against the counter strategy CS_i. This resulted in two sets of games per new team with exactly the same number of instances, the baseline games (set 1) and the counter strategy games (set 2). Note that in the course of experiments some games had to be removed from consideration because of server crashes or connection errors (see N_1 and N_2 in table 2). The results show that in five out of six cases the classification and the related counter-strategy yielded significantly better scores than the baseline.

Table 2. Goal-difference mean and number of games of the new teams against baseline and selected counter-strategies. p is depicted if difference is significant (1-tailed t).

Team	Mean of counter-strat.	N_1	Baseline mean	N_2	p
Cyberoos 2000	-4.694	172	-5.125	183	/
FC DrWeb 2001	-5.46	126	-6.89	128	0.05
Helli Respina 2001	-13.84	72	-18.19	83	0.05
Virtual Werder 01	-0.839	30	-1.375	31	0.05
Mainz Rolling Brains	-13.21	23	-21.08	23	0.025
FC Portugal 2001	-34.27	21	-44.58	23	0.005

Experiment 3. While the significant better goal-differences against the classified counter-strategies in experiment 2 are a necessary condition for showing that the method is successful, they are not a sufficient condition. It might still be the case that all counter-strategies are better in general than the baseline strategy. Especially since the scores are partly very negative, the improvements might be due to a floor effect. In such a case, the classification would be obsolete, because any choice between the counter-strategies would yield better results than using the baseline strategy. So another experiment was run, in which several games were run against the new teams in which the strategy was randomly selected. That is, DD played against team N_i by using randomly selected counter-strategies CS_j. If the means of these games are less than the means of experiment 2 in which the classified counter-strategy was used, it can be assumed that the optimal strategy was used in the classification & selection-runs. The outcome of this experiment was non-uniform (see table 3). In three cases there was no significant difference between the random selection and the selection based on classification. One of those teams had not yielded significantly better results in experiment 2, so this case was not surprising. At least in the three other cases, the selection based on classification performed significantly better than the random selection. This means that in these cases the most suitable opponent model

Table 3. Goal-difference mean of games in which the strategy was selected randomly and of games, in which classified strategy was used. Differences to values in previous tables are due to an increased number of games.

Team	random	N_1	classif.	N_2	α
Cyberoos 2000	-4.77	126	-4.69	172	/
FC DrWeb 2001	-5.50	168	-5.85	163	/
Helli Respina 2001	-16.68	173	-14.25	173	0.05
Virtual Werder 01	-1.352	54	-0.8197	61	0.025
Mainz Rolling Brains	-15.65	336	-14.11	343	0.1
FC Portugal 2001	-32.50	42	-33.96	45	/

and the corresponding counter-strategies were selected in experiment 2. It also means that the opponent-models and counter-strategies did indeed generalize over the new teams, making the FBDOM approach successful.

However, the three cases in which the approach was not better than random selection have to be discussed. One possible explanation is that the strategies of the involved teams are so similar, that the related counter-strategies perform similarly well. Interestingly, two of these three cases were classified as ATTCMU or FCPortugal most of the times (see [9] for details), which are very similar to each other anyway, even for human observers. From this it can be concluded, that also the counter-strategies for ATTCMU and FCPortugal might perform similarly, which would contribute to the lack of significant difference. Another reason for the outcome that several counter-strategies performed similar might be that they are only different in the defensive parts, because the opponent models focussed on offensive behaviors. So, whenever the team was in a defense situation, there was no difference between any of the counter-strategies. Anyway, in the three cases which did not achieve significant improvements, the counter-strategy that was selected by the classification was not better than the others. This might also be due to the fact that the five created opponent-model/counter-strategy pairs cannot be assumed to cover all existing teams. As of now there is no evidence how well the six new teams are covered by the opponent-models. Based on this last thought it is strong evidence for the quality of FBDOM, that three cases were nevertheless significantly better than the baselines.

Experiment 4. In order to test the amount of data needed for the classification, six observation lengths were tested. The same recorded games and settings as in experiment 1 were used, with the difference that only the best parameter setting was used and that the observation length was variied. All observations started at kick-off. The results show that the classification performs very well even for very short observation windows (see table 4). After 100 cycles the classification is significantly ($p < 0.05$) better than random selection (which is 20%). After 250 cycles the classification is already correct in more than 50% of the cases. The accuracy gets better the more data is acquired.

Table 4. Accuracy of selection for different lengths of observation window.

Intervall length	accuracy
100	42.5 %
250	57.5 %
500	62.5 %
1000	62.5 %
3000	67.5 %
6000	82.5 %

This means that the classification cannot only be done offline by analysing logfiles, but also online. This also renders the idea applicable to select rather general models in the beginning of the game, and select more detailed models when more data is acquired.

4 Conclusion

A method for representing opponent models in multi-agent-systems was introduced and its performance was experimentally evaluated in the RoboCup domain. It was claimed that for classifying an opponent it is sufficient to focus on distinct and stable features instead of processing the complete behavior for all situations. The assumption was that a set of opponent models covers a great amount of existing opponent behaviors. The experiments showed that the identification accuracy was high for the modelled teams, so the claim can be supported that features are a well-suited method to describe opponent behaviors.

Regarding the coverage of new teams, the experiments were non-uniform, but hint in a promising direction. There are some methodological difficulties to measure the impact of counter-strategies. In a perfect experimentation setting, each created counter-strategy would perform well against only one opponent model, and bad against all other models. Yet, this can only be achieved in restricted toy-domains or against manually created opponents, but not under realistic conditions with using real teams. Obviously in the experiments the five created opponent models were not enough to cover all new teams. In five of six cases, the selected counter-strategy performed better than the baseline, and three of these five cases were also better than random selection. More work is needed to verify that the cause for the unsuccessful cases was the similarity between the opponent models and the small number of models which cannot be expected to generalize over all new teams. Creating more elaborate models that also contain information about defensive situations or in-depth analysis of the existing offensive behaviors could be helpful for this further work.

However, features form compact opponent models which successfully generalized over several new teams, so that the related counter-strategies were effective against previously unknown opponents. This also revealed that tactics can be identified by certain typical features, which are at this state of RoboCup still independent of the opponent, as the experiments suggest. Because of this, opponent models can easily be created for a team by observing arbitrary opponents playing against that team.

Acknowledgements

Thanks to Claus Rollinger and Wilfried Teiken for many useful comments.

References

1. S. Buttinger, M. Diedrich, L. Hennig, A. Hoenemann, P. Huegelmeyer, A. Nie, A. Pegam, C. Rogowski, C. Rollinger, T. Steffens, W. Teiken: The Dirty Dozen Team and Coach Description. In: A. Birk, S. Coradeschi, S. Tadokoro, editors: RoboCup-2001: Robot Soccer World Cup V, Springer, Berlin, 2002
2. R.B. Calder, J.E. Smith, A.J. Courtemarche, J.M.F. Mar, A.Z. Ceranowicz: Modsaf behavior simulation and control. In Proceedings of the 2nd Conference on Computer Generated Forces and Behavioral Representation, STRICOM-DMSO, 1993
3. David Carmel, Shaul Markovitch: Incorporating Opponent Models into Adversary Search. Proceedings of the Thirteenth National Conference on Artificial Intelligence, AAAI Press, Portland, Oregon (1996)
4. Gal Kaminka, D. V. Pynadath, Milind Tambe: Monitoring Deployed Agent Teams. Proceedings of the Fifth International Conference on Autonomous Agents (Agents-2001), Montreal, Canada (2001)
5. A. Kautz, F. Allen: Generalized plan recognition. In Proceedings of the National Conference on Artificial Intelligence, pp. 32-37, AAAI Press, Menlo Park, CA, 1986
6. M. Chen, E. Foroughi, F. Heintz, Z. X. Huang, S. Kapetanakis, K. Kostiadis, J. Kummeneje, I. Noda, O. Obst, P. Riley, T. Steffens, Yi Wang, and Xiang Yin. Soccerserver Manual v7. RoboCup Federation, 2001.
7. Patrick Riley, Manuela Veloso: On Behavior Classification in Adversarial Environments. In Lynne E. Parker, George Bekey, Jacob Barhen (eds.): Distributed Autonomous Robotic Systems 4, Springer, Heidelberg, Germany, pp. 371-380 (2000)
8. Patrick Riley, Manuela Veloso: Planning for distributed execution through use of probabilistic opponent models. In Proceedings of the IJCAI-2001 Workshop PRO-2: Planning under Uncertainty and Incomplete Information, 2001
9. Timo Steffens: Feature-based Declarative Opponent-Modelling in Multi-Agent-Systems. Master thesis, University of Osnabrueck (2002)
10. Milind Tambe, Paul S. Rosenbloom: Architectures for Agents that Track Other Agents in Multi-Agent Worlds. In M. Wooldridge, Joerg P. Mueller, M. Tambe (eds.): Proceedings on the IJCAI Workshop on Intelligent Agents II : Agent Theories, Architectures, and Languages, Springer, Heidelberg, Germany, pp. 156-170 (1996)
11. A. Tversky: Features of similarity, Psych. Review, 84(4), July 1977, pp. 327-352
12. Paul Viola: Complex feature recognition: A bayesian approach for learning to recognize objects. Technical report AIM-1591, AI Lab, MIT, 11, 1996

Scenario-Based Teamworking, How to Learn, Create, and Teach Complex Plans?

Ali Ajdari Rad[1], Navid Qaragozlou[1], and Maryam Zaheri[2]

[1] Computer Engineering Faculty, Amirkabir University of Technology, Tehran, Iran
{alirad,navidq}@aut.ac.ir
[2] Computer Science Department, University of North Carolina at Charlotte, USA
mzaheri@uncc.edu

Abstract. This paper presents the application of a novel method in the multi-agent teamwork field called Scenario-based Teamworking (SBT). In SBT method a team of cooperative intelligent agents could be able to execute complex plans in nondeterministic, adversary, and dynamic environments which communication cost is high. The base idea of this method is to define Scenario for different situations. With a graph of scenarios, a team of agents can execute, learn, adapt, and create team plans automatically. This method has implemented in a soccer team of intelligent agents (players and coach) and evaluated in the standard RoboCup simulator environment [1] and results show a significant improvement.

1 Introduction

One of the most important and complicated problems in designing multi-agent systems is the agents teamwork. A team of intelligent agents without cooperation will not going to act as well as a team of agents with less individual intelligence but teamwork understanding. The more complex environment and the higher cost for communication among agents make it harder to design a method for teamwork and managing agents.

The creation of the robotic soccer, the robot world cup initiative (RoboCup), is an attempt to foster AI and intelligent robotics research by providing a standard problem where wide range of technologies can be integrated and examined [2]. Some of the fields covered include multi-agent collaboration, strategy decision making, intelligent robot control and machine learning. In the RoboCup simulation league, there is a soccer simulator that simulates a soccer game. Each team should introduce 11 players and an optional coach to this simulator. The simulator sends the sensory information to players and players should declare their actions to the simulator via a network connection in a standard protocol. Like a real match, each player can not see the entire field (just watch what is in front of him) and has limited stamina. Players should communicate via the soccer simulator (with sending say commands). In this situation, a coach with global and noiseless view of the field can dynamically improve the teamwork of this team.

Similar to real soccer games, the simulated teams can have a coach. The duty of the coach is to employ appropriate tactics based on abilities of teammates and also the

D. Polani et al. (Eds.): RoboCup 2003, LNAI 3020, pp. 137–144, 2004.

strategy of the opponents. Furthermore, the coach is responsible to finding the weakness of its own team and improves their teamwork by applying appropriate strategies [3].

In this paper we present a powerful method to define plans for a team of soccer player agents called Scenario-based Teamworking. The main idea in this method was introduced by the Canned Plans concept of Essex Wizard team [4]. Using scenario – based approach, player agents are able to learn, adopt and execute complex team plans against their opponent, and the coach agent is able to modify and teach plans to its players. Another advantage of SBT is opponent modeling and the capability of automatically creating new plans.

2 Teamwork Based on Scenarios

The SBT method is based on the concept of defining scenarios. In this method a scenarios is defined for each team plan. Each scenario includes:

- Triggers: conditions that explain the current situation of the environment and internal state of the agent. These conditions are divided as follows:
  - Data Triggers: facts, produced by direct information about the environment. For example, agent is in area #1, or ball is on opponent side.
  - Time Triggers: concepts which are dependent on time, such as playing modes.
  - Communication Triggers: situations that are affected by agents' communications, such as "Agent #1 said …" or "Agent #2 sent a pass request."
  - Action Triggers: situations, which are related to an agent's action, such as owning the ball, or shooting the ball.
  - Situation Triggers: Conditions that are brought about by what happens in the game. Usually these conditions are related to the high level concepts in soccer such as attack mode, or crowding around the ball.

  According to the above classifications, we can define different and complex situations. For example:
  T1: Agent1 is ball owner & Agent2 is near Area3 & Agent3 sends a pass request
- Goal: describes the final goal of the plan. In simulated soccer field, we categorize the final goal as *Scoring*, *Clearing*, and *Possession,* when the team is ball owner; and *Blockade, Close goal*, and *Close pass,* when the opponent is the ball owner. Scoring scenarios occur near the opponent's goal and describe a plan that its final action is shooting to opponent's goal and scoring. Clearing scenarios occur near home goal and describe a plan that aims at kicking the ball away from home goal. Possession scenarios occur in the middle of the field and aim art keeping the ball and creating a chance to achieve Scoring situation. Blockade scenarios are selected when some agents want to obtain the ball from opponents (make pressure on ball). In Close goal scenarios, agents will close home goal so that the opponent ball owner will not be able to score. The final goal of Close pass is to force the opponent ball owner to keep the ball or make a bad pass by closing its useful pass lines.
- Abort Conditions: describes the conditions that abort the plan. They are defined just like Triggers (but in negative meaning). With separating these two set of con-

ditions, we make the concept simple. In addition, we can benefit from some features, such as approximate matching and risk management involving Triggers or Abort conditions.

- Evaluation: each scenario has its own evaluation parameters. We can classify evaluation parameters in to two groups: Cost and Score parameters. Cost is a real number that is determined by the designer in the beginning. The designer can describe the amount of cost as a function of time, power of the agent, number of agents participating in the scenario, the effects of the incomplete scenario (if this scenario fails) and other concepts. The score is also a real number and shows the rewards that are obtained if the scenario finishes successfully. These parameters can be learned and changed during a game.
- Side Effects: doing a scenario will change the environment and create new situations. For example "playing in the field width" scenario makes the play wide, and a scenario based on fast and long passing will increase the speed of the game.

Considering above descriptions, we define the main plan in the scenario model. Each scenario includes a set of sub-plans that are performed step-by-step. Each step includes actions, whose its main goal is scoring is shown in figure 1.

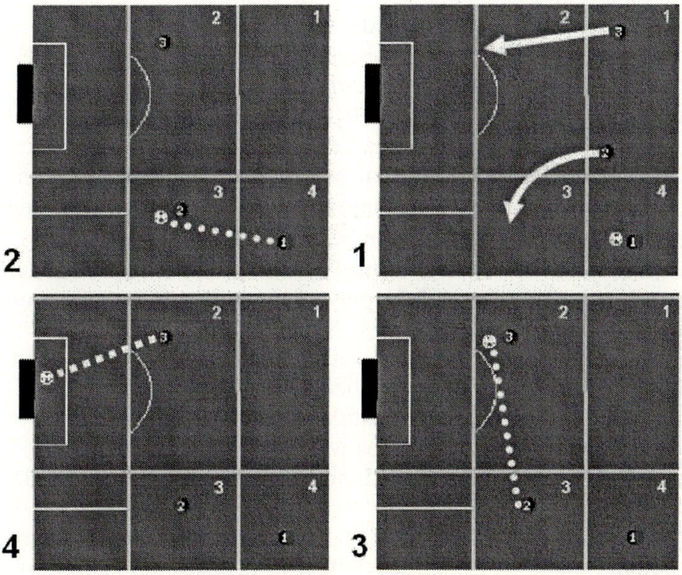

Fig. 1. Sample Scoring scenario.

```
Goal: Scoring   Cost: 3.8   Score: 17.3
Step 1:
Triggers: Agent1 is ball owner
          Agent1 is in Area3
          Agent2 is near Area4
          Agent3 is near Area2
```

```
Actions:  Agent1 passes ball to agent2 in Area4
          Agent2 receives pass from Agent1 in Area4
          Agent3 moves to Area2
Step 2:
Triggers: Agent2 is ball owner
          Agent2 is in Area4
          Agent 3 is in Area2
Actions:  Agent2 passes ball to agent 3 in Area2
          Agent3 receives pass from Agent2 in Area2
Step 3:
Triggers: Agent3 is ball owner in Area2
Actions:  Agent3 shoots to goal
Side Effects: Speed up the game
```

In the real game three players in the above condition can choose this scenario even with the lowest and least accuracy information. To make sure that agents are in the same scenario, each agent can have a priority to suggest a scenario. For example the agent who owns the ball has more priority to suggest the scenario (selecting a scenario could be done in voting or contract-net techniques). Choosing a scenario identifies the agent's roles and then there will be a matching procedure to match players with agents in the scenario, such as Player#7→agent2 (This assignment shows the player number 7 should accept agent number 2 role in selected scenario). This matching also could be done central or distributed (We implement central matching). In this case each player knows its role, and the player can do it without extra communication (in fact with a few communications).

A major problem in the other methods (like Canned Planes) was incompatibility of plans with the new situation or new opponent's plan. In SBT model, while a scenario is performed in a game and it is successful, its score increases with a coefficient. If the scenario fails, then its score decreases with a coefficient. In this way we can have a kind of adaptation during the game. Later we define a more flexible method for adaptation.

In the above, we have described how to define and implement a scenario. There is one more problem remaining which it is effect of past scenarios on current one. In a system like RoboCup, the current situation is affected by previous situations, such as environment and agents' behaviors; therefore one of the most important decision factors is the previous situations. For example in playing soccer, the percentage of success for the scenario shown in fig 1 after a scenario that its main goal is *Possession* is more than a scenario that its main goal is *Scoring*. So the success or fail of a scenario is related to its previous scenarios.

We implement this fact with making a graph by scenarios. Each node of the graph represents a scenario. Presence of an edge between each two nodes shows the probability of happening two scenarios sequentially. The weight of each edge represents the probability of success of the destination scenario after playing the departure scenario. The above idea makes the team to learn and perform a sequence of scenarios. For example a graph which starts with "Clearing" scenario and ends with a Scoring scenario is a complete plan.

For implementing this part, we use a graph of scenarios. In the first phase, scenarios are not related to each other. The first scenario is chosen in an experimental game based on matching of triggers. Agents execute the selected scenario and when it finishes; another scenario is selected based on triggers matching. In this case an edge is created between the first scenario and the second one in the graph. We continue the process as long as nodes of the graph are connected together with an acceptable threshold. In the second phase, we randomly choose a path of the graph that begins with the clear scenario and end with scoring scenario. Now agents execute scenario sequence without any evaluation. When a scenario wins we give positive score to the edge that leads us to the node, and if the scenario fails we give negative score to that edge and the sequence is failed. If the total path is successful we can give positive score to the total path.

In this manner, meaningful weights would be assigned to the edges of the graph in different games. Finally, in a real game, agents select a scenario based on the previous parameters and weight of the output edges. Also we can continue the learning phase during the real game. In this order, we can use the effect of more performing scenarios. At last scenarios are chosen with respect to these parameters:

1. Matching triggers with current conditions: Scenarios are evaluated based on matching with the current situation of the game. We can obtain a matching score for each trigger set (just like fuzzy rule based systems). For example nearness of a player to an area could be used as a fuzzy variable in this case.
2. Matching abort conditions with current conditions: Matching abort conditions are evaluated like triggers. This parameter has negative effect in the decision process.
3. Matching the scenario goal with the local team goal: The Scenario goal is compared with the team goal. For example, all teams desire scoring near opponent goal area. So scenarios whose goal is clearing should not be chosen and vice versa for scoring scenarios near home goal.
4. Matching side effect of the scenario with general strategy of the team. If a team wants to speed up the game, the scenarios having speed up side effect should be chosen. So the side effect of the scenario should be considered in the decision process.
5. We take into consideration costs and scores. In general, scenarios that have good score and low cost should be chosen. For example, a scenario that takes long time (high cost), or has lost many times (poor score) should be omitted.

In implementation, we mapped all above parameters to a real number between 0 and 1. Finally the optimum scenario is determined by using a weighted average of all the above-mentioned parameters. The designer can adjust the effect of each parameter with its coefficient.

Another benefit of SBT approach is the ability of the coach to automatically create new scenarios. This can be done in two ways:

– Watching the opponent's playing method, the coach can model the opponents' scenarios by determining current trigger, opponent's selected actions, side effects, and other features of a scenario (some features may not be determined exactly). The opponent modeling procedure can be done in a similar way [6, 7].

- Creating new scenario with evolutionary methods by the coach. New scenarios are built with random triggers and actions and then an evaluation phase determines the usability of it. Because of complexity of the environment, some limited rules should be considered to reach a better performance.

Then these new scenarios are added to team's scenario bank and are used during games.

In a real game, based on the opponent's conditions (prior knowledge about its pervious plays), one of the scenario graphs will be selected by human, before the beginning of the game (a team may have different scenario graphs for different strategies). Players select and execute scenarios and observe the result. Results will modify the graph so the behavior of team will be changed based on its online experiments. The modification of graph could be done distributed (each player modifies its graph and some decision sharing is done) or done by the coach and broadcast to players periodically. So two levels of adjustable autonomy are seen here: Human-Coach (if online modification is allowed) and Coach-Player.

3 Evaluation

In this part, we explain the results which are gained by performing and evaluating SBT method. The scenario bank of Pasargad team has around 50 scenarios. 50% of the scenarios are with Possession goal, 20% with Scoring, and the rest have other main goals. Usually, in each time of game two scenarios are performing, and the average time of any scenario is 100 cycles. For evaluating the effect of SBT method in a team, the evaluation part has implemented by two version of Pasargad team with different decision methods (Individual techniques are same).

1. Pasargad team plays based on decision tree.
2. Pasargad team plays based on scenario model.

We should mention that there are 10 games held for evaluation, and then the average result is rounded by 0.5. Each win scored 3 and each draw scored 1.

3.1 Decision Tree Version

Pasargad team has a high capability for individual actions, because of its specific architecture and optimum algorithms, which are using for technical actions in the player's code. So this team can show good performance with using simple ways to organize team working (Table 1).

Table 1. Results that obtained by Pasargad team that uses decision tree.

	1st half	2nd half	Final	Defeat	Draw	Score
FCPortugal2000	2-1	2-0	4-1	6	3	6
ATTUnited2000	1-0	2-1	3-1	3	4	13
Dirty Dozen	0-4	0-4	0-8	0	0	30
Pasargad	3-2	2-2	5-4	2	7	10

3.2 SBT

We conducted games between the same teams and Pasargad that play based on the scenario model. The result changed remarkably (Table 2).

Table 2. Results that obtained by Pasargad team that uses SBT method.

	1st half	2nd half	Final	Defeat	Draw	Score
FCPortugal2000	1-1	1-2	2-3	1	4	21
ATTUnited2000	0-1	0-3	0-4	0	2	26
Dirty Dozen	0-5	0-9	0-15	0	0	30
Pasargad (DTree)	1-2	1-4	2-6	0	1	28

In the decision tree version, Pasargad scored 1.4 average score and obtained an average result 3-3.5 for a game. Using SBT method, Pasargad reached 2.6 average score and 1-7 average results per game. So SBT could increase scoring about 40% and increased results about 6 goals per game.

Another interesting conclusion regarding the behavior is convergence of SBT approach. The team performance improves during the game and better results are always obtained in the 2nd half. Successful sequences of scenarios are repeated during the game. This fact means that the weaknesses of opponent strategy are found and team uses this knowledge to select appropriate scenarios to defeat the opponent. Also success of selected scenarios increases about 50% in the second half.

The benefit of using the SBT approach is shown in the game between the two versions of Pasargad shown in the last row of table 2. Both teams use the same base team, and the individual player actions are the same. A 60% growth of scoring is observed here just due to the SBT approach.

In a different experiment, the power of the SBT approach is seen more clearly. We made the duration of games five times longer than standard game duration (30'000 cycles), so the team had more time to learn. Table 3 shows the scoring results for two games with 6000 cycle window.

Table 3. Performance growing during a long game.

	1st	2nd	3rd	4th	5th
FCPortugal2000	2-3	2-4	2-7	1-4	0-5
ATTUnited2000	0-3	0-4	0-8	0-9	0-8
Dirty Dozen	0-16	0-18	0-24	0-26	0-26
Pasargad (DTree)	2-5	1-8	1-12	0-13	1-14

It seen that there is a fast growth of success until the 3rd segment of the game. Then this growing is slowed down, and reaches to the highest result that Pasargad with SBT could reach against its opponent (if the game was continued the result approximately would be unchanged). According to this, using offline training and the SBT approach (for example using past games against a specific opponent), the team can achieve a result twice better than the ordinary result.

4 Conclusion and Future Works

The Scenario-based approach was used in this paper for analyzing and arranging. The results show improvement of teamwork. Also implementing this team and analyzing situations are easier.

The SBT approach could be used in other areas of multi agent systems. In simulated soccer games, implementation of SBT with standard coach language [6], representing an architecture that adapts this concept [7], using efficient methods for automatically creating new scenarios, and using more parameters to implement efficient adjustable autonomy for coach and players are suggestion for future works.

Acknowledgements

We hereby appreciate Dr. Reza Safabakhsh our advisor and Dr. Ahmed Abdollahzadeh the professors of computer engineering department of Amirkabir University for their technical advices.

References

1. Noda, H. Matsubara, K. Hiraki, and Frank, 1. "Soccer server: A tool for research on multi agent systems", Applied Artificial Intelligence, 12:233-250, 1998.
2. Riley, P., and Manuela Veloso, M., "Coaching a Simulated Soccer Team by Opponent Model Recognition", In Proceedings of the Fifth International Conference on Autonomous Agents (Agents-2001), 2001.
3. Pourazin, S., Ajdari Rad, A., Atashbar, H., "Pardis, Our team in RoboCup simulation league", Linkoping University, Electronic press, pp 95-97, 1999.
4. Kalyviotis N. and Hu H., "A Co-operative Framework for Strategic Planning", Proceedings Towards Intelligent Mobile Robots (TIMR) '01, Manchester, 2001.
5. Riley, P., and Manuela Veloso, M., "Towards Behavior Classification: A Case Study in Robotic Soccer", In Proceedings of the Seventeenth National Conference on Artificial Intelligence (AAAI-2000), AAAI Press, Menlo Park, CA, 2000.
6. Ajdari Rad, A., "Design and implementation of a soccer coach; Dynamic leading of a team of intelligent agents", BE project, Computer engineering faculty, Amirkabir University of Technology, Feb. 2001.
7. Qaragozlou, N., "Design and implementation of soccer player architecture; an architecture for multi agent systems, dynamic and real-time environments", BE project, Computer engineering faculty, Amirkabir University of Technology, Feb. 2001.

Specifying Agent Behaviors with UML Statecharts and StatEdit

Jan Murray*

Institut für Informatik
Universität Koblenz-Landau, Campus Koblenz
Universitätsstraße 1, D-56070 Koblenz, Germany
murray@uni-koblenz.de

Abstract. The use of agents and multiagent systems is widespread in computer science nowadays. Thus the need for methods to specify agents in a clear and simple manner arises. One way of achieving this is by means of a graphical formalism. For using such a formalism the availability of tools, that support a developer, is of great importance. In this paper we present an approach to specifying agent behaviors on different levels of abstraction with the help of UML statecharts. Cooperation between different agents can explicitly be modeled. To help a developer with applying this formalism to the specification of agent behaviors the statechart editor StatEdit is presented. This development tool supports not only the modelling of an agent but a simple form of code generation as well.

1 Introduction

The use of agent technologies and multiagent systems has gained entrance into almost all branches of computer science. Thus the need for standards and design techniques has arisen. In order to gain wide acceptance an agent specification and design procedure must fulfill several constraints. It has to be as precise as possible to avoid ambiguities in the design of an agent. Nevertheless, the formalism must be easy to understand and use. It would also be desirable, that the application of formal methods is supported by the specification mechanism. This calls for a formal semantics.

All of these requirements can be met by a graphical formalism. Such a formalism is usually easy to understand. The definition of the individual graphical elements that make up the formalism can be done in an unambiguous way. Defining a formal semantics is also possible. Another thing that is always desirable for any formalism, especially graphical ones, is the availability of development tools which help a developer model and implement a system. In this paper we present both, a graphical way of modelling agent behaviors and an editor for working with this formalism.

Our approach is based on UML statecharts [11]. With this approach it is possible to specify not only behaviors for single agents on different levels of abstraction, but multiagent plans as well. Statecharts are a means for describing the behavior of a system in response to external events. Their graphical notation is intuitive and easy to understand. In addition to that the use of UML as a specification and modelling language is already

* Supported by the grant *263/8-1* from the German research foundation *DFG*.

D. Polani et al. (Eds.): RoboCup 2003, LNAI 3020, pp. 145–156, 2004.

widely accepted. Although the specification of UML statecharts does not provide a formal semantics, work on this has already been done, mostly with the aim of verifying properties of UML models, e.g. in [8, 14]. A tool for editing statecharts is also presented. With the StatEdit editor an agent designer can easily create statecharts describing agent behaviors. The resulting statecharts can also be exported to a number of other formats, which helps a developer with creating running code from the specification.

The rest of the paper is organized as follows. Section 2 briefly summarizes the relevant parts of the UML statechart formalism. Section 3 introduces our approach to agent specification with UML. The specification is described on different levels of abstraction. Special attention is laid on the specification of multiagent plans (Section 3.3). In Section 4 the statechart editor StatEdit, which is being developed at the University of Koblenz, is presented. This editor allows for the graphical design of agent behaviors and exporting the resulting statechart to different formats, e.g. Prolog or XML. Section 5 finally concludes the paper with an overview over related work, some final remarks and an outlook to future work.

2 UML Statecharts

The behavior of a system – like a program or an agent – can be described as a sequence of states the system is in. Depending on (external) events the system changes from one state to another. Such a change may be accompanied by the execution of an action.

In the Unified Modelling Language (UML) [11] *statecharts* are used to model behavioral aspects of systems. Statecharts are basically directed graphs, where different kinds of nodes represent states and pseudostates and edges stand for transitions. Labels on the edges describe properties of the transitions. Statecharts are *hierarchical* state transition diagrams, i.e. states in a statechart can contain other states or even whole state machines. In the following we will briefly summarize those parts of the UML statechart formalism that are employed by us for the design of agents.

In UML a *state* (represented as a box with rounded corners) is considered a period in the life of a system/agent during which a certain condition holds or an activity is performed. An agent may for example remain in a state while it waits for some external event to occur. In UML three different types of states are distinguished. *Simple states* are atomic in the sense that they do not possess any internal structure. *Composite states* can be further decomposed. They contain submachines which describe the activity associated with the composite state. *Concurrent states* are special types of composite states. A concurrent state contains two or more composite substates, which are called *regions*. If an agent is in a concurrent state, it is in *all* regions simultaneously. Thus concurrent states are used to model concurrent activities in a system or an agent.

The state of a system can be changed in reaction to external events. Such a *transition* is shown as a directed edge from state s_1 to state s_2, which is labeled with $e[c]/a$, where e is an event, c a boolean expression, and a an action. The (informal) semantics of t is "if the system is in state s_1 and event e occurs and the condition c holds, then the system executes action a and changes to state s_2".

Pseudostates may be seen as transient simple states, i.e. a system cannot remain in a pseudostate – it is left without delay. In other respects a pseudostate behaves just

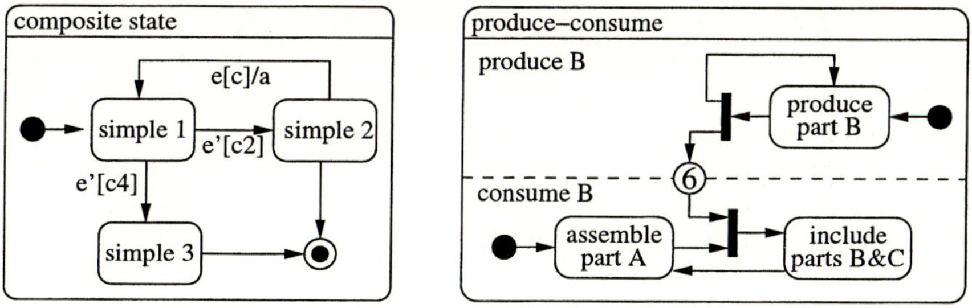

Fig. 1. Statechart examples. On the left a composite state with simple substates is shown. The right statechart shows a concurrent state modelling a producer consumer scenario.

like a simple state. With the help of *fork* and *join pseudostates* concurrent states may be entered and left. The *initial pseudostate* points to a designated start state of a state machine. The *final state* of a state machine is entered, if the activity modeled by this machine is finished. In this case a *completion event* is generated. Figure 1 shows two examples of statecharts.

Sometimes it is necessary to synchronize the concurrent activities in a statechart. This can be done with a *synch state*, which is shown as a circle residing on the dashed line separating different regions. The transition leaving a synch state may only be enabled if the transition entering the synch state has fired at least once. A synch state is labeled either with a positive integer giving an upper bound on the number of times the incoming and outgoing transitions have fired or with an asterisk, if there is no such upper bound.

In a typical producer-consumer scenario, for example, goods are produced by an agent and consumed by another. But the latter can only consume something that has been produced beforehand, so the need for synchronization is obvious. Figure 1 shows such a scenario on the right. Parts of type B are produced by one agent and used by another to assemble components ABC. The upper bound on the synch state is 6, so only six parts of type B may be produced, before one has to be consumed.

3 Designing Agents with UML

The behaviors of an agent can be seen as a sequence of states the agent is in. Each state corresponds to an activity or indicates that he agent is idly waiting for something to happen in its environment. Furthermore the behaviors of an agent can be specified on different levels of abstraction. This makes it possible, for example, to design an agent in a top-down manner by first specifying its tasks and behaviors on a very abstract level and then refining the abstract behaviors more and more.

One important feature of multiagent systems is the (explicit) cooperation of several agents to achieve a common goal or solve a problem. The agents play different *roles* in a shared (or multiagent) plan, i.e. they execute behaviors that solve parts of the problem or support other agents. Thus a role in a multiagent plan can once again be modeled as a sequence of states an agent passes through, partially in response to (external) events.

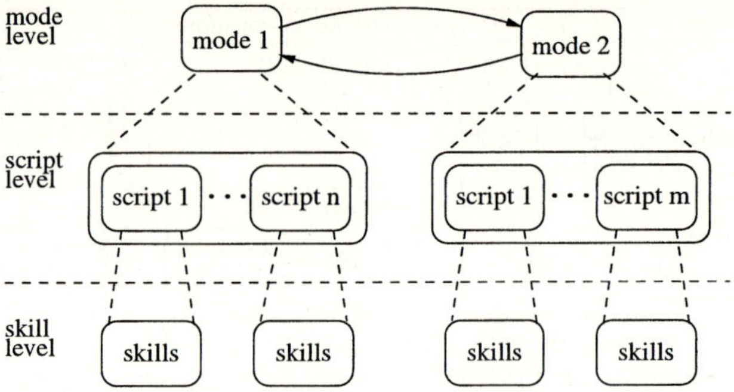

Fig. 2. Layered Architecture of an Agent.

But there is an important difference between the design of a single agent behavior and the role of an agent in a multiagent plan. When different agents work together to achieve a goal, their behaviors are not completely independent of each other, although most of the time they can be executed concurrently. But at several points of the multiagent plan *synchronization* of the individual behaviors of the agents is necessary, because sometimes agents have to work together on a subtask, or one agent has to wait for another agent to finish a subtask.

We present a layered approach to designing agents for the RoboCup Simulation League. For the specification and implementation of an agent three levels are distinguished, each of which is more global than the layer below (cf. Fig. 2).

- On the highest level – the *mode level* – global patterns of behavior are specified. They can be thought of as the most abstract desires an agent has, e.g. attacking or handling standard situations like corner kicks. These abstract desires correspond to different states an agent can be in. We will refer to them as *modes*.
- For each mode an agent has a repertoire of skeleton plans that it can use as long as it does not change its mode. The specification of these plans or *scripts* and their assignment to the global states constitute the second level of the agent design. On this level *explicit* specification of cooperation and multi agent behaviors can be realized.
- On the third and lowest level of the hierarchy the simple and complex actions the agents can execute are described. These actions, the *skills* of an agent, are used in the scripts.

So each level shows details of a higher level. The result of the design process is a layered specification of an agent. The connections between the three levels are not predefined in a rigid manner. Thus a developer can adapt the modelling technique to the kind of agent that ist to be designed. If, for example the statecharts created on all levels are merged into *one* chart, the resulting statechart models a hysteretic agent as described in [13]. If an agent for a special architecture (e.g. SOAR) is modeled, the statecharts are mapped to components of the target architecture.

Throughout the rest of the section we will describe, how UML statecharts are employed in the specification of agents. Section 3.1 explains the high level specification of

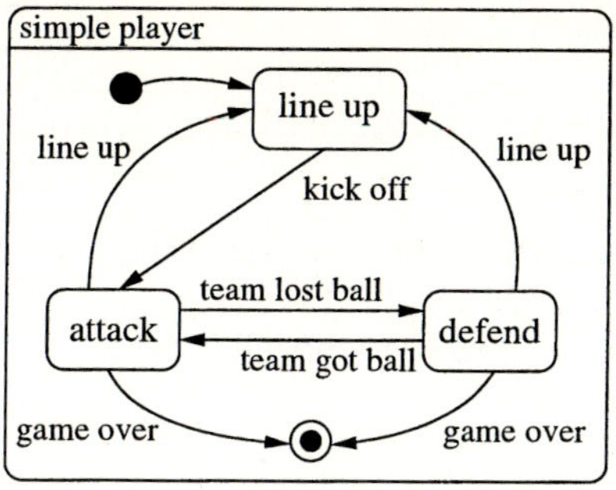

Fig. 3. Modechart for a simple player. The line up event is generated after a goal and at the beginning of each half time.

an agent, while the subsequent sections deal with the design of single agent behaviors (Section 3.2) and multiagent plans (Section 3.3). The skill level will only be addressed very briefly in Section 3.4.

3.1 Mode Level

In robotic soccer an agent frequently switches its behaviors on a very abstract level. For example, an agent may either be *defending* or *attacking*, depending on which team is controlling the ball. All changes of such global behaviors happen in response to one or more external events. If a state is associated with each of the behaviors and the events and conditions that lead to a change from one state to another are determined, the agent can already be modeled by a statechart, called *modechart*, on this very abstract level.

Consider a very simple soccer playing agent with only three such modes. Whenever the agent's team controls the ball the agent is *attacking*. If the opponent team gains control of the ball, the agent switches to a *defensive* behavior. Before each half of the game, as well as after a goal, there is a period in which the teams *line up*. Figure 3 shows the resulting modechart. As the soccer teams line up on the field before the game is started, the line-up state was chosen as the initial state. When the game is over, the agent enters its final state and ceases its activities.

The transitions between different modes are usually triggered by easily observable events and guided by guards, whose truth values can easily be determined. As a lot of behaviors an agent performs only make sense in certain situations, its action selection mechanism is guided on an abstract level by the current mode, and so the search space for selecting an action is pruned. For example, in the line-up mode the agent only has to consider a single action, namely moving to its home positon.

3.2 Script Level

Up to now only very abstract desires of an agent have been specified with the help of modecharts. But nothing has been said about how to achieve those abstract desires.

Each agent is equipped with a repertoire of *scripts*, which are short local skeleton plans for handling particular situations. In each mode an agent can access a subset of

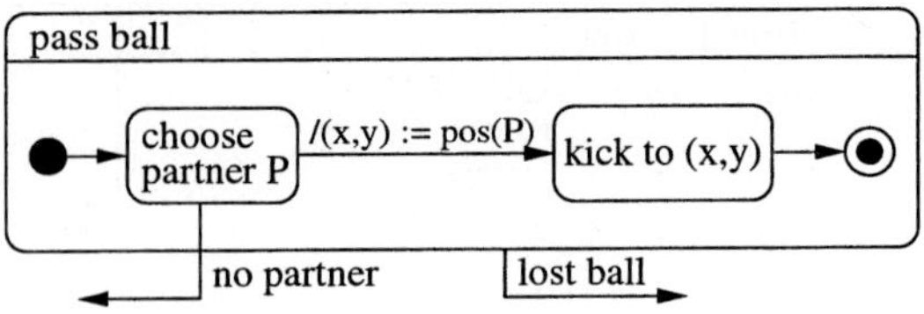

Fig. 4. A script for passing.

all scripts, depending on the situations that can arise in the particular mode. If no script is applicable, the agent has to fall back to a (possibly purely reactive) default behavior. One of the main problems of a past approach [15] to specifying such scripts lay in the rigidity of the plans. Once a script was selected for execution, it was hard to interrupt or abort it. UML statecharts, however, are very well suited for this kind of specifications.

For an agent, the execution of a script means executing a sequence of activities, some of which depend on the outcome of previous activities. As they can be associated with (simple) states in UML statecharts, the whole script can easily be represented by a composite state. Such a *script state* is entered at a designated state representing the beginning of the activity and can be left at a variety of points according to the outcome of the script. Interruption of a script in response to changes in the world are modeled by transitions originating from the edge of the composite state.

Consider the example of a *passing script* in Fig. 4. The agent selects a teammate to kick the ball to, gets its position on the field and finally kicks the ball to those coordinates. If the agent cannot find a suitable teammate or loses the ball, the script is aborted. The agent can lose the ball during either activity, so the transition handling this event originates from the state representing the whole script. But only the *choose partner* activity may fail because the agent cannot find a partner, so the corresponding transition starts from the substate modelling this activity.

3.3 Multiagent Plans

Up to now we have shown how to model scripts or behaviors of a single agent with the help of UML statecharts. But what about multiagent plans? In a multiagent plan or script several agents act *simultaneously* in order to achieve a common goal. At certain points their activities have to be synchronized. Those two additional requirements – concurrency and synchronization – are modeled with the help of concurrent states. A multiagent script is specified as a concurrent state with a region for each role hat has to be played by an agent. If the activities carried out by different agents have to be synchronized, this is modeled with the help of a synch state.

An example may help to clarify this. Consider a typical double passing situation. An agent *A* controlling the ball wants to get past an opponent *O* by playing a double pass. The agent *passes* the ball to a teammate *B* and *runs* past the opponent. The teammate *B* *dribbles* a little with the ball and *passes* it back to *A* as soon as possible. To handle this situation the agents can be equipped with a multiagent script with two roles that correspond to the behaviors of the agents *A* and *B*. The concurrent state modelling this script is shown by Figure 5. The simple states in the script correspond to the activities

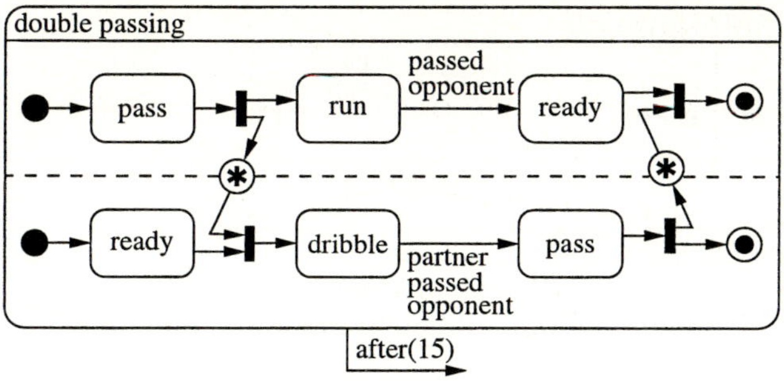

Fig. 5. A multiagent script with two roles for double passing.

of the agents. Synchronization is necessary twice in this script, namely for passing the ball. As an object (the ball) is passed between the agents and both agents can only continue their role at the respective positions if they are in possession of the ball, the need for synchronization is evident. Finally, a timeout for the execution is modeled by the transition labeled *after(15)*, which means that the execution of the script is terminated after 15 simulation steps. In this case the script has failed. Additional error transitions, e.g. modelling loss of the ball, may be added.

So far we have modeled the script as a whole. But some aspects have still been omitted. First of all, we only specified *where* synchronization has to take place, but we did not clarify *how* the activities are synchronized. As the different roles are played by different agents and their internal states are usually not known to teammates, means have to be provided that enable an agent to determine whether its partner has already reached a synchronization point or not. In addition to that an agent can only play one of the roles in a script at a time. Therefore the specification should model not only the script on the whole, but also the individual roles.

So we add a second step to the specification of a multiagent script, in which the behaviors corresponding to each role are derived from the script state to yield an *agent state*. This is done by "cutting along the dashed lines". As each role in the script is modeled by a region in the concurrent state, it is easy to see that the specification of an agent's behavior must be based upon the corresponding region. So the substates of each region are simply copied to the corresponding agent state. This process is quite straightforward, since most of the time the agents' behaviors are independent of each other.

The only spots that require special attention are the synch states. As we said before, a synch state only models the need for synchronization but says nothing about how this synchronization is realized. Unfortunately there is no unique way of handling synchronization in the derivation of an agent state, so the designer has to tackle this issue as the case arises. The required synchronization may, for example, be indicated by the change of a guard condition or the occurrence of an event. It may, however, be necessary for one of the agents to explicitly generate an event, for example by communicating its internal state.

Transitions modelling interruptions or errors are just copied from the script state. If such a transition starts on the edge of the composite state, it has to be copied to the edges

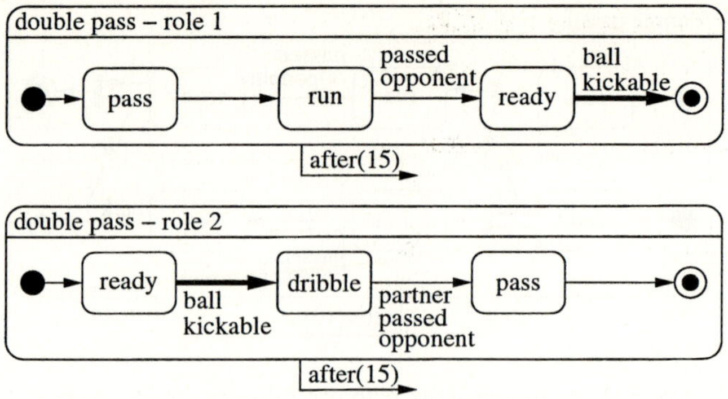

Fig. 6. The derived agent states for double passing. Bold transitions indicate synchronization.

of *all* agent states representing a role in the script. Finally, some events and guards have to be chosen, which enable the agent to notice that a situation has arisen in which the execution of a certain script is appropriate, and to determine its role in the script.

Let us now continue our example from above. The double passing script consists of two roles, so two agent states have to be generated, which are shown in Figure 6. In this example synchronization is needed, because the ball has to be transported (kicked) from one player to another.

As the ball entering the kick range of a player is an observable event and does not involve the knowledge of internal states of the teammate, synchronization can easily be handled by using a change event or by putting a guard on the respective transition edges, which prevents the transition from firing unless the ball has become kickable for the recipient of the pass. Last but not least, the transition modelling the timeout of the script has been copied to the edges of both agent states, indicating that both agents terminate the double pass after 15 cycles as a failure. The determination of the roles the agents play in the script is handled by the possession of the ball.

3.4 Skill Level

With the use of the server commands (*dash, turn, kick, turn_neck,...*) alone, the ability of an agent to interact with its environment is very limited. Therefore there are procedures or functions that provide more sophisticated *skills* for an agent at the bottom level of almost all RoboCup simulation teams.

Dribbling, for example, is modeled as a sequence of controlled kicks and dashes. The *dribbling skill* of an agent is responsible for generating the sequence of *dash, turn* and *kick* commands needed to keep the ball under control while running to a particular position. Figure 7 shows a statechart describing a simplified dribbling skill.

From the viewpoint of agent modelling the skill level is very similar to the script level. Skills are modeled like single agent behaviors. There are, however, no analogues to multiagent scripts, as skills are abilities of *one* agent only.

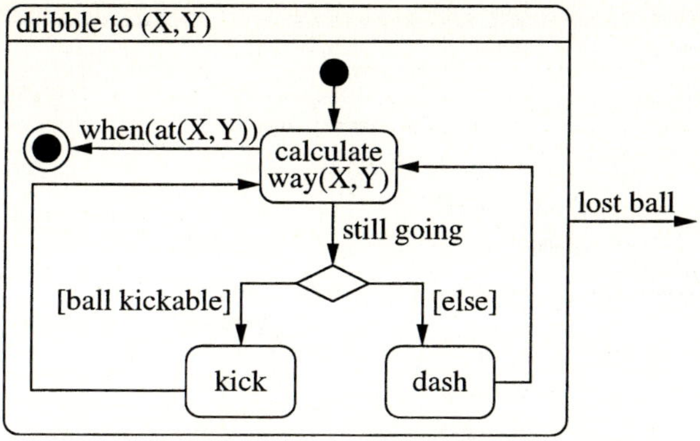

Fig. 7. A statechart modelling a simple dribbling skill.

4 StatEdit – A Statechart Editor

In order to assist an agent designer in modelling the behaviors of an agent with the help of the presented formalism the statechart editor StatEdit (Fig. 8) has been developed at the University of Koblenz [4]. With this editor the designer can easily create statecharts that specify the desired agent behaviors. Apart from the usual functionality provided by an editor for graphical elements, such as drawing, moving, erasing, or grouping objects, StatEdit offers a number of functions to support the modelling tasks of an agent developer. Two of them will be presented in greater detail, namely splitting (Sec. 4.1) and exporting (Sec. 4.2) of statecharts.

States and transitions are created and edited via pop up dialogues. The syntax of a statechart or a selection can automatically be checked, and a selected statechart can mechanically be beautified. In order to enhance the readability of a diagram, not all labels are shown at the respective elements. Transitions are only labeled with a unique identifier. The corresponding transition string is shown in a separate window when the pointer is over the transition (cf. Fig. 8). The same holds for the entry and exit actions of composite states.

4.1 Splitting Concurrent States

As we explained in Sec. 3.3 the specification of a multiagent script is done in two steps. First a concurrent script state is created. In a second step the script state is split into agent states and means of synchronization are specified.

StatEdit offers the functionality to split a concurrent state c between two regions, simply by clicking on the border between them. The result of this process are two (possibly concurrent) composite states c_1 and c_2, each corresponding to one (group) of the regions comprising the original state c. Transitions originating on the border of c are copied to the borders of both c_1 and c_2, while transitions from a substate are only moved with the corresponding state.

If a concurrent state is split between synchronized regions, synchronization is resolved with the help of events. Whenever this has to be done a unique signal event e_{sync}^{ID}

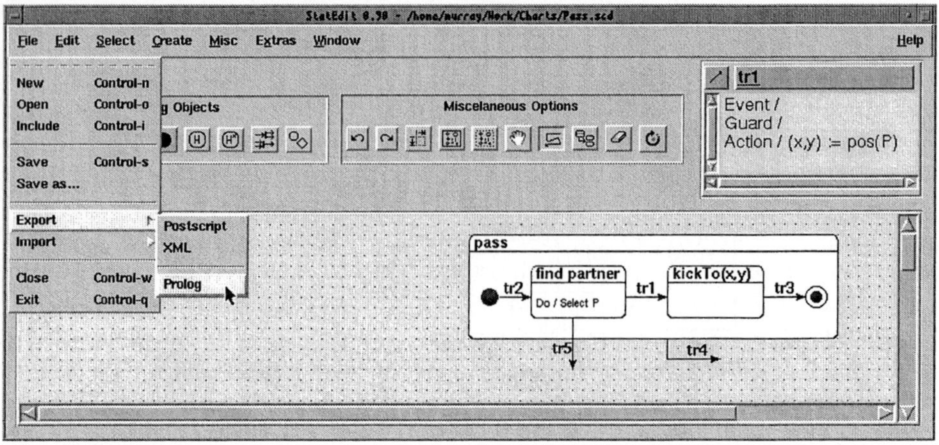

Fig. 8. The StatEdit main window with opened export menu. Prolog is selected as export format. The showinfo window on the right shows the transition string of tr_1.

is generated by StatEdit. The appropriate transitions in the resulting statecharts are labeled with transition strings, that either create e_{sync}^{ID} or are triggered by this event. Bounded sync states are resolved by putting additional guards on the transitions and incrementing resp. decrementing a global variable. Of course the user has the possibility to avoid the use of these generic events and resolve synchronization interactively.

4.2 Exporting a Statechart

Implementing a system that fulfills a given specification is a difficult and error prone process, especially for agent systems, which interact with their environment and have a certain degree of autonomy. In order to simplify this task StatEdit offers functions for exporting the created statechart to several other formats. This includes graphical formats for easy integration of figures into documentation, XML, and functions or function skeletons in different programming languages for (semi-)automatic code generation. Currently StatEdit supports conversion to EPS, XML, and a subset of Prolog, which can be interpreted by a statemachine built into our RoboCup team *RoboLog* [9, 10].

Exporting a statechart is done in two steps. First the internal representation of a statechart is converted to an XML structure which is then translated to the desired target format. This two step translation makes it easy to add further export modules, as there are already lots of XML based development tools available, which can easily be adapted for working as export functions.

5 Conclusion

This section closes the paper with some final remarks. First an overview over related work is briefly discussed. Then a short summary and an outlook to some future work will be presented.

Bergenti and Poggi [3] state that agent oriented software engineering adds another level of abstraction to the process of modelling software systems. This level, the *agent level*, treats agents as atomic units and models multiagent systems as interactions between agents. They present four agent oriented diagrams using standard UML notation. With these diagrams ontologies, agent classes and protocols are described. With this approach only aspects of the interactions among agents can be modeled. The agents themselves are treated as atomic entities. In contrast to this, our approach allows for the modelling of interactions between agents as well as describing the behavior of a single agent, i.e. the internals of an agent with only one formalism.

In [12] Odell et. al propose a number of extensions to UML for modelling multiagent systems or interactions between agents under the name of *AUML (Agent UML)*. A layered approach to specifying interaction protocols for agents is presented. At the top agent interaction is specified in different levels of detail with (extended) *sequence diagrams* and *collaboration diagrams*. *Activity diagrams* and *statecharts* are also used on this levels to emphasize certain aspects of the specification. On the lowest level intra-agent processes are specified with extensions of *activity diagrams* and *statecharts*. This approach is continued in [2] with the introduction of *protocol diagrams* into AUML, which extend the semantics of agent messages and improve inter-agent protocols. In contrast to the proposed Agent UML, our approach needs only one formalism to describe both the inter-agent and intra-agent behaviors in a multiagent system. In addition no extensions to the existing formalisms of UML have to be made.

In [7] an approach to agent specification and modelling of multiagent-systems based on object oriented formalisms is presented. Kinny and Georgeff use several object oriented modelling techniques, e.g. *class diagrams* and *statecharts*, for specifying agents. The plans that an agent may apply to reach a certain goal are described by *plan diagrams* which are based on statecharts. In contrast to the method presented in this paper plan diagrams are not used to express multiagent plans or explicit cooperation among agents. Multiagent systems are rather modeled by a class diagram, which describes the different classes of agents and the relationships among them. Object oriented mechanisms like inheritance are then used to distribute attributes between agent classes.

5.1 Summary and Outlook

We presented a graphical formalism for the layered specification of agent behaviors and multiagent systems. The formalism is based on UML statechart diagrams, which are used for modelling the behavior of systems in the object oriented software development paradigm. The approach allows for the explicit modelling of cooperation between two or more agents to achieve a common goal. From the specification of such a behavior the individual roles of the participating agents can be derived in a straightforward manner. To support a developer using our modelling technique the statechart editor StatEdit has been developed. With this editor statecharts can be created and exported to a variety of formats for further processing.

Future work includes applying the presented methods to other applications as well. First steps in this direction have already been taken [1]. The statechart editor StatEdit will be extended by further export modules. In the near future modules for integrating StatEdit with Golog [6] and the double pass architecture [5] will be implemented.

References

1. Toshiaki Arai and Frieder Stolzenburg. Multiagent systems specification by UML statecharts aiming at intelligent manufacturing. In Cristiano Castelfranchi and W. Lewis Johnson, editors, *Proceedings of the first international joint conference on Autonomous Agents and Multiagent Systems (AAMAS 2002)*. ACM Press, 2002.
2. Bernhard Bauer, Jörg P. Müller, and James Odell. Agent UML: A Formalism for Specifying Multiagent Interaction. In Paolo Cinacarini and Michael Wooldridge, editors, *Agent-Oriented Software Engineering*, 2001.
3. Frederico Bergenti and Agostino Poggi. Exploiting UML in the Design of Multi-Agent Systems. In *Proceedings of Engineering Societies in the Agents' World*, pages 96–103, 2000.
4. Michael C. Bruhn. StatEdit – a state chart editor. Diplomarbeit, Universität Koblenz-Landau, Campus Koblenz, Germany, 2003. to appear (German only).
5. Hans-Dieter Burkhard. Mental models for robot control. In *Advances in Plan-Based Control of Robotic Agents*, volume 2466 of *Lecture Notes in Artificial Intelligence*. Springer, Heidelberg, 2002. Postproceedings of Dagstuhl Seminar 01431 (Oct 21-26, 2001).
6. Frank Dylla, Alexander Ferrein, and Gerhard Lakemeyer. Acting and deliberating using GOLOG in robotic soccer — a hybrid architecture. In *Proceedings of the 2002 Workshop on "Cognitive Agents"*, September 2002. (Workshop during KI 2002, Germany).
7. David Kinny and Michael Georgeff. Modelling and design of multi-agent systems. In J. P. Müller, Michael Wooldridge, and Nicholas R. Jennings, editors, *Intelligent Agents III: Proceedings of the Third International Workshop on Agent Theories, Architectures, and Languages (ATAL-96)*, volume 1193 of *Lecture Notes in Artificial Intelligence*. Springer-Verlag: Heidelberg, Germany, 1997.
8. Johan Lilius and Iván Porres Paltor. The semantics of UML state machines. Technical Report 273, TUCS - Turku Centre for Computer Science, 1999.
9. Jan Murray, Oliver Obst, and Frieder Stolzenburg. RoboLog Koblenz 2001. In Andreas Birk, Silvia Coradeschi, and Satoshi Tadokoro, editors, *RoboCup 2001: Robot Soccer World Cup V*, volume 2377 of *Lecture Notes in Artificial Intelligence*. Springer, Berlin, Heidelberg, New York, 2002. Team description.
10. Jan Murray, Oliver Obst, and Frieder Stolzenburg. RoboLog Koblenz 2002 – short team description. In Gal A. Kaminka, Pedro U. Lima, and Raul Rojas, editors, *RoboCup 2002: Robot Soccer World Cup VI*, Fukuoka, Japan, 2002. Pre-Proceedings.
11. Object Management Group, Inc. *OMG Unified Modeling Language Specification*, September 2001. Version 1.4.
12. James Odell, H. Van Dyke Parunak, and Bernhard Bauer. Extending UML for Agents. In *Proc. of the Agent-Oriented Information Systems Workshop at the 17th National conference on Artificial Intelligence (AOIS Worshop at AAAI 2000)*, 2000.
13. Mikhail Prokopenko. Situated reasoning in multi-agent systems. Working Notes of the AAAI-99 Spring Symposium on Hybrid Systems and AI., 1999.
14. Frieder Stolzenburg. Reasoning about cognitive robotics systems. In Reinhard Moratz and Bernhard Nebel, editors, *Themenkolloquium Kognitive Robotik und Raumrepräsentation des DFG-Schwerpunktprogramms Raumkognition*, Hamburg, 2001.
15. Frieder Stolzenburg, Oliver Obst, Jan Murray, and Björn Bremer. Spatial agents implemented in a logical expressible language. In Manuela Veloso, Enrico Pagello, and Hiroaki Kitano, editors, *RoboCup-99: Robot Soccer WorldCup III*, volume 1856 of *Lecture Notes in Artificial Intelligence*. Springer, 2000.

Echo State Networks for Mobile Robot Modeling and Control

Paul G. Plöger, Adriana Arghir, Tobias Günther, and Ramin Hosseiny

FHG Institute of Autonomous Intelligent Systems
Schloss Birlinghoven
53754 St. Augustin, Germany
{paul-gerhard.ploeger,adriana.arghir,tobias.guenther,
ramin.hosseiny}@ais.fraunhofer.de

Abstract. Applications of recurrent neural networks (RNNs) tend to be rare because training is difficult. A recent theoretical breakthrough [Jae01b] called Echo State Networks (ESNs) has made RNN training easy and fast and makes RNNs a versatile tool for many problems. The key idea is training the output weights *only* of an otherwise topologically unrestricted but contractive network. After outlining the mathematical basics, we apply ESNs to two examples namely to the generation of a dynamical model for a differential drive robot using supervised learning and secondly to the training of a respective motor controller.

1 Introduction

Neural networks can serve as universal dynamical system representations, thus they constitute very powerful way of modeling [MNM02]. Simultaneously they are versatile in the tasks they can solve and without doubt neuronal networks represent an extremely successful biologically inspired solution concept. Consequently researchers begin to recast technical problems in ways amenable for solving them by the help of neural networks. Topics cover system modeling, nonlinear control, pattern classification or anticipation and prediction. Examples are found in form of feed-forward networks i.e. multi-layer perceptrons which allow autonomous driving of cars [Pom93] or the silicon retinas of Carver Mead [Mea89] which produce instantaneous optical flow very similar to natural processes found in the visual perception of animals. When it comes to even more interesting recurrent neural networks (RNN), users face some major problems. They find that using RNNs is in principle possible but mostly too difficult to be really applicable. Main problems are:

1. What is the 'right' structure for a RNN: i.e. which topology fits to the given problem best?
2. The convergence of teaching: i.e. which method will converge fast enough? There is a very pronounced desire for efficiency of training.
3. Over-fitting and exactness: i.e how to avoid too literal reproduction yet assure convergent behavior with respect to the teacher signal?

D. Polani et al. (Eds.): RoboCup 2003, LNAI 3020, pp. 157–168, 2004.

There is a proliferation of different approaches for point 1. e.g. Ellman nets, Jordan nets or Hopfield RNNS to name a few. Yet item 2. severely hinders to apply RNNs to larger class of problems. Known supervised training techniques comprise Back Propagation Through Time (BPTT), Real Time Recurrent Learning (RTRL) or Extended Kalman Filtering (EKF) all of which have some major drawbacks. Application of BPTT to RNNs requires stacking identical copies of the network thus unfolding the cyclic paths in the synaptic connections. Unlike back-propagation used in feed-forward nets, BPTT is not guaranteed to converge to a local error minimum, computational cost is $O(TN^2)$ per time step where N is the number of nodes, T the number of epochs [BSF94]. In contrast RTRL needs $O((N+L)^4)$ (L denotes number of output units), which makes this algorithm only applicable for small nets. The algorithm complexity of EKF is $O(LN^2)$. EKF is mathematically very elaborate and only a few experts have trained predefined dynamical system behaviors successfully [SV98]. In this article we approach items 1. and 2. from a different point of view and give a stunningly simple solution. We introduce the notion of Echo State Networks (ESNs) and apply this concept successfully in two problem domains, namely nonlinear system identification and nonlinear control. At this time, it seems that ESNs are also applicable to many others of the problems generally known to be solvable by RNNs such as filtering sensor data streams [Hou03] or classification of multiple sensory inputs [Sch02]. Thus they can be applied to many other every day problems of roboticists and their use is in no way restricted to the covered examples. See [Jae01b], [Jae01c], [Jae02] for an in-depth coverage of already investigated examples.

This article makes the following contribution to ESN related research: for the first time we successfully apply this technique to system modeling and controller generation for mobile robots. Using well known error norms from control theory we demonstrate that ESN based well-trained controller can compete with and even outperforms a classical handwritten one.

The remainder of this article is structured as follows: in section 2 we define basic notation and mathematics of ESNs. In the core part section 3 we describe in depth the process of teacher signal generation, training of system model and motor controller and give results on the soundness of the application of ESNs in the chosen application scenario. We close with a summary and give references.

2 Recurrent Neural Networks and the Echo State Property

In general, a discrete time recurrent neural network can be described as a graph with three sets of nodes, namely K input nodes $\mathbf{u}$, N internal network nodes $\mathbf{x}$ and L output nodes $\mathbf{y}$. We use the terms nodes, units and neurons interchangeably. Activation vectors at time point n are denoted by $\mathbf{u}(n) = (u_1(n), \ldots, u_K(n))$, $\mathbf{x}(n) = (x_1(n), \ldots, x_N(n))$ and $\mathbf{y}(n) = (y_1(n), \ldots, y_L(n))$ respectively. The interconnect edges are represented by weights $w_{ij} \in I\!R$, which are collected in adjacency matrices, such that $w_{ij} \neq 0$ implies there is an edge from node $j \to i$. We define $\mathbf{W}^{in}_{N\times K} = (w^{in}_{ij})$ for the input weights, $\mathbf{W}_{N\times N} = (w_{ij})$ for the internal connections, $\mathbf{W}^{out}_{L\times(K+N+L)} = (w^{out}_{ij})$ for the output weights and finally $\mathbf{W}^{back}_{N\times L} = (w^{back}_{ij})$ for the weights which project back from output nodes into the net, subscripts denote dimensions. Observe that direct impact from an input node to an output node or from one output node to another output node is possible. Evolution of the internal activation vector given by

$$\mathbf{x}(n+1) = \mathbf{f}(\mathbf{W}^{in}\mathbf{u}(n+1) + \mathbf{W}\mathbf{x}(n) + \mathbf{W}^{back}\mathbf{y}(n)) \tag{1}$$

where $\mathbf{f} = (f_1, \ldots, f_N)$ are the internal activation functions. Calculation of the new internal node vector from the current inputs, given old activation and old output according to equation (1) is called *evaluation.* The neural network computes its output activations according to

$$\mathbf{y}(n+1) = \mathbf{f}^{out}(\mathbf{W}^{out}(\mathbf{u}(n+1), \mathbf{x}(n+1), \mathbf{y}(n)) \tag{2}$$

where $\mathbf{f^{out}} = (f^{out}_1, \ldots, f^{out}_L)$ are the output activation functions and $(\mathbf{u}(n+1), \mathbf{x}(n+1), \mathbf{y}(n))$ denotes concatenation of input, internal and previous output activation vectors. Equation (2) is called *exploitation.* Observe that we do not *require* recurrent pathways between internal units, although we expect them to exist, and that there is *no* restriction on the topology. In our case of echo state networks we usually have full matrices for $\mathbf{W}^{in}, \mathbf{W}^{out}$ and if needed at all also for $\mathbf{W}^{back}$. $\mathbf{W}$ is a sparse matrixes with typical densities ranging from 5-20%. Successive evaluation and exploitation of the net according to equations (1), (2) might show a chaotic, unbounded behavior. Thus it is necessary to damp the system. This can be achieved by a proper global scaling of $\mathbf{W}$ (see below). With the given notation the Echo State Property (ESP) can be stated as follows [Jae02].

Definition 1. *Assume that an RNN with* $(\mathbf{W}^{in}, \mathbf{W}, \mathbf{W}^{back})$ *defined like above is driven by a predefined teacher input signal* $\mathbf{u}(n)$ *and teacher-forced by an expected teacher output signal* $\mathbf{y}(n)$ *both contained in compact intervals* U^K *and* Y^L *respectively. Then the network* $(\mathbf{W}^{in}, \mathbf{W}, \mathbf{W}^{back})$ *has the* echo state property (ESP) *with respect to* U^K *and* Y^L *iff for every left infinite sequence* $(\mathbf{u}(n), \mathbf{y}(n-1)), n = \ldots, -2, -1, 0$ *and all state sequences* $\mathbf{x}(n), \mathbf{x}'(n)$ *which are generated according to*

$$\mathbf{x}(n+1) = \mathbf{f}(\mathbf{W}^{in}\mathbf{u}(n+1) + \mathbf{W}\mathbf{x}(n) + \mathbf{W}^{back}\mathbf{y}(n)) \tag{3}$$

$$\mathbf{x}'(n+1) = \mathbf{f}(\mathbf{W}^{in}\mathbf{u}(n+1) + \mathbf{W}\mathbf{x}'(n) + \mathbf{W}^{back}\mathbf{y}(n)) \tag{4}$$

it holds true that $x(n) = x'(n)$ *for all* $n \leq 0$

Intuitively this means that if the network has been running long enough, its internal state is uniquely determined by the history of the input signal and the teacher forced output (see [Jae02] for details). The ESP is connected to certain algebraic properties of the weight matrix $\mathbf{W}$. There are sufficient conditions for RNNs to either proof or to disproof ESP. Since it eases the further presentation and the results do not depend on it, we assume $f_i(x), f^{out}_i = tanh(x)$ from now on.

Theorem 1. *Define σ_{max} as largest singular value of $\mathbf{W}$, λ_{max} as largest absolute eigenvalue of $\mathbf{W}$.*

(a) *If $\sigma_{max} < 1$ then ESP holds for the network $(\mathbf{W}^{in}, \mathbf{W}, \mathbf{W}^{back})$.*
(b) *If $|\lambda_{max}| > 1$ then the network $(\mathbf{W}^{in}, \mathbf{W}, \mathbf{W}^{back})$ has no echo states for any input/output interval $U^K \times Y^L$ which contains the zero input/output tuple $(\mathbf{0}, \mathbf{0})$.*

In practical experiments it was found that condition (a) is much too strong and that on the contrary the negation of (b) appeared to be sufficient in the sense that it always produced RNNs with ESP though this is not provable up to now. More precisely we apply the following algorithm to produce RNNs with ESP:

Algorithm 1 *1. Randomly generate a sparse matrix $\mathbf{W}_0, w_{ij}^0 \in [-1, 1]$ with a low density (e.g. 5-20% weights $\neq 0$).*
2. Normalize $\mathbf{W}_1 = 1/|\lambda_{max}|\mathbf{W}_0$, where λ_{max} is eigenvalue of $\mathbf{W}_0$ with maximal absolute value.
3. Scale $\mathbf{W} = \alpha\mathbf{W}_1$, where $\alpha < 1$, so α is the spectral radius of $\mathbf{W}$.
4. Then ESP holds for the network $(\mathbf{W}^{in}, \mathbf{W}, \mathbf{W}^{back})$

Observe that the ESP prevails regardless of the choice of ($\mathbf{W}^{in}$ or $\mathbf{W}^{back}$). Thus the open parameters of the network which require tuning are the dimensionality N of $\mathbf{W}$, spectral radius α, and scaling, sign and topology of input weights $\mathbf{W}^{in}$ and back-propagation weights $\mathbf{W}^{back}$, all of which have be adapted to the time series data of the given problem domain. The parameter α can be interpreted as the intrinsic time scale of the ESN where small α means fast reacting network and α close to one implies slow reactions. The number of inner nodes, N, relates to the short term memory capacity of the net [Jae01c]. Algorithm 1 gives a surprising answer to problem 1 from section 1: the exact interconnect topology of the RNN can be arbitrary and a condition taming the largest eigenvalues will suffice. As another convenient consequence teaching of echo state networks becomes easy and user friendly. Specifically for RNNs with ESP, *only* the matrix of output weights $\mathbf{W}^{\mathbf{out}}$ needs to be adjusted. In detail, one applies the following steps:

Algorithm 2 *Let $D = \{d_n | n = 1, \ldots, T\}$ be a set of T elements of training data d_n each consisting of a teach input vector $\mathbf{u}_{teach}(n)$ and a desired (to be taught) output vector $\mathbf{y}_{teach}(n)$. Set $\mathbf{x}(0) = \mathbf{0}$ and $\mathbf{y}_{teach}(0) = \mathbf{0}$.*

1. *Calculate the current network state $\mathbf{x}(n+1), i = 0, \ldots, T_0 - 1$ according to equation (1).*
2. *For $n = T_0, \ldots, T$ concatenate $(\mathbf{u}^T_{teach}(n+1), \mathbf{x}^T(n+1), \mathbf{y}^T_{teach}(n))$ in rows and store it in a state collecting matrix $\mathbf{M}_{(T-T_0+1)\times(K+N+L)}$*
3. *Similarly collect $(\tanh^{-1} \mathbf{y}^T_{teach}(n))$ in rows into the teacher collection matrix $\mathbf{C}_{(T-T_0+1)\times L}$*
4. *solve for $\mathbf{W}' = \mathbf{M}^{-1}\mathbf{C}$ where $\mathbf{M}^{-1}$ denotes the pseudo (Moore-Penrose) inverse of $\mathbf{M}$ and set $\mathbf{W}^{\mathbf{out}} = \mathbf{W}'^T$.*

Usually $T - T_0 + 1 >> (K + N + L)$ so step 4 amounts to the solution of an over-determined set of equations by regression, which can be accomplished by virtually any numerical SW packages in little cpu time. Usually there is no unique solution but there is unique one shortest in length. The trained network $(\mathbf{W}^{in}, \mathbf{W}, \mathbf{W}^{out}, \mathbf{W}^{back})$ is now ready to use by application of equations (1)+(2). In cases necessary arising stability problems can be cured by adding a vector term of small white noise in equation (1) during step 3. Algorithm 2 addresses item 2 - the difficult teaching problem- from section 1. ESNs teach the $L \times (K+N+L)$ coefficients of $\mathbf{W}^{out}$ only instead of the whole topology, all other parts remain untouched.

Now the mathematical basics of ESN can be summarized as follows: we can heuristically characterize them as RNNs with sparse internal interconnect topology and some restriction on the size of its maximal singular (or most of the time: eigen) value. In ESNs, only output weights are trained. Thus they ease topology and teaching related problems of classical RNNs. They have a low algorithmic complexity, allow fast teaching and are highly adaptable to AI tasks like filtering and classification. To train them, a designer still needs to fix some parameters like dimension, density and spectral radius. Signals have to be properly filtered, scaled and offset. For each of these operations there exist single or combined heuristics for an educated guess of the initial values. The most time consuming part deals with scaling of input and output ranges. Here one needs to find parameter sets specially adapted to the given problem. Speaking in terms of electrical circuit analysis shifting and scaling input/output data amounts to fixing an operating point of the ESN. Like for many other RNN models stability, data over fitting or lack of generalization abilities remains an issue also for ESNs. In the next chapter we apply ESNs to real world data stored during a game to find system models and nonlinear controllers for mobile robots.

3 Applications of ESNs to Control

Our mid size league robots have a standard differential drive with passive caster-type front wheels, so only nonholonomic movement is possible. As such the dynamics of the robot forms a nonlinear system which requires expert knowledge to be modeled in an analytic way. Consequently we preferred a black-box approach to modeling based on RNNs. They are especially powerful when approximating fast changing dynamics which is frequently the case in our behavior programming approach called dual dynamics (DD). In DD different behaviors run simultaneously on each robot [BK01].

Schematically Figure 1 displays the data flow in our robot architecture. Required left and right wheel speeds are calculated by the DD behavior program which runs on a LINUX notebook. The two values are send to the PIDs approximately every 33 msec. The PIDs convert the required speed (cm/s) into a pulse width modulation signal (PWM) in percent thus effectively controlling the voltage of the motors. At the two motors actual odometric speed values are measured. The velocity (cm/s) is feedback to the PID to close the low level control

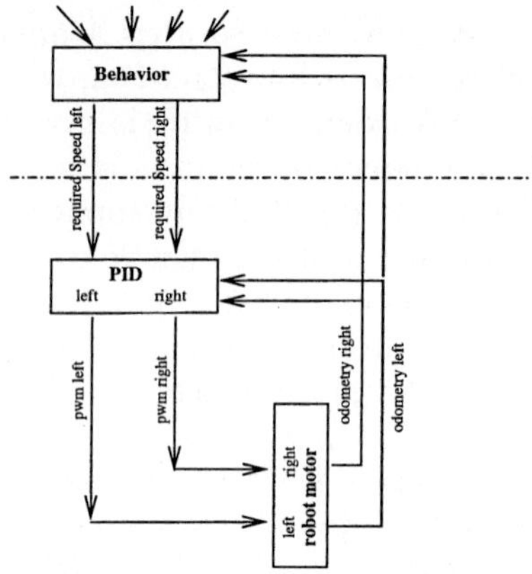

Fig. 1. Interface to low level robot architecture. Above the dashed line, we have the non-real-time PC-level. Below it there is a closed fast reactive real-time loop on μ-controller-level. Here physical modeling is mandatory.

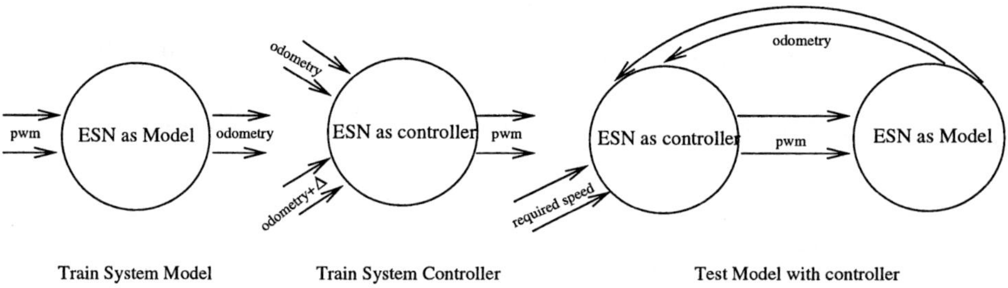

Fig. 2. Left: training of system model, middle: training of controller, right: combined simulation of controller and model, acting as a model of the physical robot.

loop and also to the behavior system running on the notebook. It operates in a non real-time way while below the dotted line the PID micro-controller operates in real-time in its feedback loop. We employ ESNs in three different situations, firstly as a system model. Inputs of the model are two PWMs and outputs are odometric velocities left and right. This can be seen as a forward model as it maps from inputs to outputs, in a context of a given state vector [Jor95]. In terms of Figure 1 this amounts to replacing the bottom box by an ESN. Secondly, as a system controller or inverse model which inverts the system by determining (i.e. by learning) the motor commands which will cause a desired modification in state. Here the trained ESN acts as controller, as it provides the necessary motor command (PWM) to achieve some desired state (i.e. the required speed). This is equivalent to the substitution of the box PID with an ESN in Figure 1. We train both replacing networks separately. Lastly in a third setup both ESNs are coupled to build an integrated model robot/low level controller pair. Then in Figure 1 the whole system below the dashed line is being modeled.

Model: Figure 2 displays the two different teaching situations and the application situation in the simulator. We begin by training the system model like depicted on the left side. According to Algorithm 1 we construct an ESN of dimension $N = 100$, 5% interconnection arcs spectral radius $\lambda = 0.8$, with two inputs and outputs (i.e. $K = 2$, $L = 2$) for left and right wheel. The inputs of the model are PWMs. Outputs are odometry and the teacher signal is set to the measured odometry from a stored trace file. The inputs were scaled down from their original domain $[-250, 250]$ to the range $[-1, 1]$. We added some mild noise of $+/-0.002$ during teaching in equation 1 as a fourth term inside the network activation function. The matrix $\mathbf{W}^{back}$ was set to zero. This is done for passive filtering tasks. In tasks involving active signal generation, the back weights are usually different from zero. The parameters and results from training are summarized in Table 1. We also computed $MSEr$ and $MSEl$ which denote the mean square error for both learned time series for the left and right wheel. Teaching took just less then a minute for a training sequence of length 6800 time steps. The evaluation time took a second on a Pentium III class machine using a MATLAB 5.30 implementation.

Table 1. Parameters used for training ESNs as Model and as Controller.

Name	Model	New Contrl.
Dimension	100	100
Density	5%	6%
λ	0.8	0.8
Inputs	2	4
Outputs	2	2
$\lvert w_{ij}^{in}\rvert$	± 0.25	0.5
Noise	± 0.002	± 0.002
MSE left	$6.5e-5$	$4e-6$
MSE right	$8.4e-5$	$4e-6$
Train steps	6800	1800

A picture of the desired and trained time series is shown in picture Figure 3. Firstly it can be stated that the trained model follows the trainer signal quite closely in general. Taking the maximum norm the overall relative error is 12%. The L^1 integral norm of the difference function between teacher and network output defined as $\int_a^b |f-g|dt/(b-a)$ is 1.1e-2 on the given time interval, the respective L^2 error norm is 4.2e-4. Taking a closer look the following observations can be made. Firstly we see how the measurement noise on top of the teacher signal is filtered away. The ESN generated signal appears to be smoother then the orignal. Secondly in the start interval [6900,7000] the ESN signal saturates and is not able to reach the desired 175 cm/sec. The explanation for this is very simple, since the used data file contained only 212 data points above 150 cm/sec and just 10 over 180cm/sec. This data set is far too small to train the network sufficiently well in this region of high speeds. By itself an ESN can neither extrapolate nor generalize for learned situations to close nearby input stimuli yet lying beyond the previously trained range. Thus it saturates at 150cm/sec. The very same explanation applies to some observed over-drives when the robot moves back. During teaching this situation prevailed for just about 10% of the time. We do not have an convincing explanation for the divergence in intervals [7250,7350] or [7750,7800] though. It might simply be mentioned that all our observations indicate that ESNs seems to behave especially well in spiking situations while in steady state situations a drift is frequently observable. The entire system model is reasonably exact to be useful in simulations and it surpasses the kinematical model in prediction quality by far. We then compared this result to a standard approach applied in the System Identification Toolbox in MATLAB. This SW supports many different methods but the default is the *prediction error method* (PEM) which is used when no special model was given by the user. Observe that PEM is still a parametric method but it chooses its parametric model all on its own using some clever heuristics. The comparison of ESN to PEM in Figure 4 proves that both models suffer from the same flaws. The training data set contained signals with extraordinary high frequencies. Consequently the fit at all rapid changes of inputs is very good, but the low frequency part is too rarely present in the teacher signal. Thus it can be concluded that the teacher stimulus set is not rich enough. More inputs sets containing rides on a straight line are needed. A final remark on the phase difference at the very end of the test data set may be noted. It is due the numerical roundoff error in the numerical calculation of the step size which PEM must use.

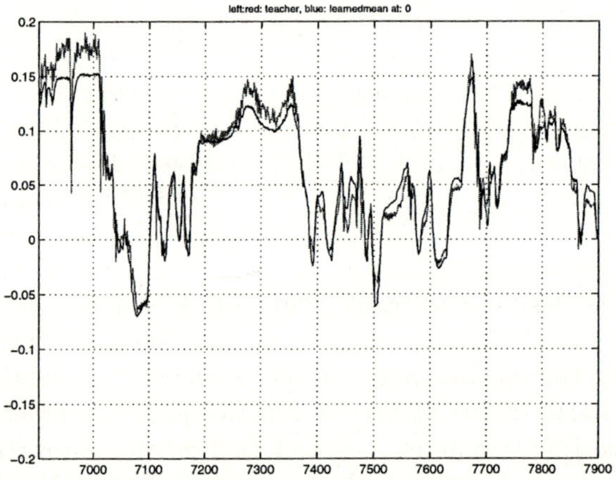

Fig. 3. Comparison of outputs of trained system model and observed robot speeds after teaching. Teacher is dashed, system is solid.

Controller: A similar teaching setup can be used to try to train a new ESN as a better controller then the given PID. If an embedded version of an ESN would be at hand this version could actually be used in the real robot replacing the

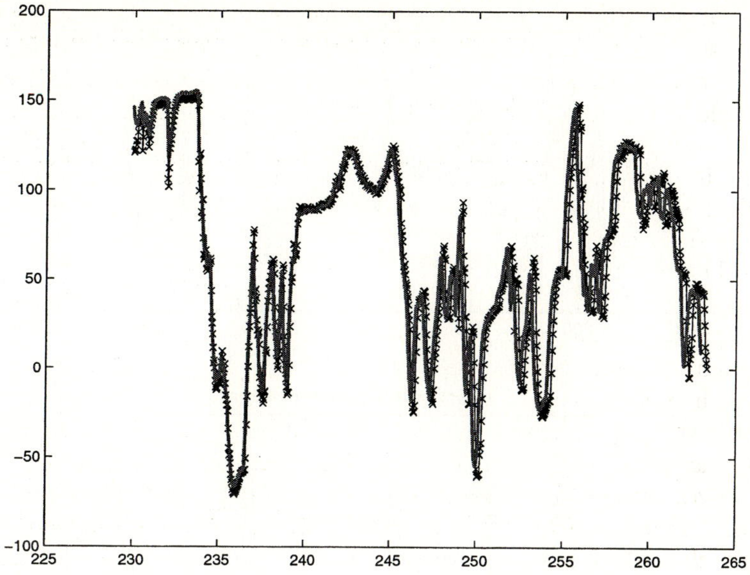

Fig. 4. Comparison of ESN method (solid line) and prediction method error (crossed line).

old PID controller. Since this hardware unit is not ready at this time we can only present a feasibility study. The ESN for this enhanced version controller has again the dimension of $N = 100$ internal units, 0.06 density, spectral radius lambda $= 0.82$, inputs $K = 4$ and outputs $L = 2$. In this case we are feeding the controller with the odometry signal itself and an incrementally delayed odometry signal (4 steps). Both are again derived from original speed data. A bigger delay it likely to enhance prediction, but would also result in damping or attenuating, which is unwanted here. In this training situation the ESN will learn to deduce how to mimic the given PWMs by using the current velocity and the desired velocity in near future. In Figure (5) we see a good fit in steady state situations on interval [50,100] as well as at rapid changes in [205,220].

Controller with System Model in the Loop: The third and last step consists of testing the new trained controller, but this time in combination the system model instead of the physical robot, see Figure 2 right frame. After initialization of start values for odometry, and PWMs, we need only to apply required speed as reference signal to the controller, while updating controller and model in a closed loop. These speeds are exactly the outputs from our DD behavior systems. Figure 5 shows the robot PID controller in the top frame and in the bottom frame it displays desired speeds, modeled odometry and PWMs. The lower part uses the new enhanced controller in combination with the system model. As it is easier, we discuss only the left motor. In the original trace on top we see a problem situation at time interval [380,460]. The desired speed is a constant but the PWM signal

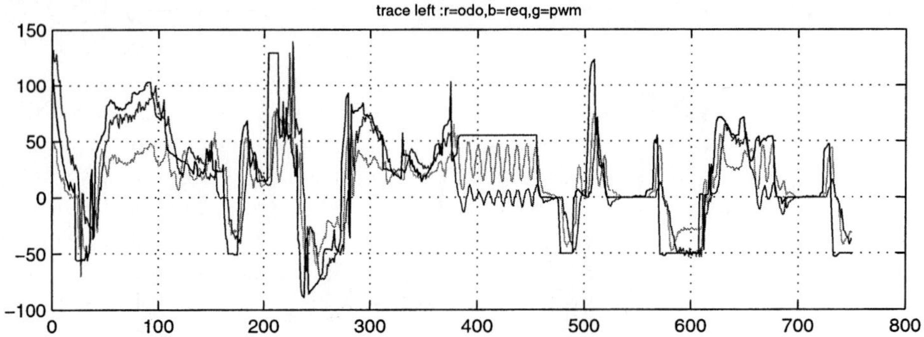

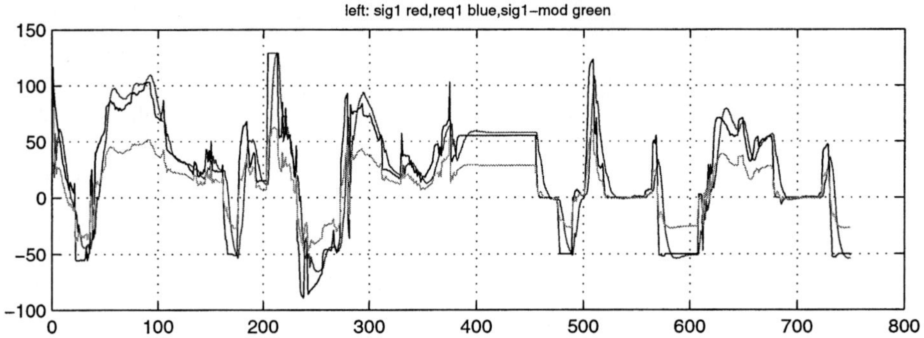

Fig. 5. Top: original trace data from real robot (black: required speed left, light grey: PWM, grey: odometry). Bottom: new enhanced controller in combination with learned physical model for an interval of 750 time steps. There are improvements at steady state situations and at points with rapid changes.

oscillates around 25 as does the measured odometry around zero. In this situation the robot was blocked by an obstacle and could not move forward as commanded by the behavior. Naturally this is an outer force not present in the simulation of the bottom frame. Instead the simulated velocity smoothly approaches the limit velocity and saturates with a significant steady state error. The situation itself was not mimicked by the trained controller instead he is able to handle it as expected with good results. This indicates that our model is well fitted. Besides that convergence in all dynamically changing situation seems to be better especially at [50,110] or at [620,660]. Another difficult situation is pictured in the time interval [200, 250]. Near step 200, required speeds are 140, but the odometry takes on this value only in the time step 225 very unstably jumping back and forth. Different from this, the bottom picture displays rectificated results for the discussed time intervals. Also the enhanced new controller can anticipate some step in future. This can be seen at time step 225. Again we computed the L^1 norm, which is one of the most popular ones in controller design, when there are fitting problems, or -as in our case- when there are big errors or "wild" points.

For the data in the original trace file the L^1 error was 2.3084 over 750 time steps. The L^1 norm for the same interval, with the optimized controller was about 0.0036.

These results clearly demonstrate the potential of our approach. Yet ESNs do share problems with other RNN based modeling approaches. For example they have to be taken as an undividable whole. This means that we can only control global network parameters like size, spectral radius etc. Changing them will impact the whole net and optimization space seems to be highly discontinuous as is it also well known from other areas like integer linear programming. Up to now we lack a systematic way to enhance convergence at one point while not sacrificing quality at others. A possible remedy might be a learnt superposition of network each being an expert in its own regime. Thus a modeling task could be decomposed and fine tuned in different independent areas. The harder or "wilder" you train the model, the better your controller will work. Another point, which is well known from learning theory, is that ESNs cannot master situations which have never been taught. The system model has to be taught "beyond limits" to be really applicable in the whole needed dynamical range. Thus in a future training sequence we want to expose the system model to dynamically wider situation by training via human operated joystick control with higher gains in comparison to the gains which the final running robot will have.

4 Summary

We introduced the notion of Echo State Networks as an easily trainable versatile variant of recurrent neural networks. We showed how these networks can be used to teach a physical system model of a differential drive robot and also how to mimic a given controller for it. Furthermore an improved controller was generated. All three applications show good agreement with observed real life data. Now an ESN can be implemented in the actual HW of the motor controller, yet the performance of this extended approach has to be studied in a future paper.

References

[A00] Christaller Th. Jaeger H Kobialka H.-U Schoell P Bredenfeld A. Robot behavior design using dual dynamics. *Tech. GMD Report*, 2000.

[Ark98] R. C. Arkin. *Behavior-based robotics.* The MIT Press, 1998.

[BK01] Ansgar Bredenfeld and Hans-Ulrich Kobialka. Team cooperation using dual dynamics. In Markus Hannebauer, editor, *Balancing reactivity and social deliberation in multi-agent systems*, Lecture notes in computer science, pages 111 – 124, 2001.

[BSF94] Yoshua Bengio, P. Simard, and P. Frasconi. Learning long-term dependencies with gradient descent is difficult. *IEEE Transactions on Neural Networks*, 5(2):157–166, 1994. (Special Issue Dynamic and Recurrent Neural Networks.).

[Gal80] C. R. Gallistel. *The Organization of Action: a New Synthesis.* Lawrence Erlbaum Associates, Inc., Hilldale, NJ., 1980.

[Hou03] Ramin Housseiny. Echo state networks used for the classification and filtering of silicon retina signals. Master's thesis, RWTH AAchen, 2003.

[Jae01a] H. Jaeger. The "echo state" approach to analysing and training recurrent neural networks. *GMD Report 148*, pages 1–43, 2001.

[Jae01b] Herbert Jaeger. The echo state approach to analysing and training recurrent neural networks. Technical report, GMD - Forschungszentrum Informationstechnik GmbH, 2001.

[Jae01c] Herbert Jaeger. Short term memory in echo state networks. Technical report, GMD Forschungszentrum Informationstechnik GmbH, 2001.

[Jae02] Herbert Jaeger. Tutorial on trainig recurrent neural networks covering bppt, rtrl, ekf and the "echo state network" approach. Technical report, GMD Forschungszentrum Informationstechnik GmbH, 2002.

[JD92] M.I. Jordan and Rumelhart D.E. Forward models: Supervised learning with a distal teacher, 1992.

[Jor95] M.I. Jordan. Computational aspects of motor control and motor learning. In H. Heuer and S. Keele, editors, *Handbook of Perception and Action: Motor Skills.* Academic Press, New York, 1995.

[KBM98] D. Kortenkamp, R. P. Bonasso, and R. Murphy. *Artificial Intelligence and Mobile Robots.* AAAI Press / The MIT Press, 1998.

[KS87] Furawaka K. Kawato, M. and R. Suzuki. A hierarchical neural network model for the control learning of voluntary movements, 1987.

[Mea89] Carver Mead. *Analog VLSI and Neural Systems.* Addison-Wesley, Reading, MA, USA, 1989.

[MNM02] W. Maass, T. Natschläger, and H. Markram. Real-time computing without stable states: A new framework for neural computation based on perturbations. *Neural Computation*, 2002.

[MW96] R.C. Miall and D.M. Wolpert. Forward models for phisiological motor control, 1996.

[Pom93] Dean A. Pomerleau. *Neural Network Perception for Mobile Robot Guidance.* Kluwer, Dordrecht, The Netherlands, 1993.

[PTVF92] William H. Press, Saul A. Teukolsky, William T. Vetterling, and Brian P. Flannery. *Numerical Recipes in C: The Art of Scientific Computing.* Cambridge University Press, Cambridge, 2 edition, 1992.

[SB81] R.S. Sutton and A.G. Barto. Toward a modern theory of adaptive networks: expectation and prediction., 1981.

[Sch02] Frank Schoenherr. Learning to ground fact symbols in behavior-based robots. In F. van Harmelen, editor, *Proceedings of the 15th European Conference on Artificial Intelligence*, pages 708–712. ECAI, IOS Press, 2002.

[SV98] Johan A.K. Suykens and Joos Vandewalle, editors. *Nonlinear modeling*, chapter Enhanced Multi-Stream Kalman Filter Training for Recurrent Networks. Kluwer, 1998.

Model and Behavior-Based Robotic Goalkeeper

Hans Lausen, Jakob Nielsen, Michael Nielsen*, and Pedro Lima

Instituto de Sistemas e Robótica – Instituto Superior Técnico
Av. Rovisco Pais, 1 – 1049-001 Lisboa, Portugal
{gruber,fisker,miguel}@control.auc.dk, pal@isr.ist.utl.pt
http://socrob.isr.ist.utl.pt

Abstract. This paper describes the design, implementation and test of a goalkeeper robot for the Middle-Size League of RoboCup. The goalkeeper task is implemented by a set of primitive tasks and behaviors, coordinated by a 2-level hierarchical state machine. The primitive tasks concerning complex motion control are implemented by a non-linear control algorithm, adapted to the different task goals (e.g., follow the ball or intercept the ball). One of the top level behaviors regularly determines the robot posture from local features extracted from images acquired by a catadioptric omni-directional vision system. Most robot parameters were designed based on simulations carried out with the Hybrid Automata Matlab/Simulink toolbox CheckMate. Results obtained with the actual goalkeeper are presented and discussed.

1 Introduction and Overview

In robotic soccer, namely in the Middle-Size League (MSL) of RoboCup, goalkeepers are interesting robots, due to the potential behavior richness they can exhibit. Indiveri [1] introduces an elegant solution for a goalkeeper based on a non-linear state-feedback control algorithm. However, little is said about the coordination of a larger set of behaviors rather than the one corresponding to ball following and blocking. Jamzad *et al* [2] describe a very efficient goalkeeper with changing shape, but they mostly concentrate on its mechanical design. Menegatti *et al* [3] were the first to introduce in RoboCup a multi-behavior goalkeeper which usually moves on an arc in front of the goal and tries to intercept the ball when a shot is headed at its goal. Nevertheless, they use an ad-hoc model for both motion control and the overall behavior coordinator.

The coordinated execution of a robotic task is one of the key features for an autonomous robot. The robot resources (e.g., sensors, actuators, shared memory, CPU) required to accomplish a given task must be properly managed and articulated with the different behaviors composing the task.

In the RoboCup MSL, one of the four allowed players is the goalkeeper. A good goalkeeper should switch among different behaviors to fulfill its role in the team.

* The three first authors are with Aalborg University and did their work at ISR/IST under a SOCRATES grant.

D. Polani et al. (Eds.): RoboCup 2003, LNAI 3020, pp. 169–180, 2004.

In this paper, the design, implementation and test of a a goalkeeper is described. The goalkeeper task is implemented by a set of primitive tasks and behaviors, coordinated by a 2-level hierarchical state machine. The primitive tasks concerning complex motion control are implemented by a non-linear control algorithm, adapted to the different task goals (e.g., follow the ball or intercept the ball), as detailed in Section 2. One of the top level behaviors, described in Section 3, regularly determines the robot posture from local features extracted from images acquired by a catadioptric omni-directional vision system. Most robot parameters were designed based on simulations carried out with the Hybrid Automata Matlab/Simulink toolbox CheckMate, as covered in Section 4. Results obtained with the actual goalkeeper are presented and discussed in Section 5. Conclusions and future work are discussed in Section 6.

The RoboCup MSL *ISocRob* team consists of four *Nomadic Scout II* robots, endowed with an omnidirectional camera and a front camera, both *Philips TouCam Pro* web-cams. The goalkeeper's kicker and cameras assembly, shown in Figure 1-a) is different from its teammates, namely due to a general 90° relative rotation of those hardware components and a larger surface for the kicking device.

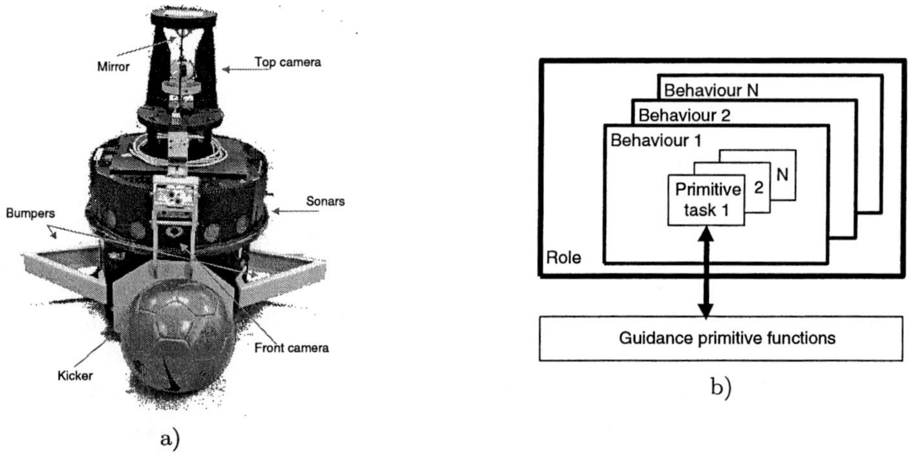

Fig. 1. ISocRob's a) goalkeeper robot; b) functional architecture

ISocRob's functional architecture is based on the concepts of roles, behaviors, primitive tasks and guidance primitive functions[4]. Figure 1-b) shows how the four elements interact with each other. Each robot is assigned a role, which consists of one or more behaviors. Behaviors are implemented using one or more primitive tasks, and each primitive task uses the guidance primitive functions to interact with the lowest level of hardware on the robot. At the top level, a role is assigned to each robot. In this case, the goalkeeper role is assigned. The role is filled by executing one of the role's behaviors, according to the current robot + environment state. Each behaviors is assigned to a state of a state machine whose arcs are traversed whenever some logical condition associated to the robot + environment state or the lower-level state machine becomes true. A behavior is executed by running a number of primitive tasks, coordinated by the lower-level state machine, where each state represents a primitive task and arcs are again traversed when logical predicates over state variables become true.

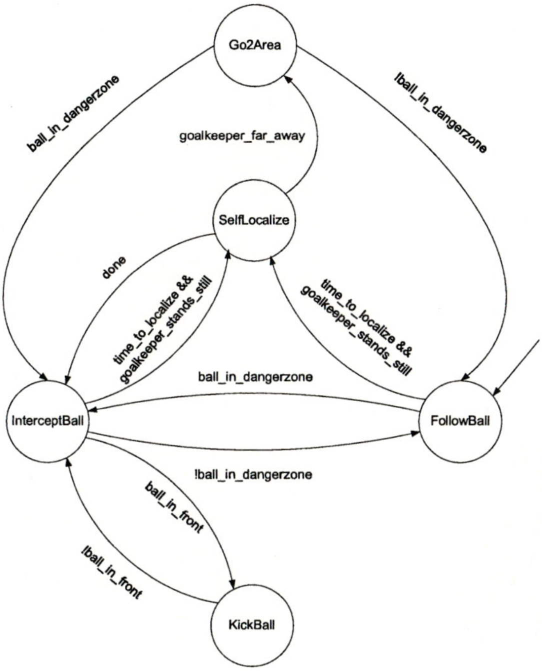

Fig. 2. The goalkeeper state machine

The state machine of the implemented goalkeeper is depicted in Figure 2. There are five behaviors in the state machine. The Go2Area behavior is implemented by one of the navigation algorithms introduced in previous papers[5]. `KickBall` is trivial. The other three behaviors (`InterceptBall`, `SelfLocalize`, and `FollowBall`) are detailed in the remaining sections of the paper.

The robot is in `FollowBall` behavior if the ball is out of the danger zone specified in Figure 3-a). The ball is not considered as a big threat, but the goalkeeper has to be able to handle a shot from the distance. The goalkeeper follows the ball while tracking an arc with adjustable radius and centered with the goal. The principle of the defensive arc is illustrated also in Figure 3-a). The `r_min` and `r_max` parameters correspond to the minimum and maximum radius of this arc. This adjustment of the radius is dependent of the distance between the ball and the goal. When the ball is near the border of the danger zone, the radius of the arc is at its minimum. When the ball is at the center line or further away, the radius is at its maximum. The robot assumes the behavior `InterceptBall`

a) b)

Fig. 3. Concepts of danger zone, maximum and minimum defensive arcs. a) `FollowBall`; b) `InterceptBall`

when the ball is in the danger zone in Figure 3-a) and moving towards the goal as seen in Figure 3-b). In this case, the goalkeeper is not following the arc anymore but rather moving on a straight line in front of the goal, trying to intercept an incoming ball.

Transitions between behaviors occur only when the associated predicates, shown in the state diagram of Figure 2, become true.

2 Motion Control

Both the `FollowBall` and the `InterceptBall` behaviors pose trajectory tracking and posture stabilization problems [6]. Nevertheless, the trajectories to be tracked will be different, therefore each behavior will have its own control algorithm, which will be derived in following subsections. Before that, we will take a brief look at the robot kinematics, required to provide the terminology for the rest of the section.

2.1 Kinematics

The goalkeeper is based on a differential-drive robot, that can be described by the same kinematic equations as the unicycle vehicle. The unicycle is a nonholonomic system with no slippage assumed. Let $q = (x, y, \phi)^T$ be the vector that describes the goalkeeper posture in configuration space. The first-order kinematic model for the unicycle is given by [7]:

$$\begin{bmatrix} \dot{x} \\ \dot{y} \\ \dot{\phi} \end{bmatrix} = \begin{bmatrix} u \cos(\phi) \\ u \; sin(\phi) \\ \omega \end{bmatrix} \tag{1}$$

Where u is the linear velocity and ω is the angular velocity. Furthermore u and ω are the control inputs.

2.2 FollowBall Behavior

Moving on an arc in front of the goal is a trajectory tracking problem. Among the possible solutions for this problem, we have chosen the algorithm described in [1], used in its original form to implement the `FollowBall`. The goal of the control design is to track and follow the arc in front of the goal until an equilibrium point (in front of the ball) is reached for the goalkeeper, as well as to achieve asymptotic stability for that equilibrium point.

Consider the kinematic model from equation 1. If the position is given in polar-coordinates an equivalent kinematic model would be:

$$\begin{bmatrix} \dot{r} \\ \dot{\alpha} \\ \dot{\phi} \end{bmatrix} = \begin{bmatrix} u \ \cos(\alpha) \\ \omega \ - \ \frac{u}{r} \ sin(\alpha) \\ \frac{u}{r} \ sin(\alpha) \end{bmatrix} \tag{2}$$

Where the angle α is given as shown on Figure 4-a).

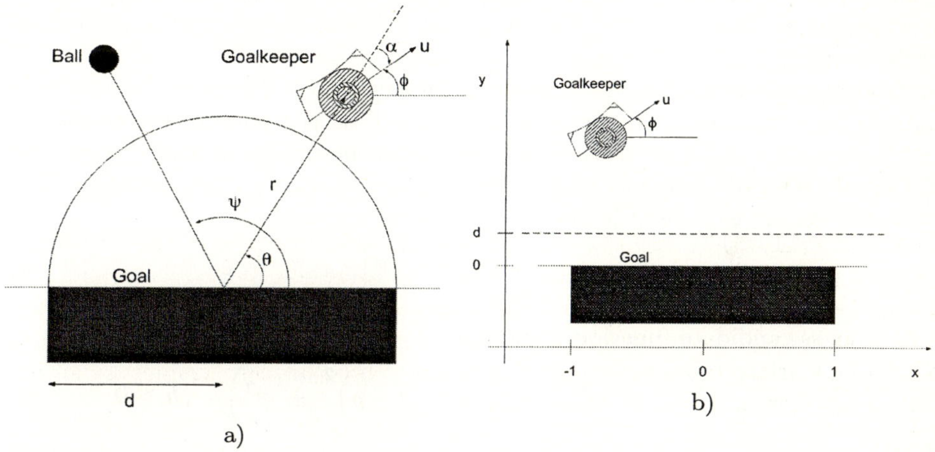

Fig. 4. The goalkeeper posture in polar coordinates: a) `FollowBall` scenario; b) `InterceptBall` scenario

In the figure, the goalkeeper is on the arc when $r = d$. Furthermore the angle α should be equal to 90° when the goalkeeper is moving on the arc in order to keep the front towards the ball. These two requirements are expressed in the following error vector:

$$e = \begin{bmatrix} e_1 \\ e_2 \end{bmatrix} = \begin{bmatrix} d - r \\ \frac{\pi}{2} - \alpha \end{bmatrix} \tag{3}$$

To achieve asymptotic stability a control law for u and ω must be derived in such a way that the time derivative of an appropriate Lyapunov function $V(e)$ becomes negative definite [8, 9]. The approach for the choice of ω is to cancel some of the undesirable terms which makes it difficult to determine the nature of $\dot{V}(e)$. The following control law for ω guarantees aymptotic stability[9]:

$$\omega = \frac{u}{r} \ \sin(\alpha) \ - \ \gamma \left(\frac{\pi}{2} - \alpha\right) \ - \ h \ u \left(d - r\right) \frac{\cos(\alpha)}{\frac{\pi}{2} - \alpha} \quad , \ h > 0 \, , \gamma > 0 \tag{4}$$

2.3 InterceptBall Behavior

If the ball moves inside the danger zone the goalkeeper switches its behavior to `InterceptBall`. When running this behavior, the goalkeeper should defend the goal on a straight line in front of the goal. It is most likely that the posture of the goalkeeper is on the arc, when the upper-level state machine switches to `InterceptBall`. The control algorithm should therefore not only be able to track the straight line, but also to make the goalkeeper able to converge to the line from any posture on the arc. This sounds similar to the trajectory tracking problem of the previous subsection, for a different trajectory. It is therefore chosen to re-derive the control algorithm used to control the angular velocity in `FollowBall` so that the equilibrium point is now on a straight line in front of the goal. Since the reasoning for the design already has been given in the previous subsection the control algorithm design will not be as detailed here. However, it is important to underline that this derivation is one of the original contributions of this work.

Consider again the kinematic model from equation 1. The straight line to be tracked, at a distance d of the goal line, as well as the posture variables x, y and ϕ are shown in Figure 4-b). When the goalkeeper moves on the straight line in front of the goal, the two requirements $y = d$ and $\phi = 0$ are satisfied. The requirements are expressed in the following error vector:

$$e = \begin{bmatrix} e_1 \\ e_2 \end{bmatrix} = \begin{bmatrix} d - y \\ \phi \end{bmatrix} \tag{5}$$

A Lyapunov candidate function is introduced:

$$V(e) = \frac{1}{2}\,\left(h\, e_1^2 \,+\, e_2^2\right) \;=\frac{1}{2}\left(h\left(d-y\right)^2 \,+\, \phi^2\right) \quad ,h > 0 \tag{6}$$

To establish if asymptotic stability is feasible the derivative of $V(e)$ is obtained:

$$\dot{V}(e) = h\,\left(d-y\right)\; u\;\sin(\phi) \,+\, \phi\omega \quad ,h > 0. \tag{7}$$

Consider the control law:

$$\omega = -\gamma\,\phi \;-\; h\,\left(d-y\right)\,u\,\frac{\sin(\phi)}{\phi} \quad ,\; h > 0\,,\gamma > 0 \tag{8}$$

Applying the control law yields $\dot{V}(e) = -\,\gamma\,\phi^2 \quad ,\gamma > 0$, which is a negative definite function. Since the control law is derived following the same line of thought as for the control algorithm of `FollowBall`, a proof for asymptotic stability will not be given here.

A control law for the linear velocity u was designed independent of the control law for ω. Based on the good results from `FollowBall`, it was chosen to use a P-controller for the linear velocity. The control law is given as $u = K_p * (\psi - \theta)$, where ψ is either the angle to the ball or the angle to the predicted interception point $x_{intercept}$, and θ is, as in `FollowBall`, the angle between the center of the goal and the goalkeeper. Depending on the direction and velocity of the ball, an interception point between the trajectory of the ball and the defending line may be used to determine the angle ψ.

3 Vision-Based Self-localization Using Local Features

The motion control algorithms of the previous section rely on good estimates of the robot posture. However, and especially for a robotic goalkeeper, frequently subject to bumps from other robots, odometry alone does not provide such a reliable estimate, as it strongly degrades over time. Therefore, one must reset it regularly with a more accurate estimate. The solution used in this work consisted of using goalkeeper local visual features (e.g., the goal and the posts) and a vision-based algorithm to provide such an estimate.

We have used ISocRob's omni-directional catadioptric vision system for this purpose. An image acquired by this system is shown in Figure 5. Since this is a norm-preserving omni-directional catadioptric system (i.e., distances on the field are preserved on the image) it is relatively simple to determine the distance from the goalkeeper to surrounding objects by processing the images. Furthermore the goal posts will always point towards the position of the robot in the (center of the) image.

Fig. 5. Omni-directional catadioptric vision image taken by the goalkeeper

Under normal operation the goalkeeper should be near its own goal at all times. So a good object to detect from the image will be the goal. Extracting the goal posts as a feature from the image, will reveal the position of the robot with respect to them. Then the posture of the robot can be estimated by simple geometry. The algorithm runs in situations while the robot is standing still.

The image processing algorithm is split in four stages, briefly detailed in the sequel.

Stage 1: Image Processing — to retrieve the contour of the goal from the image two steps are required (see Figure 6): i) segmentation of the goal in the image (identifying the pixels with the same color as the defending goal and disregarding all other pixels); ii) edge-detection revealing the contour of the goal in the image.

Stage 2: Feature Detection Using the Hough Transform — to detect the goal posts in the image containing the contour of the goal the following three steps are performed (see Figure 7): iii) Hough Transform of the image; iv) line filtering: all straight lines obtained in the previous step which do not point to the center of the image will be filtered out; v) goal post detection:

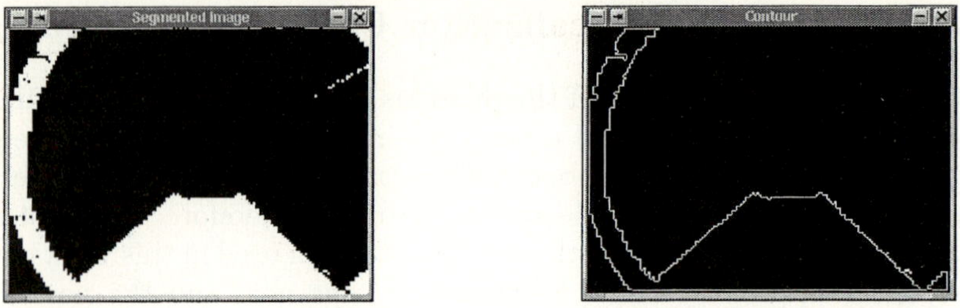

i) ii)

Fig. 6. Steps of image processing stage 1: i) Goal segmentation; ii) edge-detection

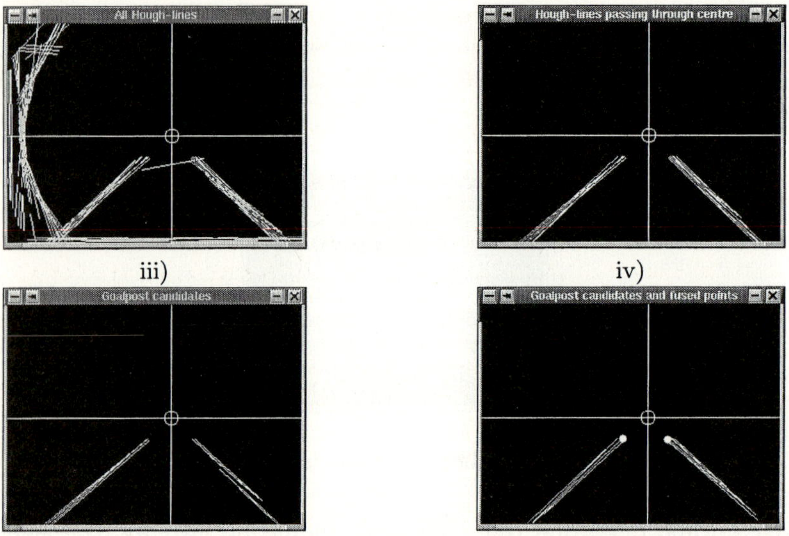

iii) iv) v) vi)

Fig. 7. Steps of image processing stage 2: iii) Hough Transform; iv) line filtering; v) goal post detection; vi) goal posts position estimates

this is accomplished by comparing the angles of all the lines in a histogram, and detecting the two angles most likely to be the angles of the goal posts. Afterwards two arrays of one or more lines are created containing the line(s) of each goal post.

Stage 3: Estimation of Goal Post Positions — If, as a result from the previous steps, there were only one line describing the goal post, the position of the goal post would be considered as the line endpoint closest to the robot. In the case there are more lines, an estimate of the line endpoint is obtained by first sorting the line endpoints, and then using a Bayesian data fusion algorithm whose details we will omit here.

Stage 4: Geometric Calculations — knowing the positions of the two goal posts with respect to the robot, and the positions of the goal posts in the field, it is possible to determine the posture of the robot using simple geometry. Again, due to limited space, the details will be omitted.

The image processing algorithm was implemented in C, using the OpenCV library developed by Intel.

4 Experimental Setup and Simulations

Simulations were designed to determine the "optimal" values for the danger zone distance from the goal line d_{dz} and the radius r of the defensive arc for the `FollowBall` behavior. The purpose is to determine the configuration that will lead to the highest number of saves by the goalkeeper when the state machine includes the `FollowBall` and `InterceptBall` behaviors only. Since both the discrete state (`FollowBall` or `InterceptBall`) and the continuous state of the robot as well as of the ball (i.e., their position and velocity) will matter for this purpose, the goalkeeper was modeled as a hybrid automaton [10], resulting from the composition of the goalkeeper and ball hybrid automata. The hybrid automaton was implemented in Simulink using CMU's CheckMate toolbox. Three tests were performed. In the first test, several values of d_{dz} used for a danger zone as wide as the field. In the second test, the best values for d_{dz} were taken and the danger zone width varied. Finally, for the best pair determined in the previous tests, several values for r were tested. In each of the three tests, shots were fired from different locations in the field, each shot aiming towards the goal. For every different setup for the danger zone or the radius of the defensive arc, 625 shots were fired against the goal. The speed and the direction of the shots differed. In five groups of 125 shots each, the speed was set to 2m/s, 2.5m/s, 3m/s, 3.5m/s, 4m/s respectively. The goalkeeper was initially at the center of the goal, on the defensive arc.

From these tests, the best results (64.6% of goals saved) were obtained for a danger zone with the field width, $d_{dz} = 4.5$ m and $r = 0.6$ m. A constant radius was also used for the defensive arc, instead of `r_min` and `r_max` referred in Section 1. The failure rate of 35.4% is mainly due to the relatively slow top speed achieveable by the goalkeeper ($1m/s$) and the high speed of some of the simulated shots.

5 Experimental Results

5.1 Self-localization

To test the performance of the self-localization algorithm, two issues were considered: the success rate of the algorithm and the error of the estimated postures. The robot was placed with the same orientation (90° in the field frame, where x points towards the opponent goal z points upwards and y is such that a right-handed frame is obtained) in 60 different positions in front of the goal,

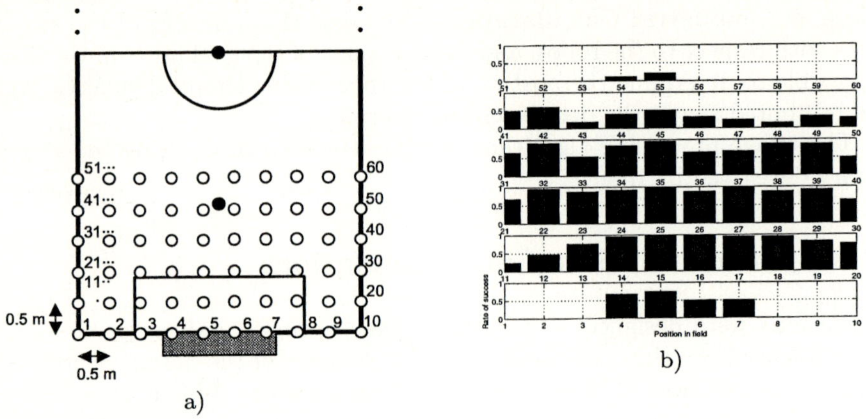

Fig. 8. Self-localization results: a) goal-keeper field positions; b) estimation error at different field positions

as illustrated in Figure 8-a). In each position the algorithm was run until 20 estimates of the posture were determined (in some it was not possible to get 20 estimates though).

As explained earlier the algorithm does not return any posture if one of several conditions is not met (e.g. if only one goal post is found in the image). With this in mind, the success rate for the algorithm is defined as the ratio between algorithm runs which are successful (a posture is found) and the total algorithm runs. The success rate was evaluated in each of the 60 positions in the field. The results are plotted in Figure 8-b. The success rate when the robot is on the goal line outside the goal posts is zero, since the goal color is not visible from here, and it is needed for posture disambiguation. It is also noticeable that the success rate is very low in positions 2 m and further away from the goal. The algorithm has the best success rate in positions 1 and 1.5 meters away from the field end line, and in general there are more successes in positions in front of the goal.

Concerning the posture estimation errors, in general errors are larger (around 20 cm) in the x direction than in the y direction (around 10 cm). Orientation errors are typically below 10°. Unfortunately, large outliers can sporadically occur, especially on the x coordinate. Therefore, a test is made which accepts an odometry reset only if the new value does not differ from the current odometry estimate for more than a given threshold. The algorithm average run time is of approximately 0.4 s.

5.2 Motion Control

The results regarding the motion control cover the the goalkeeper performance while tracking the defensive arc, shown for two different speeds in Figure 9-a), under the `FollowBall` behavior, and when attempting to intercept a ball heading

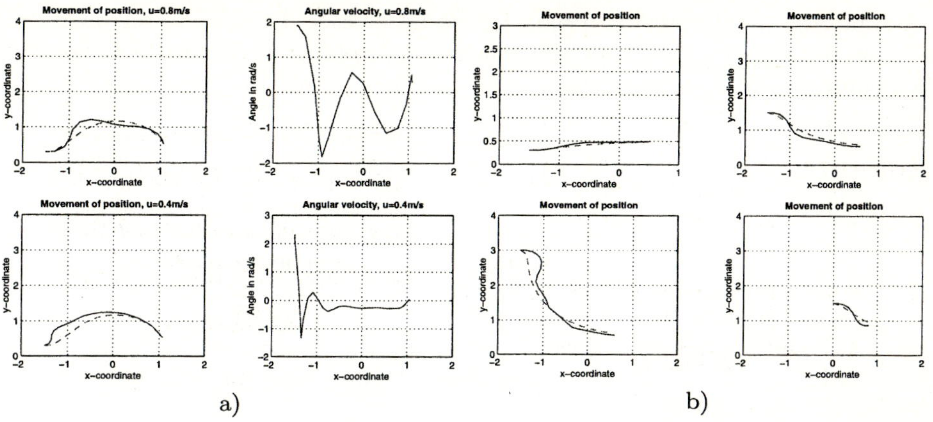

Fig. 9. a) Results of simulated and actual robot motion for the `FollowBall` behavior. The goalkeeper moves from its starting position to the arc and stops as it reaches the line between the ball and the center of the goal. The solid line shows the movement in goal coordinates logged from odometry and the dashed line is the simulated motion. The radius for the defending arc is 1.2 m. The parameters used for the control algorithm were $d = 1.2$, $h = 4$, $\gamma = 2$, $K_p = 1.5$ and $u_{max} = 0.8\frac{m}{s}$. b) Results of simulated and actual robot motion for the `InterceptBall` behavior under four situations where the goalkeeper has different start positions. The circle in (2, 1) represents the ball. The line from the ball to the origin illustrates a shot towards the goal. The defending line is placed at $d_{dz} = 0.5$. Starting position for the goalkeeper in the figures from the left to right is (-1.5, 0.3), (-1.5, 1.5), (-1.5, 3) and (0, 1.5).The dashed line is the simulated trajectory and solid is the logged trajectory from odometry

towards the goal, starting from four initial positions, under the `InterceptBall` behavior, as shown in Figure 9-b).

In general, the tests made show a good performance when following the ball, even though some differences from the simulated results were found, essentially due to unmodeled dynamics in the simulations. Among those, the critical problem is the almost unstable convergence to the arc at higher speeds. Regarding ball interception, a problem arises due to the usage of a constant value for d_{dz}, instead of a variable one, e.g., equal to the current x coordinate of the goalkeeper. Due to speed limitation, for high ball speeds the robot tends to "open" the goal while moving towards the intercept line, since it first rotates of 90°.

6 Conclusions and Future Work

This paper presented an integrated design, implementation and test of a robotic goalkeeper which included hybrid automata modeling of the behavior coordinator, non-linear motion control for trajectory tracking and posture stabilization and vision-based self-localization. The results are very promising, but further work needs to be done, namely to improve the reaction speed, to use a variable intercept line and include further behaviors, such as leaving the goal to face an opponent robot carrying a ball. In the latter case, inspiration on work concerning behavior coordination with smooth and conflict-free transitions between behaviors [11] will be considered.

References

1. Indiveri, G.: *An Introduction to Wheeled Mobile Robot Kinematics and Dynamics.* Slides for lecture given at RoboCamp, Paderborn (Germany) (2002)
2. Jamzad, M., Chitsaz, H., et al: A goalkeeper for middle size robocup. RobCup-2000: Robot Soccer World Cup IV (2001)
3. Menegatti, E., Nori, F., Pagello, E., Pellizzari, C., Spagnoli, D.: Designing an omnidirectional vision system for a goalkeeper robot. RobCup-2001: Robot Soccer World Cup V (2001)
4. Lima, P., Custódio, L., Ventura, R., Aparício, P.: A functional architecture for a team of fully autonomous cooperative robots. RobCup-99: Robot Soccer World Cup III (1999)
5. Marques, C., Lima, P.: A multi-sensor navigation system for soccer robots. RobCup-2001: Robot Soccer World Cup V (2002)
6. de Wit, C.C., Siciliano, B., (Eds)., G.B.: Theory of Robot Control. Springer-Verlag, London, UK (1996)
7. Oriolo, G., Luca, A.D., Vendittelli, M.: *WMR Control via Dynamic Feedback Linearization: Design, Implementation and Experimental Validation.* IEEE Transactions on Control Systems Technology (2002)
8. Khalil, H.K.: Nonlinear Systems. Prentice-Hall (2002)
9. Indiveri, G.: *On the Motion Control of a Nonholonomic Soccer Playing Robot.* RobCup-2001: Robot Soccer World Cup V (2002)
10. van der Schaft; Hans Schumacher, A.: An Introduction to Hybrid Dynamical Systems. Springer (1999)
11. Uchibe, E., Kato, T., Asada, M., Hosoda, K.: Dynamic task assignment in a multi-agent/multitask environment based on module conflict resolution. In: IEEE Intern. Conf. on Roboticas and Automation. (2001) 3987–3992

Evolving Visual Object Recognition for Legged Robots

Juan Cristóbal Zagal, Javier Ruiz-del-Solar, Pablo Guerrero, and Rodrigo Palma

Department of Electrical Engineering, Universidad de Chile
{jzagal,jruizd,pguerrer,ropalma}@ing.uchile.cl

Abstract. Recognition of relevant game field objects, such as the ball and landmarks, is usually based upon the application of a set of decision rules over candidate image regions. Rule selection and parameters tuning are often arbitrarily done. We propose a method for evolving the selection of these rules as well as their parameters with basis on real game field images, and a supervised learning approach. The learning approach is implemented using genetic algorithms. Results of the application of our method are presented.

1 Introduction

Teams of the four legged league have generally reported good vision strategies for the recognition of objects on the game field [3,14], such as the ball, landmarks and goals. Approaches often rely on the sequential application of a set of recognition rules, such as making comparisons of sizes and distances between regions of connected pixels of certain colors. These rules are pre-engineered and their parameters are manually adjusted until they become useful at recognizing objects under different locations, illuminations, and poses into the image. After applying theory and models to the problem, the engineering task often falls into a trying and error optimization process. Some teams [14] have even reported their uncertainty with respect to the application of some rules.

We believe that this engineering process can be supported or even automated by the use of a supervised learning approach, for which a large set of real pre-classified images can be used as a training data. These real images are expected to cover as much as the interesting examples that one might imagine. A main advantage of this approach is that such system will gain its knowledge from its own experience rather than being product of an arbitrary design.

Machine learning methods like evolutionary computation provides optimization tools which are used in robotics for learning behaviors [5], such as the case of evolutionary robotics, and also for the adaptation of perceptual systems [6,10,11,12]. A main idea is that genetic algorithms can search for solutions on highly dimensional spaces contaminated with natural noise. Evolved systems are expected to be more robust to unseen data sets than those resulting from simulated evolution.

In this work we will explore a method for evolving the selection and tuning of a group of visual object recognition rules, intended for recognizing objects in the context of the RoboCup four legged league. The proposed evolutionary learning approach is based on the use of a large set of pre-classified real images. In this paper are presented results for the detection of ball, goals, and landmarks. We are currently work-

D. Polani et al. (Eds.): RoboCup 2003, LNAI 3020, pp. 181–191, 2004.

ing towards the final goal of this approach, which is to derive rules for the recognition of other robot players into the game field. Given the complexity of this problem, it has not yet received sufficient attention from the research community.

In section 2 the related work is presented. Section 3 describes our implemented vision module which is used for extracting candidate object regions, and section 4 describes our proposed object recognition approach. Section 5 describes our results, and finally section 6 presents the conclusions and projections of our work.

2 Related Work

Cliff et al [4,5] uses genetic algorithms for evolving neural-network based controllers for visually guided robots. They use a computer graphics based model for simulating the robot vision, their model considers the introduction of certain amount of noise for preventing it from being entirely deterministic. The resulting approach was computationally expensive and with poor image resolution.

The approach of using evolutionary computation for computer vision problems has been widely explored, for example Köppen et al [9] proposes framework for the automated generation of texture filters using both, genetic algorithms (GA), and genetic programming (GP).

The object recognition problem, addressed with evolutionary computation, has been first attacked for the character recognition task. Koza [7] shows an experiment using GP for the classification of just four characters on small bitmaps, this approach relies on using a computationally expensive attention marker method. He also proposes a system which uses Automatically Defined Functions (ADFs) which were successful at finding solutions [8], but required populations of extremely large size (8000 individuals). Andre [1] uses both GP and GA simultaneously, first a GA determines feature templates and then a GP is used for classifying character bitmaps.

Teller and Veloso [13] used genetic programming for their proposed Parallel Algorithm Discovery and Orchestration (PADO) system. This system performs object recognition on real gray scale images. Genetic programming is used to induce programs which operate on pixel values in the image and return a confidence value that the given image corresponds to the class which is intended to be recognized.

3 Our Vision Module

The software architecture of our UChile1 four legged team is divided in task oriented modules. One of them, the vision module, is in charge of recognizing relevant objects from the images captured with the robot cameras. This module in particular, is mainly inspired on the large experience showed by the UNSW and CMPack teams [3,14]. This module is decomposed into four processing sub-modules: color segmentation, run-length encoding, labeling of connected regions, and finally object recognition.

For the color segmentation sub-module we use a look-up table of 64 levels in each YUV dimension, the table is generated by taking a large number of color samples (about 5000) from images of the game field. Once all samples have been collected, a median filter is operated over the look-up table values having the effect of clearing the interfaces between clusters of different colors and filling empty elements inside clusters which were not assigned during data collection. This process is particularly useful for solving ambiguities between red and orange clusters for example. A main consideration is that we train not just our seven color classes, but also a class for the set of non-relevant colors.

The output of the labeling sub-module is a set of connected regions of certain color, or blobs. Each blob can be characterized with a set of descriptors such as the size in pixels, the integer color index which in this case might take the values $\{0,1,2,3,4,5,6\}$, a set of coordinates describing the bounding box, and the coordinates of its center of mass.

The task of the object recognition sub-module is to identify image regions which are related to the relevant game field objects. The recognition of objects is performed by evaluating the response of a set of rules. For example, the detection of a ball usually requires that the related blob has the color of the ball, and if this is not the case one might expect to reject this candidate blob. These rules operate over all the image blobs or over combinations of them, such as pairs of blobs.

4 Learning Visual Object Recognition

4.1 General Approach

We propose to evolve the visual object recognition sub-module by first collecting reference region descriptors of objects which are present on a large set of real images; this stage is performed by an expert user. Then candidate regions are defined as those automatically extracted with the vision system, or combinations of them, see Figure 3 (left). Then, under a supervised rule learning process, candidate regions are compared with corresponding reference regions on each image, and the overall degree of correspondence serves as fitness for a genetic algorithm which learns the system recognition rules. Clearly, the effectiveness of the recognition sub-module is directly related to the degree of correspondence between candidate regions and reference regions.

We have used in our experiments a set of 180 real images for reference accumulation; these images contain objects such as the ball, landmarks and goals, as well as non relevant objects on the surroundings of the game field. The images consider a broad range of viewpoints, rotations, non canonical poses, and even variations on the illumination conditions. Figure 1 shows examples of these images.

In order to generate a database containing object identifiers for each reference image blob, or the so called references, we have developed a software which allows an expert user to define image regions related to relevant objects in terms of their bounding rectangles, and linking them to their corresponding identifier by just pressing on the corresponding object button. Figure 2 shows a screenshot of this software tool.

During the learning process a genetic algorithm evolves a population of recognition rules intended for detecting a particular object. These rules operate over region descriptors which are automatically extracted from each image with our vision module. In case a region, or a combination of them, is regarded as an object, its degree of correspondence with the reference is calculated by means of a correspondence quality function. The overall degree of correspondence between detected regions and references is then used as fitness for each individual generated with the genetic algorithm.

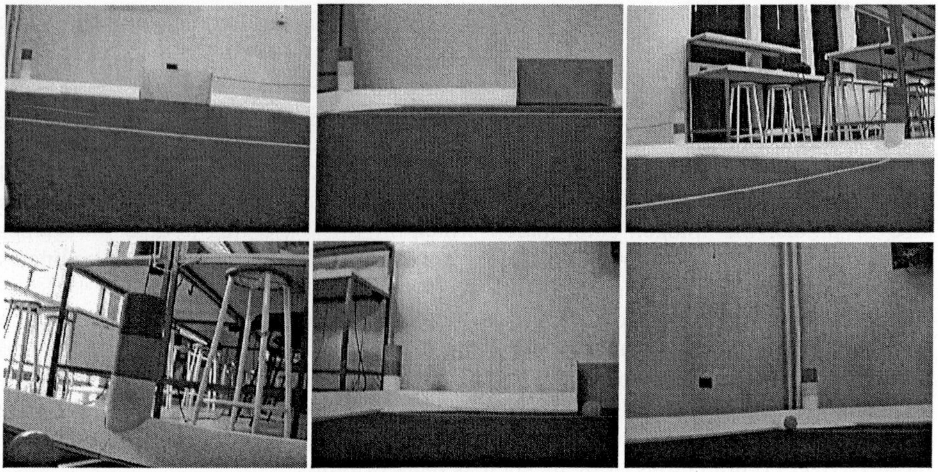

Fig. 1. Examples of images collected from the game field. Each image is subject to inspection from both an expert and our visual system under adaptation.

4.2 Fitness Function

Assigning a good fitness function is not trivial in this case. This measure should take its maximum when there is a perfect overlap between reference and candidate image regions, but it is not necessarily clear how to handle partial overlaps between them. Köppen et al [9] has proposed a quality function which well fits on our problem. It consist on measuring the area A of the reference region which does not overlaps the candidate, the area B of the overlapping region, and the area C of the candidate region which does not overlaps the reference, see Figure 3 (right). This results in the following three measures:

$r_1 = B / (A+B)$, the relative amount of correct overlapping pixels within the reference,
$r_2 = 1 - (C / (Q - A - B))$, the relative amount of correct empty pixels within the image, where Q is the total number of image pixels, and
$r_3 = B / (B+C)$, the relative amount of correct overlapping pixels within the candidate.

The intention is that genetic search increases all these measures, but we can identify some priorities among them. For example it is desired that the correspondence degree counts better for subsets of the reference, as well as for subsets of the reference which are supersets of other subsets of the reference. We also would like to refuse to assign good correspondence degrees to false positives, i.e. empty regions.

The following weighted correspondence degree, as proposed by Köppen, accounts for these requirements:

$$CD = 0.1r_1 + 0.5r_2 + 0.4r_3 \tag{1}$$

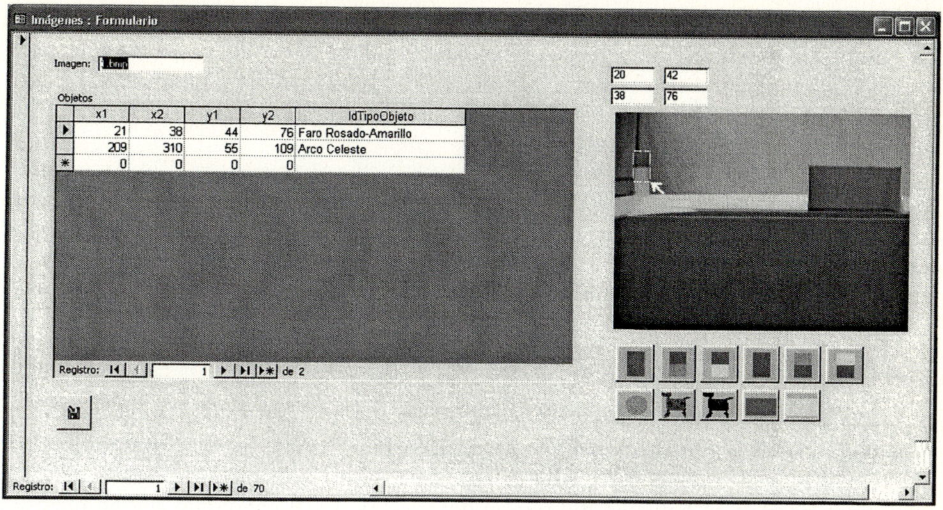

Fig. 2. A screenshot of the software developed for labeling image regions which are related to game objects. It can be seen how coordinates (left) defining rectangles, and object identifiers, are related to each object which is being selected. This task is performed by just pressing the corresponding object button (right).

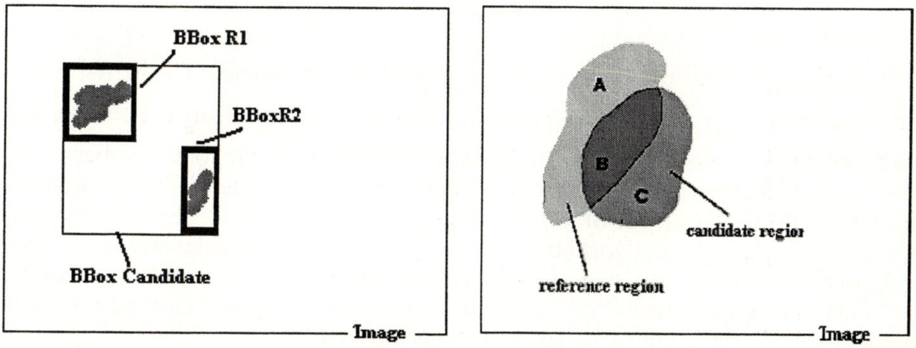

Fig. 3. Left: Illustration of how the candidate image region for the beacon detection is derived from two image regions. It can be seen how the resulting rectangle is defined in terms of the region bounding boxes. Right: An example of partial overlap between a reference region and a candidate region. The sub regions A, B and C are defined as the figure shows.

Using this measure, we compute the fitness for each individual as the sum of the correspondence degrees over the whole set of images, only when at least one candidate region is being selected:

$$fitness = \sum_i CD_i \tag{2}$$

As a consequence, genetic search evolves the population towards the higher weighted objective first, in this case r_2, rejecting false positives, and then towards r_3, allocating correct portions of the reference. Once candidates correspond to subsets of the reference, the fitness is increased by expanding them to cover the whole reference.

4.3 Genetic Rule Representation

For each candidate region, detected with the vision process, a set of *N* binary rules $R_i=f(\{p_{ij}\},\{Region\_Descriptors_k\})$ is evaluated. Each rule is described in terms of *M* parameters p_{ij}. The rules have as argument the set of candidate region descriptors. In general the binary rules have the following structure:

$$R_i = \begin{cases} 1 & if \quad p_{i1} \geq COND \geq p_{i2} \\ 0 & if \quad not \end{cases} \tag{3}$$

Where *COND* might correspond to a value, as for example the size of a region, the quotient between regions sizes, and in general to the result of logical or arithmetic operations performed between the region descriptors. Each candidate region receives a score computed as the weighted sum of the rule outputs. The region having a maximum score is regarded as an object if and only if its score is greater than a certain threshold. This score is computed as follows:

$$Score = \sum_i w_i R_i \geq T \tag{4}$$

In our implementation, the weights $w_i \in [0,1]$, the thresholds $T \in [0, N]$ and all the *M* parameters $p_{ij} \in [0,1]$ are represented as 16 bit strings. Thus the chromosome which encodes a rule has length *16x(N+M+1)*, where *M* is the total number of rule parameters. In some cases, as it will be indicated, the parameters are re-scaled or discretized to a reduced set of values. These chromosomes will be evolved with a genetic algorithm. This algorithm uses fitness-proportionate selection with linear scaling, no elitism scheme, two-point crossover with a crossover probability Pc=0.75 and mutation with a mutation rate of Pm=0.015 per bit. The population size is 8 individuals evolved over a course of 100 to 150 generations.

5 Results

5.1 Ball Recognition Experiment

We have chosen a group of six rules for the ball recognition experiment. The shapes of these rules, as well as the range of their parameters are indicated on Table 1. Figure 4 shows results of this experiment. We can first notice from these results, that the trivial color test implicit on rule R6, is satisfied, i.e. the system learns to choose the right color of the ball. The parameter P61 is discretized as the integer number 7 which correspond to exactly the index of the orange color. It is also recognized as the most important rule, since it has the maximum weight. The second rule on importance is R1 the low value of its parameter P11 indicates that the candidate region should have a minimum width of two pixels. The third rule in importance, R4 indicates that the bounding box of the region should be relatively square with a minimum quotient between width and height, or vise versa, of P41=0.5754. The fourth rule, R5 indicates that the quotient between the region size and the bounding box size should fall between P51=0.027 and P52=0.553, which is quite logical for our implementation. The lower bound accounts for those cases in which the ball correspond to a silhouette of segmented orange pixels, i.e. a region small in size but with a large bounding box. The upper bound serves for rejecting orange regions which are too close to a square. The fifth rule, R2 indicates that the region height should be at least a quarter of the image height. The last rule, R3 establishes that the region size should be at least one third of the image size. However, we should notice that the weight of R3 is quite low therefore it is not a relevant rule. The resulting threshold scaled by 6 (the number of rules) correspond to T=0.99, which can be compared to the maximum theoretical score of 3.69 (the sum of weights). This means that the threshold is set at the 26% of the maximum score.

Table 1. The six rules for the ball recognition experiment, described in terms of their shape, and parameters range. The region descriptors correspond to **width_reg**, the region width; **height_reg**, the region height; **area_reg**, the region area; $\mathbf{h_{bb}}$, the height of the region bounding box; $\mathbf{w_{bb}}$, the width of the region bounding box; and **color_reg**, the color of the region.

Rules for Ball Detection		
Rule	Activation Condition	Parameters Range
R_1	width_reg $> P_{11}$	integer [0,image_width]
R_2	height_reg $> P_{21}$	integer [0,image_height]
R_3	area_reg $> P_{31}$	integer [0,image_size]
R_4	$\min(w_{bb}/h_{bb}, h_{bb}/w_{bb}) > P_{41}$	double [0,1]
R_5	$P_{51} <$ area_reg/area_bbox $< P_{52}$	double [0,1]
R_6	color_reg $= P_{61}$	integer[1,num_colors=7]

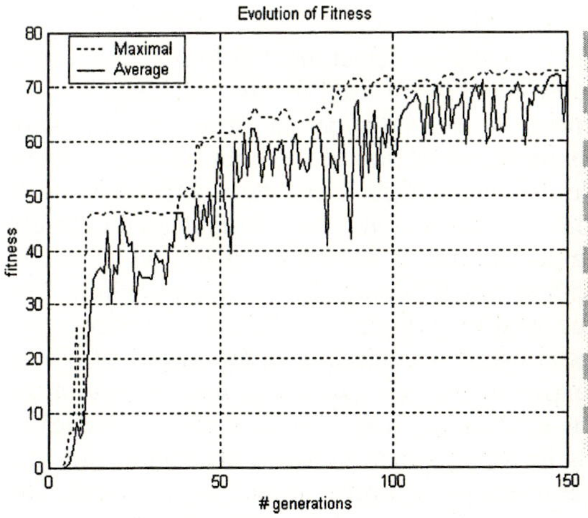

Best Evolved Individual		
Pr.	Value	Interpretation
w1	0.952347	Ranking 2
w2	0.45491	Ranking 5
w3	0.07095	Ranking 6
w4	0.805679	Ranking 3
w5	0.459213	Ranking 4
w6	0.961151	Ranking 1
P11	0.00372314	Minimum width 2 pix.
P21	0.253232	1/4 image height
P31	0.308493	1/3 image size
P41	0.575424	Relatively square
P51	0.0270081	Counts for silhouettes
P52	0.552963	Rejects false balls
P61	0.98053	Orange color index.
T	0.166499	Equivalent to 0.99

Fig. 4. Resulting evolution of fitness (left), and the resulting weights, parameters, and threshold for the best evolved individual (right), for the ball recognition experiment.

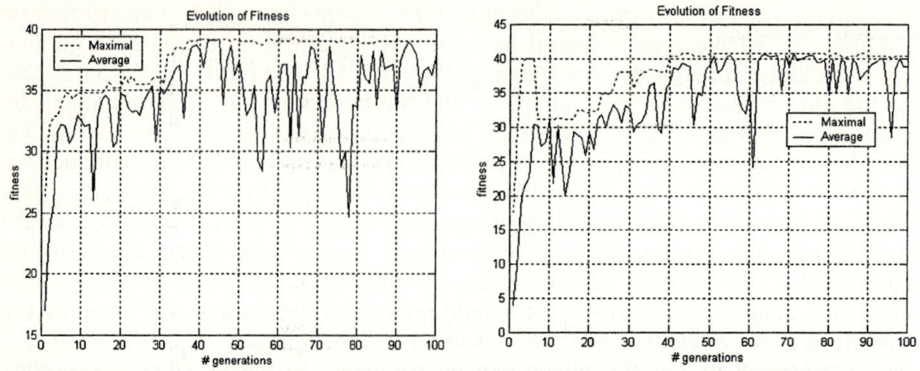

Best Evolved Individual Light Blue Goal		
Pr.	**Value**	**Interpretation**
W1	0.802551	Ranking 1
W2	0.784454	Ranking 2
W3	0.348038	Ranking 5
W4	0.366501	Ranking 6
W5	0.606674	Ranking 3
W6	0.661896	Ranking 4
P11	0.0901642	Min width 9%
P21	0.0627899	Min height 6%
P31	0.647183	Min size 6%
P41	0.65889	Square-like ratio
P51	0.400696	Lower area bound
P52	0.864853	Upper area bound
P61	0.779999	L-blue. color indx.
T	0.0565796	Equivalent to 0.34

Best Evolved Individual Yellow Goal		
Pr.	**Value**	**Interpretation**
W1	0.999039	Ranking 1
W2	0.524292	Ranking 5
W3	0.859741	Ranking 2
W4	0.0393372	Ranking 6
W5	0.83844	Ranking 3
W6	0.743652	Ranking 4
P11	0.0493469	Min width 5%
P21	0.589996	Min height 6%
P31	0.503348	Min size 5%
P41	0.0301056	Square-like ratio
P51	0.173755	Lower area bound
P52	0.9658875	Upper area bound
P61	0.134277	Yellow color indx.
T	0.0552063	Equivalent to 0.33

Fig. 5. Results for the light blue goal experiment (left), and for the yellow goal experiment (right). The evolution of fitness (top), and resulting weights, parameters, and thresholds for best evolved individuals (bottom), are indicated.

5.2 Goal Recognition Experiment

In general there are fewer rules reported for the goal detection than for the case of the ball. We will use the same group of rules which were used for the ball detection experiment, see Table 1. This group seems to be a good super set of relevant rules for our analysis. Figure 5 shows the results for the detection of the light blue and yellow goals. We have again that the color rule R6 obtains right parameters corresponding to light blue and yellow for each corresponding experiment. This rule is ranked in fourth place in both experiments. The most important rule is in both cases R1, which establishes a 9% and a 5% of the image width as lower bounds for candidate regions. The second place is for R2 in the light blue goal experiment and R3 in the yellow goal experiment. In both experiment it is established a minimum region height of 6% of the image height. Similarly, in both experiments, R3 establishes that the minimum region size should be 6% and 5% of the total image size. The third place is for R5 in both experiments, it establishes bounds for the quotient between the region size and its bounding box sizes, the resulting parameters are similar in both experiments. The fifth place is for R3 in the light blue goal experiment and for R2 in the yellow goal experiment. The less important rule as indicated in both experiments is R4, taking quite different parameters on each experiment. One interpretation of this is that genetic search concentrates on optimizing the parameter of relevant rules, leaving the irrelevant ones with arbitrary parameters. The thresholds are quite similar in both experiments; their corresponding scaled values are 0.33 and 0.34. The sum of weights is 3.569 for the light blue goal experiment, and 4 for the yellow goal experiment. Therefore the threshold is established at the 10% of the maximum score.

Table 2. The six rules for the ball recognition experiments, described in terms of their shape, and parameters range. The region descriptors correspond to **distX**, the distance between regions in the x axis; **distY**, the distance between regions in the y axis; **Sreg**, the region size; and **color_reg**, the color of the region.

Rules for Bacon Detection		
Rule	Activation Condition	Parameters Range
R_1	distX(reg1,reg2) < P_{11}	integer [0,image_width]
R_2	distY(reg1,reg2) < P_{21}	integer [0,image_height]
R_3	min(Sreg1/Sreg2,Sreg2/Sreg1) > P_{31}	double [0,1]
R_4	1/2 (Sreg1+Sreg2)/sqrt(dist(reg1,reg2)) > P_{41}	double [0,1]
R_5	color reg1 = P_{51} OR color reg2 = P_{51}	integer[1,num_colors=7]
R_6	Color_reg2 = P_{61} OR color reg2 = P_{61}	integer[1,num_colors=7]

5.3 Beacon Recognition Experiment

In this experiment candidate regions are generated as a combination of the bounding boxes of two image regions, see Figure 3 (left). In this case it is necessary not to just evaluate the rules over each region extracted from the image, but also to evaluate these rules over all the possible pairs of them, clearly the rules of this experiment have two region descriptors as input. The pair of regions which obtains the maximum score is selected if satisfies equation 4, and a candidate region is derived from them as indicated in Figure 3 (left), this region is used for calculating the correspondence degree as indicated in equation 1. For this experiment we have selected a group of six rules, presented in Table 2. We will explore the particular case of the category beacon of colors pink-light-blue and light-blue-pink without distinguishing between the vertical orders of the colors.

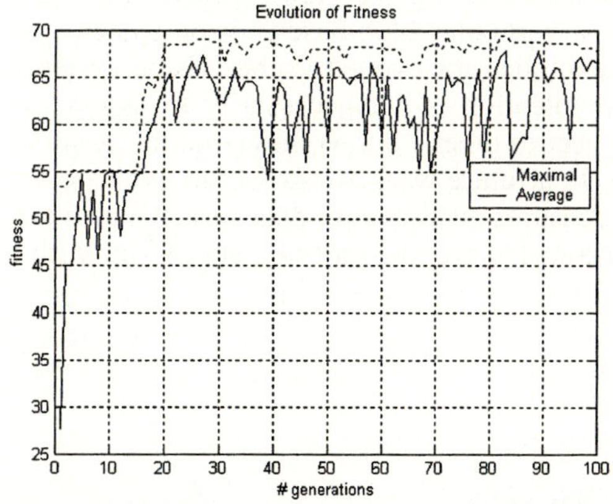

Best Evolved Individual		
Pr.	**Value**	**Interpretation**
W1	0.549803	Ranking 3
W2	0.40321	Ranking 5
W3	0.101231	Ranking 6
W4	0.538501	Ranking 4
W5	0.633870	Ranking 2
W6	0.875532	Ranking 1
P11	0.032041	Min distance in X 3%
P21	0.083218	Min distance in Y 8%
P31	0.0031345	Quotient sizes.
P41	0.832451	Relative distances
P51	0.896420	Color is pink
P61	0.7559361	Color is light blue.
T	0.138913	Equivalent to 0.83

Fig. 6. Resulting evolution of fitness (left), and the resulting weights, parameters, and thresholds for the best evolved individual (right), for the beacon recognition experiment.

It can be seen form these results that the two color rules R6 and R5 are regarded as the more important ones, their corresponding parameters fit exactly to the expected colors. The third rule on importance correspond to W1 which performs a minimal check for the horizontal distance between the blobs, the corresponding parameter establishes a threshold of 3% of the image width. The fourth rule in importance is R4 which checks for the distances between blobs with invariance to the scale of the objects, this rule was proposed in [14]. The fifth rule is R2 establishing a minimum vertical distance between regions of 8% of the image height. Finally R3 is regarded as the less important rule.

The maximum theoretical score is in this case 3.1 which means that the threshold was established at the 27% of the maximum score.

6 Conclusions and Projections

We have presented a method for automating and aiding the selection and tuning of visual object recognition rules in the domain of the RoboCup four legged league. The system shows to be consistent with the training data sets, and it allows the extraction of interesting parameters for different rules as well as the identification of the more relevant ones from a given set. It was particularly explored the case of ball, goal and landmark detection. We aim at extending this research, first we will explore rules for the detection of other robots into the game field, and then we will explore the application of a similar learning method for aiding the visual estimation of robot pose.

In the presented approach, the resulting parameters are dependent on the color calibration stage. If the color detection is poor or noisy, the resulting recognition system will be adapted to these specific conditions, with the consequence of having to retrain the system for each different lighting condition. This inconvenient is solved by ensuring accurate color detection. Our color detection system is accurate under different lighting conditions, and its calibration is performed in just about 15 minutes. In practice, we haven't had to perform the rule training when changing the lighting conditions.

The intention of this work is to present a method for the improvement of a vision system. Although we have just analyzed a particular blob-based vision system, we believe that a similar methodology can be applied for evolving other vision systems, such as grid based or corner based. Our intention is not necessarily to assess improvements of our vision system with respect to others, but to show that the result of this learning platform performs similarly. In a future work we expect to evaluate our system in terms of standard quantitative measures, using larger data sets. We also aim at comparing our visual system with those which are known to be successful within the RoboCup domain.

Acknowledgements

This research was partially supported by FONDECYT (Chile) under Project Number 1030500. We thank the reviewers for valuable comments on how to improve this paper.

References

1. Andre, D.: Automatically Defined Features – the Simultaneous Evolution of 2-Dimensional Feature Detectors and an Algorithm for Using Them. In: Kinnear, K. (eds.): Advances in Genetic Programming. MIT Press, Cambridge, Massachusetts (1994) 477-494.
2. Chalup, S., Creek, N., Freeston, L., Lovell, N., Marshall, J., Middleton, R., Murch, C., Quinlan, M., Shanks, G.: When NUbots Attack! The 2002 NUbots Team Report. Newcastle Robotics Laboratory. The University of New Castle. NSW 2308, Australia (2002) http://robots.newcastle.edu.au/NUbotFinalReport.pdf
3. Chen, S., Siu, M., Vogelgesang, T., Fai Yik, T., Hengst, B., Bao Pham, S., Sammut, C.: The UNSW RoboCup 2001 Sony Legged Robot League Team. In: Birk, A., Coradeschi, S., Tadokoro, S, (eds.): RoboCup 2001: Robot Soccer World Cup V. Springer-Verlag Berlin, Germany (2002) 39-48.
4. Cliff, D., Husbands, P., Harvey, I.: Evolving Visually Guided Robots. In: Meyer, J. A., Roitblat, H., Wilson, S. (eds.): Proceedings of the Second International Conference on Simulation of Adaptive Behaviour (SAB92). MIT Press Bradford Books, Cambridge, MA (1993) 374-383.
5. Cliff, D., Harvey, I., Husbands, P.: General Visual Robot Controller Networks via Artificial Evolution. In: Casement, D. (eds.): Proceedings of the Society of Photo-Optical Instrumentation Engineers Conference (SPIE-93), Session on Intelligent Robots and Computer Vision XII: Algorithms and Techniques, Boston, MA (1993)
6. Johnson, M., Maes, P., Darrell, T.: Evolving Visual Routines. In: Brooks, R., Maes, P. (eds.): Artificial Life IV. MIT Press (1994) 198-209.
7. Koza, J.: Genetic *Programming*. MIT Press, Cambridge, Massachusetts (1992)
8. Koza, J.: *Genetic Programming II: Automatic Discovery of Reusable Programs*. MIT Press, Cambridge, Massachusetts (1994)
9. Köppen, M., Teunis, M., Nickolay, N.:A Framework for the Evolutionary Generation of 2D-Lookup Based Texture Filters. In: Proceeding of Lizuka Conference, Lizuka, Japan (1998) 965-970
10. Martin, M. C.: Genetic Programming for Robot Vision. In: SAB 2002, the Seventh International Conference on the Simulation of Adaptive Behaviour (2002).
11. Martin, M.C.: Genetic Programming for Real World Robot Vision. In: Proceedings of the 2002 IEEE/RSJ International Conference on Intelligent Robots and Systems (IROS), EPFL, Lausanne, Switzerland, September-October (2002) 67-72
12. Nguyen, T., Huang, T.: Evolvable 3d Modeling for Model-Based Object Recognition Systems, In: Kenneth, E. (eds.): Advances In Genetic Programming. MIT Press (1994) 459-475
13. Teller, A., Veloso, M.: PADO: A New Learning Architecture for Object Recognition. In: Ikeuchi, K., Veloso, M. (eds.): Symbolic Visual Learning, Oxford University Press (1996) 81-116
14. Veloso, M., Lenser, S., Vail, D., Roth, M., Stroupe, A., Chernova, S.: CMPack-02: CMU's Legged Robot Soccer Team. Carnegie Mellon University Report (2002) https://www.openr.org/page1_2003/code2002SDK/CMU/cmu_teamdesc.pdf

Coaching Advice and Adaptation

Patrick Riley and Manuela Veloso

Carnegie Mellon University, Computer Science Department, Pittsburgh, PA 15232
{pfr,mmv}@cs.cmu.edu
http://www.cs.cmu.edu/~{pfr,mmv}

Abstract. Our research on coaching refers to one autonomous agent providing advice to another autonomous agent about how to act. In past work, we dealt with advice-receiving agents with fixed strategies, and we now consider agents which are learning. Further, we consider agents which have various limitations, with the hypothesis that if the coach adapts its advice to those limitations, more effective learning will result. In this work, we systematically explore the effect of various limitations upon the effectiveness of the coach's advice. We state the two learning problems faced by the coach and the coached agents, and empirically study these problems in a predator-prey environment. The coach has access to optimal policies for the environment, and advises the predator on which actions to take. We experiment with limitations on the predator agent's actions, the bandwidth between the coach and agent, and the memory size of the agent. We analyze the results which show that coaching can improve agent performance in the face of all these limitations.

1 Introduction

In spite of the increasing complexity of the relationships among autonomous agents, one agent coaching or advising another is still somewhat uncommon. We frequently see this relationship among human beings, yet further computational understanding of how this relationship extends to autonomous agents is needed.

As we start to move beyond agent systems of short, limited duration, we believe that advising relationships will become more important. Many current agent systems implicitly allow a complete transfer of knowledge of an agent's behavior structure, both because of the size of the domains and the homogeneity of the agent's representations of the world. In other words, if one agent knows a good way to behave in the world, all agents can easily copy that knowledge. A coach relationship does not assume this is possible. The goal is to obtain knowledge transfer in spite of limitations in communication bandwidth or differences in behavior representation. If the coach wants the agent to act differently, the coach must still consider the best way to *guide* the agents to this behavior, rather than simply transferring the knowledge. Additionally, a coach should adapt its idea of optimal behavior to the strengths and abilities of the agent being coached.

Specifically, our research on coaching refers to one autonomous agent providing advice to another about how to act in the world. Through research on a complex simulated robot soccer domain [1, 2], we have been exploring how one

D. Polani et al. (Eds.): RoboCup 2003, LNAI 3020, pp. 192–204, 2004.

agent can generate advice for teams. With this experience, we have now undertaken a series of systematic experiments in more controlled environments. We present a general scheme for how the coach and agent interact in Figure 1. The coach can only affect the world through communication with the agent. Both the coach and the agent are learning. This interaction between the coach and agent is challenging, as the coach has only an indirect effect on the world and does not get to see the working of the agent's decision mechanism. We empirically investigate varying the limitations of the agent as well as limitations of the communication between the coach and agent.

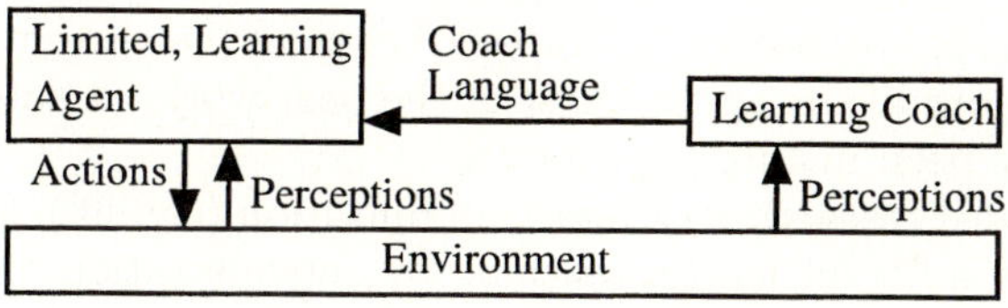

Fig. 1. The high-level view of coach-agent-environment interaction

While our previous work has focused on adapting to an adversary and compiling past experience, the research here deals primarily with adapting advice to the agent being coached. One hypothesis that underlies our current work is that in order for a coach to be effective coaching multiple different agents, it needs to adapt its advice to the peculiarities of each agent. This is supported by the results of the coach competition at RoboCup 2001 as discussed in [1,3] where the ability of a coach to improve a team's performance varied vastly across different teams. This paper contributes interesting interaction modes of a coach and coached agent and a thorough empirical exploration of the ramifications of various limitations on that interaction. We start from the idea that advice is recommending an action for a particular state and explore the effects in different interaction modes. We choose basic reinforcement learning algorithms as a starting point since our interest is in the comparative performance of different algorithms and what that suggests for the general coaching problem.

2 Related Work

This work focuses primarily on how an advice giver can interact with an advice taker, and how the characteristics of that interaction affect the performance of the agent. One important component of this is how the advice-taker will function. A variety of work has been done on using advice (often from humans) to improve autonomous agent performance.

A significant challenge is to operationalize "fuzzy" advice. For example, [4] describes a multi-step method to incorporate advice into a reinforcement learner which uses a neural network to represent the value function. [5] takes a similar approach. High level advice from a human being is translated to if-then rules describing intermediate goals for the agents. By combining these with qualitative descriptions of the domain dynamics, these rules can then be refined with a genetic algorithm. [6] presents a model for advice taking in a recurrent connectionist network. Advice changes the current activation state rather than the weights of the network.

Clouse and Utgoff [7,8] take a similar approach to ours to study the effect of advice on a reinforcement learner. Their focus is primarily on the advice-taker and how to incorporate advice into the value updates, and little attention is paid on how the advice was generated in the first place. They do empirically explore how the rate of advice affects performance. However, they do not allow the training agent to give advice for states other than the current state or explore limiting the memory of the agent. Most of the past work on autonomous agents *giving* advice has been in tutoring systems.

A closely related research area is that of imitation (e.g. [9]). Similar work has gone under many different names: learning by demonstration. behavior cloning, learning by watching. and behavior or agent imitation. In all the cases, the robot or agent is given examples of some task being done successfully (often from a human demonstration), and the agent's goal is to perform the same task. The demonstration of the task can be seen as a set of advised actions for the agent.

3 Continual Advice

We first consider an agent without limitation and where the coach can communicate some advice every step with the agent. The first question which must be answered is how an advice-receiving agent incorporates advice into its behavior.

3.1 Learning Algorithms

We begin with a basic Q-learning agent. The state space is assumed to be represented explicitly.

During learning, the agent has a fixed exploration probability ϵ; with probability ϵ the agent takes a random action and with probability $(1 - \epsilon)$ it chooses uniformly randomly between the actions with the highest Q-value. This is commonly known as ϵ-greedy exploration.

For all Q-learning done throughout, the learning rate decays over time. For each Q-table cell, we keep track of the number of times that cell has been updated (call it v), and calculate the learning rate for each update as $\frac{1}{1+v}$.

Then we add a coach agent. Every step, the coach recommends a single action. Table 1 shows the algorithm used by the coached agent. The only difference with a basic Q-learner is in choosing an action. The new parameter β controls the probability that the agent will follow the coach's advice. Otherwise, the agent reverts to the normal exploration mode. Except where noted, we use $\beta = 0.9$.

The coach is also a Q learner, but its actions are to advise *optimal* actions. The algorithm for the coach is shown in Table 2. While this algorithm is apparently too complex for this particular task, we will continue to use the same

Table 1. Algorithm for an agent Q-learning and receiving continual advice

```
ContinualAdviceTaker(ϵ, β)
  a := recommended action
  with prob β
    do a
  with prob 1 − β
    Choose action (ϵ greedy)
  Q-update for action performed
```

Table 2. Algorithm for the coach Q-learning and providing advice

```
CoachContinualAdvice(ϵ)
  g := last rec. (a.k.a guidance)
  if (see agent action (call it a))
    if (a is optimal)
      Q-update for a
    else
      no Q-update
  else
    Q-update for g
  recommend action (ϵ greedy)
```

algorithm later in more complex settings. There are two primary cases (which we vary as an independent variable): seeing the agent's actions and not seeing the agent's action. If the coach sees the actions and the agent takes an optimal action, the coach performs a Q-update with that action. This is based on the assumption that if the coach had advised that action, the agent probably would have done it. Note that in general, there may be multiple optimal actions in a particular state. However, if the action is not optimal, the coach does nothing; the coach's Q-table does not even include non-optimal actions.

Note that this algorithm requires that the coach know all optimal actions. For these experiments, we precompute the optimals beforehand.

What the coach is learning here (and will be learning throughout) is not what the optimal actions are (the coach already knows this), but rather learning about what actions the agent is taking. The coach restricts its Q-table to just the optimal actions and then pessimistically initializes the table. The Q-table then provide an estimate of the value achieved by the agent when the coach *recommends* an action (which is not the same as an agent *taking* the action). This will have important consequences when the advisee agent has limitations in the following sections.

3.2 Experimental Results in Predator-Prey

We use a predator-prey environment in order to test all the algorithms. The simulation is built upon the SPADES simulation system [10]. The agents operate on a discrete 6x6 grid world. The agents have 5 actions, stay still or move north, south, east, or west. There are virtual walls on the outside of the grid so if an agent tries to move outside the grid, it stays where it is.

There are *two* prey agents which, for these experiments, move randomly. The predator agent's goal is to capture at least one prey agent by being on the same square as it. The coach's job is to advise the predator agent. The world is discrete, all agents have a global view of the world, and all moves take place simultaneously.

The world is operated episodically, with an episode ending when at least one prey is captured; the predator can simultaneously capture both prey if all three agents occupy the same square. The predator and coach receive a reward of 100 for each prey captured and a reward of -1 every step that is taken. The agents try to maximize *undiscounted* reward per episode.

The state space of the world is $36^3 = 46656$ (the predator location, prey 1 location, prey 2 location), though for any state which is a capture for the predator (2556 states), there are no actions so no learning need take place. After a capture, all agents are placed randomly on the grid. Further note that approximately 20% of the states in this environment have multiple optimal actions.

The optimal average reward per step is approximately 17, which means on average between 5 and 6 steps to capture a prey. The average reward per step reflects the value of a policy since we are using episodic undiscounted reward. In all cases, the coach has access to the complete true values of all actions in all states and therefore optimal policies of the environment.

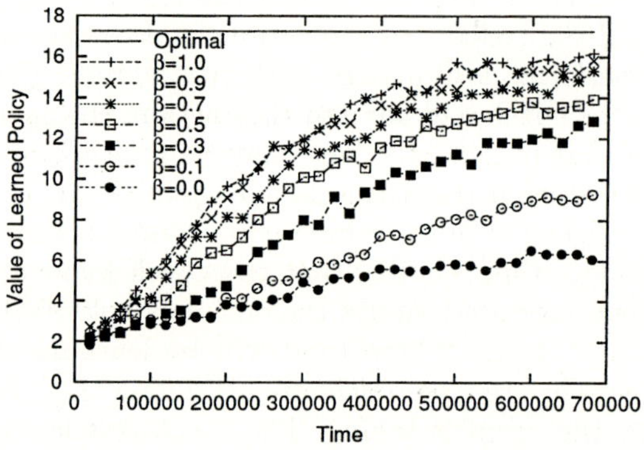

Fig. 2. Data for a predator agent learning with a coach advising the agent every cycle

We alternated periods of learning and evaluation. Every 5000 steps, 5000 steps of policy evaluation were done. The results throughout are the average values per step obtained during the policy evaluation periods. In order to smooth the curves and make them more readable, the values of two periods of policy evaluation were averaged together before plotting.

Throughout, we used $\epsilon = 0.1$ as the exploration parameter. The Q table was initialized to 0 everywhere, and the optimal Q values are positive everywhere.

The $\beta = 0.0$ line in Figure 2 shows the results for the predator agent learning without the coach. The agent is learning, though over 700,000 steps (half of which are learning, half are evaluation), the agent achieves only about 40% of optimal.

The rest of Figure 2 presents the results of the coach advising (with the algorithm CoachContinualAdvice as shown in Table 2) and the predator taking advice (with the algorithm ContinualAdviceTaker as shown in Table 1), varying the value of β to the ContinualAdviceTaker algorithm. This value β controls how often the predator listens to the coach, and how often the predator ignores the advice. As expected, with the coach, the predator agent learns much more quickly and reaches nearly an optimal policy. The coach is effectively guiding the agent to the right actions, which improves the speed at which the predator's Q table reflects an optimal policy. Also, the more often the predator listens to the coach (i.e. as β increases), the faster the learning and the better the performance. However, there are diminishing returns with the performance difference between $\beta = 0.7$ and $\beta = 0.9$ rather slight and the difference between $\beta = 0.9$ and $\beta = 1.0$ not significantly different.

4 Limited Agent Actions

We now consider cases where the action space of the coached agent is limited. For our purposes, this will simply mean that the agent is not allowed/able to perform some actions.

4.1 Learning Algorithms

A revised algorithm for the coached agent is shown in Table 3. The only difference from ContinualAdviceTaker (Table 1) is that if the coach recommends an action the agent can not perform, a random action is performed. This is intended to simulate an agent which is not fully aware of its own limitations. By trying to do something which it can't do, some other action will result.

Table 3. Algorithm for a predator with limited actions. Note that for the random action done, the same action is chosen every time that state is visited

```
LimitedContinualAdviceTaker(ε, β)
  a := recommended action
  with prob β
    if (a is enabled)
      do a
    else
      do random enabled action
  with prob 1 − β
    Choose action with ε greedy exploration
  Q-update based on action performed
```

The coach can follow the same CoachContinualAdvice algorithm as before (Table 2) and should be able to learn which optimal policy the agent can follow. If the coach can see the agent's actions, then only those Q-values for actions which the agent can perform will have their values increased. If the coach can not see the actions, then recommending an action that the agent can perform will tend to lead to a higher value state than recommending an action the agent can not perform (since the agent then acts randomly). In this manner, the coach's Q-table should reflect the optimal policy that the agent can actually perform, and over time the coach should recommend less actions which the predator agent can not do. Note that this algorithm implicitly assumes that the agent can perform some optimal policy.

4.2 Experimental Results in Predator-Prey

We once again use the predator-prey world as described in Section 3.2.

We assume that the limitations on the agent still allow some optimal policy. In particular, in this environment, 9392 of the 46656 states have more than one optimal action. Every choice of optimal action for each of those states gives an optimal policy, and we restrict the predator agent to be able to perform exactly one of the optimal policies. The same restriction was used for all experiments that are directly compared, but different experiments have different randomly chosen restrictions.

Further, for a given state action pair, the same random action always results when the agent is told to do an action it can not do. We also chose to use $\beta = 1.0$ to emphasize the effects of the coach's advice.

Results of this learning experiment are shown in Figure 3. An important issue to consider when analyzing the results is how much our limitation algorithm actually limits the agent. Only 20% of the states have multiple optimal actions, and of those, almost all have exactly 2 optimal actions. If an agent performs optimally on the 80% of the state space with one optimal action and randomly anytime there is more than one optimal action, the average reward per step of this policy is approximately 14.2 (optimal is 17.3).

In order to provide a reasonable point of comparison, we also ran an experiment with the coach providing intentionally bad advice. That is, for every state where the predator has an optimal action disabled, the coach would always recommend that action, and the predator would always take the same random action in a given state. This is the data line "Always Rec. Disabled" in Figure 3. The data line for the predator learning without limitations and with the coach (with $\beta = 0.9$ from Section 3) is shown labeled "No Limitation." Whether or not the coach sees the actions, the learning coach achieves better performance than the baseline of bad advice and approaches the performance without limitations at all.

A natural question to ask at this point is whether the coach is really learning anything about the agent's abilities. One measure of what the coach has learned about the agent is to examine what percentage of the coach's recommended actions are ones that the agent can not do. Figure 4 shows this value as the simulation progresses. The three lines represent the case where the coach is not learning (basically flat), learning and not seeing the agents actions, and learning and seeing the agent's actions. As one would expect, with learning, the percentage of bad actions recommended goes down over time, and seeing the actions enables faster learning.

5 Limited Bandwidth

Up to this point, the coach was giving advice to the agent every cycle. We see this as an unrealistic interaction mode with the coach, and this set of experiments

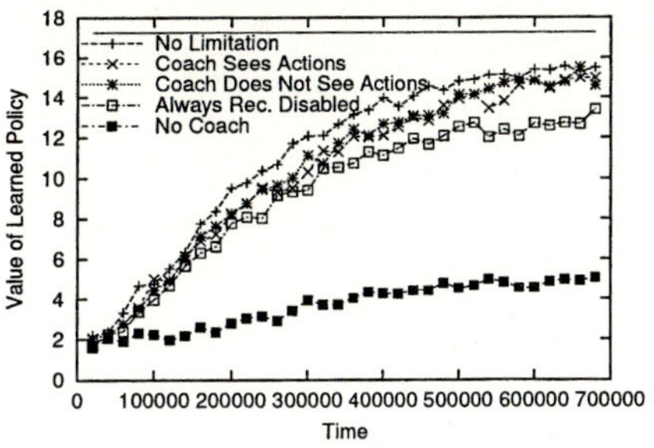

Fig. 3. The value of the policy learned by a limited predator agent under various coach conditions

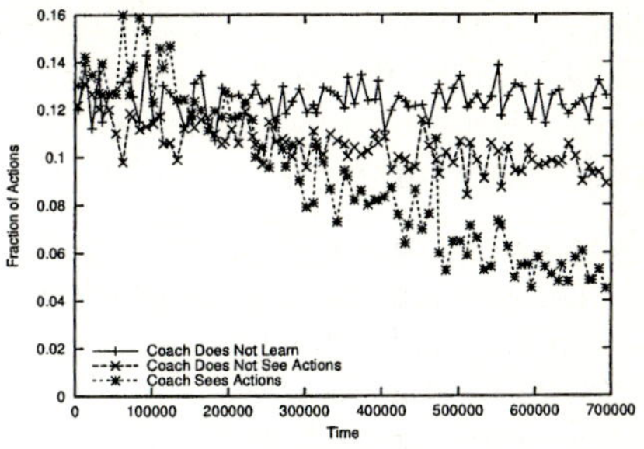

Fig. 4. As a function of time, the percentage of coach recommended actions which the predator can not do

Table 4. RandomState strategy for coach giving advice with limited bandwidth

```
CoachRandomState()
  if (time for K more advice)
    do K times
      s := random state
      a := random optimal action for s
      advise (s, a)
```

deals with limiting the amount of advice provided, while still using an advisee agent with a limited action space (as described in Section 4).

5.1 Learning Algorithms

First, to allow the coach to talk about states that are not the current state, a single piece of advice is now a state-action pair. The coach is advising an agent to perform a particular action in a particular state. Further, the coach has limitations on how much advice it can give. We use two parameters: I is the interval for communication and K is the number of states that the coach can advise about. Every I cycles, the coach can send K pieces of advice.

The agent stores all advice received from the coach and consults that table in each state. The new algorithm is shown in Table 5. For this experiment, the table T simply stores all advice received and if multiple actions have been advised for a given state, the table returns the first one received which the agent is capable of doing.

We propose two strategies for sending advice. The first strategy is mostly random; the coach randomly chooses K states and sends advice for an *optimal* action for each of those states (see Table 4). Note that while we call this a random strategy, it is still providing optimal advice about the states it chooses. If the bandwidth between the coach and advice taker were large enough, RandomState would send the entire optimal policy to the agent.

Table 5. Algorithm for a predator Q-learner with limited bandwidth with the coach; T is a table which stores past advice from the coach

```
LimitedBWAdviceTaker(ε, β, T)
  if (advice received from coach)
    add advice to T
  if (current state is in T)
    a := rec. action from T
    with prob β
      do a
    with prob 1 − β
      Choose action (ε greedy)
  else
    Choose action (ε greedy)
  Q-update for action performed
```

Table 6. OptQ strategy for coach giving advice with limited bandwidth

```
CoachOptQ(ε)
  Q* := optimal Q-table
  V* := optimal value function
  every cycle
    s := current state
    a := action predator took
    put (s, V*(s) − Q*(s, a)) in W
    if (a was an advised action for s)
      perform Q-update
    if (time to send K advice pieces)
      W′ = (smallest K values of W)
      for each (s′, x) in W′
        if (x ≈ 0)
          s′ := random state
        a′ := action for s′ (ε greedy)
        advise (s′, a′)
```

The other strategy, which we call "OptQ" is more knowledge intensive. It requires the entire optimal Q table and always seeing the coached agent's actions. The algorithm is given in Table 6. Like RandomState, OptQ always provides optimal advice for the states it chooses. The difference is that OptQ attempts to choose better states about which to advise. The basic idea is to advise about the states in the last interval for which the agent performed the least optimal actions. Note that the *smallest* values are chosen first since all of the values in W are negative. If the algorithm runs out of states about which to advise, it simply chooses random ones rather than wasting the bandwidth.

While neither RandomState nor OptQ may be good algorithms to implements in a real world setting, they provide good bounds on the range of performance that might be observed. RandomState puts no intelligence into choosing what states to advise about, and OptQ uses more information than would likely be available in any real world setting. The improvement that OptQ achieves over RandomState indicates how much benefit could be achieved by doing smarter state selection in an advice giving framework like this one.

5.2 Experimental Results in Predator-Prey

We once again use the predator prey world as described in Section 3.2. The predator has limited actions as described in Section 4.2. For the parameters limiting advice bandwidth, we chose I to be 500 and K to vary between 1 and 50. Recall that every I cycles, the coach can send K pieces of advice.

The results are shown in Figure 5 for different values of K. Note first the that as the amount of information the coach can send gets larger (i.e. K gets

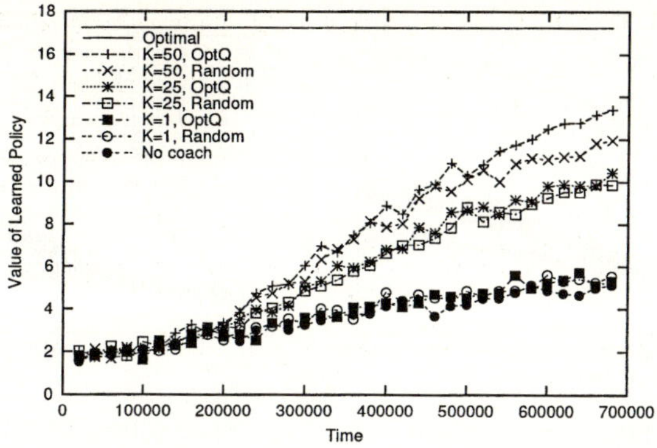

Fig. 5. Policy evaluation for the predator for various coach strategies and values of K

larger), the agent learns more quickly. While the same performance as the coach advising every cycle is not achieved, we do approach that performance. However, it should be noted that the agent is remembering all advice given.

The surprising result is that the OptQ algorithm with its much higher information requirements does not perform much better than the RandomState strategy. There are two things to consider. First, by the end of the simulation, the RandomState strategy achieves fairly large coverage of the state space. For example, with $K = 25$, the random strategy is expected to provide advice about 47% of the states[1]. Since the agent remembers all of the advice, near the end of the run it is essentially getting optimal advice for one in two states. Secondly, the OptQ strategy only advises about states about which the agent has already been and taken an action. The chance of encountering one of these states again is about the same as encountering any other given state.

Differences between RandomState and OptQ performance only emerge at the $K = 50$ level. This suggests as the bandwidth of the advice interaction is increased, the first benefit obtained by the agent is simply by obtaining some coverage of the state space with optimal advice (see the $K = 25$ line). Only after the bandwidth is increased more do we see *how* the advice is given start to make a difference.

6 Limited Bandwidth and Memory

In Section 5, the coached agent has a limited action space and there is limited bandwidth between the agents. However, the coached agent remembers all advice which the coach has sent. This is probably unrealistic once the world becomes

[1] With basic probability theory, one can calculate that if T is the number of pieces of advice and S the size of the state space, the expected number of distinct states about which the random strategy advises is $S(1 - (\frac{S-1}{S})^T)$. At the end of 700,000 steps with $K = 25$, 35000 pieces of advice have been given. With $S = 46656$, this is approximately 22035 states.

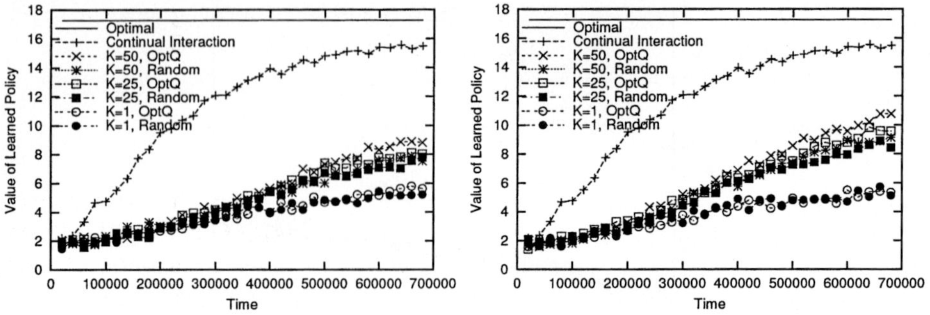

$M = 5000$; 11% of state space $M = 10000$; 21% of state space

Fig. 6. Results of learning with limited predator, limited bandwidth, and limited memory, M is the number of pieces of advice the predator can remember. The $M = 1000$ case is not shown as all data lines perform at approximately the $K = 1$ level

large enough that the advised agent is not using a full state/action table. Therefore, we explore the additional limitation of varying the amount of advice which can be remembered.

6.1 Learning Algorithms

We consider a straightforward FIFO model of memory. The coached agent has a fixed memory size, and when it is full, the agent forgets the oldest advice.

The coach strategies are the same as before: RandomState (Table 4) and OptQ (Table 6). The coached agent can still uses LimitedBWAdviceTaker (Table 5), but now the state advice table T only stores the last M pieces of advice heard, where M is an independent variable which we experimentally vary.

6.2 Experimental Results in Predator-Prey

We once again use the predator prey world as described in Section 3.2. The predator has limited actions as described in Section 4.2. Figure 6 shows the results. With the smallest memory of $M = 1000$, the predator does not improve significantly over having no coach at all. This can be explained simply because the memory can only hold approximately 2% of the state space. As the amount of memory is increased, the agent's performance improves. It should be noted that for $K = 1$, throughout the entire simulation the coach only sends 1400 pieces of advice, just barely more than the smallest memory. Therefore, we can not expect increasing the memory of the to improve performance for $K = 1$ since memory is not really the limiting resource for most of the simulation.

We see the same general effects of the differences between the RandomState and OptQ strategies here as when there was no limited memory (Section 5.2). The difference is that the size of memory effectively lowers the absolute performance of all techniques. It is interesting to note that absolute performance improves (in the $M = 5000$ and $M = 10000$ cases) when moving from $K = 25$ to $K = 50$ only when using the OptQ strategy and not RandomState.

7 Conclusion

This paper examines the problem of how one automated agent (the coach) can give effective advice to another learning agent (the advisee). Our focus has been on proposing and exploring modes of interaction between the agents, as well as various limitations on the agents. We explore both problems of giving and receiving advice.

A thorough experimental analysis was done in a controlled predator-prey environment. The predator agent is a Q-learning agent that takes advice by probabilistically following the advised action. The coach agent was given a large amount of information about the world, such as the full dynamics and full information about the value and optimality of all actions.

The results consistently show that advice indeed improves the advisee's performance under all forms of limitation tested. For constant advice, our results show that the more an agent listens to coach advice, the better it performs, with the expected diminishing returns. Further when the bandwidth between coach and agent is limited, two effects are observed. First, for smaller bandwidths the presence of advice helps regardless of how that advice is chosen. Only when the bandwidth is increased further does the choice of which states to advise about begin to matter. Lastly, limiting the memory of the agent lowers overall performance, but the same general trends still hold.

Further, we have shown how the coach can learn to adapt its advice to the limitations of the agent. We show empirically that the coach is effectively learning about the actions that the limited agent can do.

The challenge of coaching involves learning about an environment, learning about the agents, and effectively communicating advice while taking into account agent abilities. We believe that this work is a core step towards completely answering that challenge.

References

1. Riley, P., Veloso, M., Kaminka, G.: An empirical study of coaching. In Asama, H., Arai, T., Fukuda, T., Hasegawa, T., eds.: Distributed Autonomous Robotic Systems 5. Springer-Verlag (2002) 215–224
2. Riley, P., Veloso, M.: Planning for distributed execution through use of probabilistic opponent models. In: AIPS-2002. (2002) 72–81 *Best Paper Award.*
3. Riley, P., Veloso, M., Kaminka, G.: Towards any-team coaching in adversarial domains. In: AAMAS-02. (2002) 1145–1146
4. Maclin, R., Shavlik, J.W.: Creating advice-taking reinforcement learners. Machine Learning **22** (1996) 251–282
5. Gordon, D., Dubramanian, D.: A multi-strategy learning scheme for knowledge assimilation in embedded agents. Informatica **17** (1993)

6. Noelle, D., Cottrell, G.: A connectionist model of instruction following. In Moore, J.D., Lehman, J.F., eds.: Proceedings of the Seventeenth Annual Conference of the Cognitive Science Society, Hillsdale, NJ, Lawrence Erlbaum Associates (1995) 369–374
7. Clouse, J.A., Utgoff, P.E.: A teaching method for reinforcement learning. In: ICML-92. (1992) 92–101
8. Clouse, J.: Learning from an automated training agent. In Gordon, D., ed.: Working Notes of the ICML '95 Workshop on Agents that Learn from Other Agents, Tahoe City, CA (1995)
9. Sammut, C., Hurst, S., Kedzier, D., Michie, D.: Learning to fly. In: ICML-92, Morgan Kaufmann (1992)
10. Riley, P.: MPADES: Middleware for parallel agent discrete event simulation. In Kaminka, G.A., Lima, P.U., Rojas, R., eds.: RoboCup-2002: The Fifth RoboCup Competitions and Conferences. Number 2752 in Lecture Notes in Artificial Intelligence. Springer Verlag, Berlin (2003) *RoboCup Engineering Award* (to appear).

Technical Solutions of TsinghuAeolus for Robotic Soccer

Yao Jinyi, Lao Ni, Yang Fan, Cai Yunpeng, and Sun Zengqi

State Key Lab of Intelligent Technology and System
Department of Computer Science and Technology
Tsinghua University , Beijing, 100084, P.R.China
{Yjy01,Noon99,Caiyp01}@mails.tsinghua.edu.cn, albertyang@263.net
szq-dcs@mail.tsinghua.edu.cn

Abstract. TsinghuAeolus is the champion team for the latest two RoboCup simulation league competitions. While our binary and nearly full source code for RoboCup 2001 had been publicly available for the entire year, we won the champion again in Fukuka, with more obvious advantage. This paper describes the key innovations that bring this improvement. They include an advice-taking mechanism which aims to improve agents' adaptability, a compact and effective option scoring policy which is crucial in the option-evaluation framework, and thorough analysis of interception problem which leads to more intelligent interception skill. Although not strongly interrelated, these innovations come together to form a set of solutions for problems across different levels.

1 Introduction

RoboCup Simulation League provides a challenging platform to promote research. By six consecutive years' effort of the whole community, the simulation match has improved amazingly, from being seemingly stupid to so delicate that some people think it's more than real. It should in great part owe to the whole community's great effort on sharing binaries and source codes. As the champion team in Seattle, TsinghuAeolus2001 released both executable and source code. Its effectiveness is convincingly proved by the fact that the newcomer Everest, who started completely from the released TsinghuAeolus2001 source code, won the runner-up in RoboCup 2002.

There was a debate in Fukuka on how to schedule simulation platform's development, alternatively speaking, how to develop the platform to help all the participators do research on it. Since the attraction of RoboCup depends partially on competition, one of the principles should be to encourage winning from better research. On the other hand, RoboCup is so complex an environment that various problems have to be solved to build a competitive team. We feel it's important to summarize what we have reached so that the following researchers can build a strong team easily, on the basis of which they can focus on what they are interested in. For this purpose, in addition to releasing source code, we

D. Polani et al. (Eds.): RoboCup 2003, LNAI 3020, pp. 205–213, 2004.

try to describe our various key innovations in this paper, especially those that make TsinghuAeolus2002 improve from TsinghuAeolus2001.

Taking external advice is an important way toward adaptability. Our first innovation is an advice-taking mechanism, which can be attached to an existing autonomous agent. It works somewhat like case-based architecture. The detail is discussed in Section 2. In RoboCup, the architecture for action selection in ball possession is the key component for a team. But few teams have explained their work on this aspect in detail. Our second innovation focus on the creation and evaluation of pass options. It is discussed in Section 3. The third innovation is further analysis on interception, on the basis of which we develop an adversarial interception skill. More detail is discussed in Section 4.

2 A Scheme Mechanism for Adaptability

Adaptability has been a desirable quality for autonomous agent all along. But in some complex environments, it's not easy for an autonomous agent to improve by itself. RoboCup simulation platform, which is a highly dynamic, adversarial multiagent environment, is such a typical case. The state space of 22 agents in a continuous soccer field seems immense, and it's hard to decide how an action, in a sequence of hundreds of joint actions, is responsible for final success or fail. Besides that, player's local and noisy eyeshot is also a big disadvantage. Recently there is an increasing focus on external advice which is expected to take a quite important role for agent's adaptability in many real applications. In RoboCup simulation, an online coach acts as an advice-giving agent. In human matches, teammates also help each other a lot by advising. Besides these, It will also be quite interesting and meaningful if a domain expert can direct an autonomous agent online by giving advice.

Given external advice, an autonomous agent needs to incorporate them into its existing reasoning process. [9] converts the advice to internal neural hidden units, which are then integrated into the existing knowledge-based neural network. In this way, advice is incorporated seamlessly and can be redressed by later learning. But it depends on a knowledge-based neural network which is not applicable in many domains including RoboCup. [3] converts coach advice to some behavior which competes with other spontaneous behaviors in the same framework of behavior manager. One of its advantages is that advice won't be followed blindly. We develop a compact and efficient mechanism to incorporate advice. It is successfully applied in TsinghuAeolus2002. There are two main roles here: scheme and scheme manager. Programmly each piece of advice is converted into an object called "scheme", which is then inserted into a corresponding scheme manager. When the agent does related reasoning, the scheme manager will be called to run in predefined point. Through the whole process, there are three aspects most concerned:

- **State of Advice**
  Advice is often given as condition-dependent. We define four states for each scheme: sleeping, waiting, working and obsolete. Three conditions are required to update a scheme's state: C_{start}, C_{end} and $C_{activate}$. Before C_{start}

is satisfied, the scheme is sleeping. When C_{end} is satisfied, it is obsolete. Between them, if $C_{activate}$ is satisfied, it's working, otherwise waiting. Only working schemes are really incorporated into the agent's decision-making.

- **Where to Consider Advice**
  This is not always a problem. But if you have already an autonomous agent and hope to incorporate advice into its existing decision process, which was not designed to be advice-compatible, you have to consider it. This problem seems domain-dependent and its solution is quite heuristic. A basic idea we use is to classify schemes. Then we heuristically select a point, somewhere in the agent's decision process, for each class instead of each piece of advice. For each class of advice, there is a corresponding scheme manager responsible to update its schemes' states and run all the working ones in some order. Alternatively speaking, the scheme managers also determine what kinds of advice may be incorporated. Those pieces of advice that can't be classified into any existing scheme manager have to be abandoned.
- **Conflicting Issues**
  How to handle conflicting issues is concerned by many research work [10] [3]. We mainly handle two kinds of conflicting cases: exclusive and competitive. For exclusive case, according to some heuristic criteria, an value of rank is given to each scheme, which determines the scheme's order of being processed in its manager. The scheme with the highest rank can be considered first and then the other kindred schemes are ignored. For example, it's reasonable to think that an advice from coach is more authoritative then one from some teammate, so we give the coach advice a higher rank which guarantee it can exclude other teammates' kindred advices. For competitive case, we give an value of advice strength S_a which means the extent that the adviser recommend this advice. A typical competitive case is option-evaluation architecture(See Section 3). When an option is advised, its priority is added by an extra value which is proportional with S_a. We'd like to address that the priority is first calculated according to evaluation, which means the detail of this advice is verified by the agent itself. This is a fundamental difference between exclusive case and competitive case. The extra value can also be negative which means this option is discouraged. Generally, we use a piece of advice to encourage a group of options. For example, "pass more to region R", all the pass options whose destinations lie in R are encouraged to same extent.

A simple scheme is a five touple $\{C_{start}, C_{end}, C_{activate}, Rank, Content\}$. The first four have been explained above. The last one is an extendable component, the instances of which often differ across classes of schemes. For example, S_a is extended in *Content* for classes in competitive cases. The explanation and execution of *Content* depends on the corresponding scheme manager.

We haven't used online coach to give any advice because we haven't worked on online learning by coach yet. Although some of other teams' online coaches are available, it's not reasonable to assume their advices more advisable than an agent's own autonomous behaviors. To experiment the capability of this mechanism, we give a variety of heuristic advices by ourselves. Some of them are listed here, described in natural language.

- When Playmode is changed, scan the field with wide visual mode in less than 7 cycles
- When Player P_a gets the ball in Region R_A, run to Region R_B and keep some distance from P_a
- Don't pass across our own forbidden area.

It proves to be quite natural and smooth to incorporate them into the existing TsinghuAeolus agent under the scheme mechanism. If using online coach to give advice, we would need to code conditions and actions in some language, for example CLang [15]. And after the interpretation, the left things are just what have been discussed above. In the TsinghuAeolus2002 champion agent, there are totally 5 scheme managers and about 20 pieces of advice that may be given according to the situation. We make a comparison between enabling advice and disabling it. The result is shown in Table 1. It proves to work very well. The improvement should owe to both good advice and the suitable mechanism to incorporate it.

Table 1. Goals in 18000 cycles(30 minutes). To suppress the randomness of heterogenous players across different matches, the parameters for all player types are forced to be same with that of normal type, so all the agents for both sides are homogenous. The another side is TsinghuAeolus2002 in all the cases, with scheme enabled or disabled

Opponent Team	Scheme Enabled	Scheme Disabled
FCP2002	0 - 24	0 - 16
USTC2002	2 - 18	3 - 13
TsinghuAeolus2002	2 - 2	4 - 0

3 Action Selection for Ball Possession

For controlling the player with the ball, the option-evaluation architecture [2] fits well. It involves two layers. The lower layer is in charge of producing options, such as dribbling, passing and shooting and so on. The upper one is to evaluate these options according to various score policies. Then the option with the highest score is to be executed. Although not called exactly as option-evaluation, the corresponding modules in [8], [7] and [12] seem to be quite similar. They differ only on the creation of options and score policies, which determine a team's competence in great part.

As introduced in [2], passing options are created at discreet angle increments and speed increments. For each angle-speed pair, it's rejected if some adversary could intercept the ball before any teammate. Our way to generate pass options was described briefly in [12]. The full implementation source code is available in [14]. We also considering passing the ball at discrete angle increments, but not at speed increments. For each angle, the speed section dominated by each agent (Any speed in which, paired with the angle, forms such a pass option that the agent can receive the ball in advance of the others.) can be calculated in a simplified model that involves only the positions of all the agents, instead of other things such as their bodyfacing and heterogenous ability. A key technique in this algorithm is reviewed in Section 4. Viewed in the polar coordinate of passing angle and speed, the whole domain is separately dominated by the agents. Here each agent's domain is exactly the set of pass options that will be received by it superior to the others. It's approximated in both dimensions, angle and speed, by discretization in [2]. Our method only needs to discretize the angle dimension, but the cost is that it's specific to the simplified model.

Option's score is best understood as expected reward in the sense of Reinforcement learning[2]. A set of options defined over an MDP constitutes an embed SMDP[11]. [11] gives theoretical insight into the related learning issues. But it's hard to be implemented in RoboCup right now. As we feel, the biggest difficulty is how to automatically give the reward for an executed option accurately. In the whole process from kick-off to goal, there are lots of options executed by different agents. It's hard, even for human, to decide whether it is good or not for each of them. Until now, we haven't known the score policy of any top team is fully dependent on learning. There is much research work that can be done on this subject in the future.

[2] calculates the score of an option by $p_s * v_s + (1 - p_s) * v_f$ where p_s is the possibility of success and v_s and v_f are the values of succeeding and failing respectively. TsinghuAeolus2001 also used this natural form and even took a nearly same way to calculate p_s and v_s. Our improvement for TsinghuAeolus2002 comes from further analysis and approximation of these factors. For p_s, there are at least two factors responsible to the failure of a pass course, the noise of kicking action and the unaccurate knowledge of other agents, such as their positions, body facing and responsiveness and so on. The danger of the first one is approximated by the margin of the option relative to the agent's domain in the polar system of passing angle and speed. Obviously, bigger is this margin, more robust it is as kicking concerned. The other one can be approximated by the margin between its and other agents' interception times. For v_s, we also divide it into two factors, direct reward v_{dr} and future reward v_{fr}. v_{dr} is calculated according to a static potential field which is built and stored in a neural network beforehand. It's a function of location that increases as the ball advances towards the opponent's goal. v_{fr} is approximated by evaluating the suppositional situation when the receiver gets the ball. If there's an open space around it or a good shooting opportunity, we say there is high future reward expected. It's not hard to heuristically combine them to get the final score. In fact, since scores are used only in competition between the options in the same cycle, we don't need to keep them consistent through cycles as what is generally required in Reinforcement Learning.

Our advantage in pass is quite obvious in RoboCup2002. According to the statistical data, the counts of TsinghuAeolus' successful pass are double as many as that of the other top teams'.

Table 2. Pass counts Statistic. The data are generated by Matrix's Team Assistant's analysis on TsinghuAeolus' logs in RoboCup2002

Opponent Team	Success Pass	Score
FCP2002	79 - 198	0 - 7
Everest	80 - 161	0 - 7
rUNSWIFTII	68 - 193	0 - 6

4 Adversarial Soccer Skill

Basic soccer skills, such as kicking, interception, dribbling and so on, are essential for a soccer team. In simulation platform, these problems seem quite fitting the classical machine learning techniques. Various algorithms, such as RL, A* and so on, have been tried and worked well here. A more challenging problem is to exert these skills in an adversarial environment. In [5], RL is used to learn dribbling against an opponent. In [13], Q learning and A star algorithm are combined to provide a fast and efficient approach for kicking in adversarial environment. One of our innovations in TsinghuAeolus2002 is an adversarial interception skill which is based on the further analysis of the interception problem.

Interception problem refers to how to turn and dash, for a given player, until getting the ball given initial physical information of the ball and itself. Supervised learning by collecting successful examples is an empirical way to obtain this skill, which was recommended in [1]. But later, analytical way seems to be more popular for the advantage of efficiency. [2] presents a numerical algorithm for computing interception times. We have done the similar work and made some improvement.

A simplified model of interception involves an arbitrary ball position B_{p0}, initial ball speed B_v and its direction, and receiver position P_r. Other factors, such as the receiver's body facing, velocity and so on, can be considered heuristically later. The ball's moving route is fixed if random noise is not concerned, The reachable area for the receiver in t cycles later is a circle with its initial position as the center. The circle's radius increases linearly by t in the receiver's maximal speed v_p . With t as the third dimension, we can get a 3D view of the interception process, as illustrated in figure 1. The ball's route C_b is a curve and the receiver's reachable area R_r is a cone. The overlay part of C_b with R_p is exactly the set of points where interception may be completed, which we call as *Sol*. [2] calculates the least time to complete interception. But it's not always the best choice to intercept in the least time. Imagine such a scene, a player lies in front of the ball when the ball is moving forward with a high speed, it's not wise for him to run backward to get the ball and then dribble forward. The better choice is to run forward to meet with the ball. This kind of scene can be found a lot in human's match, especially in anti-offside case. So it's useful to figure out *Sol* completely. In the 3D coordinate, C_b lies in an vertical plane which intersects the surface of R_r in a hyperbola C_p. As has already been pointed out in [2], there are at most three intersection points between C_b and C_p.

To look into this problem, assume C_p and the direction of the ball's velocity is fixed while B_v is alterable. With varying B_v, C_b may be tangent to C_p in

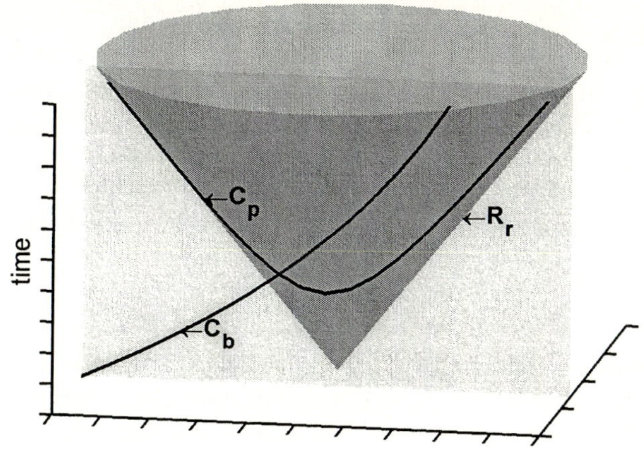

Fig. 1. Interception

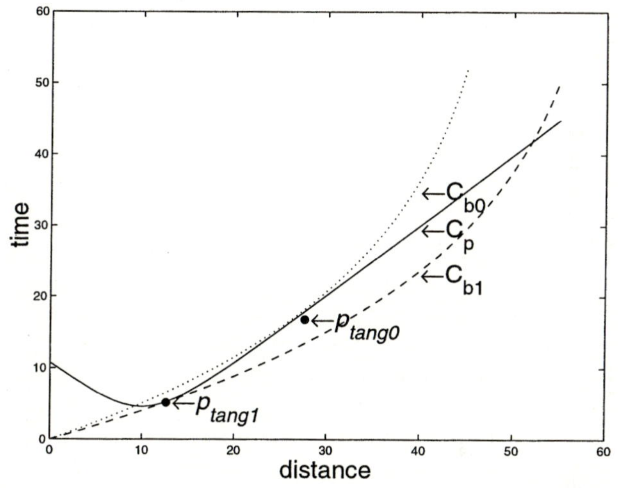

Fig. 2. Two tangent cases

two cases, as illustrated in figure 2. The corresponding values of B_v are b_{tangv0} and b_{tangv1}. The corresponding tangent points are p_{tang0} and p_{tang1}. It's easy to know that there is only one intersection point between C_b and C_p when B_{v0} is in $(0, b_{tangv0})$ or in $(b_{tangv1}, +\infty)$, two when B_v equals b_{tangv0} or b_{tangv1}, and three when B_v is in (b_{tangv0}, b_{tangv1}). Given C_p, p_{tang0} and p_{tang1} can be approximately figured out by Newton methods with appropriate initial values for update. Then the corresponding values of B_v are easy to figure out.

Back to the original problem where both C_p and C_b are given, we use Newton Method to calculate the intersection points, plus some preprocess and tricky treatment. C_p is first divided into two symmetrical parts at its apex p_s. It's obvious that there is not more than one intersection point between the left part with C_b, which, if exists, can be figured out with standard Newton Method. For the right part, assume there are three intersection points p_0, p_1 and p_2. We only discuss this case in the following. The other cases can be handled in a similar and simpler way. It's obvious that p_0 lies between p_s and p_{tang1}, p_1 between p_{tang1} and p_{tang0}, p_1 between p_{tang0} and $p_{+\infty}$(a point in C_p whose x value is big enough). We use $f_{cb}(x)$ to represent the function of C_b, and $f_{cp}(x)$ for the right half of C_p. The standard Newton update equation is as following.

$$x_{i+1} = x_i - \frac{f_{cp}(x_i) - f_{cb}(x_i)}{f'_{cp}(x_i) - f'_{cb}(x_i)} \tag{1}$$

But it's possible to overshoot. $f'_{cp}(x)$ lies always in $(0, v_p)$. With the maximal and minimal value replacing $f'_{cp}(x)$ in 1, We get the following two modified update rules.

$$x_{i+1} = x_i - \frac{f_{cp}(x_i) - f_{cb}(x_i)}{0 - f'_{cb}(x_i)} \tag{2}$$

$$x_{i+1} = x_i - \frac{f_{cp}(x_i) - f_{cb}(x_i)}{v_p - f'_{cb}(x_i)} \tag{3}$$

To take p_s as the initial point and (2) as update rule, x_i converges to $x(p_0)$. Because in the iterations, $x_{i+1} < x_i, x_i > x(p_0)$, which is not hard to prove and we don't intend to detail here. Similarly, we can reach p_1 by taking p_s as the initial point and (3) as update rule, p_2 by taking $p_{+\infty}$ as the initial point and (2) as update rule.

There is another related problem which is crucial for creating pass options as introduced in Section 3. It originates from the need to know the minimal pass speed to pass across some point before some player can get it. Here C_p, a point p on it, and the direction of the ball's velocity are given, and the problem equals to figure out the maximum B_v when there is at least an intersection point before p between C_b and C_p. The critical case must be that C_p and C_b intersect in either p_{tang0} or p. Assume p_{tang0} and p is the intersection point separately, it's easy to get the corresponding values of B_v. The bigger one is exactly the answer.

5 Conclusion

The interception problem is a basic but important one. We present a thorough view of it in analytical way, which leads to a stronger adversarial skill.

TsinghuAeolus have showed amazing passing skill in Fukuka. The crucial component responsible for that includes the creation and evaluation of pass options. It's the accurate analysis of this problem that leads to our advantage.

An integral advice-taking system for an advised agent involves several steps: receiving, parsing and taking. The scheme mechanism addresses the last step. It can be attached to an existing complex agent without too much work. And it forces few requirements on the existing agent's framework. We also demonstrated its effectiveness. With this mechanism as the basis, we're working on agents' adaptability in the hope of improving the on-line team's behavior.

References

1. Peter Stone. Layered Learning in Multi-Agent Systems. PhD Thesis, Computer Science Dep.,Carnegie Mellon University, December 1998
2. Peter Stone and David McAllester. An architecture for action selection in Robotic Soccer. In Fifth International Conference on Autonomous Agents(Agents'2001)
3. Paul Carpenter, Patrick Riley, Manuela Veloso and Gal Kaminka. Integration of Advice in an Action-Selection Architecture
4. Sebastian Buck and Martin Riedmiller. Learning Situation Dependent Success Rates of Action in A RoboCup Scenario. In PRICAI 2000, Melbourne, published in Lecture Notes in Artificial Intelligence, Springer 2000, p809
5. Martin Riedmiller and Artur Merke. Using machine learning techniques in complex multi-agent domains. In I. Stamatescu, W. Menzel, M. Richter and U. Ratsch (eds.), Perspectives on Adaptivity and Learning (2002), LNCS, Springer
6. Remco de Boer and Jelle R. Kok. The Incremental Development of a Synthetic Multi-Agent System: The UvA Trilearn 2001 Robotic Soccer Simulation Team. Master's thesis, University of Amsterdam, The Netherlands, February 2002.
7. Jelle R. Kok, Remco de Boer, Nikos Vlassis, and F.C.A. Groen. UvA Trilearn 2002 team description. In G. Kaminka, P.U. Lima, and R. Rojas, editors, RoboCup 2002: Robot Soccer World Cup VI, page 549, Fukuoka, Japan, 2002. Springer-Verlag.
8. Nuno Lau and Luis Paulo Reis. FC Portugal 2001 Team Description: Flexible Teamwork and Configurable Strategy. RoboCup-2001: Robot Soccer World Cup V, Andreas Birk, Silvia Coradeshi, Satoshi Tadokoro editors, LNAI, Springer Verlag, 2002
9. Richard Maclin and Jude W.Shavlik. Incorporating Advice into Agents that Learn from Reinforcements. Proceeding of the Twelfth National Conference on Artificial Intelligence.
10. Benjamin N. Grosof. Conflict handling in advice taking and instruction. Technical report, IBM Research Report 20123, 1995.
11. Sutton, R.S., Precup, D., Singh, S. Between MDPs and semi-MDPs: A Framework for Temporal Abstraction in Reinforcement Learning. Artificial Intelligence 112:181-211. 1999
12. Jinyi Yao, Jiang Chen, Yunpeng Cai and Shi Li. Architecture of TsinghuAeolus. RoboCup-2002: Robot Soccer World Cup V, Andreas Birk, Silvia Coradeshi, Satoshi Tadokoro editors, LNAI, Springer Verlag, 2002
13. Jinyi Yao, Jiang Chen, and Zengqi Sun. An application in RoboCup combining Q-learning with Adversarial Planning . In The 4th World Congress on Intelligent Control and Automation(WCICA'2002).
14. TsinghuAeolus2001 Souce Code. Available in http://www.lits.tsinghua.edu.cn/RoboCup.
15. Noda et al, RoboCup Soccer Server Users Manual, RoboCup Federation, June 2001.

A Real-Time Auto-Adjusting Vision System for Robotic Soccer*

Matthias Jüngel, Jan Hoffmann, and Martin Lötzsch

Institut für Informatik, LFG Künstliche Intelligenz, Humboldt-Universität zu Berlin
Rudower Chaussee 25, 12489 Berlin, Germany
{juengel,jan.hoffmann,loetzsch}@informatik.hu-berlin.de

Abstract. This paper presents a real-time approach for object recognition in robotic soccer. The vision system does not need any calibration and adapts to changing lighting conditions during run time. The adaptation is based on statistics which are computed when recognizing objects and leads to a segmentation of the color space to different color classes. Based on attention, scan lines are distributed over the image ensuring that all objects of interest intersect with the number of lines necessary for recognition. The object recognition checks the scan lines for characteristic edges and for typical groupings of color classes to find and classify points on the outlines of objects. These points are used to calculate size and position of the objects in the image. Experiments on Sony's four-legged robot Aibo show that the method is able to recognize and distinguish objects under a wide range of different lighting conditions.

1 Introduction

The RoboCup real robot soccer leagues (middle-size, small-size, Sony four-legged league) all take place in a color coded environment. The method described in this paper was implemented for the Sony four-legged league. The objects the robot needs to see are two-colored flags for localization (pink and either yellow, green, or sky-blue), two goals (yellow and sky-blue), the ball (orange), the robots of the two teams (wearing red and blue tricots), and the white field lines (center circle, middle line, penalty lines).

A very common preprocessing step for vision-based object recognition in color coded scenarios is color segmentation using color tables, e. g. [2, 7]. Such methods directly map colors to color classes on a pixel by pixel basis, which has some crucial drawbacks. On one hand, the color mapping has to be adapted when the lighting conditions change, on the other hand, the mapping results in a loss of information, because the membership of a pixel in a certain class is a yes/no decision, ignoring the influences of the surrounding pixels. Some researchers try to overcome these limitations [4], but the solutions are too slow to work under real-time conditions on a robot with limited computational power.

* The Deutsche Forschungsgemeinschaft supports this work through the priority program "Cooperating teams of mobile robots in dynamic environments".

D. Polani et al. (Eds.): RoboCup 2003, LNAI 3020, pp. 214–225, 2004.

A method for off-line self-calibration on sample images that first filters the images to improve color invariance and then applies k-means clustering for the adaptation of color classes is described in [6]. Lighting conditions can differ depending on the position on the field or the situation (the ball can be in the shadow of a player, the carpet can appear in different shades when seen from different viewing angles). In this paper a method is presented that adapts color classes during run time, so it is not necessary to perform any kind of pre-run calibration.

Although it is not necessary to use the full resolution of images to detect objects, it helps in determining their sizes and positions precisely. A very fast method for object finding which uses the concept of perspective view is suggested in [5]. Several additional methods restricting the number of pixels being processed during object finding are presented in this paper.

2 Image Processing

The key ideas of the image processing method presented in this paper are that speed can be achieved by avoiding to process all pixels of an image, and independence of the lighting conditions can be reached by focusing on contrast patterns in three different color channels and auto-adapting color classification.

2.1 Guiding Attention

Usually objects in the image are larger than a single pixel. Thus for feature extraction, a high resolution is only needed for small objects. The performance and robustness of image processing can be improved by guiding more attention to areas within the image where small objects are expected. Instead of processing all pixels in the image, a grid of scan lines is used to reduce the number of pixels processed. Different types of additional information about the image and the environment can be incorporated:

1. Image sequences: a robot usually processes a continuous sequence of images. Low level information (e.g. color clusters) gained in the previous image can be used to speed up object detection (e.g. a ball will not have moved too far from one image frame to the next, therefore it is in most cases beneficial to start searching for the ball in the area of the image where it was previously detected).
2. Iterative processing: a part of the image is processed and the information gained is used in the further processing of the image. For example, the image can first be scanned for more prominent features; once these are found, their location in the image can hint to the whereabouts of other features.
3. Other Sensors: readings from other sensors (such as distance or tilt sensors) can be analyzed and used to guide visual attention. For example, far away objects may or may not be of importance and therefore only parts of the image need to be scanned.

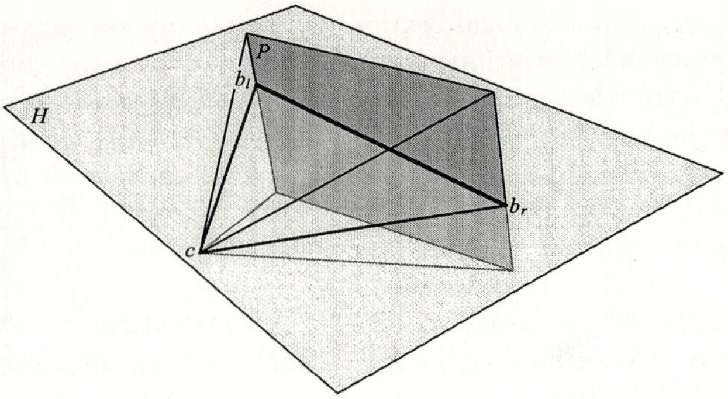

Fig. 1. *Calculating the horizon.* The horizon is the line of intersection of the projection plane P and the plane parallel to the ground H. $b_{l/r}$ is the intersection of the horizon with the left and the right border of the image, c is the focal point of the camera.

4. Knowledge about the environment: domain specific knowledge can be used to simplify image processing. Heuristics can be derived from this, e. g. in the RoboCup domain, robots and the ball are in contact with the field at all times. Therefore the larger parts of these objects will be below the horizon in the image.
5. Explicit feedback of (high level) world model information: The robot's knowledge (or hypothesis) of where it is in the environment can be used to simplify the search for relevant features. Only areas in the image that, according to the hypothesis, contain information are being scanned.

2.2 Arranging Scan Lines

First the position of the horizon in the image is calculated from the rotation of the head and the rotation of the body. The roll and tilt of the body are estimated from the readings of the robot's acceleration sensors (indicating the direction of the gravity), while the rotation of the head is determined from the angles of the three head joints (tilt, pan and roll). Then the areas below and above the horizon are scanned. For each of the areas an optimized grid layout is used that assures that relevant features cannot escape detection.

Calculating the Horizon. The horizon is the line of intersection between the two planes P and H where

P is the projection plane of the camera, and H is the plane parallel to the ground through the origin c of the camera coordinate system. To calculate the horizon line pixels $\vec{b}_{l/r}$ on the left/right border of the image and points $\vec{h}$ on the horizon H are examined, see Fig. 1:

$$\vec{b}_{l/r} = \begin{pmatrix} \frac{s}{\tan\alpha} \\ \pm s \\ z_{l/r} \end{pmatrix} \qquad \vec{h} = \begin{pmatrix} x \\ y \\ 0 \end{pmatrix} \tag{1}$$

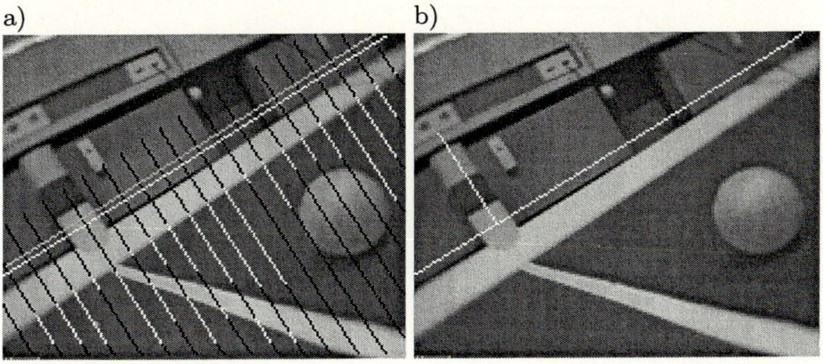

Fig. 2. *Scan Lines.* a) Gray horizontal line: horizon calculated from sensor readings, white horizontal line: horizon calculated from the white border. Vertical scan lines extend from the horizon to the lower half of the image. The white parts are used for auto-adaptation. b) A scan along the horizon determines the foot of the flag and its width. A second scan line vertical to the horizon through the center of the foot is used to determine the height of the flag.

with opening angle 2α and the screen resolution $2s$. Pixels that are both on the horizon and on the border of the image satisfy

$$R \cdot \overrightarrow{h} = \overrightarrow{b}_{l/r} \tag{2}$$

with R being the rotation matrix of the camera. This can easily be solved for the pixel coordinates of the horizon in the camera image:

$$z_{l/r} = \frac{\mp r_{32}s - r_{31}s \cdot \cot\alpha}{r_{33}} \tag{3}$$

with the entries r_{nm} of the rotation matrix.

Grid Lines below the Horizon. From the horizon, vertical scan lines (i.e. perpendicular to the horizon) extend into the lower half of the image, see Fig. 2a). The lines are evenly spaced. The distance of the lines depends on the expected distance to the ball. It is chosen such that at least two lines will intersect with the ball. Since the ball is the smallest object encountered on a RoboCup field, all other objects (robots) will also be hit. In addition to the dynamic objects mentioned, some static features of the field are found below the horizon: field lines, lower regions of the goals, and field borders.

Scan Lines above the Horizon. Color coded flags are placed at the corners and at the side of the field above borders to offer additional help for localizing the robot. Scan lines to find the flags are positioned in three steps: First based on the robots position and the direction of the camera a prediction of the positions of all visible flags in the image is calculated and vertical scan lines are arranged through the predicted flags. If the predicted flags are not found in the image a second attempt to find the flags is started on a scan line parallel to the horizon. If

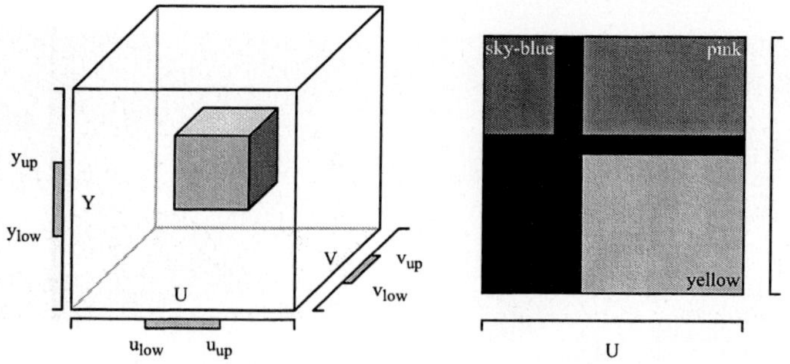

Fig. 3. *Left:* The reference color is a sub cube in the YUV color space. Other colors are defined relative to this cube. *Right:* The ranges for sky-blue, pink and yellow in the color cube.

this also fails the sensor readings are assumed to be disturbed by the influence of other robots and the position of the horizon is recalculated based on the position of the white border found in the image. If the foot of a flag is found on the scan line parallel to the horizon (based on sensors or image) a vertical scan line is positioned there, see Fig. 2b).

2.3 Color Classification and Adaptation

Each pixel at the grid lines is assigned to a color class. The color class depends on the position of the color of the pixel in the color cube. How the color cube is subdivided into color classes and how this subdivision is adapted automatically to changing lighting conditions is described in this section.

Color Classification. The camera of the Aibo provides YUV-images. In YUV, the luminosity of a pixel is stored in the Y channel, its color information in the chrominance channels U and V. To classify the color a very simple model is used. Within the YUV color cube, a reference color is defined. In the soccer application this reference color is the green of the carpet. This color is defined by upper and lower bounds in the three dimensions, thus producing a sub cube. Other colors are defined in relation to this cube. For example, *sky-blue* is defined as having a V value greater than the upper bound of the reference cube and a U value lower than the lower bound of the reference cube. Other colors are defined in a similar fashion. Only a limited number of colors is classified. Colors not necessary for object recognition are omitted, see Fig. 3.

As this method is so simple it is not possible to separate colors that are close to each other in color space, e.g. yellow and orange. Such colors are grouped into one color class. This disadvantage is accepted due to the simplicity of auto-adaptation of the reference color and the segmentation of the color cube. Ambiguities that arise from the bounded number of distinguishable colors are resolved by using additional information from the image and the environment (see sections 2.4 and 2.5).

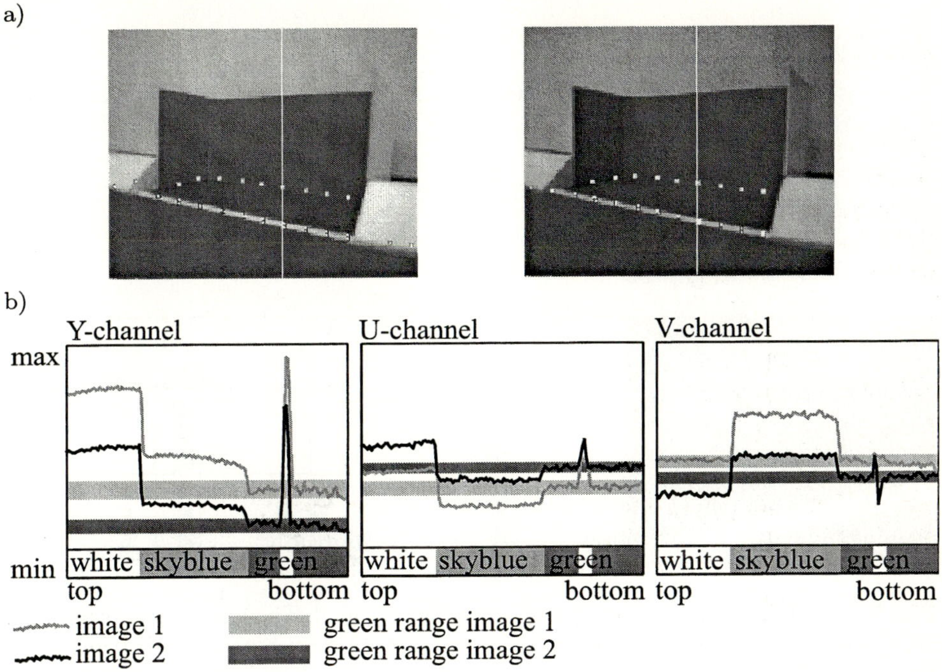

Fig. 4. a) Two images of the same scene recorded under different lighting conditions with a highlighted scan line and the recognized points on lines (border, field line, goal). Left: day light, right: artificial light. b) The intensities of the three channels along the scan line for both images. Left: image 1, right: image 2. The light and dark gray bars show the range of green after auto-adaptation for the left and the right image. For each image the values for sky-blue in the U channel are lower than the values for green. But the values for sky-blue in the U channel in image 2 are greater than the values for green in image 1 which shows the need for auto-adaptation.

Adaptation to Lighting Conditions. The effect of changing lighting conditions to the position of the colors in the color space is illustrated in Fig. 4. The positions of all colors move along the three axes and the intensities are amplified or weakened. But the relative position of the colors to each other remains. Thus it is sufficient to change position and dimensions of the reference cube with the lighting conditions.

To update the reference cube, from each image a set of green pixels is extracted. This extraction is not based on the previous reference cube but on a different heuristic: The changes in the YUV channels on a scan lines have certain characteristic properties which are used to identify pixels that are considered to be green. A scan line starting near the horizon will show a characteristic drop in the luminosity channel Y where the color changes from the white of the border to the green of the field. Starting from this point, all pixels on the rest of the scan line are considered to be green. If there is a colored object on the field,

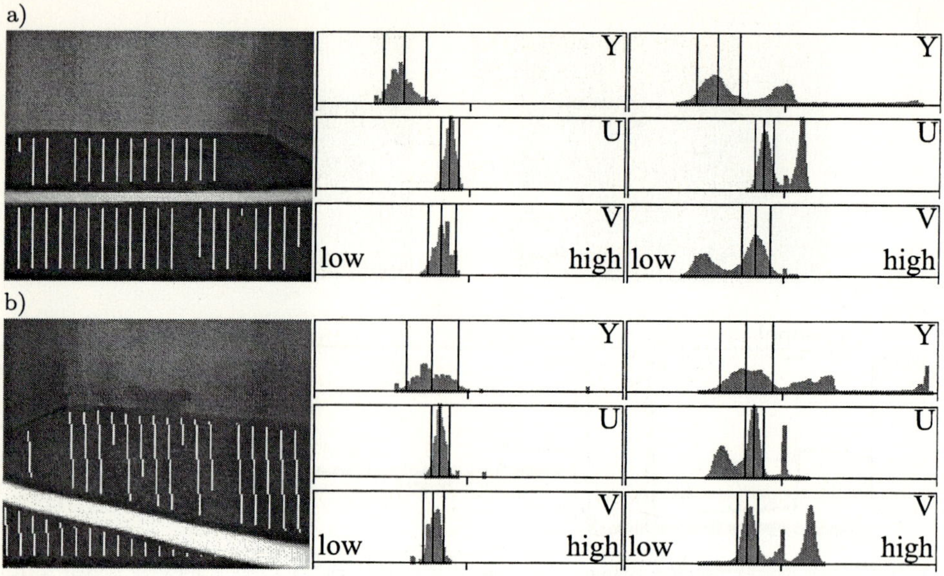

Fig. 5. Left: Image of a yellow (a) and a sky-blue (b) goal. The highlighted pixels have been selected by the heuristic for green extraction. Center: Frequency distribution of the intensities of the extracted pixels. The vertical lines show the 98% ranges. Right: Frequency distribution of the intensities of each pixel in the image. The vertical lines show the 98% ranges obtained from the extracted pixels.

there is a change in one of the color channels and the rest of that scan line is omitted (see the white parts of the vertical scan lines in Fig. 2a). This condition is rather stringent to assure that no pixels are wrongly added to the set of pixels used for color-adaptation.

This method is very robust and simple. No classification of clusters in color space or the frequency distribution of intensities for all pixels is needed.

The reference cube is updated with every image with enough green in it. Most of the images taken during a RoboCup game satisfy this condition. The YUV values of the extracted green pixels are used to recalculate the reference cube such that the new cube contains almost all green pixels. A typical distribution of the intensities for each channel of the selected pixels is shown in Fig 5. Note that the distribution of green pixels cannot be assumed as Gaussian but is dependent on scene illumination, carpet texture, camera noise, etc. The lower and the upper bound of the cube in each channel are chosen such that they enclose the most frequent 98% of the intensities.

2.4 Edge Detection and Classification

The edge detection finds characteristic changes in the three color channels. The points in the image where edges are found are classified using two criteria: the three dimensional contrast patterns and the surrounding color classes.

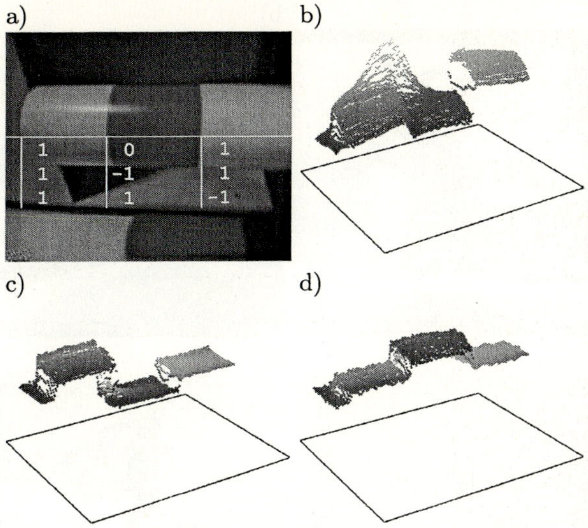

Fig. 6. *Scan Lines:* a) An image of a pink/sky-blue/white (from left) flag with a scan line and Y-U-V contrast patterns for each edge. b) the Y-channel of the image of the flag as height map. There is an up-edge (1) at the begin of the flag, the intensity increases near the highlight but there is no edge at the change from pink to sky-blue (0), there is an up-edge (1) from sky-blue to white. The first row of values in the contrast patterns is therefore (1,0,1). c, d) The height maps of the U- and V-channel visualize the second and the third row of values in the contrast patterns. Up-Down-Up (1,-1,1) and Up-Up-Down (1,1,-1).

Contrast Patterns. The contrast pattern for a pixel p_i on a scan line is defined as

$$C(p_i) = \begin{pmatrix} c_1 \\ c_2 \\ c_3 \end{pmatrix}$$

with

$$c_j = \begin{cases} 1 & , value_j(p_i) - value_j(p_{i-1}) > t_j \\ -1 & , value_j(p_i) - value_j(p_{i-1}) < -t_j \\ 0 & , else \end{cases}$$

for $1 \leq j \leq 3$, where $value_j(p_i)$, denotes the values of the color channels of the ith pixel of the scan line and t_j denotes the thresholds for large changes in the three color channels. These thresholds are chosen once so that only edges and no noise inside objects are detected. The thresholds do not need to be recalibrated when lighting conditions change, see Fig. 6.

Each pixel p_i with $C(p_i) \neq (0, 0, 0)$ is assigned to one or more edge classes. For example $(0, -1, -1)$ is assigned to *yellow-black*, *yellow-green*, *orange-black*, and *orange-green*, because a significant drop in the U and V channel is an indication for one of these edges.

Fig. 7. *Scan Lines:* a) Three types of lines: field/border, field/goal, and field/line. Points on the outline of the ball. Circle calculated from the points. b) The ball on the field and a flag. c) The ball inside the yellow goal. d) A partly occluded ball and a flag.

Surrounding Color Classes. The color classes of the pixels surrounding the detected edges at the scan lines are taken into account to resolve ambiguities appearing during the classification of contrast patterns and to filter edges caused by noise.

For example the set of edge classes { *yellow-black*, *yellow-green*, *orange-black*, *orange-green* } can be reduced to { *yellow-green*, *orange-green* } if the preceding pixels belong to the color class orange/yellow and the following pixels are green.

The contrast pattern $(0, 1, 1)$ occurs at the top edge of the orange ball in a yellow goal but sometimes also if there is a reflection of a spotlight on a robot. Such contrast patterns are assigned to *yellow-orange* only if the preceding and the following color class is orange/yellow, see Fig. 7c).

2.5 Object Recognition

The object recognition combines several classified edges to objects and determines the size and position of the objects in the image. Different combining methods are applied to edges from different types of the scan lines.

Flags. To determine size and position of the flags in the image, the sequence of classified edges on each vertical grid line above the horizon is considered.

Each flag is characterized by a triple of classified edges containing the expected color changes at the upper edge of the flag, at the color change inside the flag, and at the lower edge of the flag. Thus a simple pattern matching algorithm is sufficient to detect all flags. Flags found in an image are represented by the four angles describing their bounding rectangle (top, bottom, left, and right edge) in a system of coordinates that is parallel to the field. The angles to the top and the bottom of the flag are determined from the positions of the first and the last edge of the matching edge pattern. To determine the angles to the left and the right edge of the flag, a new scan parallel to the horizon is started from the center of the section with higher contrast to the background, i. e. the pink one, in both directions. The first contrast pattern with a large decrease for the U-channel on this scan line marks the "end" of the flag and provides a point on a vertical edge.

Goals and Field Lines. Four different types of lines can be detected on the RoboCup field: edges between the sky-blue goal and the field, edges between the yellow goal and the field, edges between the border and the field, and edges between the field lines and the field, see Fig. 7a). The object detection has not to extract *lines* from the image, but *pixels on lines* instead. This approach is faster and more robust against misinterpretations, because lines are often partially hidden either by other robots or due to the limited opening angle of the camera. The *pixels on lines* are the input for a Monte-Carlo localization that is described in [8].

Ball. The edges at the bottom of the ball and at the bottom of the yellow goal have the same contrast patterns and the same color classes above and below the edges and thus are in the same edge class. But the top edge of the ball is unique because the top of the yellow goal is always about the horizon and does never intersect with the scan lines. Thus bottom ball edges are only accepted if there is a top ball edge above. As described in section 2.4 even a ball in a yellow goal can be detected, although yellow and orange are in the same color class.

The points found on the border of the ball are measured values and thus will not always lie on one and the same circle. First for each triple of points the resulting circle is calculated. In a second step each of the circles obtains a vote from each pixel having a distance of less than 3 pixels to the circle. The circle with most votes is assumed to be the border of the ball. This method assures that even partly occluded balls are detected, see Fig. 7.

3 Experimental Results

3.1 Lighting Independence

The approach described in this paper was tested in robot soccer games. The robots were able to localize themselves on the field and to handle the ball in an accurate manner. Approximately 80% of all taken images could be used to re-calibrate the color classification. Independence of lighting conditions was tested by manually changing the white balance modes of the camera (indoor, outdoor, fluorescent light) and changing the lighting conditions itself. The objects on the first image taken under new conditions are not detected as auto-adaptation is always done for the next image. This does not matter, because each of the 25 images per second provided by the robot is processed. If enough green is in the first image with new conditions the objects in the second image are recognized properly. The need for auto-adaptation was shown by changing lighting conditions while keeping old color calibration which lead to misclassifications. The system fails when there is very little contrast in the image. This is the case in bright daylight (overexposed camera images) and under very little light (underexposure). In both cases colors become indistinguishable even for the human observer. If the scenario is lighted up with a very yellow light and the white balance of the camera is set to fluorescent light mode, the orange of the white of the borders and the yellow of the goal become indistinguishable, too. Under conditions similar to those at competition sites, the selected white balance mode had no influence on the quality of object recognition.

3.2 Performance

On an Aibo which is equipped with a 400 MHz processor the whole image processing (calculating and scanning the lines, finding edges, classifying colors and recognizing object) takes 7 ms if the grid lines are dense and 4 ms if the grid lines are wide spaced (ball is expected near the robot). There is no difference in running time if the auto-adaptation is switched off. Thus the approach is faster than classifying the color of each pixel of the image and then finding regions as described in [2]. It is as fast as the method that was used by the GermanTeam for the RoboCup 2002 competitions in Fukuoka that is described in [1, 3] and which requires manual color calibration.

4 Conclusion

This paper presents a real-time auto-adjusting vision system for robotic soccer. The vision system extracts features without processing whole images by guiding attention to the areas of high interest. Independence of lighting conditions is reached by an auto-adaptation of color classes and a matching of contrast patterns. The object recognition employs environmental constraints to determine size and position of the objects. This results in a fast, auto-calibrating, and precise feature extraction.

Acknowledgments

The authors would like to thank Hans-Dieter Burkhard, Joscha Bach, Uwe Düffert and Thomas Röfer for useful discussions.

References

1. J. Bach and M. Jüngel. Using pattern matching on a flexible, horizon-aligned grid for robotic vision. *Concurrency, Specification and Programming - CSP'2002*, 1:11–19, 2002.
2. J. Bruce, T. Balch, and M. Veloso. Fast and inexpensive color image segmentation for interactive robots. In *Proceedings of the 2000 IEEE/RSJ International Conference on Intelligent Robots and Systems (IROS '00)*, volume 3, pages 2061–2066, 2000.
3. U. Düffert, M. Jüngel, T. Laue, M. Lötzsch, M. Risler, and T. Röfer. GermanTeam 2002. In *RoboCup 2002 Robot Soccer World Cup VI, G. Kaminka, P. Lima, R. Rojas (Eds.)*, Lecture Notes in Computer Science, 2003. to appear. more detailed in http://www.tzi.de/kogrob/papers/GermanTeam2002.pdf.
4. R. Hanek, T. Schmitt, S. Buck, and M. Beetz. Towards RoboCup without color labeling. *RoboCup International Symposium 2002*, 2002.
5. M. Jamzad, B. Sadjad, V. Mirrokni, M. Kazemi, H. Chitsaz, A. Heydarnoori, M. Hajiaghai, and E. Chiniforooshan. A fast vision system for middle size robots in RoboCup. In *RoboCup 2001 Robot Soccer World Cup V, A. Birk, S. Coradeschi, S. Tadokoro (Eds.)*, number 2377 in Lecture Notes in Artificial Intelligence, pages 71–80, 2002.
6. G. Mayer, H. Utz, and G. Kraetzschmar. Towards autonomous vision self-calibration for soccer robots. In *Proceedings of the 2002 IEEE/RSJ Intl. Conference on Intelligent Robots and Systems*, 2002.
7. F. K. H. Quek. An algorithm for the rapid computation of boundaries of run length encoded regions. *Pattern Recognition Journal*, 33:1637–1649, 2000.
8. T. Röfer and M. Jüngel. Vision-based fast and reactive Monte-Carlo localization. *to appear: IEEE International Conference on Robotics and Automation*, 2003.

Knowledge-Based Autonomous Dynamic Colour Calibration

Daniel Cameron and Nick Barnes

Department of Computer Science and Software Engineering, University of Melbourne
camerond@students.cs.mu.oz.au, nmb@cs.mu.oz.au

Abstract. Colour labeling is critical to the real-time performance of colour-based vision systems and is used for low-level vision by most RoboCup 2002 physically based teams. Unfortunately, colour labeling is sensitive to changes in illumination and manual calibration is both time consuming and error prone.
In this paper, we present KADC, a robust method for Knowledge-based Autonomous Dynamic Colour Calibration. By utilising the known geometry of the environment, landmarks are identified independent of colour classifications. Colour information from these landmarks is used to construct colour clusters of arbitrary shape. Clusters are dynamically updated through actions and by the use of a similarity metric, the Earth Mover's Distance (EMD). We apply KADC to the RoboCup Legged League, generating a colourtable purely from geometrical knowledge of the environment and dynamically update this colortable to compensate for non-uniform changes in lighting conditions.

1 Introduction

Image segmentation based on colour labeling is used extensively in real-time vision. Colour labeling requires colourspace to be partitioned into a set of colour classes, each corresponding to a symbolic colour such as 'blue' or 'green'. Each pixel is labeled with a symbolic colour according to location in colourspace and from this, objects of a distinct colour can be efficiently and reliably identified. For optimal effectiveness, colour labeling requires distinctive colour partitions that are not generally available in real world systems. In such circumstances, colour labeling can still serve as a useful component of multi-modal systems.

RoboCup playing environments have been constructed such that landmarks can be easily identified by their distinct colour scheme. By constructing a mapping between colourspace and colour classes of interest such as green (field), yellow (goal & beacons) and orange (ball), landmarks can be tracked in real-time. Unfortunately, this mapping degrades rapidly with any significant change in lighting conditions. As a result, RoboCup competitions are held under controlled lighting conditions designed to minimise colour misclassification. Colour classification poses a significant hurdle for the evolution of RoboCup to a more realistic environment. The lack of robust autonomous dynamic colour classification restricts RoboCup venues, requires an idealised playing environment, and contributes to long setup times.

D. Polani et al. (Eds.): RoboCup 2003, LNAI 3020, pp. 226–237, 2004.

In Robocup, determining which colourspace coordinates map to which colour classes has been commonly achieved through linear thresholding. The simplest form of this is achieved by thresholding each colour component, either in the native camera colourspace [1], or in a transformed colourspace [2]. Unfortunatly, colour classes may not correspond to the rectangular boxes in colourspace defined by these thresholds. To reduce misclassification, some RoboCup teams have used learning algorithms such as neural networks [3] and decision trees [3, 4] to semi-autonomously generate arbitrary concave colour classes. Lookup tables in colourspace allow for the most accurate representation of colour classes as they handle arbitrary distributions. As with [1], colourspace fidelity is reduced due to the prohibitive cost of storing lookup values for the entire colourspace.

Current autonomous calibration techniques vary widely in their method of calibration and domain of application. Austermeier *et al.*, [5] presents a calibration technique that compensates for colour distortions from a reference illumination. Self-organising feature maps are used to compute a vector field for transforming pixels to their original colour under the reference illumination. This method is computationally expensive and requires a physical colour calibration chart to be available under all conditions to be calibrated for. Legenstein *et al.*, [6] presents a less powerful method suited for finding coloured objects that calibrates from a set of coloured strips carried by the robot.

Mayer *et al.*, [7] presents a method of semi-autonomous colour self-calibration using Mid-Sized League robots. The robot collects sample images from the known environment using colour classifications generated for previous lighting conditions. A filter to improve colour consistency is applied to these images which then undergoe k-means clustering. From each of these clusters, HSV thresholds are found that determine each colour class. These colour classes are then manually mapped to symbolic colours.

In this paper, Section 2 presents an overview of the system developed. The system is composed of three components: actions which allow for the rejection of landmarks that do not appear in plausible locations, colour independant landmark identification to eliminate classification divergence due to feedback, and a probabilistic colour clustering model which is formally defined in in Section 3. Section 4 outlines the Earth Mover's Distance (EMD), a metric used in image databases [8] adapted for use as a cluster similarity metric. Important implementation considerations as well as Legged League specific details are given in Section 5, with experimental results in Section 6.

The probabilistic cluster representation used makes no *a priori* assumptions about the distribution of the cluster in colourspace. Notably, concave and disjoint clusters can be represented without loss of information.

The approach we take considers a broader problem in colour calibration than most papers. Rather than model uniform lighting changes across the environment, we also consider non-uniform shifts in colourspace. If a light fails at one end of the field (as occurred during the 2002 Legged League competition) the illumination of the yellow landmarks drops, whilest the blue remain constant, and the pink is stretched in colourspace. Further, we consider the problem where the colour of landmarks is not known at all prior to calibration. Our results demonstrate the learning of landmark colours given no a priori information other than the geometry of the environment.

2 System Overview

In order to generate colour classifications, landmarks must be identified and the colour information from the landmarks used to generate the classifications. By taking actions based on geometric knowledge, these landmarks can be identified with greater certainty than a given set of random images. The system presented in this paper can be broadly decomposed into three components: actions, landmark identification, and colour classification.

Actions: A key difference between robot vision and computer vision in general is the ability to take actions. By using knowledge of the environment, actions can be used to increase the probability of particular landmarks being visible. By only considering landmarks when they are expected to be visible, false positives can be reduced, leading to more robust classification. By removing human interaction the overhead of classification is reduced such that it is practical to recalibrate extensively. This calibration can be performed dynamically during a game in parallel with existing systems.

Landmark Identification: Once the robot is positioned such that a landmark is expected to be visible, the image is segmented into 4-connected regions of self-similar colours using a segmentation algorithm independent of colour classifications (e.g., edges / regions). A landmark is composed of one or more sections of a single colour. Each section corresponds to a 4-connected region in the image. To identify a landmark, sets of regions are evaluated as to the probability that they correspond to the landmark. Spatial constraints known from the geometry of the environment (such as goals immediately above the field or beacon sections of similar size and vertically aligned) are used to evaluate the probability of a set of regions corresponding to a landmark.

Colour Classification: Once a landmark has been identified, the colour information from each section is used to construct a candidate cluster for each section. EMD is used to determine whether candidate clusters for the landmarks are sufficiently close to the current clusters of the colours associated with a known colour scheme (such as the current blue cluster and current yellow cluster for the goals in the Legged League). If a match is found, the current cluster is updated using the candidate cluster. For all updates, the probability that the region set corresponds to the landmark determines the magnitude of the update.

3 Clustering Framework

Cluster Representation: To handle the uncertainty present within classifications, cluster are represented within a Bayesian inference framework. Occupancy grids [9] are used in mobile robotics to model obstacles. Here we apply occupancy grids over colourspace to form the basis of clusters. An occupancy grid is used for each colour to map the distribution of that colour in colourspace. Colourspace is trivially divided into cells, each mapping to a distinct colourspace coordinate. Each cell C has an associated state variable $s(C)$ representing whether that cell maps to the colour $s(C) = \text{COL}$, or to non-colour $s(C) = \text{NON}$. These two discrete states are mutually exclusive and exhaustive thus the probability of a cell being in either state sums to 1 $P[s(C) = \text{COL}] + P[s(C) = \text{NON}] = 1$. Since $P[s(C) = \text{NON}] = 1 - P[s(C) = \text{COL}]$, the probability mass function $P[s(C) = \text{COL}]$ defined over all cells defines the colour cluster for a given symbolic colour.

If no information is available, non-informative priors are used for initialisation:

$$P[s(C) = \text{COL}] = P[s(C) = \text{NON}] = 0.5 \tag{1}$$

Cluster Updating: Since the probability of each cell mapping to a symbolic colour is treated as an independent random variable, we can use Bayes theorem to calculate the change in probability of cell C due to the classification M:

$$P[s(C) = \text{COL}] \leftarrow \frac{P[M \mid s(C) = \text{COL}]}{P[M]} P[s(C) = \text{COL}], \tag{2}$$

Similarly, Bayes theorem also applies to NON:

$$P[s(C) = \text{NON}] \leftarrow \frac{P[M \mid s(C) = \text{NON}]}{P[M]} P[s(C) = \text{NON}], \tag{3}$$

where

$$P[M] = \begin{array}{l} P[M \mid s(C) = \text{COL}] \cdot P[s(C) = \text{COL}] \\ +P[M \mid s(C) = \text{NON}] \cdot P[s(C) = \text{NON}]. \end{array} \tag{4}$$

Equations (2) and (3) can be combined and rewritten as:

$$\frac{P[s(C) = \text{COL}]}{P[s(C) = \text{NON}]} \leftarrow \frac{P[M \mid s(C) = \text{COL}] \cdot P[s(C) = \text{COL}]}{P[M \mid s(C) = \text{NON}] \cdot P[s(C) = \text{NON}]}. \tag{5}$$

For efficiency reasons, the log-likelihood of the cell probability is stored. The likelihood of a hypothesis H given the probability of that hypothesis $P(H)$ is:

$$L(H) = \frac{P(H)}{P(\neg H)}, \tag{6}$$

taking the log of this results in:

$$\log L(H) = \log \left(\frac{P(H)}{P(\neg H)} \right). \tag{7}$$

Expressed in terms of likelihoods, (5) becomes:

$$L[s(C) = \text{COL}] \leftarrow L[M \mid s(C) = \text{COL}] \cdot L[s(C) = \text{COL}], \tag{8}$$

taking the log of both sides gives:

$$\log L[s(C) = \text{COL}] \leftarrow \log L[M \mid s(C) = \text{COL}] + \log L[s(C) = \text{COL}]. \tag{9}$$

Which yields a simple update strategy: the cluster value $\log L[M \mid s(C) = \text{COL}]$ due to M is added to the previous cluster value for each cell C.

Classification Representation: For each classification, a multiset of colourspace coordinates (one for each pixel in the landmark section) representing the colour information of the landmark is used to generate a new cluster which is subsequently merged with the appropiate symbolic colour cluster. Initially, the new cluster is uninformative with $\log L[s(C) = \text{COL}] = 0$ for all cells C (corresponding to (1)). For each element of the multiset, the corresponding cell is updated as per (9).

Unlike sensors such as laser range-finders used in robot navigation, accurate sensor models are not available for landmark classification. In this paper we have modelled these probabilities as

$$L[M \mid s(C) = \text{COL}] = \frac{f}{1-f} \exp\left(\frac{1}{A}\right) \tag{10}$$

where f is the fitness of the landmark and A is the image area of the associated region (both of which are defined below).

A colourspace coordinate close to a coordinate identified as part of a landmark section is more likely to correspond to the colour of the landmark. For example, if a pixel is identified as blue, pixels which are similar to the identified pixel are also likely to be blue. We model this by blurring updates over colourspace using Gaussian blur.

Cluster Decay: In uncontrolled environments, lighting conditions change over time and the clusters representing landmarks drift from their initial positions. To incorporate this into (9), a decay factor γ indicating the rate of depreciation of previous classifications has been introduced resulting in:

$$\log L[s(C) = \text{COL}] \leftarrow \log L[M \mid s(C) = \text{COL}] + \gamma \log L[s(C) = \text{COL}] \tag{11}$$

Cluster Discretisation: Since each cluster is treated independently, multiple symbolic colours can map to a single colourspace coordinate. If the colour labeling algorithm can utilise this ambiguity [10] then each cluster can be thresholded independently as given by (12).

$$S \in m(C) \leftrightarrow \log L_S[s(C) = \text{COL}] > t_S \tag{12}$$

where $m(C)$ is the mapping from colourspace cell C to a set of symbolic colours, L_S is the likelihood for the cluster of symbolic colour S and t is the inclusion threshold of log-likelihood for S.

In many colour labeling applications, symbolic colours are mutually exclusive and each colourspace coordinate must be mapped to a unique symbolic colour. In order to support such applications (which includes the Legged League), the cluster merging strategy is augmented to include a negative classification. That is, all clusters which are not being added to are subtracted from. For example, classifying a landmark as green implicitly classifies the associated candidate cluster as not red and not blue etc. Equation (9) could be augmented to modify all other clusters, however, this information is only needed during the discretisation process, a more efficient solution is to subtract when discretisation is required:

$$S \in m(C) \leftrightarrow \log L_S[s(C) = \text{COL}] - \sum_{S_i \neq S} \log L_{S_i}[s(C) = \text{COL}] > t_S \quad (13)$$

which guarantees mutual exclusion for all thresholds $t_S \geq 0$.

4 Cluster Similarity Metric – EMD

A cluster similarity metric is required in the cases where landmark identification cannot unambiguously determine which symbolic colour a region should be labeled and to prevent misclassified landmarks from being added to clusters. The former occurs in the Legged League for the goals and the beacons. These landmarks have identical geometry and differ only by colour and position.

This section briefly presents a overview of EMD (for a full description see [8]), and its application to dynamic colour classification. EMD was first proposed as a similarity metric between colour images and used for image retrieval and display in image databases. EMD derives it's name from the analogy that is the normalised minimum work required to fill a set of holes with earth by moving this earth from another set of mounds.

More formally, given a set of producers X each producing x_i units, a set of consumers Y each demanding y_j units, and a cost c_{ij} per unit of transportation from $i \in X$ to $j \in Y$ determined by the distance between i and j, the total cost of transportation is given by:

$$\sum_{i \in X} \sum_{j \in Y} c_{ij} f_{ij}, \quad (14)$$

where f_{ij} is the amount transported from i to j. EMD is defined as the normalised minimum total cost of transportation:

$$\text{EMD}(X, Y) = \min \left(\frac{\sum_{i \in X} \sum_{j \in Y} c_{ij} f_{ij}}{\sum_{j \in Y} y_j} \right) \quad (15)$$

This minimisation problem is an instance of the classical transportation problem from linear programming, to which efficient solutions exist. Thus, EMD takes a two weighted sets of coordinates in some space, a ground distance in that coordinate space, and computes the minimum average distance required to move the first set to the second. The weight of each element of the set determines the capacity of that producer/consumer or in the earth moving scenario, the amount of earth required to empty/fill the mound/hole. Rubner *et al.* [8] refers to these weighted sets of coordinates as 'signatures'.

Application to Dynamic Colour Classification: In the context of cluster similarity the signature of a cluster is determined directly from cell values. The signature of a cluster is composed of all colourspace coordinates with the weight of each coordinate equal to the normalised cell value. Cells with a value of zero may be omitted since they have no effect on the EMD. By normalising by the total sum of cell values, EMD is unaffected by cluster size and exists for all informative clusters. The ground distance used for calculating the cost of transportation is given by the straight-line (Euclidian) distance (L2 norm) between the coordinates in native colourspace. This ground distance does not correspond to the difference perceived between two colours but the colour distance as determined by the capture hardware. For applications requiring human interaction such as image databases, colourspaces that closely resemble the human perception of colour have been used [8].

5 Implementation

The system has been implemented on a Sony ERS-210 entertainment robot. A full size 2002 RoboCup Legged League field within a laboratory environment with partial natural lighting is used. The field comforms to Robocup 2002 Legged League requirements with exception that the white field lines are not present.
Cluster Compression: A naive implementation of probabilistic clustering using 24-bit colour, 32-bit floating-point precision and 10 colour classes requires $2^{24} \cdot 4 \cdot 10 = 640$Mb of storage. To reduce the memory footprint the least significant bits of each channels are ignored. By ignoring the lowest 4 bits of Y, 2 bits of U and V, the total memory required to store the clusters is reduced to 2.56Mb. Note that this corresponds to a reduced resolution occupancy grid over colourspace in which probabilistic formulation given in Section 3 also holds. To further compress clusters, clusters are tiled according to their most significant bits: 2 for Y and 3 for U and V. Colourspace thus is divided into 256 tiling blocks, each containing 256 units. Since clusters only occupy a small portion of colourspace, most blocks will have a value of zero in all cells and these are stored implicitly. It was found that on average, 8-16 blocks were occupied thus reducing the memory footprint from 2.56Mb to around 120Kb.

Calculating EMD directly from the cluster signature is computationally prohibitive so an approximation is used. For each tiling block, the centre of mass and total weight are used as the cluster signature. This information is cached for each tile for efficient EMD computation.
Actions: The action sequence undertaken by the robot attempts to maximise the probability of a landmark being visible when that landmark is being considered for cluster updating. The following sequence of actions was used for classification:

1. walk back 0.5m, look down
2. look up, turn around until a goal visible
3. walk to 1m from goal, look left, look right
4. turn around, goto step 2.

Initially, the robot must be on the field, when the robot looks down, the only landmark present will be the field, except is when the robot is at the edge of the field facing the white barrier. (1) reduces the probability of seeing the white

barrier. Using (2), visible goals will be vertically centred in the image, simplify identification. Since there is one beacon either side of the goal, beacons should be visible and bounds can be placed not only on the expected image location but also on the expected size after (3). As the robot cycles between the goals, and its localisation converges to a line running between the two goals and the probability of finding beacons at the precalculated angles continually increases.
Landmark Identification: All landmarks in the Four Legged League are composed of one, or in the case of beacons, two homogeneous sections, each of a single colour. Since colour within a section varies only due to lighting conditions, and the colour of adjacent sections are widely separated in colourspace, both edge-based and region-based image segmentation techniques should be able to identify the regions of the image that correspond to landmark sections robustly. For this paper, an edge-based approach based on Canny edge detection [11] has been used. Canny edge detection is performed on each of the three input channels separately and the result combined in a boolean OR operation with an edge being present if it is present in at least one input channel with the result run-length encoded for efficiency. A region merging algorithm similar to [1] is used to identify all the 4-connected regions of non-edge in the image.
Landmark Fitness: Since the robot pose is known from the actions taken, simple heuristics can be used for landmark identification. The green field must occupy most of the lower part of the image. The goals must be vertically centred and should be twice as wide as they are high. Beacon sections must be of similar size and aligned vertically. Beacons and goals should be of the estimated size and approximately convex (the convex hull area is compared to the actual area).

6 Experimental Results

Static Illumination: To highlight the differences between calibration techniques, three colour classifications were generated. The first classification was generated on-board, updating only for landmarks that expected to be visible. The second was generated on a PC by transmitting images over a wireless LAN. Unlike the on-board classification, the PC-based autonomous classification updated all clusters without regard to expected image contents. The third classification was generated manually from the transmitted images. Due to limited bandwidth, the on-board autonomous classification was able to process more images and hence the image sequence used for this classification differs from the image set used for the autonomous PC-based classification and the manual classification.

To evaluate the three different methods, a second set of images was taken with the robot manually positioned in front of each landmark. These images served as a test set from which a reference classification was established by manually labelling regions of each image with appropriate colours. Misclassification rates were determined by comparing the classifications generated to the reference classification for each image in the test set. Due to the ambiguity in determining which edge pixels are considered to correspond to a landmark, the reference classification was conservative in its classification of landmarks and any false

Table 1. Misclassification rate (%) after self-calibration under constant illumination.

	Manual	AIBO	PC
Beacon - Blue	1.0	2.2	1.1
Blue Beacon - Pink	0.3	0.1	0.6
Beacon - Yellow	9.4	3.3	4.2
Yellow Beacon - Pink	2.3	6.2	4.1
Goal - Blue	1.3	1.5	0.4
Goal - Yellow	8.6	6.1	4.0
Field - Green	2.0	0.3	2.5

positives connected to the reference classification were assumed to correspond to the landmark.

As can be seen in Table 1, all three classification techniques generated classifications of comparable quality. The high misclassification rate associated with the yellow clusters can be partially attributed to the small sample size as the robot spent most of its time near the blue goal and beacons.

Dynamic Illumination: A second experiment was conducted to evaluate the sensitivity to changes in illumination of the technique presented in this paper. For this experiment, four beacons were placed next to the blue goal as shown in Figure 1. Initially, the room was illuminated by two pairs of floodlights, and standard office neon lighting with the illumination gradually decreased to zero lighting, then increased to full lighting at a faster rate.

The first row of images shown in Figure 1 demonstrates the classifications after one floodlight is turned off. The manual calibration shown on the right exhibits some degree of degradation in classification of the blue goal. Note that since the autonomous classification only classifies the blue beacon (since no yellow goal has been seen hence the pink-yellow and pink-green beacons cannot be disambiguated), the pink section of the yellow beacon in the left of the image has been misclassified due to differences in illumination between the beacons.

In the second to fourth rows, illumination continually decreases. The autonomous classification shows little change whereas the static classfication completely misclassifies. The fourth row displays an instance of misclassification in the left yellow beacon due to poor image segmentation under the darker lighting conditions. The edge between the pink and the green was incomplete in previous frames and as such, the region classified as field also included the pink area of the beacon. Note however that this misclassification was rectified in subsequent frames and by the fifth row, the classification returned to pink.

To correctly re-identify landmarks under modified illumination, the change in illumination must be small enough that there is no ambiguity between the landmark under the new illumination, and other landmarks of the same shape of different colour. The fifth row shows classifications after an abrupt increase in illumination. Note that unlike the field and goal, the classification of the blue beacon was unable to recover from the abrupt change.

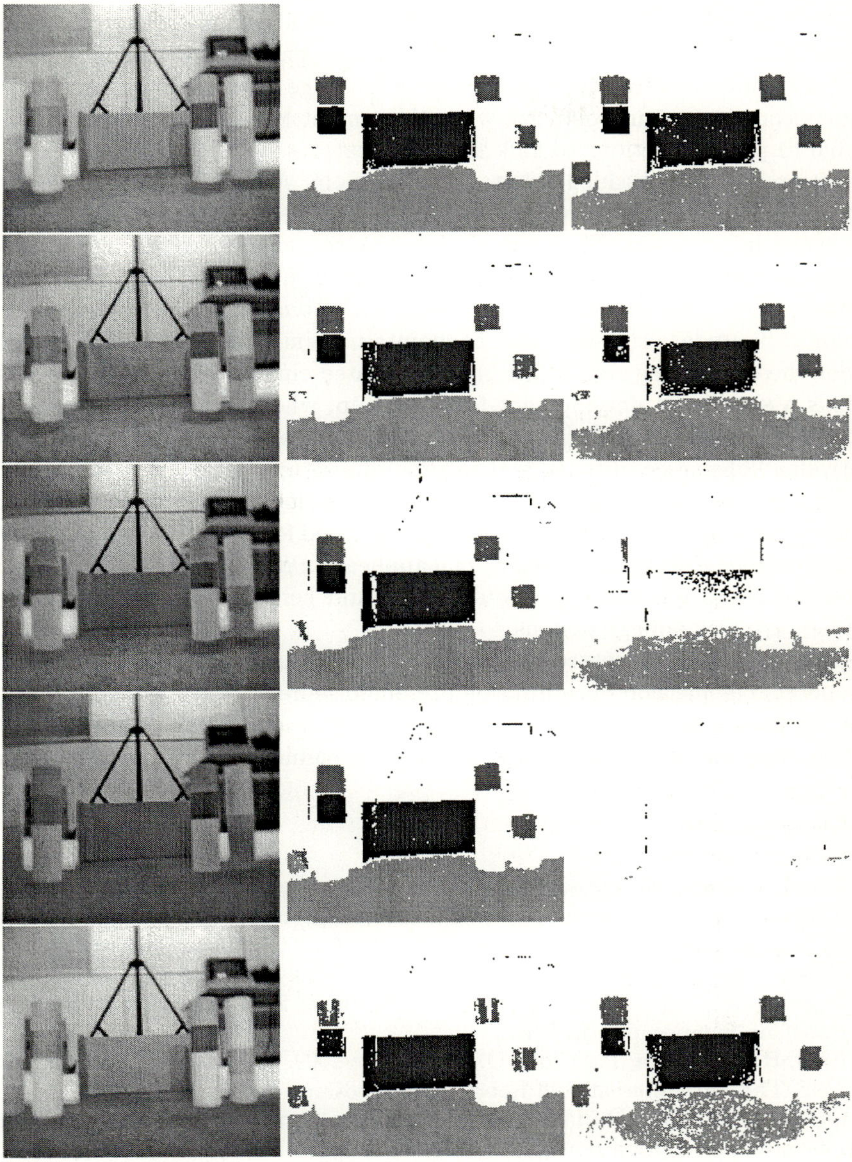

Fig. 1. Lighting conditions are dynamically changed during the experiment. On the left is the actual image, the dynamically calibrated classification in the middle, and a static classification done at the start of the experiment on the right.

7 Conclusion

In this paper, we presented KADC, a new method for Knowledge-based Autonomous Dynamic Colour Calibration. This method and algorithm facilitates the generation of new colour classifications, and dynamically updates existing colour classifications under changing lighting conditions given certain assumptions. Specifically, that a method is available for segmenting the regions independent of colour classification, that we have knowledge of where important features might occur in the environment, what they might look like, and a robot which is able to perform actions to look for these features. We have implemented this algorithm on an embedded robot platform (Sony AIBO), and have used it to generate partial colour tables for the Legged League competition given no apriori colour information. The method was able to generate colour classifications that were comparible to hand segmented clusters under constant illumination. Further, under changing lighting conditions, the technique presented is able to dynamically update classifications to maintain a similar quality of classification. Under these changing conditions hand generated clusters have high error rates. This is a significant development for RoboCup, where in the physical leagues, teams have issues of classification robustness, are only able to compete under controlled lighting conditions, and require long setup times.

The use of a probabilistic interpretation of colour classes allows for uncertainty in classifications to be modeled explicitly and for colour classes to be represented without loss of information. The application of EMD to cluster similarity allows for the comparison of colour cluster and hence a measure of similarity between arbitrary image regions.

The techniques presented are applicable to colour-based vision applications in general. The method of calibration presented represents a significant improvement in the robustness of colour classification and allows colour labeling to be used in a more general class of problems with dynamic non-uniform illumination.

References

1. Bruce, J., Balch, T., Veloso, M.: Fast and inexpensive color image segmentation for interactive robots. Proceedings of the 2000 IEEE/RSJ International Conference on Intelligent Robots and Systems (IROS '00) **3** (2000) 2061–2066
2. Oda, K., Kato, T., Ishimura, T., Katsumi, Y.: The kyushu united team in the four legged robot league. In: Proceedings of RoboCup 2002, The 2002 International RoboCup Symposium. (In Press)
3. Akm, H.L., Pavlova, P., Yildiz, O.T.: Ceberus 2002. In: Proceedings of RoboCup 2002, The 2002 International RoboCup Symposium. (In Press)
4. Cayouette, F., Sud, D., Patel, K., Sarikaya, D., Cooperstock, J.: Mcgill reddogs 2002 team description report. In: Proceedings of RoboCup 2002, The 2002 International RoboCup Symposium. (In Press)
5. Austermeier, H., Hartmann, G., Hilker, R.: Color-calibration of a robot vision system using self-organizing feature maps. Artificial Neural Networks - ICANN 96. 1996 International Conference Proceedings (1996) 257–62
6. Legenstein, D., Vincze, M., Chroust, S.: Finding colored objects under different illumination conditions in robotic applications. In: Proceedings of SPIE Vol. 4197 Intelligent Robots and Computer Vision XIX: Algorithms, Techniques, and Active Vision. (2000)
7. Mayer, G., Utz, H., Kraetzschmar, G.K.: Towards autonomous vision self-calibration for soccer robots. IEEE/RSJ International Conference on Intelligent Robots and Systems (2002) 214–219

8. Rubner, Y., Tomasi, C., Guibas, L.: A metric for distributions with applications to image databases. IEEE International Conference on Computer Vision (1998) 59–66
9. Elfes, A.: Occupancy grids: a stochastic spatial representation for active robot perception. In: Proceedings of the Sixth Conference of Uncertainty in AI. (1990) 60–70
10. Akm, H.L., Pavlova, P., Yildiz, O.T.: Ceberus 2002. In: Proceedings of RoboCup 2002, The 2002 International RoboCup Symposium. (In Press)
11. Canny, J.: A computational approach to edge detection. IEEE Trans. Pattern Analysis and Machine Intelligence **PAMI-8(6)** (1986) 679–698

Playing Robot Soccer under Natural Light: A Case Study

Gerd Mayer, Hans Utz, and Gerhard K. Kraetzschmar

University of Ulm, James-Franck-Ring, 89069 Ulm, Germany

Abstract. The recent debate in the RoboCup middle-size community about natural light conditions shows that a more in-depth analysis of the problems incurred by this is necessary in order to draft out a focused and realistic roadmap for research. Based on real-world images taken under varying lighting conditions, we performed descriptive and statistical analysis of the effects on color-based vision routines. The results show that pure color-based image processing is not likely to perform well under varying lighting conditions, even if the vision system is calibrated on a per-game base. We conclude that color-based vision has to be combined with other methods and algorithms in order to work robustly in more difficult environments with varying illumination.

1 Introduction

One of the long-term goals of the RoboCup initiative is to construct soccer robots capable of playing in typical soccer playgrounds, including outdoor soccer fields. As a consequence, soccer robots must be able to play under varying lighting conditions, ranging from temporally stable artificial illumination (fluorescent or floodlight light) to natural illumination with varying temporal dynamics (slow changes during the day, fast changes by clouds). An open challenge for the community is to draft a roadmap for research, which ultimately yields soccer robots playing well in all these environments.

Currently, the middle-size league plays robot soccer in an environment with strictly controlled artificial lighting conditions. Providing this environment is laborious and expensive for tournament organizers. For this reason, the RoboCup middle-size community discussed on its mailing list, whether competitions can be held under daylight conditions in the near future or not. Some teams have already tried to play under natural lighting conditions, e.g. at fairs or exhibition games, and problems seemed to be not so bad. Nevertheless, permitting daylight constitutes a significant change in the environment and is expected to severely affect game play and robot performance. Although everyone seems to intuitively agree on that, the roadmap discussion would greatly benefit from more detailed scientific investigations on this topic. This paper contributes to this discussion by providing an analysis of the effects of various lighting conditions on color-based image processing for soccer robots.

The paper is organized as follows: Section 2 discusses various lighting conditions, and their effect on image processing and vision routines. The image data

D. Polani et al. (Eds.): RoboCup 2003, LNAI 3020, pp. 238–249, 2004.

used and the methods applied for further analysis are described in sections 3 and 4. Experiments and results are presented in Section 5. Section 6 draws some conclusions and discusses potential solutions.

2 Survey of Lighting Conditions

In order to gain a better understanding of the causes and effects of varying lighting conditions, we analyze several different scenarios for playing (robot) soccer.

The first scenario is the one currently in use for RoboCup tournaments: An indoor soccer field is illuminated by *artificial light*. The current rules constrain light intensity to lie between 800 and 1200 lux, but do not impose additional constraints, e.g. on color temperature. Special care is often taken to provide diffuse lighting. Once calibrated, the lighting system remains unchanged during the whole tournament. Although there is no a-priori specification of lighting conditions, teams can perform an on-site, one-time calibration of their vision systems and optimize their performance for the lighting situation on a particular field. The calibration should account for the remaining small variations of lighting across the field. This process needs some care, but can be done with reasonable effort. It results in robots with vision systems well tuned to the lighting situation on site, but which cannot deal with unexpected disturbances like those caused by flashlight.

The next scenario is characterized by *small changes over time.* An example would be the slow movement of the sun during the day. In addition, the shadows thrown by objects in the environment, e.g. from trees, pillars, or even the robot itself (see the rightmost image[1] in Figure 2), will wander over time, and dramatically change local perception of other objects like the goal or the ball. This effect can be even stronger in indoor environments illuminated by natural light through windows, which are often located on one side of the room. Assuming that lighting changes remain small within the relevant match period, robot teams can handle this scenario with suitable technology for fast on-site, per-game vision calibration. Due to the presence of shadows, the remaining variations in lighting conditions within the field are harder to deal with.

Fig. 1. The computer science faculty building at TU Munich, where most test images were taken.

A third scenario is characterized by *sudden, discrete variations* of illumination. This happens especially on days with partially cloudy skies, when a cloud temporarily covers the sun. Very sudden lighting changes also occur, if something shadows the sun for a short moment only, like people bending over a robot's camera. As these changes are hardly predictable, pre-game calibration to cope with such effects should be very difficult, if not impossible.

[1] All color images of this paper can be found at
`http://smart.informatik.uni-ulm.de/DataSets/RoboCup/NatLight/`

3 Image Data

A set of soccer robot camera images covering all previously described scenarios was collected in two different locations in order to allow for further analysis. One location was the new computer science faculty building of Technical University of Munich, where we were invited for a few friendly games against the Agilo robot soccer team. The soccer field was set up in the large central hall (see Figure 1), the outer hull of which consists almost completely of windows. The images were made during a period of about 5 hours around midday and early afternoon, and at the late evening during setup. The other location was a lab room at the University of Ulm with large windows on one side of the room. In both locations, the same carpet, goals, and balls were used.

Image set I (Munich, evening) The first set of images were taken during setup in the late evening. So, no natural light was present. The field was illuminated only by the fluorescent lights. Thus, illumination consisted of constant, very diffuse light, so neither hard shadows nor strong reflections posed a major problem. Although light intensity was probably at the lower end of the range, this setting was certainly closest to the RoboCup setting currently specified.

Image set II (Munich, daytime) The second set of images were taken over a period of 5 hours around midday and early afternoon of the next day. The weather was sunny with occasional clouds. Illumination of the field was both with the indoor lighting structure and heavy natural light influence through the dyed windows. Because the sun was very low and surrounding buildings shaded her from throwing hard shadows on the field, the overall light was still rather diffuse, although the dyed windows influenced the color spectrum significantly (see the center image in Figure 2). In addition, illumination changed slightly over time; see Section 5 for examples on this.

Image set III (Ulm, daytime) A third series of images was made in our laboratory with windows on only one side of the room. Because they were collected at a sunny day, illumination consisted of very directed light, which led to hard shadows and non-uniform lighting.

Fig. 2. Examples of the three different lighting conditions used for the experiments. Neon light in Munich in the leftmost picture, pure daylight in Munich in the middle one, and again natural but directed light in our lab in the right image.

Some image examples are shown in Figure 2. Note the differences in color and illumination uniformity, although exactly the same carpet and the same goals were used.

4 Method

In order to investigate the effects of varying lighting conditions, it would be best to take a number of test images for precisely the same situation on the playground, but under different, carefully controlled lighting conditions. A direct comparison of these images would then be possible *without* assuming or implying any particular method for image processing. Unfortunately, such a data set is virtually impossible to obtain, because neither can the weather be controlled nor would it be possible to guarantee comparability of images taken under motion, so need to use other investigative means for evaluation.

As color-based image processing still is the dominating approach in RoboCup middle-size league, we evaluate the effects on the color information in sets of real-world images taken under different lighting conditions. Furthermore, we investigate how these variations affect common color-based image processing methods. In our team, the early visual processing consists of three steps: *Transformation* (map image from RGB to HSV color space), *Classification* (Assign a color class to each pixel) and *Blob detection* (find regions of contiguous pixels with the same color class). Most further processing, like object recognition or field line detection for self-localization, make somehow use of the color-segmented regions image delivered by this process. Many teams use a similar approach, so that the results presented here should apply to them as well.

The second step is the one of interest here, because the quality of the classification process has a strong influence on any further processing. In our case, colors are mapped from the HSV color space to a set of target color classes, containing all colors relevant to RoboCup, and a catch-all color class *gray*, to which all colors not of interest are mapped. The mapping is simply defined by manually specifying for each target color class a lower and upper bound in each color space dimension. If the target color classes can be sufficiently narrow defined and are far away from each other in the color space (high separability), then the calibration process trades off misclassifications between a target color class and the catch-all color class. Often, however, target classes are bordering each other or would even have to overlap (i.e. there exist particular pixel values belonging to different classes, sometimes even in the same image), so calibration must directly trade off misclassifications between two target color classes.

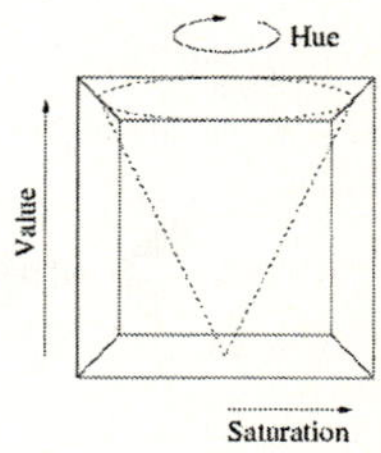

Fig. 3. Illustration of the HSV color space structure.

The HSV color space is a circular, cone-like representation of the dimensions hue, saturation and brightness. Black is at the cone peak on the bottom, white at the innermost part of the upper plate and the colors arranged orbitally around the white color (illustrated in Figure 3).

For small variations, like those even present on the current RoboCup field and lighting setup, the calibration process can make cuboids of target colors large enough to compensate for these variations and still obtain good classification results. The open question is whether this can also be done for larger variations, like those present in scenarios involving daylight. In order to answer this question, we perform the following experiments:

1. We first compare the target class boundaries for several scenarios. For each scenario, the boundaries are the result of a manual calibration process based on a representative set of images.
2. Using the manually calibrated color classificator for each scenario, we color-classify a larger set of images. The spatial distribution of pixel values in each color class is characterized for each scenario by computing and comparing their mean and standard deviation.
3. In order to show that the results are not significantly influenced by the chosen color classification procedure and its calibration process, the previous results are compared with equivalent distributions of pixel values in HSV space obtained from idealized color classifications, which were generated by manually masking in some images areas with objects of interest and mapping their pixel values to HSV space.
4. Finally, a few special examples were selected to illustrate some situations, whose effects are insufficiently described by the above statistics. Even though these effects may be marginal problems or side-issues, they cannot be ignored and, under different circumstances, they may occur more frequently.

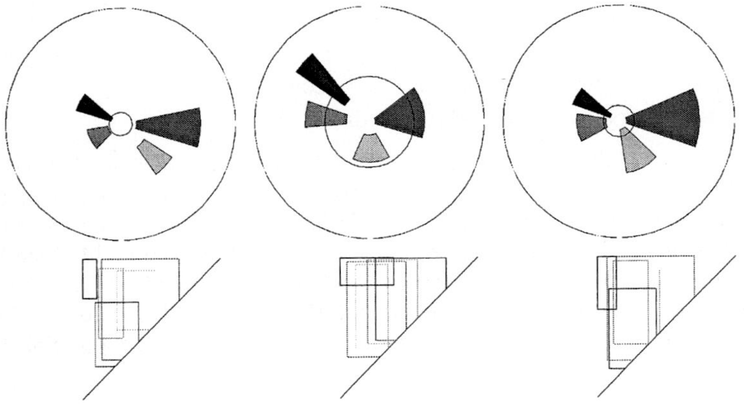

Fig. 4. Results of manually calibrating color classificators for different scenarios. The cuboids for the target color classes are projected onto the hue/saturation plane in the upper row, while the lower row contains projections onto the saturation/brightness half-plane. Figures on the left relate to image set I, figures in the middle to image set II and figures on the right relate to image set III.

5 Experiments and Results

5.1 Comparing Manual Calibration Results for Different Scenarios

Figure 4 illustrates the resulting calibrations for the different scenarios. The conflict resulting from the overlap between the white cuboid and other color classes is resolved with a priority rule, which gives any other target color class priority over white.

The figures show that the color cubes for the midday and lab settings are larger and closer to each other (e.g. the blue and green cone on the rightmost calibration for the lab scenario). This means an increased risk to misclassify pixels from different color classes. The nearer the color descriptions come together, the smaller gets the catch-all class *gray* in between them.

5.2 Variations in Spatial Distribution of Color Classes

For this step, over one hundred images were processed with the appropriately calibrated color classificator for a particular image set. Then, for each color class the frequency and spatial distribution of HSV color values were collected from the original images. Figure 5 illustrates the blobs within the HSV color space resulting from this processing step for the fluorescent light scenario.

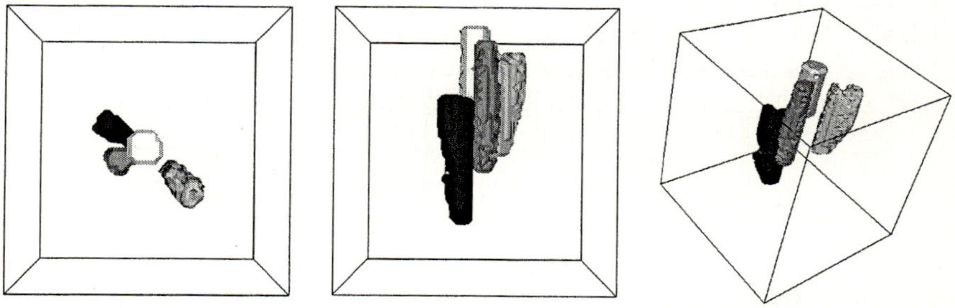

Fig. 5. Three-dimensional plots of the distribution of pixel values in HSV color space for several target color classes, as obtained by color-classifying a single test image taken from image set I. Colors in the figures represent the target color classes (e.g. yellow for the yellow goal). The left image show a top-down view, the middle one the view from one side and the right an isometric version of the same plot. The leftmost plot corresponds with the top-left image in Figure 4, the middle with the bottom-left image in Figure 4. The plot on the right gives a better view of the three-dimensional structure.

For the set of images of each scenario, the standard deviation and the weighted arithmetical mean are calculated for the distributions of pixel values for each color class. Table 1 show these values for the three different scenarios and three different colors.

The standard deviation for the different colors increases the more difficult lighting conditions become. The standard deviation for the hue values remains relatively constant for all illuminations, but the standard deviation values for saturation and even more for brightness increase significantly. Note f.i. the values for the green floor: During the night in Munich under constant fluorescent light, the whole playing field was illuminated quite uniformly. As a result, standard deviation for saturation was very low. The same location during midday shows an increased standard deviation for saturation. The increased value is possibly due to slowly varying illumination over time rather than a non-uniform lighting. Standard deviations increase also for the laboratory setting. Because these images are recorded during a short period, the increased standard deviation is the result of the non-uniform, very directed light, that throws hard shadows and leaves parts of the playing field darker. The weighted mean also strengthens the previous insight. Finally, an effect of the colored windows is that while the hue value for the blue goal remains almost unaffected by daylight influence during midday, the centers of the other colors change significantly.

Table 1. The standard deviation and the arithmetical weighted mean over the three dimensions in the HSV color space (hue, saturation and brightness). Each value is specified for four different (G=green, B=blue, Y=yellow and O=orange) color blobs.

	Hue				Saturation				Brightness			
	G	B	Y	O	G	B	Y	O	G	B	Y	O
Standard Deviation												
Evening	6.3	4.6	3.2	6.4	7.1	10.0	12.4	47.2	22.8	28.4	15.2	59.1
Midday	3.3	2.8	9.8	11.4	17.6	34.9	13.3	25.7	30.6	37.4	26.4	45.1
Laboratory	6.9	6.6	8.3	8.3	16.0	13.7	12.1	46.2	31.7	49.7	22.8	64.2
Weighted Mean												
Evening	151	210	44	4	49	62	112	96	157	116	181	167
Midday	193	218	78	352	94	119	62	64	169	175	160	155
Laboratory	176	206	64	5	67	37	37	99	160	116	85	174

Spatial Distributions of Idealized Color Classifications

To exclude the possibility that the results from the previous section are too strongly influenced by the quality of the manual calibration procedure, we compute spatial distributions of idealized color classifications and compare them with the previous results. For each scenario and each target color class, several images with different views of the object are selected. In each image, the object represented by the target color class is manually masked and assigned to the color class. Then, for each color class the spatial distribution of HSV color values were collected from the original images, just like in the previous section.

Figure 6 illustrates the spatial distributions in the three different setups for the target color classes green (on top-left, representing the carpet on the floor), blue (top-right, representing the blue goal), and yellow (bottom, representing the yellow goal). The three-dimensional structure is visualized using an isometric perspective. Note, that the colors used in these images do not represent target color classes, but different lighting situations. The green blob describes the object under fluorescent light, the blue cluster the same object under midday natural light conditions, and the red blob the same object under directed daylight in our laboratory. The different level of transparency in the plots describe different iso-levels of the density distribution, with a convex hull drawn over all points with the same histogram value (i.e. the opaque part has a higher density then the transparent part, but are the same cluster from the same color).

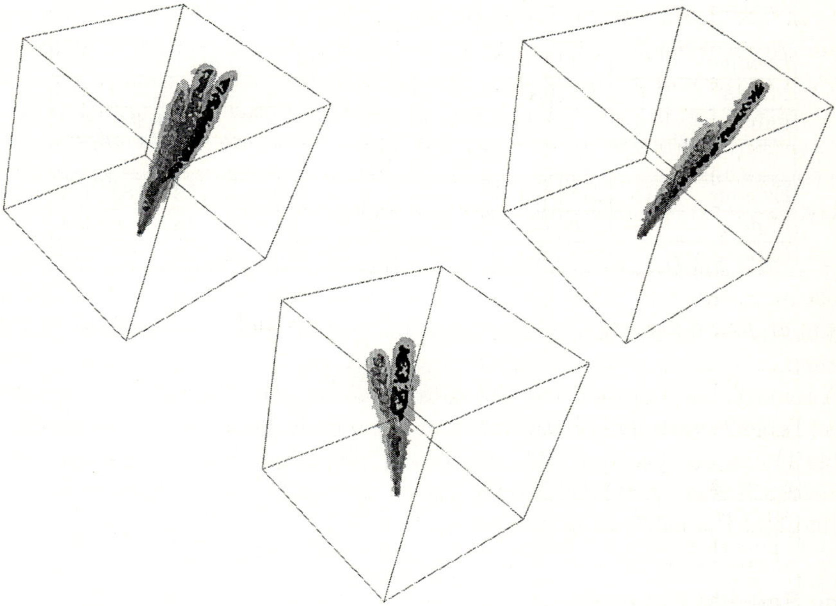

Fig. 6. Three-dimensional plots of the arrangement of the same object respectively under different illuminations within the HSV color space. The first image describe the green color, the top-right plot the blue goal and the bottom image the colors of the yellow goal.

The plots in Figure 6 seem to confirm the values from Table 1. For example, in the first image in Figure 6, displaying the spatial distribution of the green floor color, the opaque part of the red distribution (describing image set III) is spatially much larger in the brightness dimension then the others. This illustrates nicely the extremely varying illumination in this case. Note also, that for the yellow target color displayed in the bottom image, the red blob remains constrained on very low brightness values. This is explained in more detail in Section 5.2.

The same statistics – standard deviations of the different color channels within the HSV color space distribution and weighted mean – are calculated for the idealized color classifications and displayed in Table 2. Because the yellow and blue corner posts were also masked out of the images for the idealized color classification, the values for these two colors decrease slightly. The brightness value for the yellow goal in the lab setting remains on a very low level. The hue value for the blue goal is virtually the same, whereas for the other colors it changes significantly. The top-right plot in Figure 6 illustrates the large stan-

Table 2. The standard deviation and the arithmetical weighted mean over the three dimensions in the HSV color space (hue, saturation and brightness). Each value is specified for four different (G=green, B=blue, Y=yellow and O=orange) color blobs.

	Hue				Saturation				Brightness			
	G	B	Y	O	G	B	Y	O	G	B	Y	O
Standard Deviation												
Evening	6.5	3.9	2.9	3.6	8.3	18.0	12.0	29.5	24.5	26.4	16.4	42.3
Midday	3.6	3.3	6.7	28.6	19.1	40.0	12.1	26.4	32.0	49.7	28.7	50.0
Lab	10.7	13.1	8.4	25.4	18.9	7.9	7.4	59.3	41.9	10.9	9.8	85.1
Weighted Mean												
Evening	151	211	44	7	49	72	115	148	152	124	183	221
Midday	194	218	74	348	92	135	68	63	162	182	166	151
Laboratory	174	208	65	2	60	25	36	92	149	58	82	173

dard deviation for the blue goals during the games under natural light (the blue cluster).

A remarkable congruence of the data obtained from idealized color classifications (Table 2) with those obtained from a manually calibrated color classificator (Table 1) can be observed. This is a strong indication that the effects described in Section 5.2 are not significantly biased by our choice of color classification method and the calibration procedure.

Some Specific Examples

Finally, a few specific examples are presented which may serve as worst-case scenarios for certain aspects. The first example illustrates the problem of distance-dependent variations of color recognition. In Figure 7 two views on the yellow goal (within our laboratory) and the appropriately segmented images are shown. The yellow goal can only be detected in the second image, where the robot is closer to the goal. However, this problem is not due to a badly adjusted calibration of the color classificator. In the right image, parts of the background are already getting classified as blue, which indicates that the calibration is at the limit for the darker boundaries. Note that the yellow part of the corner post, which is lightened directly by the daylight through the window, can be detected in both images, while the yellow goal, whose side panel shadows the back part of the goal, cannot.

Another example illustrates the influence of the colored windows during our friendly game. Figure 8 shows two images with exactly the same robot with identical cyan color markers in front of the same white border. The left image is taken from image set I, the right image from image set II. The magnified details illustrate, how dramatically the color values are changing and especially the distance between the cyan and white values decreases. Under fluorescent light, the (spatial) arithmetical distance between the two magnified RGB values (255,255,255) and (199,255,255) is 56 (assuming all values are within a range of [0..255]). Under natural daylight conditions with dyed windows, the distance

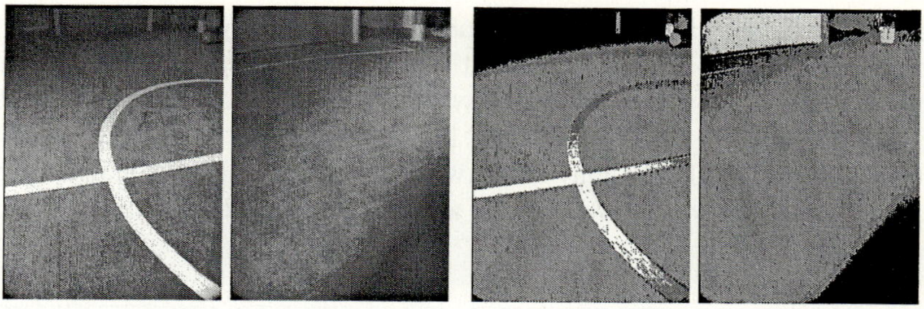

Fig. 7. Examples on how the same color can be detected differently dependent on the distance. The left two images shows the original recording, the right the color-segmented image.

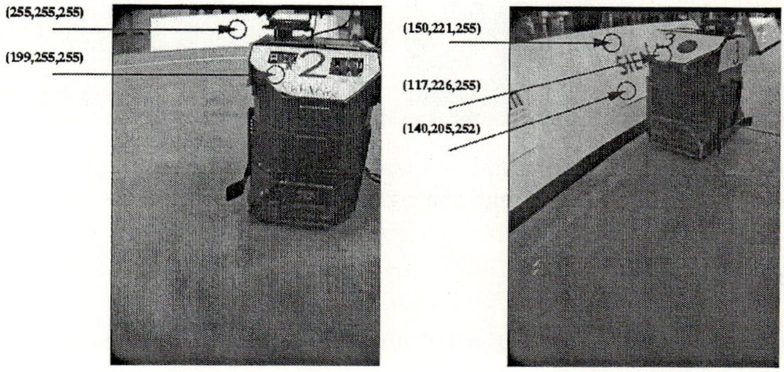

Fig. 8. Examples on how colored panes can influence the color. The left image shows the robot with color marker in front of a white border under fluorescent light, the right image the same robot in front of the same border, this time under daylight conditions.

between the upper detail (150,221,255) and the color marker detail (117,226,255) decreases to 33 or even to 31 between the color marker and the lower wall-detail. As an immediate consequence, we were not able to create a manual calibration, that allowed to separate the color markers reliably from other objects.

The last example illustrates the effects of quick changes in natural daylight illumination within a short time period. The images in Figure 9 were recorded within 14 minutes. The leftmost image looks much like the example in Figure 2 for the natural daylight scenario. Note the bluish coloring of the floor, caused by dyed windows. The second image shows almost the same view on the field a short time later on, when the sun was briefly covered by clouds, and the field was more or less illuminated by the additional fluorescent light. The RGB values for the green color range within each image illustrates this effect nicely: In the first image the weighted mean for all green values within the RGB color space is (81,190,168), whereas the weighted mean for the green color in the second image is (88,149,155). Note the strong shift for the green (middle) value. During such situations, a fixed calibration of the color classificator causes classification quality to deter significantly, as the right images of Figure 9 demonstrates. The second image from right segments the floor almost perfectly, assigning green color to most parts of the carpet, whereas the rightmost image shows a significant increase of the catch-all color class gray within the floor. Remember, that this effect happens in our case during only 14 minutes – and could happen in even shorter intervals, so that it may occur within a single half of a match, with no opportunity of team coaches to perform any kind of manual or manually initiated re-calibration.

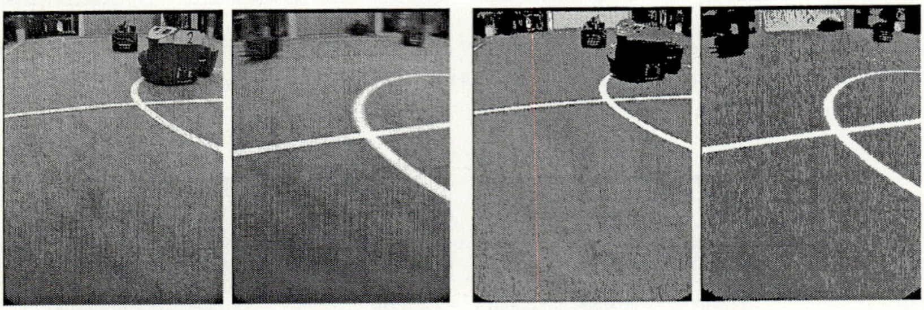

Fig. 9. Example on how the illumination may change within a short time. The left two images shows the original recording, the right the segmented images.

6 Conclusions

From the experiments and statistical analysis performed we can draw the following conclusions:

- Strong variations in field illumination have a strong effect on the spatial distribution of target color classes. Between certain situations differences are so strong that the spatial distribution of certain color classes do not even overlap any more (see Figure 4).
- Due to these strong variations of the target color classes in color space it will not be possible to obtain (neither manually nor by some automated procedure) a color classificator performing well under strongly varying lighting conditions. Making color class cuboids large enough to account for lighting variations causes too much noise and too many misclassifications for any particular, locally stable lighting setup.
- Because of most teams still being largely dependent on working color-based vision methods, introducing natural light to the field setup is likely to have dramatic effects on game performance.

We suggest to undertake the following steps in order to obtain robot vision systems which are more robust against lighting variations:

- As lighting will always be different on different playgrounds, teams should develop and use automated on-site vision calibration routines (see e.g. [1]), which the robot can perform autonomously when put into a new field.
- Another possible path to improve color-based image processing in RoboCup is to enhance and apply algorithms for improving color constancy. These

algorithms try to "stabilize" color recognition, so that the same colors look almost the same under different illuminations. Exemplary work in this area was performed by Forsyth [2,3], or by Jobson and Rahman [4,5].
– A third thread of work necessary for RoboCup is on using additional visual features aside of color for object detection and tracking, and possibly integrating them with color-based methods. The realtime processing requirements implied by the RoboCup setup pose a particular challenge, especially for many already existing methods in this area. A notable example for work in this area which has already been proven in several RoboCup competitions is by Hanek et. al. [6,7]. They use deformable models (snakes), which are fitted to known objects within the images by an iterative refining process based on local image statistics.

New results and methods in any of these areas would be of interest to everyone working on sophisticated applications in robotics, e.g. service robotics in normal, naturally lighted habitations or any variant of outdoor robotics. The RoboCup middle-size league should foster research in this area by defining appropriate technical challenges involving variations in lighting conditions.

Acknowledgment

The work described in this paper was done in project *Adaptivity and Learning in Teams of Cooperating Mobile Robots* supported by DFG SPP-1125. We thank Günther Palm and the University of Ulm for supporting our robot team, and the Agilo team from TU Munich (Michael Beetz, Sebastian Buck, Robert Hanek, Thorsten Schmitt and Freek Stulp) for inviting us to the adventure of holding a friendly game under daylight conditions and the pleasant teamwork we enjoyed with them.

References

1. Mayer, G., Utz, H., Kraetzschmar, G.K.: Towards autonomous vision self-calibration for soccer robots. Proceeding of the IEEE/RSJ International Conference on Intelligent Robots and Systems (IROS-2002) **1** (2002) 214–219
2. Forsyth, D.: A novel approach for color constancy. International Journal of Computer Vision **5** (1990) 5–36
3. Forsyth, D.: Sampling, resampling and colour constancy. In: Proceedings Computer Vision and Pattern Recognition. (1999) 300–305
4. Rahman, Z., Jobson, D.J., Woodell, G.A.: Multiscale retinex for color image enhancement. In: International Conference on Image Processing. (1996)
5. Jobson, D.J., Rahman, Z., Woodell, G.A.: Properties and performance of a center/surround retinex. (1997)
6. Hanek, R.: The contracting curve density algorithm and its application to model-based image segmentation. IEEE Conference Computer Vision and Pattern Recognition **1** (2001) 797–804
7. Hanek, R., Schmitt, T., Buck, S., Beetz, M.: Towards robocup without color labeling. RoboCup International Symposium 2002 (2002)

Tracking Regions

Felix von Hundelshausen and Raúl Rojas

Free University of Berlin, Institute of Computer Science
Takustr. 9, 14195 Berlin, Germany
{rojas,hundelsh}@inf.fu-berlin.de

Abstract. In this paper we present a simple and new algorithm that tracks the contour of several homogenous regions in a sequence of images. The method exploits the fact that, when i.e. observing a moving object (exposing a homogenous region), the regions in two consecutive frames often overlap. We show that the method is valuable for the RoboCup domain: It allows to track the green playing field and the goals very efficiently, to detect the white marking lines precisely, enabling us to recognize features in them (the center circle, the quatre circles, corners, the rectangle of the penalty area,...). It is also useful to find the ball and the obstacles. Furthermore, it provides data for path planning based on potential field methods without further computation. We compared the algorithm with the fastest existing method and measured a speed enhancement of 30 percent. In contrast to other methods, our algorithm not only tracks the center of blobs but yields the precise boundary shape of the objects as a set of point sequences. First tests with real world data have confirmed the applicability for other domains than RoboCup.

1 Introduction

The algorithm presented in this paper emerged from our aim to localize our middle size robot by detecting the marking lines on the playing field. Our robot is equipped with an omnidirectional catadioptric vision setup, with the camera looking upwards into a convex mirror. In order to recognize features like the center circle, the quatre circles, corners, rectangles, T-junctions, etc., it is important to detected the lines precisely in the images, without false positives and last but not least connected. The latter issue is important for fast feature detection. Only if the order of line points is known, we can efficiently calculate curvature measures and tangent directions, which is important for both, relative matching as described in [14], and feature recognition.

To detect the marking lines the straight forward way is to use an edge detector. Possible solutions could be based on the Roberts Cross Edge Detector[19], the Sobel Edge Detector [18], the Canny edge detector[6], the compass operator [20], edge detectors using the Laplacian of Gaussian, Gabor filters[16] or wavelet based solutions [15]. However, applying such a scheme to the whole image is time consuming. In particular, when processing 15 up to 30 frames per second for real-time vision, it is important to restrict the area within the image to which

D. Polani et al. (Eds.): RoboCup 2003, LNAI 3020, pp. 250–261, 2004.

the detector has to be applied. Furthermore, the problem of connecting the detected edge points remains [8]. Another approach, which can be found in [11], uses the Hough transform [10] for grouping, but this is not efficient.

To overcome the problem of linking model based prediction of edges has been used extensively over the last years [3]. However, the problem is that the model has to be known, and that the initial pose must be given. Although the determination of the initial pose is possible by propagation of a probability distribution [2], it is time consuming. Moreover, problems with matching under larger translations are known [7], since the models converge to local minima.

We approached the line detection problem differently: The idea is to detect the regions between the lines (see figure 1a). Since their boundaries neighbor the lines, they can easily be recognized: Just step around the boundaries, calculate the normal of the boundary curve at the actual position and apply a direction specific detector. For instance, search for a green-white-green transition.

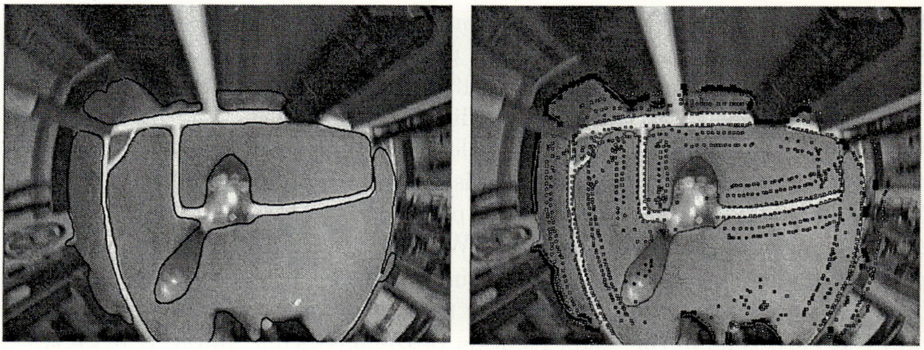

(a) (b)

Fig. 1. (a) The green regions are tracked. The boundaries of the regions are visualized by black lines. They are represented as sequences of points. (b) Objects are searched along the boundaries. Black points mark detected obstacles, for the marking lines three points are investigated along the normal. One on the line, and two at each side of the line to verify a green-white-green transition. Detected ball points are painted black for the sake of visibilitiy.

The method is also useful for detecting the obstacles and the ball: Just look for something orange or black next to the boundaries of the green regions (figure 1b). Doing so, false balls outside the playing field are not detected since they are not next to a green region in the image.

In this paper, we will show how the regions and their boundaries can be tracked very efficiently. We use the results of the last image to calculate the solution for the next, and on average only 10 percent of the pixels have to be accessed per frame. Our tracking algorithm is based on the region growing paradigm [4][22] [21][1][9][17]. However, we extend the paradigm to track regions over time.

The remainder of this paper is organized as follows: Section 2 reviews the basic region growing algorithm and describes the key observation that leads to the extension of the algorithm. In section 3 we describe the boundary extraction. Section 4 illustrates the generality of the algorithm. Section 5 shows results of the algorithm applying it within the RoboCup domain and comparing it with the currently most used tracking algorithm by robotic soccer vision systems[5]. Finally, section 6 concludes the paper.

2 Extending the Region Growing Paradigm

We first give a short review of a specific region growing algorithm which will be extended afterwards.

2.1 Region Growing by Pixel Aggregation

In *Region Growing by Pixel Aggregation* one starts with a small region (i.e. a single pixel) and consecutively adjoins pixels of the region's neighborhood as long as some homogeneity criterion is fulfilled. This method can be implemented in the following way: We reserve a two-dimensional array V which has the same size as the image and which has a boolean entry for each pixel that indicates whether the respective pixel has been visited yet. Further, we maintain a queue Q that stores the spreading boundary pixels (their coordinates) of the region during the execution of the algorithm. We refer to the stored elements in Q as *drops*, following the idea that we pour out a glass of water at the seed pixel and that the drops of this water spread over the region. Q just stores the boundary drops at each time during the execution of the algorithm. Initially, if we start with a single pixel, Q stores a single drop corresponding to the pixel which is also marked as visited in V. The algorithm continues with a loop which is performed as long as any drops are stored in Q. In each pass of the loop one drop is extracted from the queue and the neighboring pixels are investigated. In the case of a 4-neighborhood we inspect the top, right, bottom and left pixels. After assuring that the pixels have not yet been visited, respectively, we determine whether they hold for a specific homogeneity criterion, color similarity for instance. For each pixel that conforms with this condition a new drop is instantiated and stored in Q. After the loop terminates Q is empty and the region is marked in V. Figure 2 shows an example of a growing process that finds the region of the yellow goal.

2.2 The Key Observation

Our goal is to efficiently track regions over time. To give a specific example, we want to track the yellow goal, while the robot moves. Assume, that the robot starts at pose P_A and takes an image I_A. Then the robot moves a little to pose P_B and takes another image I_B. Assume further that we have determined the region A of the yellow goal in I_A and the corresponding region B in I_B by the above

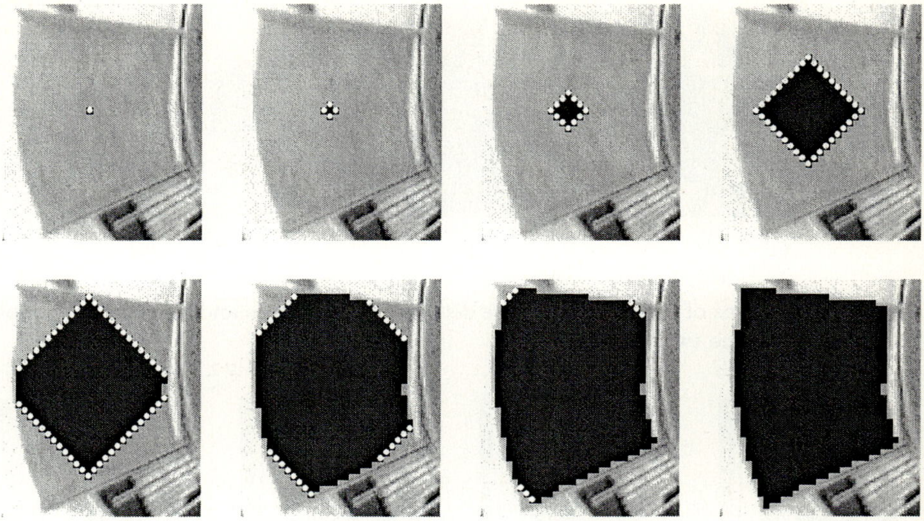

Fig. 2. The array V is visualized by the dark gray surface (visited elements). The white circles represent the drops. Here, the algorithm does not work on single pixels, but on blocks of 4×4 pixels.

region growing algorithm. If the video frequency is high enough with respect to the movement of the robot, then the two regions will overlap as depicted in figure 3. If we apply the region growing algorithm separately to image A and B, then the running time for each is linear in the number of pixels (or blocks of pixels) within the respective regions. We want to develop a more efficient method. Assume that we have extracted region A, then roughly speaking, we want to use the drops of the region growing of A to somehow flow to the region of B by making use of the overlap. The drops of region A should first shrink to the intersection S of A and B and then grow to the boundary of B. Thus, in order to find region B we don't start with a seed pixel, but with a whole seed region, the intersection of A and B. To realize the algorithm a method of how to shrink region A to the intersection of A and B has to be developed. This will be done in the following subsection.

2.3 Shrinking Regions

We will develop the shrinking method in two steps. First we consider the case of shrinking a region without stopping criterion. That is, the region shrinks until it vanishes. Next, we modify the method so that shrinking stops at the intersection of A and B.

In the first step, we don't need any image information but just the array V in which region A is marked and a queue of drops at the boundary of that region. As we will need two different queues for growing and shrinking later, we denote the queue used here as Q' to avoid confusion. We apply the same operations as in region growing with one exception: We reverse the meaning of V. As a

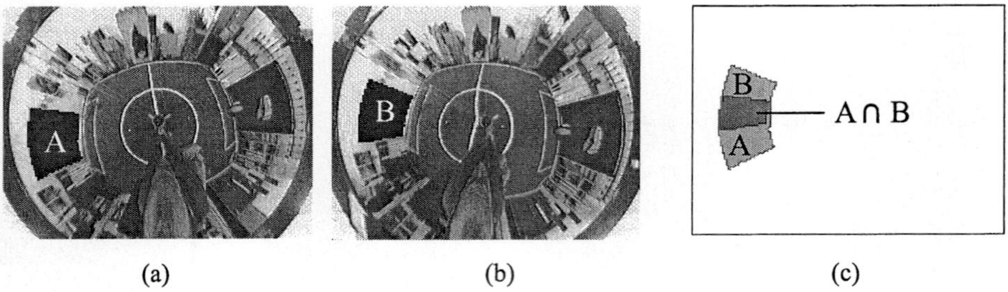

Fig. 3. The regions of the yellow goal are determined (a) before and (b) after the robot has moved. (c) The two regions overlap.

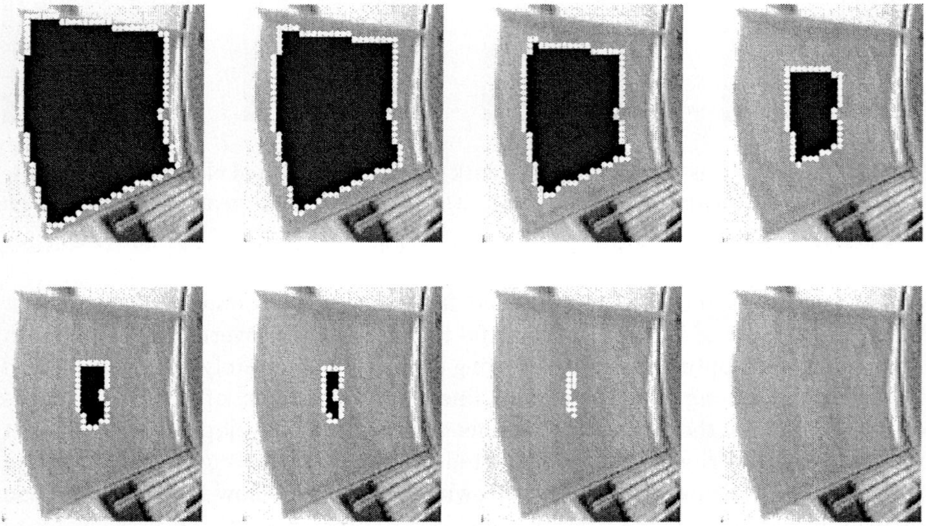

Fig. 4. The region shrinks until it vanishes. For this process no image information is used.

result, the drops can only spread within the region marked in V and since they initially are placed at the boundary their only means of escape is to move from the outer to the inner of the region. Instead of marking elements in V, as in region growing, entries in V are cleared while the drops spread. At the end, Q' is empty and V is cleared completely. This process is illustrated in figure 4.

In the second step we use image I_B to determine when shrinking should stop. We emphasize that the initial region (marked in V) is due to image I_A while only image I_B is referenced during shrinking. Each time after a drop has been extracted from Q' we verify the homogeneity criterion for the corresponding pixel in image I_B. If the pixel belongs to the region, then the drop is not allowed to spread anymore. As a result the region shrinks to the intersection of region A and B as depicted in figure 5. This is exactly inverse to the growing, where drops only spread to neighbors that fulfill the homogeneity criterion.

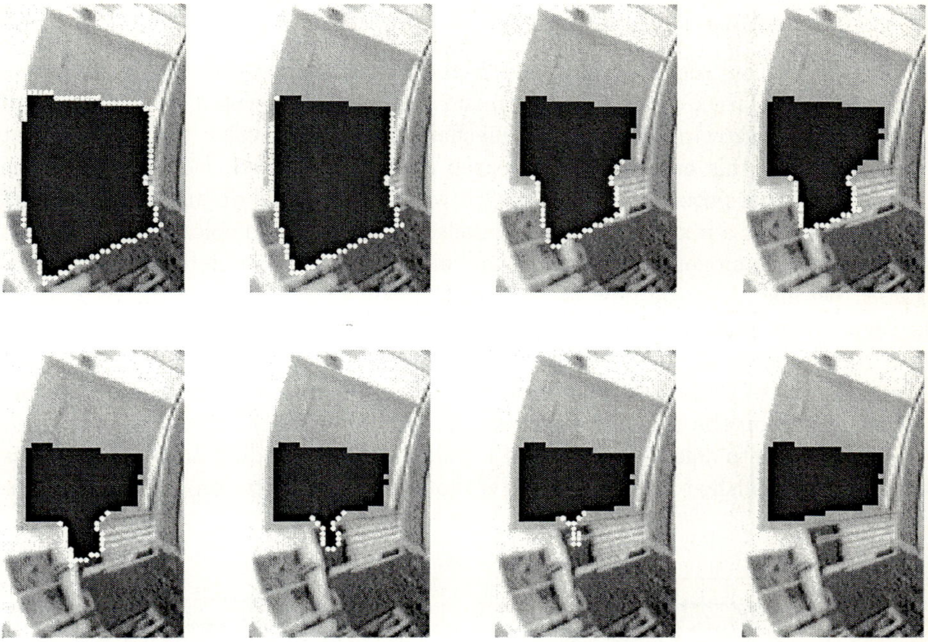

Fig. 5. The region shrinks up to the intersection area, that means, until all drops are within the region of the new image.

2.4 Alternating Shrinking and Growing

Region growing is a well known technique which was extensively studied 20 years ago. However, our new contribution is that region growing can also be used to shrink regions and that alternation of shrinking and growing allows to track large regions extremely efficiently.

To alternate shrinking and growing in order to track regions some problems concerning the interface between the two stages must be solved. The first problem is that after growing the queue of drops is empty but shrinking initially requires a list of drops at the boundary of the region. In the same way shrinking ends with an empty list of drops, but growing requires seed drops. To solve this problem each of the two processes, growing and shrinking, has to build the initial queue for the other procedure. We accomplish this by using two queues, Q and Q'. Growing assumes the initial drops to be in Q and after execution Q is empty and Q' has been built up which stores the initial drops for shrinking. Shrinking runs with these drops and initializes Q for the next growing.

During growing, when the neighbors of an extracted drop are inspected, the drop is inserted into Q' as initial drop for shrinking, if any of its neighbors does not belong to the region. Vice versa, a drop that is extracted from Q' during shrinking is inserted to Q as initial drop for growing, if shrinking stops for this drop.

2.5 Controlling the Tracking

Since the tracked regions can be lost if the movement between two consecutive images is too large, a control mechanism has to be integrated into the loop of shrinking and growing. The control mechanism checks whether tracking has lost the region. In this case an initial search has to be started. Depending on the application this procedure might search within the whole or just within a part of the image for a pixel or a block of pixels having a certain color or texture that satisfies the homogeneity criterion and maybe some other detection criterion. Later, we also describe how to extract the polygonal shape of the regions. If a region gets lost, and an initial search is started, then first several regions can be tracked and by using the information about the size and shape of the regions some of them can be discarded. In this way the computational power is concentrated on the regions of interest. To exclude a region from being tracked one simply has to delete its entries in V and the corresponding drop queues. This can be accomplished by a shrinking without stopping criterion as illustrated in figure 4.

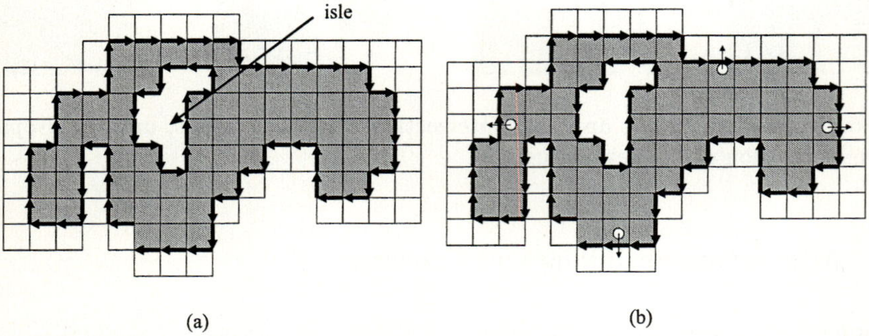

Fig. 6. (a) The boundary of a region is composed of a sequence of edge vectors. (b) The direction of the edge vectors depend on the direction of the spreading at which they are detected during the growing phase.

3 Boundary Extraction

The boundary of each region consists of a chain of small edge vectors (see figure 6a). Each edge vector represents one of the four sides of a pixel and the last edge vector ends at the beginning of the first. During the growing phase, when a drop reaches a border, the corresponding edge vectors are inserted into a special data structure, the *connectivity grid*. It is a two-dimensional array, one wider and one higher than the image. The cells do not correspond to the pixels in the image but to the inter-pixel positions (corners of pixels), respectively. Each cell has 4 bits, marking whether an edge vector starts at the corresponding position, directed up, right, down or left, respectively. Initially, the grid is cleared. After growing the grid contains the edges. Now, starting at any edge, one can follow the edges through the grid and clear them in the same pass. Since situations appear where an edge has two possible successors, it is important to apply a rule: With respect to the current direction, always choose the most left/right turning possibility. In this way, it is guaranteed that all chains of edges are closed. After having extracted the edges, the connectivity grid again is cleared and ready for the next insertion process. This approach is similar to [4], but our method occupies less memory, since [4] employs a supergrid of size $(2n + 1) \times (2m + 1)$ and we just use a grid of size $(n + 1) \times (m + 1)$, where $n \times m$ is the size of the image. The result of boundary extraction is as set of point sequences, each forming a closed curve.

4 Homogeneity Criterion

It is important to understand, that the described algorithm works with any homogeneity criterion. We will give some examples. For RoboCup, our images are in YUV 422 format. The YUV bytes of a pixel (24 Bit) are an index into a 16 MB big lookup table. Each entry in the LUT is one byte big, and each of the 8 bits represent a color class. We refer to a pixel's LUT-entry as its class mask. Verifying whether the pixel supports a certain class or collection of classes is then possible with a single AND-operation. To be robust against image noise, we do not run our region-tracking algorithm on single pixels, but on blocks of 4×4 pixels, in case of an image resolution of 640×480. Then, for tracking the green field, we define a block to belong to the green region, if the number of "green" pixels exceed a certain threshold (12 is a typical value). We have tested this with extremely noisy images and we can assert that it works well.

However, the choice of homogeneity criterion is free. For instance, one could define that two blocks of pixels are homogeneous, if there texture is similar (assuming a texture classifier). One could also define, that two pixels are homogeneous, when an edge detector at the respective position yields low response. The region tracking algorithm will run with any criterion. However, it is the task of future research, to find the best criteria, and more important, how the criteria can be calibrated automatically and adapt to changing lighting conditions.

5 Results

The advantage of our region tracking algorithm is that only the non-overlapping parts of corresponding regions of successive images are processed. However, if the regions don't overlap, the algorithm has to lunch initial searches in each frame and degenerates. Therefore the question is, how often do the regions overlap? Of course, the answer depends on the application. In the following, we present results from our RoboCup application. Here, we have tracked three different types of regions: the green playing field and the blue and yellow regions (goal and post markings). While the robot moved through the environment, we computed the fraction of the processed area with respect to the size of the image and the percentage of overlapping of the tracked regions. We determined these values for each pair of successive frames and built the respective average over a longer time

Table 1. The average fraction over a sequence of frames of the processed area with respect to the size of the image and the percentage of the overlapping area with respect to the area of the tracked regions has been determined for different movements and speeds. The last two columns indicate the number of frames and the number of initial searches that had to be started.

Rotation	processed fraction of the image	% overlapping	frames	initial searches
$56^\circ/s$	8.0%	80%	181	11
$74^\circ/s$	8.4%	77%	142	11
$110^\circ/s$	8.8%	71%	97	13
$145^\circ/s$	10.0%	65%	74	13
$200^\circ/s$	11.0%	60%	62	25
Translation	processed fraction of the image	% overlapping	frames	initial searches
$0m/s$	6.3%	93%	138	0
$0.26m/s$	5.9%	89%	114	0
$0.38m/s$	6.8%	84%	82	2
$0.52m/s$	7.0%	82%	61	0
$0.65m/s$	7.2%	80%	48	0
$0.74m/s$	6.9%	81%	42	1
$0.83m/s$	7.9%	75%	40	0
$1.0m/s$	7.8%	73%	31	1
$1.0m/s$	7.7%	79%	24	2

span (see table 1). We also counted the number of initial searches and repeated the experiment for different movements (rotation/translation) and speeds of the robot. Here, we should mention that our robots have omnidirectional wheels and are able to rotate and translate into any direction at the same time.

The tracked regions overlap greatly and only a small fraction of the image data is accessed (below 10%). As expected, the rotation is more disadvantages then the translation. This is because objects that are far away from the robot (as the goals) have a high speed in the image, if the robot rotates. Therefore the regions might not overlap, and initial searches have to be executed. With a rotational speed of $200^\circ/s$ 25 initials searches had to be done within 82 frames. Assuming that an initial search requires accessing all the pixels in the image, the algorithm yet performs well, compared with common methods, which access all the pixels in each frame.

Although the theoretic running time of our tracking algorithm is evidently faster than any algorithm that touches all the pixels in each image, the question is whether this is also true for the practical running time, since the maintenance of the drops, the queues and the connectivity grid might be more time consuming than a straightforward implementation.

Therefore, we decided to compare the algorithm with the color segmentation algorithm proposed in [5], which is believed to be the fastest color tracking algorithm at the moment and which is applied by most teams in RoboCup. The algorithm is based on classifying each pixel into one or more of up to 32 color classes using logical *AND* operations and it employs a tree-based *union find* with path compression to group runs of pixels having the same color class. We calibrated both algorithms to track green, blue and yellow regions. We did not track the orange ball and black obstacles, because our system uses different methods than color class tracking to recognize them. The following table gives the absolute running times of the algorithms over 1000 frames with the robot moving. Each image consist of 640 × 480 pixels (*YUV 4:2:2*) and we applied a Pentium III 800 Mhz processor.

CMU Algorithm	Our Algorithm
37.47 s	22.67 s

Thus, our algorithm is significantly faster. Moreover, our algorithm also extracts the contour curves of all regions at each frame.

However, we do not claim that our algorithm performs better in all possible cases. There might be applications where the other algorithm performs better. This will typically be in cases, where many small and fast moving regions are to be tracked.

6 Conclusion

We have proposed a new method that is able to efficiently track and extract the boundaries of several homogeneous regions. The efficiency is accomplished by not processing the entire images but concentrating on parts where regions are expected. The algorithm exploits the fact that corresponding regions in successive images often overlap and it extends the region growing paradigm: Tracking is accomplished by alternating shrinking and growing. We provide a homepage for the algorithm, with source code available in *C++* and demo videos[1].

The algorithm is also very useful for edge extraction and tracking (also for other features). The advantage is that feature detectors can be very selective and that they have to be applied at a few locations only. For instance, when searching for edges, a detector which responds to edges having a predefined direction can be used. This is possible, because the boundary contour of each region and their corresponding normal directions are computed by our method.

We have also demonstrated the application of the algorithm in a practical problem. In our real-time vision system for RoboCup we use the algorithm to track regions like the goals, in order to locate and recognize these objects. However, we also use the method to detect the marking lines on the playing field, which are matched to a predefined model for the purpose of robot self-localization. What makes the algorithm practical is, that both, region and edge tracking, can be accomplished in the same run.

[1] http://page.inf.fu-berlin.de/~hundelsh/research/RegionTracking/index.htm

The method also exposes a useful interface for higher level algorithms. For instance, visual attention can be implemented by controlling which regions should be tracked. Here the primary issue is that when excluding a region from tracking, the algorithm is really faster, since less pixels are accessed in the images. Another interface is the extracted boundary contour. Since the algorithm is based on region growing connected boundaries are guaranteed. The boundary can serve as a base for recognition of objects and movements or as a reference into the images. In RoboCup for instance, we track the regions of the green playing field and the boundary helps us to find other objects on the playing field. This is because all objects like the ball and other robots being on the playing field are next to the green regions in the images. Therefore their boundaries can be used as a reference into the image where to search for the objects and they can be rapidly detected. There is another advantage of the method, concerning the interface between vision and path planning. When using potential field methods for path planning [12], defining the field is an expensive operation. The occupancy grid for the tracked green field can be used as potential field. Obstacles must not be inserted, since they are not part the green region. However, the marking lines must be inserted to the grid to allow the robot to pass over them. When using omnidirectional vision, there is also the advantage that the resolution of the path planning will be high for near but rough for distant positions.

There are three open questions future work should concentrate on. The first is related to the homogeneity criterion. In this paper we have assumed a predefined criterion, such as certain color classes for instance. However, rigidly defining a homogeneity criterion will result in an inflexible algorithm. How can the homogeneity criterion be automatically defined? This is not just a question of finding some thresholds but also of the kind of criterion to be used (color, texture,...). Related to the homogeneity criterion is the question of scale [13]. How many pixels should constitute a drop? If, for instance, there was a region whose pixels had all the same color, then the algorithm could work on single pixels and the homogeneity criterion could be based on color difference. But, if texture is present, more than a single pixels has to serve as a unit. The last question concerns the correspondence of regions over time. Can the fact of overlap be exploited for correspondence definition?

References

1. Rolf Adams and Leanne Bischof. Seeded region growing. *IEEE Transactions on Pattern Analysis and Machine Intelligence*, 16(6):641–647, 1994.
2. Neils Gordon Arnaud Doucet, Nando de Freitas, editor. *Sequential Monte Carlo Methods in Practice.* Statistics for Engineering and Information Science. Springer, 2001. ISBN 0-387-95146-6.
3. A. Blake and M. Isard. *Active Contours.* Springer-Verlag, 1998.
4. Claude R. Brice and Claude L. Fennema. Scene analysis using regions. *Artificial Intelligence*, 1:205–226, 1970.
5. James Bruce, Tucker Balch, and Manuela Veloso. Fast and inexpensive color image segmentation for interactive robots. In *Proceedings of IROS-2000, Japan*, October 2000.

6. John Canny. A computational approach to edge detection. *IEEE Transactions on Pattern Analysis and Machine Intelligence (PAMI)*, 8:679–698, 1986.
7. Laurent D. Cohen. On active contour models and balloons. *CVGIP:IU*, 1991.
8. G. W Cook and E J Delp. Multiresolution sequential edge linking. *Proceedings of the IEEE International Conference on Image Processing*, pages 41–44, October 1995.
9. S. A. Hojjatoleslami and J. Kittler. Region growing: a new approach. *IEEE Transactions of Image Processing*, 7(7):1079–1084, 1998.
10. P. V. C. Hough. Method and means for recognizing complex patterns. US Patent 3,069,654, December 1962.
11. P Jonker, J Caarls, and W Bokhove. Fast and accurate robot vision for vision based motion. *Lecture Notes in Computer Science*, 2019:149–155, 2001.
12. J Latombe. *Robot Motion Planning.* Kluwer Academic Publishers, 1991. ISBN 0-7923-9206-X.
13. T. Lindeberg. *Scale-Space Theory In Computer Vision.* Kluwer Academic Publishers, 1994.
14. Feng Lu and Evangelos Milios. Robot pose estimation in unknown environments by matching 2d range scans. *CVPR94*, pages 935–938, 1994.
15. S. Mallat and S. Zhong. Characterization of signals from multiscale edges. *IEEE Transactions on Pattern Analysis and Machine Intelligence (PAMI)*, 14(7):710–732, 1992.
16. R. Mehrotra, K. R. Namudura, and N. Ranganathan. Gabor filter-based edge detection. *Pattern Recognition*, 25:1479–1494, 1992.
17. Theo Pavlidis. Integrating region growing and edge detection. *IEEE Transactions on Pattern Analysis and Machine Intelligence*, 12(3):225–233, 1990.
18. K. K. Pingle. Visual perception by computer. In A. Grasselli, editor, *Automatic Interpretation and Classification of Images*, pages 277–284. Academic Press, New York, 1969.
19. L. G. Roberts. Machine perception of three-dimensional solids. In J. T. Tippet et al., editor, *Optical and Electro-Optical Information Processing*, pages 159–197. MIT Press, Cambridge, 1965.
20. Mark A. Ruzon and Carlo Tomasi. Color edge detection with the compass operatro. *IEEE Conference on Computer Vision and Pattern Recognition*, 2:160–166, 1999.
21. A. Siebert. Dynamic region growing, 1997.
22. S. W. Zucker. Region growing: Childhood and adolescence. *Computer Graphics and Image Processing*, 5:382–399, 1976.

Fast and Robust Edge-Based Localization in the Sony Four-Legged Robot League*

Thomas Röfer[1] and Matthias Jüngel[2]

[1] Bremer Institut für Sichere Systeme, Technologie-Zentrum Informatik, FB 3
Universität Bremen
roefer@tzi.de
[2] Institut für Informatik, LFG Künstliche Intelligenz
Humboldt-Universität zu Berlin
juengel@informatik.hu-berlin.de

Abstract. This paper presents a fast approach for edge-based self-localization in RoboCup. The vision system extracts edges between the field and field lines, borders, and goals following a grid-based approach without processing whole images. These edges are employed for the self-localization of the robot. Both image processing and self-localization work in real-time on a Sony Aibo, i. e. at the frame rate of the camera. The localization method was evaluated using a laser range sensor at the field border as a reference system.

1 Introduction

The Sony Four-Legged Robot League (SFRL) is one of the official leagues in RoboCup. Besides the use of four-legged robots, there are some other specialties in that league. The first one is that the robot platform is standardized, i. e. the Sony Aibo ERS-210 and ERS-210A (cf. Fig. 1a) are the only permitted systems, and they can only be used without any modification. Therefore, in some sense the SFRL can be seen as a *software league*, because it is neither possible nor required to construct robots. Another characteristic is that the robots are completely autonomous, i. e. there is no external computer beside the field (except from one running the so-called *game manager* for the referee) that can help the players in their calculations. The main sensor of the Sony Aibo is the camera located in its head. The head can be turned around three axes (tilt, pan, and roll), and the camera has a field of view of 58° by 48°. Thus, all teams in the league have to tackle the problem of *directed vision* (in contrast to *omni-vision* as often used in the middle-sized league or in the first approaches of the small-sized league to local vision systems). With 20 degrees of freedom, the color camera, and more than 30 further sensors, the movements and the sensor equipment of the robots are the most complex in RoboCup so far, and they have to be controlled by a single 200 MHz MIPS processor (400 MHz in the ERS-210A), i. e. all algorithms

* The Deutsche Forschungsgemeinschaft supports this work through the priority program "Cooperating teams of mobile robots in dynamic environments".

D. Polani et al. (Eds.): RoboCup 2003, LNAI 3020, pp. 262–273, 2004.

a) b)

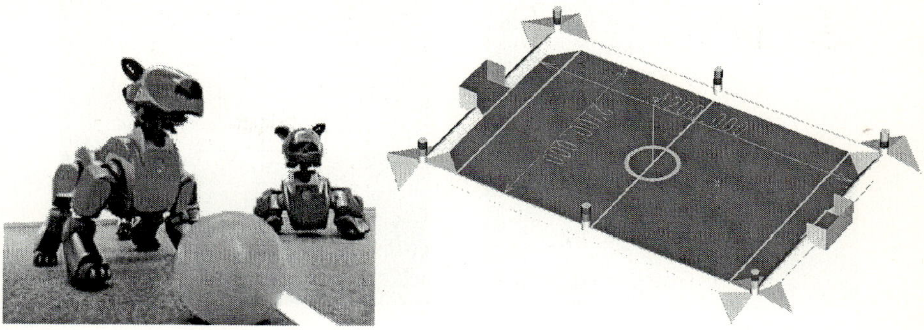

Fig. 1. a) Two Sony Aibo robots and a ball. b) The field used in the SFRL.

used, e. g. for image-processing or self-localization, have to be highly efficient to run in real-time.

The soccer field in the SFRL has a size of approximately 5m×3m (cf. Fig. 1b). As the main sensor of the robot is a camera, all objects on the RoboCup field are color coded. There are two-colored flags for localization (pink and either yellow, green, or skyblue), the two goals are of different color (yellow and skyblue), the ball is orange (as in all RoboCup leagues), and the robots of the two teams wear tricots in different colors (red and blue). However, there are no flags on a real soccer field, and as it is the goal of the RoboCup initiative to compete with the human world champion in 2050, it seems to be a natural thing to develop techniques for self-localization that do not depend on artificial clues. In the SFRL, all teams have to participate in three technical challenges as part of the RoboCup championship. In 2003, self-localization without the six two-colored flags around the field is one of these challenges. This challenge can be seen as a preparation to remove the flags in the soccer games in 2004.

2 Grid-Based Line Detection

The localization method presented in this paper relies on the detection of edges between differently colored objects on the field: the edges between the skyblue goal and the field, the edges between the yellow goal and the field, the edges between the border and the field, and the edges between the field lines and the field (cf. Fig. 2a). The key idea of the method presented here is not to actually extract *lines* from the image, but *pixels on lines* instead. This approach is faster and more robust against misinterpretations, because lines are often partially hidden either by other robots or due to the limited opening angle of the camera.

A very common preprocessing step for vision-based object recognition is color segmentation using color tables, e. g. [1, 9]. Such methods directly map colors to color classes on a pixel by pixel basis, which has some crucial drawbacks. On the one hand, the color mapping has to be adapted when the lighting conditions change, on the other hand, the mapping results in a loss of information, because

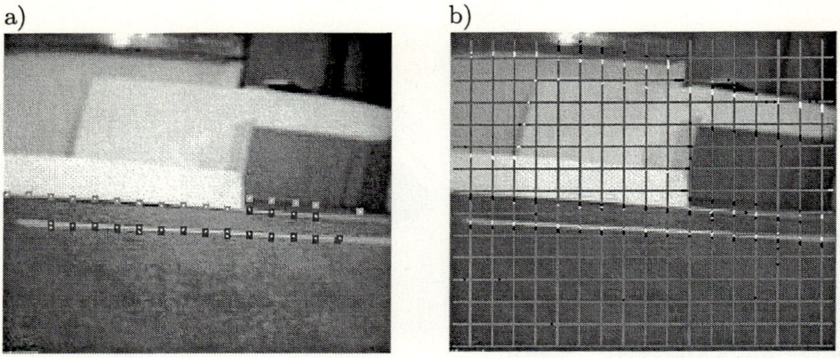

Fig. 2. Detection of lines. a) Three types of lines: field/goal, field/border, and field/line. b) The scan lines are scanned from top to bottom and from left to right. White pixels: increase in Y-channel, black pixels: decrease in Y-channel.

the membership of a pixel in a certain class is a yes/no decision, ignoring the influences of the surrounding pixels. Some researchers try to overcome these limitations [5], but the solutions are too slow to work under real-time conditions on a robot such as the Aibo.

The key ideas of the image-processing method used in this paper are that speed can be achieved by avoiding processing all pixels of an image, and a certain independence of the lighting conditions can be reached by focusing on contrast patterns in the three different color channels. In case of the Aibo, these channels are Y, U, and V.

To find pixels on edges, in the image horizontal and vertical lines having a distance of ten pixels to each other are scanned from left to right and from top to bottom following the method described in [6] (cf. Fig. 2b). In contrast to this method color classification is only applied when a significant decrease in the Y-channel is recognized, because the field is darker then the adjacent surfaces of the field lines, the border, and the goals. If such a decrease in brightness has been detected, the colors above and below are this edge are checked for being green, white, skyblue, or yellow using a color table (cf. [7] for a solution to this problem without using color tables).

If the color above the decrease in the Y-channel is skyblue or yellow, the pixel lies on an edge between a goal and the field. The differentiation between a field line and the border is a bit more complicated. In most of the cases the border has a bigger size in the image than a field line. But a far distant border might be smaller than a very close field line. For that reason the pixel where the decrease in the Y-channel was found is assumed to lie on the ground. With the known height and rotation of the camera the distance to that point is calculated by projecting it to the ground plane. The distance leads to expected sizes of the border and the field line in the image. For the classification these sizes are compared to the distance between the increase and the decrease of the Y-channel in the image. The projection of the pixels on the field plane is also used to determine their relative position to the robot (cf. Fig. 3).

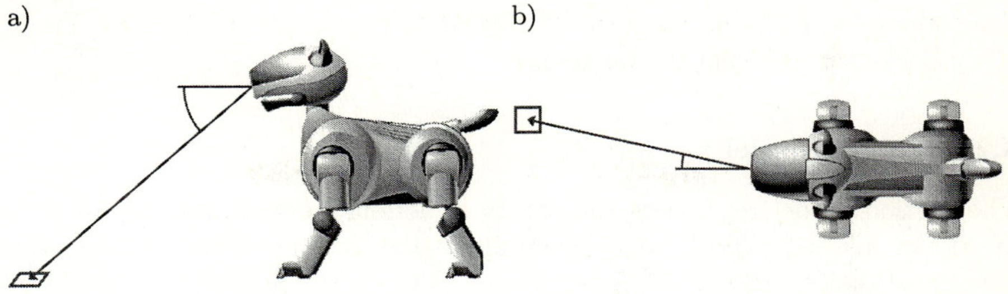

Fig. 3. The projection of edge points to the field plane. a) Vertically. b) Horizontally.

3 Self-localization Based on Edge Points

An approach to self-localization is the so-called Monte-Carlo localization (MCL) by Fox *et al.* [3]. It is a probabilistic method, in which the current location of the robot is modeled as the density of a set of particles. Each particle can be seen as the hypothesis of the robot being located at that position. Therefore, such particles mainly consist of a robot pose (x, y, θ), i. e. a vector representing the robot's x/y-coordinates and its rotation θ.

In many implementations, MCL was used on robots equipped with distance sensors such as laser scanners or sonar sensors, e. g. in the original one [3]. Only in a few approaches, vision is used for self-localization [2, 11]. Self-localization in RoboCup is different, because the area the robots can be located at is relatively small, i. e. the field, but in that area the position of the robots has to be determined quite precisely to allow different robots of the same team to communicate about objects on the field, and to follow some location-based rules of the game. Odometry is very unreliable, because the robots walk, and they tend to push each other around. As the Aibo is equipped with a sensor with a narrow opening angle of 58°, only a few objects usable for self-localization can be seen at once, and sometimes misreadings are in the majority. The method presented here takes these circumstances into account.

3.1 Monte-Carlo Localization

A Markov-localization method requires both a *motion model* and an *observation model.* The motion model expresses the probability for certain actions to move the robot to certain relative positions. The observation model describes the probability for taking certain measurements at certain locations.

The localization approach works as follows: first, all particles are moved according to the motion model of the previous action of the robot. Then, the probabilities q_i are determined for all particles on the basis of the observation model for the current sensor readings. Based on these probabilities, the so-called *resampling* is performed, i. e. moving more particles to the locations of samples with a high probability. Afterwards, the average of the probability distribution is determined, representing the best estimation of the current robot pose. Finally, the process repeats from the beginning.

3.2 Motion Model

The motion model represents the effects of actions on the robot's pose. First of all, an odometry position is maintained that is derived from the motions performed (gaits, kicks, etc.). As this value is only a rough estimate, in addition a random error Δ_{error} is assumed that depends on the distance traveled and the rotation performed since the last self-localization. For each sample, the new pose is determined as $pose_{new} = pose_{old} + \Delta_{odometry} + \Delta_{error}$. Note that the operation + involves coordinate transformations based on the rotational components of the poses.

3.3 Observation Model

The localization is based on the points on edges determined by the image-processing system (cf. Sect. 2). Each pixel has an edge type (field, border, yellow goal, or blue goal), and by projecting it on the field, a relative offset from the body center of the robot is determined. Note that the calculation of the offsets is prone to errors because the pose of the camera cannot be determined precisely. In fact, the farther away a point is, the less precise the distance can be determined. However, the precision of the direction to a certain point is not dependent on the distance of that point.

Information Provided by Edge Points. The four edge types provide very different information: *The field lines* are mostly oriented across the field. As lines only provide localization information perpendicular to their orientation, the field lines can only help the robot to find its position along the field. The field lines are seen less often than the border. *The border* is surrounding the field. Therefore it provides information in both Cartesian directions, but it is often quite far away from the robot. Therefore, the distance information is less precise than the one provided by the field lines. The border is seen from nearly any location on the field. *Goals* are the only means to determine the orientation on the field, because the field lines and the border are mirror symmetric. The goals are seen only rarely.

If the probability distribution for the pose of the robot had been modeled by a large set of particles, the fact that different edges provide different information and that they are seen in different frequency would not be a problem. However, to reach real-time performance on an Aibo robot, only a small set of samples can be employed to approximate the probability distribution. In such a small set, the samples sometimes behave more like individuals than as a part of joint distribution. To clarify this issue, let us assume the following situation: as the field is mirror symmetric, only the recognition of the goals can determine the correct orientation on the field. Many samples will be located at the actual

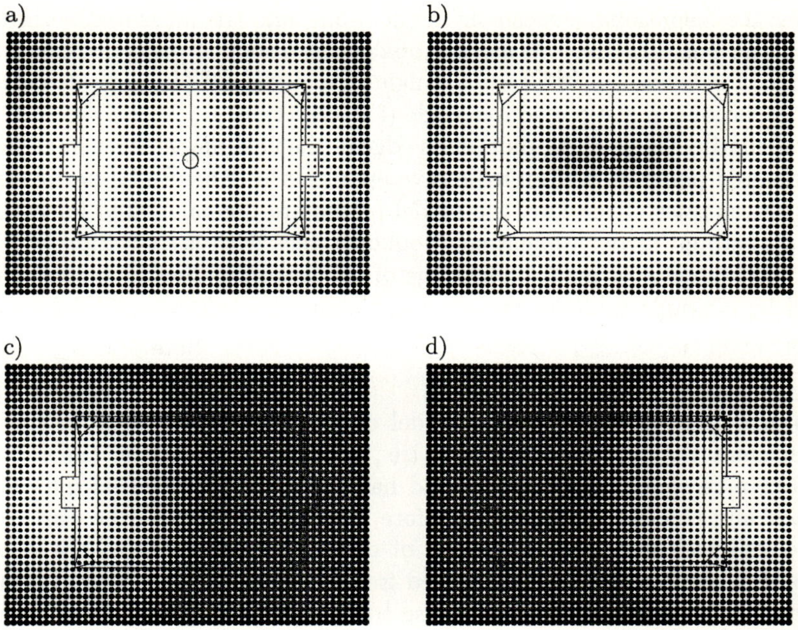

Fig. 4. Distances from edges. Distance is visualized as thickness of dots. a) Field lines. b) Border. c) One goal. d) The other goal.

location of the robot, but several others are placed at the mirror symmetric variant, because only the recognition of the goals can discriminate between the two possibilities. For a longer period of time, no goal is detected, but the border and the field lines are seen. Under these conditions, it is possible that the samples on the wrong side of the field better match the measurements of the border and the field lines than the correctly located ones, resulting in a higher probability for the wrong position. So the estimated pose of the robot will flip from one orientation alternative to the other without ever seeing a goal. This is not the desired behavior, and it would be quite risky in actual soccer games.

To avoid this problem, separate probabilities for field lines, borders, and goals are maintained for each particle.

Closest Model Points. The projections of the pixels are used to determine the three probabilities of each sample in the Monte-Carlo distribution. As the positions of the samples on the field are known, it can be determined for each measurement and each sample, where the measured points would be located on the field if the position of the sample was correct. For each of these measured points in field coordinates, it can be calculated, where the closest point on a real field line of the corresponding type is located. Then, the horizontal and vertical angles from the camera to this model point are determined. These two angles of the model point are compared to the two angles of the measured point. The smaller the deviations between the model point and the measured point from a hypothetic position are, the more probable the robot is really located at that position. Deviations in the vertical angle (i. e. distance) are judged less rigidly than deviations in the horizontal angle (i. e. direction).

Calculating the closest point on an edge in the field model for a small number of measured points is still an expensive operation if it has to be performed for, e. g., 100 samples. Therefore, the model points are pre-calculated for each edge type and stored in two-dimensional lookup tables with a resolution of 2.5 cm. That way, the closest point on an edge of the corresponding type can be determined by a simple table lookup. Figure 4 visualizes the distances of measured points to the closest model point for the four different edge types.

Probabilities. The observation model only takes the bearings on the edges into account that are actually seen, i. e., it is ignored whether the robot has *not* seen a certain edge that it should have seen according to its hypothetical position and the camera pose. Therefore, the probabilities of the particles are only calculated from the similarities of the measured angles to the expected angles. Each similarity s is determined from the measured angle ω_{seen} and the expected angle ω_{exp} for a certain pose by applying a sigmoid function to the difference of both angles weighted by a constant σ:

$$s(\omega_{seen}, \omega_{exp}, \sigma) = \mathrm{e}^{-\sigma(\omega_{seen}-\omega_{exp})^2} \qquad (1)$$

If α_{seen} and α_{exp} are vertical angles and β_{seen} and β_{exp} are horizontal angles, the overall similarity of a sample for a certain edge type is calculated as:

$$p = s(\alpha_{seen}, \beta_{seen}, \alpha_{exp}, \beta_{exp}) = s(\alpha_{seen}, \alpha_{exp}, 4) \cdot s(\beta_{seen}, \beta_{exp}, 100) \qquad (2)$$

Calculating the probability for all points on edges found and for all samples in the Monte-Carlo distribution would be a costly operation. Therefore, only a single point of each edge type (if detected) is selected per image by random. To achieve stability against misreadings, resulting either from image processing problems or from the bad synchronization between receiving an image and the corresponding joint angles of the head, the change of the probability of each sample for each edge type is limited to a certain maximum. Thus misreadings will not immediately affect the probability distribution. Instead, several readings are required to lower the probability, resulting in a higher stability of the distribution. However, if the position of the robot was changed externally, the measurements will constantly be inconsistent with the current distribution of the samples, and therefore the probabilities will fall rapidly, and resampling (cf. Sect. 3.4) will take place.

The filtered probability p' for a certain edge type is updated ($p'_{old} \rightarrow p'_{new}$) for each point of that type:

$$p'_{new} = \begin{cases} p'_{old} + 0.01 & \text{if } p > p'_{old} + 0.01 \\ p'_{old} - 0.005 & \text{if } p < p'_{old} - 0.005 \\ p & \text{otherwise.} \end{cases} \qquad (3)$$

The probability q of a certain particle is the product of the three separate probabilities for edges of field lines, the border, and goals:

$$q = p'_{field\ lines} \cdot p'_{border} \cdot p'_{goals} \tag{4}$$

3.4 Resampling

In the resampling step, the samples are moved according to their probabilities. This is done in two steps: First, the samples are copied from the old distribution to a new distribution. Their frequency in the new distribution depends on the probability q of each sample, so more probable samples are copied more often than less probable ones, and improbable samples are removed. In a second step that is in fact part of the next motion update, the particles are moved locally according to their probability. The more probable a sample is, the less it is moved. This can be seen as a probabilistic random search for the best position, because the samples that are randomly moved closer to the real position of the robot will be rewarded by better probabilities during the next observation update steps, and they will therefore be more frequent in future distributions. The samples are moved according to the following equation:

$$pose_{new} = pose_{old} + \begin{pmatrix} \Delta_{trans}(1-q) \times rnd \\ \Delta_{trans}(1-q) \times rnd \\ \Delta_{rot}(1-q) \times rnd \end{pmatrix} \tag{5}$$

rnd returns random numbers in the range $[-1 \dots 1]$. Typical values used for Δ_{trans} and Δ_{rot} are 20 cm and 30°.

3.5 Drawing from Observations

So far, the observation of edge points has only been used to determine the probability of the robot for being at a certain location. However, observations can also be used to generate candidate positions for the localization, i. e. to place samples at certain positions on the field. This approach follows the *sensor resetting* idea of Lenser and Veloso [8], and it can be seen as the small-scale version of the Mixture MCL by Thrun *et al.* [10]. As a single observation cannot uniquely determine the location of the robot, candidate positions are drawn from all locations from which a certain measurement could have been made. To realize this, the robot is equipped with a table for each edge type that contains a large number of poses on the field indexed by the distance to the edge of the corresponding type that would be measured from that location in forward direction. Thus for each measurement, a candidate position can be drawn in constant time from a set of locations that would all provide similar measurements. As all entries in the table only assume measurements in forward direction, the resulting poses have to be rotated to compensate for the direction of the actual measurement.

Such candidate positions are used to replace samples with a low probability. Whether a sample j is replaced or not is also drawn, based on the probability of that sample in relation to the average probability of all samples, i. e. if the following condition is satisfied:

$$\frac{rnd}{n}\sum_{i}^{n} q_i > q_j \tag{6}$$

In this case, *rnd* provides a random number between 0 and 1. If a sample is replaced, the new sample has probabilities p' that are a little bit below the average. Therefore, they have to be acknowledged by further measurements before they are seen as real candidates for the position of the robot.

3.6 Estimating the Pose of the Robot

The pose of the robot is calculated from the sample distribution in two steps: first, the largest cluster is determined, and then the current pose is calculated as the average of all samples belonging to that cluster. To calculate the largest cluster, all samples are assigned to a grid that discretizes the x-, y-, and θ-space into $10 \times 10 \times 10$ cells. Then, it is searched for the $2 \times 2 \times 2$ sub-cube that contains the maximum number of samples. All samples belonging to that sub-cube are used to estimate the current pose of the robot. Whereas the mean x- and y-components can directly be determined, averaging the angles is not straightforward, because of their circularity. Instead, the mean angle θ_{robot} is calculated as the orientation of the sum of all direction vectors:

$$\theta_{robot} = \text{atan2}\left(\sum_i \sin\theta_i, \sum_i \cos\theta_i\right) \tag{7}$$

4 Experiments

To judge the performance of the localization approach, two different experiments were conducted. The first one measures the localization error when the robot is continuously moving. The second one evaluates the precision in reaching certain goal points.

4.1 Experimental Setup

To be able to evaluate the precision of an approach for self-localization, a reference method for localization is required. Gutmann and Fox [4] have analyzed different localization approaches using the Aibo by manually controlling the robot around using a joystick, and whenever it reached a position that was previously marked, they stored the position of that marker and the position as calculated by the robot in a log file. They also stored all perceptions of the robot, allowing them to test different localization approaches based on the same data.

The setup used for the experiments presented in this paper is a little bit different. To be able to continuously track the position of the robot, a laser

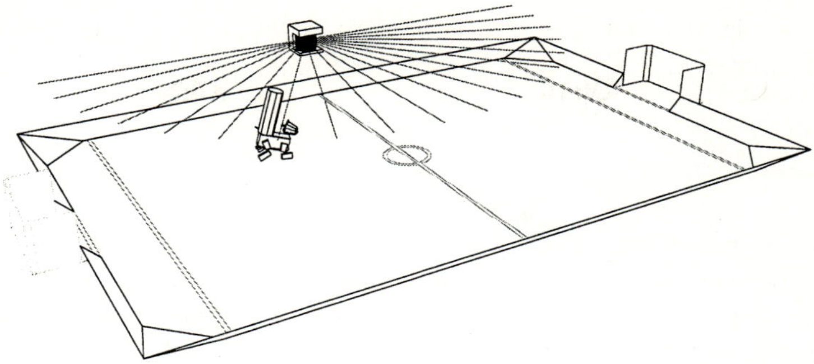

Fig. 5. The experimental setup. The laser scanner is fixed to the border of the field. The robot carries a vertical paper tube on its back that is measured by the laser sensor.

range finder was placed at the border of the field. Within its opening angle of 180°, it measured distances in a height of 35 cm, i. e. above the goals. The robot used for the experiment was equipped with a paper tube on its back that was high enough to be detected by the laser range finder (cf. Fig. 5). This way, the position of the robot could easily be determined by searching for an area that was significantly closer to the laser scanner than the neighboring areas. The shortest distance within that area plus the radius of the tube was used as distance to the robot. Together with the angle under which the robot was measured, the exact location of the robot was determined.

In both experiments, the robot was continuously turning its head from left to right and vice versa. The Monte-Carlo localization method used 100 samples.

4.2 Experiment 1

The goal of the first experiment was to judge the precision of the localization approach when the robot is continuously moving. To accomplish this, the robot was randomly moved around on the field with a maximum speed of 15 cm/s using a joystick. The positions of the robot as calculated by the robot itself and as measured by the laser scanner were stored in a file. The experiment took about 18 minutes, resulting in approximately 5300 measurements.

The result was an average error of 10.5 cm, i. e. less than 4% of the width of the soccer field and less than 2.2% of its length. 60% of the measurements had an error less than this average. Figure 6a shows the path traveled and the errors made. Please note that this outcome is similar to the results presented in [4], with the two exceptions that Gutmann and Fox used color marks for localization, and that they performed their experiments on a small 3m×2m field. In addition, they worked on a log file, allowing them to optimally adjust the parameters of their algorithms, e. g. the Monte-Carlo localization approach used needed only 30 samples.

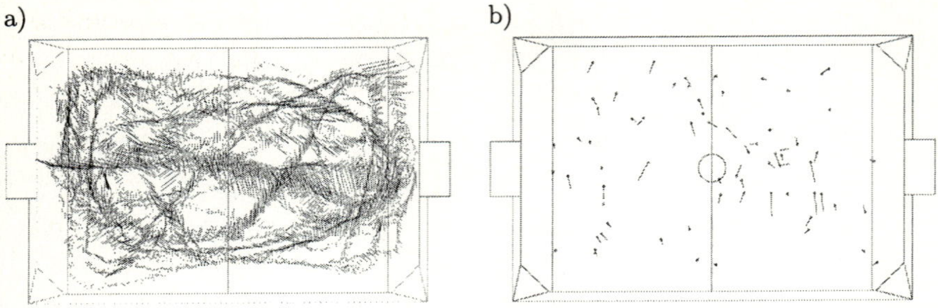

Fig. 6. Experimental results. Each line connects a position calculated by the robot with one determined by the laser scanner. a) First experiment. b) Second experiment.

4.3 Experiment 2

The goal of the second experiment was to evaluate the precision in reaching certain goal points. In this experiment, random goal positions were given to the robot. The system then performed the so-called *go-to-point* skill to reach the specified location. When the robot did not move anymore, the coordinates of the goal position, the position calculated by the robot, and the position measured by the laser scanner were stored in a file. In the experiment, 68 positions had to be reached.

There were two results: the average error between the goal position and the position reported by the laser scanner was 9.4 cm. 66% of the goals were reached with smaller deviations. However, the go-to-point skill does not reach the goal position precisely. It often stops one or two cm too early. Therefore, the average error between the position measured by the robot and the position measured by the laser sensor is smaller, namely 8.4 cm. 60% of the goals were even reached with a smaller error. Figure 6b shows the 68 goal positions and the positions reached by the robot.

5 Conclusions and Future Work

This paper presents an approach for edge-based self-localization in the SFRL. It is based on a vision system that extracts edges without processing whole images. The localization method is a variant of the well known Monte-Carlo localization. While using only a small number of samples, it increases the stability of the localization by maintaining separate probabilities for different edge types for each sample. These probabilities are only adapted slowly. This results in a fast, robust, and precise self-localization of the robot, and it can be seen as a milestone for the SFRL, because it shows that a self-localization without the color beacons is possible.

However, the results presented in this paper only show that edge-based localization is possible for a robot that is alone on the field. Further experiments have to show whether the localization method will also work during actual RoboCup games. While the recognition of the edge points is quite robust, the head cannot swing from left to right and back during actual soccer games, because the robot has to track the ball. Therefore, the situation is different, and it requires for a suitable control strategy for the head posture.

Acknowledgments

The authors would like to thank all members of the GermanTeam for providing the basis for their research, namely Martin Lötzsch, Uwe Düffert, and Max Risler.

References

1. J. Bruce, T. Balch, and M. Veloso. Fast and inexpensive color image segmentation for interactive robots. In *Proceedings of the 2000 IEEE/RSJ International Conference on Intelligent Robots and Systems (IROS '00)*, volume 3, pages 2061–2066, 2000.
2. F. Dellaert, W. Burgard, D. Fox, , and S. Thrun. Using the condensation algorithm for robust, vision-based mobile robot localization. In *Proc. of the IEEE Computer Society Conference on Computer Vision and Pattern Recognition (CVPR)*, 1999.
3. D. Fox, W. Burgard, F. Dellaert, and S. Thrun. Monte Carlo localization: Efficient position estimation for mobile robots. In *Proc. of the National Conference on Artificial Intelligence*, 1999.
4. D. Gutmann, J.-S. Fox. An experimental comparison of localization methods continued. In *Proc. of the IEEE/RSJ International Conference on Intelligent Robots and Systems*, Lausanne, Switzerland, 2002. EPFL.
5. Robert Hanek, Thorsten Schmitt, Sebastian Buck, and Michael Beetz. Fast image-based object localization in natural scenes. In *IEEE/RSJ International Conference on Intelligent Robots and Systems (IROS) 2002, Lausanne*, 2002.
6. M. Jamzad, B. Sadjad, V. Mirrokni, M. Kazemi, H. Chitsaz, A. Heydarnoori, M. Hajiaghai, and E. Chiniforooshan. A fast vision system for middle size robots in robocup. In *5th International Workshop on RoboCup 2001 (Robot World Cup Soccer Games and Conferences)*, number 2377 in Lecture Notes in Computer Science, pages 71–80. Springer, 2002.
7. M. Jüngel, J. Hoffmann, and M. Lötzsch. A real-time auto-adjusting vision system for robotic soccer. In *7th International Workshop on RoboCup 2003 (Robot World Cup Soccer Games and Conferences)*, Lecture Notes in Artificial Intelligence. Springer, 2004. to appear.
8. S. Lenser and M. Veloso. Sensor resetting localization for poorly modeled mobile robots. In *Proc. of the IEEE International Conference on Robotics and Automation (ICRA)*, 2000.
9. F. K. H. Quek. An algorithm for the rapid computation of boundaries of run length encoded regions. *Pattern Recognition Journal*, 33:1637–1649, 2000.
10. S. Thrun, D. Fox, and W. Burgard. Monte carlo localization with mixture proposal distribution. In *Proc. of the National Conference on Artificial Intelligence*, pages 859–865. AAAI, 2000.
11. J. Wolf, W. Burgard, and H. Burkhardt. Robust vision-based localization for mobile robots using an image retrieval system based on invariant features. In *Proc. of the IEEE International Conference on Robotics and Automation (ICRA)*, 2002.

A Symmetry Operator and Its Application to the RoboCup

Kai Huebner

Bremen Institute of Safe Systems, TZI, FB3
Universität Bremen, Postfach 330440, 28334 Bremen, Germany
khuebner@tzi.de

Abstract. At present, visual localization of soccer playing robots taking part in the RoboCup contest is mainly achieved by using colored artificial landmarks. As known, this method causes further vision problems like color classification and segmentation under variable light conditions. Additionally, robots confined to use visual sensor information from common cameras usually waste time in switching between the modi of playing soccer and searching landmarks for localization. An upcoming approach to solve these problems is the detection of field lines. Motivated by our research in using a compact symmetry operator for natural feature extraction in mobile robot applications, we propose its application to the RoboCup contest. Symmetry is a structural feature and as results show, it is highly independent of illumination changes and very compliant to the task of line detection. We will motivate symmetry as a natural feature, discuss the symmetry operator and finally present results of the field line extraction.

1 Introduction

For mobile robot tasks it is preferable to concentrate on visual sensor information only. In contrast to other types of sensor data, camera images offer additional information for range estimation, object classification, localization and navigation. Human vision proves that it is possible to realize these tasks without using laser scanners or ultrasonic sensors. In this case, performance is relying on robust image features to be recognized, tracked or classified. In structured workspaces, extraction of those features may be simple if constant light conditions, color markings and maps can be used. In arbitrary environments, however, it is necessary to extract natural and cognitive features.

One of those is symmetry. Most objects in our world have a high degree of symmetry, maybe because it is appropriate to a certain kind of beauty, simplicity or usefulness. Just like animals or plants are quite symmetrical in shape, humans are inclined to use symmetry in art, architecture and artifacts. Therefore, symmetry has been tested in psychological experiments to examine how and how effective human vision explores symmetric objects and scenes [5, 6].

Especially reflective symmetry and its orientation seem of high importance for human vision. Eye-tracking experiments pointed out that there are differences in quality and speed of detecting several types of symmetry, e.g. vertical

D. Polani et al. (Eds.): RoboCup 2003, LNAI 3020, pp. 274–283, 2004.

mirror symmetry (reflective symmetry about a vertical axis) features very quick and accurate detection in most cases. Assuming the presence of horizontal or vertical reflective symmetry in a scene, people are able to initially detect and take advantage of the symmetry axis for further visual exploration.

Because of its relevance in human perception, symmetry has been widely studied in psychological context. Palmer and Hemenway [6] studied the latency for detecting different types of reflective symmetry in a set of arbitrary oriented polygon shapes. Their results show that detection is fastest for vertical symmetry, next fastest for horizontal symmetry and slowest for skewed symmetries. Locher and Nodine [5] made experiments on visual detection and attention of symmetry in composed pictures and showed that the axis of symmetry is used as a perceptual landmark for visual exploration. Ferguson [2] presents the detection of symmetry using visual relations, adjustment of an object's reference frame using its symmetry axes and analyzation of orientation effects.

In computer vision tasks, symmetry has also been used in different applications. Sun [8] proposes a fast symmetry detection algorithm to detect the main symmetry axis of an image. Reisfeld et al. [7] define a generalized symmetry transform including reflective symmetry to extract regions of interest in arbitrary images and to determine an object's value of symmetry. They use a square mask detecting symmetry and gradient-based interest to establish a symmetry picture. Similar results are achieved by Kovesi [4] by analyzing the frequency components of an image. This is based on the idea that if most components have their minimum or maximum at a point of interest, it will correspond to a point of huge symmetry. Chetverikov [1] computes symmetry in order to find orientations of faces in portraits or to detect structural defects in industrial applications. Therefore, he defines a regularity value based on the symmetrical regularity of a pattern. Face detecting is done by Zabrodsky et al. [10] by utilizing symmetry. By using explicitely defined rotational symmetry, they are able to reconstruct partially occluded objects, in the case that they are roughly rotational symmetric (e.g. in an image with partially occluded asymmetric flowers).

2 The Symmetry Operator

Related approaches use symmetry in a global sense. Some use symmetry as a feature of the image itself [1, 8], detect reflective symmetries of any direction [1, 4, 7, 8] or additionally incorporate rotational symmetries [10].

Our line detection method is based on a compact 1-dimensional symmetry operator for arbitrary images [3]. For each pixel of the image, a qualitative value of reflective symmetry in horizontal or vertical direction is determined. *Vertical symmetry* is defined as symmetry about a vertical axis, thus only pixels in the same image row $R = [p_0, p_{w-1}]$ have to be considered for the detection of vertical symmetry about a pixel $p_i \in R$, where w is the width of the image. The same is applied for horizontal symmetry regarding only one column of the image. Furthermore, robot vision requires processing of real images. Because of the common image distortion in real images, an operator detecting exact, mathematic symmetry fails and offers erroneous symmetry images. Therefore, we propose the following qualitative symmetry operator based on a normalized mean square error function:

$$S(p_i, m) = 1 - \frac{1}{C \cdot m} \sum_{j=1}^{m} \sigma(j, m) \cdot g(p_{i-j}, p_{i+j})^2 \tag{1}$$

where $m > 0$ is the size of the surrounding of p_i in which its value of symmetry shall be detected. Thus, the complete number of pixels considered is $2m$. C is a normalization constant depending on the used color space and on $\sigma(j, m)$, which is a radial weighting function. The difference between two opposing points p_{i-j} and p_{i+j} is determined by a gradient function $g(p_{i-j}, p_{i+j})$, which usually is the Euclidian distance of the corresponding color vectors $\overline{p}_{i-j}$ and $\overline{p}_{i+j}$. For all presented experiments, we used 8-bit gray-scale representation with

$$g(p_{i-j}, p_{i+j}) = \begin{cases} \|\overline{p}_{i-j} - \overline{p}_{i+j}\| & \text{if } p_{i-j} \in R \wedge p_{i+j} \in R \\ c & \text{otherwise} \end{cases} \tag{2}$$

where c is the maximum error available (depending on color space), and a linear weighting function additionally depending on m

$$\sigma(j, m) = 1 - \frac{|j|}{m+1} \tag{3}$$

The larger m the more it may exceed the visible region R and the more the error function gets burdened with the maximum error c (see Eq. 2). In this case, image border regions (left and right for vertical symmetry, see Fig. 1b and 1d; top and bottom for horizontal symmetry, see Fig. 1c and 1e) get more influenced by the effect of fading.

Important symmetry axes can be found at places where not necessarily high symmetry values but symmetry peaks can be detected. Though the extraction of maxima and minima of a symmetry image causes more distortion in resulting binary images, it is more significant than using a threshold value. Thresholds may vary from application to application or even from image to image. Additionally, appropriate thresholds are difficult to find for normalized symmetry. A symmetry value of 0 corresponds to hard black-white transitions between each pair of opposing points p_{i-j} and p_{i+j}, while a value of 1 corresponds to exact parity. Thus, high symmetry values are more frequent and much more dense, which makes threshold setting very ineffective. Symmetry is more adapted for the application of local extrema, since it is a regional feature characterizing the local environment (in contrast to local features like edges). Since calculation of vertical symmetry in one row is independent of those in other rows or columns, maxima and minima can be detected line by line, respectively column by column for horizontal symmetry. Results of this symmetry axes detection are presented in Fig. 2. Note that each result has been achieved by only using the symmetry operator and maximum detection, without any kind of pre- or post-processing like Gauss filtering, segmentation or related techniques.

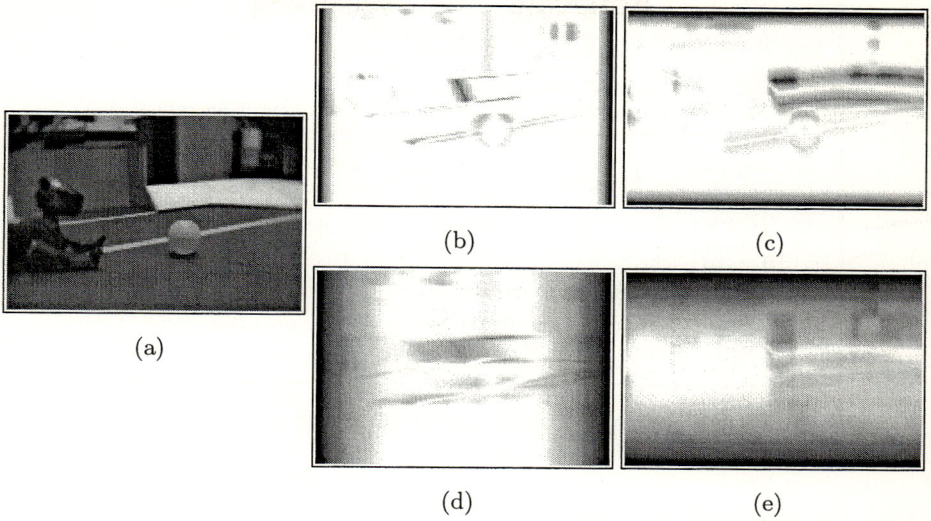

Fig. 1. Symmetry images of a RoboCup image (174×114) (a): vertical and horizontal symmetry using $m = 10$ (b,c) and $m = 50$ (d,e).

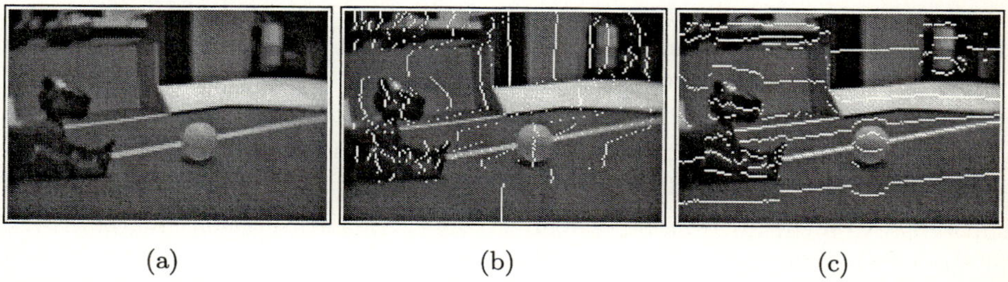

Fig. 2. Symmetry maxima of the RoboCup image (a) using $m = 5$ for vertical (b) and horizontal (c) symmetry axes extraction.

3 Experimental Results

In addition to the extraction of symmetry axes, symmetry feature points can be extracted as crossings of vertical and horizontal symmetry axes. Zhang and Huebner [11] used this feature type to track and classify points of interest with a mobile robot using an omnidirectional vision sensor. In the context of panoramic images, another application is the usage of histograms of vertical symmetry axes as a feature to recognize doors or the direction of the hallway. The several methods of feature extraction are presented in Fig. 3, but surely offer further expansion. Choosing the best feature method mainly depends on the specific application. Vertical symmetry histograms and symmetry feature points were useful in panoramic images for mobile robot tasks like range estimation. In this section, we will demonstrate our RoboCup line detection experiments based on symmetry axes extraction.

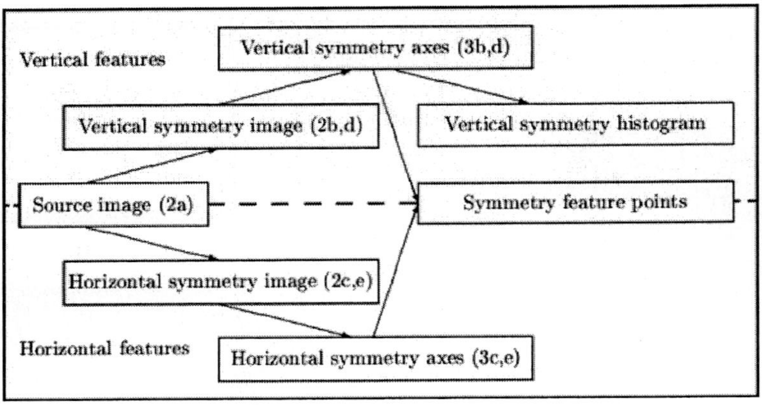

Fig. 3. Several methods of symmetry feature extraction.

3.1 Symmetry Line Filter

The task of line extraction was motivated by visual mobile robot localization in the RoboCup contest. In this context, localization highly depends on robust recognition of color markings that have to be explored around the field. For robots with common cameras, searching those marks deters from concentrating on game objects like the ball. Thus, it is only possible to either localize the robot or to capture the ball at a time, accordingly it would be more efficient to likewise concentrate on the field for localization. A possible and more intelligent solution could be offered by the extraction of field lines.

Line extraction techniques usually need some preprocessing, let edge detection, thresholding or thinning only be a few examples. Using symmetry, we can detect lines as a structure from arbitrary images. For example, a horizontal line is a structure where we should continuously detect a given A_1SA_2-pattern (= Asymmetry$_1$–Symmetry–Asymmetry$_2$) in each small vertical neighborhood along the line. Actually, A_1S is sufficient, because the symmetry axis S implies that there is another A_2 symmetric to A_1. An example for detection of this structure is shown in Fig. 4, where only horizontal symmetry axes using $m = 3$ were detected. If the specific AS-pattern can be found in the same environment, we can assume it is part of a line.

3.2 Line Detection

In the following, two approaches are presented to extract lines from the images resulting from the proposed symmetry line filter. The first one is a modified Hough approach using the Wallace Muff space [9], which represents a line by its start and end point on the image border rectangle.

The Muff parameter space (Fig. 5b) shows that there are two lines that lead from the left to the right side of the symmetry line image (Fig. 5a). The result shown in Fig. 5c seems quite acceptable, but it needed several adaptation steps

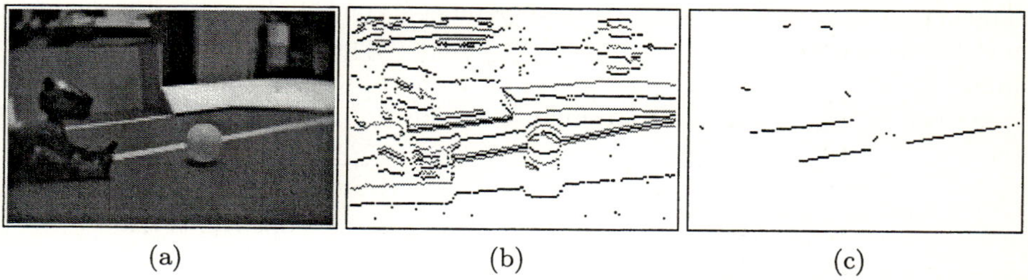

Fig. 4. (a) RoboCup image. (b) Horizontal symmetry maxima and minima (gray = A, black = S) using $m = 3$. (c) Filtered line points by AS pattern.

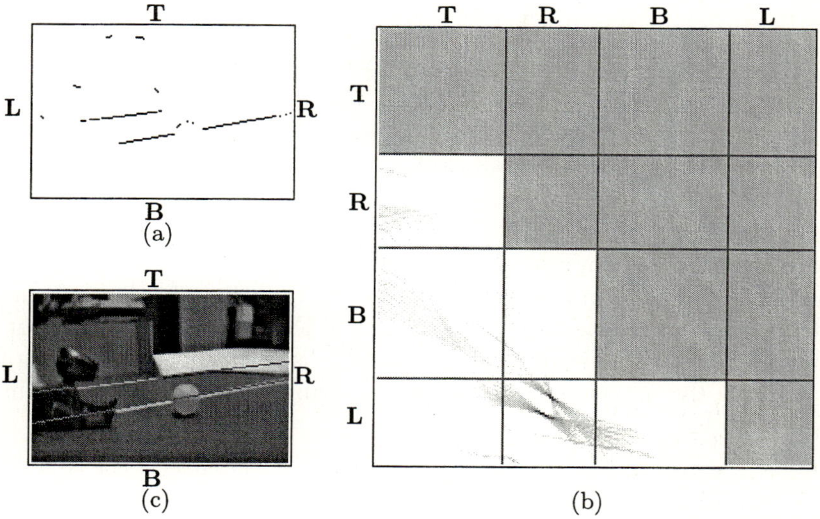

Fig. 5. (a) Symmetry line image. (b) Muff parameter space. (c) Lines found in Muff space.

for this result, mainly because of the difficulty to extract maxima in Muff space. There are further disadvantages of this approach, for example, a line now is represented by its image border points, thus information about line segments is lost. Additionally, curve segments may also be detected as lines with this method, that yet is complex enough without a modified Hough transform for circle detection.

Because of these disadvantages, we developed another approach that takes advantage of the fact that most feature points of the symmetry line image only have one or two neighboring feature points. Simply using the number of feature points in the 3×3-neighborhood of a point p, each feature point can be classified as follows:

- Type A: if p has **no** neighbor, it is not interesting for line extraction.
- Type B: if p has **one** neighbor, it is start or end point of a line.
- Type C: if p has **more** neighbors, it is part of a line.

Thus, we only have to search for feature points of type B (a line's start point) and recursively search the next neighboring point of type C, until we find another point of type B (the line's end point). Therefore, we use the search patterns described in Fig. 6 that are rotational invariant:

Fig. 6. The two search patterns for line segmentation.

Suppose X and Y are points of type B or C, and Y has been detected as the neighbor of X. Now we can start searching the neighboring fields as proposed, until we find a new feature point. If no feature point is found, Y is the end point of the current line, otherwise we proceed at Y in the same manner. Note that the fields left empty can not be occupied by feature points because of the symmetry maxima detection.

Each line segment can now be represented as the list of feature points found by this method. Based on this representation, we can access further information about the line, e.g. the variance of each point to the line described by start and end point. This measure is very useful to easily distinguish curves from straight lines, because the maximum variance will probably exceed a few pixels in the first case (see Fig. 7).

Some results of arbitrary RoboCup images are presented in Fig. 8. In each case, we had to search for thin white horizontal lines. Thus, we applied the hor-

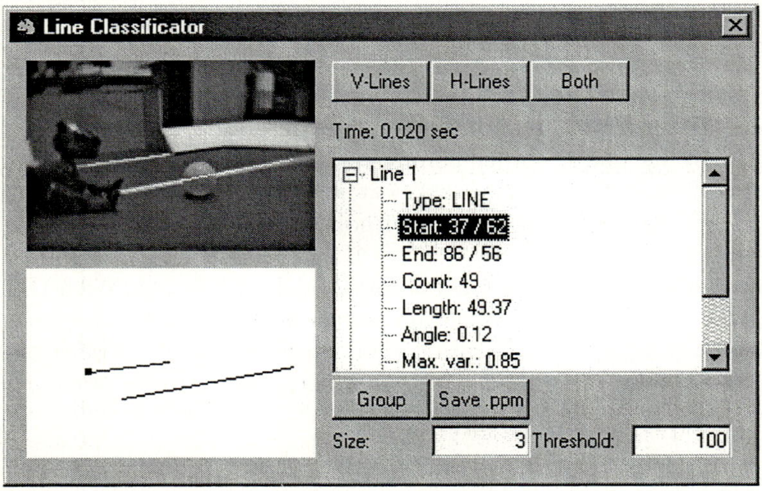

Fig. 7. Screenshot of the Line Classificator Dialog.

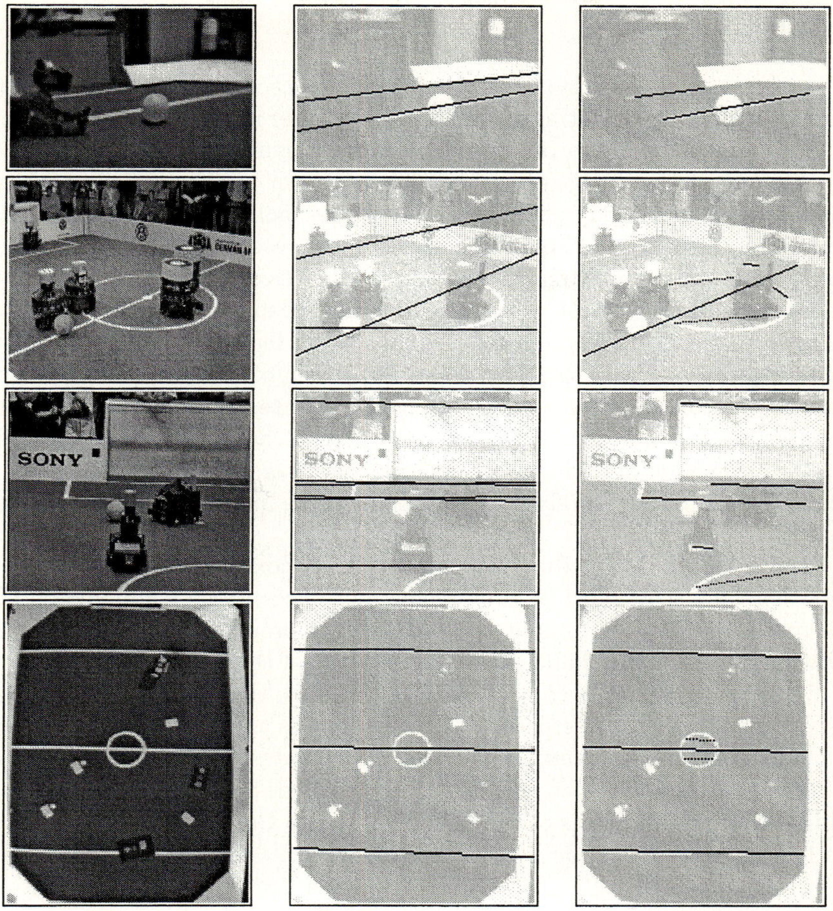

Fig. 8. Results of the horizontal symmetry line detection and classification. In each line: source image, lines detected by Muff space approach, lines and curves (dotted) detected by own approach.

izontal symmetry operator using $m = 3$ and included an illumination threshold neglecting those feature points having a gray-scale value smaller than 100 in the source image. Additionally, we did not care about lines shorter than a given threshold and implemented a heuristic to combine line segments, in the case that they seem to belong to the same field line, but are disrupted by occluding objects.

As proposed, the performance of this method is quite acceptable. Additionally, it is more compact and faster than the Muff space approach. It needs less adaptation, but offers extraction of line segments and classification of curves. In Fig. 8, all line segments are simply plotted as lines from start point to end point, but those identified as curve segments are dotted.

4 Conclusions

We proposed a compact symmetry operator detecting horizontal and vertical reflective symmetry. Resulting symmetry images offer multiple feature extraction methods, especially binary images derived from symmetry axis detection are interesting for further image processing. The operator can be applied to arbitrary images without prior adaptation and without thresholds, the only parameters to specify are the size of the operator mask and the resolution of symmetry data. As a structural feature, symmetry is additionally less dependant on illumination changes, thus no color table or classification is needed.

The operator and the proposed line extraction technique are able to detect lines in a promising and fast way, which is a basic condition for further applications like localization and navigation. As presented, horizontal and vertical symmetry are not purely restricted to detect exact horizontal or vertical lines. For all experiments, only the horizontal operator was used for extracting nearly horizontal lines, but for surely finding all lines in the image, both operators must be used.

Considering the algorithm in comparison to competitive approaches will be another aim. The effort of time seems quite acceptable compared to other gradient-based operators, but it is worse than approaches based on color segmentation. On the other hand, it is an advantage of the proposed operator that it is highly independent of color. In addition to this, the quality of the operator is very convincing with reference to its compactness and line detection ability.

Future work in the context of the RoboCup contest will also consist of supporting localization methods with the lines detected. Some possible approaches would be the use of lines as observations for Monte Carlo Localization or using them for position estimation by Spatial Reasoning.

References

1. D. Chetverikov. Fundamental Structural Features in the Visual World. In *Proceedings of the International Workshop on Fundamental Structural Properties in Image and Pattern Analysis*, pages 47–58, Budapest, 1999.
2. R. W. Ferguson. Modeling Orientation Effects in Symmetry Detection: The Role of Visual Structure. In *Proceedings of the 22nd Conference of the Cognitive Science Society*, Hillsdale, New Jersey: Erlbaum, 2000.
3. K. Huebner. A 1-Dimensional Symmetry Operator for Image Feature Extraction in Robot Applications. *The 16th International Conference on Vision Interface*, June 2003.
4. P. D. Kovesi. Symmetry and Asymmetry From Local Phase. In *The 10th Australian Joint Conference on Artificial Intelligence, Poster Proceedings*, pages 185–190, Perth, 1997.
5. P. Locher and C. Nodine. The perceptual value of symmetry. *Comput. Math. Applic.*, 17:475–484, 1989.
6. S. E. Palmer and K. Hemenway. Orientation and Symmetry: Effects of Multiple, Rotational, and Near Symmetries. *Journal of Experimental Psychology: Human Perception and Performance*, 4(4):691–702, 1978.

7. D. Reisfeld, H. Wolfson, and Y. Yeshurun. Context Free Attentional Operators: the Generalized Symmetry Transform. *International Journal of Computer Vision*, 14:119–130, 1995.
8. C. Sun. Symmetry detection using gradient information. *Pattern Recognition Letters*, 16:987–996, 1995.
9. R.S. Wallace. A Modified Hough Transform For Lines. *IEEE Intrenational Conference on Computer Vision and Pattern Recognition*, 85:665–667, 1985.
10. H. Zabrodsky, S. Peleg, and D. Avnir. Symmetry as a Continuous Feature. *IEEE Transactions on Pattern Analysis and Machine Intelligence*, 17(12), 1995.
11. J. Zhang and K. Huebner. Using Symmetry as a Feature in Panoramic Images for Mobile Robot Applications. In *Proceedings of Robotik 2002*, volume 1679 of *VDI-Berichte*, pages 263–268, Ludwigsburg, 2002.

RoboCup as an Introduction to CS Research

Peter Stone

Department of Computer Sciences, The University of Texas at Austin
1 University Station C0500, Austin, Texas 78712-1188
pstone@cs.utexas.edu
http://www.cs.utexas.edu/~pstone

Abstract. This paper proposes using topics central to RoboCup, particularly autonomous agents and multiagent systems, as the subject-matter for a course designed to introduce undergraduate students to all facets of computer science research. Experiences are presented from the design and implementation of such a course. The course is structured around an ongoing incremental programming project that culminates in a class tournament in the RoboCup Soccer Server, an open-source infrastructure built to support multiagent systems research and education.

1 Introduction

Most upper-division computer science (CS) majors have determined that they enjoy taking CS classes, or at least that they are relatively good at it. However, this determination may not be indicative of a propensity for computer science *research*. Indeed, many CS Ph.D. candidates have discovered that they are *not* cut out for research only after investing several years in a graduate program. Conversely, there are presumably those students who could have enjoyed quite successful research careers had they only thought to give it a shot.

One obvious explanation for these phenomena is that traditional undergraduate computer science courses demand very different skills from those required of researchers[1]. In particular, traditional coursework typically requires students to:

- read textbooks;
- sit (often silently) in large lectures;
- execute programming tasks with correct and complete answers;
- work alone; and
- take exams.

On the other hand, researchers typically must:

- read about and critically assess original research;
- speak in public;

[1] A similar argument likely holds in other areas of scientific research. This paper speaks specifically from a computer science perspective due to the fact that the author's experience lies mainly in that field.

D. Polani et al. (Eds.): RoboCup 2003, LNAI 3020, pp. 284–295, 2004.

– collaborate effectively with peers;
– devise solutions and/or approaches to open-ended problems; and
– write about these solutions and/or approaches.

Given these differences, it is not surprising that proficiency at coursework does not correlate perfectly with proficiency at research. In response to this observation, many Ph.D. programs now encourage and/or require their students to engage in research activities from the outset of their graduate studies with the aim of helping students to determine quickly whether they are truly interested in pursuing a career in research.

This paper advocates giving students an opportunity to make this determination before even entering graduate school. One method for doing so is to encourage students to engage in individual undergraduate research projects. However such opportunities are quite varied in scope and limited in availability. Another method for doing so – the one put forth in this paper – is to offer undergraduate coursework specifically designed to require the same skills required of professional researchers.

This paper argues that RoboCup is an ideal topic around which to build such a course. Experiences are presented from the design and implementation of such a course focusing in particular on *autonomous agents and multiagent systems.* The course is structured around an ongoing incremental programming project culminating in a class tournament in the RoboCup Soccer Server, an open-source infrastructure built to support multiagent systems research and education [11].

The remainder of this paper is organized as follows. Section 2 presents some general principles to be followed by a course designed to introduce research to undergraduates. Section 3 details a specific implementation of these principles in the form of a course entitled *Autonomous Multiagent Systems.* Section 4 presents anecdotal results from this course and concludes.

2 General Principles

This section proposes a set of five principles to be followed when creating a course for the purpose of introducing undergraduates to scientific research. These principles are illustrated in the context of an AI-based course that makes extensive use of the RoboCup domain in Section 3.

Open-Ended Project. The most crucial element of such a course is a long-term open-ended assignment that leaves as much room as possible for creativity and innovation. For practical purposes and to avoid mass frustration on the parts of the students, it must be possible to get *some* result fairly easily. However it is very important that there be no natural stopping point. Instead, students should propose an approach and a goal for their projects on their own before beginning.

Students are used to assignments with the characteristic that they can recognize when they have reached a correct or sufficient result and thus can safely quit. However, researchers often work on long-term projects with no clear answers. Thus it is important to have some such challenging project in the class. The difficulty then becomes motivating the students to meet the challenge.

One motivational technique is to have the project culminate in a class competition. Even if the competition has no impact on the student's grade, it can be a strong impetus to going above and beyond the effort typically reserved for a class project. AI is rich in domains that are appropriate for such competitions, with the RoboCup soccer domain [11] being perhaps the most popular example. Experiences with this domain are reported in Section 3.

Read Research Papers. Students are used to reading secondary sources, such as textbooks, in which many ideas are synthesized and put into a coherent perspective. While these sources may be preferable to primary sources from a pedagogical perspective, they do not prepare students for the activity of reading and assessing cutting-edge research papers.

Consequently, course readings should be predominantly primary sources. They need not be the most current articles in the field. But they should be assigned in the form in which they were originally published. Whenever possible, readings with opposing views should be selected so that the students must form their own opinions about their relative merits.

Finally, the articles must be chosen so as to be accessible to students without extensive background (other than previous course readings). This requirement is perhaps the most difficult to meet for many course topics.

Require Class Participation. It is not sufficient to merely *assign* the research papers. There must be some incentive for the students to read them from a critical perspective. For example, students can be required to come to class prepared with questions about the readings, or brief written reactions to the readings can be required.

Whatever the method, the goal should be to have every student verbally express reactions to the readings. Especially in its more philosophical guise, AI is rife with topics that excite stimulating class discussions.

Foster Improved Scientific Writing. By all accounts, writing is perhaps the skill most lacking in students today, perhaps particularly so for science majors. As such, it is essential that the course place a heavy emphasis on scientific writing. In particular, the students should at least be required to write a report of their project in the form of a research paper. This report should count for a significant part of their final grades.

In addition to the final report, there should be at least one prior writing assignment so that students can receive, and act on feedback from the instructors. Ideally, this feedback should concern style and organization as well as content.

Encourage Collaboration. Students are often made to complete their work entirely on their own. They may ask their instructors for hints or help, but they are usually not allowed to work with their classmates. In contrast, a glance at the bylines in the literature shows that most research papers are the result of collaborations. Thus, especially for the open-ended project, students should be given the opportunity to work in small groups.

Surely there are many ways in which the above principles can be implemented in the class. Indeed, the details are necessarily dependent upon the subject being taught. Section 3 illustrates these principles in concrete form within a computer science curriculum.

3 Implementation

The main thesis of this paper is that RoboCup is an ideal topic around which to build a course that implements the principles laid out in Section 2. The main advantages of RoboCup are that it it is an appealing domain for students of both genders; it is a popular testbed domain used by active researchers in the field with many recent papers published pertaining to it; it admits for a wide variety of research foci; and it lends itself well to exciting, visual competitions. No other domain known to the author combines all of these aspects as well as does RoboCup.

This section details a specific implementation of the general principles in the form of a course focusing on autonomous agents and multiagent systems (AAMAS). AAMAS is one of the fields to which RoboCup participants have contributed consistently and prominently over the years. Despite being the basis for a large subfield of AI, there is no generally accepted definition of artificial intelligence *agents.* In loose terms, agents are programs that (i) sense their environment, (ii) make decisions about how to act based on these sensations, and (iii) then execute these actions. Autonomous agents do all three of these steps on their own, i.e. without a human in the loop. Multiagent systems are collections of multiple agents that interact with one another. The field of AAMAS covers a wide variety of research foci and applications, including software-based information processing, robotic control of multiple agents, entertainment agents, and tutoring agents [6].

The course described in this section is called *Autonomous Multiagent Systems.* It has been taught twice, most recently during the fall of 2002 [2]. It will be offered again during the spring of 2004 [3]. The course provides a broad introduction to autonomous agents with an emphasis on multiagent systems. Topics include:

- agent architectures;
- inter-agent communication;
- teamwork;
- distributed rational decision making;
- agent modeling;
- multiagent learning; and
- entertainment agents.

This paper will not go into detail about the subject matter of the course except as far as is necessary for the purpose of clarity. The subject matter is

[2] `http://www.cs.utexas.edu/~pstone/Courses/378fall02`

[3] `http://www.cs.utexas.edu/~pstone/Courses/378spring04`

in a sense subordinate to the main purpose of the course. Students are told explicitly that in order to succeed they will need to attain a mastery of the subject. However evaluation is based primarily on their ability to engage in the full range of activities required of researchers.

The remainder of this section describes how each of the principles presented in Section 2 is implemented in the course.

3.1 Project

The central focus of the course is a semester-long build-up towards a class robotic soccer competition in the RoboCup Soccer Server [10]. The Robot Soccer World Cup, or RoboCup, is an international research initiative that uses the game of soccer as a domain for artificial intelligence and robotics research. The RoboCup Soccer Server, coupled with the large body of available client code, is an infrastructure that is designed to be appropriate for both scientific research and education [11].

Soccer Server is a multiagent environment that supports 22 independent agents interacting in a dynamic, real-time environment. The server embodies many real-world complexities, such as noisy, limited sensing; noisy action and object movement; limited agent stamina; and limited inter-agent communication bandwidth. AI researchers have been using the Soccer Server to pursue research in a wide variety of areas, including real-time multiagent planning, real-time communication methods, collaborative sensing, agent/opponent modeling, and multiagent learning [2].

In addition to the server itself being publicly available under an open-source paradigm, users have contributed several clients that can be used as starting points for newcomers to the domain. One example is the CMUnited simulated soccer team [15], champion of the RoboCup-98 and RoboCup-99 robotic soccer competitions. After winning the competitions, much of the CMUnited source code became publicly available, and several groups used it as a resource to help them create new clients for research and as entries in subsequent competitions. Figure 1 shows a screen shot of Soccer Server.

Soccer Server and the CMUnited client code are *widely* and *freely* available over the internet using an *open source* paradigm. The software is *packaged* for easy installation, *supported* both by the developers and by the large *community* of current users.

This infrastructure is a *comprehensive, implemented* MAS designed for *simulation* experiments. It consists of several independent *components*, including visualization, sample client, and coach modules. The coach module is often used as a tool for *experiment construction.* The most natural and compelling form of *measurement* is game results in tournaments with multiple teams, but the infrastructure also includes *data collection and analysis* tools for more rigorous scientific measurement.

Judging by the large user community (over 1000 researchers worldwide), this infrastructure is very *usable*; the fact that it has been successfully used for multiple international competitions is a testament to its *robustness.* New

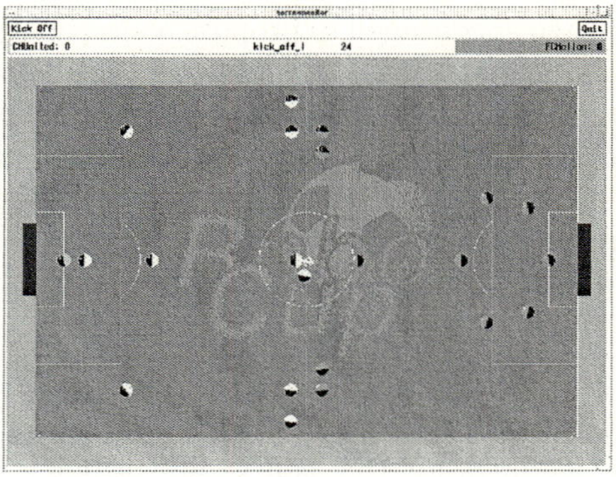

Fig. 1. Window image of Soccer Server

users can take advantage of its *progressive complexity* by starting with a single agent and gradually increasing the size of teams and their communicative and organizational capabilities. A recent addition to the infrastructure is the ability to induce *intentional failures* by disabling selected players.

The italicized words above are all characteristics identified by Gasser [5] as essential or desirable for MAS infrastructures that support science and education.

Two additional favorable properties of Soccer Server are that it is both an appealing domain for students of both genders [14] and a popular testbed domain used by active researchers in the field. An IJCAI-97 challenge paper [8] identified three general research challenges that can be addressed within Soccer Server as being

- multiagent learning;
- teamwork structures; and
- agent/opponent modeling.

As laid out in [15], other relevant research issues include inter-agent communication in single-channel, low-bandwidth environments; coordination with limited communication, collaboration in a dynamic real-time environment; organizational structures; distributed sensing/sensor fusion; resource management; agent monitoring; and multiagent planning. These research topics are all addressed by various researchers in the continuing series of RoboCup books [1, 2, 7, 16, 17].

In addition to being the substrate domain for much original research, Soccer Server has been used previously as a basis for a few other university courses (e.g. [4, 19]). These courses have previously focussed primarily on teaching AI and/or multiagent systems rather than introducing CS research in general.

Students in the course described here are assigned a series of four preliminary programming assignments designed to get them familiar with Soccer Server and the CMUnited client code. By the end of these preliminary assignments, they have created a fully functional team, but not one that is particularly competent.

The students are then encouraged to propose an improvement on this team as the topic for their final projects[4]. The improvement need not be for the purpose of creating a winning team in the final tournament. For example, one student proposed to use machine learning techniques to learn a good goaltender without paying attention to the rest of the team. In practice, about half of the students make earnest attempts to create top quality teams, while the remainder focus on interesting side issues. Both types of projects can yield (and have yielded) quality results.

3.2 Readings

The majority of the readings for this course are primary sources chosen both to introduce particular topics relevant to the course and to engender some controversy. For example, during one of the early weeks, the students learn about "agent architectures" by reading both an article espousing completely reactive agents (e.g. [3]) and an article that argues in favor of using more deliberative agents (e.g. [13]).

In order to encourage the students to complete the readings in a timely fashion, they are required to submit a brief written answer to a single question pertaining to the readings at least 2 hours before the class starts. For example, a question associated with the above readings has been:

> Part one of your current programming assignment can be implemented reactively. Describe a non-reactive soccer-playing behavior.

Associated with readings on "agent modeling" was the following:

> Describe a domain not mentioned in the readings in which agent modeling could provide a benefit.

Note that these questions can be answered briefly and have no right answers. But in order to respond, students must complete and understand the readings. The fact that the responses are due two hours before class allows the instructor to incorporate them into the class discussion.

3.3 Class Participation

Another effect of the questions pertaining to the readings is that the students *do* come to class prepared to discuss the readings. There have been many extended and heated class discussions pertaining to specific aspects of agents and multiagent systems.

Another important component of class participation is that each student is required to moderate at least one class discussion pertaining to that week's readings. The student discussion is specifically *not* a presentation of the readings: the students are encouraged to assume that the class members have all done the readings. Rather, they are instructed to either defend a controversial statement

[4] They are also given the option to propose a programming project in a multiagent domain of their choice, but typically few students choose to do so.

or pose a question and be prepared to defend either side depending on how the class reacts. Their explicit goal is to moderate an interesting and extended discussion.

This activity turns out to be one of the most difficult for the students to complete. Many of them are not used to speaking in front of a class, and they have rarely been put in the position of *facilitating* discussions as opposed to defending specific positions. Some examples of successful questions generated by the students are:

- "Is it better to build in teamwork at the architecture level, or to first build individually functioning agents and then add on teamwork constructs?"
- "We've seen some examples of machine learning agents that spend lots of time and effort learning, but don't do as well as hand-coded agents. Is it worth all the trouble?"
- "Do we want machines (programs, appliances) to have emotions and personality?"
- "Nash Equilibrium is limited in its applicability due to its exponential complexity and the inability to give a deterministic solution. Agree or Disagree?"

3.4 Writing

The course requires a good deal of writing from the students. As already mentioned, the students are required to provide weekly written responses to questions related to the readings. They receive feedback pertaining to the clarity and soundness of their responses.

Much more significantly, the students are required to write three written documents pertaining to their final projects. First, they write project proposals defining their goals for their projects as well as the proposed methods for achieving them. Second, they revise their proposals and add a section on their work in progress to create progress reports. Finally, they write final reports in the format of conference papers.

In practice, these writing assignments often show an evolution of the students' ideas. For example, a common proposal is to use some machine learning technique that has been mentioned in the readings to improve some aspect of the soccer team's performance. However, based on feedback from the instructors and their initial attempts, the students often realize that their initial proposal was too ambitious and scale it back to more realistic levels given the time available. As one student who was attempting to use machine learning in conjunction with a simple general agent architecture wrote in the final report: the "machine learning part has proven difficult and elusive, and has turned into a research project over suitable data, choice of representation, and machine learning algorithms." Nonetheless, there have been many modestly successful uses of machine learning in the student teams.

3.5 Collaboration

Productive collaboration cannot be forced. However it should be encouraged. In this course, students are given the opportunity, but are not required, to work in pairs on the final project. Teams of two must still write their proposals and reports individually, with clear indications of what role each person played in the collaboration; and as such, more is expected from them as a final product.

The robotic soccer project lends itself to such collaboration nicely since there are many different ways in which the students can divide up the work. For example, one can focus on offense while the other focuses on defense; one can focus on low-level skills while the other focuses on team strategy; one can work on agent development tools while the other focuses on using them to create the team; etc. Each of the these three task division strategies has been used by students in the class.

3.6 The Tournament

The class culminates in the class tournament. The students are told at the outset that performance in the tournament will have no negative impact on their grades (A strong performance *can* have a positive impact.). Nonetheless, the tournament is a strong motivational factor for the students. Visitors are invited to the event and the students present their approaches orally and field questions as their teams are playing.

The performance spread among the teams is often very large, especially given the fact that some students do not focus on creating winning teams. In the end, all class champions have been tested against a mid-range RoboCup entry and lost significantly: despite starting with a fairly detailed client code base, the students are not able to attain competitive world-class levels. However, given their time limitations this fact is neither surprising nor discouraging.

4 Results and Conclusion

The true measure of this course is the degree to which it has influenced students' decisions to pursue a career in computer science research, either positively or negatively. Perhaps it would be possible to devise a controlled study to measure a relevant statistic of this form. However, such a study has not been done and is beyond the scope of this paper.

There is, however, anecdotal evidence available, at least from the positive perspective. Both times the course has been run, at least one student has described the course in graduate school applications as a primary motivation for going on to do research. In addition, two students from the Fall 2002 version of the course actively contributed to the UT Austin entry in the RoboCup 2003 competition. In one case, the student played a key role in the winning entry in the on-line coach competition. The research involved may lead to this student's senior thesis.

Various course evaluations and surveys have also provided evidence that the students appreciate the opportunity to be exposed to the various components of scientific research. Some of the comments from students have included:

- "I really like reading from research literature. Just the fact that it's actual research provokes curiosity;"
- "The discussions we have in class are quite unique, I haven't had such involving discussions in any class before;"
- "Format of the class is perfect. I've waited through three years of college for a class like this.... I like the class so much that my other classes now disappoint me;" and
- "I loved this course, and it's the best possible way to be exposed to AI."

Another indication that the course has been successful is that many of the students' final projects have addressed interesting research topics. Of the 25 students who have successfully completed the course to date, 24 have done their final projects in the Soccer Server domain[5].

A rough breakdown by topic follows. The list does not account for exactly 24 projects because some projects (4) were done by pairs of students, and some are counted under more than one topic.

- Five projects made use of *machine learning* in some way. Two of these used reinforcement learning to learn aspects of the team, in one case just the goalie, and in another case, the entire defensive strategy. A third project learned models of player and ball motion so as to predict their positions when they are not visible to the player. One project aimed towards learning to recognize and predict breakaway situations, and another used on-line adaptive formations via opponent modeling. This last was the winner of the class tournament in 2002. During the competition, it displayed an ability to adjust its players' positions on the field appropriately during play.
- Four projects examined the efficacy of some general *agent architectures* in the robotic soccer domain. Specifically, they used a BDI model, cooperative behavior nets, a subsumption architecture, and a decision tree represenation.
- Three projects incorporated aspects of *ant-based* system design as presented by Parunak [12]. One modeled the entropy flow through the system. One built an ant-like system using the subsumption architecture and then attempted to improve the result via machine learning. One focused on finding the minimal reactive rules necessary to create a functioning team.
- Three projects focussed on inter-player *communication*. One of these developed a paradigm for communincation of high-level strategies. Another enabled players to communicate their world states. The other focussed on the challenge of quickly propagating information from a decision-maker to the rest of the team despite the limitations on message bandwidth enforced by the simulator.
- Three projects focussed on improving high-level general soccer *strategies*, including players' roles, teams' formations, etc.
- Three projects were geared primarily towards creating *winning teams*. In these cases, a wide range of challenges were addressed, often with a good

[5] The other did an interesting study of a multiagent-based approach to neural network formation using the Swarm package [9].

deal of heuristics and manual tuning of parameters. In general, low-level behaviors were the main emphasis. The teams resulting from these projects did indeed do well in the class tournaments, winning in 2001 and finishing 2nd and 3rd in 2002.
- Two projects implemented *planning* approaches to the Soccer Server domain. In both cases, planning was done dynamically during the course of the game. In one case, each individual constructed entire team plans and then only executed its own part of the best team plan.
- One ambitious project focussed on implementing the team from scratch (i.e. without the benefit of any code base). Whereas most projects have been programmed in C++ or Java due to the existing code bases in those languages, this project was programmed in Perl.
- One project facilitated short-term real-time coordination among a few agents for the purpose of executing certain *combination plays*, such as the "give-and-go" and the centering cross.
- One project addressed the challenge of maintaining an accurate world state despite only seeing a part of the environment at any given time. The approach was to predict the movements of other agents and the ball, and in the absence of visual information to update the player's world state accordingly.

Several of these projects were well-written and interesting enough to warrant follow-up experiments and (usually with a fair amount of additional effort) could possibly have led to eventual publications. However, so far, none of the reports has been extended in such a way.

All of the course materials are available on-line, including the RoboCup soccer server system and the CMUnited client code. As such, the course should be fully repeatable at other institutions. One goal of this paper is to fully describe the motivations behind the course so as to encourage such repetition. The primary goal has been to illustrate the successful use of a course on autonomous agents and multiagent systems, with project assignments in the RoboCup Soccer Server, as an introduction to CS research.

In the future, the course may be extended to include exposure to real robots as well as the Soccer Server. During the spring of 2003, the author taught a graduate level course[6] that culminated in a class entry in the RoboCup Sony four-legged robot league [18]. The opportunity to work with these legged robots may be extended to undergraduates in future iterations of the course described in this paper.

Acknowledgements

Thanks to Tucker Balch, Jeremy Cooperstock, Gal Kaminka, Manuela Veloso, and José Vidal for ideas from their related courses. The author also wishes to thank Vinay Sampath Kumar, a TA for the 2002 version of the course, as well as the NYU and UT Austin students who have participated in the course.

[6] `http://www.cs.utexas.edu/~pstone/Courses/395Tspring03`

References

1. Minoru Asada and Hiroaki Kitano, editors. *RoboCup-98: Robot Soccer World Cup II.* Lecture Notes in Artificial Intelligence 1604. Springer Verlag, Berlin, 1999.
2. Andreas Birk, Silvia Coradeschi, and Satoshi Tadokoro, editors. *RoboCup-2001: Robot Soccer World Cup V.* Springer Verlag, Berlin, 2002.
3. Rodney A. Brooks. Intelligence without representation. *Artificial Intelligence*, 47:139–59, 1991.
4. Silvia Coradeschi and Jacek Malec. How to make a challenging AI course enjoyable using the RoboCup soccer simulation system. In Minoru Asada and Hiroaki Kitano, editors, *RoboCup-98: Robot Soccer World Cup II.* Springer Verlag, Berlin, 1999.
5. Les Gasser. Mas infrastructure definitions, needs, and prospects. In *Proceedings of the Autonomous Agnets 2000 Workshop on Infrastructure for Scalable Multi-Agent Systems*, Barcelona, Spain, June 2000.
6. Maria Gini, Toru Ishida, Cristiano Castelfranchi, and W. Lewis Johnson, editors. *Proceedings of the first international joint conference on Autonomous agents and multiagent systems.* ACM Press, 2002.
7. Hiroaki Kitano, editor. *RoboCup-97: Robot Soccer World Cup I.* Springer Verlag, Berlin, 1998.
8. Hiroaki Kitano, Milind Tambe, Peter Stone, Manuela Veloso, Silvia Coradeschi, Eiichi Osawa, Hitoshi Matsubara, Itsuki Noda, and Minoru Asada. The RoboCup synthetic agent challenge 97. In *Proceedings of the Fifteenth International Joint Conference on Artificial Intelligence*, pages 24–29, San Francisco, CA, 1997. Morgan Kaufmann.
9. N. Minar, R. Burkhart, C. Langton, and M. Askenazi. The swarm simulation system: A toolkit for building multi-agent simulations, 1996. available at `http://www.santafe.edu/projects/swarm/overview/overview.html`
10. Itsuki Noda, Hitoshi Matsubara, Kazuo Hiraki, and Ian Frank. Soccer server: A tool for research on multiagent systems. *Applied Artificial Intelligence*, 12:233–250, 1998.
11. Itsuki Noda and Peter Stone. The RoboCup soccer server and CMUnited clients: Implemented infrastructure for MAS research. *Autonomous Agents and Multi-Agent Systems*, 7(1&2), July 2003. To appear.
12. H. Van Dyke Parunak. "go to the ant": Engineering principles from natural agent systems. *Annals of Operations Research*, 75:69–101, 1997.
13. Reid Simmons. Structured control for autonomous robots. *IEEE Transactions on Robotics and Automation*, 10(1):34–43, February 1994.
14. Elizabeth Sklar, Amy Eguchi, and Jeffrey Johnson. Robocupjunior: learning with educational robotics. In Gal A. Kaminka, Pedro U. Lima, and Raul Rojas, editors, *RoboCup-2002: Robot Soccer World Cup VI.* Springer Verlag, Berlin, 2003.
15. Peter Stone. *Layered Learning in Multiagent Systems: A Winning Approach to Robotic Soccer.* MIT Press, 2000.
16. Peter Stone, Tucker Balch, and Gerhard Kraetszchmar, editors. *RoboCup-2000: Robot Soccer World Cup IV.* Lecture Notes in Artificial Intelligence 2019. Springer Verlag, Berlin, 2001.
17. Manuela Veloso, Enrico Pagello, and Hiroaki Kitano, editors. *RoboCup-99: Robot Soccer World Cup III.* Springer Verlag, Berlin, 2000.
18. Manuela Veloso, William Uther, Masahiro Fujita, Minoru Asada, and Hiroaki Kitano. Playing soccer with legged robots. In *Proceedings of IROS-98, Intelligent Robots and Systems Conference*, Victoria, Canada, October 1998.
19. José M. Vidal and Paul Buhler. Teaching multiagent systems using RoboCup and biter. *The IMEJ of Computer-Enhanced Learning*, 4(2), 2002.

RoboCup in Higher Education: A Preliminary Report

Elizabeth Sklar[1], Simon Parsons[2], and Peter Stone[3]

[1] Department of Computer Science, Columbia University
1214 Amsterdam Avenue, New York, NY 10027, USA
sklar@cs.columbia.edu

[2] Department of Computer and Information Science
Brooklyn College, City University of New York
2900 Bedford Avenue, Brooklyn, NY 11210, USA
parsons@sci.brooklyn.cuny.edu

[3] Department of Computer Sciences, The University of Texas at Austin
1 University Station C0500, Austin, TX 78712-1188, USA
pstone@cs.utexas.edu

Abstract. Since team-based projects have been proven to be an effective pedagogical tool, we have been using RoboCup challenges as the basis for class projects in undergraduate courses. This paper unifies several independent efforts in this direction and presents early work in the development of shared resources and evaluation. We outline three courses and describe the related class projects in order to make the context of our investigation clear and to make it possible for others to replicate or extend our work, and contribute to the shared resource.

1 Introduction

Educational robotics, the use of robotics as a form of hands-on learning environment, is becoming increasingly common as robot kits are becoming more accessible and affordable [31]. Creative instructors are finding ways to teach science topics using these technologies, organizing tournaments around the robots; the energy, enthusiasm and motivation displayed by students of all ages is unsurpassed. We have found RoboCup – especially Soccer Simulation and the RoboCupJunior challenges – to be particularly conducive to college-level classroom use. The ability to demonstrate theoretical models and complex algorithms with a hands-on, accessible medium strengthens the learning experience.

In this paper, we document our experiences incorporating RoboCup activities into undergraduate courses with the idea of uniting others who are doing the same. Our aim is two-fold: one, to create a repository for related curricular materials; and two, to build a common instrument and database for evaluating the RoboCup learning environment. While we have found the link between RoboCup and traditional coursework in Introductory Robotics, Artificial Intelligence and Multiagent Systems to be a natural one, we presume that this arises out of our familiarity with RoboCup through longterm involvement with the initiative. In developing our repository, we are hoping to make the notion of incorporating RoboCup into such coursework a relatively easy task for uninitiated instructors, by providing syllabi, reading lists and project descriptions.

D. Polani et al. (Eds.): RoboCup 2003, LNAI 3020, pp. 296–307, 2004.

Others have experimented with the RoboCup paradigm in undergraduate classrooms. Coradeschi and Malec used the RoboCup soccer simulator in a course on Artificial Intelligence Programming[1] [11]. Birk has developed a course on Autonomous Systems that uses the Small-Size RoboCup League for practical exercises[2]. Vidal and Buhler [34] have developed a series of graduate level courses on multiagent systems using the RoboCup Simulation league.

We are also, of course, not the first to use robot kits in an undergraduate classroom as a hands-on learning environment. In 1989, Martin created the MIT Robot Design project course (6.270) [16, 18]. Yanco [36] has adopted this course, ending the term with a Botball[3] tournament. Mataric's "Introduction to Robotics" [20] takes a hands-on approach to the introduction of the basic concepts in the field of robotics and concludes with a contest where robots play a ball game in a hexagonal field. There are also courses using hands-on robotics that do not focus on teaching robotics as the main subject. Littman's "Programming Under Uncertainty" [15] teaches about methods for programming under uncertainty and a variety of machine learning techniques.

Aside from constructing a shared repository of course materials, we are interested in conducting a comprehensive evaluation of the pedagogical value of educational robotics in general and RoboCup activities in particular. Following on the work of Sklar et al. [30], we are interested in trying to pinpoint the educational value of robotics and the RoboCup initiative at the undergraduate level. Uniting multiple instructors means that we can not only share experiences, but also collect course evaluation data on a grander scale, allowing us to perform analysis across a broader cohort, with a range of academic as well as cultural backgrounds.

2 Robotics in Undergraduate Education

Here, we describe three classes where we have successfully used RoboCup challenges as term projects.

2.1 Introduction to Robotics

This introductory course looks at robotics from several aspects: technically, historically and socially. Many of the technical aspects are based on Mataric's course described above. The course is designed for non-engineering students to gain a hands-on experience with technology, as well as a basic understanding of the field of robotics and the challenges facing the field today. Part of the course is spent reading and discussing classic material that relates to robots – including non-technical aspects such as science fiction, psychology, cognitive science and education. The remainder of the course takes a hands-on approach to introducing the basic concepts in robotics, focusing on autonomous mobile robots. LEGO Mindstorms robots[4] are used, and students must complete two projects with them. First, they must build robots to execute a line-following

[1] `http://www.ida.liu.se/~silco/AIP/`
[2] `http://www.faculty.iu-bremen.de/birk/lectures/COURSES/autosys.html`
[3] `http://www.botball.org`
[4] `http://www.legomindstorms.com`

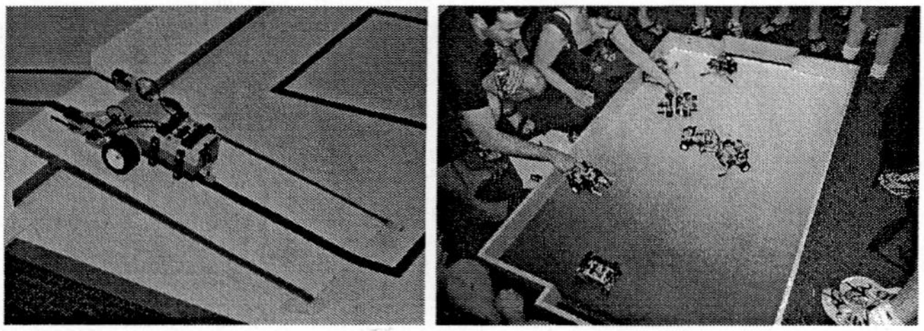

(a) Line-following maze (b) Soccer

Fig. 1. Robot contests.

task culminating in a maze contest. Second, they construct robots to play soccer and perform in a RoboCupJunior style two-on-two tournament.

The course begins with an introduction to robotics and agent-based artificial intelligence [28]. Then we cover some history and the basics of building and programming with LEGO Mindstorms [17, 19], using Not Quite C[5] [4, 5]. The robots are used as examples for the remainder of the topics, which introduce the general areas in robotics: effectors, sensors and control [18, 23]. The area of control is covered in more depth, discussing various architectures including deliberative, reactive, hybrid and behavior-based [1, 6, 8, 22]. Learning is also discussed [13, 21, 35]. Other topics presented include artificial life [3, 10], edutainment [14, 32], cognitive science and psychology [7, 24] and science fiction [2].

The course is taught over a 14-week semester. There is one 75-minute lecture and one 75-minute lab per week. There are two exams, and students submit written lab reports documenting their software and hardware developments. They are encouraged to record results of tests made and changes to their designs. Students also prepare a term project, presented both written and orally.

Sklar taught this course in Spring 2001 at Boston College[6]. Twenty-seven students were enrolled, three of whom were female. All were undergraduates, and there was a mix of ages: first year (1 student), second year (3), third year (10) and fourth year (13). The Computer Science Department at Boston College is in the School of Management and there is no engineering school in the university, so the hands-on technical experience of these students was limited. Sixteen members of the class were Computer Science or Information Technology majors. The rest came from Biochemistry (1 student), Communication (1), Economics (4), History (2), Marketing (1), Mathematics (1) and Physics (1).

The students were placed in groups of three for working on the robotics projects. Since the experience levels of the class was so diverse, Sklar assigned the groups, attempting to balance each group with an equal number of beginning and advanced students. Students were given some lab time during the scheduled course period in order to work on the projects. However, this was not enough time to perfect robots to perform well in the contests, so many of the students met outside of class time to work on the

[5] `http://www.baumfamily.org/nqc/index.html`

[6] `http://www.cs.columbia.edu/~sklar/teaching/spring2001/mc375/default.html`

robots. Each student was required to submit a lab report individually, which included an assessment of the contribution of their teammates. The efforts of team members are never balanced, however the inequities were obvious in reviewing the lab reports, even without the peer assessment component. Students' grades were based on the lab reports, not on their robots' performance in the contests.

The two contests were held in a public space and students were encouraged to invite their friends to watch. The excitement of the crowd and the visibility of the event motivated students to work harder after the first (maze) contest in preparing for the second (soccer) contest.

The term projects presented a major challenge for these students, who were not typically asked to do any writing in Computer Science classes. They were required to submit a brief project proposal several weeks prior to the final due date, in order to get them started and also to provide feedback about the appropriate nature of the topic. The range of topics chosen was quite broad, from the use of nanotechnology in surgical robots to the history of robots dating back to ancient Greece. Each student gave a ten-minute oral presentation on their chosen topic. This was difficult for many students who were not used to speaking in front of a class. Although discussion following the presentations was encouraged, very little actually occurred and several students skipped class on presentation days when they were not speaking. Course evaluation results (below) confirmed that the motivation surrounding the term project was minimal.

Students were given a survey at the end of the course. Forty-four percent of the class responded. The survey collected demographic data and also queried the students about their learning experience. They were asked to identify which elements of the course were helpful in learning the material and which elements of the assessment were valuable in helping them to solidify and demonstrate their knowledge of the subject. The results are shown in Figure 2.

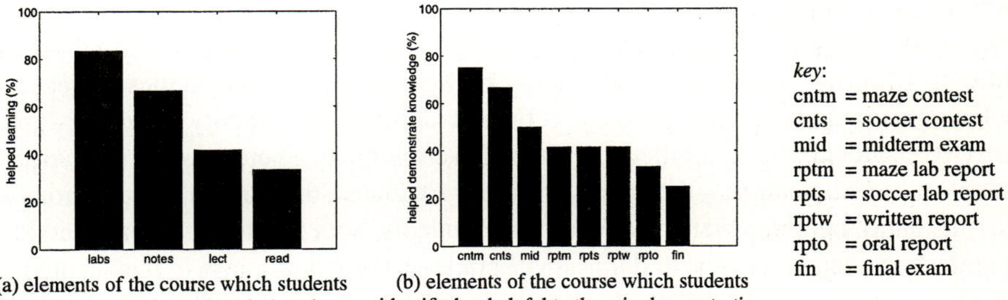

(a) elements of the course which students identified as helpful to them in learning the material

(b) elements of the course which students identified as helpful to them in demonstrating their knowledge of the material

Fig. 2. Survey Results from Introduction to Robotics course, Spring 2001.

Overwhelmingly (83%), the students felt that the labs (i.e., building and programming the robots) was helpful for learning the material, whereas only 33% said that the reading was helpful. Seventy-five and sixty-seven percent responded that the two contests (maze and soccer, respectively) were valuable in helping them solidify and demonstrate their knowledge of the material. This confirms our intuition that the hands-on components provide more effective learning experiences than other aspects of coursework, particularly at the introductory level. We speculate that the readings chosen were perhaps too advanced for most of the class.

Student comments were overall quite positive, including the following statements:

- "Great course... loved the lax atmosphere and hands-on experience. I'd recommend the course to any CS major."
- "I think the class idea is great. It is a great hands-on experience to try out. The labs were very fun times."

There were many comments that the mixed age group was helpful for all students, as the inexperienced students learned from the more advanced, and in assisting others, the advanced also learned more themselves. Negative remarks centered around requests for more lab time and less time spent on oral presentations.

2.2 Artificial Intelligence

The modern view of Artificial Intelligence (AI) [28] is that it is the study of intelligent agents – autonomous computing systems that perceive their environments and act upon them in a way that both responds to changes in their environments and works towards underlying goals. Robots are prototypical agents that have to move around, and react to, their environments in pursuit of their goals. Thus it is highly appropriate to explore areas of a typical artificial intelligence syllabus using robotics projects.

This course is designed to give a broad understanding of the basic techniques in use today for building intelligent computer systems. The syllabus broadly follows the outline of Nilsson's *Artificial Intelligence: A new synthesis* [25]. Students learn about state-space representations, problem reduction, means-end analysis, and reinforcement learning. They study search methods including depth-first, breadth-first and best-first search, as well as hill-climbing and alpha-beta pruning. Predicate calculus is introduced, along with various methods of theorem proving. The course is taught over a 14-week semester, with two 75-minute lectures a week, two exams and two robotics projects.

Parsons taught the course for the first time at Columbia University in Spring 2002[7]. Thirty-five students were enrolled, of whom 6 were female. There was a wide range of students taking the course – the bulk were undergraduates (68% of the 19 for whom we have this data), but there were also 6 graduate students, both Master's students and PhD students, and even within the undergraduate students there was a mix of ages from first year (1 student), second year (1), third year (8) and fourth year (4). The majority of the undergraduates were Computer Science majors (74%). The rest were from biology (1 student), economics (1), electrical engineering (1) and mechanical engineering (2).

The students formed themselves into groups of three to four and had to program LEGO robots in the Not Quite C programming language to perform two tasks – a RoboCupJunior style line-following rescue task (in which the robot had to follow a convoluted line, detect an obstacle and back-up, climb and descend a gradient, and finally detect and head towards a light source) and to play a simplified version of the RoboCupJunior soccer task (the robot started at one end of a standard RoboCupJunior

[7] `http://www.cs.columbia.edu/~sp/4701-2.html`

two-on-two soccer pitch, with the ball at the halfway point, and the robot had to manoeuvre the ball into the opposite goal, rather like a penalty kick into an empty goal). The culmination of the project was a contest in which the basis for competition was the cumulative time taken to complete both tasks, and the students also wrote a report on the project. Students were also given the option of an extra-credit project of building a dancing robot exactly as in the RoboCupJunior dance competition.

Two course evaluations were administered. One was an official evaluation done by the engineering school. The other was an informal paper-and-pencil survey given out in class. The results of the engineering school's evaluation showed that 55% of the 33 students who responded gave the robotics project they undertook a rating of 5 (on a 5-point scale) for interest, and two-thirds gave it a rating of 4 or 5. Twenty-one percent of the same cohort of students gave the project a rating of 5 for the amount learned during the project, and 58% rated it 4 or 5.

The informal survey probed more into the students' perception of the value of the project as opposed to other aspects of the course. In particular, students were asked to identify which aspects of the course most contributed to helping them learn the material, and which aspects were most helpful to them in demonstrating knowledge of the material. The results are given in Figure 3. These show that the students felt that the project work was not as helpful in learning as some of the more traditional aspects of Computer Science courses, but was more useful than the textbook and additional reading material (which no students felt were useful). The picture is much the same for the demonstration of knowledge, with students rating the contest as more helpful than the final, but less helpful than the homework and midterm. The report was rated least useful of all (they really hated having to write a report). Despite the rather unencouraging figures from this second survey, the very obvious enjoyment that the majority of the students took in the projects encouraged us to repeat the experiment the following semester.

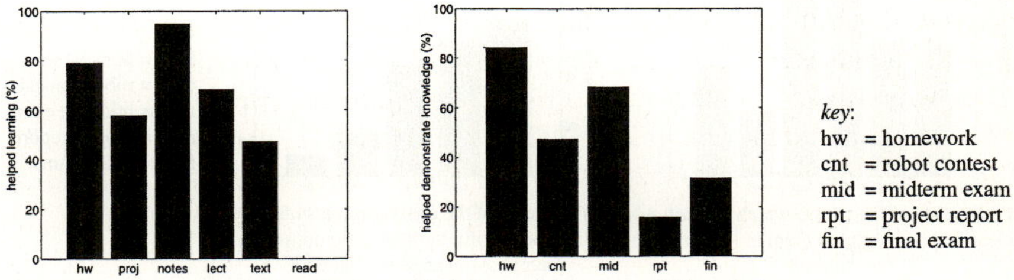

(a) elements of the course which students identified as helpful to them in learning the material

(b) elements of the course which students identified as helpful to them in demonstrating their knowledge of the material

Fig. 3. Survey Results from Artificial Intelligence, Spring 2002.

The second offering of the course was given by Parsons at Brooklyn College, City University of New York (CUNY) in Fall 2002[8]. With the exception of the absence of Masters and PhD students, the cohort was broadly similar to that at Columbia in terms of the factors we measured. Eighteen students were enrolled, of whom 7 were female.

[8] `http://www.sci.brooklyn.cuny.edu/~parsons/courses/cis32-fall-2002/`

Again there was a wide range of students taking the course – all were undergraduate students but there was a considerable mix of ages with 2 second year students, 3 third year, 10 fourth year and 2 fifth year students[9]. The majority of the undergraduates were Computer Science majors (16 students). The remaining students were from Political Science and Business.

In this offering, the project was given much later (due to problems with access to the robots) and while the teams were the same size (3 students), the project took the form of just the line-following exercise described above. Again the project ended with a contest and there was an extra-credit option to build a dancing robot.

Once again, an unofficial survey was administered, and the results are presented in Figure 4. These results are a little more encouraging than those from Columbia. This time the project was still felt to be less helpful in learning than lectures or lecture notes, but on a par with homework and more helpful than additional readings or the textbook. In terms of demonstrating knowledge, the students felt that the project was more helpful than either midterm or final. While gratifying, these figures should be viewed with some suspicion. First of all, these students had self-selected to do robot projects (students were allowed to do a non-robotics project instead if they preferred). Second, as a result of the other project, these figures are based on a very small sample of just 11 students. Finally, it seems that some of the reaction is because it is so unusual for students at Brooklyn College to get to do project work (in the open comment part of the survey several confessed that this was the only project they had ever done). The effect of this influence is supported by the fact that broadly similar results were generated by students who did the non-robotics project (though the very small number of students in this category makes the results extremely unreliable and so they are not presented here).

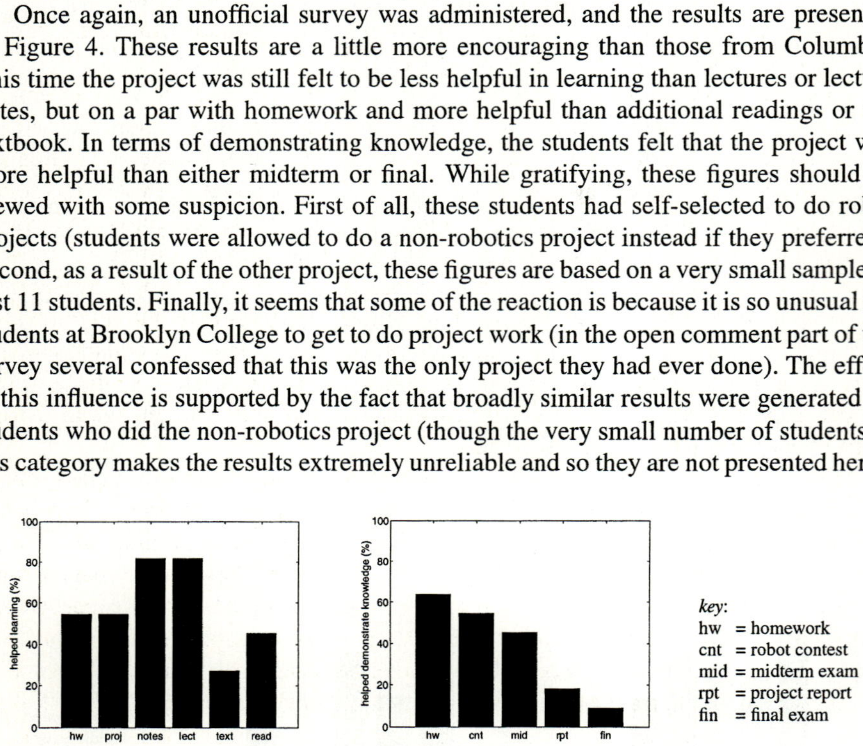

(a) elements of the course which students identified as helpful to them in learning the material

(b) elements of the course which students identified as helpful to them in demonstrating their knowledge of the material

Fig. 4. Survey Results from Artificial Intelligence, Fall 2002.

Comments from the student surveys from both offerings of the course include:

- "When working with the robot, I learnt that nothing is perfect in the real world. A lot of times the outcome is very unexpected."
- "It reminded me of why I want to stay away from hardware as much as possible."
- "It helped immensely! It helped me understand some of the concepts covered in the lecture."

[9] It is common for CUNY students to take more than four years to complete since many study part-time.

2.3 Autonomous Multiagent Systems

Autonomous agents and multiagent systems (AAMAS) is one of the fields to which RoboCup participants have contributed consistently and prominently over the years. Despite being the basis for a large subfield of AI, there is no generally accepted definition of artificial intelligence *agents*. In loose terms, agents are programs that (i) sense their environment, (ii) make decisions about how to act based on these sensations, and (iii) then execute these actions. Autonomous agents do all three of these steps on their own, i.e., without a human in the loop. Multiagent systems are collections of multiple agents that interact with one another. The field of AAMAS covers a wide variety of research foci and applications, including software-based information processing, robotic control of multiple agents, entertainment agents and tutoring agents [12].

This course provides a broad introduction to autonomous agents with an emphasis on multiagent systems. Topics include agent architectures, inter-agent communication, agent teamwork, distributed rational decision making, agent modeling, multiagent learning and entertainment agents. In addition to teaching about AAMAS, the course aims to introduce undergraduates to the full spectrum of research activities engaged in by professional computer science researchers, emphasizing the difference between these activities and the activities of a typical undergraduate student [33]. As such, the course includes an open-ended programming project, readings from the research literature, public speaking and writing requirements, and opportunities to collaborate with peers. In order for students to succeed, they need to attain a mastery of the AAMAS subject. However assessment is based primarily on their ability to engage in the full range of activities required of researchers.

The central focus of the course is a semester-long build-up towards a class robotic soccer competition in the RoboCup Soccer Server [26]. Students are assigned a series of four preliminary programming assignments designed to get them familiar with Soccer Server and the CMUnited client code [27]. By the end of these preliminary assignments, they have created a fully functional team (although not one that is particularly competent). The students are then encouraged to propose an improvement on this team as the topic for their final projects[10]. For example, one student proposed to use machine learning techniques to train a good goaltender without paying attention to the rest of the team.

The majority of the readings for this course are primary sources chosen both to introduce particular topics and to engender some controversy (e.g., reactive [9]) versus deliberative [29] agent architectures). To encourage the students to complete the readings in a timely fashion, they are required to submit a brief written answer to a single question pertaining to the readings at least 2 hours before the class starts. The fact that the responses are due two hours before class allows the instructor to incorporate them into the class discussion. Another effect of the questions is that the students *do* come to class prepared to discuss the readings. As a result there have been many extended and heated class discussions.

An important component of class participation is that each student is required to moderate at least one class discussion pertaining to that week's readings. They are in-

[10] They are also given the option to propose a programming project in a multiagent domain of their choice, but typically few students choose to do so.

structed to either defend a controversial statement or pose a question and be prepared to defend either side depending on how the class reacts. This activity turns out to be one of the most difficult for the students to complete. Many of them are not used to speaking in front of a class, and they have rarely been put in the position of *facilitating* discussions as opposed to defending specific positions.

The course requires a good deal of writing from the students. As above, the students are required to provide weekly written responses to questions related to the readings. They receive feedback pertaining to the clarity and soundness of their responses. More significantly, the students are required to write three documents pertaining to their final projects. First, they write project proposals defining their goals for their projects as well as the proposed methods for achieving them. Second, they revise their proposals and add a section on their work in progress to create progress reports. Finally, they write final reports in the format of conference papers.

Students optionally work in pairs on the final project. Teams must write their proposals and reports individually, with clear indications of what role each person played in the collaboration; and as such, more is expected from them as a final product. The robotic soccer project lends itself to such collaboration nicely since there are many different ways in which the students can divide up the work.

The class culminates in a simulated soccer tournament. The students are told at the outset that performance in the tournament will have no negative impact on their grades (while a strong performance *can* have a positive impact). Nonetheless, the tournament is a strong motivational factor for the students. Visitors are invited to the event and the students present their approaches orally and field questions as their teams are playing. The performance spread among the teams is often very large, especially given the fact that some students do not focus on creating winning teams. All class champions have been tested against a mid-range RoboCup entry and lost significantly: despite starting with a fairly detailed client code base, the students are not able to attain competitive world-class levels. However, given their time limitations this fact is neither surprising nor discouraging.

Stone has taught this course twice, first at New York University in the Fall of 2001. Fifteen students were enrolled, only one of whom was female. All students were Computer Science majors. All were graduate students: 12 masters and 3 Ph.D. The second offering of the course was at the University of Texas at Austin during Fall of 2002[11]. Again, fifteen students were enrolled, however none were female. This time, the cohort were undergraduates. Fourteen were seniors (fourth year) and one was a junior (third year). Most students were Computer Science majors, with the remainder majoring in Computer Engineering.

Course evaluations and surveys were administered at the conclusion of both courses. At NYU, the course was rated 4.5 out of a possible 5 (highest rating). At UT Austin, the course was rated 4.6 out of a possible 5. Student comments have also provided evidence that the students appreciate the opportunity to be exposed to the various components of scientific research. Both times the course has been run, at least one student has described the course in graduate school applications as a primary motivation for going on to do research. In addition, two students from the Fall 2002 offering are actively contributing

---

[11] `http://www.cs.utexas.edu/~pstone/Courses/378fall02`

to the UT Austin entry in the RoboCup 2003 competition. In one case, the research is leading up to the student's senior thesis.

An informal survey was administered in the middle of the term to the UT Austin cohort. Students were asked to rate the programming assignment on a scale of 1 to 5. Thirty-three percent gave it the highest rating; 57% gave it the second highest, while 10% scored it average and no students entered low marks. Students also rated the reading assignments on the same scale. Twenty percent gave the highest rating; 60% gave the second highest, 13% scored it average and 7% gave the lowest rating.

Some of the comments from students have included:

- "The discussions we have in class are quite unique, I haven't had such involving discussions in any class before;"
- "Format of the class is perfect. I've waited through three years of college for a class like this.... I like the class so much that my other classes now disappoint me;" and
- "The only thing I dislike about the class is that we are limited in our application of our knowledge. Our education in AI is directed at implementing a RoboCup soccer agent. I feel that if we were able to apply our knowledge to other aspects, we would gain an even better understanding of artificial intelligence."
- "The simulator code was kinda tricky to understand."

3 Discussion

The three courses described above offer an interesting comparison, not only in terms of content and presentation but also in regard to the cohorts of students enrolled. Collectively, the courses have been offered five times at five different universities, providing a broad range of backgrounds, demographics and experience levels – from first-year undergraduates at a private, non-engineering college to fourth-year undergraduates at a large state university, and including graduate students from both a public city university and two large private universities. Thus the positive feedback across the board in regard to the robotics projects is an encouraging and significant factor.

Comments about the reading materials were typically less enthusiastic, as the figures presented in the previous section indicate. Stone's course, where students were required to respond to the reading in short written assignments and were then given the opportunity to discuss the readings formally in class, fared better than the other courses, where reading was assigned in a more traditional manner – without written reading responses and primarily the material was presented by the instructors in lectures where questions were encouraged (but infrequent) and discussion was not the central theme. Comparatively, Stone's classes had fewer students, so more effective discussion was possible. Nonetheless, several students from all the courses commented that they wished there had been better connections between the readings and the project work. This type of feedback is valuable to us and our colleagues in improving the existing courses as well as designing new ones.

The challenge presented to students who were required to make oral presentations to the classes (in both Sklar's and Stone's courses) is also notable. No matter what career path is ultimately taken, students need to know how to communicate their ideas. The development of oral presentation skills is important and should be encouraged, despite students' dislike of this aspect of the courses. Perhaps more creative ways of oral reporting can be incorporated into all the courses.

The evaluations performed on Stone's course so far is limited to generalized questions about whether students liked the course and standard questions about the instructor and workload. While this level of information is useful to administrators, we are interested in gathering more specific data on the students' learning experiences. The discrepancy in evaluation methodology from one course offering to another is one of the factors that has spurred us to create the repository mentioned here. This will include a standard instrument for measuring the effectiveness of specific coursework and the general RoboCup learning environment.

4 Summary

We have presented our development of a repository for educational robotics activities, particularly focused on RoboCup challenges. Our goal is to build an on-line space for sharing curricular materials and to develop a unified instrument and database for evaluating the RoboCup learning environment. As examples of the type of information we are seeking and archiving, we have presented an account of our collective experiences incorporating RoboCup activities into undergraduate courses.

We will continue our efforts with the existing courses described above as well as development of new courses. The Autonomous Multiagent Systems course will be offered in Fall 2003 by Parsons at the CUNY Graduate Center and in Spring 2004 by Stone at the University of Texas at Austin[12]. Sklar is currently adapting her course to an introductory computer science curriculum, using the robotics as a basis for demonstration. The survey instrument will be adapted to both courses and administered at the end of each term.

We hope to encourage others to join in this community venture. Our on-line repository can be found evolving at `http://agents.cs.columbia.edu/er`. We welcome contributions and participants.

References

1. R. C. Arkin. *Behavior-Based Robotics*. MIT Press, 1998.
2. I. Asimov. *I, Robot*. Doubleday, Garden City, NY, 1950.
3. T. Balch, Z. Khan, and M. Veloso. Automatically Tracking and Analyzing the Behavior of Live Insect Colonies. In *Proc of Agents-01*, Montreal, 2001.
4. D. Baum. *Dave Baum's Definitive Guide to LEGO Mindstorms*. APress, 2000.
5. D. Baum, M. Gasperi, R. Hempel, and L. Villa. *Extreme MINDSTORMS: An Advanced Guide to LEGO MINDSTORMS*. Apress, 2000.
6. A. Birk. Behavior-based robotics, its scope and its prospects. In *Proc of 24th IEEE Industrial Electronics Society Conf.* IEEE Press, 1998.
7. V. Braitenberg. *Vehicles: Experiments in Synthetic Psychology*. MIT Press, 1984.
8. R. Brooks. A robust layered control system for a mobile robot. *IEEE Journal of Robotics and Automation*, 2(1), 1986.
9. Rodney A. Brooks. Intelligence without representation. *Artificial Intelligence*, 47, 1991.
10. A. Colorni, M. Dorigo, and V. Maniezzo. An Investigation of some Properties of an Ant Algorithm. In *Proc of PPSN-92*. Elsevier Publishing, 1992.

[12] `http://www.cs.utexas.edu/~pstone/Courses/378spring04`

11. S. Coradeschi and J. Malec. How to make a challenging AI course enjoyable using the RoboCup soccer simulation system. In *Proc of RoboCup-98*. Springer Verlag, LNAI 1604, 1998.
12. Maria Gini, Toru Ishida, Cristiano Castelfranchi, and W. Lewis Johnson, editors. *Proc of AAMAS-02*. ACM Press, 2002.
13. I. Harvey, P. Husbands, and D. Cliff. Issues in Evolutionary Robotics, Cognitive Science Research Paper Serial No. CSRP 219. Technical report, Univ of Sussex, Brighton, UK, 1992.
14. H. Kitano, M. Asada, Y. Kuniyoshi, I. Noda, and E. Osawa. RoboCup: The Robot World Cup Initiative. In *Proc of Agents-97*, 1997.
15. M. Littman. CPS196: Programming Under Uncertainty. `http://www.cs.duke.edu/~mlittman/courses/cps196/`
16. F. Martin. 6.270: The MIT LEGO Robot Design Project Competition. `http://fredm.www.media.mit.edu/people/fredm/projects/6270/`
17. F. Martin. Kids Learning Engineering Science Using LEGO and the Programmable Brick. In *Proc of AERA-96*, 1996.
18. F. Martin. *Robotic Explorations: A Hands-On Introduction to Engineering*. Prentice Hall, 2000.
19. F. Martin, B. Mikhak, M. Resnick, B. Silverman, and R. Berg. To Mindstorms and Beyond: Evolution of a Construction Kit for Magical Machines. *Robots for Kids: Exploring New Technologies for Learning*, 2000.
20. M. Mataric. CS 445 Introduction to Robotics, a LEGO-kit-based hands-on lab course. `http://www-scf.usc.edu/~csci445/`
21. M. J. Mataric. Reward Functions for Accelerated Learning. In W. W. Cohen and H. Hirsh, editors, *Proc of Machine Learning Conf.* Morgan Kaufmann Publishers, 1994.
22. M. J. Mataric. Behavior-Based Control: Examples from Navigation, Learning, and Group Behavior. *Journal of Experimental and Theoretical Artificial Intelligence, special issue on Software Architectures for Physical Agents*, 9(2-3), 1997.
23. P. J. McKerrow. *Introduction to Robotics*. Addison-Wesley Publishing Co., 1991.
24. M. Minsky. *Society of Mind*. Picador, 1987.
25. N. J. Nilsson. *Artificial Intelligence: A New Synthesis*. Morgan Kaufmann, 1998.
26. Itsuki Noda, Hitoshi Matsubara, Kazuo Hiraki, and Ian Frank. Soccer server: A tool for research on multiagent systems. *Applied Artificial Intelligence*, 12, 1998.
27. Itsuki Noda and Peter Stone. The RoboCup soccer server and CMUnited clients: Implemented infrastructure for MAS research. *Autonomous Agents and Multi-Agent Systems*, 7(1&2), July 2003. To appear.
28. S. J. Russell and P. Norvig. *Artificial Intelligence: a Modern Approach*. Prentice Hall, 1995.
29. Reid Simmons. Structured control for autonomous robots. *IEEE Transactions on Robotics and Automation*, 10(1), Feb 1994.
30. E. Sklar, A. Eguchi, and J. Johnson. Robocupjunior: learning with educational robotics. In *Proc of RoboCup-2002*, 2002.
31. E. Sklar and S. Parsons. Robocupjunior: a vehicle for enhancing technical literacy. In *Proc of AAAI-02 Mobile Robot Workshop*, 2002.
32. E.I. Sklar, J. Johnson, and H.H. Lund. Children Learning from Team Robotics: RoboCup Junior 2000 Educational Research Report. Technical report, The Open Univ, UK, 2000.
33. Peter Stone. RoboCup as an introduction to CS research. In *Proceedings of the RoboCup-2003 Symposium*, Padova, Italy, July 2003.
34. José M. Vidal and Paul Buhler. Teaching multiagent systems using RoboCup and Biter. *The IMEJ of Computer-Enhanced Learning*, 4(2), 2002.
35. R. A. Watson, S. G. Ficici, and J. B. Pollack. Embodied Evolution: Embodying an Evolutionary Algorithm in a Population of Robots. In *Proc of CEC-99*, 1999.
36. H. Yanco. 91.450: Robotics I. `http://www.cs.uml.edu/~holly/91.450/`

Scaffolding Children's Robot Building and Programming Activities

Peta Wyeth, Mark Venz, and Gordon Wyeth

School of Information Technology and Electrical Engineering
University of Queensland
Brisbane, Australia
{peta,mvenz,wyeth}@itee.uq.edu.au

Abstract. Since 2001 the School of Information Technology and Electrical Engineering (ITEE) at the University of Queensland has been involved in RoboCupJunior activities aimed at providing children with the Robot building and programming knowledge they need to succeed in RoboCupJunior competitions. These activities include robotics workshops, the organization of the State-wide RoboCupJunior competition, and consultation on all matters robotic with schools and government organizations. The activities initiated by ITEE have succeeded in providing children with the scaffolding necessary to become competent, independent robot builders and programmers. Results from state, national and international competitions suggest that many of the children who participate in the activities supported by ITEE are subsequently able to purpose-build robots to effectively compete in RoboCupJunior competitions. As a result of the scaffolding received within workshops children are able to think deeply and creatively about their designs, and to critique their designs in order to make the best possible creation in an effort to win.

1 Introduction

This paper describes the RoboCupJunior activities implemented by the School of Information Technology and Electrical Engineering (ITEE) at the University of Queensland. RoboCupJunior aims to engage 10 to 17 year old children in robot building and robot programming through structured challenges and competitions. The University of Queensland has organized two competitions (in 2001 and 2002) featuring the three RoboCupJunior challenges of Dance, Rescue and Soccer. The competitions themselves have been structured to account for age and opportunity differences, and to provide an environment where learning continues to take place.

The most significant efforts, however, have been placed in the development of appropriate workshops to introduce children and teachers to the RoboCupJunior program and the associated technology – most notably the LEGO® RCX™ and ROBOLAB™ products. In order for the workshop initiative to achieve this goal extensive efforts have been made to ensure that workshops teach children the fundamentals of robot building and programming in an engaging and meaningful way. The teaching methods incorporated in ITEE robotics workshops are outlined in this paper.

The robotics workshops provided by the School of Information Technology and Electrical Engineering have proven to be incredibly successful with over 2200 chil-

D. Polani et al. (Eds.): RoboCup 2003, LNAI 3020, pp. 308–319, 2004.

dren across 60 schools in Queensland participating. Teachers, children, school administrators and government bodies have embraced the initiative. The success of the workshops is primarily due to their ability to provide the necessary robotics education to support children's robot building activities. Queensland students are the current (2002) world champions in the soccer challenge. Furthermore, Queensland students were winners in every challenge at both age levels in the Australian championships in 2002. Queensland's successes in both national and international RoboCupJunior competitions are testament to the success of the initiative.

The development of the workshop program has been guided by the knowledge that many children require scaffolding in their robot building endeavors. Scaffolding allows children to acquire the knowledge they need to become independent robot builders and programmers. This paper discusses the workshop development process, the underlying theoretical model, the practical implementation of this model, and the improvements which have been made based on student and teacher feedback. Empirical data from both workshops and RoboCupJunior competitions is used to explore the extent to which the initiative has provided children with the scaffolding they require to become successful robot builders and programmers.

2 Background

The ITEE Robotics Workshop initiative is guided by the belief that the most powerful way to learn about technology is to become a creator of technology. One of the most effective means by which children can create technology is to develop an understanding of the dynamic and programmable properties of that technology. This is an idea that has been advocated by many in the past 20 years. Papert, in his landmark work Mindstorms [9] recognized that computer programming as an educational activity had great potential as a vehicle for the acquisition of useful cognitive skills such as problem solving and reflective thinking. Other researchers have also identified the importance of allowing children to experience the unique dynamic and programming properties of computers and in doing so allowing children to become creators, not just consumers, of computing activities [11], [5], [15].

Researchers at MIT continued the work of Papert [9], continuing his vision of computing in which children explore ideas by constructing their own computer programs. Resnick and his group at the MIT media lab based their research on this philosophy. They started with the development of LEGO/Logo [12] which combined the LEGO® Technic™ product with the Logo programming language providing children with an environment where they could build and program robots.

Robot building and programming is a natural – and exciting – extension of computer programming activities. Through building and subsequently programming robots, children are building agents which they can program to perform a wide variety of different behaviors. This process allows children to directly see the consequences of their programming activities – the resultant robot behaviors. In this robot building process children have become empowered as they purposefully create robots to achieve a specific function. Researchers are currently exploring classroom technologies that enable children to learn from construction. Many researchers have identified that technology which supports children becoming involved in design projects provides rich opportunities for learning [4], [6], [8].

The final assumption that underpins the Robot Workshop initiative revolves around the importance of scaffolding for children in their robot building endeavors. Vygotsky, a prominent development psychology, was the first to advocate that complex forms of thinking have their origins *social* interactions [2]. Scaffolding – a process whereby important activities are modeled through cooperative dialogues between a skillful tutor and a novice – is an important feature of social collaboration that fosters cognitive growth [2], [14]. Subsequently, guided participation model form the foundation of the ITEE robotics workshops. Within this model, the workshops provide structured learning activities that are carefully tailored to the children's abilities. Tutors are available to provide helpful hints and instructions, to monitor the children's progress. The tutors gradually reduce their levels of support as the children become more confident and competent.

RoboCupJunior is an excellent context within which children can be introduced to the field of robotics [7], [16]. The RoboCupJunior robotics competitions provide an additional level of importance to the robot building activities of children. Through the competition children are able to work in teams to create competitive robots. This competitive environment motivates children to work of creating robots that are skillfully able to complete specific tasks.

2.1 RoboCupJunior Australia

RoboCupJunior is a project-oriented educational initiative that organizes local, regional and international robotic events for young students [13]. Within RoboCupJunior three team challenges have been developed:

- **Soccer:** 2-on-2 teams of autonomous mobile robots play games in an 1800mm x 1200mm field. The soccer challenge in the Queensland RoboCupJunior competition is open to the senior participants, aged between 14 and 18 years.
- **Rescue:** Robots race to rescue victims from artificial disaster scenarios, varying in complexity from line-following on a flat surface to negotiating paths through obstacles on uneven terrain. For the Australian competition, the robot is required to find its way to a hazardous area – following a contrasted line – to rescue the victim. The challenge is open to the middle school entrants, aged from 10 to 15.
- **Dance:** One or more robots perform to music, in a display that emphasizes creativity of costume and movement. Within the Australian RoboCupJunior competition the Dance challenge is split into two age categories, junior (10 - 12yrs) and senior (13 - 18yrs).

Children between the ages of 10 and 18 produce a robot or robots to compete in one or more of these three challenges. Australian RoboCupJunior competitors primarily use, but are not limited to, the LEGO® MINDSTORMS™ robot construction environment to create their robots. The LEGO® MINDSTORMS™ Kits provide children with an environment in which they can create and program robots. The LEGO® products are comparatively inexpensive and most importantly reusable, allowing children to easily work through create-improve-demolish processes.

2.2 Robot Building with LEGO

At the core of the LEGO® MINDSTORMS™ Kit is the RCX™ brick. The RCX brick is an autonomous microcomputer embedded in a LEGO brick (seen in Figure 1) that can be programmed to serve as the "brain" of any LEGO construction [1]. The RCX is programmable, microcontroller-based brick that can simultaneously operate three motors, three sensors, and has an infra-red serial communications interface [10].

Fig. 1. The LEGO RCX Brick, a light sensor and the motor that children use to build robots.

The key elements necessary for using the RCX are the RCX brick itself, an infra-red transceiver, and a personal computer. Additional components, such as motors, sensors, and other building elements, in combination with this base system allow the creation of functional autonomous robotic devices [10]. LEGO provides an array of analog sensors capable of measuring light intensity, rotation and touch as well as a DC motor (see Figure 1). Within a LEGO MINDSTORMS kit there are also gears, wheels, axles and bricks which in combination with the other elements provide a comprehensive robot-building environment. All of the LEGO parts are self contained units allowing users the opportunity to create robots without the having to machine their own structures or design electronics components. LEGO MINDSTORMS provides both children and adults with opportunities to develop robots that move, think, and react.

2.3 ROBOLAB

ROBOLAB is a software development environment designed for use in the programming of RCX-based creations. The programs created using ROBOLAB can be downloaded to the RCX using the infrared transceiver. The RCX can then run the program independent of the computer.

The ROBOLAB software development environment is predominantly used within the ITEE robotics workshops. ROBOLAB is an iconic programming environment. The icons represent actions that the robot may perform as well as programming structures such as loops and decision statements, and commands. Users construct programs by selecting, placing and connecting icons in a ROBOLAB diagram.

ROBOLAB has a number of levels to accommodate the varying abilities of students. Pilot is the basic elementary section and Inventor is designed for use by more advanced students. While both use icons to represent commands or structures, within the Pilot section the number and order of icon options is restricted to ensure the success of the user.

The second category is Inventor. Inventor has been designed to meet the needs of students in the middle and upper grades of school. This category allows the users access to all of the ROBOLAB programming icons. As a result, students have the freedom to design programs of their choice.

3 Robotics Activities Coordinated by ITEE

The School of Information Technology and Electrical Engineering offers a number of robotics activities to schools. Three-hour robotics workshops are conducted during school terms to provide children interested in robotics the knowledge necessary to independently create their own robots. Due to demand, the workshops have been expanded to include groups of teachers and student teachers. For the past two years ITEE has also offered three-day summer camps designed to provide children with a greater understanding of the LEGO robot building and programming process. These workshop and camp activities culminate with the annual RoboCupJunior Queensland competitions. In addition to these activities, teachers, school administrators and government bodies have actively sought the advice of ITEE robotics staff with respect to robotics curriculum issues. ITEE has provided assistance though a consultation process. Each of these activities is discussed in detail in the following sections.

3.1 Robotics Workshops

The University of Queensland's School of Information Technology and Electrical Engineering has been running robotics workshops since 1995 [17]. In 2001, these workshops were redesigned specifically for the RoboCupJunior initiative. They are open to school children and operate during the school terms. The workshops are three hours long. During the workshops students build a robot and then spend time programming the robot to perform simple tasks. Due to demand, the workshop program has been extended recently to include teachers and student teachers.

The workshops are designed to allow students to work at their own pace through the building and programming processes. Students work in pairs for most workshops, however when workshop numbers are large they may work in groups of three. Each pair or group of three are provided with a computer with the ROBOLAB software installed, an RCX brick and an infra-red transceiver, as well as motors, sensors, and general LEGO bricks. Experienced tutors are available to answer questions and provide support. There are usually two tutors who participate in each of the workshops.

There are two levels of workshops: beginner and intermediate. Beginners are defined as those students who have not used ROBOLAB. These students work through activities which outline the fundaments of the ROBOLAB environment and guide them through the creation of simple programs. Intermediate students, those who have already participated in a beginner's workshop or who have used the ROBOLAB programming environment elsewhere, are given more complex activities to complete.

3.1.1 Workshop Participants

Workshop and summer camp participants are school children between the ages of 10 and 17. Children usually come to the workshops as a school excursion. Over the last year approximately 2200 students from 60 different schools have attended the robotics workshops. Participants are predominantly from schools in the Brisbane metropolitan area, but participants have also attended from Northern NSW, North Queensland and some rural areas. Both public and private schools have attended the workshops; however the majority has been public schools.

Approximately 60% of students who attend the robotics workshops are from primary schools and are aged between 10 and 12. For the primary schools workshops generally include between 25 and 30 students. Workshops for older children who attend secondary schools usually comprise between 10 and 15 students. Students are predominately volunteers who have a keen interest in robotics. They are accompanied by teachers and mentors and these adults are encouraged participate as well.

Although both boys and girls attend the robotics workshops generally there are a greater number of boys in attendance. Boys make up approximately 60% of participants from the co-educational schools that participate.

3.1.2 Robot Building

During the ITEE robotics workshops participants are required to build the specified robots and subsequently program it using specially designed worksheets. The robot that the participants use in their programming exercises is a differential drive robot (wheel-chair configuration) has two motors (a motor driving each wheel), two light sensors which are capable of reading levels of light intensity. Both of these sensors are trained onto the ground and the light intensity measured is that reflected from their own light source.

Students are initially directed to the website ITEE RoboCupJunior website http://www.itee.uq.edu.au/~robocup/junior/, which contains the build instructions for the LEGO™ robot that is used in the workshops. This robot takes between 45 to 90 minutes to build depending on the robot building skills and experiences of the participants. Workshop attendees construct a robot that has been designed by the workshop tutors making this process one of "build-by-numbers". This strategy has been put in place for two reasons:

1. Evidence from early workshops which allowed children to construct their own robots suggested that children could easily spend the three hours playing with the LEGO™. By providing the children with a robot "recipe", they move on to programming tasks more readily.
2. The robot used in the workshops is a structurally sound design. The robot is robust enough to survive falls and collisions. In addition, from building such a robust LEGO structure it is intended that children learn some of the principles of sound LEGO construction.

Fig. 2. The robot that participants build in the ITEE robotics workshop.

Figure 2 depicts the robot that children construct in the workshop. The build instructions are pictorial. This enables children with limited LEGO™ construction experience to successfully construct a robust robot. Each step in the build instructions shows the LEGO™ piece as it is about to be placed, with an arrow indicating its destination. The next image shows the piece in position. Where possible an additional illustration depicts the intended destination of those pieces just connected. Colored backgrounds are used to separate differential the required parts, the steps, and sub-assemblies. There is no text describing new parts.

On completion of the workshop robot, children then move on to robot programming activities. Children generally create programs for their robots for the remainder of the workshop. They generally spend approximately 90 minutes creating programs for the robots. These activities are described to children in a series of worksheets.

3.1.3 Worksheets

All of these worksheets use ROBOLAB at the Inventor 4 level. This level provides access to all available icons, but requires the users to manually connect added icons. The worksheets are designed so that by progressing through the worksheets the children gradually build up knowledge of how to get their robot to produce certain behaviors. The purpose of the worksheets is twofold:

1. to describe the development environment and its associated syntax; and
2. to scaffold the novice programmer by guiding through the creation of simple programs.

Each worksheet introduces a concept or a syntax requirement. The format of the worksheets is such that they provide step-by-step instructions of how to construct a particular program. The description is text-based and is supported by an image of what the program should look like at that step. Figure 3 below provides an example of a worksheet activity.

Children who are new to the ROBOLAB development environment participate in the introductory workshop. While there are nine worksheets in all, during an introductory workshop students have only to complete the first five worksheets. By the end of these five worksheets, the students have been shown how to use ROBOLAB to program their robot to turn a motor on and off, travel in a straight line, turn 180 degrees, stop when a dark color is detected and consistently follow a dark line. The children in Figure 4 are attendees at introductory workshops.

A second intermediate workshop is available for children who have completed the beginner's workshop or who have had previous experience with ROBOLAB. During this second workshop children are given the opportunity to explore programming concepts related to structured programming. They complete worksheets 6 to 9. By the end of the intermediate workshop children are able to create loops, program robots to deal with decisions, use variables and apply their knowledge in construction of their own line following program.

3.1.4 The Evolution of ITEE Robotics Workshops

The ITEE Robotics workshops have evolved over the last two years. Improvements made have been in response to issues identified by teachers, students and tutors. There has been a gradual refinement of workshops to meet the needs of participants. The processes outlined above are the result of this refinement.

Over the past two years the robot construction instructions have improved considerably. These improvements are based on feedback received from children. The first series of robot build instructions used both pictures and text. While using these instructions the children were often observed asking questions that they would have

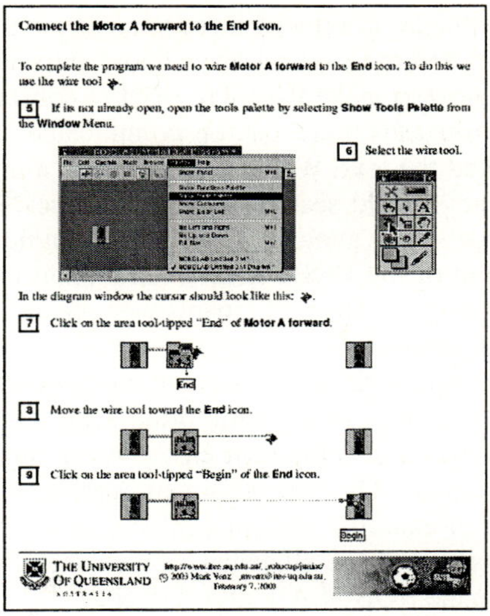

Connect the Motor A forward to the End Icon.

To complete the program we need to wire **Motor A forward** to the **End** icon. To do this we use the wire tool.

5 If its not already open, open the tools palette by selecting **Show Tools Palette** from the **Window** Menu.

6 Select the wire tool.

In the diagram window the cursor should look like this:.

7 Click on the area tool-tipped "End" of **Motor A forward**.

End

8 Move the wire tool toward the **End** icon.

9 Click on the area tool-tipped "Begin" of the **End** icon.

Begin

THE UNIVERSITY OF QUEENSLAND

Fig. 3. The robotics worksheets provide step-by-step instructions on how to complete a particular programming task.

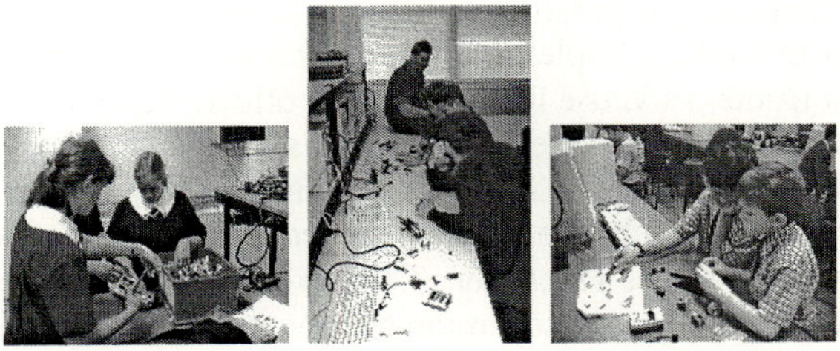

Fig. 4. Robotics workshops engage children aged between 10 and 17 in robot building and programming activities.

known the answers to if they had read the text. In response the children's failure to read the important text associated with images, the second generation robot construction instructions primarily contained images. An image of the parts needed for each step was shown in a table above each step. A textual description of the parts and how many were needed was also given. Questions arising from these build notes where primarily about which part was required. These questions were answered by tutors referring children to the text accompanying the parts table.

The final version of the build instructions contains images only. The only text in these instructions is sequence numbers and size information. As mentioned, these notes have added a step which shows the LEGO™ piece as it is about to be placed, with an arrow indicating its destination. The number of queries during the construction of the robot has decreased dramatically. The most common question asked is "How did you make those pictures?"!

The programming worksheets also went through a development cycle. The first version only contained one worksheet. This worksheet asked the students to program the robot with a number of exercises. The worksheet familiarized participants with the development environment and introduced them to concepts of actions states, program looping, binary decisions and message passing. It also introduced the concept of multitasking and described how to make the robot react to sensor readings. Observations suggested that these worksheets were far too complicated for the children. For the most part students ignored the text. When challenged by a problem, they would consult the tutors, before they would search the text for clues. Exercises which had no example program were generally avoided. Observations indicated that the participants had no real understanding of the ROBOLAB environment or programming on completion of the workshop session. The participants were generally unable to create their own programs without assistance.

Based on these observations, the programming worksheets were changed dramatically. The second worksheet was greatly simplified. Children were given more detail on how to use the ROBOLAB development environment and provided greater detail about the ROBOLAB syntax. The second set of worksheets provided ROBOLAB iconic solutions to each of exercise, and an explanation of the principles underlying the solution. Observations of these students showed that the worksheets had only partially helped them use the ROBOLAB environment.

The third set of worksheets reverted to a more text based approach, but covered the description of the environment and solutions to the exercises by progressing step by step through the exercises with supporting images. Again the bulk of the text in the worksheets was ignored. Information hidden in the text was not found. Empirical evidence suggests that after completing the worksheets in these early workshops only about 1/3 of participants could use the light sensor effectively.

Within the current worksheets each exercise is now a separate worksheet. Each worksheet introduces a concept or a syntax requirement. Again the worksheets provide a step per action, but each step is accompanied by an image of what the program should look like at this stage. The order of the exercises was also changed with the introduction of using the light sensor in the third worksheet. In earlier iterations, use of the light sensor was the final concept covered by students in the workshop. This final set of worksheets covers less programming information than earlier worksheets, instead concentrating on the functionality of key icons, and focusing on how to use the ROBOLAB development environment. The new worksheets are based on observations across many workshops. They have been designed to prevent students from encountering the programming difficulties common in earlier workshops.

3.2 RoboCupJunior Queensland Competitions

The RoboCupJunior Competitions provide a strong motivation for children to build robots, and to critically think about their robot creations. The competitive aspects encourage the children to think deeply and creatively about their designs, and to critique their designs in order to make the best possible creation in an effort to win.

However, the competition is more than motivation – it is a great educational opportunity. The competition is a gathering of student minds; an opportunity for students to share their ideas. Students are keen to share, and are usually ready to offer advice. This was first observed during the 2001 competition, where some of the teams that fielded non-functional robots at the start of the competition had robots that functioned well by the end. When asked how this had come about, the students responded that they had received assistance from other students: their competitors.

In 2002, inter-team interaction was encouraged further by the complete exclusion of adults from the team setup and practice areas. Teachers and parents were invited to sit in the stands and observe, rather than actively participating in the setup of the robots. The students quickly tired of climbing the stairs to the stands, and started asking other students nearby. Social interactions built quickly, and inter-team sharing flourished. Figure 5, a photograph taken at the RoboCup 2001 competition, gives a feeling for the interest and excitement generated during the competition.

Fig. 5. The soccer quarter finals of RoboCupJunior Queensland 2001 held at the University of Queensland.

3.2.1 Observations of Robots Built for Competition

All but one of the teams used LEGO™ and the RCX™ brick to build their robots. The exception was the team from Brisbane Grammar School with the custom design that won the 2002 championships. Despite the explicit nature of the building and programming instructions provided in the workshops, no teams arrived with the workshop design for the construction or programming. All of the teams had taken it upon themselves to come up with completely new designs, or to significantly customize the workshop design. There was a notable contrast between younger and older robot builders. The younger team (aged 10 to 13) would typically build an initial base, and then add components to fix problems rather than re-designing the whole robot. Older teams were more able to see the benefit in going back to the design phase and re-considering their first steps.

Programming the RCX was predominately performed using ROBOLAB, with a number of teams also using NQC [13]. Dance robots, in both primary and secondary divisions, used highly linear programs. Rather than using a loop with a counter, students would cut and paste long strings of commands to form a long line of motor control elements with timer elements to set the duration of each motion. There was almost no use of sensors or sensor programming. Rescue robots, on the other hand, tended to be programmed with a linear sequence of behaviors. The students would first run a line following behavior, followed by a search behavior. Soccer produced the most diverse range of programming styles, and included the most examples of the use of NQC. Many students used a state based style, where a single behavior would execute for a fixed amount of time, or until certain sensor conditions were met, before moving to another behavior. Others used multi-tasking to execute parallel behaviors that would compete for control of the robot based on competency measures, in a similar style to the subsumption architecture [3]. A robot programmed in this manner was the runner-up to the world champions in the 2002 competition.

4 Future Improvements

Further research is being conducted into the impact on robotics activities on children's ability to independently build and program robots to achieve particular goals. A full usability study of the ROBOLAB™ development environment is being undertaken to highlight usability issues which both support and hinder the programming efforts of students. In addition, in the next year workshop participants will be observed and recorded as they work through programming tasks in an effort to gain an in-depth understanding of the concepts which cause the most difficulties. While observations of children using the current worksheets indicate that children are having success programming their robots during the workshops, further studies will be undertaken to evaluate the degree to which children are able to use this knowledge at a later date.

In the coming year minor improvement may be made to both the robot construction notes and the programming worksheets. The tutors are currently exploring ways in which the build notes could include three dimensional vector models of the robot and animated steps in the construction process. Such improvements are possible in an interactive web-based environment. While the worksheets are fundamentally sound, minor improvements in the ways certain processes are explained may be made. The tutors are also in the process of developing three additional intermediate worksheets which cover GOTO statements, as well as the programming concepts of multitasking and event handling.

5 Conclusions

Over the past two years, the School of Information Technology and Electrical Engineering at the University of Queensland has been involved in delivering robotics tuition and providing a competitive format to further this educational process, to school communities across Queensland. The robotics workshops and summer camps initiated by ITEE have provided opportunities, which may not have otherwise been available, for a wide range of children to develop knowledge and skills in robot building and programming. This scaffolding has helped students who have gone on to compete in RoboCupJunior Dance, Rescue and Soccer competitions. These competitions have been successful in creating a community of children avidly interested in building robots, sharing ideas and striving to "do it all better next year".

References

1. Apple Computer: Web site: ROBOLAB. http://www.apple.com/education/LTReview/spring99/robolab/
2. Bee, H.: The Developing Child, 9th Ed. Allyn & Bacon: Boston, MA (2002)
3. Brooks, R.A.: Elephants don't play chess. Autonomous Robots, 6, 3-15 (1990)
4. Harel, I.: Children designers. Norwood: Ablex (1991)
5. Kay, A.: Observations about children and computers. Advanced Technology Group, Learning Concepts Group, Apple Research Laboratory Research Note No. 31. [Online]. Available: http://www.atg.apple.com/technology/reports/RN31.html. (1994)

6. Lehrer, R: Authors of Knowledge: Patterns of Hypermedia Design. In S. P. Lajoie, and Derry, S. J. (Eds.) Computers as Cognitive Tools, Hillsdale, New Jersey: Lawrence Erlbaum Associates, Inc., 197-227. (1993)
7. Lund, H. H.: Robot Soccer in Education. In Advanced Robotics Journal, (2000), 13(8).
8. Marx, R., Blumenfeld, P., Krajcik, J., Soloway, E.: The Growth of Wisdom, Elementary School Journal, University of Chicago Press. (1994)
9. Papert, S.: Mindstorms: Children, computers, and powerful ideas. New York: Basic Books. (1980)
10. Proudfoot, K.: Web site: RCX Internals. http://graphics.stanford.edu/~kekoa/rcx/
11. Resnick, M., Bruckman, A., & Martin, F.: Pianos not stereos: Creating computational construction kits. Interactions, (1996) 3 (5) , 41-50.
12. Resnick, M. and Ocko, S.: LEGO/Logo: Learning Through and About Design. In Constructionism, edited by I. Harel & S. Papert. Norwood, NJ: Ablex Publishing. (1991)
13. NQC: Web site: Not Quite C. http://www.baumfamily.org/nqc/
14. Shaffer D. R. Developmental Psychology: Childhood and adolescence. 6th Edition. Wadsworth Group: Belmont CA. (2000)
15. Sheingold, K.: The microcomputer as a symbolic medium. In R. D. Pea, & K. Sheingold (Eds.), Mirrors of minds: Patterns of experience in educational computing, pp 198-208. Norwood, NJ: Ablex Publishing Corporation. (1987)
16. Sklar, E., Eguchi A., and Johnson, J.: RoboCupJunior: learning with educational robotics. In Proceedings of RoboCup-2002: Robot Soccer World Cup VI. 2002
17. Wyeth, G.F.: An Introductory Course in Mechatronics: Robo-Cricket. Mechatronics and Machine Vision in Practice, IEEE Computer Society Press, pp. 20-25. (1997)

Planning Trajectories in Dynamic Environments Using a Gradient Method

Alessandro Farinelli and Luca Iocchi

Dipartimento di Informatica e Sistemistica
Università di Roma "La Sapienza"
Via Salaria 113, 00198, Roma, Italy
{farinelli,iocchi}@dis.uniroma1.it

Abstract. In this article we propose an extension for a path planning method based on the LPN-algorithm to have better performance in a very dynamic environment. The path planning method builds a navigation function that drives the robot toward the goal avoiding the obstacles. The basic method is very fast and efficient for a robot with few degrees of freedom. The proposed extension integrates the obstacle dynamics in the planning method to have better performances in very dynamic environments. Experiments have shown the effectiveness of the proposed extension in a very dynamic environment, given by RoboCup soccer matches.

1 Introduction

Path planning is a fundamental task for autonomous mobile robots. Every application that involves the use of an autonomous mobile robot has to deal with path planning and obstacle avoidance. Examples of such applications are: air or navy traffic control, exploration and work in hostile environment (such as sub-sea or space), people rescue during disaster or dangerous situation, office work, robotic soccer, etc. Path planning is a critical task for the success of those applications and many different approaches can be found in literature to tackle the problem [16]. Well known techniques based on road-map construction, cell decomposition, or artificial potential fields, are widely used.

In many applications the robots have to cope with a dynamic environment, in which the problem of path planning and obstacle avoidance becomes much harder, because the robots have to take into account that configuration of the work space changes as time flows. Among other application domains, RoboCup [13] is a very good experimental field for such applications because the RoboCup competitions are characterized by a highly dynamic and hostile environment.

Different solutions for path planning in dynamic environments have been proposed in literature. A first group of methods do not take into account any explicit representation of time and are mostly focussed on extending or adapting a standard path planning method proposed for static environments. The goal is to have a very fast algorithm to plan trajectories in a static obstacle configuration and replan the trajectories at a fixed time interval to take into account

D. Polani et al. (Eds.): RoboCup 2003, LNAI 3020, pp. 320–331, 2004.

environmental changes [14, 18, 3, 15, 10]. Another group of works is based on explicitly considering the time dimension during the path planning process. Some of those works rely on the assumption of knowing in advance the dynamic evolution of the environment [9, 12]. As the complete knowledge of the dynamics of the environment is not available in most cases, other authors have studied the possibility of predicting the future evolution of the environment to plan the trajectories [20, 17, 21, 22].

In this paper an approach to the problem of path planning in dynamic environments is proposed[1]. The general idea is to integrate the obstacle's dynamics characteristic into the planning method, by changing their representation in the robot's configuration space. This is a general approach already used in literature, [9, 20], and can be used with different trajectory planning methods. The effectiveness of the approach depends on the specific planning method, and relies on how the information about the obstacles dynamics are taken into account.

The basic approach, from which the present work has started, is described in [14] and provides for an efficient implementation of a path planning method based on numerical artificial potential field similar to [23, 11]. The method in [14] (referred in this paper as LPN) has very good performance for a robot with few degrees of freedom. In particular this method computes a navigation function that avoids local minima, resulting in very effective trajectories. However by not considering the dynamics of the obstacles often the method results in undesired behavior of the robot when facing moving obstacles. Our extended method (LPN for Dynamic Environment, or LPN-DE) gives us better results in highly dynamic environments as compared with the LPN, but still keeps a low computational time and avoids local minima. The LPN and the LPN-DE algorithms have been implemented and compared with several experiments both in a simulated environment and with real robots in the RoboCup environment.

The LPN method is presented in Section 2 while its extension LPN-DE is described in Section 3. In Section 4 experimental results are given while in Section 5 an analysis of the related works is presented. Finally, conclusions are drawn in the last section.

2 LPN Gradient Method

In this section we describe the method presented in [14] (Linear Programming Navigation gradient method or LPN), while in the next section we describe the extension that we propose to improve its performances when the dynamics of the environment are taken into account.

The LPN gradient method is based on a numerical artificial potential field approach. The method samples the configuration space and assigns a value to every sampling point using a linear programming algorithm. The values for the sampling points are the sampled values of a navigation function; this means that

[1] This research is partially supported by MIUR (Italian Ministry of Education, University and Research) under project RoboCare (A Multi-Agent System with Intelligent Fixed and Mobile Robotic Components).

the robot can find an optimal path, just following the descendent gradient of the navigation function to reach the goal. The method can take as input a set of goal points, the final point of the computed path will be the best one with respect to the cost function discussed below. In order to build the navigation function a path cost must be defined. A path is defined as an ordered set of sampling points

$$P_n = \{p_n, p_{n-1}, ..., p_0\} \quad (1)$$

such that:

- $p_i \in \mathcal{R}^2$
- $\forall i = n, ..., 1 \; p_i$ must be adjacent to p_{i-1} or along the axis of the work space or in diagonal
- $\forall i, j$ if $i \neq j$ then $p_i \neq pj$
- p_0 must be in the set of the final configurations
- $\forall i = n, ..., 1 \; p_i$ must not be in the set of the final configurations

Given a point p_k, a path that starts from p_k and reaches one of the final configurations p_0 will be represented as $P_k = \{p_k, ..., p_0\}$. A cost function for a path P is an arbitrary function $F : \mathcal{P} \mapsto \mathcal{R}$, where $\mathcal{P}$ is the set of paths. This function can be divided into the sum of an *intrinsic cost* due to the fact that the robot is in a certain configuration, and an *adjacency cost* due to the cost of moving from one point to the next one:

$$F(P_k) = \sum_{i=0}^{i=k} I(p_i) + \sum_{i=0}^{i=k-1} A(p_i, p_{i+1}) \quad (2)$$

where A and I can be arbitrary functions.

Normally, I depends on how close the robot is to obstacles or to "dangerous" regions, while A is proportional to the Euclidean distance between two points.

The value of the navigation function N_k, in a point p_k, is the cost of the minimum cost path that starts from that point:

$$N_k = \min_{j=1,...,m} F(P_k^j) \quad (3)$$

where P_k^j is the j-th path starting from point p_k and reaching one of the final destinations and m is the number of such paths.

Calculating the navigation function N_k for every point in the configuration space directly, would require a very high computational time, even for a small configuration space. The `LPN` algorithm [14] is used to efficiently compute the navigation function. It is a generalization of the `wavefront algorithm` [19, 6] that is based into three main steps:

- assign value 0 to every point in the final configuration and an infinite value to every other point;
- put the goal set points in an *active list*;
- at each iteration of the algorithm, operate on each point of the active list, removing it from the list and updating its 8-neighbors (Expansion phase).

The expansion phase is repeated until the *active list* is not empty. To update a point p we operate as follow:

- for every point q in the 8-neighbors of p compute its value, adding to the value of p the moving cost from p to q and the intrinsic cost of q.
- if the new value for q is less than the previous one, update the value for q and put it into the *active list*.

In Figure 1 an update step is shown.

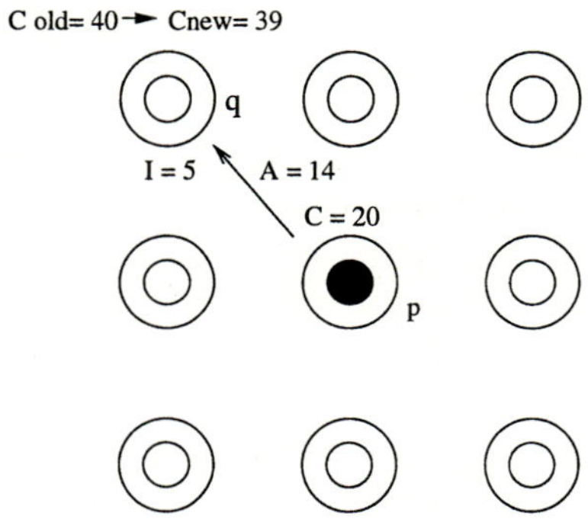

Fig. 1. Update step

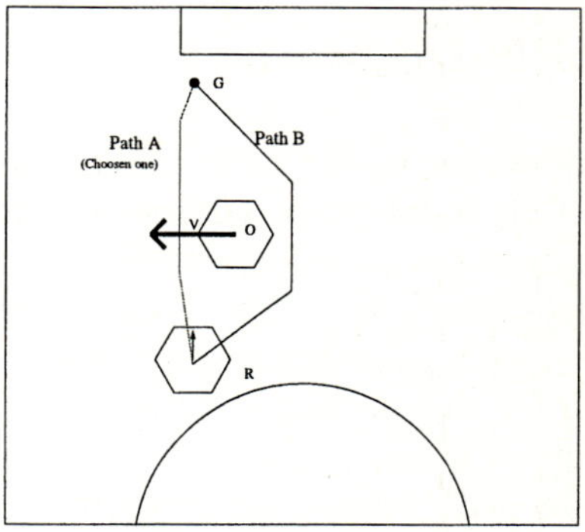

Fig. 2. Problems due to dynamic obstacle

The navigation function is computed according to the intrinsic cost of the sampling points in the configuration space. Suppose the obstacles in the workspace are given by a set of obstacle sampling points. Let $Q(\cdot)$ be a generic function and $d(p)$ the Euclidean distance for the sampling point p from the closest sample point representing an obstacle, then $I(p) = Q(d(p))$. In order to compute $d(p)$ for every sampling point we may apply the LPN algorithm giving the obstacle sampling points as the final configuration set, and assigning in the initialization phase a value 0 to the intrinsic cost for all the sampling points. Once $d(p)$ (and then $I(p)$) is computed for every sampling point p, we can execute again the LPN algorithm to compute the navigation function.

This method has been chosen for our extension for several reasons. First of all the method is very fast: it is possible to apply the trajectory planning every 100 ms on a grid of 100×100 sampling points on a modest CPU (Pentium 266 MHz)[14]. Moreover, the method finds an optimal path with respect to the cost function used. Those properties are very interesting for the RoboCup environment, which is the environment that we used for our experiments. Thanks to the low computational requirement the sampling of the configuration space can be exploited in order to reach a satisfying precision (10 cm in our case). On the other hand, the known information about the environment (such as field shape and position of fixed obstacles) can be easily and effectively taken into account. Moreover, the intrinsic cost function can be tailored in a very direct and simple manner, to obtain different trajectories, and so different behaviors of the robot.

However, the LPN method described in this section does not take into account the dynamics of the environment. This is a limitation when coping with very dynamic obstacles and it can result in non-optimal paths in some critical situations. As an example, in Figure 2 we can see that for the robot R to reach the goal G, path B would be better than path A, when the obstacle dynamics is considered, but the LPN method would choose path A, that is the optimal trajectory with respect to the current time frame.

3 LPN-DE Method

The extension to the LPN method for application in Dynamic Environments (LPN-DE) is based on additional information about moving obstacles, which are represented by a velocity vector. Observe that we do not require to have complete knowledge of the dynamics of the environment in advance (that would not be possible in many cases), but only assume to know an estimation of velocity vectors for the obstacles. Notice also that this requirement can be obtained by an analysis of sensor data, and its precision is not critical for the method, that is robust with respect to errors in evaluating such velocity vectors. Moreover, the proposed extension does not depend on the technique used to compute the velocity vectors for the obstacles. How those information are obtained, mainly depends on specific sensors of the robot and our particular implementation will be described later. By suitably taking into consideration the information about the velocity vectors in the planning method we show that our extension gives better result than the basic LPN algorithm.

In order to clarify how the LPN-DE integrate the obstacle dynamics information into the planning process, we introduce the concept of *influence region* of an obstacle. This definition is based on the fact that the intrinsic cost $I(p)$ is generally limited, i.e. it is 0 for points p such that $d(p) > \delta$, for a threshold δ.

Definition 1 *The* influence region *for an obstacle O_i is the set of sample points p whose intrinsic cost $I(p)$ is modified by the obstacle O_i.*

Basically the influence region of an obstacle O_i is the area surrounding the obstacle that we want to avoid, i.e. the set of point around O_i for which $I(p) > 0$. Considering the obstacle pointwise, an intrinsic cost function depending only on the distance $d(p)$, will result in circular *influence regions*, as it can be seen in Figure 3. The LPN-DE extension that we propose here defines a function $I(p)$ resulting in *influence regions* as shown in Figure 4.

As intuitively shown in Figures 3 and 4, the LPN-DE extension is designed in such a way to prefer trajectories for which the probability of collisions (or trajectory intersections) with moving obstacles is minimum. In fact, the function $I(p)$ in LPN-DE modifies the shape of the *influence region* according to the information about the obstacle dynamics, as shown in Figure 4. Consequently

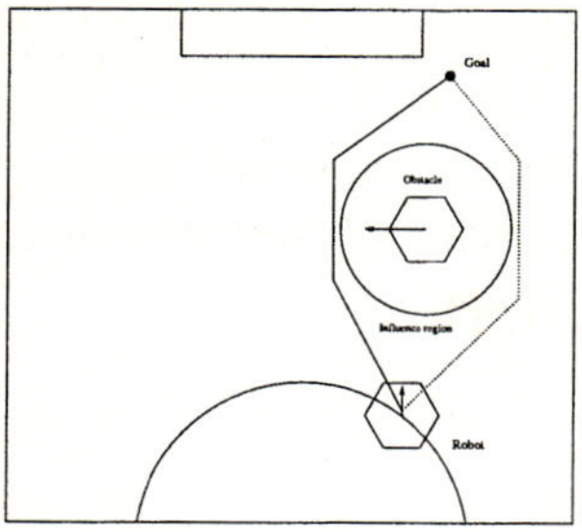

Fig. 3. Simple influence region (LPN)

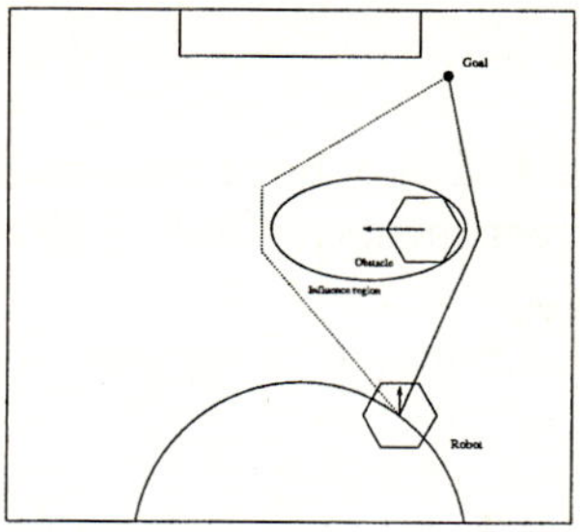

Fig. 4. Extended influence region (LPN-DE)

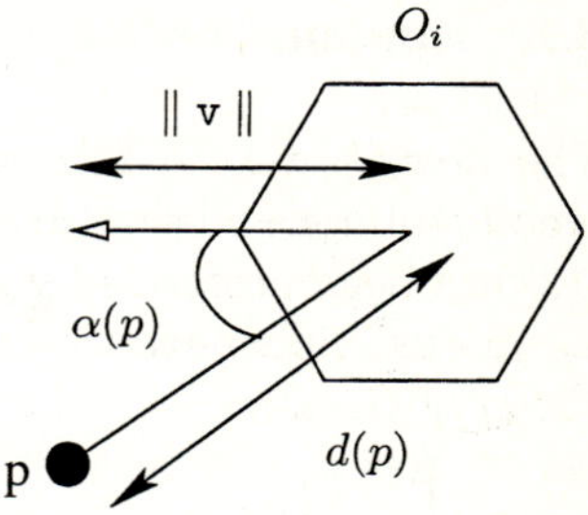

Fig. 5. Values for computing $I(p)$

the behavior of the robot is more adequate to the situation, since the chosen path will be the one passing behind the other robot, which is shorter an safer.

If a given point p belongs to the influence region of an obstacle O_i the intrinsic cost function for p will depend not only on the distance from O_i, but also on the position of p with respect to the velocity vector of O_i.

Therefore the function $I(p)$ will depend on (see Figure 5):

- the distance $d(p)$ of the point p with respect to O_i
- the angle $\alpha(p)$ between the velocity vector and the line reaching the center of O_i,
- the velocity's module $\| \mathrm{v} \|$ of O_i

So $I(p)$ in LPN-DE is given by a function $Q(d(p), \alpha(p), \| \mathrm{v} \|)$.

To compute the intrinsic cost function $I(p)$, depending on the values described above, the LPN algorithm has been slightly changed. We represent each moving obstacle as a circle and we use the robot sensors for obtaining, at every new planning cycle, (an estimation of) the center, the radius and the velocity vector of each obstacle. The basic structure of the algorithm is the same as described in Section 2, but the intrinsic cost function, the initialization and update steps for computing the intrinsic cost change. Specifically:

1. $I(p) = Q(d(p), \alpha(p), \| v \|)$.
2. In the initialization step we have to represent the velocity vector (module and heading) of each obstacle in each of the sampling points representing the center of an obstacle; we give to each of those point a high value and put to zero the values of all the other sampling points.
3. In the update step, if the new computed cost for a point q (I(q)) adjacent to a point p is greater then the actual cost:
   - update the cost of the point q,
   - propagate the velocity vector and position of the obstacle center,
   - put q in the *active list*.

In particular at each step we need to propagate the velocity vector and position information in all the points that are part of the influence region of the obstacle because those information are needed in the computation of the intrinsic cost function I(p).

3.1 Implementation

The described LPN-DE method has been implemented on different kinds of mobile robots, and several experiments in the RoboCup Middle-Size environment [13] has been performed.

In our implementation the function $Q(d(p), \alpha(p), \| v \|)$ is:

$$\begin{cases} C & if\ d(p) \leq D \\ C - F(\alpha(p), \| v \|)d(p) & if\ d(p) > D \text{ and } F(\alpha(p), \| v \|)d(p) < C \\ 0 & if\ d(p) > D \text{ and } F(\alpha(p), \| v \|)d(p) \geq C \end{cases}$$

where

- C is the intrinsic cost of a sample point inside an obstacle.
- D is the radius of the augmented obstacle.
- $F(\alpha(p), \| v \|) = \alpha(p) * \beta * \left(\frac{1}{\|v\|} + \gamma\right)$

For higher value of $\alpha(p)$ the value of $F(\alpha(p), \| v \|)$ becomes higher, so that the influence region results stretched along the axes of the obstacles velocity vector. For lower value of $\| v \|$ the value of the function $F(\alpha(p), \| v \|)$ becomes higher, so that the influence region is greater for greater values of the obstacles velocity module. The function has been chosen because it is a very simple function that gives us a reasonable shape for the influence region (similar to the one represented in figure 4). The value of the parameters have been tuned for the specific application.

4 Experiments and Results

In this section we present some experiments made in order to show the improvement in the performance of the LPN-DE method. The experimental environment has been the RoboCup Middle-Size field installed in our laboratory, that is slightly reduced with respect to current rules. For the implementation of the method we have used sampling rate of 100 mm. The main robot sensor is a color camera and the experiments have been performed on different kinds of robots, with different kinematics (unicycle-like and holonomous ones). The method has also been used by the SPQR middle-size team during the German Open 2002.

We have performed two kinds of experiments: the first one is based on a simulation environment, in which we have collected a set of quantitative data; the second kind of experiments has been done on real robots, substantially confirming the behaviors obtained in simulation.

In the simulation environment we made the assumption that the robots know in every instant the current positions and the velocity vector of the obstacles, in the experiments with real robots those information have been extracted by software vision modules from images acquired by the color camera.

4.1 Experimental Setting

Figures 6 and 7 present a typical situation occurred in all the experiments we have performed. The robot should reach a goal point, while a moving obstacle crosses the planned trajectory. In Figure 6 the robot uses the LPN method to plan the trajectories while in Figure 7 the LPN-DE one. The black dots in the figures represent sample points influenced by the moving obstacle, the lines starting from the robot represent the planned trajectory at that instant. The trajectory planned in Figure 7 using the LPN-DE method is better than that planned in Figure 6 as the following experiments show.

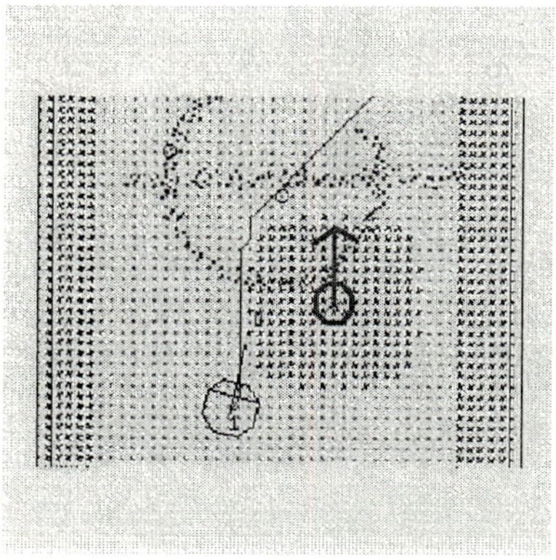

Fig. 6. LPN method

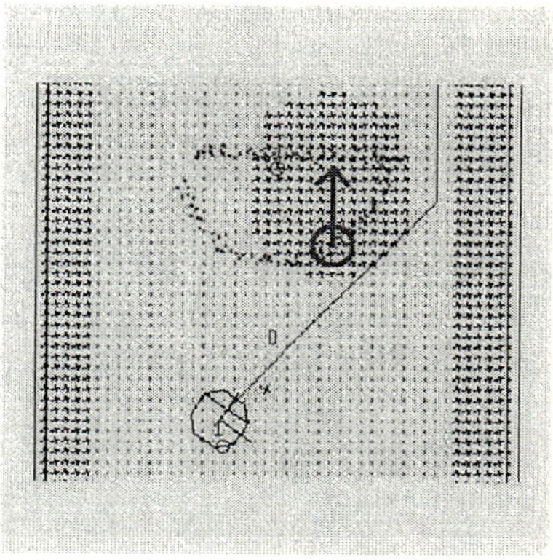

Fig. 7. LPN-DE method

4.2 Experiments with a Simulator

In this set of experiments the robot has to reach a predefined set of check points in sequence, while a simulated obstacle follows trajectories that intersect many times the standard path of the robot. An example of such experiments is given in Figure 8, in which the robot has to reach iteratively the check points highlighted with circles, while a moving obstacle moves along the bold path.

This kind of experiments have been performed by measuring the average time needed for the robot to reach the next point in the sequence, considering different velocities for the robot and the obstacles and comparing the LPN method and its extension LPN-DE. The table in Figure 9 shows the average time (from a 15 minutes execution of the experiment) needed to reach the next check point, for different velocities of the robot and of the obstacle and comparing the values of LPN (upper cells) with LPN-DE (lower cells). The values in the table shows that LPN-DE generally performs better than LPN, since it is generally able to avoid critical situations in which the robot trajectory intersects the obstacle path.

4.3 Experiments with Real Robots

The second set of experiments has been done on real robots by considering two robots that, in order to reach two different target positions, must plan trajec-

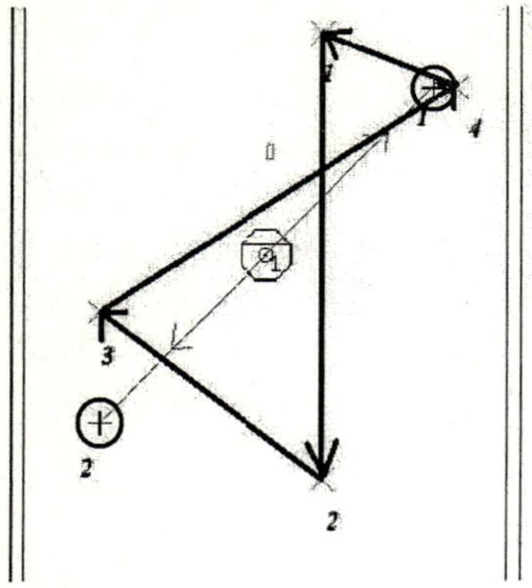

Fig. 8. Simulation experimental setting

	Vel Max Robot 800 mm/sec	Vel Max Robot 1600 mm/sec
Vel Max Obst. 1000 mm/sec	6.3 sec	5.9 sec
	5.8 sec	5.6 sec
Vel Max Obst. 2000 mm/sec	6.5 sec	6.2 sec
	6.3 sec	5.9 sec

Fig. 9. Simulation experimental results

tories that intersect each other. Also in this case we have evaluated the average time for reaching the two target points and we have obtained results similar to the ones presented in the previous section, in which the LPN-DE algorithm generally performs better.

In fact a typical situation arising in this experimental setting is reported in Figures 10 and 11, that represent the actual trajectories followed by the two robots, whose initial positions S_1 and S_2 are shown in the left side of the figures and the target points G_1 and G_2 are in the right side.

In the first case both the robots use the LPN method, and their behavior is not optimal since one robot (robot 1) always tries to pass in front of the other one. The second case, in which both the robots use the LPN-DE method, shows instead that the additional information about the obstacle dynamics has been properly exploited in the LPN-DE method for planning a better trajectory in presence of moving obstacles. Notice also that in any case the trajectories in Figure 11 are not optimal, as they would be if any robot would know in advance the trajectory planned by the other. Thus the robots execute a small portion of their path in parallel as in the previous case. However, in this case, at a certain point one of the robots (specifically robot 1) is able to detect such situation and decides to pass behind the other robot.

5 Related Work

Among the several techniques proposed in the literature for path planning in dynamic environments, we have mainly compared our approach to approaches that have been experimented in the RoboCup domain, both because we choose it as our testbed, and because in this domain a solution to the problem of obstacle avoidance and path planning among moving obstacle is particularly needed.

The work described in [8] presents an approach based on modified potential field. The method focuses on the generation of low level commands that enable the robot to dribble the ball amongst moving obstacle, but do not address the

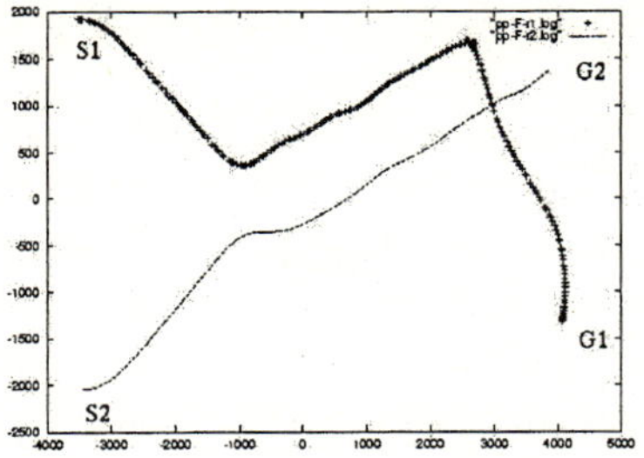

Fig. 10. LPN method

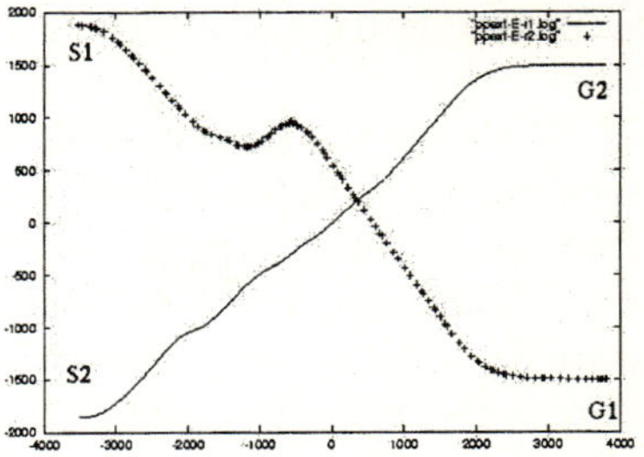

Fig. 11. LPN-DE method

problem of local minima, that is very important in a crowded dynamic environment, like RoboCup soccer; where often several robots are close to the ball. Our approach on the other hand, as already pointed out, generates navigation functions that does not contains local minima.

The problem of local minima is instead addressed in [4]. where the authors propose a method based on a combination of random exploration and biased motion towards the goal configuration. This method do not solve completely the problem of local minima, but reduces their occurrence.

A different approach is presented in [5], this work is focused on an hybrid path planning method that choose the best path planning algorithm depending on the specific situation. Simulation experiments show that the hybrid method outperform all the simple methods, but no guarantee is given on the optimality of the chosen trajectory.

Finally, [1] presents an adaptive planner, based on a static planner (developed in [2]), that tries to slightly change the plan while the environment configuration evolves. The experiments reported show good results for this approach; however, the time required for the trajectory generation is not constant with respect to obstacle configuration, and in particular becomes quite high in particular situations (e.g. when the adaptation of the planned trajectory fails).

With respect to all the discussed approaches the present work is a novel attempt to exploit the obstacles dynamic inferred from the environment during the task execution. In fact, the LPN method is guaranteed to plan optimal trajectories with low computational time requirements in real-time and by avoiding local minima. Our LPN-DE extension maintains all the above advantages of the LPN method (including low computational time), but performs much better in highly dynamic situations.

The general approach of integrating the obstacles dynamic information into the path planning procedure, changing the obstacle configurations accordingly, could be also extended to other path planning methods; as an example a suitable method for an effective extension based on this approach is the one described in [7]. In this work harmonic functions are used in order to compute a free path for a robot, using a potential field approach. Those functions contain very interesting properties, such as completeness in the path calculation, robustness to unanticipated obstacles and rapid computation.

6 Conclusions

In this article we have presented an approach to cope with dynamic environments extending a very efficient method for path planning presented in [14]. This method has been chosen because it is appropriated for our testing environment (RoboCup). The basic idea of the LPN-DE relies on the integration of the information about the velocity of moving obstacles into the path planning algorithm. In particular, in the present implementation, the dynamics of the obstacles have been integrated in the LPN method, modifying the cost function used for computing the global navigation function. Although this extension does not guarantee optimal trajectories along the time dimension, the performed experiments show that the LPN-DE method performed better than LPN, when facing fast moving obstacles, keeping the same computational requirement.

The LPN method and the extension have been implemented on robotic platforms with different kinematic models (unicycle-like and holonomous). As the present work is focused on the generation of effective trajectories for dynamic environments, we did not investigate the issues related to the low level control of the robotic platform. We rather provided a simple implementation of a low level robot controller that given the desired trajectory and the robot kinematic model, is able to drive the robot along the trajectory fulfilling the application constraints, that in the RoboCup case are high speed and reasonable precision.

The reported experiments show that better results can be obtained by integrating into the path planning method the information on the obstacles dynamics, and that the proposed extension in highly dynamic environments turned out to be effective and it has actually improved the performance of the robotic platform. As future work, we are currently investigating the possibility of applying a similar extension for dealing with moving obstacles to other path planning methods.

References

1. J. Baltes and N. Hildreth. Adaptive path planner for high dynamic environment. *RoboCup 2000: Robot Soccer World Cup IV*, 2000.
2. A. Bicchi, G. C. Casalino, and Santilli. Planning shortest bounded-curvature paths for a class of nonholonomic vehicles among obstacles. *Proocedings of the IEEE Int. Conference on Robotics and Automation*, 1995.
3. J. Borenstein. The vector field histogram-fast obstacle avoidance for mobile robots. *Robotics and Automation, IEEE Transactions on*, Volume: 7 Issue: 3:278 –288, 1991.
4. J. Bruce and M. Veloso. Real-time randomized path planning for robot navigation. In *In Proceedings of IROS-2002*, Switzerland, October 2002.
5. S. Buck, T. Schmitt, and M Beetz. Planning and executing joint navigation tasks in autonomous robot soccer. In *In RoboCup Int. Symposium*, Seattle, 2001.
6. Arkin R. C. Integrating behavioral,perceptual and world knowlodge in reactive navigation. In *Robotics and Autonomus Systems 6*, pages 105–122, 1990.
7. C. I. Connolly and R. A. Grupen. On the application of harmonic functions to robotics. *Journal of Robotic Systems*, 10(7):931–946, 1993.

8. B. Damas, L. Custodio, and P. Lima. A modified potential fields method for robot navigation applied to dribbling in robotic soccer. In *Proc. of RoboCup 2002 Symposium*, Fukuoka, Japan, 2002.
9. P. Fiorini and Z. Shiller. Motion planning in dynamic environments using the relative velocity paradigm. *IEEE Int. Conference on Robotics and Automation*, vol.1:560 –565, 1993.
10. A. Fujimori, P. N. Nikiforuk, and M. M. Gupta. Adaptive navigation of mobile robots with obstacle avoidance. *Robotics and Automation, IEEE Transactions on*, Volume: 13 Issue: 4:596 –601, 1997.
11. R. A. Jarvis. Collision-free trajectory planning using the distance transforms. *Mechanical Engineering Transaction of the Institution of Engineers*, ME10(number 3):187–191, 1985.
12. R. Kindel, D. Hsu, J.C. Latombe, and S. Rock. Kinodynamic motion planning admist moving obstacle. *Proocedings of the 2000 IEEE Int. Conference on Robotics and Automation*, 2000.
13. H. Kitano, M. Asada, Y. Kuniyoshi, I. Noda, E. Osawa, and H. Matsubara. Robocup: A challenge problem for AI and robotics. In *Lecture Note in Artificial Intelligence*, volume 1395, pages 1–19, 1998.
14. Kurt Konolige. A gradient method for realtime robot control. *AIROS*, 2000.
15. B. Kreczmer. Application of parameter space discretization for local navigation among moving obstacles. *Robot Motion and Control, 1999. RoMoCo '99. Proceedings of the First Workshop on*, pages 193 –198, 1999.
16. J. C. Latombe. *Robot Motion Planning.* Kluwer Academic Publisher, 1991.
17. J. Miura and Y. Shirai. Modelling motion uncertainty of moving obstacles for robot motion planning. In *IEEE Int. Conference on Robotics and Automation, Proceedings. ICRA '00*, volume Volume: 3, pages 2258 –2263, 2000.
18. G. Oriolo, G. Ulivi, and M. Vendittelli. Path planning for mobile robots via skeleton on fuzzy maps. *Intelligent Automation and Soft Computing*, 1996.
19. S. Thrun, A. Buken, W. Burgarg, D. Fox, T. Frohlinghaus D. Hennig, T. Hofmann, M. Krell, and T. Schmidt. Map learning and high-speed navigation in RHINO. In *AI-based Mobile Robots : Case studies of successful robot systems.* MIT Press, Cambridge,MA, D. Kortenkamp, R.P. Bonasso, and R. R. Murphy editors, 1998.
20. M. Yamamoto, M. Shimada, and A. Mohri. On-line navigation of mobile robot under the existence of dynamically moving multiple obstacle. In *Assembly and Task Planning, 2001, Proceedings of the IEEE Int. Symposium on*, 2001.
21. Huiming Yu and Tong su. A destination driven navigator with dynamic obstacle motion prediction. *Proceedings of the IEEE Int. Conference on Robotics and Automation*, 2001.
22. N. H. C. Yung and C. Ye. Avoidance of moving obstacles through behaviour fusion and motion prediction. In *Systems, Man, and Cybernetics, 1998. 1998 IEEE Int. Conference on*, volume Volume: 4, pages 3424 –3429, Oct 1998.
23. A. Zelinsky. Using path transforms to guide the search for findpath in 2D. *IJRR*, 13:315–325, 1994.

Local Multiresolution Path Planning*

Sven Behnke

International Computer Science Institute
1947 Center St., Berkeley, CA, 94704, USA
behnke@icsi.berkeley.edu

Abstract. Grid-based methods for finding cost optimal robot paths around obstacles are popular because of their flexibility and simple implementation. However, their computational complexity becomes unfeasible for real-time path planning if the resolution of the grid is high.
These methods assume complete knowledge about the world, but in dynamic environments where sensing is done on board the robot, less is known about far-away obstacles than about the ones in close proximity. The paper proposes to utilize this observation by employing a grid of variable resolution. The resolution is high next to the robot and becomes lower with increasing distance. This results in huge savings in computational costs while the initial parts of the paths are still planned with high accuracy. The same principle is applied to the time-axis, allowing for planning paths around moving obstacles with only a moderate increase in computational costs.

1 Introduction

Path planning is an important subtask of the robot navigation problem, which is to find a path from a start configuration to a target state and to traverse it without collision. The navigation problem can be decomposed into three subtasks: mapping and modeling the environment, path planning, and path traversal with collision avoidance.

Many approaches to path planning have been described in the literature [8, 9]. They can be grouped into local and global methods. Local path planning methods do not attempt to solve the problem in its full generality, but use only the information available at the moving robot to determine the next motion command. One well known local path planning technique is the potential field method [7]. Here, the robot follows the gradient of a force field. The field is generated by attractive potentials pointing towards a target and by repulsive potentials that point away from obstacles. The potential field method has a low computational load and generates smooth paths that stay away from obstacles. However, the greedy gradient descent may get trapped in local minima. It is hence most useful in environments where local minima are unlikely. Furthermore, it can be used for fast reactive obstacle avoidance.

* This work was supported by a fellowship within the postdoc program of the German Academic Exchange Service (DAAD).

D. Polani et al. (Eds.): RoboCup 2003, LNAI 3020, pp. 332–343, 2004.

In contrast, global methods assume complete knowledge about the world. They frequently rely on the concept of free space, the configurations the robot can take without collision [10]. It is convenient to shrink the robot to a point while growing the obstacles accordingly to obtain the free space.

Roadmap path planning methods inscribe a graph into the free space that contains all possible paths. For instance, a roadmap defined by a visibility graph [11] can be used to find the shortest path around polygonal obstacles. Another possibility to define a roadmap is to use Voronoi borders [12] as graph edges in order to find a path that stays far away from obstacles. To rapidly explore high-dimensional configuration spaces planners have been proposed that randomly sample configurations and connect samples in free space by simple local paths, thereby creating probabilistic roadmaps [6]. One disadvantage of these methods is that only a binary representation (occupied/free) of the configuration space is possible.

Another class of global planning methods decompose the free space into cells. Exact cell decomposition results in cells of different simple shapes as required by the shape of obstacles. Approximate cell decomposition methods use predetermined cell shapes, sizes, and positions to approximate the free space [1]. Popular approximate cell decompositions include uniform coverage with square cells and quadtree representations [5] that use smaller cells next to the obstacle borders.

Once the decomposition is determined, dynamic programming can be used to find an optimal path. This requires to fill out a data structure, e.g. a multidimensional table, that contains solutions for all possible subproblems. If the resolution of a decomposition is high or the state space has many dimensions this can still be computationally demanding.

The computational efficiency of path planning is essential for online-problems [2], where paths are planned and executed in real time, for on board planning, where the computational resources are limited, and for planning in dynamic environments, where frequent replanning is required. All three of the above constraints are present in many leagues of the RoboCup competition.

On the other hand, in dynamic environments a detailed path from the start to the target has little chance of execution. Obstacles move unexpectedly as the robot traverses the path and hence continuous replanning is required. Furthermore, due to limited local sensing capabilities, far-away obstacles can be determined only with reduced accuracy. These observations motivate the local multiresolution path planning method proposed in this paper. The idea is to use high resolution to represent the configuration space in close proximity to the robot and to lower the resolution with increasing distance from the robot. This concentrates the planning resources to the begin of the path, the part that must be traversed first and where the information about obstacles is most reliable. While the computational load is reduced dramatically, the immediate movement of the robot can still be planned with high accuracy.

The paper is organized as follows. The next section describes grid-based path planning and details its extension to the multiresolutional case. In Section 3, the traversal of planned paths and the effects of the initial robot motion are covered.

```
PlanPath(target, obstacles, 𝒩){
    grid = eval = ClearGrid();
    q = InitPriorityQueue();
    while (!q.empty()){
        p = q.pop();
        if (target == p)
            return previous;
        for(n∈ 𝒩(p)){
            if(eval(n) == 0)
                best = ∞;
            else
                best = eval(n);
            new = p.cost + grid(p)·n.l0 + getGrid(n, obstacles, grid)·n.l1;
            if(new<best){
                eval(n) = new;
                q.push([n, new, getHeuristic(n, target)]);
                previous(n) = p;
            }
        }
    }
}
```

Fig. 1. A^* search for the least cost path to the target. The cost of a configuration is computed by getGrid and stored in the grid array. A heuristic function is used to guide the search. The cost of the shortest path to a configuration is maintained in the eval array. The reverse path is produced by traversing the previous list starting at target.

Section 4 applies the multiresolution idea to the time axis to find paths around moving obstacles. The paper concludes with a discussion of the experimental results and indicates possibilities for future work.

2 Grid Based Path Planning

Grid-based path planning methods decompose the configuration space into an array of cells. Costs are associated with the cells to represent the occupancy by obstacles. Neighboring cells are connected by edges. The cost of an edge can be derived from the cost of the two cells it connects. An minimal cost path can be found by searching this graph.

2.1 Basic Method

The basic algorithm used to find the least cost path in such a graph is summarized in Figure 1. It maintains in the eval array the cost of the best path seen so far to each of the configurations reached. The costs of grid cells are computed by getGrid as new cells are explored and stored in the grid array for future use. The edges of the graph are given as adjacency list $\mathcal{N}$ for each grid point. Each directed edge contains two lengths, l_0 and l_1, that describe the distance to the cell border from its start and its end node, respectively. The cost of an edge is

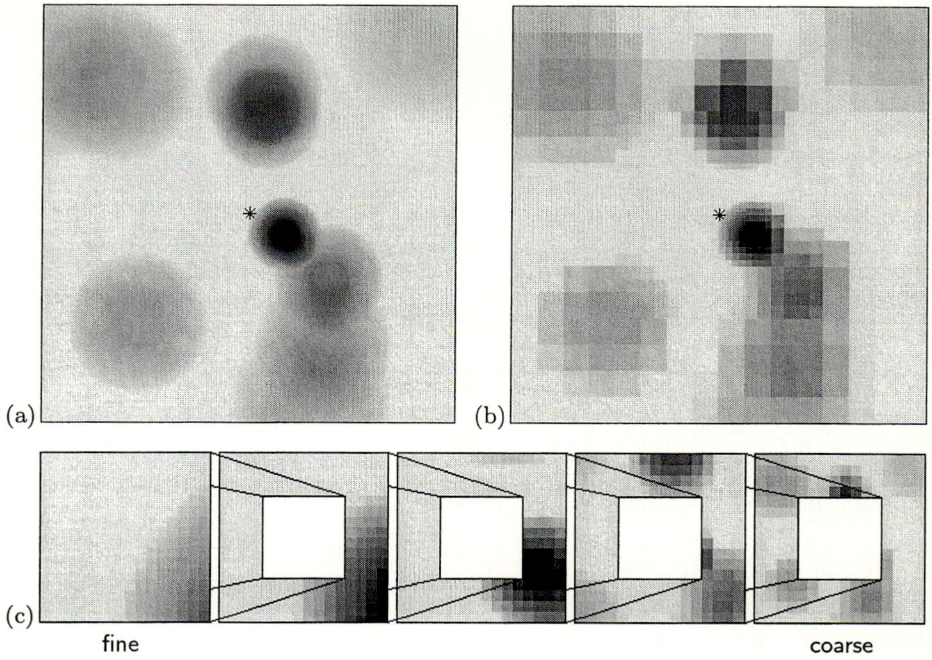

Fig. 2. Cost function used for the experiments. The robot is located in the center of the grid, as indicated by the star. Obstacles are represented by fuzzy discs. With distance from the center, the discs become larger and their height (indicated by darkness) decreases. (a) single resolution 256×256; (b) multiple resolutions 16×16×5 shown overlaid; (c) multiple resolutions 16×16×5 shown side by side.

computed as weighted sum of both grid values. The cost of a path is the sum of its edges.

InitPriorityQueue initializes the search with the start nodes. Since the algorithm is used here for local search around the current robot position, the search always starts at the center of the grid. The algorithm expands the nodes first that have the lowest accumulated cost until the best path cannot be improved any more.

2.2 Cost Function

The cost function that describes the occupancy of a grid cell can be chosen arbitrarily. Here, simple disk-like obstacles are modeled, as illustrated in Figure 2(a). 10 obstacles are placed at random positions. Their radius $r = r_o + r_r + r_d$ consists of a fixed component, which represents the radius of the obstacle r_o plus the radius of the robot r_r, and a variable component r_d that increases linearly with distance from the grid center. The far-away obstacles are modeled larger, because their position can be sensed with less accuracy from the perspective of the robot and because they might move before the robot gets close to them.

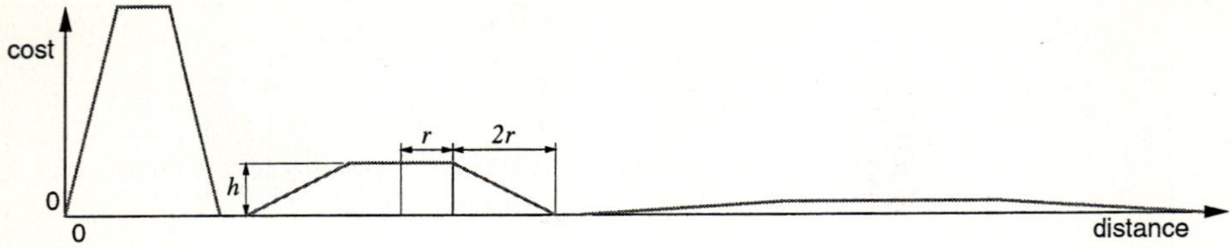

Fig. 3. Cut through cost function for obstacles with different distances from the robot.

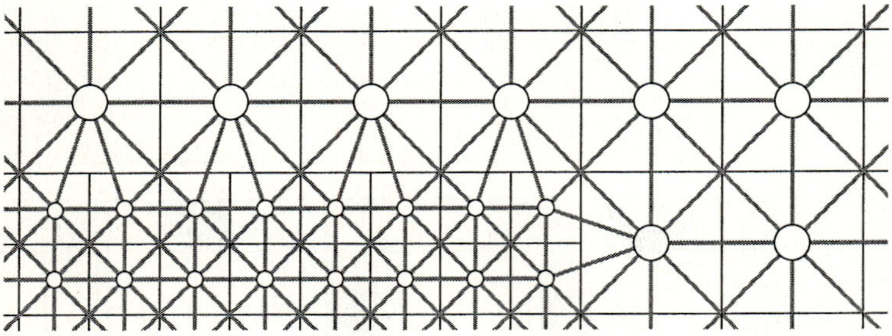

Fig. 4. Connectivity of the multiresolutional cell grid. A detail of the border between two resolution levels is shown.

Each obstacle is also characterized by a height h which is inversely proportional to the squared radius to keep its integral constant. The cost increase of a grid cell that is caused by an obstacle depends on their distance, as illustrated in Figure 3. It is constant if the distance is smaller than the radius and decreases linearly to zero at three times the radius. To compute the cost of a grid cell, the contributions from all obstacles are added to a uniform base cost.

The cost function is a simple and flexible way to express uncertainty. Obstacles with noncircular shapes could be included into the cost function in an analogous way.

2.3 Non-uniform Resolution

It is not necessary to represent the entire grid with a high uniform resolution. Since far-away obstacles cover a larger area, a coarser resolution suffices there to approximate them. This is illustrated in Figure 2(b). Here, the resolution is high in the center of the grid and decreases towards the outside. This corresponds to the situation shown in Figure 2(c). Multiple low-resolution grids of size $M \times M$ are stacked concentrically. The inner part $([\frac{1}{4}M \ldots \frac{3}{4}M][\frac{1}{4}M \ldots \frac{3}{4}M])$ of a grid is not used, but the next grid level is placed there, until the highest resolution is reached. To cover the same area as a uniform $N \times N$ grid with the same inner resolution, only $K = \log_2(N/M) + 1$ levels of size $M \times M$ are needed. If N is large compared to M this lowers the number of grid cells substantially. In the following experiments, I use $N = 256$, $M = 16$, and $K = 5$. Hence, the flat grid has 64 times as many cells as the multiresolutional grid.

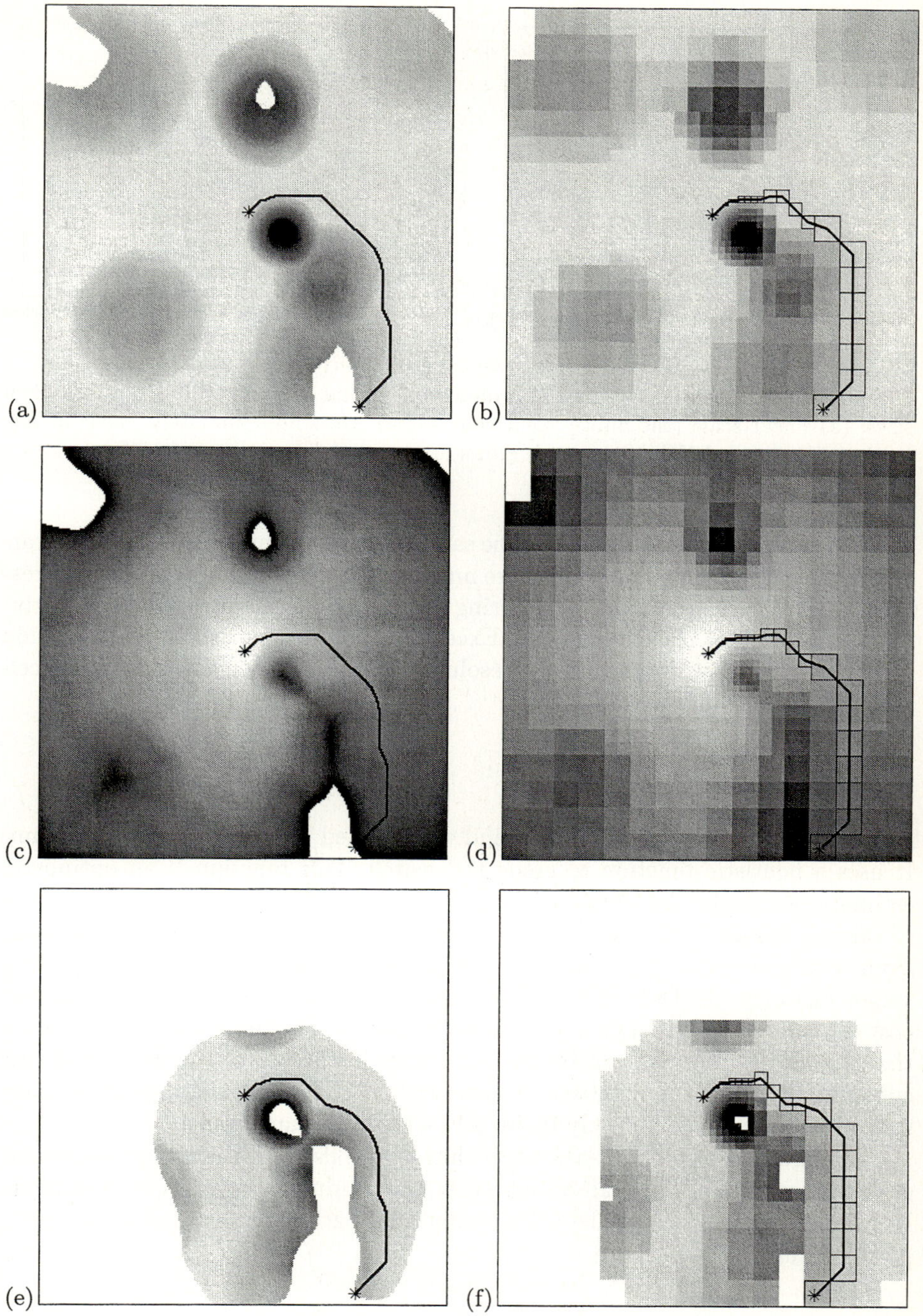

Fig. 5. Path planning without (a-d) and with (e-f) heuristics using a flat (a,c,e) and a multiresolutional (b,d,f) grid. (a,b,e,f) show the status of the grid and (c,d) show the status of the of the eval array after the search terminated at the target (lower star).

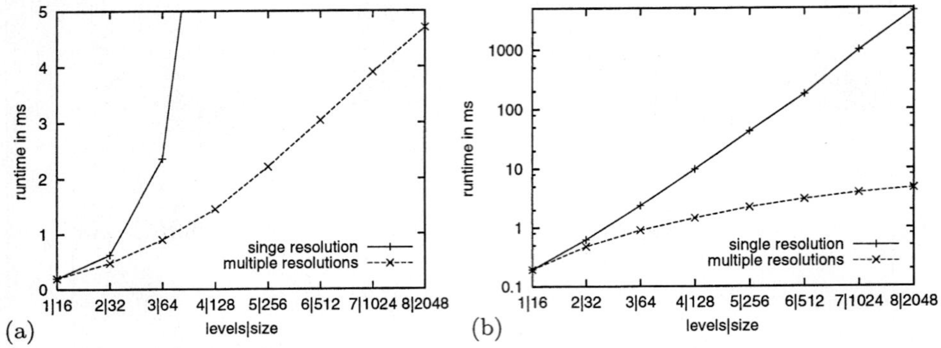

Fig. 6. Runtime of the flat and the multiresolutional search for different problem sizes: (a) linear scale (the multiresolutional search grows approximately linear in K); (b) logarithmic scale (the flat search grows approximately exponential in K).

The connectivity within cells of the same level of this multiresolutional hierarchy is set to the 8-neighborhood. Care must be taken at the borders between resolution levels to connect the neighboring cells. Figure 4 illustrates the connectivity that is used for the experiments. Except for the corners, each high-resolution cell connects to two adjacent low-resolution cells and each low-resolution cell connects to four high-resolution cells.

2.4 Heuristics

The A^* algorithm [4] is an efficient and well studied best-first search algorithm. It uses a heuristic function to guide the search. This function is an optimistic estimate of a path's total cost.

Since each grid cell has at least the base cost, the remaining part of the path from a grid point to the target cannot be less expansive than the Euclidean target distance weighted by the base cost. Hence, the sum of the accumulated cost of the best path to a grid point plus this heuristics can be used to determine the expansion order. As can be seen in Figure 5, the use of this heuristics can substantially lower the number of visited grid cells. The altered expansion order may alter the path found only if two paths have the same costs.

The figure also compares the algorithm for the flat and the multiresolutional grid representation. One can observe that the produced paths are very similar. In particular, the start of the multiresolutional path is as detailed as the path produced using the flat grid.

2.5 Runtime

The different cell numbers between the flat and the multiresolutional grid result in different runtimes. Figure 6 displays how this difference grows with the problem size. The runtimes represent the measured average running time of the path

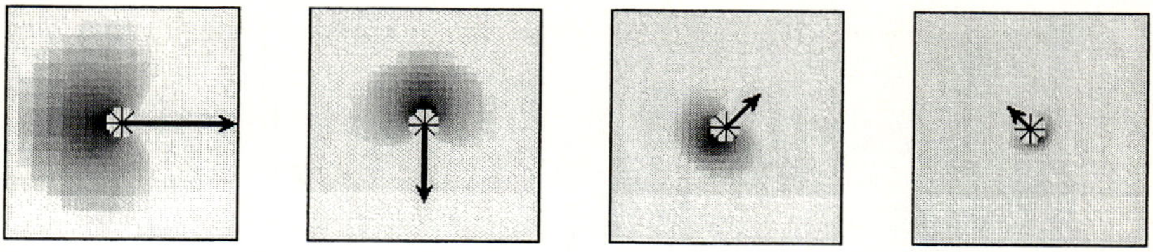

Fig. 7. Obstacle placed behind a moving robot to account for its initial velocity. It discourages sudden changes in direction.

planner to random targets with random obstacles. A 1 GHz Athlon processor has been used for the measurement. The algorithm has been implemented in C++. At the leftmost data point, where $K = 1$ and $N = M = 16$, both representations are identical. As N gets larger, K is adjusted accordingly. One can observe that the runtime grows approximately exponential with K when a single resolution is used and grows, after some cache effects, approximately linear when multiple resolutions are used. This corresponds well to the growth of the cell numbers. For $K = 8$ and $N = 2,048$, the flat search needs on average 4.55s while the multiresolutional search needs only 4.70ms on average.

3 Continuous Planning and Execution

In a dynamic environment, a planned path cannot simply be executed. Since the obstacles move, the plan must be updated as the robot follows its trajectory. Furthermore, in order to make consecutive plans compatible, the initial robot motion must be taken into account.

3.1 Initial Condition

One simple way to account for the initial velocity of the robot is to place an additional obstacle behind it, as shown in Figure 7. This obstacle favors paths that initially continue in a similar direction the robot is already moving. The larger the robot's initial velocity, the more severe a sudden change in direction would be and hence the more pronounced this obstacle must be.

3.2 Partial Execution and Replanning

Figure 8 illustrates how two different initial conditions lead to two different paths. The figure also shows, how the robot generates a trajectory by moving along the initial segment of the path. The path is continuously updated. As the robot comes closer to initially far-away obstacles, their radius decreases, since their position can now be determined with greater precision and they are less likely to move. Hence, the robot passes these obstacles closer than originally planned.

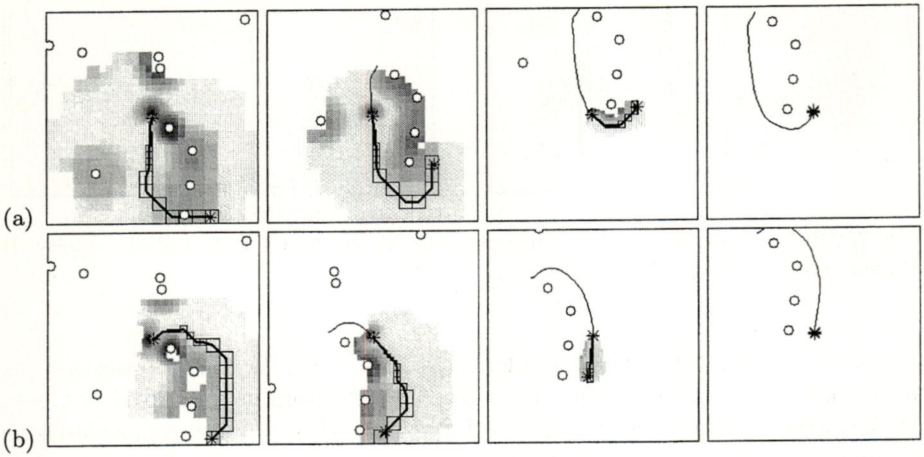

Fig. 8. Different initial conditions lead to different paths. As the path is executed, the robot is followed by the obstacle representing its velocity. (a) initial downward movement; (b) initial rightward movement. The thin line indicates the generated trajectory.

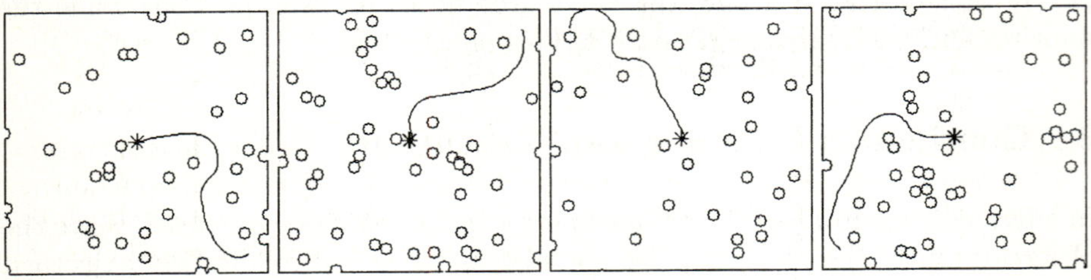

Fig. 9. Paths executed in an environment with many obstacles. The robot started in one of the corners and generated the trajectory while driving to the target (star).

Figure 9 shows some additional trajectories that have been generated in an environment with more obstacles. The trajectories are smooth, relatively short, and stay away from obstacles. Hence, they are suitable to reach the target fast while avoiding the chance of collisions.

4 Dynamic Planning

If the movement of obstacles can be estimated, a dynamic path can be planned by extending the dimensionality of the configuration space [3]. Figure 10 shows how the time axis can be represented in a multiresolutional fashion. For illustrative purposes the robot's position has been reduced to a single dimension.

Since time advances only in one direction, the higher-resolution arrays are not centered in the middle of the time-axis, but are located at its start. Hence, the first time-steps of the path are modeled with high precision while later time-steps are longer. This leads only to a moderate increase in computational complexity.

As can be seen, obstacles are not circular any more, but look like a line that becomes wider and flatter with distance from the origin. The two obstacles

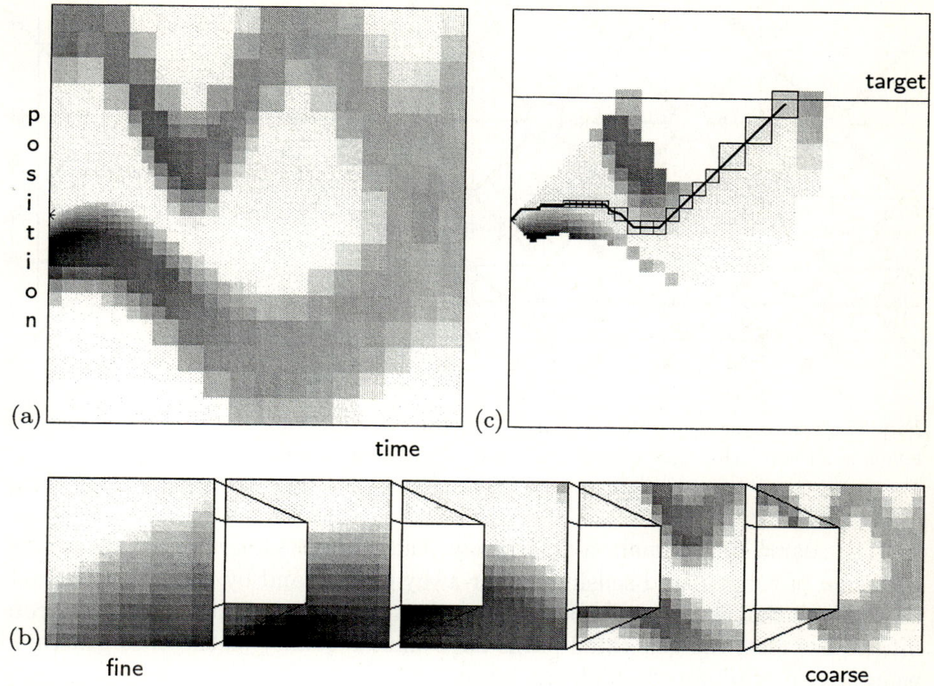

Fig. 10. Dynamic planning. The horizontal axis corresponds to time while the vertical axis represents space. The robot is located at the center of the left edge. (a) multiresolutional cost function 16×16 ×5 shown overlaid; (b) cost function 16×16 ×5 shown side by side; (c) planned path to the target position, represented by a horizontal line.

shown move along sinusoidal trajectories. To plan a minimal time path to a target-position not a single target cell, but a line of cells at this position and all points of time must be considered as search target.

Part (c) of the figure shows a planned path. One can see that the robot first moves upwards to avoid the lower obstacle, then moves downwards to avoid the upper obstacle and finally moves straight to the target. A direct motion to the target is not possible, since the maximum speed of the robot has been set to one. This is reflected in the connectivity shown in Figure 11. The edges are a subset of the edges from Fig. 4. Only edges that advance in time and do not exceed the maximum speed are included.

5 Conclusions

The paper proposed a local multiresolutional path planning algorithm. In contrast to quadtree algorithms [5] that focus the computational resources at the obstacle borders, this algorithm represents the configuration space next to the robot with higher resolution than far-away from it. This leads to the use of fewer grid cells, as compared to a representation that is based on a uniform grid. These savings result in substantially lower runtimes.

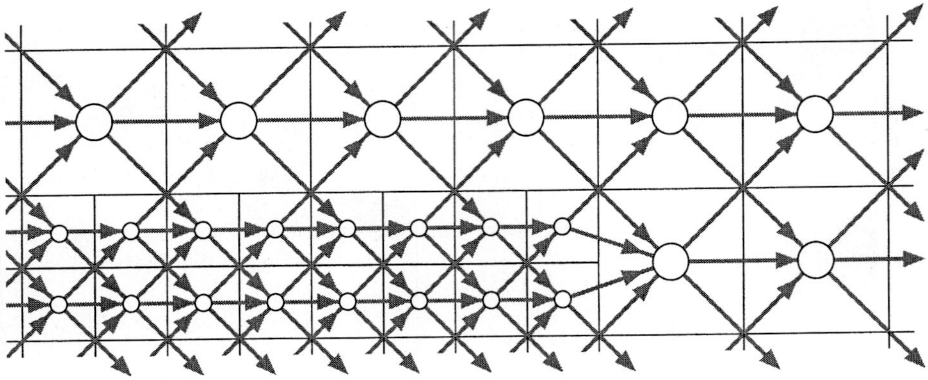

Fig. 11. Edges for dynamic planning. A detail of the border between two resolution levels is shown. All edges advance in time. The maximal speed is one.

The coarse approximation of far-away obstacles was motivated by the limited precision of robot-based sensing for far-away objects and by the larger obstacle movements that are possible before the robot comes close to them. If these conditions are met, the generated paths have similar quality as the ones generated using a grid of uniformly high resolution.

Since the runtime of the multiresolutional path planner is very short, it can be used for continuous replanning. This is not wasteful, since only the initial part of the path that is executed immediately after planning is planned in detail.

An example with a two-dimensional configuration space has been presented. The generated trajectories facilitated the fast movement towards the target while at the same time minimizing the chances of collisions.

Furthermore, it has been shown, how to include time into the configuration space. This makes planning with moving obstacles possible. The non-uniform sampling of the time-dimension leads only to a moderate increase in computational costs.

So far, the kinematics of the robot has not been included in the configuration space. Since the running time of the planner is only a few milliseconds long, it would be feasible to increase the dimensionality of the configuration space and still replan at a high rate. One could e.g. explicitly model the orientation or the velocity of the robot. This will be subject to future research.

References

1. R. Brooks and T. Lozano-Pérez. A subdivision algorithm in configuration space for findpath with rotation. In *Proceedings of the 8th International Conference on Artificial Intelligence (ICAI)*, pages 799–806, 1983.
2. James Bruce and Maria Manuela Veloso. Real-time randomized path planning for robot navigation. In *Proceedings of the 2002 IEEE/RSJ International Conference on Intelligent Robots and Systems (IROS '02)*, October 2002.

3. B. Donald, P. Xavier, J. Canny, and J. Reif. Kinodynamic motion planning. *Journal of the ACM*, 40(5):1048–1066, 1993.
4. P.E. Hart, N.J. Nilsson, and B. Raphael. A formal basis for the heuristic determination of minimum cost paths in graphs. *IEEE Transactions on Systems Science and Cybernetics*, SSC-4(2):100–107, 1968.
5. S. Kambhampati and L.S. Davis. Multiresolution path planning for mobile robots. *IEEE Journal of Robotics and Automation*, RA-2(3):135–145, 1986.
6. L. Kavraki, P. Svestka, J.C. Latombe, and M. Overmars. Probabilistic road maps for path planning in high-dimensional configuration spaces. *IEEE Transactions on Robotics and Automation*, 12(4):566–580, 1996.
7. O. Khatib. Real-time obstacle avoidance for manipulators and mobile robots. *International Journal of Robotic Research*, 5(1):90–98, 1986.
8. J.-C. Latombe. *Robot Motion Planning.* Kluwer Academic Publishers, Boston, MA, 1991.
9. J.-C. Latombe. Motion planning: A journey of robots, molecules, digital actors, and other artifacts. *International Journal of Robotics Research*, 18(11):1119–1128, 1999.
10. T. Lozano-Pérez. Spatial planning: A configuration space approach. *IEEE Transactions on Computers*, C-32(2):108–120, 1983.
11. N.J. Nilsson. A mobile automaton: An application of artificial intelligence techniques. In *Proceedings of the 1st International Joint Conference on Artificial Intelligence*, pages 509–520, Washington, DC, 1969.
12. C. O'Dunlaing and C. K. Yap. A 'retraction' method for planning the motion of a disc. *Journal of Algorithms*, 6:104–111, 1986.

A Humanoid Approaches to the Goal – Reinforcement Learning Based on Rhythmic Walking Parameters

Minoru Asada[1,2], Yutaka Katoh[1], Masaki Ogino[1], and Koh Hosoda[1,2]

[1] Dept. of Adaptive Machine Systems
[2] HANDAI Frontier Research Center
Graduate School of Engineering, Osaka University
{ogino,yutaka}@er.ams.eng.osaka-u.ac.jp
{asada,hosoda}@ams.eng.osaka-u.ac.jp

Abstract. This paper presents a method for generating vision-based humanoid behaviors by reinforcement learning with rhythmic walking parameters. The walking is stabilized by a rhythmic motion controller such as CPG or neural oscillator. The learning process consists of two stages: the first one is building an action space with two parameters (a forward step length and a turning angle) so that infeasible combinations of them are inhibited. The second one is reinforcement learning with the constructed action space and the state space consisting of visual features and posture parameters to find feasible action. The method is applied to a situation of the RoboCupSoccer Humanoid league, that is, to reach the ball and to shoot it into the goal. Instructions by human are given to start up the learning process and the rest is completely self-learning in real situations.

1 Introduction

Since the debut of Honda humanoid [3], the research community for biped walking has been growing and various approaches have been introduced. Among them, there are two major trends in the biped walking. One is model based approach with ZMP (zero moment point) principle [4] or the inverted pendulum model [14] both of which plan the desired trajectories and control their bipeds to follow them. In order to stabilize the walking, these methods need very precise dynamics parameters for both the robot and its environment.

The other one is inspired by the findings [2] in neurophysiology that most animals generate their walking motions based on the central pattern generator (hereafter, CPG) or neural oscillator. CPG is a cluster of neural structures that oscillate each other under the constraint of the relationships in their phase spaces, and generates rhythmic motions that interact with the external environment and the observed motion can be regarded as a result of the entrainment between robot motion and the environment. This sort of approach does not need model parameters that are as precicise as ZMP or the inverted pendulum.

D. Polani et al. (Eds.): RoboCup 2003, LNAI 3020, pp. 344–354, 2004.

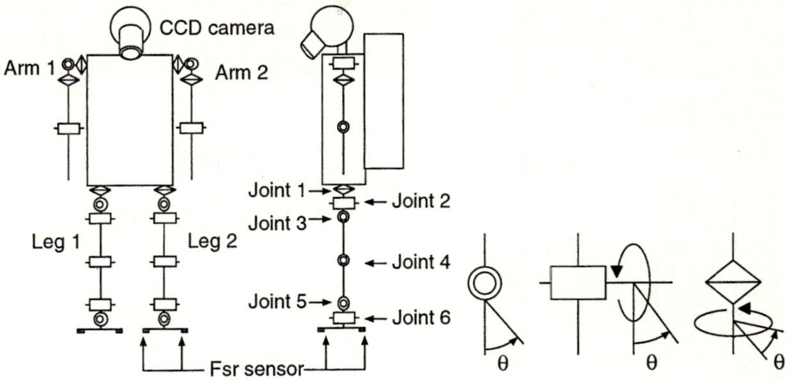

(a) Model of robot (b) Mark of joint

Fig. 1. A model of biped locomotion robot

Taga et al. [11] gave the mathematical formulation for neural oscillator, constructed a dynamic controller for biped walking on the sagittal plane, and showed the simulation results which indicated that his method could generate stable biped motions similar to human walking. Others extended his method to three dimensions [8] and adaptive motion on the slope by adjusting the neural oscillator [1].

The second approach seems promising for adaptation against changes in the environment. To handle more complicated situations, the visual information has been involved. Taga [12] studied how the robot can avoid an obstacle by adjusting the walking pattern assuming that the object height and the distance to it can be measured by the visual information. Fukuoka et al. [5] also adjusted CPG input so that a quadruped can climb over a step through the visual information. In these methods, however, the adjusting parameters were given by the designer in advance. Therefore, it seems difficult to apply to more dynamic situations, and learning method seems necessary.

This paper presents a method for generating vision-based humanoid behaviors by reinforcement learning with rhythmic walking parameters. A rhythmic motion controller such as CPG or neural oscillator stabilizes the walking [13]. The learning process consists of two stages: first one is building an action space with two parameters (a forward step width and a turning angle) so that infeasible combinations of them are inhibited. The second one is reinforcement learning with the constructed action space and the state space consisting of visual features and posture parameters to find feasible action. The method is applied to a situation of the RoboCupSoccer Humanoid league [6], that is, to approach the ball and to shoot it into the goal. Instructions by human are given to start up the learning process and the rest is solely self-learning in real situations.

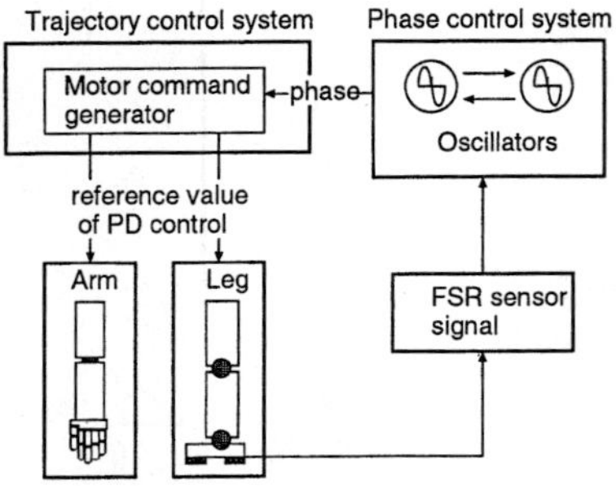

Fig. 2. A walking control system

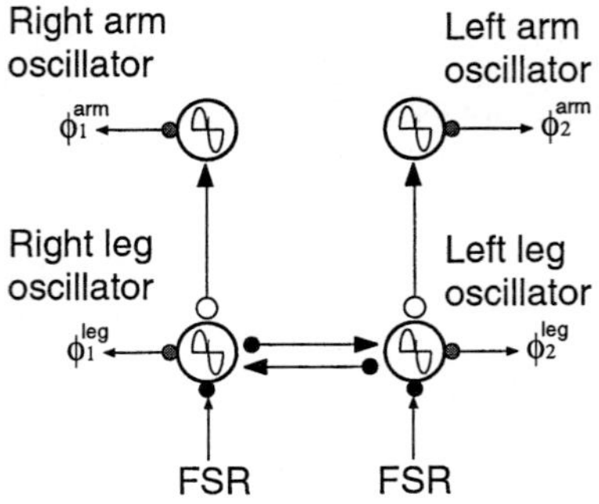

Fig. 3. A phase control system

2 Rhythmic Walking Controller

2.1 Biped Robot Model

Fig. 1 shows a biped robot model used in the experiment which has a one-link torso, two four-link arms, and two six-link legs. All joints rotate with a single DoF rotation. Each foot has four FSRs to detect reaction force from the floor and a CCD camera with a fish-eye lens is attached at the top of the torso.

2.2 Rhythmic Walking Controller Based on CPG Principle

Tsujita and Tsuchiya [13] designed a rhythmic walking controller based on CPG principle and generated walking motions adaptive to the environment through the mutual entrainment between non-linear neural oscillators. Following their design principle, we build a controller which consists of trajectory and phase ones to control reciprocation of each leg. (see Fig. 2). The former drives motors attached to joints according to the motor command from the latter which consists of two oscillators. The phase controller receives the feedback signal of reaction force from the floor through the FSRs attached at the soles.

The stable walking motion is realized as follows.

1. Each leg motion has two kinds of modes: a free leg mode and a support leg one both of which trajectories are specified by the designer in advance (see Fig. 4).
2. In each mode, the joint trajectories are given as a phase function in terms of the corresponding neural oscillators.
3. Mode switching is triggered by phase shift of the free leg caused by the ground contact information from the FSRs. That is, if the free leg contacts with the floor, the phase of the free leg (the support leg, too) is accelerated, and mode switch (free leg $\longleftrightarrow$ support leg) happens.
4. Various kinds of walking are generated with two parameters: a forward step length β and a turning angle α (see Fig. 5).

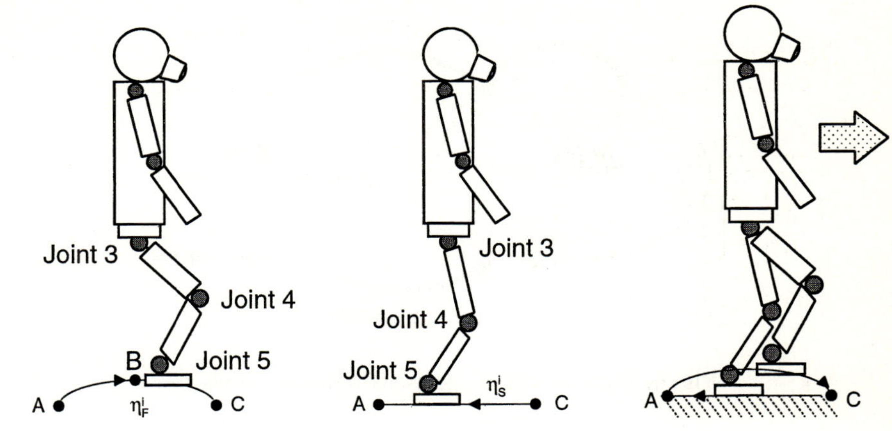

(a) Trajectory of free leg

(b) Trajectory of support leg

(c) Trajectory of leg

Fig. 4. Trajectories of the leg

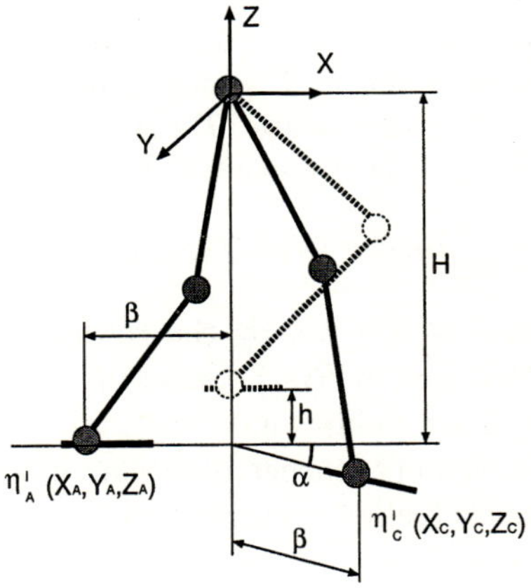

Fig. 5. Waking parameters

3 Reinforcement Learning with Rhythmic Walking Parameters

3.1 Principle of Reinforcement Learning

Recently, reinforcement learning has been receiving increased attention as a method for robot learning with little or no *a priori* knowledge and higher capability of reactive and adaptive behaviors. Fig. 6 shows the basic model of robot-environment interaction [10], where a robot and an environment are modelled by two synchronized finite state automatons interacting in a discrete time

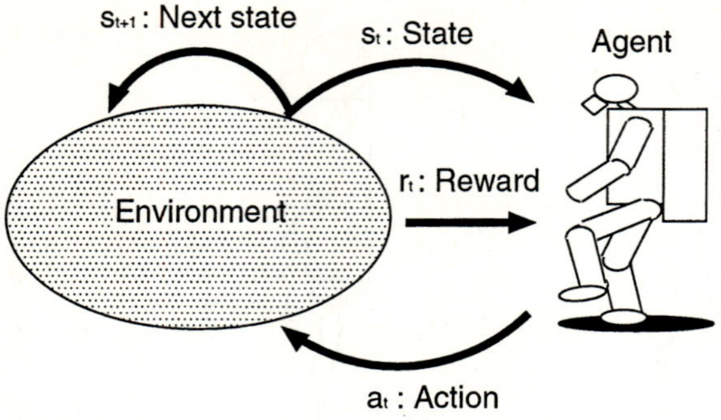

Fig. 6. Basic model of agent-environment interaction

cyclical processes. The robot senses the current state $s_t \in \boldsymbol{S}$ of the environment and selects an action $a_t \in A$. Based on the state and action, the environment makes a transition to a new state s_{t+1} and generates a reward r_{t+1} that is passed back to the robot. Through these interactions, the robot learns a purposive behavior to achieve a given goal. In order for the learning to converge correctly, the environment should satisfy the Markovian assumption that the state transition depends on only the current state and the taken action. The state transition is modelled by a stochastic function $\boldsymbol{T}$ which maps a pair of the current state and the action to take to the next state ($\boldsymbol{T} : \boldsymbol{S} \times \boldsymbol{A} \rightarrow \boldsymbol{S}$). Using $\boldsymbol{T}$, the state transition probability $P_{s_t,s_{t+1}}(a_t)$ is given by

$$P_{s_t,s_{t+1}}(a_t) = Prob(\boldsymbol{T}(s_t, a_t) = s_{t+1}). \tag{1}$$

The immediate reward r_t is given by the reward function in terms of the current state by $R(s_t)$, that is $r_t = R(s_t)$. Generally, $P_{s_t,s_{t+1}}(a_t)$ (hereafter $\mathcal{P}^a_{ss'}$) and $R(s_t)$ (hereafter $\mathcal{R}^a_{ss'}$) are unknown.

The aim of the reinforcement learner is to maximize the accumulated summation of the given rewards (called *return*) given by

$$return(t) = \sum_{n=0}^{\infty} \gamma^n r_{t+n}, \tag{2}$$

where $\gamma\,(0 \leq \gamma \leq 1)$ denotes a discounting factor to give the temporal weight to the reward.

If the state transition probability is known, the optimal policy which maximizes the expected *return* is given by finding the optimal value function $V^*(s)$ or the optimal action value function $Q^*(s,a)$ as follows. The derivation of them can be found elsewhere [10].

$$\begin{aligned} V^*(s) &= \max_a E\{r_{t+1} + \gamma V^*(s_{t+1}) | s_t = s, a_t = a\} \\ &= \max_a \sum_{s'} \mathcal{P}^a_{ss'} \left[\mathcal{R}^a_{ss'} + \gamma V^*(s')\right] \end{aligned} \tag{3}$$

$$Q^*(s,a) = E\{r_{t+1} + \gamma \max_{a'} Q^*(s_{t+1}, a') | s_t = s, a_t = a\}$$
$$= \sum_{s'} \mathcal{P}^a_{ss'} \left[\mathcal{R}^a_{ss'} + \gamma \max_{a'} Q^*(s', a')\right] \quad (4)$$

In this paper, the learning module examines the state transition when both feet contact with the ground. The state space, **S**, consists of the visual information s_v and the robot posture s_p, and the action space consists of two parameters of rhythmic walking. Details are explained in the following subsections.

3.2 Construction of Action Space Based on Rhythmic Parameters

The learning process has two stages. The first one is to construct the action space consisting of feasible combinations of two rhythmic walking parameters (α, β). To do that, we prepared the three-dimensional posture space s_p in terms of the forward length β (quantized into four lengths: 0, 10, 35 60 [mm]), the turning angle α (quantized into three angles: -10, 0, 10 [deg].) both of which correspond to the result of the execution of the previous action command, and the leg side (left or right). Therefore, we have 24 kinds of postures. First, we have constructed the action space of the feasible combinations of (α, β) excluding the infeasible ones which cause collisions with its own body. Then, various combinations of actions are examined for stable walking in the real robot. Fig. 7 shows the feasible actions (empty boxes) for each leg corresponding to the action, which determines the resultant posture of the next step. Due to the differences in physical properties between two legs, the constructed action space was not symmetric although it should be theoretically.

3.3 Reinforcement Learning with Visual Information

Fig. 8 shows an overview of the whole system which consists of two layers: adjusting walking based on the visual information and generating walking based on neural oscillators. The state space consists of the visual information s_v and the robot posture s_p, and adjusted action a is learned by dynamic programming method based on the rhythmic walking parameters (α, β). In a case of ball shooting task, s_v consists of ball substates and goal substates both of which are quantized as shown in Fig. 9. In addition to these substates, we add two more substates, that is, "the ball is missing" and "the goal is missing" because they are necessary to recover from loosing their sight.

Learning module consists of a planner which determines an action a based on the current state s, a state transition model which estimates the state transition probability $\mathcal{P}^a_{ss'}$ through the interactions, and a reward model (see Fig. 10). Based on DP, the action value function $Q(s,a)$ is updated and the learning stops when no more changes in the summation of action values.

$$Q(s,a) = \sum_{s'} \mathcal{P}^a_{ss'} [\mathcal{R}_s + \gamma \max_{a'} Q(s', a')], \quad (5)$$

where $\mathcal{R}_s$ denotes the expected reward at the state s.

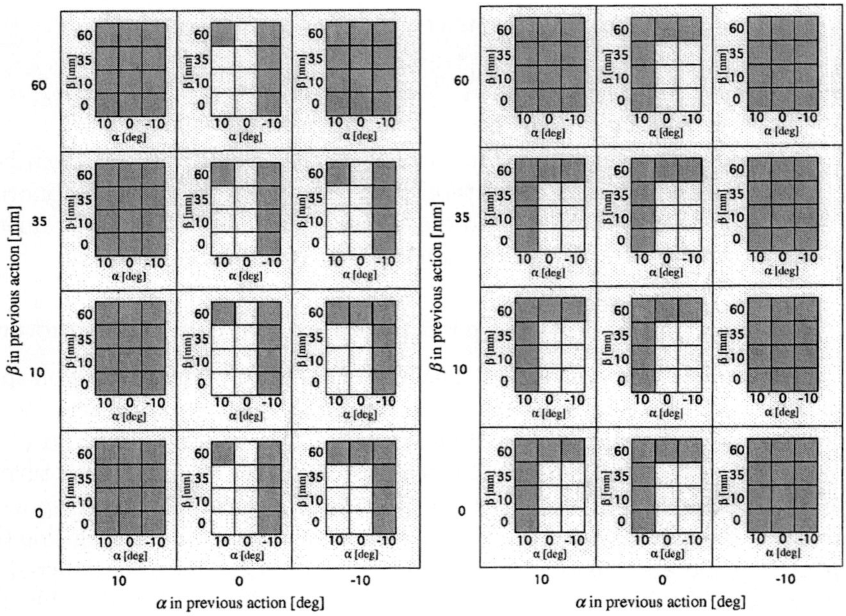

(a) left leg (b) right leg

Fig. 7. Experimental result of action rule

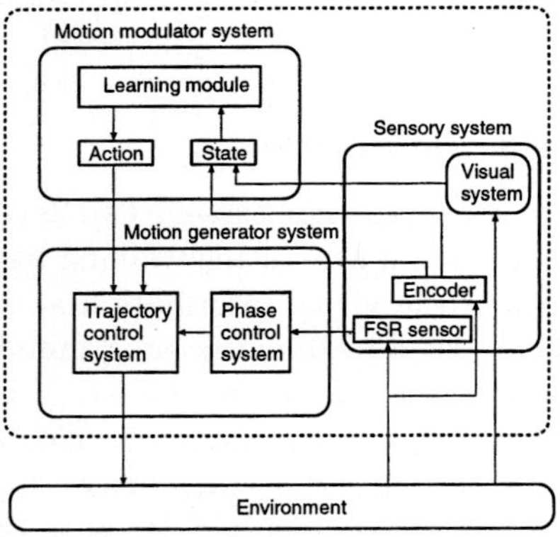

Fig. 8. Biped walking system with visual perception

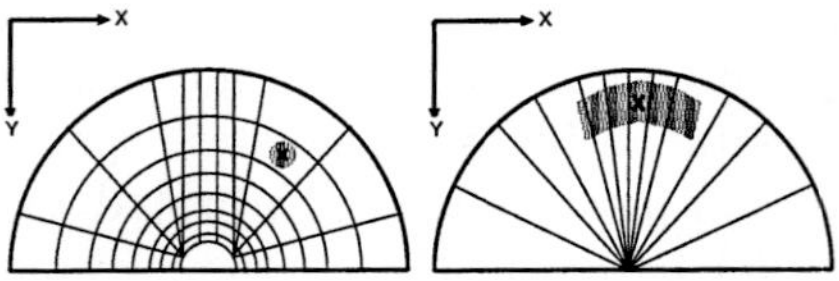

(a) State space of ball (b) State space of goal

Fig. 9. State space of ball and goal

4 Experiments

4.1 Robot Platform and Environment Set-Up

Here, we use a humanoid platform HOAP-1 by Fujitsu Automation LTD. [9] attaching a CCD camera with a fish-eye lens at the head. Figs. 11 and 12 show a

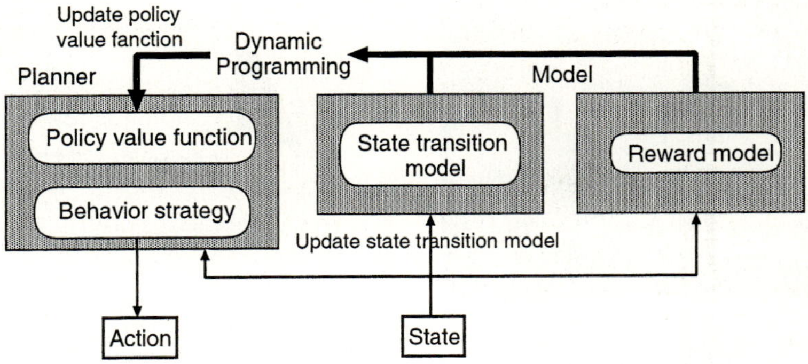

Fig. 10. A learning module

Fig. 11. HOAP-1

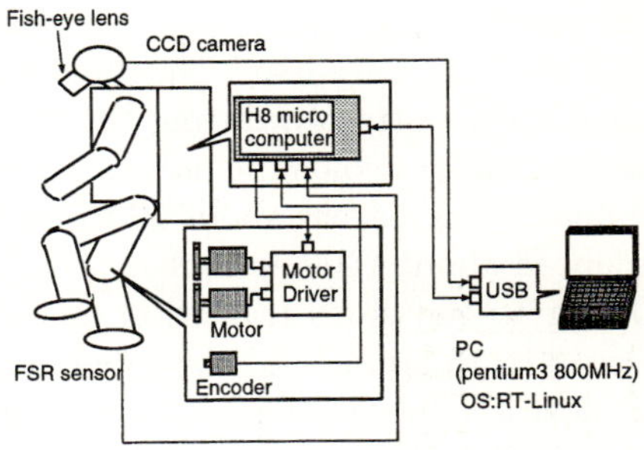

Fig. 12. An overview of the robot system

picture and a system configuration, respectively. The height and the weight are about 480[mm] and 6[kg], and each leg (arm) has six (four) DoFs. Joint encoders have high resolution of 0.001[deg/pulse] and reaction force sensors (FSRs) are attached at soles. The colour image processing to detect an orange ball and a blue goal is performed on the CPU (Pentium3 800MHz) under RT-Linux. Fig. 13 shows the on-board image.

The experimental set-up is shown in Fig. 14 where the initial robot position is inside the circle whose center and radius are the ball position and 1000 [mm], respectively, and the initial ball position is located less than 1500 [mm] from the goal of which width and height are 1800 [mm] and 900 [mm], respectively. The task is to take a position just before the ball so that the robot can shoot a ball into the goal. Each episode ends when the robot succeeds in getting such positions or fails (touches the ball or the pre-specified time period expires).

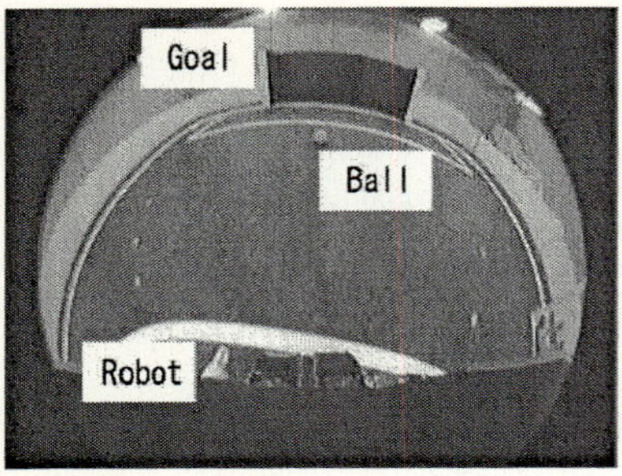

Fig. 13. Robot's view (CCD camera image through fish-lens)

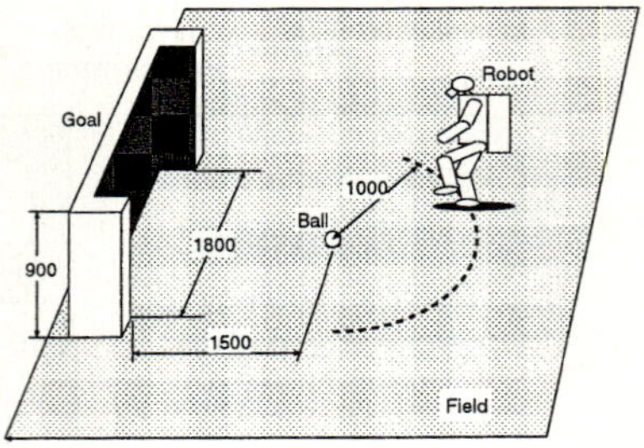

Fig. 14. Experimental environment

4.2 Experimental Results

One of the most serious issues in applying the reinforcement learning method to real robot tasks is how to accelerate the learning process. Instead of using Q-learning that is most typically used in many applications, we use a DP approach based on the state transition model $\mathcal{P}^a_{ss'}$ that is obtained separately from the behavior learning itself. Further, we give the instructions to start up the learning, more correctly, during the first 50 episodes (about a half hour), the human instructor avoids the useless exploration by directly specifying the action command to the learner about 10 times per one episode. After that, the learner experienced about 1500 episodes.

Owing to the state transition model and initial instructions, the learning converged in 15 hours, and the robot learned to get the right position from any initial positions inside the half field. Fig. 15 shows the learned behaviors from different initial positions. In Fig. 15, the robot can capture the image including both the ball and the goal from the initial position while in Fig. 15 (f) the robot cannot see the ball or the goal from the initial position.

5 Concluding Remarks

A vision-based behavior of humanoid was generated by reinforcement learning with rhythmic walking parameters. Since the humanoid generally has many DoFs, it is very hard to control all of them. Instead of using these DoFs as action space, we adopted rhythmic walking parameters, which drastically reduces the search space and therefore the real robot learning was enabled in reasonable time. In this study, the designer specified the state space consisting of visual features and robot postures. State space construction by learning is one of the future issues.

Acknowledgments

This study was performed through the Advanced and Innovational Research program in Life Sciences from the Ministry of Education, Culture, Sports, Science and Technology, the Japanese Government.

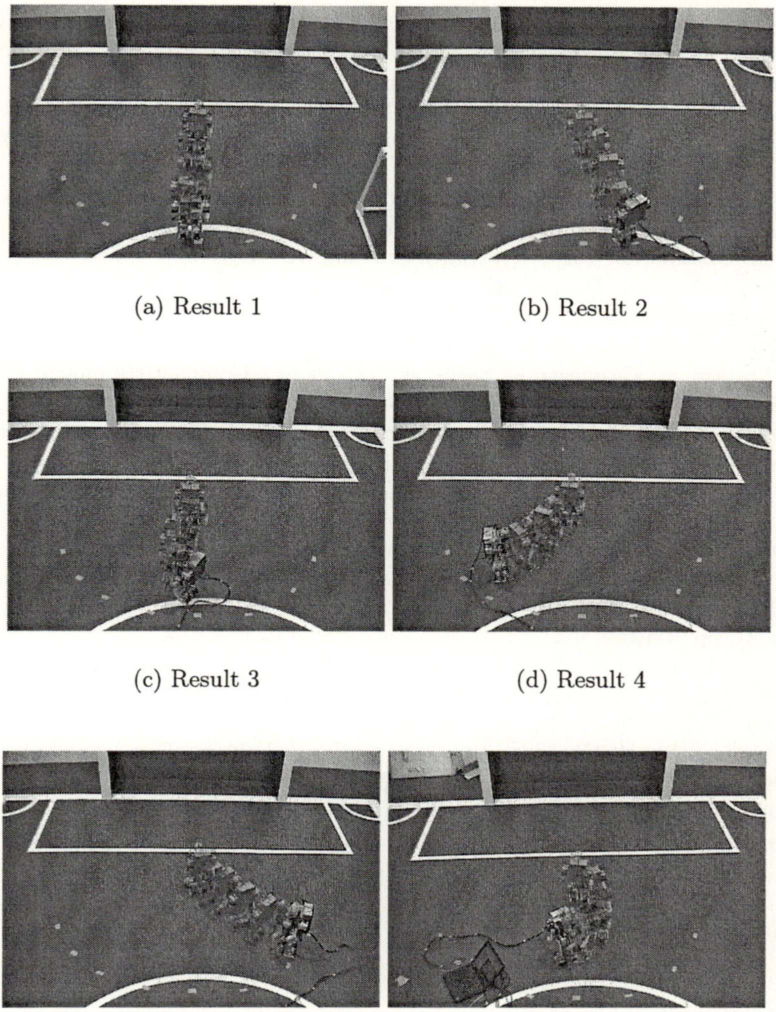

(a) Result 1 (b) Result 2

(c) Result 3 (d) Result 4

(e) Result 5 (f) Result 6

Fig. 15. Experimental results

References

1. A. Fujii, A. Ishiguro. Evolving a cpg controller for a biped robot with neuromodulation. In *Climbing and Walking Robots* (2002), pp. 17–24.
2. S. Grillner. Neurobiological Bases of Rhythmic Motor Acts in Vertebrates. *Science*, Vol. 228, (1985), pp. 143-149.
3. K. Hirai, M. Hirose, Y. Haikawa, T. Takenaka. The Development of Honda Humanoid Robot. In *ICRA* (1998), pp. 1321–1326.
4. S. Kajita and K. Tani. Adaptive Gait Control of a Biped Robot based on Realtime Sensing of the Ground Profile. In *Proc. of IEEE/RSJ International Conference on Intelligent Robots and Systems* (1996), pp. 570-577.

5. H. Kimura, Y. Fukuoka, and H. Nakamura. Biologically inspired adaptive dynamic walking of the quadruped on irregular terrain. In *Proc. of 9th International Symposium of Robotics Research* (1999), pp. 271-278.
6. H. Kitano and M. Asada, The RoboCup humanoid challenge as the millennium challenge for advanced robotics, *Advanced Robotics*, vol. 13, no. 8, 2000, pp. 723-736.
7. M. Laurent and J. A. Thomson. The role of visual information in control of a constrained locomotor task. *J. Mot. Behav*, Vol. 20, (1988), pp. 17–37.
8. S. Miyakoshi, G. Taga, Y. Kuniyoshi, and A. Nagakubo. Three Dimensional Bipedal Stepping Motion using Neural Oscillators –Towards Humanoid Motion in the Real World–. In *Proc. of IEEE/RSJ International Conference on Intelligent Robots and Systems* (1998), pp. 84-89.
9. Y. Murase, Y. Yasukawa, K. Sakai, etc. Design of a Compact Humanoid Robot as a Platform. In *19th conf. of Robotics Society of Japan*, (2001), pp. 789-790.
10. Richard S.Sutton and Andrew G.Barto. "Reinforcement learning:An Introduction", MIT Press/Bradford Books, March, (1998).
11. G. Taga, Y. Yamaguchi, H. Shimizu. Self-organized control of bipedal locomotion by neural oscillators in unpredictable environment. *Biological Cybernetics*, Vol. 65, (1991), pp. 147–159.
12. G. Taga. A model of the neuro-musculo-skeletal system for anticipatory adjustment of human locomotion during obstacle avoidance. *Biological Cybernetics*, Vol. 78, (1998), pp. 9–17.
13. K. Tsuchiya and K. Tsujita Locomotion Control of a Quadruped Locomotion Robot based on Central Pattern Generator Model. *J. Robotics Society of Japan*, (2002), Vol. 20, pp. 21-24. (in Japanese)
14. J. Yamaguchi, N. Kinoshita, A. Takanishi, and I. Kato. Development of a Dynamic Biped Walking System for Humanoid –Development of a Biped Walking Robot Adapting to the Human's Living Floor–. In *Proc. of IEEE/RSJ International Conference on Intelligent Robots and Systems* (1996), pp. 232-239.

Design of Walking Gaits for TAO-PIE-PIE, a Small Humanoid Robot

Jacky Baltes and Patrick Lam

Department of Computer Science, University of Manitoba
Winnipeg, Canada
jacky@cs.umanitoba.ca
http://www.cs.umanitoba.ca/~jacky

Abstract. This paper describes the methodology that we used to design and implement balancing and walking gaits for TAO-PIE-PIE, a small 30cm tall humanoid robot. TAO-PIE-PIE is a fully autonomous robot with all power, sensing, and processing done on-board. It is also a minimalistic design with only six degrees of freedom. Nevertheless, its performance is comparable to that of other more complex designs. The paper describes three patterns: (a) a straight walk, (b) a turn on the spot, and (c) a kicking pattern. Sensor feedback is provided by two gyroscopes that provide angular velocity in the left-right and forward-backward plane and a CMOS camera providing vision information. The feedback from the gyroscopes is not used to actively control the walking gait, because the signal is noisy and it would be computationally to expensive for the current processor hardware. Instead, coarse feedback from the gyroscopes is used to monitor the transition from one phase of the pattern to the next. TAO-PIE-PIE proved to be a successful design winning a number of honors at international competitions.

1 Introduction

This paper describes the design and implementation of a stable walking gait, a turning motion, and a kicking motion for TAO-PIE-PIE, the third generation humanoid robot developed in our laboratory.

Recent advances in material science, control engineering, robotics, and Artificial Intelligence has enabled researchers to build fully autonomous humanoid robots that can walk, dance, climb stairs, and other functions.

For the first time, these robots are not limited to academia and research laboratories, but have been developed as commercial products. Recently, several companies have developed commercial humanoid robotic platforms, for example Honda, Fujitsu, Mitsubishi, and Sony.

These designs have many degrees of freedom and are very complex mechanical and electronic systems. Correspondingly, they are expensive.

Nevertheless, many research questions about humanoid robots remain unanswered. Apart from the general problems of localization, computer vision, path planning, motion planning, and task planning, these also include some problems that are specific to the control of humanoid robots. For example, what is the minimum set of actuators needed for stable walking, what sensor information is necessary to walk over uneven

D. Polani et al. (Eds.): RoboCup 2003, LNAI 3020, pp. 355–365, 2004.

terrain, how to minimize the energy required in walking, or what are the trade-offs between walking speed and stability.

This paper describes our work on designing TAO-PIE-PIE, a small humanoid robot. One of the design goals of TAO-PIE-PIEwas a minimal design; using only as many degrees of freedom as necessary to achieve a stable walk.

The paper describes related work in section 2. The design requirements for our humanoid robot are shown in section 3. Section 4 gives details of the mechanical and electronic design of TAO-PIE-PIE. The methodology used to develop and details of the implementation of the walking gaits are given in section 5. We implemented three different walking patterns: a straight walk, a turn, and a kick. The paper concludes with section 7. This section also critically evaluates the performance of TAO-PIE-PIE against that of other robots.

2 Related Work

Many teams worldwide have also developed small humanoid robots. This section gives an overview over other small humanoid robot designs. This section attempts to give an overview over the different types of designs rather than trying to be exhaustive.

2.1 Viki

Viki was developed by Prof. Lund's team from the University of Southern Denmark [7]. Similar to TAO-PIE-PIE, Vicki is a minimalistic design approach. It embodies a bottom up approach focusing on the interaction between physical properties and control.

Viki uses only five motors. Two motors are used to turn the legs sideways, one motor moves the hip, and one motor moves the upper body. Another motor is used to control the arms of the robot.

Viki only uses four motors for the walking motion compared to TAO-PIE-PIE's six RC servos. However, Viki's mechanical design is significantly more complex including a gear box and timing belts. It is difficult to compare the kinematic abilities of the two robots. Viki has the ability to turn either leg sideways, but can not kick a ball straight forward. TAO-PIE-PIE can not turn its hips sideways, but it can kick with either the right or left foot.

It is also interesting to note that TAO-PIE-PIE Viki (both based on minimalistic design philosophies) were by far the smallest robots in the competitions. TAO-PIE-PIE is 28cm tall, whereas Viki is about 25cm tall.

Viki does not include any sensors for balancing and walking. Rather it seems to have been developed primarily for the RoboCup Junior competition, since it has a set of infrared sensors that can detect an infrared emitting ball over long distances.

2.2 Pino and Morph

Pino and Morph are a humanoid robots developed by the Kitano Symbiotic Systems group [9]. Pino is approximately 50cm tall. It uses high torque servo motors to control 26 DOF (6 DOF in each leg, 5 DOF in each arm, 2 DOF in the trunk, and 2 DOF in the head). Pino includes a vision sensor, posture sensors, joint angle sensors (using the potentiometers of the servos), and force sensors in the feet.

Morph is a next generation humanoid robot. It is similar to Pino's design, but smaller.

2.3 Robo Erectus

Robo Erectus is a series of small humanoid robots developed by Changjiu Zhou at Singapore Politechnic, Singapore [10]. It is a approximately 40cm tall and has 12 DOFs. It uses standard RC servos as actuators.

Although its walking gait is not as stable as that of other robots, it can move at an amazing speed of approximately 10cm/sec. Robo Erectus was the fastest of the small sized robots in the RoboCup competition achieving a second place finish in the robot walk even against taller robots.

2.4 Kaist Robot

Prof. Kim from KAIST, Daejon, Korea developed a 50cm tall robot which uses DC motors and timing belts. The robot performed superbly during the 2002 FIRA HuroSot competition and won first prize. [5].

2.5 Johnny Walker

Johnny Walker is a 40cm tall humanoid robot developed by Dr. Thomas Bräunl from the University of Western Australia [2]. It also uses the Eyebot controller. It has four DOFs in each leg. Feedback is provided by a vision cameras as well as two accelerometers, two gyroscopes, and force sensors in the feet.

2.6 Development of Walking Gaits

There is a rich literature on the design of walking gaits for specific humanoid robots (see for example [8, 6]). However, much of this work was done in simulation only, and those using real robots succeed only by constraining the range of motion of the robot (e.g., through the use of oversize feet, or by restricting the motion in the saggital plane).

The representation of the walking gait is also important. A smooth control curve is needed, since otherwise the motion of the robot will be too jerky. Jerking introduces unwanted forces into the motion of the robot, is not energy efficient, and introduces even more noise into the output of gyroscopes and accelerometers. Ijspeert describes the walking gaits with a set of differential equations [4]. The differential equations describe a set of attractors using Gaussian kernel functions. A walking gait is represented as a set of weights on the kernel functions, and thus an attractor landscape.

3 Design Requirements

TAO-PIE-PIE was intended as a research vehicle to investigate methods for deriving control methods for stable walking patterns for humanoid robots. Stable walking, especially over uneven terrain is a difficult problem. One problem is that current actuator technology (RC Servos, DC motors) generate less torque in comparison to their weight than human muscle. Another problem is that feedback from gyroscopes and actuators is very noisy. The necessary smoothing of the input signals makes it hard to use it in actively controlling the walking motion.

Another research direction was the investigation of computer vision based methods for balancing and walking. Humans use vision when balancing. This can be demonstrated convincingly by having a person balance on one leg and then ask them to close their eyes, which makes balancing much harder for people. The idea is to use the optical flow field of the camera to control the robot's walking motion. This requires fundamental scene interpretation since to extract a motion vector (in general, a six dimensional vector representing translation in the X, Y, and Z plane as well as pan, tilt and swing angles) from an image sequence requires knowledge of the geometry of the scene. This problem can be simplified by making assumptions about the world and limiting the amount of

Furthermore, TAO-PIE-PIE was intended to compete at international humanoid robotic competitions such as RoboCup and FIRA HuroSot. Among other, this meant that TAO-PIE-PIE had to be able to balance, walk, run an obstacle course, dance, and kick a ball.

Cost was an important design criteria in TAO-PIE-PIE's development. Previous experience has shown us that the use of commonly available cheap components does not only help to keep the cost of a project down and the Head of Department happy, but it also has lead to the development of novel, versatile, and robust approaches to problems in robotics.

For example, most teams in the small sized league use a camera mounted directly overhead. Since the viewing field is limited with a standard lens, most teams purchased wide angle lenses and expensive cameras. In contrast, our small size league team, the All-Botz [1], mounted the camera with a side view. This made the vision problem harder but resulted in the development of more complex and robust camera calibration routines. However, this effort is now paying off since the development system is flexible and robust enough to handle the newer larger playing fields introduced in 2002 as well as even larger playing fields planned for the future.

Another design goal was to reduce the number of degrees of freedom (DOF) of the robot. This reduces the cost of the humanoid robot as well as increases its robustness. Each DOF adds extra complexity in the mechanical design and the design of the control electronics. Furthermore, reducing the number of DOFs in the robot allows us to exploit the dimensions of the humanoid walking problem. The minimum set of DOFs that allow a humanoid robot to walk is of interest since it leads to energy efficient designs.

4 Mechanical Design of TAO-PIE-PIE

TAO-PIE-PIE is the third generation of humanoid robots developed in our lab. The first two humanoid robots RX-78 and Zaku were both based on commercially available toy model figures. Both model provided important stepping stones and insights into the design of a small humanoid robot.

A number of lessons were learned from RX-78 and Zaku and included in the mechanical design of TAO-PIE-PIE.

Firstly, the plastic model kits were not strong enough to withstand the pressures and forces during walking and thus would often break. Therefore, TAO-PIE-PIE was constructed from aluminum.

Secondly, Zaku proved the need for a joint in the left-right (so-called sagittal) plane to shift the center of gravity from right to left. Although Zaku was able to shuffle in a straight line, the absence of a joint in the saggital plane meant that it could not lift its feet of the ground and thus had a very short stride length. The only way for Zaku to shift the center of gravity from left to right was by bending its knees. TAO-PIE-PIE therefore had an extra DOF in each ankle which allows it to shift in the saggital plane.

Thirdly, the importance of feedback in the control of humanoid walking was demonstrated was demonstrated by RX-78 which used a modified mouse as make-shift pitch and roll sensor. TAO-PIE-PIE includes two gyroscopes to measure the angular velocity in the left-right and forward-backward plane.

Figure 1 shows the mechanical construction of TAO-PIE-PIE.

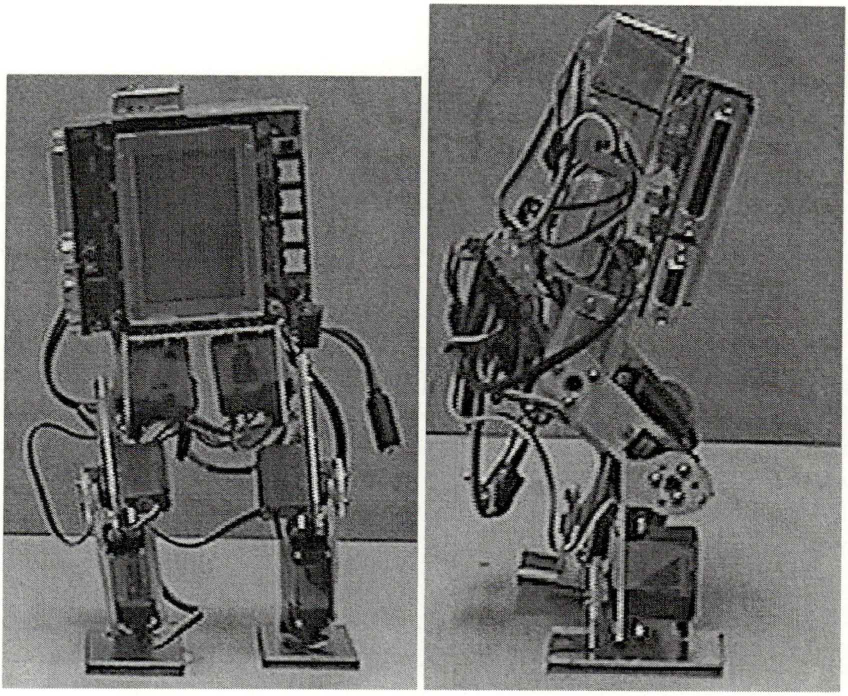

Fig. 1. Front and side view of TAO-PIE-PIE.

The actuators and sensors consist of widely available RC servos and RC gyroscopes for remote controlled cars and helicopters.

The Eyebot controller ([3]) was chosen as embedded processor, since it is relatively inexpensive, yet powerful enough to provide vision information. A small CMOS camera provides visual feedback for the robot.

The details of the mechanical construction of TAO-PIE-PIE are shown in Figure 2. The mechanical design was done in conjunction with Dr. Nadir Ould Kheddal's robotics group at Temasek Politechnic, Singapore.

TAO-PIE-PIE is constructed out of 0.5mm aluminum. The RC servos are used as structural components in the design.

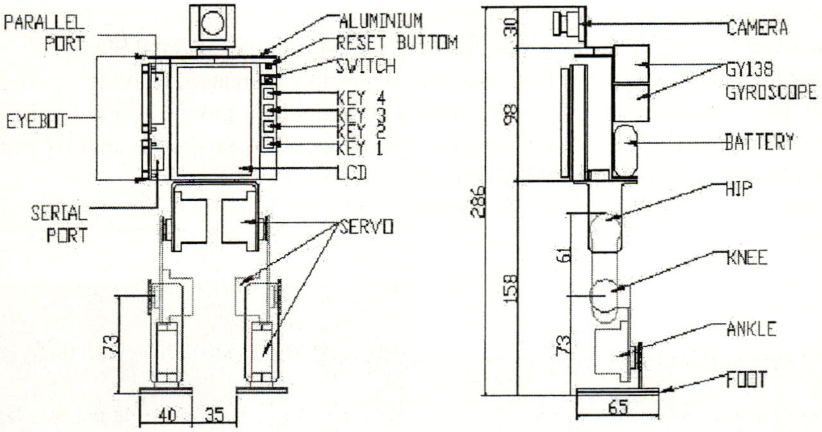

Fig. 2. CAD Drawing of TAO-PIE-PIE.

5 Development of Walking Gaits

One of the fundamental problems in humanoid robots is the development of stable walking patterns. A walking pattern is dynamically stable if the COP is within the supporting area. A statically stable walking pattern also has the COM within the supporting area.

We used a divide and conquer approach and divided the walking gait into six phases: three for the right and three for the left leg.

The phases were selected in such a way that the robot is statically stable at the end of each phase.

5.1 Inverse Kinematics

Inverse Kinematics allow us to compute the correct joint angle to position a robotic link at a target position. After determining the desired location for the COM, we compute joint angles for all RC servos in the leg. By controlling the joints angles, we can control the stability during the motions.

To keep the humanoid balanced, we must keep the center of mass within the supporting region during the termination of all phases of the walking motion.

To solve the inverse kinematics model of the humanoid robot we use a simplified two link model of the robot.

Solving the equations for θ_1 first and then for θ_2, we can find a solution to the inverse kinematics problem. The solution for θ_1 and θ_2 are shown below.

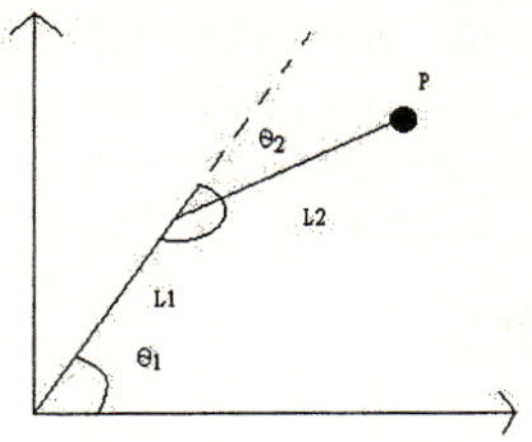

Fig. 3. Inverse Kinematics model of the robot. P, represent center of mass. $L2$ - hip joint link, $L1$ - knee joint link.

$$\theta_2 = \frac{\cos(x^2 + y^2 - {L_1}^2 - {L_2}^2)}{2L_1L_2}$$

$$\theta_1 = \frac{-(L_2 \sin(\theta_2)x + (L_1 + L_2 \cos(\theta_2)y)}{(L_2 \sin(\theta_2)y + (L_1 + L_2 \cos(\theta_2)x)}$$

5.2 Pattern for Straight Walk

The walking pattern for a straight walk is shown in Figure 4. The pattern is based on the control curves shown in the previous section.

The walking pattern consists of six phases. The walking pattern repeats itself after the sixth phase. The bottom row of images shows the approximate position of the COM.

TAO-PIE-PIE starts in phase 1 — "Two Leg Stand" — where the right leg is in front and the left leg is behind. Both legs are on the ground and the COM is between the two legs.

From phase 1, TAO-PIE-PIE moves to phase 2 — "One Leg Stand" —. In this phase, the ankle servo generates a torque which moves the COM to the inside edge of the right leg. This also results in the back (left) leg to lift off the ground.

During the transition from phase 2 to phase 3 — "Ready for Landing" — is in static balance. TAO-PIE-PIE moves the free left leg forward and positions it so that it is ready for land. The COM moves to the front of the supporting leg. This stabilizes the transition to phase 4.

During the transition from phase 3 to phase 4 — "Two Leg Stand Inverse" — the robot is in dynamic balance. The supporting leg extends its knee joint to shift the COM over the front edge of the supporting leg. The ankle servo of the supporting leg generates a torque to move the COM over the right side. The left leg will touch the ground in front of the right leg.

Phases 5 to 6 are the mirror images of phases 2 to 3. After phase 6, the motion continues with a transition to phase 1.

5.3 Pattern for Turning Walk

TAO-PIE-PIE possesses two different patterns for changing the direction of its walk from a straight line walk: (a) varying the stride length, and (b) lower body twist.

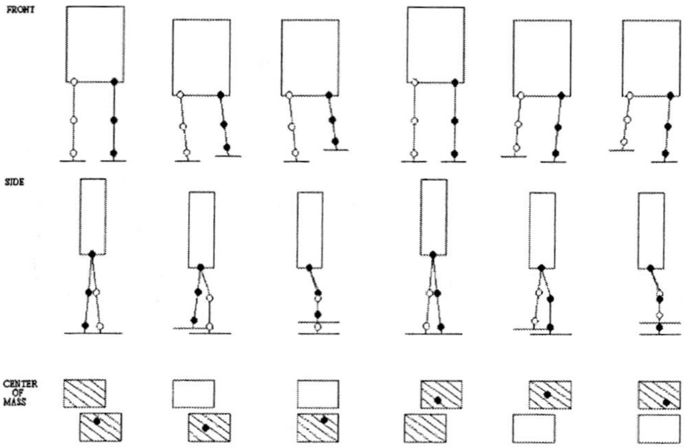

Fig. 4. Walking Pattern of TAO-PIE-PIE.

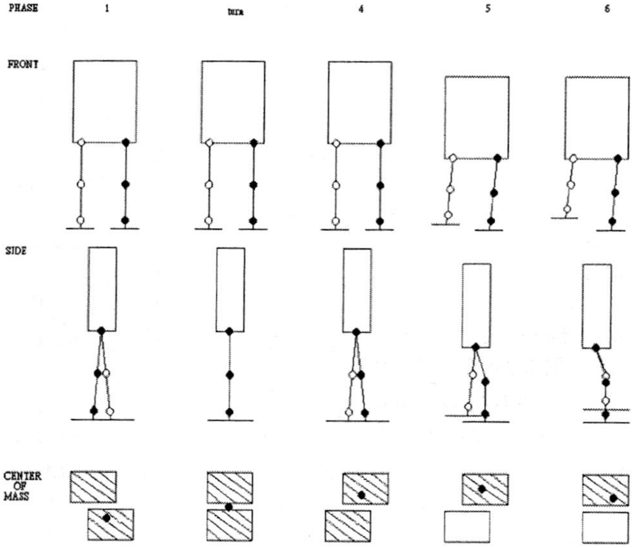

Fig. 5. Turning Pattern of TAO-PIE-PIE.

By changing the speed at which TAO-PIE-PIE moves through phases 2 and 5 respectively, it can vary the stride length of the left and right part of the walking pattern which turns the robot into this direction. However, the turning rate is slow and requires a lot of distance for TAO-PIE-PIE to turn noticeably.

TAO-PIE-PIE can turn on the spot by twisting its lower body, which is shown in Figure 5. The turn occurs in phase "turn." In this phase, the front and back legs will swap position. During the turn, the COM and the COP are in the center between the two feet.

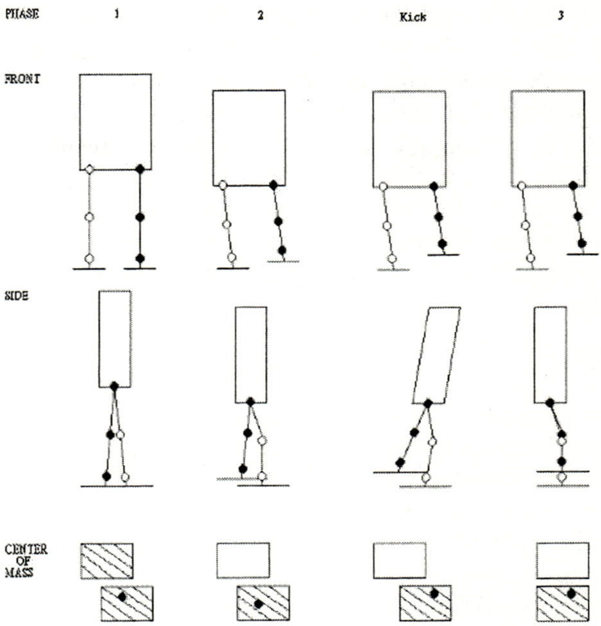

Fig. 6. Kicking Pattern of TAO-PIE-PIE.

5.4 Kicking Pattern

The RoboCup competition also required our robots to kick a ball. Therefore, we developed a kicking pattern for TAO-PIE-PIE. The kicking pattern shown in Figure 6 is similar to the walking pattern.

The difference is that in phase 2, the rear leg is moved back as far as possible. This increases the range of motion of the kick, which results in more energy for the kick. To keep the robot balanced, TAO-PIE-PIE leans the upper body forward by moving both hip joints.

This is necessary to keep the COM over the supporting area of the front leg.

TAO-PIE-PIE then snaps the back leg forward as quickly as possible. At the same time, it straightens out the upper body, which readies it for landing on the kicking foot.

5.5 Sensor Feedback

The only feedback about the motion of TAO-PIE-PIE is provided by two gyroscopes that provide information about the angular velocity in left-right and forward-backward plane respectively.

However, the computational requirements of on-board computer vision are large for the Eyebot embedded processor (MC 68332). Furthermore, the data from the gyroscopes is noisy and without significant pre-processing unsuitable to control the motion during the transition from one phase to another.

Instead, only large changes in the output of the gyroscopes are used to recognize when a certain phase has been entered successfully. For example, the robot listens for a sudden change in the angular velocity in the left-right plane to determine when the free leg has landed in phases 1 and phases 4 of the straight walking pattern.

The feedback from the gyroscopes is also used to detect abnormal behavior. For example, if the robot's foot is caught on the carpet and instead of moving the leg forward, the robot will fall onto the leg too early. If this abnormal feedback is detected the robot attempts to stabilize itself by putting both feet on the ground as quickly as possible and straighten up its upper body. The motion will then stop until both gyroscopes show little angular velocity.

6 Evaluation

It is difficult to quantitatively evaluate the performance of different walking gaits and humanoid robots. Many factors that are not directly related to the walking gaits (e.g., size of the feet of the robot) can influence the performance of a humanoid robot. Furthermore, all humanoid robot designers attempt to find a balance between the stability and speed of the walking pattern. A robot that walks quicker usually has a higher chance of falling. Therefore, not only the maximum speed, but also the probability that the robot will fall at any given speed needs to be considered.

Therefore, this section will provide anecdotal evidence based on observations during the 2002 RoboCup and FIRA HuroSot competitions to highlight TAO-PIE-PIE's strength and weaknesses.

TAO-PIE-PIE competed at two international competitions for humanoid robots in 2002: FIRA HuroSot and RoboCup. It performed well in the competitions winning a technical merit award for fully autonomous operation at FIRA HuroSot and second and third places at RoboCup.

More importantly, TAO-PIE-PIE's walking gaits have proved themselves to be very robust. TAO-PIE-PIE walked for a combined time of over an hour and only fell ones during the FIRA HuroSot competition. A similar result was achieved during RoboCup.

TAO-PIE-PIE also was the only robot in the small robot class that demonstrated during the RoboCup 2002 penalty kick competition, that it could quickly turn on the spot and kick a ball into either the right or left part of the goal. This strategy was successful up until the final against Footprints. After having been scored on by TAO-PIE-PIE in the preliminary rounds, the Footprints goal keeper moved quickly out of the goal in a straight line and thus covered the angle.

This highlights the biggest shortcoming of TAO-PIE-PIE at the moment. Its walking gait is slowed compared to that of other robots. It moved at an average speed of only 1cm/s. This is comparable to the performance of Vicki, the only other very small robot in the competition, but is very slow compared to robots in the 40cm class. The fastest robot in this class was Robo Erectus which walked at about 10cm/secs.

7 Conclusion

TAO-PIE-PIE has shown itself to be a powerful and flexible platform for research into humanoid robotics. It has proven itself during international competitions winning a second place in the RoboCup and a technical merit award in the FIRA 2002 competitions.

We have learned important lessons in the design of humanoid robots from TAO-PIE-PIE, which we will use in the design of the next generation humanoid robot HIRO. HIRO will use four additional DOFs (two in the hip and a pan and tilt camera). It also features a faster embedded processor (Intel Stayton), which allows us to implement better on-board computer vision algorithms. One of the main goals of the HIRO platform will be to investigate methods for augmenting the balancing of the robot using visual feedback.

Currently, only limited feedback from the gyroscopes is used to control the motion of the robot. With the addition of a more powerful embedded systems, finer grained control over the motion is possible.

Acknowledgements

The authors would like to thank Nadir Ould Kheddal's and his students at Temasek Politechnic, Singapore. We further would like to gratefully acknowledge the help of Jitsin Tin and Yan Naing also from Temasek Politechnic, Singapore.

References

1. Jacky Baltes, Nicholas Hildreth, and David Maplesdon. All botz. In Manuela Veloso, Enrico Pagello, and Hiroaki Kitano, editors, *RoboCup-99: Robot Soccer World Cup III*, pages 653–656, New York, 2000. Springer.
2. Thomas Bräunl. Eyebot android. WWW, August 2000. http://robotics.ee.uwa.edu.au/eyebot/robots/android.html
3. Thomas Bräunl. Thomas bräunl's homepage. WWW, November 2002. http://robotics.ele.uwa.edu
4. A. J. Ijspeert, J. Nakanishi, and S. Schaal. Movement imitation with non-linear dynamical systems in humanoid robots. In *Proceedings of the IEEE International Conference on Robotics and Automation (ICRA)*, pages 1398 – 1403, 2002.
5. Jung-Hoon Kim, Ill-Woo Park, and Jun-Ho Oh. Design of a humanoid biped robot lower body. In *Proceedings of the 3rd International Workshop on Human-friendly Welfare Robotic Systems*. KAIST, January 20 - 22 2002.
6. Andrew L. Kun and W. Thomas Miller. Control of variable speed gaits for a biped robot. *IEEE Robotics and Automation Magazine*, 6(3):19–29, September 1999.
7. Leonid Paramonov and Henrik Hautop Lund. A minimalistic approach to humanoids. In *Proceedings of the IEEE-RAS International Conference on Humanoid Robots*, Nov 22 - 24 2001 2001.
8. Jerry Pratt and Gill Pratt. Intuitive control of a planar bipedal walking robot. In *Proceedings of the IEEE International Conference on Robotics and Automation (ICRA '98)*, Leuven, Belgium, 1998.
9. Fuminori Yamasaki, Tatsuya Matsui, Takahiro Miyashita, , and Hiroaki Kitano. Pino the humanoid: A basic architecture. In Peter Stone, Tucker Balch, and Gerhard Kraetszchmar, editors, *RoboCup-2000: Robot Soccer World Cup IV*, pages 269–278. Springer Verlag, Berlin, 2001.
10. Changjiu Zhou. Linguistic and numeral heterogenous data integration with reinforcement learning for humanoid robots. In *Proceedings of the 1st IEEE Conference on Humanoid Robots*, 2000.

ProRobot – Predicting the Future of Humanoid Robots

Ralf Regele, Paul Levi, and Wolfgang Bott

FZI Forschungszentrum Informatik
Mobility Management & Robotics
Haid-und-Neu-Str. 10-14
76131 Karlsruhe, Germany
{regele,levi,bott}@fzi.de

Abstract. Humanoid robots are without question a hot topic in research today. But will they really be the next break-through invention that changes the face of the world, or are they just another over-hyped research toy? ProRobot is a study funded by the European Commission that will have a close look on the future of humanoid robots and their economic and social impact. The complete study will be published in summer 2003, and will especially concentrate on the prospects of research efforts and the differences of research activities throughout the world.

Keywords: Humanoid robots, study, roadmap, future predictions, socio-economic analysis

1 Introduction

The development of humanoid robots is in an interesting phase at the moment. The technological preconditions seem to be met, the first prototype developments look very promising, commercial interest is awakening. However, the future of humanoid robots is not clear, as it is not yet known if humanoid robots can manage the leap from the research stage to the general usage by non-researchers. There seem to be as many promising application fields as there are doubts about them. To help clarify the situation, the ProRobot study was set up. The study is funded by the European Commission, and will not only show the current state-of-the art of humanoid robot technology, but will give a full socio-economic analysis of this emerging area. The analysis will be based on a cost-benefit model comprising of five layers:

Social Benefit. What is the benefit to the end user? Measures the value of a new technology in terms of its usefulness to an end user or group of users in their everyday life.

Usability. What is needed to make the technology user-friendly? Measures the effort required by a user or group of users to be able to use a given technology.

Technical Viability. What technological opportunities make the project valid? Are new component technologies, infrastructures and methods available to implement the idea?

Feasibility. Can a production of the robots be assumed realistically? Are there obstacles or side-effects that make the technology too challenging?

D. Polani et al. (Eds.): RoboCup 2003, LNAI 3020, pp. 366–373, 2004.

Exploitability. Can the product be sold? What is the market value and competition for the technology, product, or service. This depends upon the added value it offers to the user (i.e. social benefit) and the cost the user must make to acquire the benefit (i.e. usability).

In this way, predictions on the potential market size should be given, including predictions on the advance of all needed technologies. Significant technological barriers and socio-economic obstacles will be identified which may motivate specific R&D or support activities. As a conclusion, a road map will be given that shows all expected technological advances, research efforts and market developments.

The complete study will be published by the European commission in August 2003, and will be freely accessible to the public. As the study is under development right now, this paper can only give an overview about the related problems and tasks, while the detailed analysis and road-maps will be found in the study itself.

In this paper, we will give now a basic overview about the tasks and problems when predicting the future of humanoid robots.

2 So What Is Humanoid?

When talking about humanoid robots, the question will soon arise what can be called a humanoid robot, and what not. There is no widely accepted definition for a humanoid robot. For most developers it's enough if anybody can recognize a robot as humanoid just by looking at him. However, as the development of humanoid robots became more and more popular, there were soon machines where it was quite unclear if they still can be called 'humanoid robots'. All in all, there are two existent points of view: The first is to call a robot a 'humanoid robot' if he looks humanoid, which means that he has a structured body similar to a human being: An upright body with two legs (preferable with knee joints and the right proportions) and two arms, hands with five fingers and a head on top of all. The problem here is that even very cheap toy gadgets, that can hardly be called robots at all, meet this definition. Nevertheless this point of view is quite popular, because it's the most natural way for normal people to decide if a robot is humanoid. The other, more academic point of view is to define a robot as 'humanoid', if he can act like a human being. The emphasis here lays more on the possible actions of a robot, and less on his outer appearance and structure. If he can achieve typical 'humanoid' tasks in a normal 'humanoid' environment, he can be called humanoid, no matter how many arm joints he has or how many legs he uses. This point of view is not very satisfying, as some robots may achieve humanoid tasks without looking humanoid at all. Both approaches work together, if the robot needs the humanoid body to achieve the humanoid tasks. For moving in a normal house, for example, he needs to climb stairs and therefore needs legs instead of wheels. This is especially true if the robot must work in environments which are especially designed for the abilities of the human body. An example would be the driver seat in a standard automobile, where all arms and legs are needed in the right spot to allow the usage of the driving controls. The conclusion is that a robot that should act totally like a human must look totally like a human. He does not need however to copy the restrictions of the human body, so knees which bend in both directions might make him look a little bit non-humanoid, though.

What really brings together both points of view is the interaction with real humans. It is a widely accepted fact that a humanoid robot does not make much sense if he cannot interact with humans in one or the other way. For this task the humanoid appearance of the robot is without question beneficial. If a real human wants to work with a robot, or maybe just wants to play with the robot, he will feel much more confident if the robot is shaped in a friendly, humanoid form. He is more aware of what the robot can do, and what he cannot do, making the over-all work much easier.

In practice, most researchers and developers do not seem to care much about a clear definition for humanoid robots. Normally a collection of tasks is given, which should be fulfilled by the robot while using the structure of a humanoid body. Popular tasks and properties for such a humanoid robot are:

- Being a mobile robot with power supply and computer control on-board
- Navigating and moving in an environment made for humans
- Biped walking in a humanoid style
- Gripping and manipulating objects designed for humans
- Cooperative working with humans
- Interacting with humans without endangering their safety
- Having autonomous behavior
- Communicating with humans in a simple and intuitive way
- Using a stereo-vision system as main sensor system
- Using learning and adaptive behavior strategies
- Using human-like intelligence
- Having a design pleasing to real humans

While a complete humanoid robot should of course be able to show all of the above behavior, the technology at the moment is not advanced enough. Most humanoid projects are busy working on one or two of these tasks at the moment.

For the ProRobot study, we will examine all projects that intend to work on a humanoid robot, and work at least on two tasks of this list.

3 The Technological View

Not all technology for normal robots is essential for humanoid robots, and not all technology for humanoid robots is needed for normal robots. Based on the list of tasks that was given in the previous section, we can have a closer look on the different technology fields that are especially interesting when developing a humanoid robot. The progression of these technologies will greatly effect the performance of future humanoid robots, so a prediction of the future developments in this field is surely interesting.

Bipedal Walking Technologies: Often considered to be the core technology for a humanoid robot, there is not only the problem of mechanics, but also the sensorial problem of keeping the balance. True dynamic walking is still a challenge for every humanoid robot, but recent developments are looking quite promising, and good progress is expected in the future if the research interest is staying high. The research of bipedal walking is closely linked to the research on humanoid robots [1].

Navigation in Human Environments: This task should not be underestimated, as the natural environment of humans is quite complex. The robot must navigate with many dynamic objects (called humans) in his workspace while guarantee extreme safety to them. Most service robots face the same problems, and the development is still in a very basic level.

Gripping and Manipulation of Objects: For humanoid arms, the large technology knowledge of industrial manipulators can be used, so the arm development is considered to be a minor problem. However, humanoid shaped hands and gripping patterns are still under research, and working together with a humanoid sensor system is no trivial task. Many service robots deal with similar problems.

Communication with Humans in a Natural Way: Understanding and speaking in natural language is a complex task, but is already heavily researched by countless developers of computer systems needing a human-machine interface. Impressive progress has been made, and the commercial interest is rather high, so it seems as if solutions already on the market can be used for the humanoid robot.

Humanoid Vision and Senses: There seems to be no need to make the sensor system of a humanoid robot especially human-like, although stereo-vision systems are preferred by most development teams. Image processing is already a hot topic in research nowadays, so most times standard systems are used for the humanoid robots.

System Integration: This is the problem of putting all of the above technologies into a single, autonomous humanoid body while keeping the weight as low as possible. The robot must have good computational power, and must carry his own power supply with him. While computers are getting smaller and faster in regular steps, the power supply is still an unsolved problem. Batteries for supplying a very power consuming robot over a long period are very heavy, a problem well known from the development of service robots.

Depending on the intended tasks for the robot, some technologies might be more important than others. For RoboCup, the goal was formulated to have a team of humanoid robots playing against human opponents in the year 2050, which would certainly need good solutions for all technology fields above. Fig. 1 gives an overview about the state of the art today with help of a score card.

4 The Economic View

No prediction of future developments can be made without regarding the commercial interest in the desired technologies. The developers of humanoid robots are quite aware of the fact that these robots should be for sale one day, not only for a few scientists, but for a wide audience of normal people. So, while examining what the robot technology can achieve, we should also think about what the robot is intended to do. There are many different application fields for humanoid robots, from the more futuristic scenarios to the markets where robots are already for sale at the moment. In all application fields the main characteristic is that the robot must interact with real humans. Without any form of human interaction, a specialized robot will always have advantages over the humanoid robot.

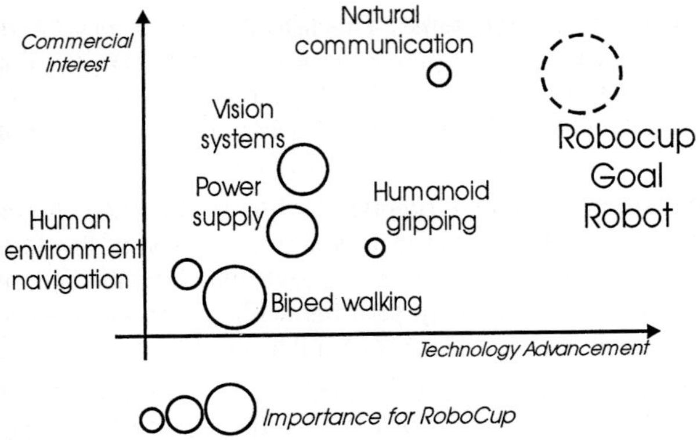

Fig. 1. An overview of the current state of the different technologies needed for a humanoid robot. Bubble sizes represent the importance of the technology for the goal of RoboCup. The RoboCup goal robot was put in there as a comparison, and was assumed to be of maximal commercial interest.

Possible fields of application are:

Industry: There are many robots working in the industrial production at the moment, but not humanoid ones. The reason is that the work is very specialized and repetitive, with human work force reduced to a minimum. The humanoid robot cannot profit from his greatest advantages, flexibility and human interaction. There are industrial fields where humanoid robots can be helpful because they can work close together with humans. The building industry is such a field, where a humanoid robot could act like a reliable co-worker, especially in unsafe environments. There are research groups that think of such an application field for their robot [2][3], but the goal seems to be very hard to reach. In addition, the cost efficiency of such work can be seriously doubted.

Military: Science fiction stories are full of super-human battle robots, which act as powerful, tireless and reliable soldiers. Chances are very high that these things remain completely fictional, as there are much better ways for robots to be used in the military. Even the U.S. military, which often favors high-sophisticated technologies, do not seem to think about this field of application. Corresponding to newspaper releases [4][5], however, they have raised a $50 million fund to research the usage of mechanized exoskeletons for human soldiers, and will share some of the problems with the developers of humanoid robots, e.g. a suitable power-supply. The U.S. defense research agency DARPA has also shown some interest in the development of autonomous, self-navigating robotic vehicles [6].

Service: This is perhaps the application field with the highest expectations for humanoid robots. Human interaction and flexibility are core characteristics of the service field, and the humanoid shape is certainly an advantage for any service robot, as the working environments doesn't need to be adjusted for such a robot. Especially working places with high personnel expenses, wide requirements and high cost pressures seem to be predestined for the usage of humanoid robots. An example is the health care field where many people are required to care for diseased, old or handicapped people. The usage of service robots seems to be not far away in the costly health care field [7].

Research: Most of the humanoid robots that are sold today are meant primarily for research purposes. While this is good for itself, it is mainly regarded as an intermediate step in the development of robots for other purposes. There are some research groups which develop robots very similar to real humans, and try in this way to learn more about the functionality of the human body and mind.

Entertainment: Last but not least, the entertainment sector is not one that should be underestimated, as its economic power might be greater than the service sector. For example, the retail figures of the latest U.S. census [8] show that over $62 billion have been spent on sporting and toy products, and $32 billion on consumer electronics like radio and television sets (not including computers), while only $10 billion have been spent on classic household appliances like refrigerators, dishwashers, vacuum cleaners and so on. Application fields for a humanoid robot in the entertainment sector seem to be unlimited, with high-sophisticated robots used as perfect toy companions. Many humanoids which are for sale today are meant primary as toys, while using impressive technology [9].

Advertisement: It might seem to be a little odd to mention the advertisement sector as an application field for humanoid robots, but in practice, many companies are raising high fundings for their humanoid robot projects mainly because of the reputation and prestige they gain with them. Humanoid robots are a good field to show the technological abilities of a company, and can get the interest of a big, world-wide audience including the media. The research community can certainly profit from these marketing issues. After all, it was also a question of prestige that brought man to moon and back.

The ProRobot study will carefully look at all of the above mentioned application fields and examine their future potential for humanoid robots. Commercial interest has always given a significant boost to the research activities, and so the future of humanoid robots will depend on how they can be used in an economically reasonable way.

5 The Social View

As mentioned in the previous section, the interaction with real human beings is one of the most important tasks for a humanoid robot. The question now is if real humans are willing to interact with a robot. The quality of cooperative work of humans and humanoids depend on the efforts by both sides, and service or entertainment robots are totally useless if people are afraid of them. Even assuming there is a perfect humanoid robot who can achieve all wanted tasks, the question remains if he is accepted by the people. How persons react on humanoid robots depends on a number of factors, some of them are shown below:

Local Society and Culture: People of different regions and nations react different on humanoid robots. It is well-known that people in Japan are quite enthusiastic about them, while people in Europe do not seem to care too much for them. The reasons for this might be based in the fundamental relationship of people towards new technology. The local differences might vanish when the new technology gets more common. After all, many new technologies are similarly popular throughout the world today.

Intended Audience: Humanoid robots are not intended to work solely with robotic specialists, researchers and academic persons. Instead, they should be able to interact with normal people who do not have any preconditional knowledge of robotics. Different audiences will react differently on a humanoid, though. It might be more difficult to establish a care robot in a home for old people than as a toy for children. It is therefore necessary to have a close look on the target audience before introducing a humanoid robot.

Safety Concerns: Absolute safety is a prerequisite for the successful introduction of humanoid robots in the human society. No human will work together with a humanoid robot if he must fear to be smashed by him. With every little safety concern the public opinion will drop noticeably. This is well-known from other technologies. Nuclear power and genetic engineering were once celebrated as wonderful technologies and are nowadays regarded with much more suspicion. There are even people who do not want to use a mobile phone because they fear the electromagnetic radiation. Robotic technology today is regarded as harmless for the most part, and great attention should be paid that it stays so.

Employment Concerns: To face the facts, robots have a slight reputation of stealing jobs. Much work once done by human workers is today accomplished by industrial robots. In most cases, this will presumably be not true for humanoid robots. They are hardly a solution for industrial work and make only sense when there are humans to interact with. However, there might be employees that seem to be replaceable by humanoid service robots. Anyhow, a long time will pass until humanoids are advanced enough (and cheap enough) to replace any human worker.

Movies and Literature: Humanoid robots are long time known from fictional stories and movies, and the image they got from this sources may influence the opinion of real humans about real humanoids. Chances are high, that the first thing a normal person on the street has in mind when he hears about 'Humanoid robots' is a Hollywood creation. For long times, robots were the bad boys of fictional stories, trying to enslave humanity for countless times. In the recent years, humanoid robots were used in a much more positive roles, and they act now mostly as comic relief ('Star Wars', Disney's 'Treasure Planet') or even as heroic sidekicks ('Star Trek'). It is good to see that a wide audience can accept a humanoid robot in this way.

Appearance: As the development of humanoid robots progresses, the question of the optical appearance and the design of the robot arises. Human users are strongly influenced by the look of the robot, and will react differently because of the emotions caused by its appearance. For example, a huge, 2-metres-sized robot will frighten many people, while they are quite comfortable with a child-sized robot with the same abilities. The realistic mimicking of the human appearance can be more frightening than a futuristic metallic look of the robot. It is therefore obvious that the role of design will increase when humanoid robots reach sale status.

6 Conclusion

This paper presented the scope of the ProRobot study. The study will concentrate on the predictions of technological advances, economic scenarios for humanoid robots and social problems concerning these robots, as presented in this paper. These factors will be combined in a socio-economic analysis of the humanoid robot market. All potential markets will be discussed, but also potential obstacles and problems. A roadmap for future research activities will be given to act as a recommendation for research efforts. As the study is funded by the European Commission, a main point of interest will be the comparison of European efforts with other activities in the world. The study will be finished in the end of August 2003 and will be made available by the European Commission.

The content of the study as well as more resources for humanoid robots are presented on the web site of the ProRobot project, which is available under http://www.aboutrobotics.net.

Acknowledgment

The ProRobot project "Paving the Way for Humanoid Robots" is funded in the Competitive and Sustainable Growth (GROWTH) Programme by the European Commission with project no. G1MA-CT-2002-00015.

References

1. Berns, K.: The walking machine catalogue. http:// www.fzi.de/ipt/WMC/preface/walking_machines_katalog.html
2. Kaneko, K., Kanehiro, F., Kajita, S.,Yokoyama, K., Akachi, K.,Kawasaki, T., Ota, S., Isozumi, T.: Design of Prototype Humanoid Robotics Platform for HRP, IEEE Int. Conf. On Robots and Systems, vol.3, pp.2431-2436, 2002
3. AIST National Institute of Advanced Industrial Science and Technology: Success in Having a Humanoid Robot drive an Industrial Vehicle Outdoors, Press Release, 2002
4. Weiss, P.: Dances with Robots. Science News, Vol. 159, No. 26, 2001
5. Main, J.: Exoskeletons for Human Performance Augmentation (EHPA). http://www.darpa.mil/DSO/thrust/matdev/ehpa.htm
6. DARPA Defense Advanced Research Projects Agency: MARS homepage. http://www.darpa.mil/ipto/research/mars/index.html
7. Hans, M., Graf, B. Schraft, R.D.: Robotic Home Assisstant Care-O-Bot: Past – Present – Future. In Proc. Of the 11th IEEE Int. Workshop on Robot and Human interactive Communication, ROMAN2002, pp. 380-385, Berlin, Germany, 2002
8. United States Department of Commerce: U.S. Census Bureau. http://www.census.gov/
9. Sony Corporation Press Release: Sony Develops Small Biped Entertainment Robot. http://www.sony.net/SonyInfo/News/Press/200203/02-0319E/, 2002

Traction Monitoring for Collision Detection with Legged Robots

Michael J. Quinlan, Craig L. Murch,
Richard H. Middleton, and Stephan K. Chalup

School of Electrical Engineering & Computer Science
The University of Newcastle, Callaghan 2308, Australia
{mquinlan,cmurch,rick,chalup}@eecs.newcastle.edu.au
http://robots.newcastle.edu.au

Abstract. With the introduction of commercially available programmable legged robots, a generic software method for detection of abnormalities in the robots' locomotion is required. Our approach is to gain satisfactory results using a bare minimum amount of hardware feedback; In most cases we are able to detect faults using only the joint angle sensors. Methods for recognising several types of collision as well as a loss of traction are examined. We are particularly interested in applying such techniques to Sony AIBO robots in the RoboCup legged league environment. This investigation provided us with a technique that enabled us to detect collisions with reliable accuracy using limited training time.

1 Introduction

An important goal of research in adaptive robotics is to develop robots which can navigate efficiently and robustly in different environments. This includes driving or walking on a variety of surfaces such as for example, sand, ice, grass or carpet, and dealing with obstacles such as stairs or rocks. These skills would be required by robots which explore changing or unknown environments, for example, disaster areas [12] or the surface of Mars [6]. The same methods would also help to improve some transport machines which currently have only a very restricted ability to move such as wheelchairs [4].

Wheeled, tracked and legged robots seem to have their advantages and disadvantages in different environments. On even and hard surfaces wheeled robots are usually faster, while legged machines would typically be better in dealing with stairs or similar obstacles. However, some six-legged robots can also walk very fast [1,5] and some tracked tank-like machines [7] have been developed which can climb stairs and jump over rocks with remarkable speed.

It is not yet completely clear what would be the practical advantages of 4–legged or biped robots over 6– or 8–legged robots and what could be useful applications other than entertainment. Nonetheless, major effort is currently put into the development of 4–legged and bipedal robots with the aim to achieve some similarity to dogs, cats, and humans [2,3].

D. Polani et al. (Eds.): RoboCup 2003, LNAI 3020, pp. 374–384, 2004.

Traction monitoring is a fundamental task in traction control and achieving collision detection for satisfactory robot movement in particular with legged or wheeled robots. If the wheels spin in sand or the legs slip on ice the machine would not be able to move forward and may even fall over. Similarly, if the robot crashes into a wall or into other robots or if its legs get entangled in some smaller obstacles, its movement could be impaired.

In the present study we investigate methods to monitor traction measures and employ them for collision detection with 4–legged AIBO robots [2] in the environment of the legged league of RoboCup [11]. The idea is that detection of an abnormal situation should be used to alter the robot's behaviour – for example, if a collision while walking forwards is detected, the robot should either walk backwards, strafe or turn to avoid the obstacle. Particular goals are to increase the speed of the robots and to find a good strategy to deal with situations where the legs of two robots get entangled (*leg-lock*), see figure 1.

Fig. 1. Leg-locked Robots.

In section 2, we give a brief overview of the Sony Legged League robot soccer environment followed by a discussion on gait control in section 3. In section 4, we describe a direct statistical approach to fault detection. Section 5 illustrates the different types of collisions which may occur and techniques to detect them, while section 6 deals with the detection of a loss of traction (*slip*). The discussion in section 7 addresses possibilities for integrating our results into behaviour control as well as future developments. The paper concludes with a summary in section 8.

2 The Legged League Robot Soccer Environment

The hardware used for this project is the Sony AIBO Entertainment Robot, model ERS-210(A). The robots are programmed in a C++ software environment. They have an internal 64-bit processor and PC Card slot allowing communication via Wireless LAN.

The legs each have three degrees of freedom to enable walking and kicking. The joints in the legs are termed the rotator, abductor and knee. The *rotator* is the shoulder joint responsible for leg movement parallel with the body (along the length axis), while the *abductor* is the shoulder joint involved in leg movement perpendicular to the body. The *knee* functions in principle like a dog's knee in nature, allowing 174° of movement in only one dimension.

The robots have a CMOS colour image sensor (camera), microphones, speakers, a temperature sensor, an acceleration sensor, pressure sensors (on the head, back, chin and legs), LEDs on the head and tail and an infrared distance sensor. The robot is powered by a lithium ion battery pack.

The dimensions of the robot (width × height × length) are 154 mm × 266 mm × 274 mm (not including the tail and ears) and the mass is approximately 1.4 kg (including battery and memory stick).

The Sony AIBO is a state of the art entertainment robot. With a good development environment and a robust design, the AIBO provides a solid platform for investigations into robotics and AI. Further information on the Sony AIBO can be found at [14].

The dimensions of the soccer field are 270 cm × 420 cm. The walls are angled at 45°. The playing surface itself is carpeted to protect the robots and to allow better grip (although the friction of the carpet seems to vary from location to location). Each team has four robots. The rules are only loosely based on soccer, but the objective of the game is identical. More detailed rules and specifications are available at the RoboCup Legged League web site [13].

3 Gait Control

Sony's 4–legged AIBO robots come equipped with a default motion system which is adequate for general use. However, it lacks the speed and versatility required to effectively play soccer.

Instead, legged league RoboCup teams typically use a system based on inverse kinematics to achieve more efficient locomotion. The motion engines used in competition tend to be highly parameterised and extremely flexible. They are mostly based on the walk developed by the legged league team of the University of New South Wales (UNSW) in Australia [8].

In the original UNSW "ParaWalk" system, each leg follows a roughly rectangular trajectory in world space. These world space coordinates are then converted to joint angles using inverse kinematics and sent to the effectors. Inclining the trajectory planes of different legs to the side allows omnidirectional motion to be achieved. Diagonally opposite legs are raised simultaneously (as in a trot).

Many variations on this general method have since been developed. The system employed for the traction monitoring described in this paper is based on the use of ellipsoidal trajectories. Note however that the fault detection methods presented below are independent of the particular locomotion system used.

4 Method for Fault Detection and Training

Current methods in legged robots use a complicated array of sensors (360° range finders, multiple cameras, sonar) [9] to minimize the chance of a robot colliding with an obstacle. In addition more sensors (torque sensors etc) [10] are used to detect a collision if one occurs. Such approaches are only possible if the developer has access to the workings of the robot. In a situation where the hardware is fixed, a technique that uses only the provided hardware is needed. We were confronted with this problem in the legged league of RoboCup. Here all teams are restricted to the use of unmodified Sony AIBO ERS-210(A) robots.

The camera and infrared distance sensor on the ERS-210(A) don't provide enough support in avoiding obstacles unless the speed of the robot is dramatically decreased. Even in the case that the robot avoids obstacles, the unpredictable movements of other robots mean that collisions are likely to occur. Again the camera and infrared distance sensor generally can't be used for detection as the majority of collisions occur outside the field of view of these sensors. For these reasons we have chosen to use the joint sensors (i.e. the angle of the joint) as the only input to our fault detection system.

The historical data needed to accurately determine the "normal" motion of the limbs is acquired using on-line training (described below). In our case the parameterised walk allows for an infinite number of parameter combinations. To minimise the number of possible combinations we have identified three key walk parameters, *backStrideLength*, *turn* and *strafe*. A fourth parameter is also important, the time parameter, t. It increases from 0 to 1 as the leg moves along its trajectory from start to finish. The parameter space was further divided into the following intervals:

Parameter	*Units*	*Min Value*	*Max Value*	*Num. of Intervals*
backStrideLength	mm	-100	100	20
turn	degrees	-25	25	12
strafe	mm	-80	80	20
t	-	0	1	20

Training involves letting the robot walk without collision or slip for a period of time, gathering sensor information from each joint. In our experiments we train the robot for about ten minutes. We then calculate the mean μ and standard deviation σ for each joint with all possible parameter combinations. To save memory while calculating the variance (1) we store only the total of the inputs, the total of the squared inputs, and the number of inputs n, rather than the complete set or sensor data $\{y_1, y_2, ..., y_n\}$. This miserly approach to memory usage allows us to perform these calculations on the limited hardware of the robot.

$$\sigma^2 = \frac{1}{(n-1)}(\sum_{i=1}^{n} y_i^2) - \frac{(\sum_{i=1}^{n} y_i)^2}{(n-1)n} \qquad (1)$$

Figure 2 shows the joint sensor as the step progresses (that is, as t moves from 0 to 1 on the horizontal axis) for six steps with the identical parameters. The parameters used for the steps shown are *backStrideLength* $\in$ [90,100], *turn* $\in$ [0,4.166) and *strafe* $\in$ [0,16).

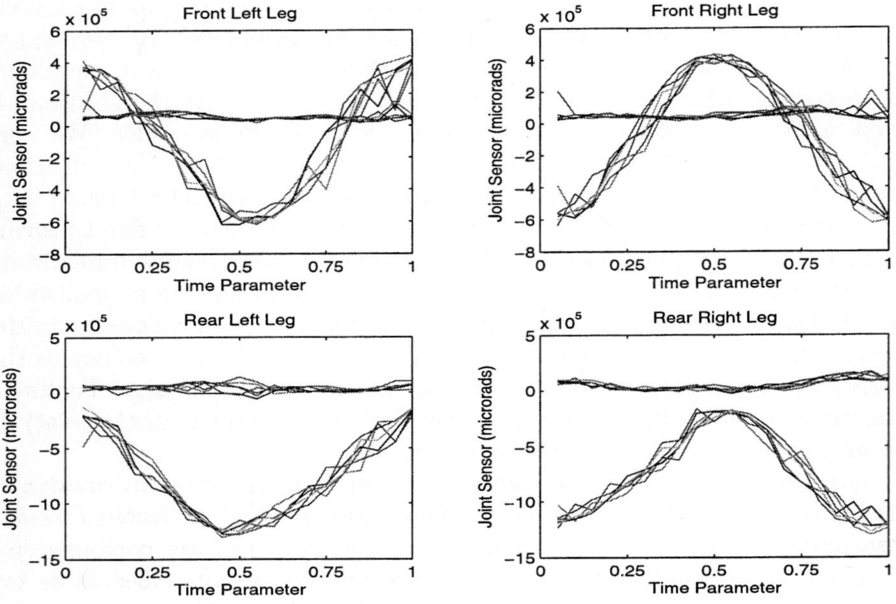

Fig. 2. Movement of legs through six forward steps with identical parameters. This shows the natural variation in "normal" motion.

Although only six steps are shown, the figure gives an indication of the natural variation in "normal" motion that occurs. The roughly horizontal lines are the abductor joints (which move very little during a forward motion), while the curved lines shown are the rotators. It can be seen that the trot gait for a forwards walk leads to the rotators on the right legs being 180° out of phase from the left legs. The unequal default positions of the rotator joints lead to the extreme values of the front and back rotators differing.

5 Collision Detection

Detection of a collision involves observing a joint position substantially differing from its expected value. In our case using an empirical study we found two standard deviations to be a practical measure. Initially we would have considered a collision to have occurred if a single error is found, but further investigation has shown that finding multiple errors (in most cases, three) in quick succession is necessary to warrant a warning that can be acted upon by the robot's behaviour system. It should be noted that if the locomotion engine was more reliable finding only one error may provide accurate results.

For RoboCup a more domain specific diagnosis system is used to provide detailed information about exactly what occurred. This section is further divided into the different types of collisions that occur and the distinguishing features that can be used to identify them.

In the experiments we collected data during different types of collisions; Walking forwards, turning and walking backwards into the field boundary. We assumed this data is representative for all other types of collisions that could occur during a soccer match, such as dog-to-dog collisions.

5.1 Walking Forwards

Our forward walking motion extracts most of its drive from the rear legs - or more precisely, their associated rotator joints. This means that collisions on the front legs result in the most noticeable change occurring on these rear rotators.

Figure 3 indicates the movement of the rear rotators. The graph plots the joint sensor values on the vertical y–axis in microrads against t on the horizontal x–axis. The bold line shows the path of a step that involved a collision. The dotted line represents the mean "normal" path of the joint (that is, during unobstructed motion), with the error bars indicating two standard deviations above and below.

It can be noticed that the clearest differences occur at roughly the midpoint of the time interval. Furthermore, the first few samples of joint positions in each step tend to be influenced by transitions between walk types. It is therefore prudent to deem a collision to have occurred only if an error occurred on the rear rotators with $t \in [0.30, 0.70)$.

Data was gathered in a two step process. First the robot was placed in a situation where no collisions would occur - this test was designed to find false positives (detecting a collision that we deemed not to have happened). False positives occurred in fewer than 1% of steps. In the second test we placed the robot directly in front of the field boundary to test the success rate of detecting a collision. In this case about 98% of collisions were detected. It should be noted that there is a level of human interpretation (and therefore human error) in the gathering of data. The output of the system is compared against what we perceived to have occurred. So if the system did not trigger an expected fault this may be the result of an incorrect human assumption.

5.2 Turning

Detecting a collision during a turn is not unlike detecting a forward collision, except a better conclusion about the impact point and therefore the position of the obstacle may be made. Our current turn motion uses only the rear legs, keeping the front legs stationary so they may be used for controlling the ball. Unfortunately, this also means that any obstruction to the rear legs severely affects the robot's ability to turn. Recognising that a collision has occurred may

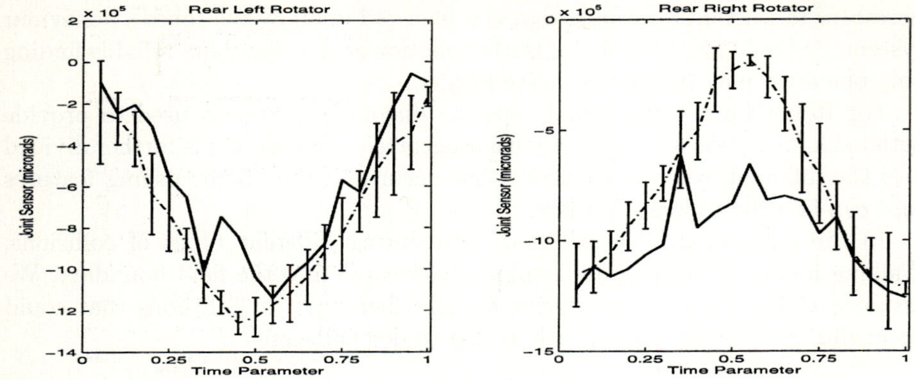

Fig. 3. Rear Rotators for a forwards walking boundary collision on both front legs, front right leg hitting first. The bold line shows the path of a collided motion. The dotted line represents the mean "normal" path of the joint (that is, during unobstructed motion), with the error bars indicating two standard deviations above and below.

therefore be very useful, as we could then choose to reemploy the front legs to assist with the turn.

Figure 4 shows the values of the rear abductors for turning to the right when the left rear leg impacts with the boundary. In this case the obstruction can easily be seen on the left rear rotator while the values for the right rear rotator remains within the standard bounds. These are reversed when turning to the left (and a collision occurs on the right rear leg).

Data was gathered in similar manner to that of a forward collision, except for the second test the robot was positioned next to the field boundary in such a way that it would collide when turning. False positives occurred on 6% of unobstructed steps while 100% of collisions were detected. These results differ from forwards collisions because the turn movement is more sensitive, thus detection of a collision is easier but the step is also more likely to generate a fault under normal operation.

5.3 Walking Backwards

Detection of a collision while walking backwards requires a slightly different mechanism compared to a collision while turning or walking forwards. Here, we found it is useful to examine the gradient of the sensor values between two specific time locations. Generally when this type of collision occurs the impact of the first leg is enough for the second leg to make little or no contact with the obstacle, meaning the motion of that leg is unaffected. Again the rear rotators give the most reliable indication that a collision has transpired.

Figure 5 shows the standard and obstructed movement of the left rear rotator when the left rear leg makes the first impact with the wall. The key time interval for the left leg is between $t \in [0.30, 0.35)$ and $t \in [0.35, 0.40)$, if a collision took place the gradient will be more than one standard deviation less than that of the

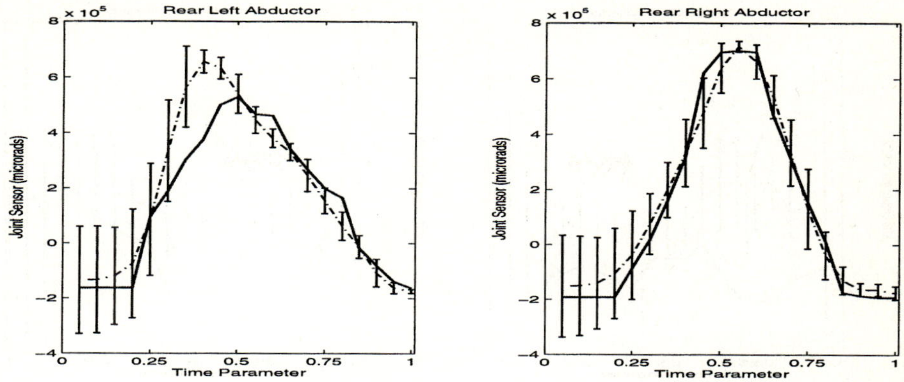

Fig. 4. Rear Abductors for turning to the right with the left rear impacting with the boundary. The bold line shows the path of a collided motion. The dotted line represents the mean "normal" path of the joint (that is, during unobstructed motion), with the error bars indicating two standard deviations above and below.

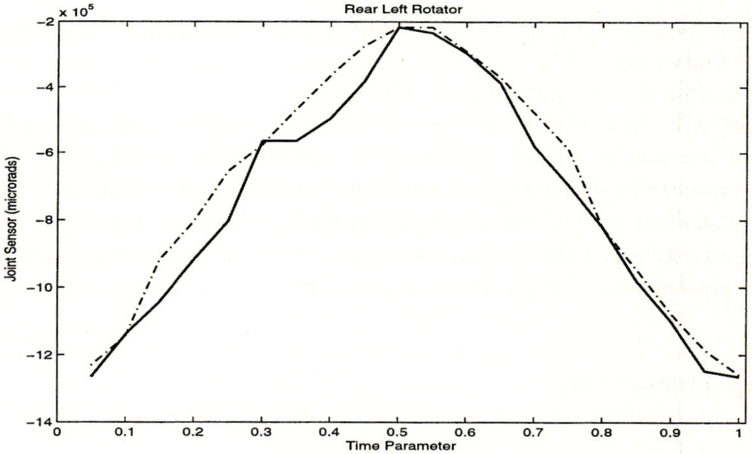

Fig. 5. Rear Left Rotator for walking backwards into the boundary, left rear leg hitting first. The bold line represents the path of a collided motion and the mean path of the joint is shown.

average step. For the right leg the key time interval is between $t \in [0.70, 0.75)$ and $t \in [0.75, 0.80)$.

We experienced no false positives while still detecting 71% of the expected collisions. Limiting detection to a single time interval decreased the accuracy of detecting a collision but ensures a low rate of false positives.

5.4 Leg-Lock

The aim of this section is to detect when the legs of two robots are locked (Fig.1). This situation occurred frequently during the 2002 RoboCup competition, and

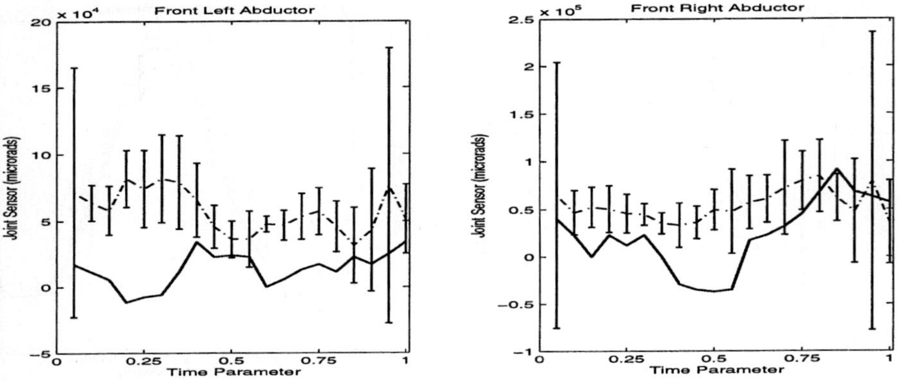

Fig. 6. Front Abductors for a leg-lock on the right front leg. The bold line shows the path of a leg-locked motion. The dotted line represents the mean "normal" path of the joint (that is, during unobstructed motion), with the error bars indicating two standard deviations above and below.

appears to be a side effect of the low forward leaning stance adopted by most teams. Robots involved in a leg-lock can take several minutes to free themselves and are useless for the game during this time.

We are able to detect when leg-lock occurs in exactly the same manner as we detect a regular forward collision (see section 5.1). Unfortunately, the two problems appear to be virtually identical as far as the joint paths are concerned. The front abductors (shown in Fig.6) do appear to show a greater deviation during leg-lock, but with our current approach it is insufficient to accurately discriminate between the two types of collision.

6 Slip Detection

We approached the detection of a slip in the exact same manner as detection of a collision. As with leg-lock, recognising a fault is simple but distinguishing a slip from a regular collision can be challenging. To simulate a slip occurring for the purposes of training, we simply allowed the robot to move around on a slippery tarpaulin on which it is almost completely unable to gain traction.

We only considered slipping during a forward walk. The key indicator of a collision is a substantial change in the rear right rotator, while the left rotator remains mostly unchanged (Fig.3). When a slip took place, the exact opposite occurred (Fig.7). We also added an additional constraint as a precaution against false diagnosis: A "slip" is only recognised if the left rear rotator is outside two standard deviations and the rear right rotator never exceeded two standard deviations anytime during the time interval, $t \in [0.30, 0.70)$. This extra constraint prevents us from detecting a slip during the middle of a step - we must instead wait until the critical part of the step has completed (that is, t must reach 0.70).

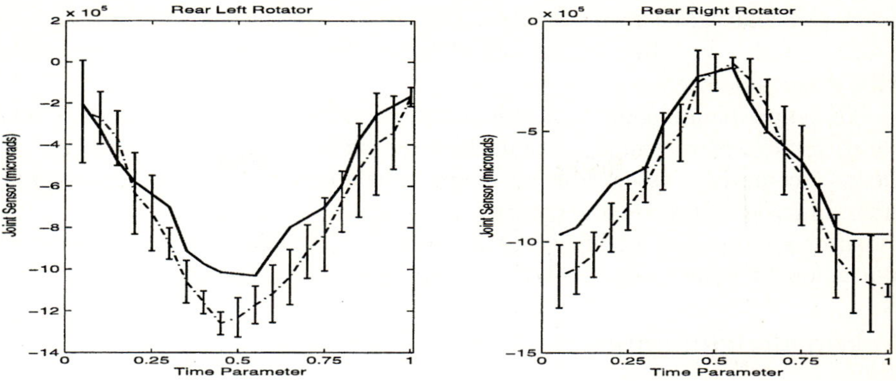

Fig. 7. Rear Rotators for a forwards step that slipped. The bold line shows the path of a slipped motion. The bold line shows the path of a collided motion. The dotted line represents the mean "normal" path of the joint (that is, during unobstructed motion), with the error bars indicating two standard deviations above and below.

7 Discussion

The techniques described above provide us with results that enable our behavior system to make previously impossible decisions. We can now detect a possible collision in locations that were once blind spots, for example while walking backwards or collisions while turning. This extra information enables us to vary our behaviour accordingly.

The most prominent use of the system is during a collision while chasing a ball. Basically our behaviour is designed so only one dog will chase at any given time. In the past the only input on whether to chase was the vision distance to the ball. You could deem that a collision occurred if you were attempted to move towards the ball yet the ball distance appears to be steady. This approach was flawed - obstruction of the ball, bad lighting or even the ball moving away at a speed equal or greater then your speed results in failure. Having the collision data enables us to quickly detect the problem and allow another un-obstructed dog to chase the ball.

In addition, we are able to improve the reliability of odometry data by taking collisions into account. This allows us to better determine the robot's location on the field.

Future improvements to the system include an unsupervised learning approach to train and classify the data. This would eliminate the human observation errors that occur and hopefully lead to a more precise and robust classification.

8 Summary

In this study, a traction monitoring system was developed in software using a minimum of hardware feedback. The techniques of the present study were developed and tested using Sony's AIBO robots. Collisions and leg-lock were detected with a very high degree of reliability. Additionally, slips are recognised with reasonable accuracy.

The experiments show that traction monitoring for a legged robot is feasible without hardware modification. Such a system provides the possibility of a future locomotion system able to adapt to changes in the environment while maintaining a high level of stability.

After appropriate modifications it should be possible to employ these techniques for different types of robots and in more general environments as well.

Acknowledgements

We would like to thank all past and present members and contributors to the University of Newcastle Legged League RoboCup team.

References

1. Clarke, J.E., Cham, J.G., Bailey, S.A., Froehlich, E.M., Nahata, P.K., Full, R.J., Cutkosky, M.R.: Biomimetic Design and Fabrication of a Hexapedal Running Robot. IEEE International Conference on Robotics and Automation (2001) 3643–3649
2. Fujita, M., Kitano, H.: Development of an Autonomous Quadruped Robot for Robot Entertainment. Autonomous Robotics **5** (1998) 7–18
3. Dario, P., Gugliemelli, E., Laschi,C: Humanoids and personal robots: Design and Experiments. Journal of Robotic Systems **18**(12) (2001) 673–690
4. Krovi, V., Kumar, V.: Optimal Traction Control in a Wheelchair with Legs and Wheels. Proceedings of the 4th National Applied Mechanisms and Robotics Conference (1995) AMR-030
5. Ritzman, R.E., Quinn, R.D., Watson, J.T., Zill, S.N.: Insect Walking and Biorobotics: A Relationship with Mutual Benefits. BioScience **50**(1) (2000) 22–33
6. Yoshida, K., Hamano, H., Watanabe, T.: Slip-Based Traction Control of a Planetarty Rover. 8th International Symposium on Experimental Robotics (ISER'02) (2002)
7. Matthies, L., Xiong, Y., Hogg, R., Zhu, D., Rankin, A., Kennedy, B., Hebert, M., Maclachlan, R., Won, C., Frost, T., Sukhatme, G., McHenry, M., Goldberg, S.: A Portable, Autonomous, Urban Reconnaissance Robot. Robotics and Autonomous Systems **40**(2-3) (2002) 163–172
8. Hengst, B., Pham, S.B., Ibbotson, D., Sammut, C.: Omnidirectional Locomotion for Quadruped Robots. RoboCup 2001: Robot Soccer World Cup V (2002) 368–373
9. Everett, R.B.: Sensors For Mobile Robots : Theory and Application. A K Peters Ltd (1995)
10. Todd, D.J.: Walking Machines - An Introduction to Legged Robots. Korgan Page Ltd (1985)
11. RoboCup web site. `http://www.robocup.org`
12. RoboCup Rescue web site. `http://www.isd.mel.nist.gov/RoboCup2003`
13. RoboCup Legged League web site. `http://www.openr.org/robocup/index.html`
14. The Sony AIBO Entertainment Robot web site. `http://www.aibo.com`

Multi-robot Control in Highly Dynamic, Competitive Environments

David Ball and Gordon Wyeth

School of Information Technology and Electrical Engineering
University of Queensland, Brisbane, Australia
{dball,wyeth}@itee.uq.edu.au

Abstract. The control and coordination of multiple mobile robots is a challenging task; particularly in environments with multiple, rapidly moving obstacles and agents. This paper describes a robust approach to multi-robot control, where robustness is gained from competency at every layer of robot control. The layers are: (i) a central coordination system (MAPS), (ii) an action system (AES), (iii) a navigation module, and (iv) a low level dynamic motion control system. The multi-robot coordination system assigns each robot a role and a sub-goal. Each robot's action execution system then assumes the assigned role and attempts to achieve the specified sub-goal. The robot's navigation system directs the robot to specific goal locations while ensuring that the robot avoids any obstacles. The motion system maps the heading and speed information from the navigation system to force-constrained motion. This multi-robot system has been extensively tested and applied in the robot soccer domain using both centralized and distributed coordination.

1 Introduction

This paper addresses multi-robot control for teams of robots that operate in uncertain, dynamic environments against unknown, competing agents. These conditions require an integrated, systematic approach that simultaneously addresses the problems of effective cooperation between robots while allowing fast, smooth reaction to ever changing conditions. This paper addresses such multi-robot control issues in dynamic, adversarial environments. The work is demonstrated within the context of robot soccer, where the task for the team is well defined and the performance is measurable, but the opposition and their reaction with the environment are highly uncertain and hard to model.

This paper specifically addresses the issues of integration between multiple layers of competency in a multi-robot system. It shows a planning system that can make short term coordinated plans based on uncertain world models; while addressing the single robot issues of behavior selection, high speed navigation and smooth motion generation. An overview of the system is presented in the next section, with the following sections explaining the detail of the individual modules.

1.1 Testing Environments

Results are given in the context of performance across three variations of the robot soccer problem, RoboCup [1]. The results have been validated by performance in the

D. Polani et al. (Eds.): RoboCup 2003, LNAI 3020, pp. 385–396, 2004.

RoboCup world championships which, in 2002, involved over 200 robot soccer teams from 30 nations. This system has been applied in the following RoboCup leagues.

- Small-Size: where the team of five real robots has a central world model from a camera mounted over the soccer field. This team is named the RoboRoos [2], [3], [4].
- Small-Size Local Vision: where the each of the five real robots on the team has its own limited view of the world from an on-board camera. This team is named the ViperRoos [5].
- Simulation: where eleven simulated robots compete with limited perception from a simulated and noisy local view. The team is named the CrocaRoos [6].

The three leagues provide significantly different testing domains for the multi-robot control system. The small-size represents the simplest domain, as all robots have the same world model. The small-size local vision league introduces the significant problem of different world models on individual robots, while the simulation league presents a further challenge in the increased number of agents and the increased size of the environment.

2 Overview of the System

The encompassing principle of the multi-robot control system used in this paper is that each module can operate competently and sensibly within its domain despite conflicting or rapidly varying information from the environment or from other modules. The domain of each module is limited by the extents of the data available to that module; a module with broad reaching data can provide direction for long term plans, while a module using only internal data can only provide assistance in generating smooth motion. The key to interfacing the various modules is that information passed between modules acts more as a series of suggestions rather than commands, respecting the need for the lower level modules to deal appropriately with their immediate task domain.

This principle is implemented in the general architecture shown in Figure 1. Each module operates with different levels of data with regard to the state of the world. The modules at the top of the diagram deal with broad, often imprecise, and always time-delayed data; while the modules at the bottom access narrow, precise and immediate data. The connections between modules show the path of resource data flowing down to the eventual control of actuation.

2.1 Types of Data

There are three types of data that provide input to the system: the global external state, the local external state, and the local internal state. The global external state represents the robot and the environment in a fixed global reference frame. This state is particularly useful for adding a priori information such as locations for goal oriented tasks. The problems with maintaining a global view often lead to this data becoming uncertain and imprecise. The local external state represents the perceived environment in a robot centered frame of reference. It is current, and mostly limited

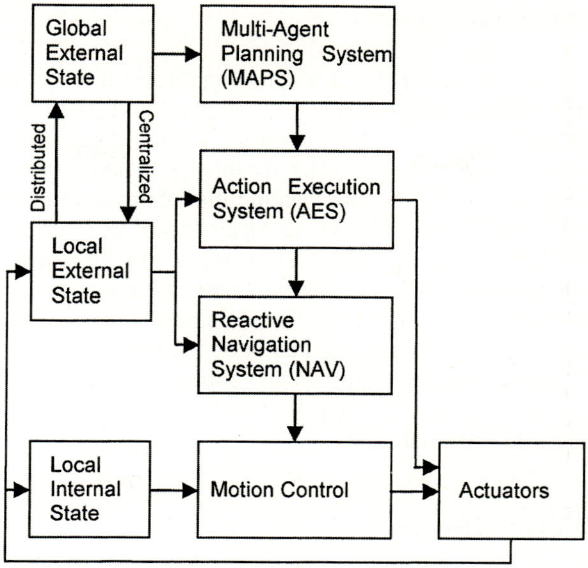

Fig. 1. The multi-robot control system. Modules at the top of the diagram use broad, imprecise data, while the modules at the bottom operate with precise, immediate data.

by the robot's external sensory abilities in terms of bandwidth and precision. The local internal state represents the dynamics and motion of the robot which can usually be measured rapidly and with high certainty.

The development of global and local external data is dependent on the nature of the multi-robot system. In the case of a system with a centralized world model generated by some external sensor, such as a camera or set of cameras with a complete view of the environment, the global data forms the basis for the local view. This is the case for the RoboRoos, in the small-size league. In a distributed system, where each robot has its own local view from on-board sensors, the global view must be formed from the local view of the robot, and communication about the local view of other robots. This is the case for both the CrocaRoos and the ViperRoos. These two distinct forms of the global data create a challenge for the planning system that relies upon it.

2.2 Modules

The Multi-Agent Planning System (MAPS) uses the global data to decompose the overall team goal into actions for the individual robots. The action from MAPS is combined with the local external state by the Action Execution System, which determines each individual robot's next immediate behavior. The Navigation system determines the robot's immediate path for the desired behavior while reactively avoiding obstacles maintained by the local external state. The motion control system smoothly maneuvers the robot in the desired direction using the local internal state for feedback.

2.3 Centralized Control System

The RoboRoos team is an example of a team using this multi-robot control system in a centralized form. The RoboRoos have been applied and extensively tested in the small size league environment. In the RoboRoos system the global external state is maintained on a central off-field PC. The RoboRoos vision system determines the global external state by identifying and locating the robots and the ball on the soccer field using an overhead camera. The Multi-Agent Planning System is also executed on this central PC. The actions that MAPS determines and the global external state of the field are sent to the robots using a broadcast radio system.

The rest of the multi-robot control system is run on the individual robots. Each robot maintains it own local external state that is derived from the common global external state. By running these systems on the robots, they are able to utilize local feedback from the actuator motion. This feedback is used to update their position in their version of the local external state.

2.4 Distributed Control System

The CrocaRoo and ViperRoo systems are examples of robot soccer teams using a distributed version of this multi-robot control system. In the CrocaRoo and ViperRoo systems the entire multi-robot control system is run on each robot.

Each ViperRoo robot has an on-board camera that is used to identify and locate other robots and the ball. The robot integrates this local view to generate a global external state for its MAPS system. The rest of the system is similar to the one described in the central control system except that the robots may communicate their local external state to their team members using a wireless network.

Each CrocaRoo agent is a network client that communicates with a network soccer server. The server sends noisy local view information to the clients. As for the ViperRoos, the CrocaRoo clients integrate this noisy local information to generate a global external state. The agent's individual MAPS modules then determines their individual sub-goals. The agent's other systems are similar to other teams, although navigation is greatly simplified. Instead of driving motors, the agents send their desired motion to the soccer server. The agents communicate with each other using a simulated low-bandwidth shouting system.

3 Multi-agent Planning System

The Multi-Agent Planning System (MAPS) is the highest level planner in the system, responsible for distributing the overall goal of the team amongst the individual robots [7]. MAPS is responsible for the multi-robot coordination and cooperation by selecting an action and an action location for each robot. MAPS determines the team's actions based on the current world model, the team goal and the currently available actions.

MAPS is a plan-by-communication system, as opposed to a plan-by-program system [8]. In a plan-by-program system plans are represented as a sequence of steps to be followed. The robots attempt to follow these steps but may be unable to cope with unforeseen contingencies. In a plan-by-communication system the robots determine how to achieve the sub-goal themselves. By using the MAPS sub-goal, but paying careful attention to local state, the robot is better able to complete its part in the plan. MAPS continually monitors the global state and does not rely on a robot to complete its action.

MAPS uses potential fields as the mechanism for determining action selection and action location. The potential fields can model the suitability of an action for the different agents, or be used to find a suitable action location. MAPS has a library of potential field functions and abstractions, where each field is a two dimensional array of values. A more positive value represents a more desirable action for an agent, or, in the case of determining action location, a more desirable location for that action. The following four types of potential fields are examples of the type of fields used.

- Basefield: This field represents favorable regions of the physical environment. The regions of the field that will always be favorable to the goals of the team are the most positive.
- Object Regions: These fields model physical objects on the field by representing an area of effect around an object. Object regions can be used to bias the positions of other team members to ensure they don't attempt to occupy the same location.
- Clear Path: This is an abstract feature that represents clear paths to objects or locations. It biases regions that offer a line-of-sight path to the point in question.
- Distance: This is another abstract feature that represents the distance from objects, thus favoring action locations close to the robot or to a goal location.

It is the overlaying of multiple fields that gives an abstract goal-biased terrain map that provides the information to determine the robots' sub-goals. By weighting the strengths and shapes of the component fields, MAPS can be tuned to give peak performance for a specific goal, or to act against specific opponent strategies.

MAPS uses potential fields to determine the location to dribble and kick the ball, where to place defending and attacking robots and which player should be kicking or dribbling the ball.

MAPS handles tasks that require coordination between multiple agents implicitly. Coordination emerges from the interaction of multiple potential fields and multiple actions. One task that requires coordination is passing. The location for the receiver is determined by overlaying clear paths to the player with the ball onto clear paths to the opponent's goal. This keeps this player in both a position to receive a pass and a position to take a shot on goal. When determining where to kick the ball a positive object region is overlaid at the receiver's location on the 'kick to' field. (The other positive object region is the goal.) By overlaying the clear paths from the ball the preference of kicking at the goal or passing to the receiver is weighed up.

3.1 MAPS in Operation

Figure 2 shows the potential field generated to determine the location to which a ball should be dribbled during a robot soccer match. The opponent's goal is on the right side and the darker robots are the opponents. There are several overlapping fields at work here.

- Basefield: The field is ramped towards the opponent's end of the field and off the walls. This encourages dribbling towards the opponents' goal.
- Clear Path: Clear paths from the opponent's goal are created. This encourages movement towards a clear shot on goal.
- Distance From: Locations further from the ball are given higher values. This encourages shorter dribble distances.
- Object Regions: The opponent's positions are given low values. This encourages a dribble to a location away from the opposition.

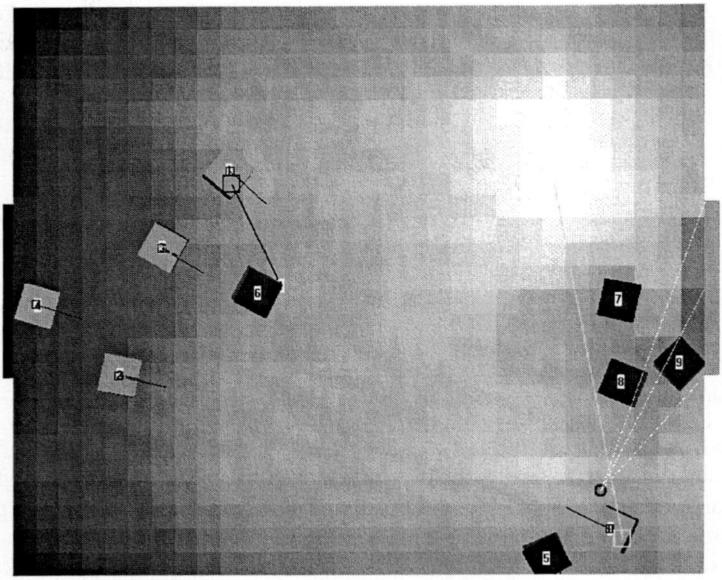

Fig. 2. An example potential field that determines where to dribble the ball to. The light colored player on the bottom right of the figure has being given a DRIBBLE action. The lightest point in the potential field represents the desired location to dribble to. Note the strong effect of the basefield and the clear path to the opponent's goal.

3.2 MAPS in Distributed Environments

In distributed environments, MAPS needs to account for each robot maintaining its own world model. The system must account for the uncertainty in the positions of objects outside the robot's local field of view. It also must account for the uncertainty in its own position relative to the global reference. Not accounting for this uncertainty could lead to individual robots attempting conflicting actions.

One method currently implemented is to distribute the robot's object region across the uncertainty in its position. This is achieved by convolving a probability distribution that represents the uncertainty in the robot's position with the potential fields that represent features of the environment. In this way the plan becomes fuzzier but maintains robustness.

3.3 Performance

Robust multi-robot control in highly dynamic, competitive environments is difficult. MAPS has shown itself to be a capable system for coordination and cooperation in such an environment. The potential field methodology of MAPS is preferred over state based approaches in highly dynamic and competitive environments. The two inherent drawback effects of using potential fields, minima and oscillations, are not persistent due to the dynamic nature of the environment. Other methods such as biasing the last selected player and the last action location further reduce these effects.

The RoboRoos system has the capability of playing against itself. A test was performed with two identical RoboRoos teams, except that one team was using MAPS, while the other used a zone-based role assignment technique. This zone strategy requires each player to maintain a fixed position unless they are the closest to the ball. In repeated tests, the team with MAPS would typically lead by 4 goals to 1 over a ten minute period. A similar test with the CrocaRoos distributed MAPS implementation showed a similar result, with the average lead being 3 goals to 1 over a similar simulated time period.

4 Action Execution System (AES)

The Action Execution System is responsible for selecting the immediate robot behavior. The local external state and the MAPS assigned action are used as resources to determine this behavior. As MAPS is a plan-by-communication style planner the AES system must fill in the details of the assigned actions. This is done by decomposing an assigned action into a series of small tasks that can be performed using simple behaviors. The AES then decides on the immediate task to be achieved.

The tasks are associated with a set of behaviors. Each behavior has a set of associated parameters and desired robot motion. Parameters associated with a robot include accelerations and top velocities, application specific actuator state and desired repulsive strength of obstacles.

By necessity, this subsystem is specific to the applied environment. This is because it is the interface between the general actions and the specifics of the environment. By using an AES system the MAPS system is removed from that level of detail. This enables the MAPS system to be portable across multiple leagues.

4.1 AES in Operation

In a real soccer match the most important and complex actions that a field player performs is kicking and dribbling the ball. As such the kick action provides a good example of the role of the AES. The kick action sequence involves acquiring the ball then dribbling until the robot is lined up for the kick to the MAPS assigned location. The AES breaks the kick action into a series of smaller tasks. These are:

1. Move to near the ball. This step moves the robot to near the ball at high accelerations and speeds. If the robot already controls the ball then skip to step 3.
2. Acquire the ball. Acquiring the ball is dependent on its position relative to the field and the opponent robots. If the ball is amongst other objects a modified version of navigation is used to acquire the ball, otherwise the robot drives directly at it.

3. Dribble until kick. Once the ball is acquired the robot must assume a pose that allows kicking to the MAPS specified location. AES chooses the appropriate motion to maintain control of the ball.
4. Kick the ball. Activate the kicking mechanism while maintaining pose.

5 Reactive Navigation (NAV)

The NAV module is responsible for achieving the desired motion behavior while avoiding obstacles. It determines the desired heading direction and distance for the motion system. It uses the local external state for object mapping. The AES system provides the relative avoidance strength for each obstacle to the NAV system. Obstacles may be physical objects such as the robots and the field boundaries. They may also be virtual obstacles. These are used to keep a robot out of a particular area for example.

NAV uses a biologically plausible reactive navigation method that is appropriate to highly dynamic environments [9]. Schema theory is a behavior based approach whereby overall robot behavior results from the interaction of many simple schemas operating in parallel. In this navigation system, multiple schemas are represented by polar mappings that are centered on the robot. Three mappings are generated: the Goal Direction (GD) Map, the Obstacle Map (OM) and the Motor Heading Map (MHM). An example of these maps and their interaction is shown in Figure 3.

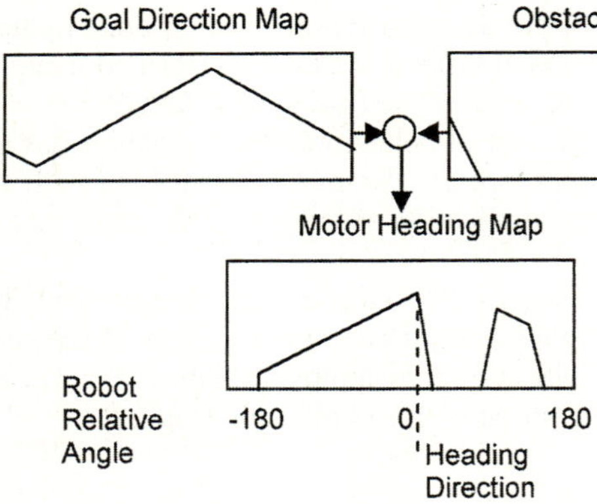

Fig. 3. Example navigation maps. The OM is showing a close obstacle just to the robots left and a more distant obstacle to the robots rear. The heading direction to avoid obstacles in marked.

The GD map represents the desired motion of the robot towards the goal location. It is a triangular map with the peak orientated towards the goal. The OM map represents the repulsive strength of each obstacle. It maps out the distance, bearing, angular width and strength of each obstacle by modeling each as a rectangular activity packet. Uncertainty is accounted by sloping the edges of this activity packet. Each obstacle is mapped separately to the OM.

The MHM map represents the best direction for the robot to head in. It is determined by a piece wise subtraction of the OM from the GD map. The peak of this map is the desired heading direction for the robot. This is determined on each run through the navigation system.

5.1 Performance

The NAV system has shown itself a capable system for achieving desired motion behavior while competently avoiding obstacles in highly dynamic environments. This is because:

- The complex problem of navigation is broken down into simple schemas that represent the different navigational influences.
- Reaction to environment change is fast as navigation is based on reactions, not on a plan.
- Navigation maintains competence even with poor behavior selection by AES due to robust avoidance of obstacles.

The deficiencies of a reactive navigation system, local minima and non-path optimal generation are generally not apparent. Local minima do not exist for long because the environment is highly dynamic and therefore states do not exist for long. Processing an optimal path is wasteful, as the environment state in which the path was planned does not exist for long.

6 Motion Control

The implementation of the motion control module is quite specific to the type of robot upon which it operates, but there are several key functions that it must perform in order for the complete multi-robot system to function correctly. The NAV module provides a desired direction of travel and a total length that remains to the goal location. The primary function of the motion control module is to ensure that the robot complies as closely as possible with that request without causing the robot to lose traction with the surface. Even the simulated agents have dynamic properties that must be carefully monitored to provide good performance.

In the case of the wheeled robots, the motion control software seeks to provide constant force acting from the wheel to the ground. Typically, this force is somewhat less than the normal force applied by the robot to the ground so that good traction is achieved. Consequently, the robot limits straight line acceleration and rotational acceleration, and must adjust speeds when maneuvering so that sufficient centripetal force can be generated by the wheels.

Maintaining good traction with the ground means that the motion control system can adequately perform its other essential role: keeping track of the robot position between external sensor updates. In high speed environments, where the robots move at over 1 m/s, significant rotation and translation can occur between external sensor updates. Furthermore, where the sensor information comes from a delayed source such as vision the motion control system can account for self motion during the delay period. By accounting self motion between updates and during sensor delays, the motion control system greatly enhances the accuracy of its interactions with the environment.

In the context of robot soccer, it is critically important to time and aim kicks correctly. Criticality in timing also applies to many robot manipulation tasks. By using the immediate feedback from local sensors and local external state that has been brought up to date with respect to self motion, the motion control software can execute timing critical functions with a precision beyond the other software modules.

Another interesting use of the motion control software is to change the point of reference for navigation commands. The RoboRoos robots have an omni-directional drive system that can simultaneously translate in a plane while rotating. In typical operations, rotation operations take place about the centre of the robot. Under direction from the AES, the motion control system can switch the rotation centre to an arbitrary point. This is applied, for example, to ball control. When the ball is directly in front of the robot, as detected by the local ball sensor, the motion control system can switch to rotating the entire robot about the ball's centre. This provides better control of the ball during dribbling operations.

7 System Performance

The RoboRoos have competed in the last five years of RoboCup with varying levels of success. During these years, the structure of the system has remained constant and has now been applied in three leagues. Even though the RoboRoos have undergone a complete mechanical redesign, the system architecture has remained as described in this paper.

In February 2003 the RoboRoos team competed in a friendly game against the RooBots team from Melbourne University, Australia. The RooBots came fourth in the 2002 RoboCup competition. The RoboRoos won the game 6-0.

The ViperRoos team is known as the first local vision team to win against a global vision team. (Final score 2-0.) In simulation the distributed MAPS system has shown the ability to coordinate multiple robots. However due to technical difficulties with the real local vision software the ViperRoos team has never reached the potential ability that is suggested by simulation results.

The CrocaRoo team has also suffered poor performance at RoboCup due to vision problems. However in lab tests, the team demonstrates the ability to play soccer competently against older teams.

7.1 System Performance against Humans

When the multi-robot control system is pitted against humans, the results are interesting. In 2001, the RoboRoos competed against humans for five hours of non-continuous play. The humans controlled their robots using game pads. Continuous running rules were adopted and the teams were limited to three players each. The final score was 196 – 24 with a convincing win for the RoboRoos. The most notable observation was the relative lack of team coordination and cooperation in the human teams compared to the multi-robot control system.

8 Next Generation Multi-robot Control

The next generation of this multi-robot control system is currently under consideration. New abilities are to be added to the system in the area of multi-agent planning. This includes integrating predictions of the opponent's future behaviors based on their current behavior and models of how their behaviors were previously executed.

The distributed MAPS system is also to be improved. Most of this work is in generating a more complete global external model through probabilistic fusion of sensed elements with information received by communication. While this model may not be more accurate, it will offer a better representation for MAPS to perform short term planning.

9 Conclusions

This paper presents a robust and layered approach to the difficult task of multi-robot control in highly dynamic, competitive environments. This approach has simultaneously addressed the need for effective robot cooperation, while allowing fast reactive response to a highly dynamic environment. System performance is high because there is:

- competency in every layer of control,
- appropriate state maintained for each subsystem,
- team goal decomposition into plan-by-communication actions by MAPS,
- behavioral approach to parameter selection by AES,
- the use of simple schemas to the reduce the complexity of navigation by NAV,
- smooth dynamic control by the motion control system.

Although work on this multi-robot control system continues, especially in the higher level control and planning systems, the overall structure of the system is to remain constant.

Acknowledgements

This multi-robot system is the hard work of many people over many years. The authors would like to acknowledge Brett Browning, Ashley Tews, Mark Chang and Mark Venz for their dedication to this system. The authors would also like to thank Chris Nolan and Jason Hirsch for their assistance in 2002 on the RoboRoos system.

References

1. H. Kitano, M. Asada, Y. Kuniyoshi, I. Noda, and E. Osawa, RoboCup: The Robot World Cup Initiative, presented at *IJCAI-95 Workshop on Entertainment and AI/ALife*, 1995.
2. G. Wyeth, B. Browning, and A. D. Tews, UQ RoboRoos: Preliminary Design of a Robot Soccer Team, *Lecture Notes in AI: RoboCup '98, 1604*, 1999.
3. G. Wyeth, A. D. Tews, and B. Browning, UQ RoboRoos: Kicking on to 2000, presented at *RoboCup-2000: Robot Soccer World Cup IV*, 2001.

4. G. Wyeth, D. Ball, D. Cusack, and A. Ratnapala, UQ RoboRoos: Achieving Power and Agility in a Small Size Robot, presented at *RoboCup 2001*, pp. 603-606, 2002.
5. M. Chang, B. Browning, and G. Wyeth, ViperRoos 2000, presented at *RoboCup-2000: Robot Soccer World Cup IV. Lecture Notes in Artificial Intelligence 2019*, Springer Verlag, Berlin, 2001.
6. G. F. Wyeth, M. Venz, H. Mayfield, J. Akiyama, and R. Heathwood, UQ CrocaRoos: An Initial Entry to the Simulation League, presented at *RoboCup-2001: Robot Soccer World Cup V. Lecture Notes in Artificial Intelligence 2377*, 2002.
7. A. D. Tews, Achieving Multi-Robot Cooperation in Highly Dynamic Environments, PhD dissertation, School of Computer Science and Electrical Engineering, University of Queensland, 2002.
8. P. Agre and D. Chapman, What are Plans For?, *Robotics and Autonomous Systems*, vol. 6, pp. 17-34, 1990.
9. B. Browning, Biologically Plausible Spatial Navigation for a Mobile Robot, PhD dissertation, Department of Computer Science and Electrical Engineering, University of Queensland, 2000.

Developing Comprehensive State Estimators for Robot Soccer

Thorsten Schmitt, Robert Hanek, and Michael Beetz

TU München, Institut für Informatik, 80290 München, Germany
{schmittt,hanek,beetzm}@in.tum.de
http://www9.in.tum.de/agilo

Abstract. This paper sketches and discusses design options for complex probabilistic state estimators and investigates their interactions and their impact on performance. We consider, as an example, the estimation of game states in autonomous robot soccer. We show that many factors other than the choice of algorithms determine the performance of the estimation systems. We propose empirical investigations and learning as necessary tools for the development of successful state estimation systems.

1 Introduction

Autonomous robots must have information about themselves and their environments that is sufficient and accurate enough for the robots to complete their tasks competently. Contrary to these needs, the information that robots receive through their sensors is inherently uncertain: typically the robots' sensors can only access parts of their environments and their sensor measurements are inaccurate and noisy. Recent longterm experiments with mobile robots [11] have shown that an impressively high level of reliability and autonomy can be reached by explicitly representing and maintaining the uncertainty inherent in the available information.

One particularly promising method for accomplishing this is *probabilistic state estimation* [10]. Probabilistic state estimation modules maintain the probability densities for the states of objects over time. The probability density of an object's state conditioned on the sensor measurements received so far contains all the information which is available about an object that is available to a robot. Successful state estimation systems have been implemented for a variety of tasks including the estimation of the robot's position in a known environment [2], the automatic learning of environment maps, the state estimation for objects with dynamic states (such as doors), for the tracking of people locations [9], and gesture recognition [11].

So far, research in state estimation for robots has focussed on the development on more reliable, more accurate, and faster algorithms [10]. With the services that autonomous robots are to provide becoming more demanding, the states that the robots have to estimate become more complex. Therefore, the choice of algorithms has become only one among many different factors that determine the performance of the state estimation systems. As a consequence, the application of state estimation algorithms becomes a difficult engineering effort. In this engineering effort, programmers

D. Polani et al. (Eds.): RoboCup 2003, LNAI 3020, pp. 397–409, 2004.

must address questions such as (1) the specification of objectives in terms of cost functions; (2) the filtering of observations; (3) the design of action models; (4) the frequency of updates; and (5) the choice and combination of algorithms. We will show that the proper design of state estimators achieves both substantial and significant performance gains.

In this paper, we argue that developing high performance state estimation systems for difficult estimation problems requires careful system design and analysis. We propose empirical investigations and learning as necessary tools for the development of successful state estimation systems. In the remainder of the paper we proceed as follows. Section 2 describes and discusses the game state estimation problem, the design of cost functions for state estimation, and different design options for state estimation systems and their possible effects on robot performance. The subsequent section 3 describes the design of a state estimator for game situations in autonomous robot soccer and investigates the impact of design decisions on the perfomance of the estimator. We end with conclusions from our results.

2 Complex State Estimation

The task of state estimation is the computation the robot's beliefs about its own state and the state of its environment. The results of the state estimation contain information such as the robot's estimated state, the accuracy and a measure of the ambiguity of the state estimate, the state of other objects.

Approached probabilistically, the state estimation problem can be considered as a density estimation problem, where a robot estimates a posterior distribution over the space of its states and the states of other objects conditioned on the available data. Denoting the state at time t by s_t and the data up to time t by d_t, we write the posterior as $p(s_t|d_t, ..., d_0; m)$. Here m is the world model (e.g., a map). We will also refer to this posterior as $b_t(s_t)$, the robot's belief state at time t.

2.1 Game State Estimation in Robot Soccer

In this paper we apply probabilistic state estimation to the assessment of game situations in autonomous robot soccer. In robot soccer (mid-size league) two teams of four autonomous robots — one goal keeper and three field players — play soccer against each other. The robots of our team are equipped with, among other components, a wireless Ethernet for communication (1), an onboard computer (2), a fixed color CCD camera (3), a guide rail for the ball (4), and a kicker (5). All sensing and all action selection is done on board of the individual robots.

The game state estimators provide the robots' action selection routines with estimates of the positions and may be even the dynamic states of each player and the ball. This estimation problem confronts probabilistic state estimation methods with several challenges: limitations of the robots as well as environmental conditions make the reasoning task difficult to perform. The camera system with an opening angle of 90° and pointed to the front gives an individual robot only a very restricted view of the game situation. In addition, relevant visual information may be occluded by other robots. Vibrations of the camera, spot light effects, specularity, and shadows cause substantial inaccuracies. Even small vibrations that cause jumps of only two or three pixel lines cause deviations of more than half a meter in the depth estimation, if the objects are several meters away. Also, the positions of the robots are uncertain and inaccurate. In addition, the robots change their direction and speed very abruptly and therefore the models of the dynamic states of the robots of the other team can only be very crude and uncertain. Finally, vast amounts of data must be processed in real-time.

In our case, the estimated game state consists of the compound state variables $Robot^1$, ..., $Robot^4$, $Ball$, $Opponent^1, \ldots, Opponent^n$. The compound state variable $Robot^i = \langle x^i, y^i, \theta^i, vt^i, vr^i \rangle$ comprises the position (x^i, y^i) and orientation θ^i of robot i and its translational (vt^i) and rotational velocity (vr^i) and $Robot^i_t$ refers to the value of these variables at time step t. Analogously, $Ball = \langle x^{ball}, y^{ball}, v^{ball} \rangle$ denotes the position and velocity of the ball, where the ball velocity is interpolated from the last ball position estimates. Finally, $Opponent^j = \langle x^j, y^j, v^j \rangle$ where v^j is again interpolated from previous estimates.

Unfortunately, the compound state variables are not independent of each others. Inaccuracies in the estimation of a robot's position causes even higher inaccuracies in the estimation of the observed robots. Also, since robots often go for the ball, the ball position often influences the movements of the robots.

2.2 Objectives of State Estimation

For the purpose of this paper, let us consider state estimation tasks that include the detection of objects, the recognition of their identity, and the inference of their states. In this setting performance aspects of state estimation include, whether the state estimation process hallucinates objects (false positive), overlooks objects (false negative), and the expected accuracy of their estimated states.

To capture these concepts we first introduce the notion of an object hypothesis. An ***object hypothesis*** h is a data structure in the robot's belief state that represents an object that the robot believes to exist. An object hypothesis h contains the most likely position $\langle x_h, y_h \rangle$ and a region of possible position r_h. The second concept that we need is the notion of ***designation*** (grounding), which is a mapping of an object hypothesis to an entity in the real world that caused the last observation supporting the hypothesis. This entity might also be a hallucination. To specify the designation function for a game episode a programmer has to step manually through the captured images and assign each observed image blob to an entity in the world.

Because this definition cannot be made operational, we will use a weaker notion of designation that can be fully automated *(this is closely related to the least square criterion)*. We say that an object hypothesis ***designates*** an object o at position $\langle x_o, y_o \rangle$ if (1) $\langle x_o, y_o \rangle$ lies within r_h and (2) there exists no other object hypothesis h' that does not already designate another object such that the distance between the most likely position $\langle x_{h'}, y_{h'} \rangle$ and $\langle x_o, y_o \rangle$ is smaller then the distance between $\langle x_h, y_h \rangle$ and $\langle x_o, y_o \rangle$.

Using our notion of designation we can now state what we mean by overlooking and hallucinating an object. We say a state estimator ***overlooks*** an object o if there exists no hypothesis h in the belief state that designates o. A state estimator ***hallucinates*** an object if its belief state contains an object hypothesis that does not designate any object in the world.

In general, the design of a state estimator is part of the design of a robot controller. Thus, the goal in the design of the robot controller is design a state estimator *se* and an action selector *as* such that taken together they will achieve the optimal performance of the robot. More formally, we intend to find a pair $\langle se, as \rangle$ such that

$$argmax_{\langle se,as\rangle}\ expected\text{-}performance(behavior(\langle se, as\rangle)).$$

The drawback of this approach is that the design of the state estimator interacts with the design of the action selector.

We can abstract away from action selection by stating the objective of state estimation as a cost function [3]. In this approach we have to design a state estimator that minimizes the given cost function $argmin_{se} cost(se)$.

So far most state estimators have been designed without specifying sophisticated cost functions. In the museum tour guide projects, some of the most successful applications of state estimation in autonomous robotics, the robot mainly estimated its own position. Therefore, the performance factors are simply whether or not the robot was lost and the average accuracy over the episodes in which it was not lost [4, 5].

In a nutshell, the cost of game state estimation in a soccer situation could be stated as *cost(belief-state,situation)* = w_1 * *hallucinations* + w_2 * *overlooked objects* + w_3 * *avg accuracy*, where w_1, w_2, and w_3 are weights that assess the relative importance of hallucinating and overlooking and the accuracy of observations. In robot soccer it is more important not to overlook objects than to hallucinate them. Overlooking opponents could result in collisions for which the robots might be sent off the field or not knowing where the ball is and therefore not being able to issue goal-directed actions. Hallucinations, on the other hand would mainly cause the robots to follow suboptimal trajectories, which is less critical.

The design of informative cost functions for game state estimation is much more subtle than suggested above and therefore the weights have to be set in situation specific ways. It is more important not to overlook the ball than the opponent players because knowing the ball position is necessary for focussed play. Overlooking team mates is not important at all because the team mates broadcast their own position estimates. It is also much more important to have accurate estimates in the area around the ball and for the ball handling robot than for objects that are not in the focus of the play or for players that only perform backup roles. Stating such informed cost functions is important because they also allow us to exploit task specific simplifications. For example, through the nature of soccer the ball handling robot is typically the one closest to the ball and facing the ball. It is therefore automatically the one that has the most accurate and reliable observations of the area around the ball.

2.3 Design Dimensions of State Estimation

When designing state estimation systems there are many design decisions to make including the ones listed below.

Decision 1: The Form of Probability Densities. One of the most critical decisions to make are the assumptions about the form of the probability densities. In many robot applications, this density is assumed to be unimodal and Gaussian distributed. Here, the distribution can be represented by the mean value and an associated covariance matrix. This has the main disadvantage that we cannot represent ambiguities in position estimates. These ambiguities can be represented and reasoned about in the Markov state estimation framework. The increased expressiveness must be paid for with higher computational cost. Particle filter are an alternative method in which probability distributions are approximated by sample sets. Particle filter have the advantage that they can be run in resource adaptive fashion. A particularity of particle filter is that they perform worse as data get very accurate.

Decision 2: State Estimation Algorithms. The choice of state estimation algorithms is constrained by the representation of the probability distributions. The most common approaches are Kalman filters, Markov localization algorithms, and particle filters. If states to be estimated involve multiple objects then observations have to be associated with object hypotheses. This is done, for example, in Reid's multiple hypothesis tracking algorithm or in Joint Probabilistic Data association filters. Recently, researchers have been proposed hybrid state estimation mechanisms, in which they combine multiple representations in parallel.

Decision 3: Decomposition and Simplification. A key problem in solving difficult state estimation problems is the complexity of the joint probability density and the huge amount of data that the probability density is conditioned on. This requires us to factorize, approximate, simplify (by making assumptions, such as the Markov assumption), and decompose the probability density [10]. State estimation problems can also be simplified based on their usage. In robot soccer, for example, instead of track exactly each opponent we can confine the estimator to track a superset of opponents. Less pressure to integrate observations of different robots. When making simplifications, assumptions, or approximations it is often possible to provide routines that monitor them.

Decision 4: Probabilistic Models. The update rules for probability densities often use parameter that are to be supplied by the programmer. These parameters can often be derived from problem specific probabilities, such as sensor or action models. There are two ways in which we can set these parameters in more informed ways. First, we can learn the values of these parameters from experience. Second, we can supply the algorithm with situation specific probabilities instead of general ones that average over all possible situations. For examples, in situations where objects become occluded, the estimator should assume a longer lifetime for object hypotheses even without supporting observations.

Decision 5: Observation Filtering. Another important design dimension is how many and which observations to take to maintain the belief state. If the estimator takes too many it might consume too much computational resources. On the other hand, it can often get away with less informative predictive models if it updates its belief with higher frequency.

Another issue is that observations are often corrupted and integrating a corrupted observation causes a less accurate belief state. For example, if a robot turns quickly the odometric data and image data do not correspond well and tracking opponents while turning quickly is therefore very unreliable. These problems can be mitigated by simply ignoring observations that are probably inaccurate and unreliable.

Decision 6: Managing Computational Resources. Yet another issue is what to spend the computational resources for. Typically, in autonomous robot control the robot has a limited amount of computational resources available that can be assigned to the different computational tasks. For example, the programmer can spend resources for a more accurate interpretation of sensory data, it can apply more sophisticated probabilistic models for predicting the resulting state (eg, ones that allow for predicting nonlinear movements), or it can use crude and cheap computation mechanisms and rather run the iterative state estimation with a higher frequency.

Discussion. We have sketched in this section a number of dimensions for the design of complex state estimators. Unfortunately, there is no general model of the influence of these design decisions on the performance of a state estimation system. Even worse, the different design decisions interact with each other in very subtle and application specific ways. Therefore, it is necessary to design state estimators specifically for the application at hand and to empirically evaluate the design decisions by comparing the expected performance with respect to the specified cost function.

3 Empirical Investigation

We will now sketch the design and the main parameters of the game state estimator and then investigate the effects of design choices on its performance.

3.1 The Game State Estimator

The game state estimation subsystem consists of the perception subsystem, the state estimator itself, and the belief state. The perception subsystem consists of a camera system with several feature detectors and a communication link that enables the robot to receive information from other robots. The belief state contains a position estimate for each dynamic task-relevant object.

The state estimation subsystem consists of three interacting estimators: the self localization system, the ball estimator, and the opponents estimator. The self localization estimates the probability density of the robot's own position based on environment features, the ball position, and its predicted position. The ball localizer estimates the probability density for the ball position given the robot's position, its observation, and the ball estimates of team mates. Finally, a robot estimates the positions of the opponents, based on its own position, the robots' appearances in the images, and the estimations of the team mates.

The decomposition of game state estimation reduces the overall complexity of the estimation problem and enables the robots to exploit the structures and assumptions underlying the different subtasks. However, as stated in the previous section, the different estimation problems are not independent.

Self Localization. State estimation is an iterative process where each iteration is triggered by the arrival of a new piece of evidence, a captured image, an odometry reading, or a partial state estimate broadcasted by another robot. These data are integrated over time to a maximum a posteriori (MAP) estimate of the robot's pose. The MAP estimate $\widehat{\mathbf{\Phi}}$ is given by $\widehat{\mathbf{\Phi}} = \text{argmax}_{\mathbf{\Phi}}\ p(\mathbf{\Phi}) \cdot p(data|\mathbf{\Phi})$ where $p(\mathbf{\Phi})$ is the prior of the pose $\mathbf{\Phi}$ summarizing all evidence gathered in the past. The prior at the current time step is obtained by predicting the pose distribution estimated for the previous time step. The second term $p(data|\mathbf{\Phi})$ is the likelihood of receiving $data$ given the robot's pose $\mathbf{\Phi}$.

Self localization runs a fast Kalman filter for tracking the position of the robot and thereby makes the assumption that the probability density is Gaussian. To detect situations where this assumption yields localization failures a particle filter with a lower frequency is run concurrently. In cases where evidence implies multi modal probability densities the particle filter detects that the robot gets lost and provides the different local maxima for reinitializing the Kalman filter.

Parameters and Assumptions. The parameters of the self localization module that can be set include the the probabilistic models $p(data|\mathbf{\Phi})$ for the different kinds of data. The programmer can also decide on the mechanisms applied to image interpretation, the frequency at which to run the algorithms, and which observations to take for state estimation.

Opponent and Ball Tracking. For opponent tracking a variant of Reid's Multiple Hypothesis Tracking (MHT) algorithm [6] is used. The objective of the MHT algorithm is to maintain a set of object hypotheses, each describing a unique real object and its position and estimate the likelihood of the individual hypotheses. The MHT algorithm deals with two kinds of uncertainties. The first one is the inaccuracy of the robot's sensors and is represented using a Gaussian probability density. The second kind of uncertainty is introduced by the data association problem, i.e. assigning feature blobs to object hypotheses. It is represented by a hypotheses tree where nodes represent the association of a feature blob with an object hypothesis. A node h_j^k is a son of the node h_i^{k-1} if h_j^k results from the assignment of an observed feature blob with a predicted state $\tilde{h}_i^{k-1}$ of the hypothesis h_i^{k-1}. Computing the association probability $P(h_{ij}^{k+1}|Z(k))$, which indicates the likelihood that observed object and object hypothesis refer to the same object given $Z(k)$, the sequence of all measurements up to time k, is the heart of the MHT algorithm [1, 7].

Parameters and Assumptions. The opponents tracking system can be parameterized as follows. First, we can set the minimum likelihood of a hypothesis to be kept in the hypotheses tree. If the threshold is set higher then the tree is pruned more aggressively. Smaller trees require fewer computational resources but have a higher risk of deleting correct hypotheses from the tree. Other parameters include the average time of detecting spurious features and of keeping hypotheses without supporting observations, which is needed to deal with occlusions. Additional parameters are the motion models for opponents, the routine for deciding whether an observation is informative enough to improve the belief state.

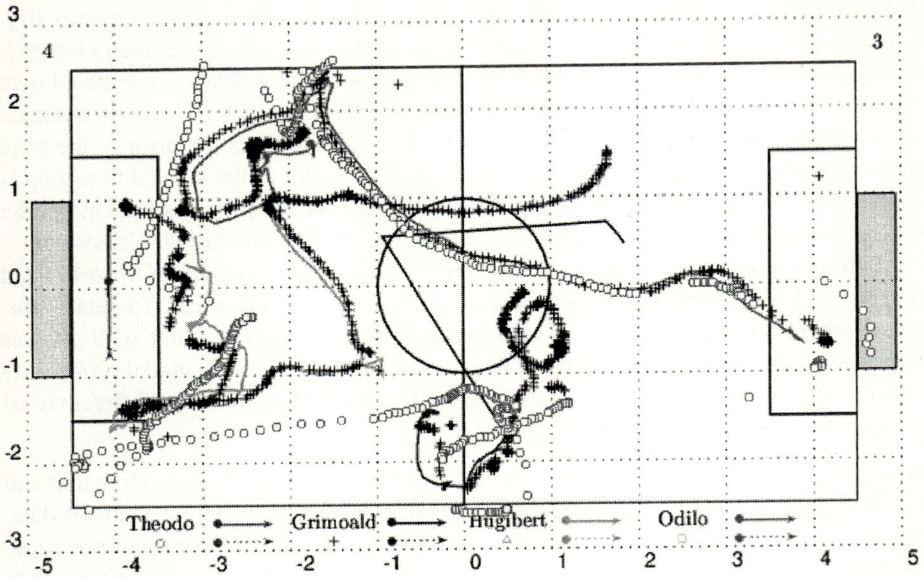

(a)

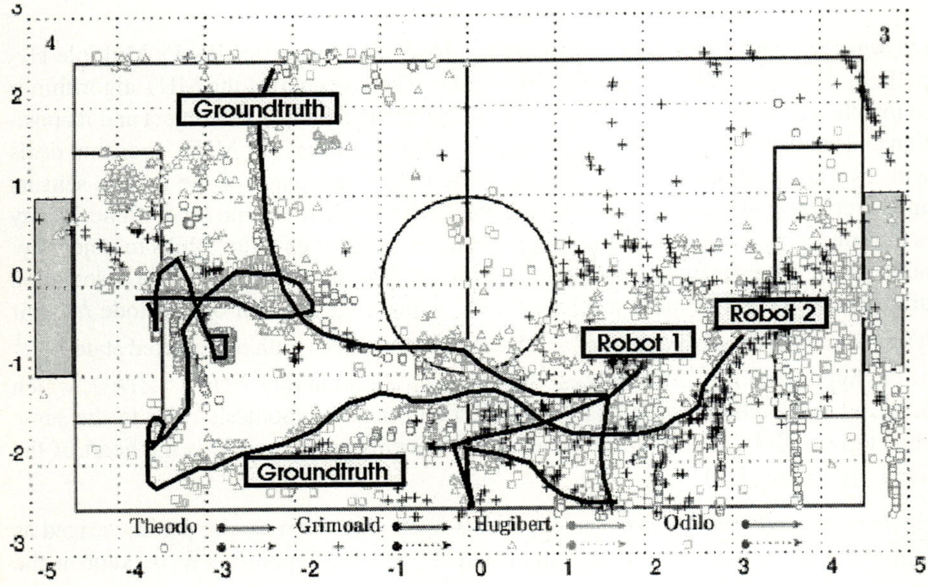

(b)

Fig. 1. Game episode (100 seconds long). Subfigure (a) shows the results of self localization. The trajectories of the robots are plotted as solid lines. Ground truth position data for robots is indicated with '+'. Ground truth position data for the ball is indicated with 'o'. Subfigure (b) shows the individual observations of an opponent where the color indicates which one of the robots made the observation. The overlying black lines are the ground truth data.

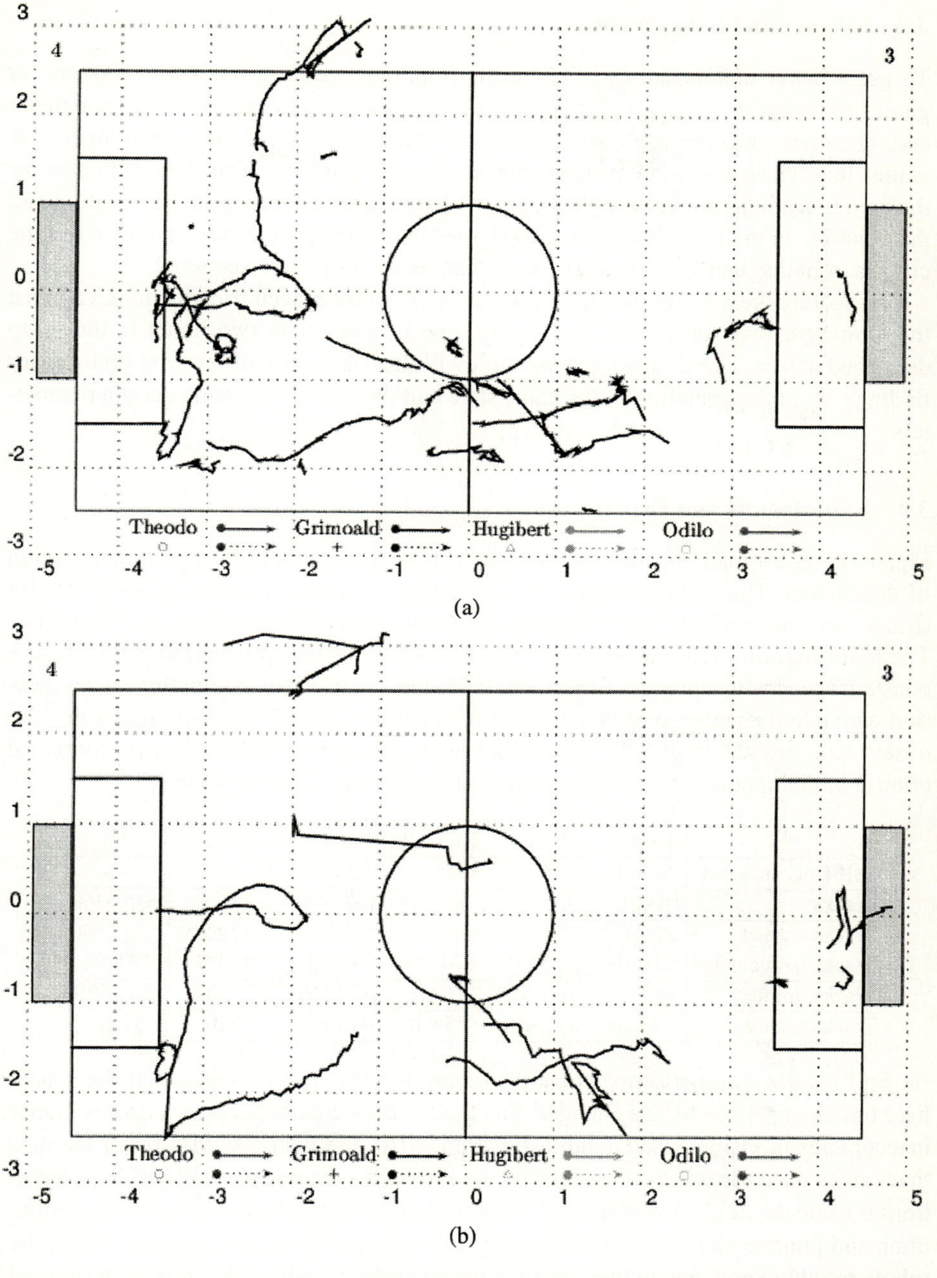

(a)

(b)

Fig. 2. Game episode (100 seconds long). Subfigure (a) shows the tracks of opponent robots exploiting observations made by all team mates. Subfigure (b) shows the tracking of opponents without cooperation.

3.2 Setup of the Experiment

To get a better understanding of the design options, we have recorded robot soccer games to evaluate system performance more carefully. To do so, we have mounted a ceiling camera into our robot lab in order to record a bird's eye view from the soccer games that will serve as the ground truth for our experiments. The state estimator for the ceiling camera has an average accuracy of 10-15cm depending on the situation and the position on the field. For evaluating the results of game state estimation we use the criteria of hallucinations and overlooking that we have stated in section 2.

To acquire the data for our empirical studies we have played friendly games against the Ulm Sparrows for a total net playing time of more than two hours in the setup described above. The following subsection will describe some of our first preliminary findings about the operation of the game state and the effects of setting certain parameters.

3.3 Experiments and Experiences

Fig. 1 shows a typical result of game state estimation with standard parameterization of the system. The Kalman filter self localization runs with a frequency of 20-30 Hz (frame rate), the particle filter self localization at about 15Hz and with 1000 samples. The MHT algorithm runs at 4*20 Hz. The action models for opponent tracking assume constant speed with a process variance of 0.03. The criterion for associating an observation with a hypothesis is a Mahalanobis distance of less than 5.9, which means that the observation lies within the 95% confidence interval of the prediction. Key performance criteria of the opponent tracking system are listed in the subsequent table.

	Robot 1	Robot 2	Robot 3	Robot 4	cooperation	1/4
Self-Loc. accuracy	11 cm	48 cm	28 cm	19 cm	—	—
Tracks	1038 51%	992 49%	1067 53%	1162 58%	1278 64%	635 31%
Tracks correct	820 41%	720 36%	893 44%	1002 50%	1087 55%	528 26%
Tracks hallucinated	218 10%	272 13%	174 9%	160 8%	191 9%	107 5%
Tracks omitted	913 59%	1008 51%	933 47%	838 42%	722 36%	1365 69%
Tracks accuracy	33 cm	37 cm	36 cm	31 cm	23 cm	37 cm

Self localization performs worse than expected [8]. The reason is that the soccer field has changed due to rule changes. The field has become larger which causes higher inaccuracies in vision-based depth estimation. Also, the field provides fewer features that can be used for self localization and the lack of a surrounding wall yields observations outside the field. Qualitatively, the robots loose track of their positions much more often and jump to alternative position estimates because of the particle filter. When the robots roughly know where they are then the accuracy is still sufficiently high (around 10-20cm).

The table lists the results for a robot with cooperation and the individual robots as columns. The rows represent the different performance measures. The maximal number of base points for the opponent tracks (given omnidirectional view and no occlusion) is 2000 for the individual robots and the cooperating robots. From those a robot with cooperation was able to observe 1278 tracks (64%), 9% of those where hallucinated track points. In the table these numbers are listed as observed tracks, correct designations, hallucinated tracks, and average accuracy.

Cooperation Helps – Does It? We can see that in our episode cooperation increases the number of detected tracks as well as increases the accuracy, which is typical. These results are depicted in Fig. 2. Fig. 2 (a) shows the opponents tracked with cooperative tracking and Fig. 2 (b) the same estimation made without cooperation. You can see that gaps in the tracks that are mainly caused by occlusions can often be closed by using the observations of team mates. Accuracy can be substantially increased by fusing the observations of different robots because the depth estimate of positions are much more inaccurate than the lateral positions in the image. This can be accomplished through the Kalman filter's property to optimally fuse observations from different robots into global hypotheses with smaller covariances.

However, cooperation is not always a good idea. We have experienced this when running the game state estimation under extremely poor lighting conditions. Under these conditions self localization could not well discriminate between some symmetric positions on the field. Therefore, integrating observations from dislocalized robots caused incoherent hypotheses and we were forced to disable the cooperation to get more stable estimates.

Specify Cost Functions Explicitly! If we add a notion of relevance to the opponent players depending how important they are for the game state then the percentage of coverage of the *relevant* opponents is much higher. This is because the robots often look into the direction of the ball and therefore see the opponents that are close to the ball more often. It would be inadvantageous to design the state estimator such that it has a uniform coverage of all positions on the field.

Tune the Parameters of Estimation Algorithms. Tuning of algorithm parameters has an enormous impact on the performance of the systems. The most promising approach seems to be to learn these parameters from the experimental data. Due to space limitation we cannot discuss the issues here.

If You Can't Do It Accurately Then Do It Fast! The column *1/4* in our table shows the performance of the system if we run state estimation at 0.25 of its normal frequency. This causes less accurate position estimation as well losing more than half of the tracks because of less redundant information and stronger dependency on action models. Recall that we assumed motion to be constant whereas motions in robot soccer are very often changed abruptly. This suggests that we can deal with worse motion models by estimating with a higher frequency.

Implement Estimators That Check Their Work! We have pointed out in section 2 that Kalman filter localization is fast but relies on the assumption that the probability distribution for the robots is Gaussian distributed. We use a particle filter in parallel that runs at a third of the frequency of the Kalman filter as a monitor that detects ambiguous position estimates and that initializes the Kalman filter localization after the position track is lost. We didn't use a particle filter on its own because we couldn't run it fast enough and because Gaussian distributions can be communicated much more compactly and used as evidences in the other state estimation problems. Our findings are that the time needed for relocalization could be reduced from about 10 seconds to about 2 seconds and that lost position tracks could be detected earlier.

Lesson. Applying multiple state estimation techniques with different strength and weaknesses in parallel is a viable design option for state estimators. In particular, if the computationally more expensive method can be used to monitor the faster but less reliable method.

Learn to Look Away! In state estimation it is often assumed that every observation provides the robot with additional information that can be used to improve the state estimate. In reality, however, this is often not true. We have used Quinlan's C4.5 decision tree learning algorithm to learn predictive rules as to whether or not to integrate an observation into state estimation considering the current situation. The robot learned rules that state, an observation is likely to be informative (within 30cm of the ground truth position) if

1. the observed distance D is less than 3.76086 and the robot has not lost track of its position
2. the robot is not turning, the angle ϕ between the observation and the direction the camera is pointed to is less than 22.91° and $D \leq 5.4253$
3. the robot is turning, $D \leq 6.87$, and $\phi \leq 43.6$

Applying these rules to filtering observations we could reduce the hallucinated tracks by one third, and further improved the track accuracy, and overlooked fewer tracks (which get otherwise corrupted by the noisy observations).

Lesson. What is interesting here is not the specific feature language and rules. They are different for different robots, environments, and cost functions. What is important is that we can try to learn predictive models as to whether or not an observation can be expected to improve the state estimate. These rules can be used to filter out worthless observations.

4 Conclusions

In our view, research in state estimation is focusing too much on algorithm design and analysis and too little on system design. As we will apply state estimation in very complex autonomous robot applications, such as household robotics, where the estimated states are extremely high dimensional we won't be successful unless we know how to parameterize the systems and how to provide them with the necessary probabilistic models. Therefore, our most important conclusion is an obvious one: the development of complex high-performance state estimation systems is a complex design problem with many design options. The choice of estimation algorithms is only one of these options but many other design dimensions have an equally large impact on system performance. The design has to be tailored to the particular application at hand and the different design option interact with each other in obscure and opaque ways. We propose empirical investigations and learning based on ground truth data as necessary tools for the development of successful state estimation systems. We have illustrated these issues using a probabilistic game state estimation in autonomous robot soccer. Our results are preliminary and a lot more has to be done to understand the design of complex state estimators properly.

References

1. Y. Bar-Shalom and T. Fortmann. Tracking and data association. Academic Press., 1988.
2. D. Fox, W. Burgard, and S. Thrun. Markov localization for mobile robots in dynamic environments. *Journal of Artificial Intelligence Research*, 11:391–427, 1999.
3. Z. Ghahramani, D. M. Wolpert, and M. I. Jordan. Computational models of sensorimotor organization. In Morasso and V. Sanguineti, editors, *Self-Organization Computational Maps and Motor Control*, Amsterdam, North-Holland, 1997.
4. J.-S. Gutmann, W. Burgard, D. Fox, and K. Konolige. An experimental comparison of localization methods. In *Proc. of the IEEE/RSJ International Conference on Intelligent Robots and Systems*, 1998.
5. J.-S. Gutmann and D. Fox. An experimental comparison of localization methods continued. In *Proc. of the IEEE/RSJ International Conference on Intelligent Robots and Systems*, 2002.
6. D. Reid. An algorithm for tracking multiple targets. IEEE Transactions on Automatic Control, 24(6):843–854, 1979.
7. T. Schmitt, M. Beetz, R. Hanek, and S. Buck. Watch their moves: Applying probabilistic multiple object tracking to autonomous robot soccer. In *AAAI National Conference on Artificial Intelligence*, pages 599–604, Edmonton, Canada, 2002.
8. T. Schmitt, R. Hanek, M. Beetz, S. Buck, and B. Radig. Cooperative probabilistic state estimation for vision-based autonomous mobile robots. *IEEE Trans. on Robotics and Automation*, 18(10):–, 2002.
9. D. Schulz, W. Burgard, D. Fox, and A.B. Cremers. Multiple object tracking with a mobile robot. In *Computer Vision and Pattern Recognition (CVPR)*, volume 1, pages 371–377, Kauai, Hawaii, 2001.
10. S. Thrun. Probabilistic algorithms in robotics. *AI Magazine*, 21(4):93–109, 2000.
11. S. Thrun, M. Beetz, M. Bennewitz, A.B. Cremers, F. Dellaert, D. Fox, D. Hähnel, C. Rosenberg, N. Roy, J. Schulte, and D. Schulz. Probabilistic algorithms and the interactive museum tour-guide robot minerva. *International Journal of Robotics Research*, 19(11):972–999, 2000.

Cooperative Soccer Play by Real Small-Size Robot

Kazuhito Murakami[1], Shinya Hibino[2], Yukiharu Kodama[2], Tomoyuki Iida[1], Kyosuke Kato[2], and Tadashi Naruse[1]

[1] Aichi Prefectural University, Nagakute-cho, Aichi, 480-1198, Japan
[2] Graduate School of Information Science and Technology
Aichi Prefectural University, Nagakute-cho, Aichi 480-1198, Japan
apurobo@ist.aichi-pu.ac.jp
http://www.aichi-pu.ac.jp/ist/lab/narulab/index.html

Abstract. One of the typical cooperative actions is the pass play in RoboCup small-size league. This paper presents three technical key features to realize robust pass play between robots. The first one is the high resolution image processing to detect the positions and orientations of the robots. The second one is the control algorithm to move and adjust the robots for the pass play. The third one is the mechanism to catch the ball moving at high speed. This paper discusses these methods and shows the effectiveness of the methods by experimental results.

1 Introduction

RoboCup gives full scope to realize a soccer game by robot [1–5]. It is important to realize a robust pass play. There are many key features to realize robust pass play between robots. From the viewpoint of software, it is important to realize the high resolution image processing algorithm to detect the positions and the orientations of the robots. If the image processing system couldn't detect moving objects correctly or it would take too much time, it would be impossible to kick and receive the ball at the promised place. A control algorithm to move and adjust both robots for the pass play is also required because the ball would not be passed at the time promised. From the viewpoint of hardware, it is necessary to design a mechanical device to catch the ball moving at high speed, because the ball collides with the robot at high speed and rebounds off to an unexpected direction.

From these considerations, we realized simple but high-speed image processing system whose processing speed is about 1.4 msec to detect and track the robots and a ball by reducing the searching range, and also implemented an algorithm to control both robots, the passing robot and the receiving robot, for the robust pass play. We improved the dribbling mechanism to catch the ball moving at high speed by intentionally adding some loose to the joint part of the dribbling device.

In the following, the image processing system and cooperative algorithm to realize a pass play are shown in sections 2 and 3, respectively. The mechanical devices to catch the ball and its modeling are shown in section 4.

D. Polani et al. (Eds.): RoboCup 2003, LNAI 3020, pp. 410–421, 2004.

2 Image Processing System

2.1 Objects to Be Detected

It is allowed to attach other submarkers to detect the identity number (ID) of each robot and/or the orientation of the robot. The number, size and layout of submarkers are unrestricted but approved colors are limited so as not to affect the robotic image processing systems used by opposing teams. Figure 1 shows submarkers of our team. A rectangle submarker (105mm× 16.5mm) has been used to determine the orientation of the robot, and small circle submarkers (1∼ 4 pieces, 8.5mm diameter) have been used to determine ID of each robot. Our image processing system aimed to detect these markers.

2.2 Image Processing System

CPU in the host computer is a Pentium Xeon 2GHz and the OS is Windows 2000. We employ a progressive scanning camera (DXC-9000, SONY) to avoid the problem of the difference between the odd frame and the even frame caused by the high-speed movement of the robot.

Global vision camera is set at 3.0m over the game field. Since the length along to the side-line including the goal area becomes about 3.2m, at least 56 degree view angle is needed. In our system, wide lens attachment (Fujinon, WCV-65, × 0.75) covers this range. The size of the global vision image grabbed by the frame grabber (Matrox, GEN/X/00/STD) is 640(W)×480(H), and the resolution is 5 mm/pixel.

Fig. 1. Top surface of robot

2.3 Image Processing Algorithm

Image processing system detects the location of each robot by using team color markers, and determines the orientation and the ID number (code) of each robot by using submarkers attached on the top of the robot. Figure 2 shows the flowchart of image processing. The outline of processing is shown below and literatures [6–9] give detailed description.

Step-1 (Input Image) Get an RGB image from frame grabber which is captured through the progressive camera. Figure 3 shows an example of input image. Figure 3(b) is the enlarged images around the robots.

Step-2 (Calculation of Search Range) Calculate the search range based on the reliability of respective objects. Since the reliability is an important parameter which characterizes the performance of image processing, the details of the calculation are explained later in this section. White rectangles around robots in Fig.4 are the search ranges. Only in these ranges, RGB color format image is converted to YUV color format image. This greatly reduces computation time.

Step-3 (Segmentation of Colored Region) In the same ranges as restricted in Step-2, the simultaneous pattern matching is executed for 9 colors (max. 32colors) in YUV color format image by using the CMU's color segmentation algorithm[10]. Since the real game is not played under the uniform light conditions, the intensity and color spectrum are not uniform over the game field. Therefore, we assign several colors as a candidate for the object. We assign 2 colors for the ball, 2 colors for each team color and 3 colors for the areas that we would like to remove from processing. This method realizes the robust color extraction under the non-uniform lighting conditions.

Step-4 (Labeling) Apply the first propagation in the labeling algorithm, which we have developed last year[9], to label the object region. Figure 5 shows the result. In this algorithm, the processing for interlaced image is not applied, since we use non-interlaced image.

Step-5 (Selection of Objects) Detect team marker region and expand the region. Then, detect ID markers and rectangle submarker in the expanded region. ID markers area and rectangle submarker area are distinguished by the size of area. Figure 6 shows the result.

Step-6 (Calculation of IDs and Directions) Calculate the ID and the rough orientation of the robot by the numbers and the positions of circle submarkers, respectively. The precision for the angle is about 8 degrees.

Step-7 (Calculation of Modified Directions) Calculate the precise orientation of the robot by detecting the long-side edges of rectangle submarker and applying the least mean square method to the detected edges. The precision goes up to less than 1 degree.

Step-8 (Record of Positions and Directions) Record the current positions and directions.

Step-9 (Output of Object IDs, Positions and Directions) Convert the objects positions from the camera coordinates to the world coordinates.

We utilize the reliability factor to decide the search range in Step-2 and Step-3. Reliability indicates how much the result of image processing is correct. Wide range should be searched if the reliability is low. On the contrary, it is enough to search in the restricted range if the reliability is high.

We defined 4 levels of reliability, i.e. non-, low-, intermediate- and high-reliability level. The reliability level is updated every frame cycle (i.e. 60 times per second). Basically, it goes up 1 level if the object detection ends in success, otherwise goes down 1 level.

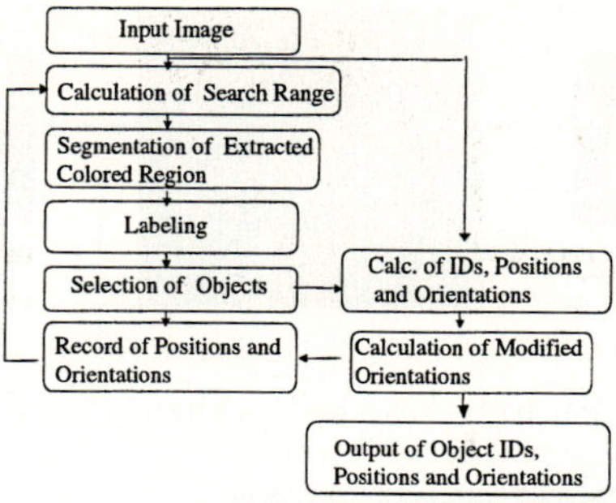

Fig. 2. Flowchart of image processing

The decreasing of the reliability level makes the search range wider and causes the increasing of computation time. However, information management system prevents the reliability level from excessive go-down. Such cases happen when the ball is occluded, since the ball detection ends in failure. The information management system keeps the reliability level at 2 (intermediate).

The search range is restricted when the reliability level is in 1–3, i.e. the range of 20 × 20 pixels, 30 × 30 pixels and 60 × 60 pixels is searched when the reliability level is 3, 2 and 1, respectively. 20 × 20 pixels range is about 10 cm × 10 cm in the real field. This search range works for the tracking of a moving object up to 3m/sec.

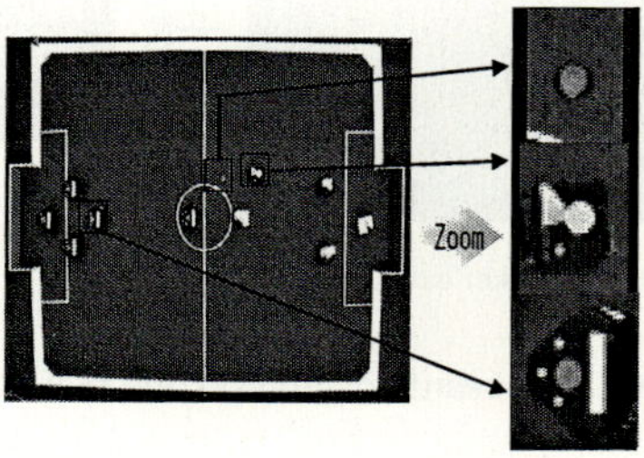

(a) global vision image (b) enlarged image

Fig. 3. An example of grabbed image from a global vision camera

3 Cooperative Control Algorithm

There are many types of cooperative play in our system, for example, pass-play, defense-play, assist-play and so on. In this paper, we discuss the pass play algorithm. Algorithms for passing robot and receiving robot are given below.

First, we define variables and flags. Let the line connecting the center of passing robot and the receiving robot be $Line_A$, and let the line on the heading direction of passing robot be $Line_B$. Let the angle between $Line_A$ and $Line_B$

Fig. 4. Restricted region to be searched (whitened area) and enlarged images

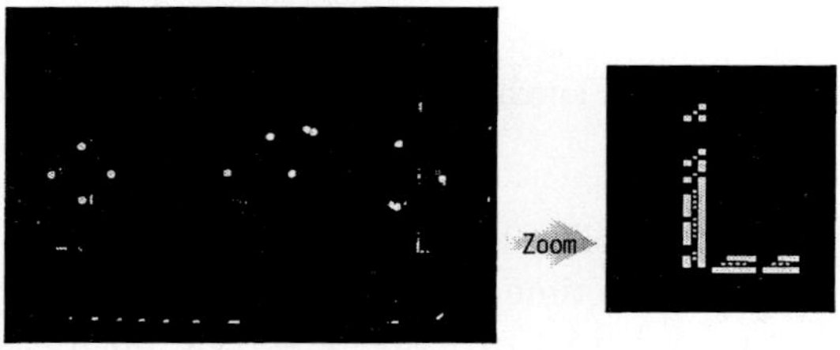

Straight lines of width 1 pixel will be recognized as separated objects. They are easily removed by area thresholding.

Fig. 5. An example of line noises to be deleted by the proposed labeling algorithm

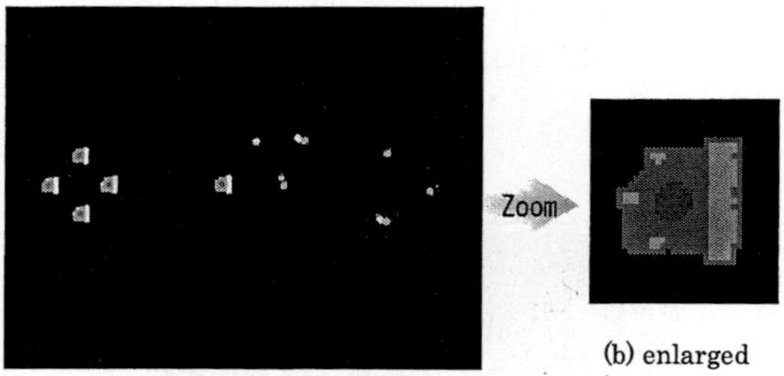

(a) the result of marker extraction

(b) enlarged image

Fig. 6. The result of marker extraction

be θ . Let the direction flag and stabilization flag be Dir and $PassCounter$, respectively.

[Initializing]

Step 1 Decide the rotation speed v_r based on the velocity profile shown in figure 7. v_r depends on θ. Start rotation.

Step 2 Start dribbling device.

Step 3 If there is an obstacle on $Line_A$, then compute which side of $Line_A$ the center of that obstacle is located. If it located on left-hand side of the passing robot, then set Dir to 1, otherwise -1. If there is no obstacle on $Line_A$, then set Dir to 0.

[Pass Algorithm]

Step 1 Do the Initializing.
Step 2 If Dir is not 0, then set $PassCounter$ to 0 and go to Step 5.
Step 3 Go along $Line_B$ at a speed of v_{ps}.
Step 4 If $|\theta|$ is less than θ_{ps} and the distance between the ball and the passing robot is less than d_{min}, then increase $PassCounter$ by 1. If $PassCounter$ is greater than n_ps and $|\theta|$ is less than θ_{min}, then start the kicking device. Reset $PassCounter$.
Step 5 Wait for the next frame cycle. Go to step 1.

In the above, $v_{ps}, \theta_{ps}, d_{min}, n_{ps}, \theta_{min}$ are constants and are determined by the experiments.

[Receive Algorithm]

Step 1 Do the Initializing.
Step 2 If Dir is not 0, then go in the direction of $Dir \times \theta_{rec}$ from $Line_B$ at a speed of $\min(v_{rec}, c_{rec} \times distance\_between\_ball\_and\_Line_A)$.
Step 3 Wait for the next frame cycle. Go to Step 1.

In the above, $v_{rec}, \theta_{rec}, c_{rec}$ are constants and are determined by the experiments.

4 Catching Mechanism and Its Analysis

4.1 Robot Mechanism

The robot must be constructed in order to execute the commands, such as a move, a shoot and a dribble. Figure 8 shows our robot. Our robot has 3 omni-wheels and can move any direction. As shown in Fig.8 (a), 3 DC motors with encoders are used (FAULHAVER). The gear ratio is 9.7:1. The kicking device that is driven by a solenoid (SINDENGEN) is mounted (Fig.8 (a) below). It can kick the ball at the maximum speed of about 3m/sec. Dribbling device utilizes the rotating roller to give backspin to the ball. Uncovered robot in Fig.8 (b) shows dribbling roller.

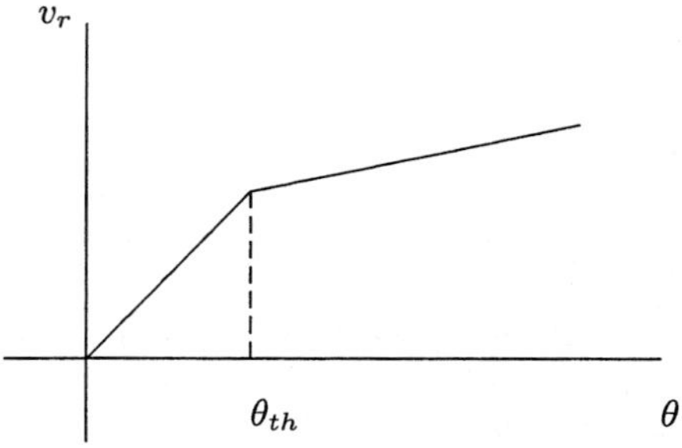

Fig. 7. Velocity profile

(a) omni-directional drive. (b) appearance of our robot.

Fig. 8. Our robot system

There are two important technical issues to realize a pass play. One is the correct recognition of the position and the orientation of each robot and the ball. The other is the mechanical improvement to catch the ball. For the former, our precise and high-speed image processing system solved the problem as described in section 2, and for the latter subject, our system solved the problem by adding shock absorption mechanism to the dribbling device. It may be natural to use a spring like the Cornell University's BigRed, but our team realizes it by another method.

The dribbling device is attached in front of the robot as shown in Fig.8 (b). The roller made of rubber rotates and gives backspin to a given ball and holds the ball. However, because the roller is hard enough not to bite into the ball which is also hard, the ball will bounce off of it if the dribbling device receives the fast ball. To solve this problem, we made the joint of the roller looser than normal and making it possible for it to move upward slightly. This simple looser joint enables the impact of the ball to be sufficiently absorbed. Still, it remains necessary to adjust the degree of looseness depending on the quality on the surface of the carpet on the robotic soccer field.

4.2 Modeling of Catching Device

There are so many factors to model real shock absorption mechanism. Since our system realizes it by adding adequate space to the joint of the dribbling roller, we analyzed it by three simplified models, (a) at the impact, (b) transition to the stable state, and (c) in the stable state.

(a) Impact Model
Let the angle between the horizontal line and the line crossing to the centers of the roller and the ball be θ, and let m_R and m_B, ν_R and ν_B be mass and velocity of the roller and the ball, respectively, in the world $X-Y$ coordinates as shown in Fig.9. Let the components of ν_R and ν_B be $(\nu_{R,x}, \nu_{R,y})$ and $(\nu_{B,x}, \nu_{B,y})$ in the local $x-y$ coordinates parallel to the tangent line and the normal line at the collision point as shown in Fig.9.

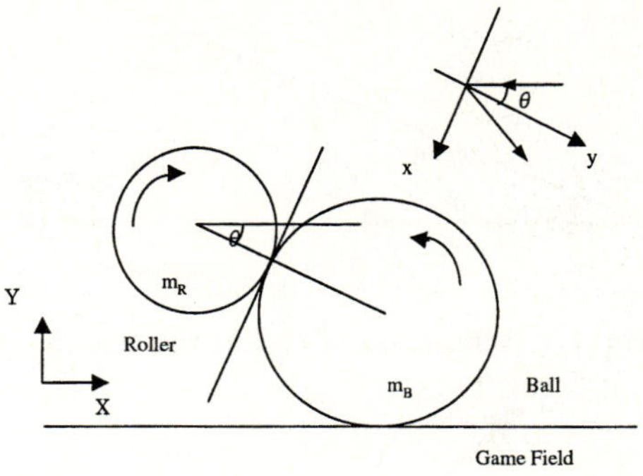

Fig. 9. Physical situation at a the impact

First, by modeling the collision without rotation of the roller and the ball by the law of conservation of moment, following equations for the y-coordinate,

$$\nu'_{B,y} - \nu'_{R,y} = -e_y(\nu_{B,y} - \nu_{R,y}) \tag{1}$$

$$m_B\nu_{B,y} + m_R\nu_{R,y} = m_B\nu'_{B,y} + m_R\nu'_{R,y} \tag{2}$$

are obtained, here dash denotes the velocity after collision and e_y is the coefficient of rebound for y-coordinate. As the same way for x- coordinate,

$$\nu'_{B,x} - \nu'_{R,x} = e_x(\nu_{B,x} - \nu_{R,x}) \tag{3}$$

$$m_B\nu_{B,x} + m_R\nu_{R,x} = m_B\nu'_{B,x} + m_R\nu'_{R,x} \tag{4}$$

are obtained, and the collision of ball and roller is modeled when they are assumed as a particle.

Since the roller and the ball are rotating, it is necessary to introduce some equations concerning to the angular momentum. Let the moment of inertia and angular velocity be I and ω, respectively. From the law of conservation of angular momentum,

$$I_B\omega_B + I_R\omega_R = I_B\omega'_B + I_R\omega'_R \tag{5}$$

and following equations,

$$r_B\omega_B = \nu_{B,x}\ ,\ r_R\omega_R = \nu_{R,x}\ ,\ r_B\omega'_B = \nu'_{B,x}\ ,\ r_R\omega'_R = \nu'_{R,x} \tag{6}$$

are realized at the collision point, here, r_B, r_R are the radius of the ball and roller, respectively. Since our system utilizes rubber of cylinder, the moments of inertia I_R and I_B are easily calculated as $I_R = \frac{1}{2}r_R^2 m_R$ and $I_B = \frac{2}{5}r_B^2 m_B$, respectively. The solutions of $\nu'_{R,x}$, $\nu'_{R,y}$, $\nu'_{B,x}$, $\nu'_{B,y}$ at the impact are calculated by solving equations (1) $\sim$ (6).

(b) Transition Model

The center of gravity of the ball repeats the reflection after the impact as shown in Fig.10. It is sufficient to consider the movement of the center if the angle θ

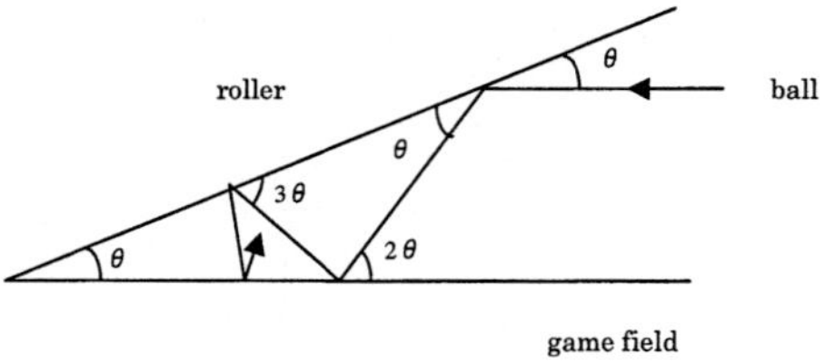

Fig. 10. Reflections of ball in the transition

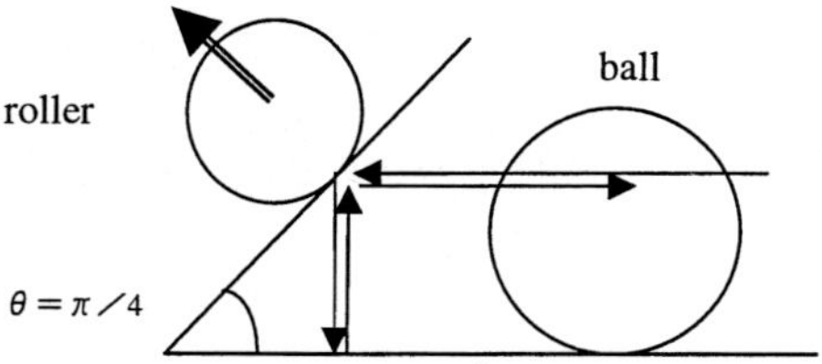

Fig. 11. An example of shock absorption at $\theta = \pi/4$

varies within the range of a few degree during the transition. Precise analysis and modeling are one of our future works, but it is possible to realize simple model as follows if the relation $e_x \equiv e_y$ is satisfied in the $x - y$ local coordinates.

Reflection angle increases reflection by reflection as shown in Fig.10, namely $\theta, 2\theta, 3\theta, ..., n\theta, ...$. After n times reflections, it exceeds $\pi/2$ and the ball returns back to the inverse direction. Figure 11 shows the case of $\theta = \pi/4$. In this case, the ball rebounds in a short time and if the roller moves to the direction illustrated with arrow in order to absorb the shock, these angular relations are satisfied during transition.

(c) Stable Model

Let the reactions be N_B, N_R and let the coefficients of kinetic friction be μ_1, μ_2 as shown in Fig.12. The equations of motion for the ball are as follows.

$$\text{For X} - \text{direction}: \quad -\mu_1 N_B + N_R \cos\theta - \mu_2 N_B \sin\theta = 0 \tag{7}$$

$$\text{For Y} - \text{direction}: \quad N_B - m_B g - N_R \sin\theta - \mu_2 N_R \cos\theta = 0 \tag{8}$$

By solving equations (7) and (8), the solutions for N_B, N_R are obtained as follows.

$$N_R = \frac{\mu_1 m_B g}{-(\mu_1 + \mu_2)\sin\theta + (1 - \mu_1\mu_2)\cos\theta} \tag{9}$$

$$N_B = \frac{(\cos\theta - \mu_2 \sin\theta) m_B g}{-(\mu_1 + \mu_2)\sin\theta + (1 - \mu_1\mu_2)\cos\theta} \tag{10}$$

So the force by backspin of the ball is derived as

$$\begin{aligned} f = \mu_1 N_B &= \frac{\mu_1(\cos\theta - \mu_2 \sin\theta) m_B g}{-(\mu_1 + \mu_2)\sin\theta + (1 - \mu_1\mu_2)\cos\theta} \\ &= \frac{\mu_1(1 - \mu_2 \tan\theta) m_B g}{1 - \mu_1\mu_2 - (\mu_1 + \mu_2)\tan\theta} \end{aligned} \tag{11}$$

From equation (11), following equations are obtained.

$$\frac{\partial f}{\partial \theta} = \frac{\mu_1^2(1+\mu_2^2)m_B g}{\{1-\mu_1\mu_2-(\mu_1+\mu_2)\tan\theta\}^2\cos^2\theta} \tag{12}$$

$$\frac{\partial f}{\partial \mu_1} = \frac{(1-\mu_2\tan\theta)^2 m_B g}{\{1-\mu_1\mu_2-(\mu_1+\mu_2)\tan\theta\}^2} \tag{13}$$

$$\frac{\partial f}{\partial \mu_2} = \frac{\mu_1^2(1+\tan^2\theta)m_B g}{\{1-\mu_1\mu_2-(\mu_1+\mu_2)\tan\theta\}^2} \tag{14}$$

It is easily known that larger force f could be obtained according to the increase of θ because $\frac{\partial f}{\partial \theta} > 0$ and $f|_{\theta=0} > 0$. It is important that f has a critical point at $\theta = \arctan\frac{1}{\mu_2}$. At this point, f becomes 0.

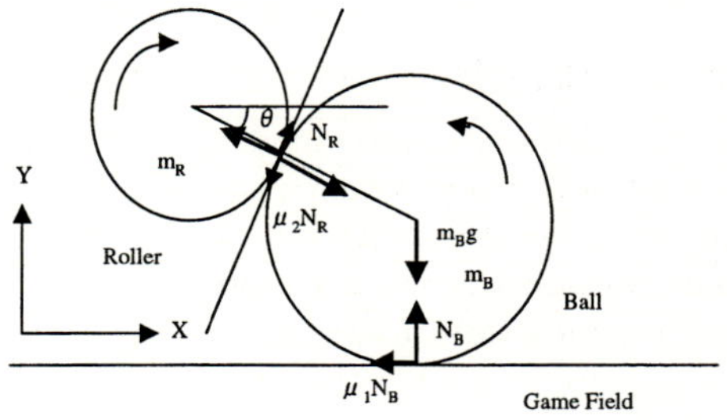

Fig. 12. Stable model by dribbling device

5 Considerations

From the viewpoint of mechanical performance, we confirmed the effectiveness to use the dribbling and kicking devices. Dribbling device worked well except for the case that the coefficient of kinetic friction of the field's carpet is high and the robot lost a ball around the wall of the game field.

For the performance of image processing system, it was experimentally confirmed that the angle accuracy is sufficient to realize cooperative play. Experimental results are shown in Table 1 under the condition that two incandescent lights are added to the fluorescent lights for making the environment near the actual game hall. As a result, the luminosity value on the game field was set to 450-800 lux. In the ID detection experiment, five robots have been placed at the free kick markers and at the center on the field, and each robot's ID was recognized 10,000 times under the condition that the robots do not move. If the distance between the recognized position of robot and its planned position is less than 10cm, the recognition is succeeded, otherwise failed.

The accuracy of angle detection is experimented on nine points on the field and the orientation of the robot was set to 0, 45, 90 and 135 degrees at each setting point. It was measured 1,000 times at each point, and the maximum angle error was calculated from the real position. There is no weak point (place) on the field to detect ID's and submarkers.

The recognition rates for the objects, ball, team color marker, submarkers, are evaluated during a real game. Although the intensity on the field is determined as 700–1000 lux by the regulation, it varies by the lighting condition. From some practical games, it is confirmed that our image processing system could extract all objects under the condition that the intensity is about 200 lux.

In our image processing system, the whole image was searched only when (1) just after the game started, (2) ball has been occluded completely for some period, and (3) shoot speed exceeds 5m/sec. Even though it takes long time for the whole image search, it is confirmed that it works in 20msec/frame. So that, our system could defend a fast ball even if it was shot from the region occluded by a robot.

Table 1. Experimental results of image processing performance

Experimental item	Planned	Result
ID recognition ratio in a static mode [%]	100.00	99.68
ID recognition ratio in a dynamic mode [%]	100.00	99.61
Maximum deviation for the position [cm]	0.50	0.097
Maximum deviation for the angle [deg]	1.00	0.78
Processing time [msec]	16.7	1.33

6 Conclusions

This paper discussed the algorithms, the robot mechanism and image processing method to realize cooperative play in RoboCup small-size league. Key features in our system are simple pass algorithms, implementation of dribbling and kicking devices, and high speed and robust image processing method. In particular, the visual feedback system of every 1/60 seconds realized high speed tracking to the ball moving at 3.0m/sec. From the viewpoint of image processing, if the light includes wide spectrum like natural light, more robust method will be required, because color blur appears around the boundary of the object to be detected by optical color aberration. The light with too wide dynamic range like a spot light bothers our image processing system, so it is also required to be more robust for the change of the intensity, exceeding the value specified in the regulation. From the view point of mechanism, further analysis and modeling of catching and shock absorption are expected. These are future works.

Acknowledgement

This paper was partially supported by The Hibi Research Grant, AI Research Promotion Foundation and the RoboCup Japanese Committee grant.

References

1. http://www.robocup2002.org/
2. M.Veloso, E.Pagello, and H.Kitano (Eds.), "RoboCup-99: Robot Soccer WorldCup III", Springer (June 2000)
3. P.Stone, T.Balch and G.Kraetzschmar (Eds.), "RoboCup 2000:Robot Soccer WorldCup IV", Springer (March 2000)
4. A.Brik, S.Coradeschi, and S.Tadokoro (Eds.), "RoboCup 2001:Robot Soccer WorldCup V", Springer (March 2002)
5. G.A.Kaminka, P.U.Lima and R.Rojas (Eds.), "RoboCup 2002:Robot Soccer WorldCup VI", The 2002 International RoboCup Symposium PreProceedings, Fukuoka, (June 2002)
6. Y.Kodama and K.Murakami, "Small-Size Robot Extraction Method by High-speed Image Processing for RoboCup", Proc. of VIEW2002, Yokohama, (Dec.2002) (In Japanese)
7. S.Hibino, Y.Kodama, T.Iida, K.Kato, S.Kondo, K.Murakami, and T.Naruse, "System configuration of RoboDragons team in RoboCup small- size league", Proc. of SI2002, Kobe, (Dec.2002) (In Japanese)
8. Y.Kodama, S.Hibino, K.Murakami, and T.Naruse, "Small Robot Detection by Using Image Processing and its Application to Action Planning and Action Analysis", Proc.of MIRU2002, Vol.1, pp.223-228, Nagoya, (July 2002) (In Japanese)
9. S.Hibino, Y.Kodama, Y.Nagasaka, T.Takahashi, K.Murakami, and T.Naruse, "Fast image processing and flexible path generation system for RoboCup small size league", The 2002 International RoboCup Symposium Pre-Proceedings, pp.45-57, Fukuoka, (June 2002)
10. Bruce, J., Balch, T. and Veloso, M.: "Fast and Inexpensive Color Image Segmentation for Interactive Robots", Proc. of IROS '00, pp. 2061-2066 (2000)

On-Board Vision Using Visual-Servoing for RoboCup F-180 League Mobile Robots

Paul Lee, Tim Dean, Andrew Yap, Dariusz Walter,
Les Kitchen, and Nick Barnes

Department of Computer Science and Software Engineering
The University of Melbourne
Melbourne, VIC 3010, Australia
http://www.cs.mu.oz.au/robocup/2003/F180/
publications/2002/localvis/index.php

Abstract. In the RoboCup F-180 league competition, vision is predominantly provided by an overhead camera which relays a global view of the field. There are inherent disadvantages in utilising this system, particularly the delays associated with the capture, transmission and processing of vision data. To minimise these delays and to equip the robots with greater autonomy, visual servoing on-board the individual robots is proposed. This paper presents evaluation of two visual servoing methods for mobile robots: position-based and image-based servoing. Traditional implementations of image-based servoing have relied on partial pose estimation, negating much of the advantage gained from using this method. This paper will present an alternative implementation of image-based servoing for approaching objects on the ground plane, which disposes of the pose estimation step and fully relies only on image features. To evaluate the suitability of both visual servoing methods to F-180, the task of docking with the ball is used as a basis of the investigation.

1 Introduction

The implementation of vision as a form of feedback for robotic tasks is a major field of research dating back to the 1970s, where much of the early investigation concentrated on pattern recognition problems [6]. Owing to the reduction in hardware costs and the increase in computing power, the focus of vision research has turned to the introduction of visual data into the control loop of a robot. Using visual data within the control loop has been termed *visual servoing.*

The classical approaches to visual servoing are position-based servoing and image-based servoing. In position-based servoing the world pose of the target is estimated from the image, generally based on a geometric model of the object, and a position-based control signal is generated accordingly. In image-based servoing, the motion is controlled directly based on information from the image plane, with the control error signal being the difference between a desired feature vector, and the current feature vector. However, there are problems with image-based servoing such that large undesired motions of the robot can occur, and

D. Polani et al. (Eds.): RoboCup 2003, LNAI 3020, pp. 422–433, 2004.

the object may move out of camera view [10, 12]. To deal with these problems, recent work has examined $2\frac{1}{2}$D visual servoing, where the camera displacement between current and desired positions is estimated in 3D coordinates without the need for a 3D model of the target [8]. Other work emphasises robot path planning in image space to avoid the problems of loosing the object from the field of view [10, 12].

In this paper, we examine visual servoing for on-board control in the RoboCup F-180 League. Currently, vision for most F-180 teams is provided by an overhead camera which relays a global view of the field and all game-play information back to a host computer. The disadvantages of a global vision system are the delays inherent in the capture, transmission and processing of the image by the host, which then issues commands to the robot. By employing vision on-board each individual robot, and using visual servoing to produce a much 'tighter' closed-loop control, decisions can be made by the robot without the intervention of the host.

The possibility of integrating both global and on-board vision provides the advantage of being able to select whichever form of vision is most appropriate for a particular task. For reflex actions such as aiming and shooting at goal, on-board vision would most likely be appropriate. Higher-level actions such as team strategies would be handled by global vision.

In F-180, the environment is well-known so a model of the object is available, facilitating position-based servoing. Further, in dealing with specific tasks such as chasing the ball, the problem is well posed for image-based servoing. We present the algorithms for both these approaches, and demonstrate through real robotic experiments using the University of Melbourne entries to the F-180 League as the platform for experimentation that both approaches are suitable for the F-180 league.

2 Theory

2.1 Position-Based Servoing

In position-based control, the task of positioning the robot is based on

1. extraction of image features during iteration of the control loop, and
2. evaluation of an estimate of the target pose with respect to the camera.

An error signal is determined from the difference between the current and desired pose. This error signal acts as an input to the system control law, as shown in Figure 1. Corke [3] states that the advantage of position-based control is that it neatly separates the computation of the feedback signal from the estimation problems involved in computing the pose from visual data. Corke, Hager and Hutchinson [7] contend that, in general, image-based control is preferable to position-based control due to factors such as positioning accuracy of the system being less sensitive to camera calibration, and the computational advantages which can be gained from a reduced number of transforms. However, Martinet

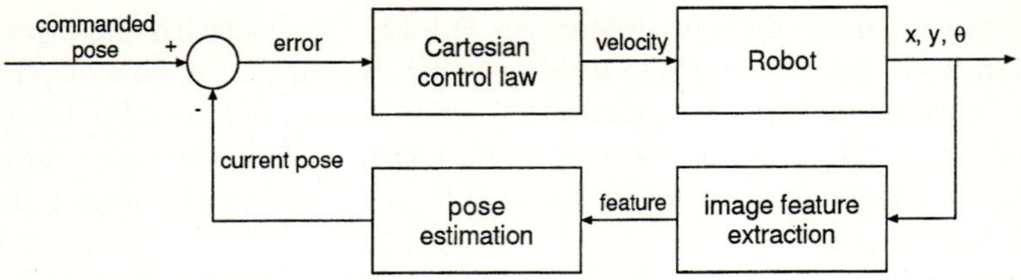

Fig. 1. Position-Based Control System

and Gallice [9] demonstrated that using non-linear state feedback, 3D visual features can still be incorporated into the control loop and achieve performance comparable to image-based servoing.

2.2 Image-Based Servoing

Image-based control consists of specifying the positioning task directly from the image without an estimation of the pose of the target. Feedback is purely from the image plane, and the error signal is computed as the difference between the desired feature vector $\mathbf{f}_\mathrm{d}$ and the current feature vector $\mathbf{f}$. A *feature vector* is a set of visual features such as the coordinates of vertices or the areas of the faces of an object. Elements of the task are therefore specified in the image space rather than the world space. i.e. in pixels rather than Cartesian coordinates.

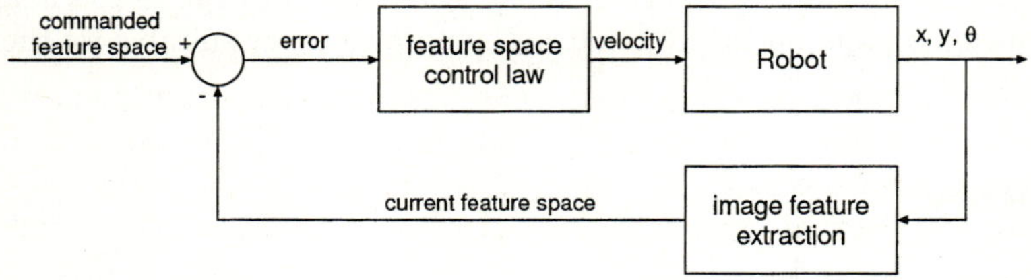

Fig. 2. Image-Based Control System

It is ideal in image-based control to reduce an appropriate error function e such that when the desired position is achieved, e is 0. While the error function is defined in the image parameter space, the input to the robot is in the task space, as shown in Figure 2. Therefore, in order to relate changes in image features to changes in the position of the robot, the concept of an image Jacobian is introduced [3, 4, 7, 11]. The following relationship is then given,

$$\dot{\mathbf{f}} = \mathbf{J}\dot{\mathbf{p}} \tag{1}$$

where $\mathbf{p} = [x, y, z]^T$ is a point in the body frame and $\dot{\mathbf{p}}$ the velocities of this point. As shown by Sharma, Sutanto and Varma [11], the Jacobian is evaluated from

$$\mathbf{J} = \frac{\partial \mathbf{f}}{\partial \mathbf{p}} = \begin{bmatrix} \frac{\partial f_1}{\partial p_1} & \frac{\partial f_1}{\partial p_2} & \cdots & \frac{\partial f_1}{\partial p_n} \\ \frac{\partial f_2}{\partial p_1} & \ddots & & \vdots \\ \vdots & & \ddots & \vdots \\ \frac{\partial f_m}{\partial p_1} & \cdots & \cdots & \frac{\partial f_m}{\partial p_n} \end{bmatrix} \tag{2}$$

Assuming that the image Jacobian is square and non-singular, a simple control law can be determined as

$$\dot{\mathbf{p}} = \mathbf{k}\mathbf{J}^{-1}\left(\mathbf{f}_\mathrm{d} - \mathbf{f}(t)\right) \tag{3}$$

where $\mathbf{k}$ is a diagonal gain matrix which will implement simple proportional control.

Image-based control possesses underlying weaknesses since visual data is interpreted using a constantly refined Jacobian matrix as part of the control loop. Chaumette and Malis [1, 2] exposed the possibility of a local minimum being reached and the image Jacobian being singular during servoing. In order to avoid these problems, $2\frac{1}{2}$D visual servoing combines visual features obtained directly from the image with position-based features [8].

3 Implementation

3.1 Position-Based Servoing

Since an image is two-dimensional, it is difficult to extract three-dimensional (3D) Cartesian coordinates, because depth is unknown. However, the 3D coordinates are required to properly implement the control system. In order to calculate the depth of the centroid of an object, an assumption was made that the object will always be on the ground and of known size, which is reasonable to assume for the ball. Since the camera height and tilt angle were known, the position of any image objects was determined by triangulation. Once the position of the object was determined, a simple proportional and derivative controller was used regulate the motion of the robot to achieve the docking position.

3.2 Image-Based Servoing

For the simple task of docking the robot with a ball, the degrees of freedom considered were translation forwards and backwards, and rotation about the current position of the robot. Translation sideways was deemed redundant, since this can be achieved through a combination of forward motion and rotation.

The pinhole camera model was selected for the perspective projection, and the feature vector $\mathbf{f} = [u, v]^T$ was utilised, where u and v represent the image coordinates of the centroid of the object. Using the coordinate system in Figure 3, the velocity of a point $\mathbf{p}$ was expressed relative to the camera frame as,

$$\dot{x} = z\omega \tag{4}$$

$$\dot{y} = 0 \tag{5}$$

$$\dot{z} = -x\omega + v_x \tag{6}$$

where v_x and ω were the chosen degrees of freedom shown in Figure 4. Note that the camera used for on-board vision had non-unit aspect ratio and was oriented on its side to achieve the greatest possible depth of vision. Thus to reflect the reorientation of the camera, the normal convention of having u and v representing the horizontal and vertical coordinates was changed (see equation 7). As well as viewing the ball when docked with the robot, it was considered more important to be able to perceive objects at a distance, rather than to have a wider viewing range close to the robot. This is reflected in the image coordinate system.

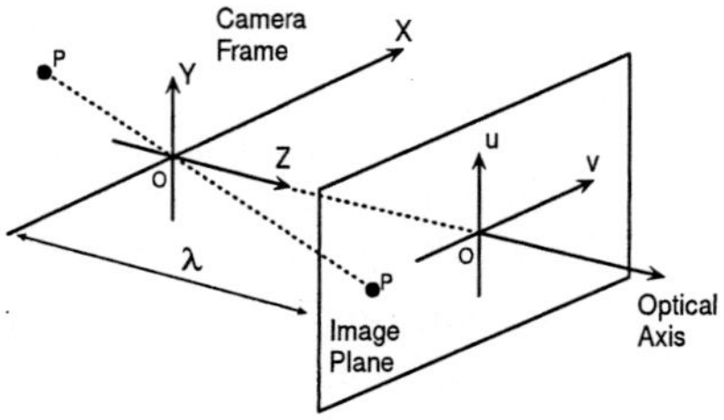

Fig. 3. Perspective projection of a pinhole camera

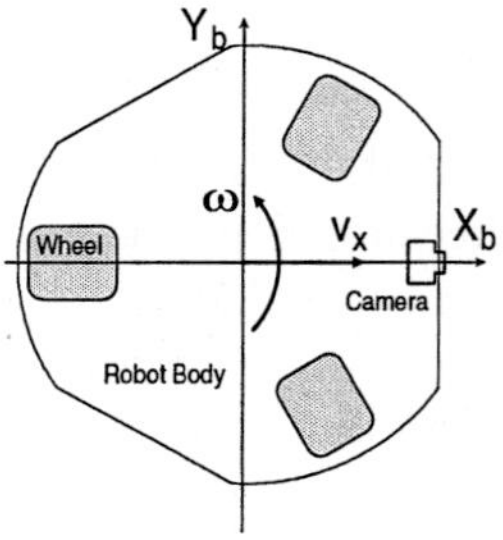

Fig. 4. Top view of configuration of omni-driven Kanga robot

The perspective projections

$$u = \lambda \frac{y}{z} \quad \text{and} \quad v = \lambda \frac{x}{z} \tag{7}$$

were used to express the velocities in (4) in terms of the feature parameters

$$\dot{x} = z\omega \tag{8}$$

$$\dot{y} = 0 \tag{9}$$

$$\dot{z} = -\frac{vz}{\lambda}\omega + v_x \tag{10}$$

Using these relationships the differential change in the image features was

$$\dot{v} = \lambda \frac{z\dot{x} - x\dot{z}}{z^2} \tag{11}$$

$$= -\frac{v}{z} v_x + \left(\frac{\lambda^2 + v^2}{\lambda} \right) \omega \tag{12}$$

$$\dot{u} = \lambda \frac{z\dot{y} - y\dot{z}}{z^2} \tag{13}$$

$$= \frac{uv}{\lambda} \omega - \frac{u}{z} v_x \tag{14}$$

Equations (12) and (14) can be written in matrix form as,

$$\begin{bmatrix} \dot{u} \\ \dot{v} \end{bmatrix} = \begin{bmatrix} -u/z & uv/\lambda \\ -v/z & (\lambda^2 + v^2)/\lambda \end{bmatrix} \begin{bmatrix} v_x \\ \omega \end{bmatrix} = \mathbf{J}\mathbf{v} \tag{15}$$

The relationship between robot velocities and changes in the image is therefore

$$\mathbf{v} = \mathbf{J}^{-1}\dot{\mathbf{f}} = \begin{bmatrix} -z(\lambda^2 + v^2)/\lambda^2 u & vz/\lambda^2 \\ -v/u\lambda & 1/\lambda \end{bmatrix} \dot{\mathbf{f}} \tag{16}$$

A simple proportional controller was used to regulate the motion of the robot for image-based servoing. The aim of this investigation was not to examine the performance of the control system, but rather to focus on the characteristics of the visual-servoing methods. The simplest control system was therefore chosen.

In order to find the inverse Jacobian, $\mathbf{J}$ must be square and non singular, that is, the determinant must be non-zero.

$$\begin{aligned} |\mathbf{J}| &= -u\lambda/z - u \cdot v^2/(z\lambda) + u \cdot v^2/(z\lambda) \\ &= -u\lambda/z \end{aligned} \tag{17}$$

According to Equation (17), the determinant can be zero only if either u is zero, or z is infinite. However, in practice these occurrences were defined as a no-ball case at the image-processing stage and handled prior to the control system. Thus, the Jacobian could be considered as always non-singular. Furthermore, in our experiments the underlying error function was monotonic and no local minima occurred in the control loop.

The derivation given for the image Jacobian considered the camera axis horizontal to the ground. However, for this implementation, the camera was tilted to enable the ball to be seen when at the base of the robot. The main effect of the tilt angle is to introduce a constant scaling factor into the robot control velocities. This scaling factor can be absorbed by the gain factor ($\mathbf{k}$) in the control law. It was found through experimentation that the appropriate selection of the controller gains results in the derived control law being effective even with camera tilt. The non-tilt derivation can therefore be used in the general case.

As discussed by Corke [3], many image-based models still require the depth of the target in the formulation of the image Jacobian. This results in a partial pose estimation, which is the basis of the position-based model. By utilising the image size of the target as an indicator of the relative distance from the camera, Zhang and Ostrowski [12] were able to remove the dependence of the image Jacobian on the depth of the target. However, for the implementation of image-based control used in this investigation, image size was not considered an accurate representation of the depth because the entire object might not be recognised. This was a major problem encountered, as specular reflections and self-shadowing under the F-180 competition overhead lighting conditions caused variable recognition of the ball. In F-180 most dominant teams rely only on overhead cameras, and the lighting design is more suitable for overhead applications. However, the centroid height was determined to be a more robust ball feature when the ball is close, as partial occlusion will only change the distance estimate slightly. Although it does cause larger errors when the ball is far away, the control behaviour will not be significant different compared with when it is close by. Depth was therefore expressed as a function of the chosen image features and the geometry of the camera placement, as shown in Figure 5.

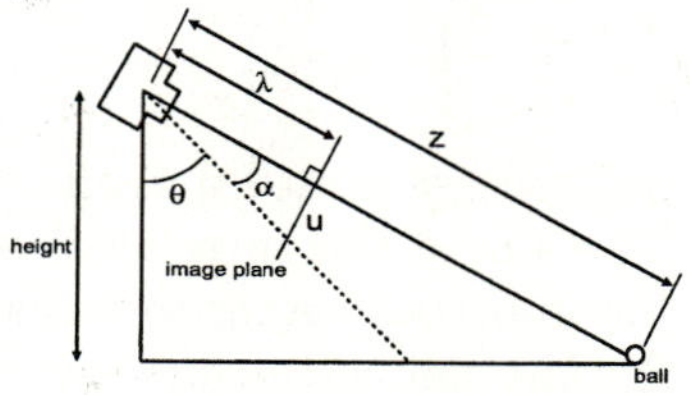

Fig. 5. Camera Placement

Using the geometry of the camera placement shown in Figure 5, the depth z and angle α were evaluated to be

$$z = \frac{h}{\sin\theta\cos\alpha + \cos\theta\sin\alpha} \tag{18}$$

$$\alpha = \arctan\left(\frac{u}{\lambda}\right) \tag{19}$$

Since the angle of elevation of the camera is known, both $\sin\theta$ and $\cos\theta$ were evaluated as constants. Furthermore, using (19) and Pythagoras' Theorem, $\cos\alpha$ and $\sin\alpha$ can be evaluated, allowing z to be determined.

4 Experimental Procedure

Two forms of experiments were designed to analyse the behaviour of each method of visual servoing: docking with a static ball, and docking with a moving ball. Both experiments performed docking with an orange golf ball, as used in RoboCup competition. The purpose of these tests was to observe the effects of system parameters such as frame rate, computation time and calibration.

In the static experiment the ball was placed at three locations in front of the robot: to the left, in the center, and to the right of the robot. The ball was also placed at differing distances from the robot: near, middle and far away from the robot respectively. Figures 6(a) and (b) illustrate where the placement locations were placed relative to the robot. The ruler seen in these images is 1m in length. For the moving target tests, the ball was released from a ramp of fixed height and rolled towards the robot. The ramp allowed reasonable repeatability of tests with a consistent velocity. The robot was required to track varying ball trajectories at 90°, 45°, and almost parallel to the robot's line of vision.

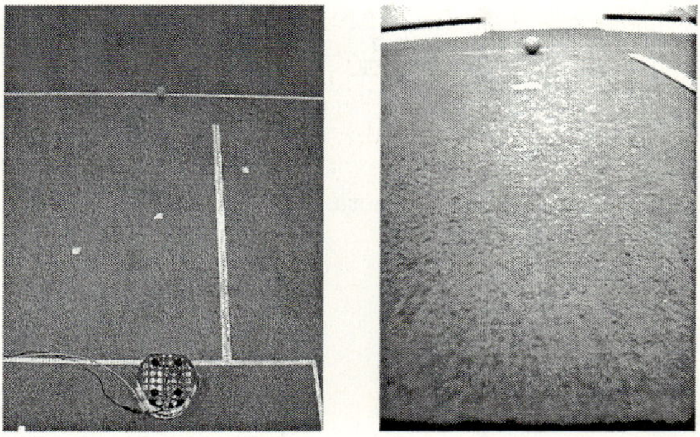

Fig. 6. Placement of the ball and test locations relative to the robot, as seen from **(a)** an overhead view and **(b)** the robot camera

5 Results and Discussion

Two sets of results for each experiment were collected for analysis. The results were of the overhead view of the test and of the on-board camera view. The overhead camera provided a world-space view of the path taken by the ball and the robot as docking occurred. The overhead view was also used to track changes in velocities as the robot approached the docking position. This was achieved by knowing the sampling rate used for the camera. The on-board view was used to observe the changes in ball position within the image frame. This data provided a better understanding of the servoing behaviour of the robot as it attempted to reach the commanded position. Note that the overhead camera was used *only* to record the motion of the robot for evaluation, *not* for servoing, which was done by on-board vision.

A representative sample of the results is given in Figures 7 to 9. Figure 7 presents the tests as seen by the on-board camera. Figures 8 and 9 present the global view of the tests seen by the overhead camera. The overhead plots contain markers at one-second intervals, while the local camera has markers at half-second intervals. The markers for image-based control are depicted by the "o", and for position-based control by the "*" marker. Complete results for all tests are given in [5].

From the results obtained there was no discernible difference in the performance of the two methods. Both servoing implementations were able to track

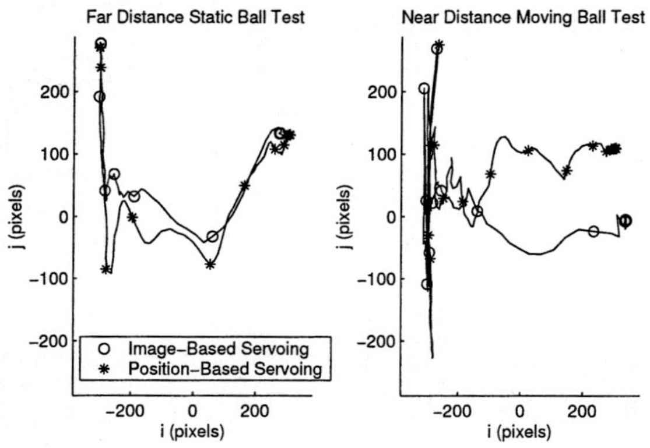

Fig. 7. Local camera view of two tests

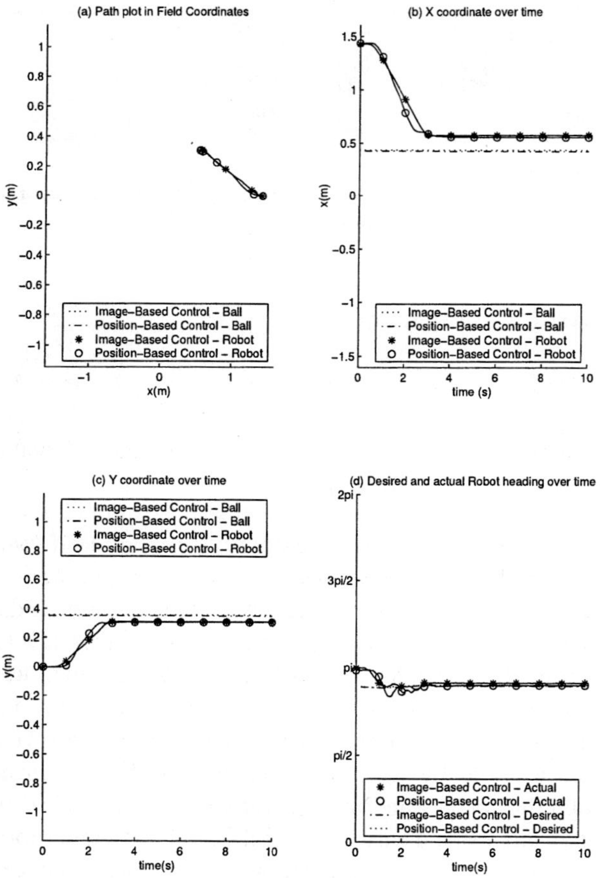

Fig. 8. Overhead camera view of far distance static ball test

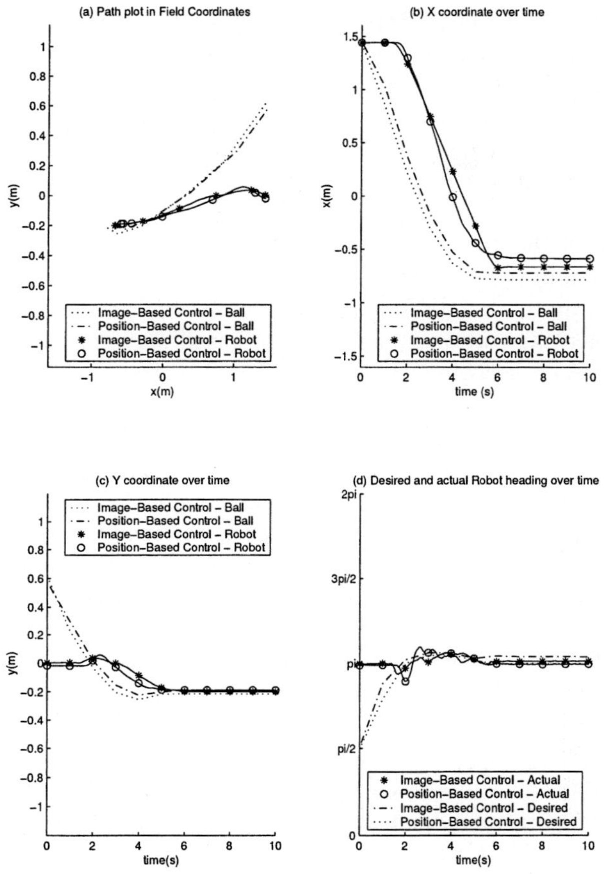

Fig. 9. Overhead camera view of near distance moving ball test

the ball reliably, and were able to successfully approach the docking position. Neither method was able to achieve the exact docking position due to the inability of the robot to move when given low-velocity commands. This was a potential control system problem which could be solved through the implementation of more advanced controllers. However, the simple controllers used for the experiments still demonstrated that visual servoing using either method can be satisfactorily achieved. Furthermore, during operation in the actual F-180 competition it would not be desirable to stop next to the ball, but rather to move to the ball at the highest speed possible in order for it to be picked up on a roller.

It was noticed that for the image-based method that the robot travelled at constant velocities to the docking position, whilst the velocities for the position-based method had a ramping effect. This is attributed to the different controllers used for each implementation. A proportional controller was used for the image-based method and accounted for the constant or linear velocities observed. For the position-based implementation, a derivative controller was added, introducing the ramping effect. The ramping effect was due to the predictive nature of a derivative controller, evidenced by the reduction in the commanded speed as the robot approached the docking position.

The position-based method was a much more intuitive means of visual servoing than the image-based method. Position-based servoing is essentially an extension of image-based servoing by the addition of a pose estimation step. Both methods required the extraction of image features in the feedback loop. The derivation of the Jacobian proved difficult to analyse for errors due to the coupling involved in the evaluation of the velocities. The position-based servoing was therefore a more straightforward method to conceptualise and implement.

In terms of computational costs, image-based servoing was much cheaper than position-based servoing. The removal of the transformation from image space to Cartesian world space allowed computation involving only addition, subtraction, multiplication, and division operations. The additional cost of using tan and arctan operations made position-based control much more computationally expensive. It needs to be considered, though, that when the complexity of the task increases, the computational cost of the image-based method would increase in proportion to the size of the Jacobian matrix. Image-based servoing could therefore potentially be more expensive than position-based methods for complex tasks.

In an application such as visual servoing calibration effects are an important issue to be considered. It was observed that for position-based servoing, external and internal calibration issues such as camera positioning and angle of field-of-view will cause errors in position estimation. However in image-based servoing, the main effect of calibration was from determination of the focal length, an integral step in the development of the image Jacobian. To minimise the erroneous influences of calibration and to achieve reliable servoing, it was therefore important for both servoing methods that the calibration be conducted offline.

6 Conclusion

In this paper, it has been shown that it is possible for visual servoing to be performed on board the F-180 robots under RoboCup conditions. The two classical methods of position-based and image-based servoing both satisfactorily achieved the task of docking a robot with static and moving targets. In conducting this investigation it was shown that both methods had advantages and disadvantages over each other and neither proved to be a superior implementation. As such, both are equally suitable to be used in the F-180 League, and the decision on using either method would be based on which features of position-based and image-based servoing are more desirable.

References

1. F Chaumette and E. Malis. $2\frac{1}{2}$D visual servoing: A possible solution to improve image-based and position-based visual servoing. In *Proc. IEEE Int. Conf. on Robotics and Automation*, volume 1, pages 630–635, 2002.

2. François Chaumette. Potential problems of stability and convergence in image-based and position-based visual servoing. In *The Confluence of Vision and Control*, number 237 in LNCIS Series, pages 66–78. Springer-Verlag, 1998.
3. P I Corke. *Visual Control of Robots: High Performance Visual Servoing*. Research Studies Press Ltd., Great Britain, 1996.
4. P I Corke and S A Hutchinson. A new hybrid image-based visual servo control scheme. In *Proc. IEEE Conf. on Decision and Control*, volume 3, pages 2521–2526, 2000.
5. T Dean, P Lee, and A Yap. Mechatronics research project: Local vision for small league robots – technical report. Technical report, The University of Melbourne, 2002. Available: `http://www.cs.mu.oz.au/robocup/2003/F180/publications/2002/localvis/Documents/Technical%20Report%202.15.pdf`.
6. B Espiau, F Chaumette, and P Rives. A new approach to visual servoing robotics. *IEEE Transactions on Robotics and Automation*, 8(3):313–326, 1992.
7. S Hutchinson, G Hager, and P I Corke. A tutorial on visual servo control. Technical report, CSIRO, Australia, 1996. Available: `http://www.cat.csiro.au/cmst/staff/pic/vservo.htm`.
8. E Malis, F Chaumette, and S Boudet. $2\frac{1}{2}$D visual servoing. *IEEE Transactions on Robotics and Automation*, 15(2):238–250, Apr. 1999.
9. P Martinet and J Gallice. Position based visual servoing using a non-linear approach. In *Proc. IEEE/RSJ Int. Conf. on Intelligent Robots and Systems*, volume 1, pages 531–536, 1999.
10. Y Mezouar and F Chaumette. Path planning in image space for robust visual servoing. In *Proc. Int. Conf. on Robotics and Automation*, pages 2759–2764, 2000.
11. H Sutanto, V Varma, and R Sharma. Image based autodocking without calibration. In *Proc. IEEE Int. Conf. on Robotics and Automation*, volume 2, pages 974–979, 1997.
12. H Zhang and J P Ostrowski. Visual motion planning for mobile robots. *IEEE Transactions on Robotics and Automation*, 18(2):199–208, 2002.

A Plugin-Based Architecture for Simulation in the F2000 League

Alexander Kleiner[1] and Thorsten Buchheim[2]

[1] Institut für Informatik, Universität Freiburg
79110 Freiburg, Germany
kleiner@informatik.uni-freiburg.de

[2] Institute of Parallel and Distributed Systems, Universität Stuttgart
70569 Stuttgart, Germany
buchheim@informatik.uni-stuttgart.de

Abstract. Simulation has become an essential part in the development process of autonomous robotic systems. In the domain of robotics, developers often are confronted with problems like noisy sensor data, hardware malfunctions and scarce or temporarily inoperable hardware resources. A solution to most of the problems can be given by tools which allow the simulation of the application scenario in varying degrees of abstraction and the suppression of unwanted features of the domain (like e.g. sensor noise). The RoboCup scenario of autonomous mobile robots playing soccer is one such domain where the above mentioned problems typically arise.

In this work we will present a flexible simulation platform for the RoboCup F2000 league developed as a joint open source project by the universities of Freiburg [13] and Stuttgart [8] which achieves a maximum degree of modularity by a plugin based architecture and allows teams to easily develop and share software modules for the simulation of different sensors, kinematics and even complete player behaviors.

Moreover we show how plugins can be utilized to implement benchmark tasks for multi robot learning and give an example that demonstrates how the generic plugin approach can be extended towards the implementation of hardware independent algorithms for robot localization.

1 Introduction

Simulation has become an essential part in the development process of autonomous robotic systems. In the domain of robotics, developers often are confronted with problems like noisy sensor data, hardware malfunctions and scarce or temporarily inoperable hardware resources. A solution to most of the problems can be given by tools which allow the simulation of the application scenario in varying degrees of abstraction and the suppression of unwanted features of the domain (like e.g. sensor noise).

The RoboCup scenario of autonomous mobile robots playing soccer is one such domain where the above mentioned problems typically arise. In contrast to the the simulation league [6] where there is a predefined type of player with fixed action and perception capabilities in the simulated environment, the F2000 league permits a great diversity of robots in shape, kinematics and sensorics. Typically, teams in the F2000

D. Polani et al. (Eds.): RoboCup 2003, LNAI 3020, pp. 434–445, 2004.

league use multi sensor approaches to gather relevant information about their environment for the task of playing soccer. Here, all kind of local sensors, like ultrasonic or infrared sensors, laser range finders, cameras or omni-vision sensors, may be used in arbitrary combinations on the robot.

Existing simulation tools for the F2000 league usually are rather specialized for a certain robot architecture of one team, restricted to a certain software language or deeply interwoven within the team's software structure [1] . Even if a certain adaptability to different kinds of robot setups or robot control properties partially was considered within some software designs [2], usually a lot of work still has to be spent to adapt the software to the individual needs of one team without any general benefit for other teams.

In this paper we present a flexible simulation platform for the RoboCup F2000 league developed as a joint open source project by the universities of Freiburg [13] and Stuttgart [8]. The simulator is based on a client/server concept that allows the simulation of diverse robots and also the simulation of team play.

The simulator achieves a maximum degree of modularity by a plugin based architecture. Software plugins are well known from the field of software engineering, particularly in the Internet context.They are basically dynamic libraries which extend the functionality of an application while loaded at runtime. We introduce a generic plugin concept for the simulation of the dynamic model and the sensors of an autonomous robot. This concept makes it possible to easily develop and share modular software components which can be re-used in various robot setups of different teams.

Furthermore we introduce plugins for robot behaviors. These are software components which implement a full player behavior in the sense of reading and reacting on sensor data while running as integrated module of the simulation. We show how they can be utilized to implement benchmark tasks for multi robot learning.

Finally we give an example that demonstrates how the generic plugin approach can be extended towards the implementation of hardware independent algorithms for robot localization. The simulator is freely available from our home page [11].

The remainder of this paper is structured as follows. Section 2 gives an overview of the server and section 3 of the client part of the simulator. The concept of plugins which is used by both parts, will be described in more detail in Section 4. In section 5 we sketch some applications like localization and learning and in section 6 we give an outlook to future developments.

2 Server Architecture

The core of the server is a 2D physics simulation that continuously computes the state of the world within discrete time intervals δ_t. At time t, the simulation calculates the subsequent world state for time $t + \delta_t$ by taking into account velocities of objects and their modified internal states, like for instance committed motion commands to a robot.

As indicated in figure 1, the physics simulation is basically influenced by robot objects existing on the server. Robots can either be controlled by clients connecting via the *message board* over a TCP/IP socket or by a *player plugin* which is started within the graphical user interface (*GUI*). In case a player plugin is created by the GUI, a robot is created and assigned to the plugin. If created by the message board, the robot will be controlled by the remote connected client.

To each robot there is a *motion plugin* and a *sensor container* associated. The motion plugin simulates the robot's dynamics, whereas the sensor container aggregates various plugins that implement the simulation of its sensors. The creation of the different plugins is done by a *plugin factory* which detects the available plugins at server startup. A more detailed description of the plugins will be given in section 4.

The individual setup of a robot is defined by a robot configuration file which contains information on its sensor and motion plugins. Furthermore the file includes a description of the robot's geometry and manipulators.

Manipulators are designed as variable parts of the robot geometry which can be switched between a set of states with differing geometric setups. Furthermore manipulator states can contain an additional parameter describing their ability to accelerate the ball which is needed for the definition of kicking devices.

Additionally, the state of the world can be influenced by the *referee module*. The intention of this module is to implement a set of rules restricting the robots actions in some way. At the current stage, however, the referee is limited to count the score, taking the ball back into the game if crossing the field boundary and resetting the game when a goal was scored. The physics simulation and the communication will be described in more detail in the following section.

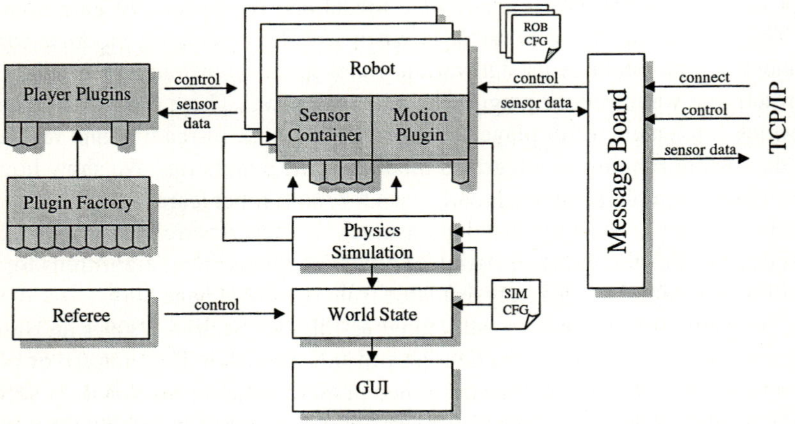

Fig. 1. Overview of the server architecture.

2.1 Physics Simulation

The physics simulation is carried out in the plane. Simulation parameters, such as friction, restitution and cycle time are configurable by the simulation configuration file. The file also defines the setup of the environment which is described by a set of geometric primitives like rectangles, ellipses and triangles which nearly any scenario can be built from. As some domains may require three-dimensional information, each primitive contains two optional parameters specifying their beginning and end within the z-plane. Robots are described by these primitives as well, but within their specific robot configuration file. With a world described by primitives, a simulation of rigid body collisions can be carried out. Unfortunately at the beginning of the simulator project there was no open source package for this simulation available. Therefore we implemented calculations based on "Dynamics", an old but enlightening book from Pestel and Thomson [10].

Given two objects with centers of mass $C_1 = (x_1, y_1), C_2 = (x_2, y_2)$, the velocities $v_{1x}, v_{1y}, v_{2x}, v_{2y}$, rotations ω_1, ω_2 and masses m_1, m_2 that collided in A, compute the velocities $c_{1x}, c_{1y}, c_{2x}, c_{2y}$ and rotations Ω_1, Ω_2 after the collision. For a solution, we need six equations to determine the six unknowns. By the law of conservation of momentum along each space axis we get:

$$m_1 v_{1x} + m_2 v_{2x} = m_1 c_{1x} + m_2 c_{2x} \tag{1}$$

$$m_1 v_{1y} + m_2 v_{2y} = m_1 c_{1y} + m_2 c_{2y} \tag{2}$$

The angular momentum of each body with respect to the origin is conserved. Due to the restriction that the simulation is carried out in the plane the angular momentum is one-dimensional and because there are two bodies, this results in two equations:

$$\omega_{1z} I_1 + m_1 (x_1 v_{1y} - y_1 v_{1x}) = \Omega_{1z} I_1 + m_1 (x_1 c_{1y} - y_1 c_{1x}) \tag{3}$$

$$\omega_{2z} I_2 + m_2 (x_2 v_{2y} - y_2 v_{2x}) = \Omega_{2z} I_2 + m_2 (x_2 c_{2y} - y_2 c_{2x}) \tag{4}$$

where I_1 and I_2 denote the moment of inertia of the two bodies. After Newton's hypothesis the ratio of the relative velocity in the X direction (if we assume the impact is in X direction) before and after the collision equals a constant $-e$, where e is the elasticity (also known as restitution):

$$c_{1x} - c_{2x} + \Omega_{1z} y_1 - \Omega_{2z} y_2 = -e (v_{1x} - v_{2x} + \omega_{1z} y_1 - \omega_{2z} y_2) \tag{5}$$

In order to get the sixth formula, we have to make assumptions about the friction between the two bodies and thus about forces in the tangent plane. In the simulator so far, we use a simple solution that interpolates between the two extremes "stickiness" and "no friction" , which has been introduced by John Henckel[5]. The shown calculation can be generalized for the three dimensional case with little effort. If there are collisions of more than two bodies at the same time, the computation seems not to be straight forward and hence is currently treated on a higher level of the simulation.

One difficulty in simulation is to determine the exact moment of collision. Usually when a collision is detected the two objects involved overlap by some degree due to a limited time scale resolution of the simulation. Thus we implemented a divide-and-conquer algorithm that searches, if an impact has been detected, until a minimal time resolution ϵ for the time of the event. So far, the accuracy of the collision detection suffices for a simulation in real-time and up to five times faster when executed on a Pentium3 600MHz computer.

2.2 Communication

Robots can be controlled by clients connecting via a TCP/IP socket to the message board. This has the advantage that clients can be programmed in any kind of language,

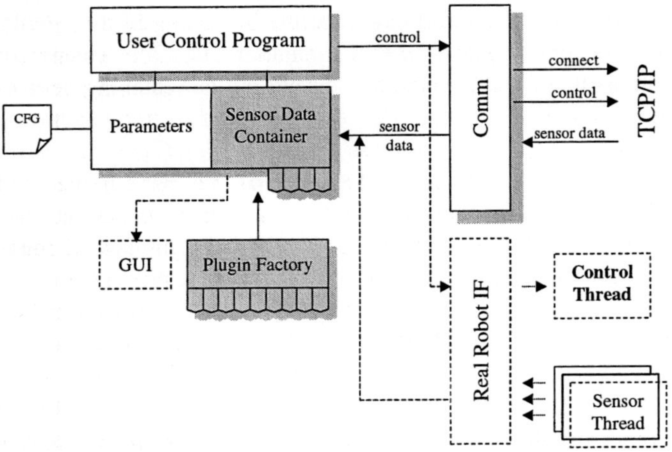

Fig. 2. Overview of the client architecture. Dashed components are optional.

as long as they comply with the protocol. The protocol for commands from the clients to the server is based on ASCII strings that start with a unique command identifier followed by a list of parameters. Due to the generic plugin architecture, the protocol between server and client does not prescribe a format for the exchange of sensor data. Thus, plugins have to implement two abstract methods, one for reading and one for writing an array of *unsigned integers* respectively. This array, which represents the sensor's current data, is then transmitted together with a unique sensor identifier and a timestamp from the server to the client. On the client side, the received array and timestamp are written into the appropriate sensor plugin.

Since computer clocks are usually running asynchronously, we had to integrate a time synchronization mechanism between server and client. This is done by exchanging timestamps between server and client during the connection phase which are used to calculate a time offset value. When transmitting sensor data to a client this offset is added to the timestamp of the package.

The time duration until the world state is completely transmitted depends strongly on the amount of data the client's sensor plugins requests. We measured on our internal network for a client requesting 180 beams of a Laser Range Finder (LRF) sensor, distance and angle to all lines, poles, and goals on the field, and odometry information, a duration between $2ms$ and $3ms$. The latency of the server, which depends on the number of simultaneously connected clients, has been measured between $6ms$ and $8ms$ in the case of 3 actively operating clients.

3 Client Architecture

Figure 2 shows an overview of the client architecture. The client consists of a module that manages the communication with the server (*Comm*), a module for parsing and accessing information from the robot configuration file (*Parameters*) and a *sensor container* holding sensor plugins which are created by the *plugin factory*.

The *Parameters* module parses the client's robot configuration file from which information is used for creating the respective sensor plugins, for example the type of feature the sensor plugin provides. The file is also transmitted to the server during the server connection phase and used there for the creation of an equivalent robot object in the simulation.

Sensor plugins created on the client side are equal to plugins created on the server, but reduced of the ability to generate data. They can be considered as data buffers which are filled by the communication module. As indicated by the dashed modules in the figure, one could also fill the container using threads that process information from real sensors. This is particularly useful when designing high-level algorithms based on the sensor container concept which work on the generated features. In this case the algorithms can easily be evaluated both in the simulation and on real robots.

The sensor plugins are automatically generated and inserted into the sensor container which provides functionality to search for and access specific sensor data objects by certain characteristics. Within the container, sensor plugins are distinguished by three identifiers:

1. The specific model of the sensor, for example DFW500[1] or LMS200[2]
2. The type of the sensor, for example camera or LRF
3. The name of the particular feature, for example *ballDistance* or *180DegreeScan*

In order to address a particular plugin, one can query the container by providing one or more of the above identifiers. This has the advantage that features can be addressed in a more abstract way by their functionality rather than by a specific sensor name. Thus plugins that provide equal features, e.g. *distanceToGoal* from a vision or LRF plugin, can be exchanged in the container without direct consequences for the user program.

Furthermore the client package provides an optional GUI module that can be used to display information from the sensors or additional information generated by algorithms on a higher level.

4 Plugin Architecture

The plugin concept was chosen since it offers the highest degree of flexibility and extensibility while maintaining an independent and stable simulator core. The plugin concept evolved in stages with the increasing need for a stronger modularity of certain parts of the simulator until reaching the current state which, we hope, meets most of the expectations and requirements for the teams of the F2000 league. Figure 3 shows the plugin architecture for the sensorics and motion simulation which will be described in the following.

Sensor Plugins. Usually the biggest discrepancies between robots of different teams apart from the robot chassis are the sensors used on the robots and the data representation used by the data processing modules, like differing coordinate frames or a camera

[1] Firewire camera used by the University of Freiburg

[2] SICK laser range finder used by the RoboCup teams of Freiburg and Stuttgart

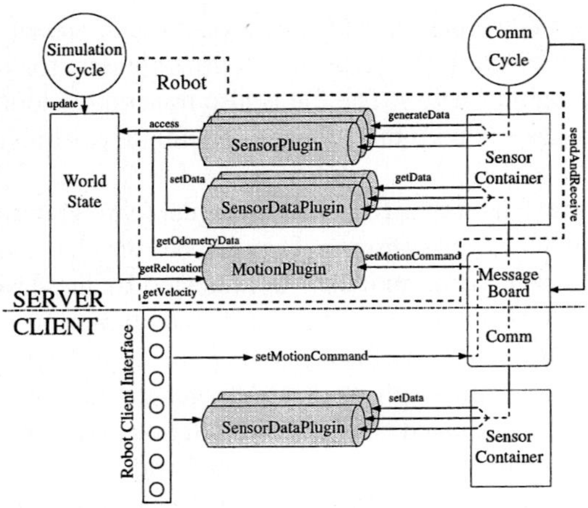

Fig. 3. Server/client view of the sensor and motion plugin concept.

image data format needed by a specific color segmentation algorithm. Consequently the sensorics play a central role within the plugin architecture of the simulator.

Although sensor plugins were referred to as one plugin type in the previous chapters for the sake of simplicity, the sensor concept consists of two separate plugin types, one for data generation and one for data storage. *Sensor plugins* access the current world state of the simulator to generate data of any desired form. This data is stored within the second plugin type named *sensor data plugin* which is needed for a transparent communication between server and client as well as for a data buffering on the client side. One functionality of the *sensor data plugins* is the transformation between their specific data representation and the data format required by the communication module.

On an update request from the client, *sensor data plugin* objects on the server side are filled by the *sensor plugins*, transformed to the data transmission format, transferred to the client and re-transformed to the original data format by the corresponding *sensor data plugin* objects on the client side. One major advantage of this concept is that sensor data processing can either be done on the client or on the server side as both share the same view on the sensor data. This can be extremely useful when the amount of data grows too large to transmit it to a client in a reasonable amount of time like e.g. a rendered high resolution camera image.

To build a new *sensor plugin* basically one method *generateData()* has to be implemented by the developer. Within this method the position of the sensor itself as well as information about all other objects concerning position, color and geometric shape are accessible by the framework within an absolute coordinate frame. This information can then be used to create the data as usually provided by the real sensor. Optionally there are auxiliary functions for simulating sensor noise or for two dimensional ray tracing. Until now there exist several sensor plugins for different kinds of sensors based on the sensor equipment used by the teams of Freiburg and Stuttgart:

– **odometry sensor** providing the current translational and rotational velocities and a position within the odometry coordinate frame.

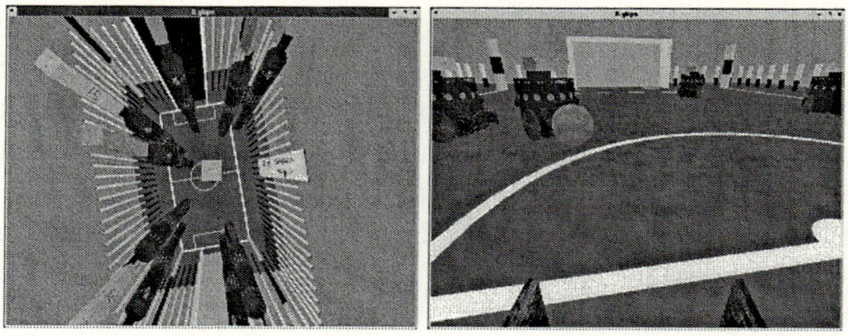

(a) (b)

Fig. 4. Two screenshots from the plugin for generating artificial camera images: (a) an omnidirectional view as seen by a perfect warp free omnidirectional camera using an isometric mirror shape as proposed by [9]. (b) a conventional camera perspective.

- **camera for iconic data** of ball, goal and corner posts within its field of view
- **iconic omnivision camera** detecting field lines and goals, corner posts and the ball in a 360° field of view
- **laser range finder** providing either raw distance data obtained by ray tracing or iconic information about obstacles and an absolute position within the field
- **ultrasonic sensor** detecting the nearest obstacle within the opening angle of the sensor
- **camera image by 3D scene reconstruction** delivers a 3D rendered view of a commonly used 2D camera or an unwarped omnivision sensor as shown in Figure 4.

Motion Plugins. The integration of different kinematics and controller types is done by a further plugin type of the architecture named *motion plugin*. Within the simulator a robot's motion is represented by a change of its pose (relocation) (x, y, θ) within a certain interval of time. This relocation of the robot is requested from the motion plugin within each update cycle of the simulation to update the robot's pose in the environment.

For the physics simulation the motion plugins must provide further information about its current velocity and the rotational speed needed for the collision calculation.

To set motion commands motion plugins can either be addressed by a set of standard commands, such as *setRotationalVelocity* or *setTranslationalVelocity*, or by a n-dimensional vector. The latter allows to address more complex motion models.

Since a motion model may require to send data about its internal state back to the client the motion plugin interface provides one further method to retrieve motion data that is automatically invoked by the default odometry sensor plugin of the server which makes the data available on the client side as well.

Currently there are three different motion plugins available. One model is based on a neural network that basically predicts the robot's state (x, y, θ, v_r, v_t) at t_1 depending on the previous state and velocity settings (v_t^{set}, v_r^{set}) at t_0. The plugin was trained for Freiburg'a Pioneer hardware base, but can easily be trained for other platforms by a small application. A similar method has also been used for state prediction by the Agilo RoboCuppers [3, 2]. A second model is a mathematical calculation of a two wheel differential drive based on a a linear acceleration assumption and parameter tuned to map the behavior of the Nomadic Scout hardware base of Stuttgart's CoPs team. A holonomic three wheel robot base was contributed by the University of Dortmund.

Player Plugins. After various experiments and applications of the simulation server, we found it a useful feature to facilitate the implementation of simple robot behaviors that run within the simulator without the overhead of a remote connected client. This turned out to be useful to simulate opponent players like e.g. goalkeepers or for gathering simulated robot sensor data to create an environment map which e.g. was used for a Monte Carlo localization approach. For this kind of application the framework offers a third type of so called *player plugins*.

Each player plugin controls one robot which is created based on its configuration file that is specified within the plugin. Its core functionality is implemented within a method *senseAndAct* that retrieves sensor information from the sensor data plugins and determines an action which usually is a command to the motion plugin or a triggering of its robot manipulators. Alternatively a player plugin can access the world state of the simulator directly and manipulate it if needed. This may either be useful when implementing a perfect player behavior without limitations by its sensors or when certain environment situations shall be created artificially.

To model coordinated or cooperative group behaviors of different player plugins the simulator provides a common communication channel for message broadcasts among player plugins. Each player can thus broadcast and receive messages from other player plugins to synchronize group actions.

Currently there are two player plugins, a goalkeeper plugin and a plugin for generating a sensor expectation model which will be described in the following chapter.

5 Applications

5.1 A Benchmark for Multiagent Learning

The proposed architecture comes with two crucial advantages that makes it valuable for robot learning: Firstly, customized sensor and motion plugins allow for different kinds of robots to be used within the same learning environment. Secondly, the usage of player plugins on the server offers a modular way of designing and distributing benchmark tasks while keeping the same setup.

Within the simulator framework, algorithms can be designed for learning on either the client or server side. In both cases, the algorithm's input is taken from a sensor data container and the output is a command to a motion plugin. Due to the abstract interfaces it is possible that an algorithm can be applied to different robot types. As already discussed in section 3, sensors are addressable by their specific model, their type or the name of a feature they provide. By addressing the feature directly, e.g. *DistanceToGoal*, the algorithm can be executed, regardless by which sensor the data was produced. Note that this works for plugins that provide features with equal identifiers and equal data format.

For the evaluation of Multiagent learning algorithms we introduced the player plugins. They allow to build unique benchmark situations that are defined by a set of robots behaving in a particular way. One plugin, which is included in the simulator package, implements the control of a goalkeeper and can be started with different policies. The plugin can be used to evaluate the performance of an opponent (or a team) that learns policies to score against the defending goalkeeper with a specific skill level. The goalkeeper plugin can be started with the policies *randomly moving, moving by a pattern, blocking the ball, intercepting the ball* and *mixed policy*. The benchmark has recently been introduced within the RoboCup Special Interest Group (SIG) on Multiagent Learning.

Our own experiences with learning a complex control task have shown that behaviors learned in the simulation can be executed successfully on a real robot as well [7]. Even tricky tasks, such as dribbling and approaching the ball, were managed well by the pre-trained robots that have been trained within games against hand-coded opponents or against themselves. Also the team from the University of Stuttgart uses the simulation for learning skills.

5.2 Plugin Based Localization

In this section we give an example that demonstrates how the plugin approach can be extended towards hardware independent implementations of robot localization algorithms. We extended the sensor plugin interface for the utilization of the well known Monte Carlo Localization method (MCL)[12]. Since the method has been well documented by several other researchers, we will not discuss it in detail any further here.

To implement MCL, one needs to build the perceptional model $p\left(s \mid l\right)$. In the case of containers, s is a vector of n features, where each feature is stored by a plugin in the container. If we assume independence among the features with respect to the pose, the perceptual model can be calculated by $p\left(s_t \mid l\right) = p\left(s_t^1 \mid l\right) p\left(s_t^2 \mid l\right) ... p\left(s_t^n \mid l\right)$ [4].

We extended the abstract interface of the sensor data plugins by the two methods *learnProbability(l_t)* and *getProbability(l_t)*. These methods are supposed for learning the sensors observation and accessing the perceptual model for the pose l_t respectively. It is defined that in case the sensor has an expectation for the queried pose, *getProbability(l_t)* returns a number between 0 and 1 and -1 otherwise. The interface also provides a method *getProbability($l_t, \mathcal{S}$)* that allows to weight a sample set $\mathcal{S}$ according to the current sensor reading at pose l_t. Note that it is assumed that there was either a call of *generateData* or *setData* beforehand to assure that the latest observations are available inside the plugin. The interface leaves open how the perceptual model has to be represented. We used a common representation that separates the perceptual model in two tables, one for the expectation of each pose and one for the sensor model itself [12]. In this case it is assumed that *learnProbability* is called with noise free data, since the data is stored in the table of expectations. The second table is pre-calculated from a Gaussian distribution that reflects the sensors noise. The return value of *getProbability* is then simply calculated by a nested lookup of the two tables. Figure 5(a) shows the expectation of the features *LRF Beam*, *PoleDistance*, *PoleAngle*, *GoalDistance*, *GoalAngle*, and 5(b) the mixed perceptual model of the latter two for the blue and yellow goal.

The perceptual model can be learned at once for all features by a special player plugin in the simulator. The plugin basically moves the robot to all poses on the field and executes *generateData* and *learnProbability* for all Markov plugins found in the container.

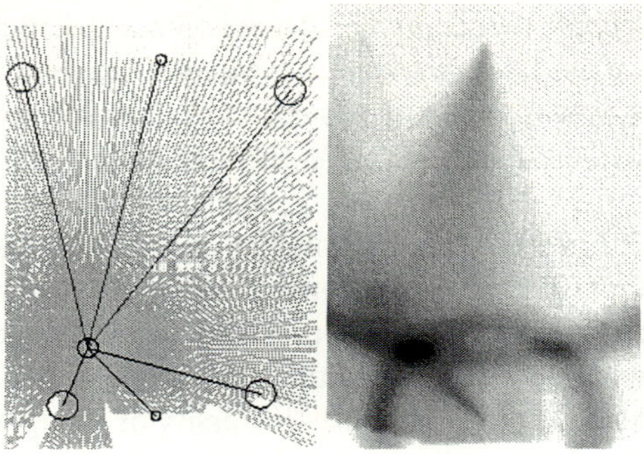

(a) (b)

Fig. 5. Calculating perceptual models: (a) the robots expectation of the features *LRF Beam*, *PoleDistance*, *PoleAngle*, *GoalDistance* and *GoalAngle* (b) the mixed perceptual model of the features *distance* and *angle* to the blue and yellow goal. Darker regions indicates a higher probability of the robot's position. Sensor noise is modeled by a Gaussian distribution with $\sigma = 20mm$ and $\sigma = 5°$ for distances and angles respectively.

During localization, the MCL implementation executes *getProbability* on the container for weighting the sample set representing the current belief. We are planning to develop a similar interface for the motion plugins. By this, the localization algorithm can be designed completely independent of the robot's hardware.

6 Summary and Outlook

In this paper we presented a modular and extendible multi-robot simulation environment for the F2000 league that is suitable for simulating sensors, robot control tasks and cooperative team play. An open plugin architecture has been introduced that permits teams to design their own robot models regarding sensor configuration and motion control. With a growing community of teams using this simulator even full test matches can be held on a regular basis without physical presence of the teams. Furthermore we showed exemplarily, how the simulator framework can be utilized for multiagent learning.

Past experiences in robotics have shown that besides a good idea for e.g. a new localization or feature extraction algorithm the concrete implementation plays an even more important role. Unfortunately, there are only a few implementations freely available and those are in turn tailored for particular hardware or data structures.

One of the key motivation of our work is to foster the exchange of software for autonomous robots. The proposed simulator architecture makes clear how this can be achieved in the case of a robot simulation. Even more beneficial, however, can be the exchange of software components for real robots. In order to accomplish this goal, one has to introduce a level of abstraction and to define standardized methods for accessing this level. We believe that the introduced concept of sensor containers provides also a good level of abstraction on real robots. In the same way, as the proposed sensor plugins generate data inside the simulation, one can implement sensor plugins that run on robots and generate feature data from real sensors. RoboCup teams using the same kind of sensors could then easily share those components.

Furthermore the example shown in section 5.2 demonstrated how algorithms can be designed to operate on generic containers rather than on specific implementations for sensor data processing. Algorithms that are based on generic containers can easily be shared among teams that use the same abstract interface.

The German research project SPP1152-RoboCup is concerned with certain aspects regarding software architecture, system integration and learning. One aim is to find a common description for sensor types and properties as well as robot geometry and physics. We will adapt and contribute our approach to the project and look forward to get one step closer to the goal of a common description of robots and more efficient software development in the RoboCup domain.

References

1. A. Bredenfeld and G. Indiveri. Robot behavior engineering using DD-designer. In *Proceedings of the IEEE/RAS International Conference on Robotics and Automation (ICRA 2001)*. IEEE, 2001.
2. S. Buck, M. Beetz, and T. Schmitt. M-ROSE: A multi robot simulation environment for learning cooperative behavior. In H. Asama, T. Arai, T. Fukuda, and T. Hasegawa, editors, *Distributed Autonomous Robotic Systems*, volume 5, Berlin, Heidelberg, New York, 2002. Springer-Verlag.
3. S. Buck, R. Hanek, M. Klupsch, and T. Schmitt. Agilo robocuppers: Robocup team description. In *RoboCup 2000: Robot Soccer World Cup IV*, Berlin, Heidelberg, New York, 2000. Springer-Verlag.
4. S. Enderle, M. Ritter, D. Fox, S. Sablatnög, G. K. Kraetzschmar, and G. Palm. Vision-Based Localization in Robocup Environments. In Peter Stone, Tucker Balch, and Gerhard K. Kraetzschmar, editors, *RoboCup-2000: Robot Soccer World Cup IV*, volume 2019 of *Lecture Notes in Artificial Intelligence*. Springer Verlag, Berlin, 2001.
5. John Henckel. Simulation of rigid bodies. `http://www.geocities.com/Paris/6502/`
6. H. Kitano, M. Tambe, P. Stone, M. Veloso, S. Coradeschi, E. Osawa, H. Matsubara, I. Noda, and M. Asada. The RoboCup synthetic agent challenge,97. In *International Joint Conference on Artificial Intelligence (IJCAI97)*, 1997.
7. A. Kleiner, M. Dietl, and B. Nebel. Towards a life-long learning soccer agent. In *Proc. Int. RoboCup Symposium '02*. Fukuoka, Japan, 2002.
8. R. Lafrenz, M. Becht, T. Buchheim, P. Burger, G. Hetzel, G. Kindermann, M. Schanz, M. Schulé, and P. Levi. Cops-team description. In A. Birk, S. Coradeschi, and S. Tadokoro, editors, *RoboCup-01: Robot Soccer World Cup V*, pages S. 616 – 619. Springer Verlag, 2002.
9. Carlos Marques and Pedro Lima. A localization method for a soccer robot using a vision-based omni-directional sensor. In *Proceedings of EuRoboCup Workshop 2000*, 2000.
10. E. C. Pestel and W. T. Thomson. *Dynamics*. McGraw-Hill, New York, 1968.
11. Simsrv. A robocup simulator for the F2000 league. `http://www.informatik.uni-freiburg.de/~simsrv`
12. S. Thrun, D. Fox, W. Burgard, and Dellaert. F. Robust monte carlo localization for mobile robots. *Artificial Intelligence*, 128(1-2), 2001.
13. T. Weigel, J.-S. Gutmann, M. Dietl, A. Kleiner, and B. Nebel. CS-Freiburg: Coordinating robots for successful soccer playing. *IEEE Transactions on Robotics and Automation*, 18(5):685–699, 2002.

Development of a Simulator of Environment and Measurement for Autonomous Mobile Robots Considering Camera Characteristics

Kazunori Asanuma[1], Kazunori Umeda[1], Ryuichi Ueda[2], and Tamio Arai[2]

[1] Chuo University, 1-13-27 Kasuga, Bunkyo-ku, Tokyo 112-8551, Japan
{asanuma,umeda}@sensor.mech.chuo-u.ac.jp
http://www.mech.chuo-u.ac.jp/umedalab/

[2] The University of Tokyo, 7-3-1 Hongo, Bunkyo-ku, Tokyo 113-8656, Japan
{ueda,arai}@prince.pe.u-tokyo.ac.jp
http://www.arai.pe.u-tokyo.ac.jp/

Abstract. In this paper, a simulator of environment and measurement that considers camera characteristics is developed mainly for RoboCup four legged robot league. The simulator introduces server/client system, and realizes separation of each robot's information, introduction of each robot's difference and distribution of processes. For producing virtual images, the simulator utilizes OpenGL and considers the effects of blur by lens aberration and so on, random noise on each pixel, lens distortion and delayed exposure for each line of CMOS device. Some experiments show that the simulator imitates the real environment well, and is a useful tool for developing algorithms effectively for real robots.

Keywords: Simulator of Environment and Measurement, Camera Model, Modeling of Measurement Error, Server/Client System

1 Introduction

So as to develop robots or a multiple robots' systems that work in real, dynamic environment, it is necessary to assume a lot of conditions and test the robots or robots' systems for the conditions. However, this requires much cost by much trials and errors. To reduce the cost, simulation is effective. Many kinds of simulators have been developed for various applications, e.g., teleoperation[2], autonomous mobile robot[3]. It is also the case in RoboCup and simulators have been developed [4, 5]. They simulate kinematics or motions, but sensing is not well considered that suffers noises, distortions, etc.

In this paper, we develop a simulator for multiple mobile robots that have the ability to produce images by considering explicitly the characteristics of a camera, and the ability to simulate the real robots' behaviors. The concrete target as the multiple mobile robots system is the RoboCup four legged robot league.

In developing the simulator, we focus on the following features:

D. Polani et al. (Eds.): RoboCup 2003, LNAI 3020, pp. 446–457, 2004.

- programs for real robots can be applied directly,
- multiple robots can share environment and interact with each other,
- images obtained by a real robot can be imitated well.

2 Outline of the Simulator

In the field, a lot of information are obtained from the ball, the landmarks, the robots, etc., and multiple tasks are executed simultaneously in real robots. The simulator is designed so that it produces virtual images precisely for each robot with arbitrary position and orientation, and that the following tasks can be executed virtually in the simulator. Consequently, debugging of programs and verification of algorithms can be effectively performed virtually on PC. Fig.1 illustrates the relation between a real robot and the simulator.

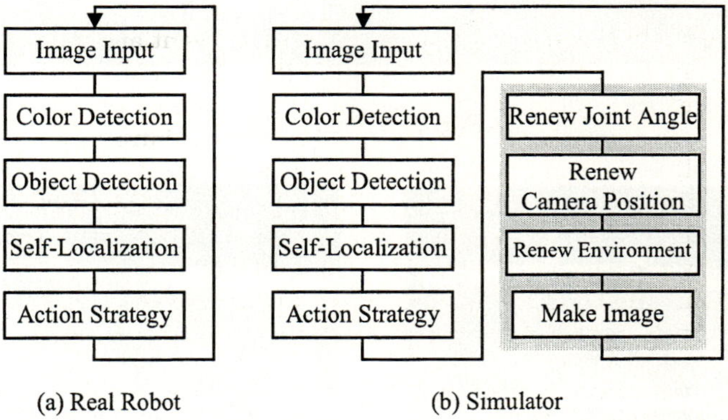

Fig. 1. Tasks in image input cycle of a real robot and of the simulator

Objects in the virtual field that is produced on the simulator are modeled with precise size by using OpenGL[7](see Fig.2.). Each robot on the virtual field produces images by evaluating its camera coordinate that is calculated using the position and orientation of the robot, the elbow angles of its legs, the pan/tilt angle of its neck(see Fig.3(a).). By considering the characteristics of its CMOS camera, the simulator can produce images that imitate real images precisely(see Fig.3(b).). Each robot applies a method of color detection for the synthesized images, and detects 8 kinds of color regions(see Fig.3(c).).

3 Server/Client System

So as to realize the requirement that multiple robots can share environment and interact with each other, server/client system is introduced in the simulator; tasks are performed cooperatively by two kinds of processes, i.e., a server process

Fig. 2. Virtual field

(a) Original OpenGL image

(b) With CMOS filter

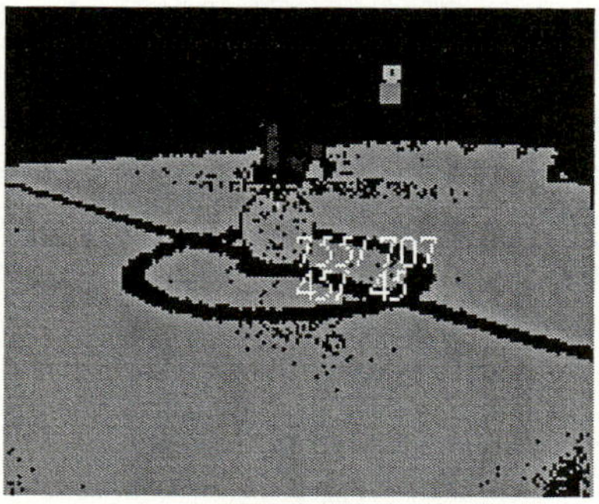

(c) Color extraction for (b)

Fig. 3. Virtual image

and client processes. The server process administrates the environment of the field totally, and the client processes execute robots' behaviors. By introducing the server/client system,

1. separation of each robot's information,
2. realization of each robot's difference,
3. distribution of processes

are realized. By 1., the condition is satisfied that each robot cannot acquire other robots' information explicitly without communication. By 2., robots' behaviors with different algorithms for each team can be implemented and a new algorithm can be easily tested. By 3., each process can be run on a different PC by using TCP/IP protocols, and calculation cost for each PC can be reduced.

Fig. 4 illustrates the structure of the server/client system. Production of a virtual image, analysis of the image, active sensing and simulation of some behavior strategy are performed by clients. Each client program is assigned to the control of each robot. By executing multiple client programs simultaneously and connecting them to the same server, multiple robots in the same field are realized. At the same time, the server administrates the environment on the field; it administrates the connected clients and the ball, recognizes the collisions, etc. Each client can recognize the information on the field through the server. Information is communicated at a constant period among the server and the clients so as to adjust the field environment.

4 Consideration of Camera Characteristics

The images are generated by OpenGL assuming an ideal pinhole camera(see Fig.3(a)), and thus they are different from real images obtained by the CMOS camera on the real robots. Therefore, we consider the camera characteristics of the CMOS camera so as to produce images that closely simulate the real images. The parameters of the ERS-2100's camera are as follows.

- CMOS 1/6inch, 178×144 pixels
- Lens F2.0, f=2.18mm
- Angle of View H:57.6deg, V:47.8deg
- White Balance: 4300K
- Frame Rate: 25fps
- Shatter Speed: 1/200sec

We consider the following characteristics.

- blur by lens aberration, focusing error, etc.
- random noise on each pixel
- lens distortion
- delayed exposure for each line of CMOS device

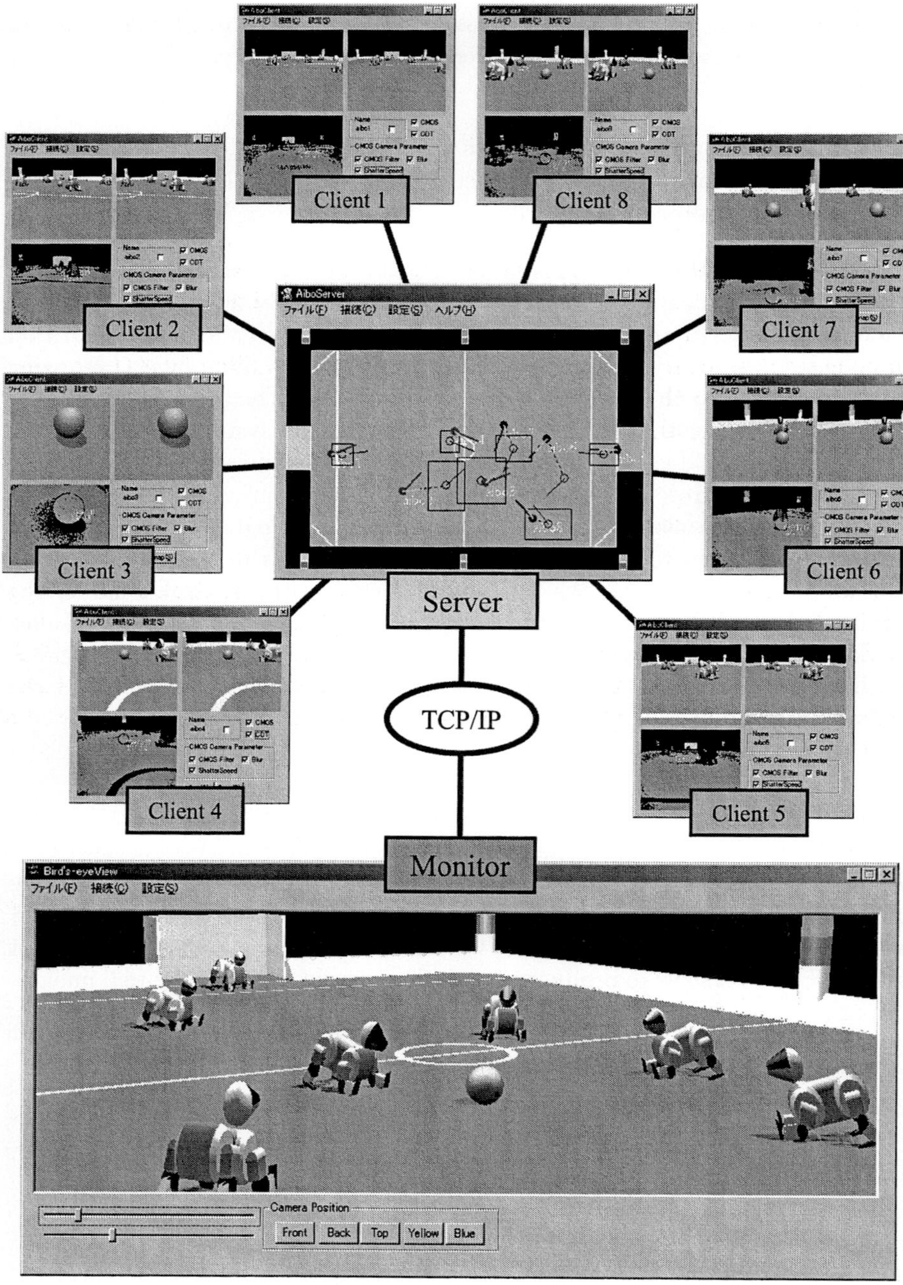

Fig. 4. Server/client system

4.1 Blur by Lens Aberration, etc.

Blur is generated by various lens aberration, focusing error, etc. Supposing the blur of each pixel is approximated by Gaussian distribution, we apply Gaussian filter to the virtual image. For an image $f(u,v)$, two dimensional Gaussian distribution function

$$G(u,v) = \frac{1}{2\pi\sigma^2}\exp(-\frac{u^2+v^2}{2\sigma}) \tag{1}$$

is convoluted, and the blurred image is obtained. In the simulator, the area of convolution is set to 5×5, and σ is set to 1.

Fig.5(a) shows an image by the real camera for the ball with the distance of 1m. Fig.5(b) is the virtual image generated by the simulator on the same condition for Fig.5(a). Fig.5(c),(d) are the close-ups of Fig.5(a),(b). Fig.5(e) is the image generated from Fig.5(d) by applying Gaussian filter. It is shown that Fig.5(e) is closer to the real image, and additionally, aliasing that appears in Fig.5(d) is not observed.

4.2 Random Noise on Each Pixel

Random noise is inevitable, and it is remarkable in principle for CMOS device. The simulator adds the random noise on each pixel with the variance that is evaluated a priori by experiments. Fig.5(f) shows the image generated from Fig.5(d) by adding random noise. Additionally, we consider limb darkening generated by the lens as shown in Fig.5(g). Parameters of limb darkening can be evaluated a priori by experiments.

4.3 Lens Distortion

Images are distorted by lens distortion. As the model of lens distortion, we apply the equation approximating a radial distortion:

$$\begin{bmatrix} \tilde{u} \\ \tilde{v} \end{bmatrix} = \frac{2}{1+\sqrt{1-4\kappa(u^2+v^2)}} \begin{bmatrix} u \\ v \end{bmatrix} \tag{2}$$

to the virtual image, where (u,v) is the coordinates on an original image and $(\tilde{u},\tilde{v})$ is that on the distorted image. κ in eq.(2) can be evaluated by camera calibration. Fig.6 shows an example of the simulation of lens distortion.

4.4 Construction of Virtual CMOS Filter

In order to make the calculation time shorter, we integrate the multiple processes for camera characteristics into one filter, called CMOS filter. By using this filter, the cycle time of producing virtual images became shorter from 208[msec] to 112[msec] by PentiumIII 600MHz. Fig.7 shows the final image by applying the CMOS filter.

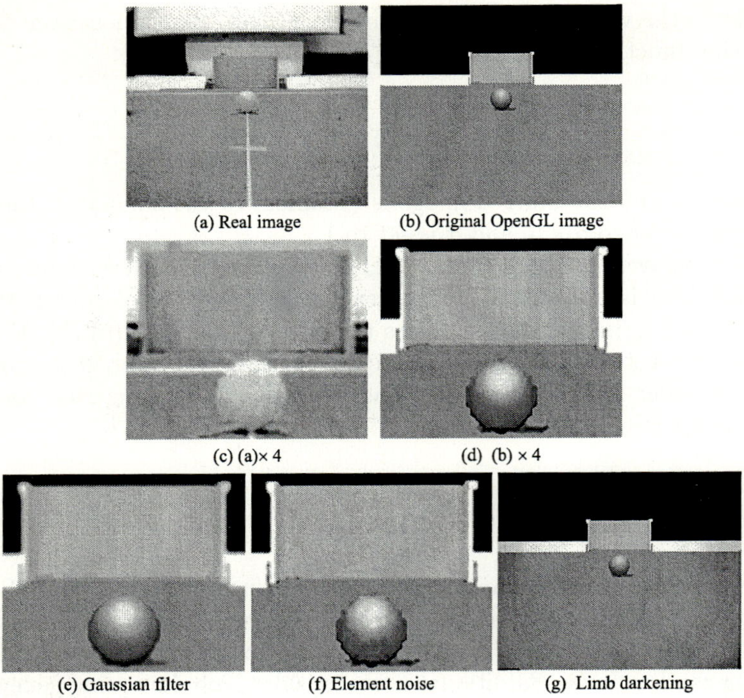

Fig. 5. Consideration of camera characteristics

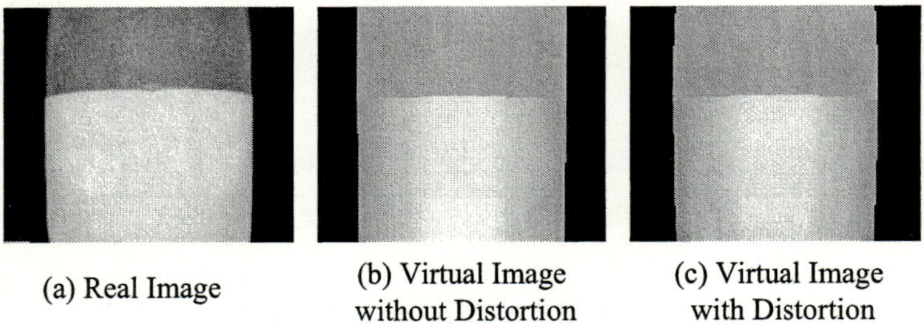

Fig. 6. Lens distortion

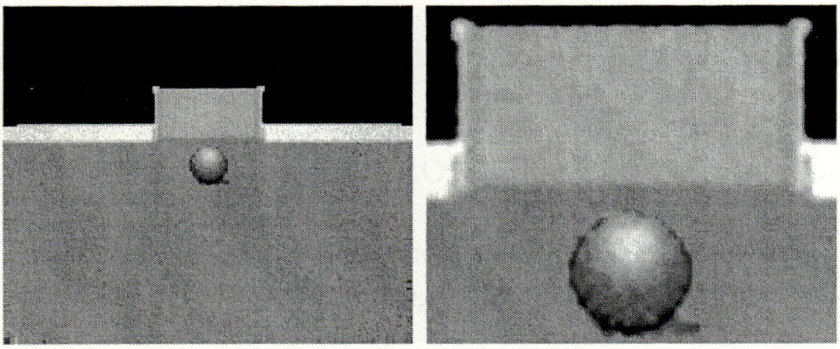

Fig. 7. CMOS filter: total consideration of camera characteristics

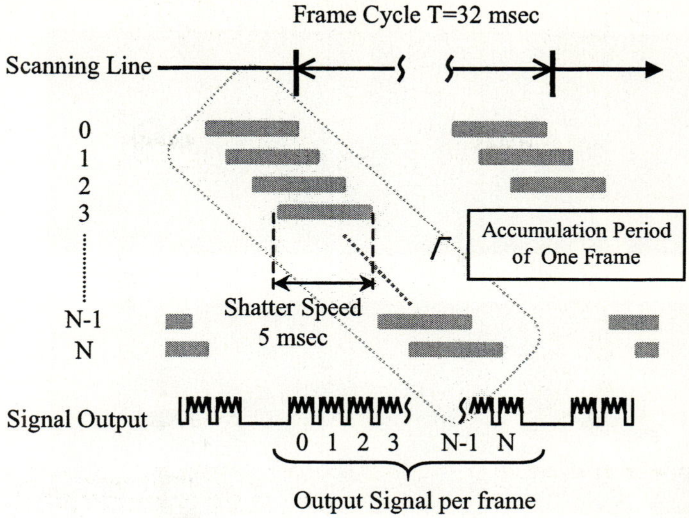

Fig. 8. Line exposure and shutter speed

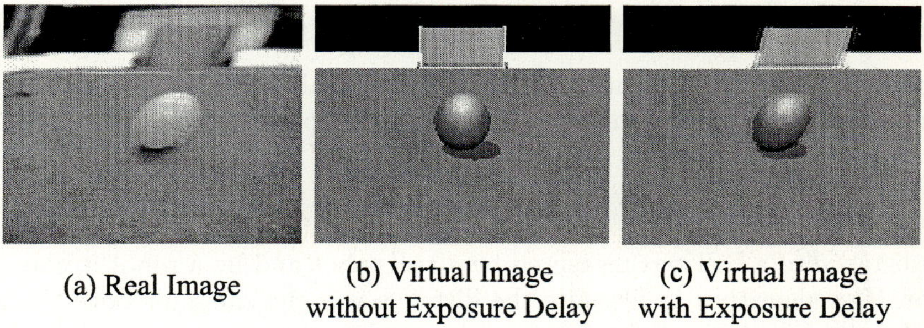

Fig. 9. Exposure delay and shutter speed

4.5 Delayed Exposure for Each Line

For motions of the robots, two more camera characteristics have to be considered. One is the motion blur. This is regardless of the kind of the camera, and depends on the exposure time. The other is specific to CMOS cameras as the ERS-2100's: exposure for each line of the CMOS device is asynchronous. There exists about one frame time delay of the timing of exposure between the first and last line of a CMOS device as illustrated in Fig.8. For ERS-2100, the phenomenon occurs especially when it fast rotates its head. Fig.9(a) shows an example for the velocity of 0.30deg/msec.

5 Evaluation of the Developed Simulator

In this section, we show some experimental results to verify that the proposed simulator well imitates the real environment. One experiment is measurement

Fig. 10. Setting of distance measurement by real robot(upper), virtual robot in the simulator(lower)

of the distance to a ball, and the other is self localization, both of which are essential and important technologies for the robocup games.

5.1 Evaluation for Ball Measurement

The distance to a ball was measured by a real robot and by a robot in the simulator. The algorithm to measure the distance was the same for both the real and the virtual robots: distance was measured from the number of pixels of the ball region, the coordinates of the center of the ball region, edge information, etc. The ball was set from constant distances from the robot as shown in Fig.10, and the distance was measured 100 times and the average and the standard deviation were evaluated for each distance. Fig.11 and Fig.12 show the results of the average and the standard deviation respectively. In Fig.11, the simulator measures almost the same distance as the real robot, especially when the camera characteristics are considered. In Fig.12, when the camera characteristics are not considered, the standard deviations are zero; and when the characteristics are considered, the deviations by the simulator show the same tendency as that by the real robot. Fig.10 shows color detection for distance measurement with/without the consideration. When considered, the color detection seems much more realistic.

5.2 Evaluation for Self-localization

We have been applying Monte Carlo Localization(MCL) that is a kind of Markov Localization methods for self localization of real robots[6]. MCL represents po-

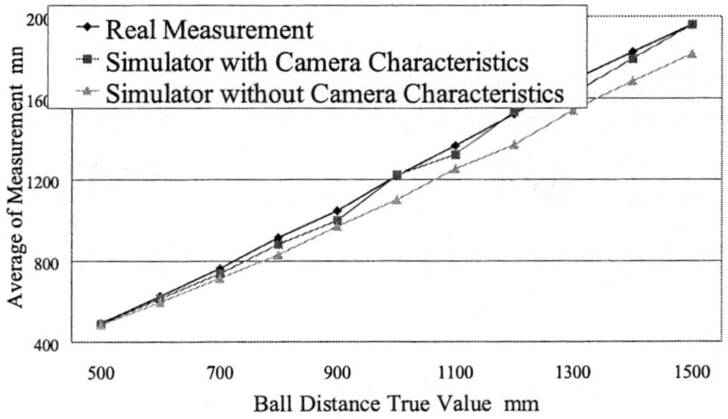

Fig. 11. Comparison of ball distance measurement: average

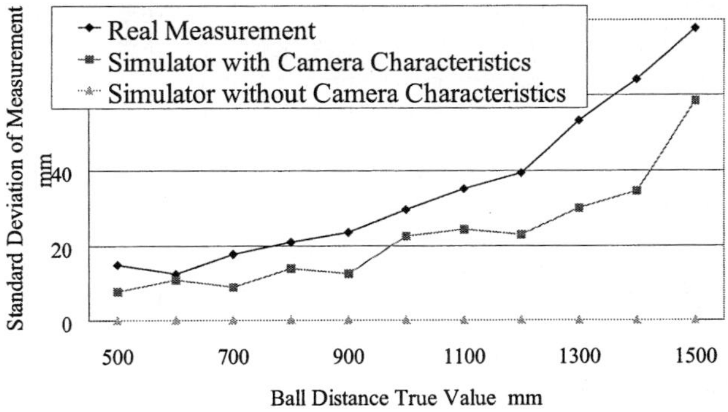

Fig. 12. Comparison of ball distance measurement: standard deviation

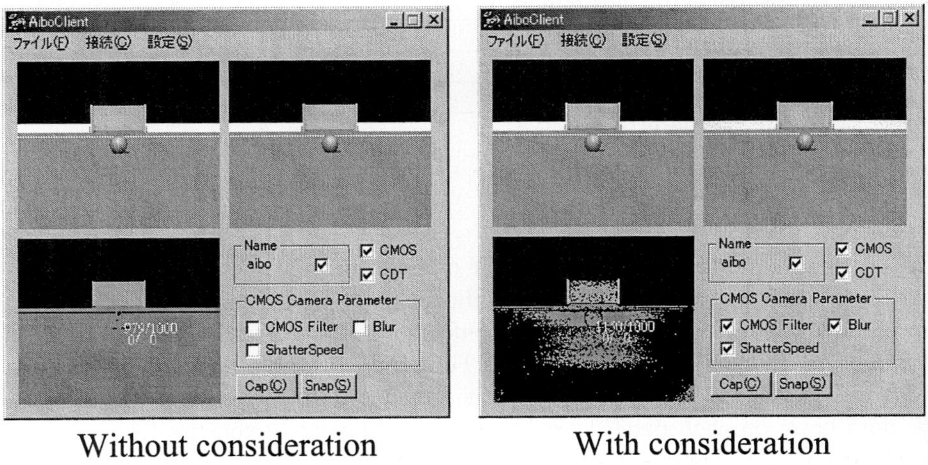

Fig. 13. Color detection for distance measurement by the simulator with/without considering camera characteristics

sition and orientation of each robot by distributed probabilistic distribution of a lot of sample points. Experiments were performed at 36 positions and orientations of a robot for 9 positions (intersection points of x=0, 1000, 1750[mm] and y=0, 500, 1000[mm]) and 4 orientations ($\theta = \pm 45, \pm 135$[deg]). Self localization was performed by rotating the camera for 30[sec]. Table 1 shows the results.

It is shown that the real robot and the simulator obtain similar results for self localization. However, the errors of the simulator are larger than of the real robot. The reason is perhaps that in the simulator, production of images and the cycle of waving head synchronized, and consequently, images had some bias.

Table 1. Comparison of self-localization

(a) Real robot

	Mean squared error	Max. error	Ave. of division width	True value in division
x	186 mm	393 mm	580 mm	61%
y	141 mm	375 mm	395 mm	58%
θ	6.5 deg	16 deg	19 deg	67%
xy	233 mm	393 mm	/	47%
$xy\theta$	/	/	/	47%

(b) Simulation without camera characteristics

	Mean squared error	Max. error	Ave. of division width	True value in division
x	272 mm	731 mm	603 mm	58%
y	218 mm	542 mm	527 mm	61%
θ	9.9 deg	23 deg	39 deg	77%
xy	349 mm	704 mm	/	42%
$xy\theta$	/	/	/	42%

(c) Simulation with camera characteristics

	Mean squared error	Max. error	Ave. of division width	True value in division
x	291 mm	731 mm	570 mm	55%
y	201 mm	564 mm	426 mm	60%
θ	9.4 deg	26 deg	26 deg	66%
xy	368 mm	807 mm	/	45%
$xy\theta$	/	/	/	45%

6 Conclusion

In this paper, we developed a simulator of environment and measurement that considers camera characteristics. The simulator introduced server/client system, and realized separation of each robot's information, introduction of each robot's difference and distribution of processes. For producing virtual images, the simulator utilized OpenGL and considered the effects of blur by lens aberration, etc., random noise on each pixel, lens distortion and delayed exposure for each line of CMOS device. Some experiments show that the simulator imitates the real environment well, and is a useful tool for developing algorithms effectively for multiple mobile robots.

References

1. Hiroaki Kitano, Minoru Asada, Yasuo Kuniyoshi, Itsuki Noda, Eiichi Osawa and Hitoshi Matsubara: "RoboCup: A Challenge Problem for AI and Robotics," Proc. of Robot Soccer World Cup I, pp.1-19, 1997.
2. N. Y. Chong, T. Kotoku, K. Ohba: "Development of a Multi-telerobot System for Remote Collaboration," Proc. IEEE/RSJ Int. Conf. on Intelligent Robots and Systems, pp. 1002-1007, 2000.
3. O. Michel: "Webots: Symbiosis Between Virtual and Real Mobile Robots," Proc. of ICVW '98 Paris France, pp.254-263, 1998.
4. T. Rofer: "An Architecture for a National RoboCup Team," Proc. of Robot Soccer World Cup VI, pp.388-395, 2002.
5. Mark M. Chang, Gordon F. Wyeth: "ViperRoos 2001," Proc. of Robot Soccer World Cup V, pp.607-610, 2001.
6. R. Ueda, T. Fukase, Y. Kobayashi, T. Arai, H. Yuasa and J. Ota: "Uniform Monte Carlo Localization – Fast and Robust Self-localization Method for Mobile Robots," Proc. of ICRA-2002, pp.1353-1358, 2002.
7. J. Neider, T. Davis and M. Woo: OpenGL Proggraming Guide, Addison-Wesley, 1992.

Simulation League: The Next Generation

Marco Kögler and Oliver Obst*

Universität Koblenz-Landau, AI Research Group
Universitätsstr. 1, D–56070 Koblenz, Germany
{koegler,fruit}@uni-koblenz.de

Abstract. We present a modular approach to model multi-agent simulations in 3D environments. Using this approach, we implemented a generic simulator which is totally decoupled from the actual simulation it performs. We believe that for Soccer Simulation League a transition to 3D states exiting new research problems and equally makes it more attractive to watch for spectators. We are proposing to use our framework as basis for a next generation Soccer Server.

1 Introduction

For eight years now Soccer Server [9] exists as a testbed for evaluating multiagent systems and has inspired a lot of researchers from Computer Science and Artificial Intelligence to compare their approaches. Since its beginning Simulation League is confined to two dimensions in order to reduce complexity. During the past years quite a number of advances have been made by the participating teams, and new features were added to the server to provide a more realistic simulation and a tougher challenge[1].

Meanwhile, the server models different types of players with distinct abilities such as speed, size, and stamina. Players can also point into directions, turn "head" and body separately from each other and perform a kind of tackling. A coach can substitute players, analyze the match, and give advice and information to players in a standardized coach language [1]. As rich as the new features appear to be, a number of caveats can be identified:

2D Simulation. A fundamental problem with the current soccer server is that it simulates the game of soccer in a 2D world, making it seem more like a game of table hockey with soccer rules. Nevertheless, the game of soccer is a three-dimensional game. The ball and players can actually leave the ground. It is absolutely necessary to move the simulation into a 3D world in order to accomplish the mission statement of the RoboCup Federation (cited from [6]): *By mid-21st century, a team of fully autonomous humanoid robot soccer players shall win the soccer game, comply with the official rule of the FIFA, against the winner of the most recent World Cup.* Small steps towards a 3D server have already been made [5]. We believe that the transition to a three-dimensional system states exiting new research problems, like for instance in the area of spatial resoning. Also applicability of approaches used in 2D for three-dimensional space has yet to be shown.

* Partially supported by the grant *Fu 263/6-1* from the German research foundation *DFG*.

[1] Thanks to Tom Howard for his great and constant effort in maintaining the Soccer Server.

D. Polani et al. (Eds.): RoboCup 2003, LNAI 3020, pp. 458–469, 2004.

Physics. The physics model used by the current RoboCup soccer server is very limited. Motion of objects is only described using acceleration, velocity and positional vectors. As objects do not have an associated mass, no real forces act upon the objects of the simulation. This also means that other effects, which usually result out of physics have to be tacked onto the simplistic model (e.g. effect of wind, or a ball slowing down). This only increases code complexity and the need for special-casing, resulting in a long-term maintenance problem. A robust physics system would handle all these effects transparently, as it is just driven by the forces which act on objects in the world. Essentially, what is now a code-driven system would become a data-driven system. Also the combination of a three-dimensional representation plus a data-driven physics system situates robots in a world that follows the same basic principles as the world of real robots. It facilitates using robot models consisting of several components connected by joints and certain degrees of freedom.

It is just soccer! This is not really a problem for the soccer server, but nevertheless it is a limitation. If the above mentioned changes were made, the soccer server would allow player programs to perceive a 3D world and perform actions within this world. The real question which should be raised at that stage is: Why restrict this system to just the game of soccer? A flexible approach would allow for various kinds of simulations including soccer or rescue so that the same platform could be used for tasks with different levels of complexities in research and education.

1.1 Resulting Goals

Now, that an overview of the subject matter at hand has been given, it should be clarified how all this relates to the system we created [7]. The goal was to create a system which can provide simulations of 3D environments. The choice to provide 3D simulations greatly simplifies the transition of real world environments to virtual simulator environments or vice versa.

Another important aspect is the integration of a robust physics system, which is responsible for producing realistic motion of the objects within the environment. This system should not be forced into the simulation, as that would limit the applicability. With "forced" we mean no general assumptions are made regarding the need for physics of the simulation. The system should remain unobtrusive and only provide services for simulations which request it.

As we are creating a simulator which can be used for multi-agent systems, we also have to think about how agents live and act in the environment. Clear interfaces for the sensing of the environment have to be presented, as well as how actions can be performed.

Also, we have to provide a means to visualize the state of our 3D environment. It is hardly possible for a human spectator to comprehend spatial relationships of complex 3D environments without having a visual reference, especially if it is not only the outcome of the simulation which is of interest to the researcher, but also the way it was achieved.

Ultimately, we want the system to be applicable to not just a single type of simulation, but an entire class of simulations, namely those which can be represented within a

3D environment. Another challenge for this system will be that it should run in realtime on a modern desktop PC.

2 World Representation

One of the biggest challenges was designing the system used for the representation of the environment being simulated. We have already seen that the environment plays a significant role in multi-agent simulations, as it controls what and how agents can perceive the virtual world that they live in. The world representation has to be accessible to a variety of different subsystems of the simulation and visualization components and be able to handle all these accesses in realtime.

In this section we will take a closer look at what is necessary to represent a complete virtual 3D world. We begin by describing the concept of an environment and then examine how the objects living within that environment have to be represented. Following that, we discuss *scene graphs* and how this concept can be used to unify everything into a single, flexible and extensible data structure.

In our simulation agents should not be the only entities which exist in the environment. It also contains passive objects, such as chairs, tables, or soccer balls. Passive objects lack the agent property, whereas agents possess sensors and effectors to perceive and interact with the environment. Aside from this, passive objects also lack the ability to reflect their current state [2]. Their state is only affected by the natural laws of the environment they live in and by agents performing actions on them (either direct or indirect). Using these ideas as a basis an environment can be defined as follows[2]:

Definition 1. *An environment E is a three-dimensional space and contains a set of objects O. Each object has a location within the environment and can be perceived and manipulated by agents. An environment contains a set of agents A, such that $A \subseteq O$.*

Definition 1 captures two important concepts. First, agents are always objects in the environment. This relationship allows agents to also be perceived and manipulated by other agents, yet it clearly illustrates that there is a semantical difference between the concept of agents and objects. Second, all objects are *situated*. This means that we are always able to tell where each object is located within the environment. This is extremely important, because without a location, the ability to perceive the environment would be ill-defined.

Definition 1 only places one real restriction on the environment, namely the three-dimensionality of the space it represents. Other than that the environment is solely defined through the objects which it contains.

3 Object Representation

An object living in our 3D environment has to meet the needs for many different subsystems of the simulation and visualization engine. For example, the visualization component has to have some information about how to display the object, whereas the physics

[2] Definition inspired by Ferber's definition of multi-agent systems [2].

component needs access to physical properties. The visualization component might require a complex 3D model, but a simple sphere which encloses the object suffices for the collision system. From now on, we will refer to these object properties as object *aspects*.

3.1 Object Aspects

In order to arrive at a scaleable and versatile system an object will be represented by modelling its aspects. Given the requirements of a 3D world and support for a physics simulation we arrive at the following categorization of aspects for an object:

Visual Aspect. This aspect captures what the object looks like. It contains all the necessary information to display the object on the screen using the graphics application programmer interface (API) that the simulator program supports (in our case *OpenGL*).

Physics Aspect. The physics aspect is used to provide an interface to the physics system. This aspect will collect different forces acting on the object, as well as its physical properties, such as mass and mass distribution (which affects how the object behaves when in motion).

Geometry Aspect. The geometry aspect is used to define the solidity of the object, its shape and size as it is used by the subsystem which detects and resolves collisions. This is usually done with simple volumes, such as spheres, capsules and boxes. We need these simple collision proxies, because the visual aspect is usually much too complex for performing real-time collision detection. Obviously, the shape and size of the geometry aspect should be chosen in a way that at least somewhat resembles the visual aspect. Otherwise one could end up with objects looking like a box (visual aspect), but behaving like a sphere (geometry aspect). When it is not possible to find a suitable collision proxy, a triangle mesh is used.

All these aspects are linked by a single property of the object, its location in the 3D world. The location determines where the visual aspect will be displayed with respect to a virtual camera. The physics aspect modifies the location of the object, as changes in location are the result of motion and motion is controlled by physics. Two objects are colliding when their respective collision proxies overlap. To determine whether this is the case or not, the geometry aspect also requires the location of the collision proxies, which is identical to the object location.

Aspects also have to be able to interact with each other. For example for resolving collisions the collision system has to resolve the situation by moving the two participating objects apart. It has to generate the forces necessary to separate the two objects and apply these forces via the physics aspect in the next simulation step.

Another difficulty which arises when trying to represent objects is, that not all of them have to posess all three aspects. Imagine a very simple environment containing two objects: A ground plane and a sphere which is floating above the plane. When the simulation starts, gravity should cause the sphere to fall down, while the plane should not. This is why the plane should not possess the physics aspect. Both the sphere and plane are solid, so the sphere should collide with the plane, bounce a few times and come to rest once its kinetic energy has been consumed.

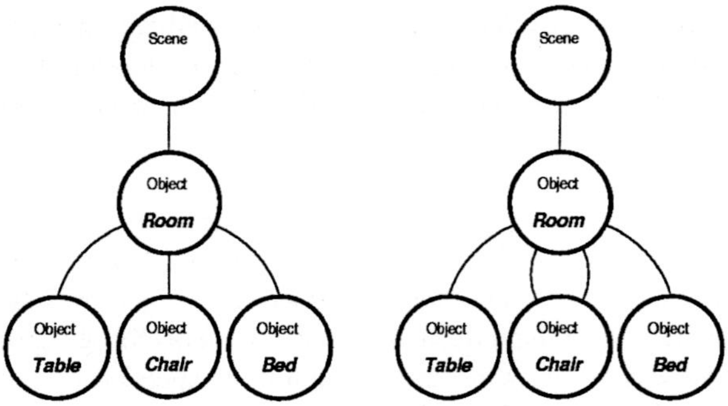

Fig. 1. a) Conceptual hierarchy b) Conceptual hierarchy with instances

The reason why some objects in the virtual world lack certain aspects comes from the fact that we are just modelling real world. Surface of the Earth in the example above is modelled by a plane. So, as our model is just an approximation to the real world, we have to bend the natural laws of the universe a bit to accurately represent it.

Now that we have looked at the concepts necessary to represent an object in a 3D environment, we have to extend these ideas to be able to represent agents within this system.

3.2 The Agent Aspect

Our definition of environment already states that an agent is an object. Therefore, it makes sense that an agent can posess the same aspects as an object. In order to represent the additional capabilities of an agent, we introduce the *agent aspect*. It has the ability to perceive the environment using perceptors, think about what actions to perform next and actually perform them using its effectors. Using this modular approach it is possible to turn almost any object into an agent and vice versa. Section 4 will deal with the agent aspect in greater depth.

3.3 Scene Graph

Scene graphs represent the de-facto standard when it comes to representing spatially-oriented data. In the early nineties, scene graphs were popularized by SGI Open Inventor as a user-friendly way to composite objects within a 3D world [10]. SGI Open Inventor was the basis for the *Virtual Reality Modelling Language* (VRML).

Conceptual Representation. Another important property of scene graphs is that they aid in structuring the 3D world by organizing the objects it contains into hierarchies, both on a conceptual and on a spatial level. For example, imagine a room which contains a table, a chair and a bed. The concept of the room *containing* these objects can be expressed through a parent-child relationship yielding the tree depicted in Fig. 1 a).

The scene node is the root of the scene graph. We see that the relationship of the chair belonging to the room manifests itself in the form of a simple link between the parent (room) and the child (chair). This form of representing such a relationship gives rise to a resource sharing scheme. When we add a second (identical) chair to the scene, the scene graph only has to create another link to the chair object to reflect that change in the scene (Fig. 1 b)).

Representing Spatial Relationships. It was stated above, that the scene graph is also capable of representing spatial relationships, but from the above descriptions we still do not know where the objects are located within the room. Before being able to tackle this problem, we have to look at the term *location* more closely.

A location in a 3D space is defined by a position vector $< t_x, t_y, t_z >$, which is the translation from the origin and the orientation of the object at that position. The orientation is usually specified in the form of a rotation matrix, see also [3].

Using a transformation matrix it is possible to convert from one coordinate space to another. For our world representation we have to cope with two different spaces: the *model space* and the *world space*. A 3D model consists of a number of points, so-called vertices. These vertices can be connected with other vertices to form triangles or polygons describing the surface of the object. Each vertex has a positional attribute which is a 3D coordinate. In order to allow this object to be independent of its location in the world, the positional attribute is specified in the objects coordinate space, the model space. The location of an object is given in the form of a transform, the object's *world transform*. Given a homogeneous position in model space in column vector form, $\mathbf{p_{model}}$, and the world transform M_{world}, we can convert it to world space simply by right-multiplying the position vector:

$$\mathbf{p_{world}} = M_{world} \cdot \mathbf{p_{model}} \tag{1}$$

The column vector form is necessary for the multiplication with the matrix to be defined. We also can concatenate several transforms. Each transform will convert from one coordinate space to another. This allows the chaining of coordinate frames and gives rise to the concept of *local transforms*. A local transform is always relative to a coordinate space. If the coordinate space is the identity space, the local transform is the world transform. The idea of local transforms allows us to add spatial relationships to a scene graph. This is done by introducing *transform nodes*. Using this new node type we can augment the scene graph for the previous example. The behavior of retrieving local and world transforms is best described inductively:

$$LocalTransform(x) = \begin{cases} \text{identity matrix} & \text{if } x \in Scene \\ \text{identity matrix} & \text{if } x \in Object \\ \text{local transform matrix} & \text{if } x \in Transform \end{cases} \tag{2}$$

$$WorldTransform(x) = \begin{cases} \text{identity matrix} & \text{if } x \in Scene \\ WorldTransform(Parent(x)) & \text{if } x \in Object \\ \text{world transform matrix} & \text{if } x \in Transform \end{cases} \tag{3}$$

The above behavior clearly shows that only transform nodes need to know about concrete local and world transforms. The corresponding matrices are data members of the actual transform objects. The world transforms are updated during a scene graph traversal which takes place every simulation step by concatenating the local transforms. However, many nodes (as we will see) do not need to specify spatial relationships.

Thus, we chose to express spatial relationships through explicit transform nodes. As world transforms are always recalculated from the local transforms, we can move entire subtrees without disturbing their spatial relationships. For example, when we modify the parent transform of the room, it would move the room and its child objects around, but the position of the objects relative to the room would remain constant. This makes assembly and reuse of complex objects very easy.

Aspect Representation. All aspects should also be reflected in the scene graph structure. The most flexible solution is to represent each aspect as a node in the scene graph. As all the aspects are (in a sense) synchronized via the object location, a node representing that location would be an ideal candidate for the aspects parent node. This role will fall to the transform node mentioned above. In the room example, we would replace the object nodes with the aspects making up that object. All edges attached to the object node would go directly to the parent transforms.

4 Simulation

In this section we add the ability to act to objects and present how all the aspects interact with each other. We will also illustrate how it is possible to add the necessary flexibility so that the simulator is able to provide more than a single specific simulation.

4.1 Agent Aspect

The agent aspect is the node in our scene graph which distinguishes agent objects from other world objects. It has to perceive the environment through perceptors, decide which actions to perform next to satisfy its goal, and finally to perform actions through effectors.

The perceptors and effectors represent the agent's interface to the environment. In order to provide a flexible solution we chose to model and implement these agent capabilities as individual classes, rather than having each agent aspect deal with them internally. Thus, the agent does not have to directly access the scene graph structure. Instead, it can receive all the necessary information through its perceptors and perform all actions through effectors. This relationship is illustrated in Fig. 2.

The approach of modeling perceptors and effectors separately is beneficial because it allows us to hide away the implementation details of the sensory and action models from the agent, which as a result only has to deal with comparatively high-level concepts. This design decision also makes the specification of agent abilities extensible. It is very easy to add new perceptors or effectors to an existing code-base. For example, the design could be exploited to simulate actual physical devices.

The agent aspect controls which perceptors and effectors it needs. When an agent aspect is attached to an object in the simulation, it performs an initialization procedure. During this procedure it can request specific instances of perceptor and effector

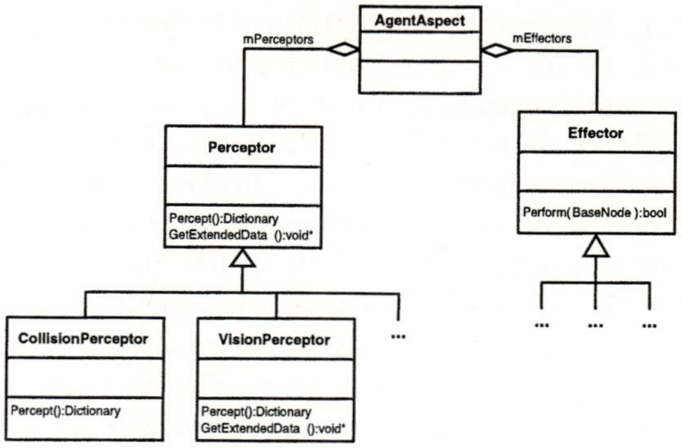

Fig. 2. UML diagram showing the relationship between agent aspects, perceptors, and effectors

classes to be added to itself. When all requested classes are accepted, the agent aspect be allowed to participate in the simulation.

Perceptors. We already mentioned above, that a very strong focus has been given to the visual aspect. This is not only the case for the scene graph, but also for the perceptors. The primary ability, which is necessary to navigate through a 3D environment is vision. A *Perceptor* class and its subclasses have the ability to process the scene graph structure to extract sensory information, which is made available to its client (the agent aspect). Every agent receives its own instance of a concrete perceptor class, so it is possible for the perceptor to store internal state. Before the agent aspect executes its thinking process, it can query all its perceptors for information about the world.

Concrete perceptors are derived from the base Perceptor class as illustrated in Fig. 2. Thus, the actual implementation of the perception algorithms can change freely without breaking the agent aspect code. This design is very similar to the *Strategy* design pattern outlined by Gamma et. al. [4]. As we would like to add arbitrary perceptor implementations to the system, the interface they expose to the agent aspect is crucial. The interface has to allow the arbitrary passing of data from the perceptor to the client. In order to strike a compromise between simplicity, safety and generality two mechanisms are made available. The first mechanism is just a dictionary of string and value pairs. This allows easy access to individual data and is sufficient for simple perceptors. We also provide a more general interface to return data like rendered images through a function which returns a C++ void-Pointer (untyped) to a memory location. As the agent aspect has knowledge about which perceptors it deals with, it also has knowledge about how to interpret the data it returns. Therefore, it is able to cast the void-Pointer back to its original type and use that data structure.

Example 1. Two implemented perceptors from our simulator framework:

CollisionPerceptor. The CollisionPerceptor allows an agent to receive information about when it has collided with another object of the world. The physics engine is responsible for detecting these collisions and will update the collision perceptor, if present, for each object involved. The only data the CollisionPerceptor of an agent will return is the path in the object hierarchy corresponding to the respective collision partner. This is done using the name-value pair return facility. The "colidee"-entry holds the above information.

VisionPerceptor. The vision perceptor is a bit more complex than the CollisionPerceptor as its functionality is not provided by an existing subsystem. The VisionPerceptor models the eye of an agent through the same kind of viewing volume we used for the camera frustum. Unfortunately, just culling objects against the VisionPerceptor's frustum is not enough, as it does not handle occlusion. Therefore, we have to perform extra visibility-checks for every object within the view frustum. This is done by tracing rays through the scene. The ray-testing is performed recursively, beginning at the root of the scene graph. The ray is always tested against the bounding-box of a node, before its interior is tested.

Effectors. An *Effector* has the ability to modify the contents of the scene graph (even create new objects). Every agent aspect can request a number of effectors during its initialization. All concrete effector implementations derive from a common Effector base class, as illustrated in Fig. 2. Thus, we also have an open design, which can be easily customized. Again, the Strategy design pattern comes to mind [4]. In a RoboCup soccer simulation, we might have (among others) a KickEffector, a MoveEffector, a DashEffector and a CatchEffector (for the goalie). For every action an agent wants to perform he needs to request its accompanying effector. This design facilitates simulation growth, being able to successively add new functions without breaking old functions. It would even be possible to add new functionality in parallel to old functions. Looking at RoboCup again, it would be possible to add experimental functionality through new effector classes. The old effectors would still be available, so this kind of experimentation would not break existing code. Existing agents could slowly migrate to the new interfaces.

4.2 Control Aspects

With all this freedom inside the agent aspects, we still have to address how the actual simulation is controlled. Basically, the legality of the agents actions somehow has to be enforced. When looking again at the game of soccer, we have an entity on the field which is responsible for making sure the rules of the game are followed: the referee. Thus, it makes perfect sense to employ a similar "construct" to watch over simulations. We call these entities *control aspects* and they are very similar to the agent aspect. In fact, on an implementation level they are equivalent, control aspects are just specialized agent aspect which posses some advanced perceptors and effectors, giving them the ability to analyze the entire scene graph.

Some simulations might be too complex to be watched over by a single control aspect. Going back to the soccer analogy, we also have two line referees which help the field referee. Thus, more than a single control aspect could be employed to monitor the simulation. This makes it easy to extend this functionality. For example, if the rules each control aspect enforces are mutually exclusive it is possible to provide several flavors of a simulation. Imagine a game of soccer with and without the offside rule, only influenced by the presence or absence of an "OffsideControlAspect".

Control aspects have the ability to register a few custom perceptors and effectors. For example, when an agent aspect is added to the simulation it is interesting for the control aspect to perceive which perceptors and effectors the agent tries to request. Based on this the control aspect also has a special effector which allows him to disqualify an agent from the simulation. Thus, aspects and effectors are much more, than just interfaces of the agent to the environment. They are also the basis for an event-like system, where perceptors act as the sinks for events.

4.3 Putting It All Together

We have looked at the different aspects of our simulation objects now in quite some detail. Some of the interactions between them have already been hinted at, but the big picture about how they all interact with each other to result in the desired simulation have yet to be presented. The scene graph unifies and triggers this simulation procedure.

Warming up. In the beginning the scene graph does not contain any world objects. It is only composed of the scene node as its root.

The simulator is initialized by registering a host of perceptor and effector classes. After this step is realized, the actual simulation can be initialized. At first the above discussed control aspects are added to the simulation. This can be one or more aspect, depending on the complexity of the task. No control aspect is needed for purely physical simulations. At this stage the world can be populated with objects and their corresponding aspects. During this step every agent aspect performs its initialization procedure, requesting perceptors and effectors. After the world is initialized, we can now start the simulation process.

Simulating. The simulation is performed in a so called *run-loop*. Every iteration of the loop corresponds to a frame being displayed by the simulator. The production of a single frame involves the interaction of all aspects to update the scene graph. At the beginning of the current iteration the world is in a legal state. This means, it contains only agent aspects which were not disqualified and the transform nodes correspond to their corresponding physics aspect's location.

The first group of aspects, which get updated by the simulator are the agent aspects. They use their perceptors to get feedback about the current world state. Each agent processes this information resulting in one or more effector being triggered to achieve its current goal. The usage of perceptors is usually monitored by a control aspect, as there often is a number of limited effector usages allowed in a simulation. For example, in the RoboCup Soccer Simulation you can only execute a single `dash` command per cycle, but you could issue a `turn_neck` command in parallel [1].

The use of effectors changes the state of the environment. A move effector will apply forces onto the physics aspect of an object, for example. This brings us to the next step, where the physics engine resolves the resulting object motion using the geometry and physics aspect. At this point, collision perceptors are also notified when applicable. Once this update pass terminates, the location of all objects already reflects the state of the next simulation iteration, bringing us to the final aspect, the visual aspect.

Updating the visual aspect of the scene graph objects begins by locating the camera. Based on this, the scene graph is culled (geometries and lights), resulting in a very conservative estimate on what needs to be displayed. Then a rendering procedure is utilized to bring the simulation on the screen and the next iteration can begin.

4.4 Parameterizing the Simulation

One of the main features of our simulator is the ability to provide more than just a single simulation. It is possible to parameterize the simulator with a *simulation description*. We have identified a simulation to be composed of two parts, the *classes* which are required to build the simulation and the *structure* in which instances of these classes are combined to form the scene graph. A simulation description contains a series of classes, which are added to the class pool of the simulator. These classes are mainly implementations of `AgentAspect`, `ControlAspect`, `Perceptor`, and `Effector` interfaces. In addition to this, the simulation description might also contain other custom classes, which have to be added to the scene graph (e.g. new visual aspects). However, this information alone would be useless, as we still need to know how the simulation is modeled using instances of these classes. This is done using a so-called *assembly script*. It describes the structure of the scene graph and the initial parameters of the objects within the scene graph.

This design requires a tremendous amount of flexibility on the implementation side, as the simulator has to be able to instantiate concrete classes without having any knowledge about them. We also need a way to be able to add new classes to a simulator that can remain static and constant. Part of this flexibility was achieved by integrating Ruby [8] as scripting language into our simulator. For the sake of brevity we omit the details here, for details on our class object system and the object hierarchy in our simulator see [7].

5 Conclusions and Future Work

In our approach we have provided a flexible way to represent 3D simulations using a scene graph architecture, modeling the aspects of objects in the virtual world and their interactions with the environment directly. By integrating a class object framework it is possible to extend the simulator at runtime, allowing it to be parameterized with simulation descriptions, ultimately giving the simulator the ability to provide more than a single simulation. This makes our simulator ideal as a platform for trying out new simulations.

When performing 3D simulations, a good visualization component is also necessary. In order to provide a flexible simulation component, we have implemented a real-time lighting solution on the basis of the *OpenGL* graphics API. The visualization component can operate in two distinct modes, ambient lighting (fast, but low quality) and local lighting (slower, but high quality), allowing it to scale based on simulation needs and hardware requirements.

We are convinced that our simulator framework would be an ideal basis for a next generation Soccer Server. Many problems have been solved with the system developed, yet a number of issues still remain.

5.1 Future Work

The current design of the simulator is very monolithic. We have a single application, which is responsible for the physics simulation, agent simulation and visualization, it is not a server system yet. The current RoboCup Simulation League Soccer Server allows for a distributed simulation. Our most favoured approach is to solve this problem at the scene graph level. The idea would be that every node in the scene graph can either be local or remote.

The way we chose to represent the simulation by modeling the different aspects of objects will greatly ease the creation of a distributed simulator, as the various parts of the simulation process are already decoupled from each other. This decoupling is one of the primary strengths of the simulator design, as it allows different aspects to be changed without disturbing the others.

A different issue is that we have only created a simple and small example simulation up to now. For a new soccer simulation, some important questions have still to be discussed, as the simulator design will encourage exploration and experimentation. How should the soccer agents be represented in 3D? Through the addition of complex joints it would also be possible to create articulated structures, such as agents with two legs. The design of the simulator allows for such changes to be made easily without breaking too much other code (if any). So there is still work to do, but the current Soccer Server was not built in a day, either.

References

1. Mao Chen, Klaus Dorer, Ehsan Foroughi, Fredrik Heintz, ZhanXiang Huang, Spiros Kapetanakis, Kostas Kostiadis, Johan Kummeneje, Jan Murray, Itsuki Noda, Oliver Obst, Pat Riley, Timo Steffens, Yi Wang, and Xiang Yin. *RoboCup Soccer Server*. The RoboCup Federation, April 2001. Manual for Soccer Server Version 7.07 and later.
2. Jacques Ferber. *Multi-Agent Systems - An Introduction to Distributed Artificial Intelligence*. Addison-Wesley, 1999.
3. James D. Foley, Andries van Dam, Steven K. Feiner, and John F. Hughes. *Computer Graphics – Principles and Practice*. Addison-Wesley, Reading, MA, 2nd edition, 1989.
4. Erich Gamma, Richard Helm, Ralph Johnson, and John Vlissides. *Design Patterns - Elements of Reusable Object-Oriented Software*. Addison-Wesley, Reading, MA, 1995.
5. Tom Howard, Artur Merke, Oliver Obst, Martin Riedmiller, and Patrick Riley. New soccer server initiative. Work in progress. Available in the Soccer Server Repository.
6. Hiroaki Kitano and Minoru Asada. RoboCup humanoid challenge: That's one small step for a robot, one giant leap for mankind. In *Proceedings of IEEE/RSJ International Conference on Intelligent Robots and Systems (IROS)*, pages 419–424, 1998.
7. Marco Kögler. Simulation and visualization of agents in 3D environments. Diploma thesis, Fachbereich Informatik, Universität Koblenz-Landau, 2003.
8. Yukihiro Matsumoto. Ruby: The object-oriented scripting language. `http://www.ruby-lang.org/en/`, 2002.
9. Itsuki Noda. Soccer server: A simulator of robocup. In *Proceedings of AI symposium '95*, pages 29–34. Japanese Society for Artificial Intelligence, December 1995.
10. Dirk Reiners, Gerrit Voß, and Johannes Behr. OpenSG: Basic concepts. In *1. OpenSG Symposium OpenSG 2002*, 2002.

Educational Features of Malaysian Robot Contest

Amir A. Shafie and Zalinda Baharum

Software and Intelligent System Development Program
Electronics and Computer Applications
SIRIM Berhad, Shah Alam, Selangor
{Akramin,zalinda}@sirim.my

Abstract. The educational experiences from robot contest of entry, junior and advance level are presented based on guided constructionism approach in education that combines hands on guidance with hands-on experience. The aim of the competition as a whole are to allow the student to (i) conceptualise the robot (ii) manage the non-deterministic characteristic of the environment and (iii) manage integrated hardware and software development projects. Indeed with this knowledge the student should be able to win a number of international robot tournaments.

1 Introduction

Technological education in Malaysia is undergoing reform in relation to its status, goals and teaching/learning strategies. This trend in reforming technological teaching/learning strategies is following the worldwide general reform process with the aim of making technology education at all levels more meaningful, intellectual and creative [1]. Real world problems, interdisciplinary approaches, project oriented learning, team cooperation and authentic assessment have become the highlights of recent curriculum innovations. The skills in focus now are used to integrate different competencies an intelligent application to construct hardware objects so that they are governed by intelligent software that continuously interacts with non-deterministic real world [2].

The goal of introducing robot games as a healthy competition in the education system is to accelerate the acquisition of general skills in problem solving and scientific concepts in experimental science domain. Robot competition involves the use of computers to acquire, analyse, control and model different worlds not reduced to screen simulation but with real device control. The educational strategies employed as the impetus for organising robot competitions are those mostly linked to theory of constructionism and refer to active pedagogy [3]. For the past two years the Software and Intelligent System Development Program of SIRIM Berhad has been actively promoting and organizing a number of robot competitions for various level of education. The aim of the competitions is to create awareness in robotic technology among the Malaysian public.

D. Polani et al. (Eds.): RoboCup 2003, LNAI 3020, pp. 470–477, 2004.

2 Competition Set-Up

The competitions have been set-up differently according to the participants' level of education. However all the competition set up have an educational purpose based on the guided constructionism approach towards technological education [4]. The set-ups of the competition were designed as a three-step educational process in which they work with imaginary robots followed by robot with modifiable body plan with complex team behaviour and lastly project coordination through robot construction and problem solving. By going through this series of robot competitions it has been the organisers hope that the students through their competitive and dedication will get a thorough understanding of robot concept in relation to the effect that the robot will have on human lives. The student also will have an understanding of the body and brain relationship with regard to the real world applications and an understanding of communication and distributed systems role in real world problem solving. In general the student should learn to manage and understand the non-deterministic characteristics of the real environment and to integrate hardware and software solutions to find the optimal set-up in solving the problem laid out for the competition.

2.1 Robot Drawing Contest

Every year in January invitations are sent through the ministry of education and mass media to invite primary school children to send in their drawings for the national robot-drawing contest. The drawings are based on the current trend of social-robot themes e.g. for the year 2002 the theme is 'Robot Maid' and for the year 2001 is 'My Friend is a Robot'.

The robot drawings are judged mainly on its originality and the ability of the participant to bring the concept in the competition theme into their drawing. The participants are from primary school (7-12 yrs old) thus their exposure to robots at this stage are mainly from televisions and toys, which has somehow hampered their effort in producing the original drawing. Therefore from the judges' observation the robots in the drawing actually depicted how the participant impression of how their favourite robot will effect their lives. Although one of the main purposes of this competition is to encourage the student to concept out an original robot, the organisers feel that their achievement in being able to concept out the theme with their favourite robot is an achievement to note. However it has to be noted that from the organisers interview with the participants, the robot are mainly thought of as the perfect being where as in real life most robots tend to break down regularly!

2.2 Learning to Manage the Non-deterministic Characteristics of the Real Environment (RoboCup Junior Soccer)

This event is adapted from the RoboCup (World Cup of Robot Soccer) and has been the penultimate event for the yearlong RoboFest Malaysia. The RoboCup is an international tournament where teams of autonomous intelligent robot compete in soccer-like games [5]. It is a serious educational event where the main purpose of the events is to give the student a hands-on experience in managing the non-deterministic characteristic of a real environment. The RoboCup Junior Soccer is a two-on-two competition i.e. there are two robots on each team. By having two robots in each team it is hoped that this rule will encourage team play because it is obvious that the match will be advantageous for the team that can develop team play. Each team has to develop the robot soccer player with the Elekit Robot Soccer M195 kit. The M195 is a miniature robot kit that includes infrared, ambient-light, touch sensors and two independent motors. The OOPic system can be programmed using a java based simulator known as 'Tile Designer' which provides the user with a graphical user interface to program the robot. The simplistic robot sensory system coupled with the low processing power of the OOPic due to high overhead caused by the java based simulator means that the robot soccer behaviour will be limited at all times.

The simplistic sensory system defines the environmental characteristic that the students have to contend with, where the ambient light and infrared strength are not uniform over different areas of the fields. Even though the students are told about the effect of the limited sensory system with regard to the information that they received several times during the training sessions, they are always amazed by changes in environmental conditions during the robot soccer player building process.

The two goal markers are equipped with different infrared signal frequency to differentiate between their own goal and the opponent goal. However the students very seldom reach the point where they have to consider the goals as they grappled with the problem of getting the robot to recognize the ball. Therefore, so far in the Malaysia RoboCup competitions the incident of own goals are frequent but it is believe that in future after the student have mastered the sensory systems the robot behaviours will be more complex.

The organisers also experienced many periods ended with the robot(s) pushing the ball into one of the four corners of the playing field. The robot(s) was unable to move the ball out from the corner, partly because of the shape of the robot. However in the second year of the competition some student tried a number of different strategies to move the ball out from the corner, e.g. turning very fast around itself while having contact with the ball, but these strategies were only successful in some particular cases with the right placement of the touch sensor.

The above situations showed that a lot of empirical tests are necessary in order for the students to make the environment suitable for the purpose that they have in mind. Often one has to manipulate the light, colours and shape of the playing field based on this many empirical tests. In this case, the interest is in teaching the future scientist about real-world application and therefore the set-up was biased by the educational purpose. For instance we chose to allow only the same kind of robot on each team. In other RoboCup and FIRA competitions, different teams are allowed to use any kind of robot that they build or buy within a specific size. Therefore as they are using one specific robot, the focus is on improving the robot behaviour through programming and creative tinkering. All teams start with the same motor characteristics, the same sensory system etc. to work with and therefore the performance can directly be compared to the properties of different robot controllers and different development process.

In a sense at this level of education, the secondary school (13-17 yrs old) there is a limit on the number of free parameters available for the students as they are using the same hardware platform for all robot soccer players. This is made as such so that the students have better focus towards solving the problem and learning how to manage real-world applications.

2.3 Learning to Integrate Hardware and Software in Distributed Systems (RoboCon)

In the RoboCon tournament the students are again given the task of making the robots, but this time they need to develop both the controller and the morphology of the robot to suit the task given. However, in this case they are confined to the budget given to them by their college and universities. This limitation requires the student to 'shop around' and opening their option to various possibilities.

The arena for this competition are large at 20m x 20 m and had different kind of colour and lines on the floor in different areas, so one could make a robot navigate around on the floor in different areas according to the colours perceived with the chosen sensor. Even though the student was able to manipulate the characteristics of the arena, it was difficult for the student to design robots that would satisfy the task requirement. The successful team however managed to use the sensor through rigorous testing phase. However more importantly the student obtained new knowledge on top of what was experienced with their entry into the competitions. At this level the knowledge is more complete as the student experience the principles for the development of controllers, mechanical aspect of the robots and integrating them through programming whilst at all time managing the project as a whole.

3 Educational Experience

In this section the use of guided constructionism will be discussed. Emphasis is put on the observation of how students learn to manage non-deterministic characteristics of real-world systems. The observation is based on our assumptions that the students' knowledge on real-world application with control of devices is minimal and in the earlier competition is none. Therefore the students have a number of ungrounded expectations of the performance of such systems. Often these expectations are based on students' previous experience with programming in deterministic environment in the computer. Also it partly arises from the whole natural science approach in which we have profound belief that the systems can be broken down into smaller systems and each of them has a deterministic functionality. Therefore, the students must go through a number of empirical experiments before they are convinced to change their unrealistic view of real-world applications [6]. The student belief in a deterministic reality can be observed mainly in their robot soccer project and from our observation that the student change this view by going through the process of building robot soccer players. Some of these experiences are documented by taking note of their questions while robot workshop are conducted.

First of all, when starting the educational process, the students have totally unrealistic beliefs of the capabilities of the robots. For the two-on-two robot soccer project with M195 robot, one of the members in the robot soccer team laid out their plan for the game as locate the ball by turning side to side and guide the ball to the opponent goal. Obviously these students had no idea about the capabilities of the simple sensory system available at their disposal. Their general idea was that they would be able to translate the human soccer player skills to the robots. After many failed experiments they admitted that the robot capabilities have to be built from the robot ability and not the ideal soccer player condition. In general most students go through the process of having to change their ideal general strategy when they achieved more experience working with the real robots in the real environment.

On the more advance level of competition (RoboCon) is that almost all students believe that they can incorporate a very precise global positioning system in order to solve the problem. The sense of locality are then established based on this global positioning system so that a relationship between the robot and the goals (the tubes). The implementations are based on a counter that keep track of how much the wheel has moved. Apparently at this stage these students do not have an idea on the role of friction, spinning etc. At the tournament many students found out that with very little interference the robot lose its orientation quickly and the method failed miserably.

Many groups realised the difficulties in making a global positioning system work in reality and teams that consider the use of environmental feedback to approximate the robot position fares much better. Obviously the students are used to having all information available in pure form in the simulation work that they performed_in their study. However the experience from this competition makes many realise that this information will not be available in a pure form in real-world applications. They change their view and start to think about how to make use of little knowledge that they might obtain via feedback from the environment. However, they also experience difficulties when trying to obtain feedback from the environment, because the more affordable sensors are almost much more primitive than the student expect.

The students initially believe that sensors give a clear and unambiguous input and that they can use abstraction. The abstraction and pseudocode is used instead of experimentation in order to overcome software complexity and only later through experimentation do the students realize the true nature of the sensors. In a sense this resemble the discussion about the classical approach to robotics in which the hardware and software tasks were believed to be distinguishable, so that the engineers could work on the hardware while AI researchers could make abstractions and work on the software only in order to create an intelligent system (robot). Nowadays the newer AI approaches on robotics begin to reject this division and focus on embodied AI.

The students found out during the test period and competition in the real field that the sensory information cannot be interpreted in a straightforward manner using fixed threshold to classify the inputs. They experience that the approach works one day but fails miserably the next day when for example it is no longer sunny weather outside of the arena. Initially the student became frustrated and blamed the hardware. However, the long process of experimentation makes the student realize that the adaptive approaches can be used to overcome this problem. In fact the winning robot in all competitions has a version of adaptive systems in their logic to overcome the changes during play. It is strongly believe that simplistic approaches in suppressing the arena information will lose out in the latter stage of the competition where the teams have better integration between hardware and software. At this stage the adaptive aspect of the robot in determining its behaviour will hold the key between winning and losing. At any level of competitions the students learn the importance of the power supply to the robots. Here they found that the due to battery power levels the logic and abstraction that they chose initially will not work as usual because the abstraction are based on ideal condition. Again the problems arise because the students believe in being able to make abstractions for example classify sensory data with fixed, pre-defined thresholds.

Therefore from an educational perspective the robot competitions put forth the importance of hands-on experiments on real world systems. This is because the students will have difficulties in believing in another view on the real world with deterministic characteristics in which abstraction is feasible. It seems like the students are only able to change this view when they are actually experiencing themselves a lot of times, that their robot will fails with a control program that depends on the abstraction. Then the student start changing their view and implement the more adaptive solutions based on their experimentations.

4 Related Work and Discussion

There exist a number of open robot competitions such as Micro Mouse and FIRST (For Inspiration and Recognition of Science and Technology) competitions. Micro Mouse has been running since the late 1970's and it consists of designing an autonomous robot known as mouse, which should navigate its way through a maze. The robot mouse has no prior knowledge of the maze configuration before its release in the maze. During the runs the robot mouse should travel from the starting point to the centre of the maze. The first two runs are used for data gathering and the final is meant for a high-speed run to obtain the fastest handicapped time. FIRST competition is an engineering contest in which high school students team up with engineers from businesses and universities. In six intense weeks students and engineers work together to brainstorm, design, construct and test their robot. The aim of the FIRST competition is to bring together businesses, schools and universities and thereby provide an exchange of resources and talent highlighting mutual needs, building cooperation and exposing students to new career choices. The similarities on the educational perspective have clear resemblance everywhere, which include the MIT series of competitions [7]. The competitions at MIT are part of hands-on, workshop like courses for undergraduates, which usually run as part of their summer course. The conclusions from these series of programs resembles our observations reported in section 3 where most of the students tend to build robots that perform only in ideal conditions. The approach taken by the Malaysian series of robot competitions is based on theoretical considerations put forward by Seymour [8], in what he terms as constructionism. Constructionism suggests that learning is achieved most effectively by participation in the construction of artefacts. The artefacts become an 'object to think with' which can be used to explore and express ideas such as In the robot competitions, the students are allowed to construct their own robots and learn about real-world applications by going through the building process.

The Malaysian RoboFest committee choose to have robotic competitions at different levels because the organizers believe that the educational process to be slightly more complex than what is often suggested in constructionism. In its most pure form constructionism theory seems to suggest that children should be allowed to play in what they find fun and they will learn by this play. However in many subject areas the students need guidance and that there exists subjects that are profound for scientific knowledge which do not lend themselves to a constructionism approach but will have to be thought in a more traditional manner especially the case when educating students with the fundamental knowledge of the subjects. Constructionism then at a higher level can be combined with other pedagogical methods to ensure that the students obtain a profound knowledge about the subject under study.

Therefore, the idea of guided constructionism uses a three step process with (i) imaginary robots, (ii) robots with modifiable body plan and team behaviour and (iii) construction of robots and complex team behaviour. Constructionism is the core of (ii) and more at (iii) while it plays a minor role in (i). With the three-step process the organizers try to ensure that the students first get the knowledge about programming in the real world and a tool to compare their different approaches. The comparison of how the robot should work is achieved at the level where the students are given the knowledge on how to modify the robot body. The students then obtain an essential knowledge about robot programming by working with the robot projects outlined in the competitions where they learn the relationship between controller and robot body plan to manipulate the environment. Hence our approaches stress on the case where the constructionism approach should be combined with other methods and that there exist essential arguments that are better acquired by the student with another approaches. However hands on experience must remain as the major role in technology education and it often facilitates the student acquisition of knowledge about an artefact.

5 Conclusions and Future Work

In this paper the Malaysia Robot Competition Program is outlined in relation to the guided constructionism ideas and the observation from it has been outlined. The test case, used now is the three-step process (i) the conceptual aspect of the artefact, (ii) the manipulation of robot behaviour in Relation to Real World Environment and (iii) the management of the integrated hardware and software project. Further, the idea can be very useful in education for a number of other objects. For instance, robot can be used as an educational tool for artificial life and biological investigation as described by Miglinio [9]. Also in this context the robot competition might be a test platform since one can for instance imagine studying the evolution of robot controllers, the evolution of communication, the evolution of suitable bodies etc. In future the difference between guided constructionism and unguided constructionism should be investigated more thoroughly in order to verify the indications presented in this paper.

Acknowledgement

Dr. Ahmad Zakaria was the chairman of the Malaysia Robofest Committee and provides a number of highly valuable comments and inspiration regarding the objectives of the Robot Competition Series funded by the Malaysian Ministry of Science and Technology.

References

1. Verner IM, Waks S and Kolberg A. Upgrading Technology Towards the Status of a High School Matriculation Subject: A Case Study, Journal of Technology Education, 1997, vol 9, no 1, Fall .
2. Leroux P, W4 : Educational Robotics, International Journal of Artificial Intelligence in Education, 1999, 10,1080-1089

3. Papert S., Constructionism: A New Opportunity for Elementary Science Education, A Proposal to National Science Foundation (1986).
4. Henril HL, Robot Soccer in Education, Advanced Robotics, 2000, vol 13, no 8, 737-752.
5. Asada M, Veloso MM, Tambe M, Noda I, Kitano H, Kraetzschmar GK, Overview of RoboCup-98, 2000, American Association of Artificial Intelligence, 0738-4602-2000.
6. Bonasso P, Thomas D., A retrospective of the AAAI Robot Competitions, 1997, American Association of Artificial Intelligence, 0738-4602-1997.
7. Dean T, Bonasso P, AAAI Robot Exhibition and Competition, AI Magazine, 1993, 14, 1, 35-48
8. Papert S., Mindstorms: Children, Computers and Powerful Ideas, Basic Books, New York, 1980.
9. Miglinio O, Lund HH, Cardaci M, Robotic as an Educational tool, J. Interactive Learning Res. 1999, 10(1), 25-48.

A Hybrid Software Platform for Sony AIBO Robots

Dragos Golubovic, Bo Li, and Huosheng Hu

Department of Computer Science, University of Essex
Wivenhoe Park, Colchester CO4 3SQ, United Kingdom
{dgolub,bli,hhu}@essex.ac.uk

Abstract. In the development of robotic systems, an interactive software platform plays an important role for control design and parameter optimisation. This paper presents a modular approach to the development of a hybrid software platform for Sony quadruped robots. Such a platform consists of an overhead vision system, a Sony AIBO robot and a desktop PC, and is designed to be interactive in order to speed up the development of various algorithms for object recognition and gait generation. Based on this platform, both the colour segmentation algorithm and the gait control algorithm are investigated. The experimental results are presented to show its operation and good performance.

1 Introduction

In the development of robotic systems, software development platforms play an important role for control design of robots, especially before a prototype system is available. A good software platform can provides simulation functions that speed up the development of different algorithms, including complex programming and huge data collection. There are mainly two types of software development platforms for robotics research. One is the pure simulation platform, in which only models of real robotic systems are presented. It is based on an assumption that the developed algorithms can be implemented in future. Another type is hybrid, which partially relies on the real robotic system. The parameters of this kind of systems are fully collected from real robots and the real environment, which is also based on real experiments. If the algorithm works well under such a hybrid platform, it can also work well on real robots. The benefit of developing a suitable software platform includes the ability to sample and store sensory data.

In this paper, we describe our research work based on Sony AIBO robots. In each AIBO, there are 20 motors for its motion control and over 30 sensors for feedback control and navigation. However, there are some difficulties in the development of control software and sensing algorithms based on AIBO robots. Firstly, programming Sony AIBO robots is complex since different algorithms and individual programs need to be developed and debugged on real robots. Such development cycle takes a long time and may have a risk to damage the robot. With a hybrid experimental platform, most of the work is done on a PC safely and efficiently. Secondly, the processor on the Sony AIBO robot is not powerful enough. Complex algorithms cannot be implemented on real robots in real time. It becomes necessary to develop complex algorithms on a PC during the development phase, then optimised in size for the real ro-

D. Polani et al. (Eds.): RoboCup 2003, LNAI 3020, pp. 478–486, 2004.

bots. Thirdly, the current interface between the Sony robot and human operators is not friendly. Each time when the experiment completed, all the results need to be downloaded from the memory stick manually, which is very time consuming. With the proposed hybrid platform, the progress of experiments is displayed on the screen directly and any problem with the experiment can be seen immediately, i.e. a convenient way to develop control and vision algorithms.

The rest of the paper is organized as follows. In section 2, a modular experimental platform is proposed, which consists of three parts: a real robot, an overhead camera and a PC. Section 3 describes the adaptive colour segmentation algorithm and some experimental results. Section 4 presents the gait generation algorithm being developed by using the proposed platform. Finally, section 5 presents a brief conclusion and future work.

2 A Hybrid Software Platform

The Sony AIBO robot is a dog-like robot mainly for entertainment. It can also be used for research on multi-agent systems and the robot soccer competition [4]. Each Sony AIBO robot has a quadruped design, approximately 30cm long and 30cm tall including the head. The neck and four legs of each robot have 3 degrees of freedom (DOFs). The neck can pan almost 90 degree to each side, allowing the robot to scan around the pitch for beacons, ball and other objects. And it has an onboard colour CCD camera and a hardware colour detection engine.

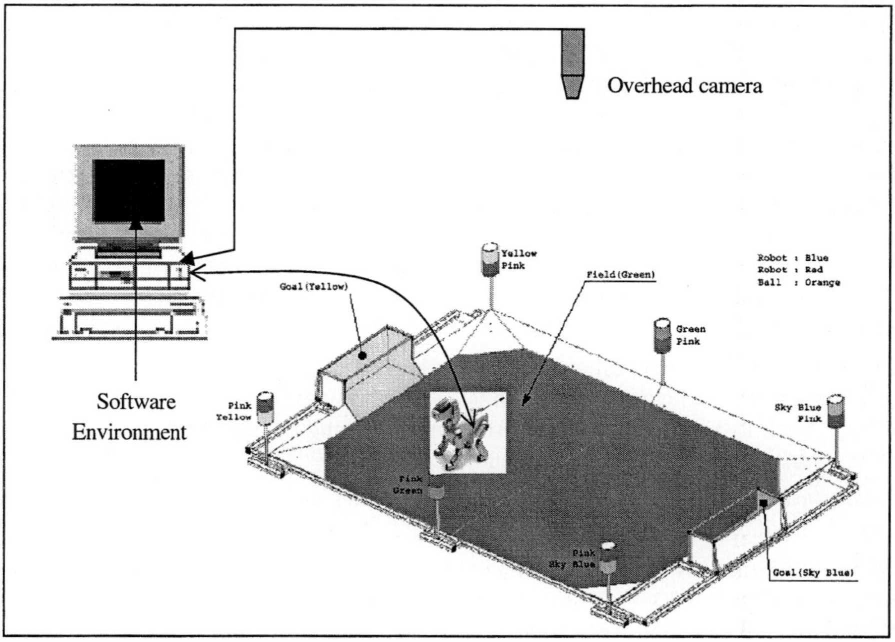

Fig. 1. System configuration of the proposed hybrid development platform

The proposed platform is based on three sub-systems: a Sony robot, a PC, and an overhead camera with one frame grab card. Figure 1 shows the system configuration in which wireless LAN is used for communication between the robot and the PC. In contrast, Figure 2 shows the basic software functions. More specifically, the program on the robot is designed to read instructions and implement basic behaviours. Its diagram is shown at the lower part of Figure 2. In contrast, the program on the PC is designed to provide a human-machine interface for the development vision and robot control algorithms. There are four main modules in it, namely,

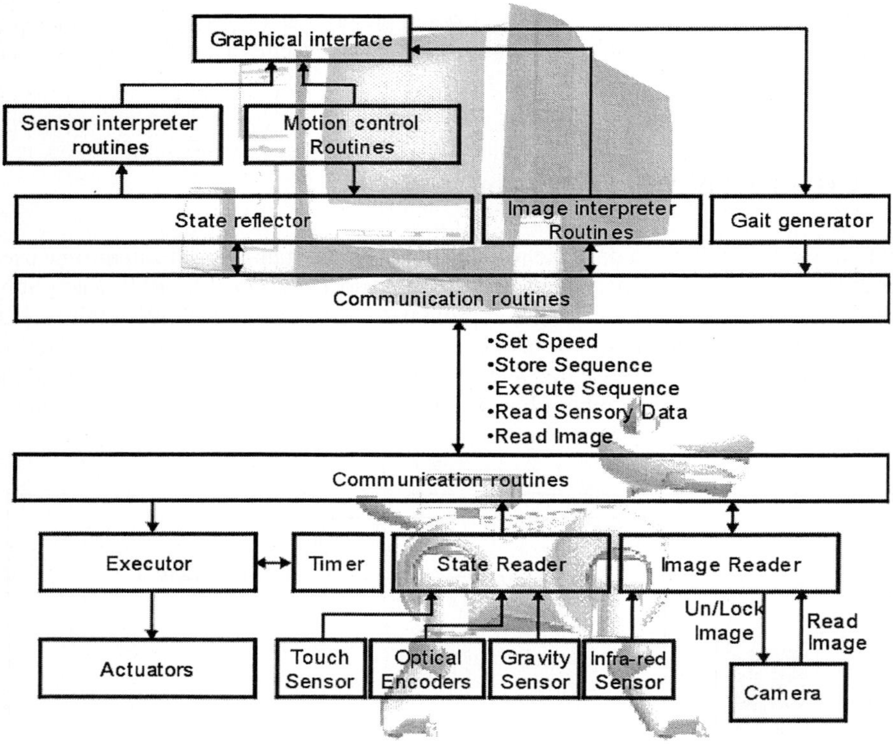

Fig. 2. Configuration of the proposed software development tool

State Reflector – An internal state reflector has been incorporated to mirror the robot's state on the host computer, which is an abstract view of the actual robot's internal state, such as sensor information from the robot and control commands from the host computer. More details can be seen in [2].

Communication Routines – The designed controller communicates with the robot using a handshake mechanism, and sends an appropriate command to the robot.

Gait Generator – The gait generator communicates with the robot by passing a sequence of arrays that are transformed into a sequence of robot movements. It creates different gaits in a form of a sequence of arrays.

Image Interpreter – Gathering image snapshots and processing images can be done completely independently from the rest of the application. The size of a captured image is 144 x 76 pixels and each pixel has three bytes for colour information. A locking mechanism has been adopted to allow the transfer of the current image to be safely completed before the new snapshot image can be grabbed.

3 Adaptive Colour Segmentation Algorithm

The task of extracting a colour object and estimating its position in images depends on colour segmentation [1][5], which needs to be robust to the varying lighting conditions and adapt to dynamic changes in the environment. The first step is to adjust the camera's shutter, aperture and amplifying rate settings, which is difficult for the Sony-legged robot. Such adjustments may cause other problems, such as change of the depth of view in the image. Therefore the implementation of different thresholds in different lighting conditions is necessary. Three main parts in our colour segmentation algorithm is shown in Figure 3.

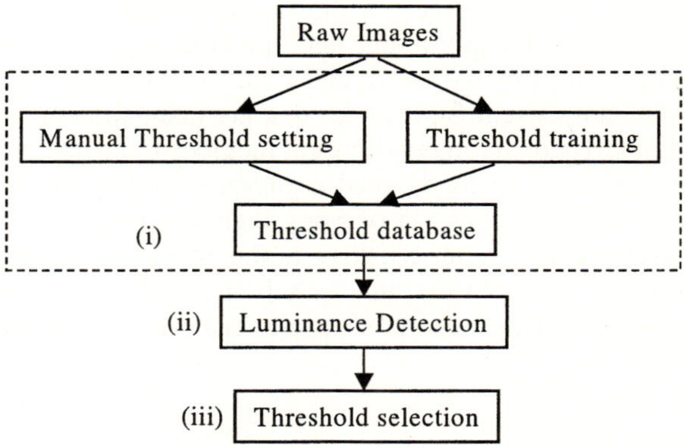

Fig. 3. The colour segmentation algorithm

3.1 Manual Threshold Setting

The threshold for the colour segmentation depends on the colour space in which the operation implemented. Commonly, HSV colour space is suitable for the colour segmentation, but the camera of a Sony AIBO robot outputs the images in YUV format and has a hardware colour detection table (CDT). The GCD method was developed by an Australia team, UNSW, to solve such a problem, which is very similar to the LUT methods presented by Schroter [8], i.e. a statistical approach. The thresholds of the GCD method can be any shapes in the UV space. The main difficulty of the GCD method is how to construct the thresholds under different lighting conditions.

Based on the proposed experimental platform, we can grab images from the on-board camera of the connected Sony AIBO. Figure 4 shows the raw image and the segmented image by using the CDT table at AGC=112. In contrast, Figure 5 presents the raw image and the segmented image by using the GCD table at AGC=112. Note that AGC represents "Auto Gain Control" here. As can be seen, the GCD method works better than the CDT method.

Fig. 4. Raw and segmented images (CDT)

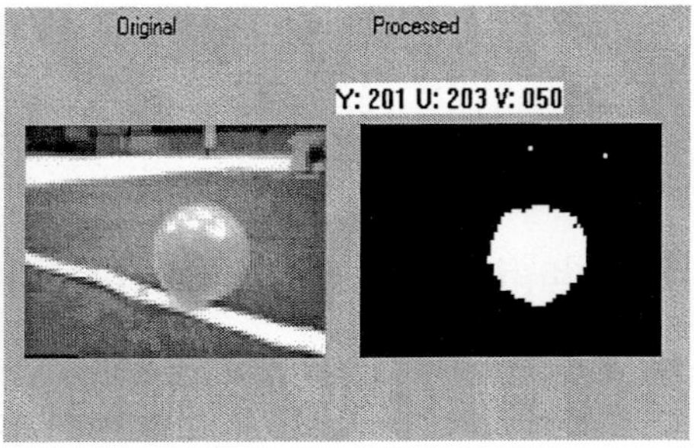

Fig. 5. Raw and segmented images (GCD)

3.2 Threshold Training Algorithm

Learning algorithms can be adopted to produce a suitable GCD or other kind of thresholds automatically [7][8]. Figure 6 shows the learning network used for GCD training, in which Y, U, and V are inputs, and output is the colour mark of the GCD. L1 is the luminance decided by the meter, and L2 is the luminance measured from the image. Training data are current GCD tables that are manually constructed with a set of lighting conditions. There are many different methods to measure the luminance of the images. Basically, most methods need a patterned object for measuring, which may be not reliable in the application of Sony legged robots. The field colour of a football pitch is green, which is not sensitive to the change of the lighting condition. However, the colour of other objects such as ball and goal is affected by light refection. With different angles, the measured results can vary dramatically under same lighting condition.

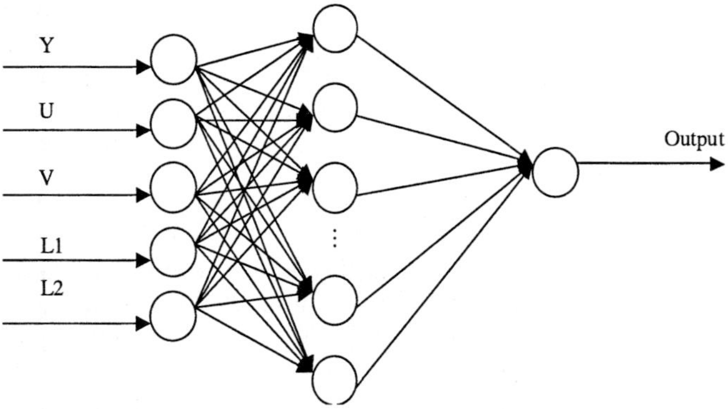

Fig. 6. A learning network for GCD training

Experiments were carried out using the real Sony robots to show the performance of the algorithm. Three different lighting conditions were adopted in the experiments: 405 Lux, 455 Lux, 535 Lux. GCD tables under 405 and 535 Lux were constructed manually. GCD tables for 455 Lux were constructed by the supervised learning. With large data sets, the speed of learning is very slow. For the learning process in a HSV colour space, the network was trained by 32MBytes test data over many hours. With the parameters such as the learning speed 5, 1 hidden layer, and 32 hidden neurons, it produced an inadequate GCD table with two many binary values 0 and the learning progress was too slow. On the other hand, with the parameters: learning speed 50, 1 hidden layer, and 8 hidden neurons, it produced a GCD table with too many binary values 1. The learning speed seemed very fast. Figure 7 shows the results obtained by using suitable GCD tables from training.

Fig. 7. Some Processed Images

4 Gait Generation Algorithm

4.1 Generation of Wheel-Like Motion

For a Sony AIBO, a trajectory refers to both the path of the movement of the tip of a limb (paw), and the velocity along the path. Thus, a trajectory has both spatial and temporal aspects. The spatial aspect is the sequence of the locations of the endpoint from the start of the movement to the goal, and the temporal aspect is the time dependence along the path. The essential conditions for stable dynamic walking on irregular terrain and on the flat ground can be itemized with physical terms:

- the swinging legs not be prevented from moving forward during the former period of the swinging phase,
- the swinging legs must be landed reliably on the ground during the latter period of the swinging phase,
- the angular velocity of the supporting legs during their pitching motion around the contact points at the moment of landing or leaving should be kept constant,
- the phase differences between the legs should be maintained in spite of delay of motion of a leg receiving disturbance from irregular terrain.

For the creation of wheel-like leg motion shown in Figure 8, we used six parameters for front and six parameters for rear legs. These twelve parameters determine the spatial aspect of a robot trajectory. The 13th parameter is bound to temporary aspect and determines the speed of paw movement.

In total, there are 13 real-valued parameters are used to determine a gait for the locomotion module. Table 1 lists some of these parameters, which are also the genes for individuals evolved by the evolutionary algorithm. These parameters specify the position and orientation of the body, the swing path and the swinging rate of the robot legs, the amplitude of oscillation of the body's location and orientation.

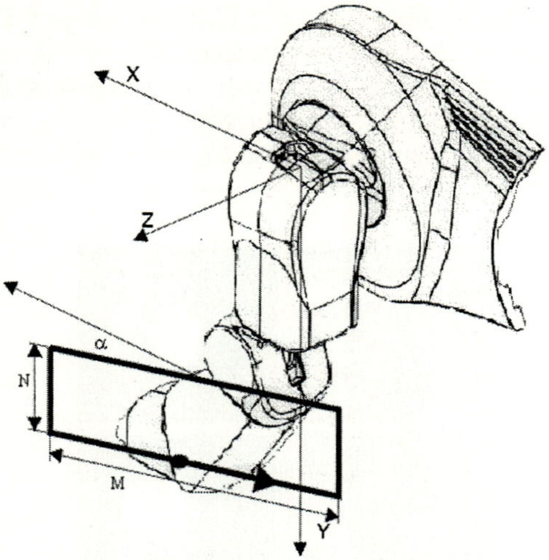

Fig. 8. Paw trajectory–Wheel-like motion

Table 1. Control Parameters for Gait Generation

	Parameter	Max value (mm)	Min value (mm)
Paw motion length	m	60	20
Paw motion width	n	50	10
X coordinate of COR	Xo	70	-20
Y coordinate of COR	Yo	140	50
Z coordinate of COR	Zo	40	-10
Paw rotation angle	α	90	-90
Paw moving speed	s	600ms/step	300ms/step

4.2 Implementation of Evolutionary Algorithms

We model natural processes, such as selection, recombination, mutation, migration, locality and neighbourhood. Figure 9 shows the structure of a genetic algorithm we implemented. Evolutionary algorithms work on populations of individuals instead of single solutions. In this way the search is performed in a parallel manner.

4.3 Evolving Results

The evaluation takes place inside of AIBO robot's football pitch. Each generation member produces a gait that runs for 5 steps. The 13th gene specifies the speed. If the robot falls over, it is fully capable of getting up and continuing its execution. After executing one member evaluation takes place and the speed and stability parameters are produced which qualitatively determines fitness values. The overhead camera records offset from the starting position and the changes when the robot to return to the starting position. In order to do that, an appropriate number of steps, e.g. forwards and backwards, are executed by the robot. We adopted a population size of 20 and run it for 50 generations, and results are presented in Figure 10.

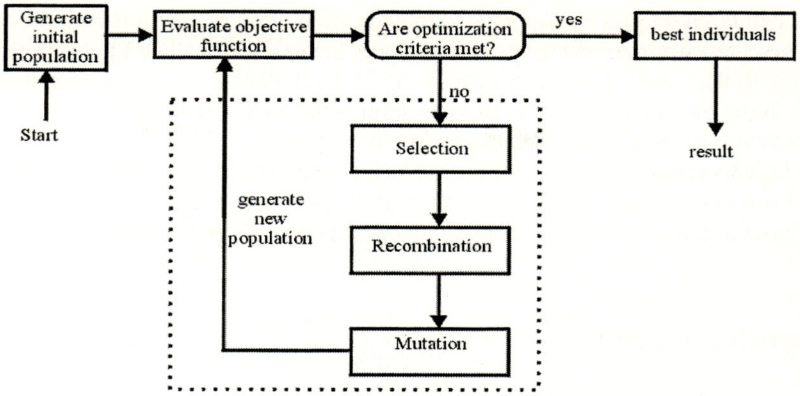

Fig. 9. Structure of a single population evolutionary algorithm

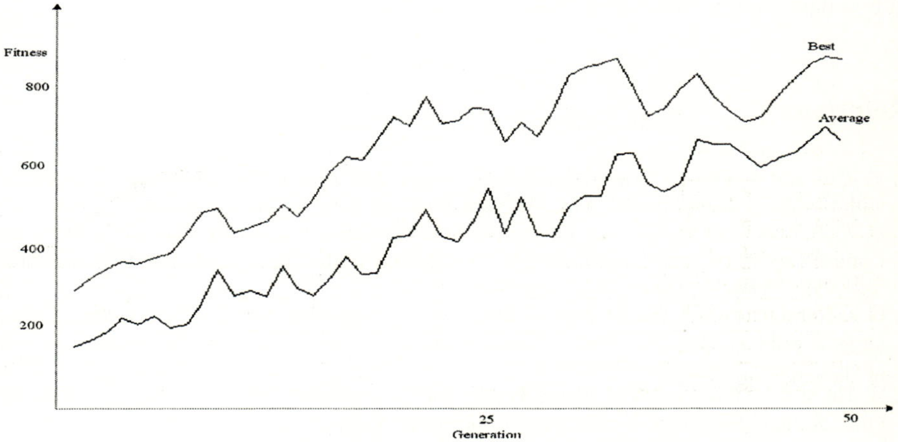

Fig. 10. The results of gait generation based on GA

At the beginning, most of the population members didn't perform very well. Some of them didn't even walk in the straight line or they even walked backward. Some of them were also causing the robot to fall. By the end of the experiment, the performance of the members significantly improved. The best member manages to walk in the straight line with good stability with the speed of 7m per minute.

We also investigated the improvement of speed and stability over a sequence of generations and the affects it has on the overall fitness functions, see [2] for details.

5 Conclusions and Future Work

A hybrid experimental platform for Sony AIBO robots is developed in this paper, which is a useful tool in the development of vision and gait generation algorithms in the RoboCup domain. Based on this platform, we can develop different vision algorithms for colour segmentation, object recognition and object tracking. Sony AIBO robots can learn good walking behaviours with little or no interaction with the designers. Once the learning method is put into place, the module can learn through its interaction with the world. The mutating and combination behaviours of the Genetic Algorithms allow the process to develop to a useful behaviour over time.

Our future work will be concentrated on three aspects:

- how to achieve the efficient operation and evaluation of tracking algorithms;
- how to improve vision system in terms of speed and accuracy;
- how to build robust and adaptive algorithms for gait generation.

Acknowledgements

We wish to thank Prof. Pierre Blazevic and his L.R.P research team for the code of the earlier version experimental platform, which gave us a starting point and motivation to develop this interactive software platform.

References

1. Y. Cui and J. Weng, "View-Based Hand Segmentation and Hand-Sequence Recognition with Complex Backgrounds", Proc. 13th Int. Conf. on Pattern Rec., pp. 617-621, 1996
2. D. Golubovic and H. Hu, An Interactive Software Environment for Gait Generation and Control Design of Sony Legged Robots, Proceedings of the 5th International Symposium on RoboCup, Fukuoka, Japan, 24-25 June 2002
3. D. Golubovic and H. Hu, A Hybrid Evolutionary Algorithm for Gait Generation of Sony Legged Robots, The 28th Annual Conference of the IEEE Industrial Electronics Society, Sevilla, Spain, November 5 - 8 2002
4. H. Hu and D. Gu, "A Multi-Agent System for Co-operative Quadruped Walking Robots", Proc. of the IASTED Int. Conference Robotics and Application, pp. 182-186, Aug. 2000
5. M. Isard and A. Blake, "Condensation -- conditional density propagation for visual tracking," Int. Journal of Computer Vision 29(1), pp. 5--28, 1998
6. B. Li, H. Hu and L. Spacek, An Adaptive Colour Segmentation Algorithm for Sony Legged Robots, Proceedings of the 21st IASTED Int. Conference on Applied Informatics, pp. 126-131, Innsbruck, Austria, 10-13 February 2003
7. T. Rendas and J. Folcher, "Image Segmentation by Unsupervised Adaptive Clustering in the Distribution Space for AUV guidance along sea-bed boundaries using Vision", 2001
8. S. Schröter, "Automatic Calibration of Lookup-Tables for Colour Image Segmentation", Proc. 3D Image Analysis and Synthesis, pp. 123-129, 1997

A Rule-Driven Autonomous Robotic System Operating in a Time-Varying Environment

Jia Jianqiang, Chen Weidong, and Xi Yugeng

Institute of Automation, Shanghai Jiaotong University
200030 Shanghai, P.R. China
mr_jia@sina.com, {wdchen,ygxi}@sjtu.edu.cn

Abstract. In this paper, the problem concerning how to coordinate concurrent behaviors, when controlling autonomous mobile robots (AMRs), is investigated. We adopt a FSM (finite state machine)-based behavior selection method to solve this problem. It is shown how a hybrid system for an AMR can be modeled as an automaton, where each node corresponds to a distinct robot state. Through transitions between states, robot can coordinate multiple behaviors easily and rapidly under dynamic environment. As an illustration, a soccer task was finished by an AMR system with this method. The robot performed well in the soccer games and won the game in the end.

1 Introduction

For an autonomous mobile robot, the ability to function in, and interact with a dynamic, changing environment is of key importance [1, 2]. A successful way of structuring the control system in order to deal with this problem is within hybrid architecture [3, 4]. This way of structuring the control system has the major advantage that it makes the system own planning ability, and at the same time, the system can react rapidly.

However, within this framework, a number of design issues need to be addressed. An important issue is how to deal with multi-behavior coordination problem. For instance, given a reactive obstacle-avoidance behavior, how should an approach–target behavior be designed and combined with it? There are two main methods to manage this problem: one is arbitration [5, 6, 7]. That is, select one behavior between several ones based on priority or through competition. This kind of method has definitude meaning and new behaviors can be added easily. But it has the major disadvantage that it both affects the performance in a negative way, not allowing for the smooth performance that concurrently active behaviors produce, and that it increases the risk of introducing chattering into the system [8]. Another method is to work with concurrently active behaviors. Different controllers affect the system simultaneously, resulting in a smooth overall performance [9, 10, 11]. However, in this case, the analysis of the system is becoming difficult as new behaviors are added. Egerstedt proposes regularization techniques to improve the first method. But he pays more attention to smoothness between behavior transitions and ignores rapidness of the system. For tasks under dynamic environment, simple, rapid, and efficient control method is necessary.

D. Polani et al. (Eds.): RoboCup 2003, LNAI 3020, pp. 487–494, 2004.

Now, autonomous soccer robot is a hot topic in AI and robot fields [12, 13]. The game has been a standard platform for theory study under dynamic, uncertain environment. In this paper, we first introduce an AMR system applicable under dynamic environment. Then, an FSM-based behavior selection method is adopted and used in the robot soccer game successfully. With this method, the robot can coordinate multiple behaviors easily and react rapidly under dynamic environment. The result of 1st CRC games (the First Chinese Robot Competition) shows the validity of the method.

2 System Description

Fig. 1. JiaoLong soccer robot

The JiaoLong robots are constructed as part of a project to build inexpensive, autonomous robots for the study of multi-robot systems. We made three robots of the same type and one goalkeeper robot as shown in Fig.1. The hardware system consists of motion platform, sensing system, communication system and control system.

The motion platform is a 45×45×60cm, differential driven car. Each robot employs a LRF (laser range finder) and two cameras as its main sensors. The whole group utilizes a wireless LAN device for communication. Every robot acts as a node in the LAN and has a wireless card that can transfer data at the speed of 11Mbps.

For software design, we propose a distributed architecture based on priority. There are one main process and three assistant processes sharing one CPU. Main process makes decision and three assistant processes dispose sensor information of LRF and two cameras. Information communication between main process and assistant process is realized through IPC (inter-process communication) mechanism. Also, this mechanism can ensure the communication between robots.

3 System Architecture and Behavior Design

3.1 Hybrid Architecture for Dynamic Environment

Robot architecture can be divided into three kinds: deliberative, reactive and hybrid architecture [3]. The first one always need precise information and is not suitable for dynamic environment. The second one makes the system react quickly and robustly, but it is not suitable for complex task. The hybrid one is suitable for dynamic environment and the system can react rapidly. Our architecture is a hybrid one, but differs from others in two aspects (Fig. 2). First, we add short-time memory in behavior-based control. Secondly, our system only needs local path-planning. The whole task will be divided into several subtasks. Robot schedules tasks according to his perception. In every time cycle, there is only one subtask executed, which generates one behavior. More details will be introduced in 3.2.

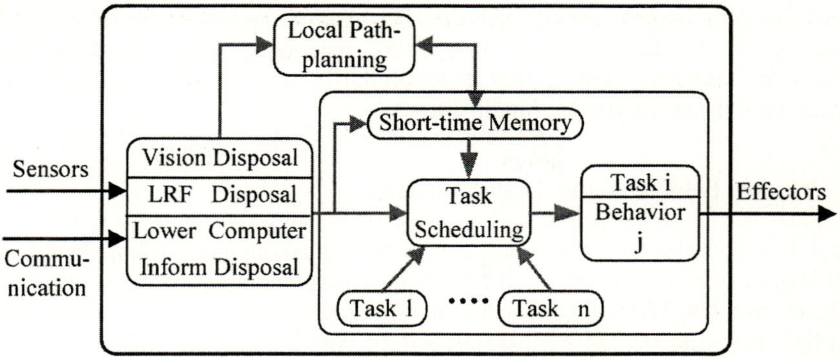

Fig. 2. Planning and control diagram

The reactive architecture characterizes having no memory and reacting rapidly. The long-time memory will ask for better hardware and lead to wrong decision due to the uncertain data. We add short-time memory in the behavior-based control according to the task. Tab. 1 shows an example of the memory content. Fig. 3 shows the meaning of the angle θ. In Fig. 3, X-axis represents robot's heading. θ means the angle between ball centre and Y-axis, and is determined by camera visual angle.

Whenever the robot sees ball, he will update the value of SIGN in Tab 1. If the ball disappears suddenly, robot could find it rapidly according to SIGN. For example, if the ball disappears at time of T, and the value of SIGN at time T-1 is positive, then the robot turns to left. Thus the ball can be found in the shortest time. If ball disappears for a long time (longer than 20 seconds), the SIGN will be set to zero until ball appears again.

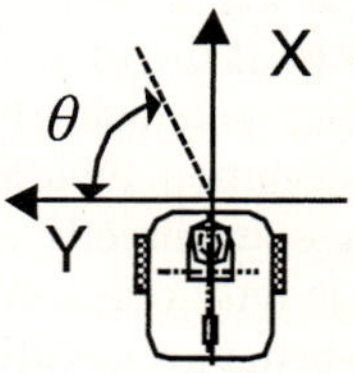

Fig. 3. Short-time memory

Table 1. Short-time memory

SIGN \ θ	$60 \le \theta \le 90$	$90 < \theta \le 120$
Ball	positive (ball on left)	negative (ball on right)
Note: Lose ball for a long time, the SIGN will be zero.		

The global path-planning has some disadvantages when using under dynamic environment. It needs precise environment model, and will make system complex. For example, in football game, the robot always be put under a dynamic environment, a proper path at this time will be improper during a short time, especially when many robots stay together. So it is not worthy of planning globally. Based on this opinion, our system only makes local path-planning. Path-planning lets robot reach a place behind the obstacle, and only needs local information about obstacles. Experiment and games verified that local path-planning could improve the robot's decision ability, and make robot react rapidly.

We also define some special behaviors. For example, when the ball stays on a corner on the playground or made by other robots, our robot will turn suddenly, thus, the ball will be moved out of the corner. Through this behavior, the ball will not be out of play, and robot can show more intelligence.

3.2 Primary Behaviors and Their Synthesis

Soccer is a complex task for robot, especially for autonomous robot. In the game, robot should search ball, move ball and shoot ball autonomously. Moreover, when he meets an obstacle, he must coordinate multiple behaviors. Generally, we decompose the whole task into several subtasks associated with ball. Behaviors are divided into two kinds: manipulating ball and avoiding obstacle. Every subtask generates a behavior associated with the ball. The final behavior should be made through arbitration or fusion. That is, we make a two-step decision: in the first step, every subtask generates a behavior, and then system decides the final behavior. From the view of human's behavior, when we do such a work, there is just a natural behaviors transition but not such a layered decision.

Thus, we decomposed the whole task into four subtasks corresponding to four states: Get_Ball, Move_To_Ball, Search_Ball and Avoid_Obstacle. If robot is in a state of the first three ones, he can manipulate ball directly and needs not to avoid obstacle. If the robot has to avoid obstacle, he will be in the forth state. In this state, a behavior generated by local path-planning will help him leave this state. Different from common avoiding behavior, robot does not care about where the ball locates, he only moves directly to the target generated by path-planning at highest speed, that is, there is only one behavior. The result is that robot escapes from the situation where multi-behavior conflicts, and can manipulates ball directly. This simple behavior makes robot transit from the forth state to one of the other three ones. Moreover, through such task decomposition, there is only one behavior generated in every state at a time, and this is the final behavior of the robot. Robot needs not to select a behavior from multi behaviors. From above, we can see that this method can simplify the control architecture, and new behaviors can be added randomly without making control system more complex.

Every robot has seven basic behaviors: MoveToBall, MoveToTarget, SearchBall, Defend, Shoot, AvoidObstacle and SearchGoal. The robot's program for accomplishing the task can be represented as a FSM diagram as shown in Fig. 4.

In this approach, each state corresponds to a suite of activated behaviors for accomplishing that step of the task. Transitions between states are initiated by real-time perception.

We will give an example to illuminate the algorithm. Suppose the robot is in the state of Move_To_Ball, at this time, an obstacle suddenly appears in the way, the robot then will transfer to the state of Avoid_Obstacle. The path-planning module generates a target for the robot as shown in Fig.5. DIST_1, DIST_2 is related to the robot's size and can be modified. During this period, if the ball moves to the place where it can be manipulated directly by the robot, robot will transfer to other state, otherwise, he will move to the target at all times. During this period, robot will not avoid obstacles except that an obstacle appears in the way and is very near. When robot reaches the target, if he cannot see the ball, he will turn according to his short-time memory. Thus the states transition is realized.

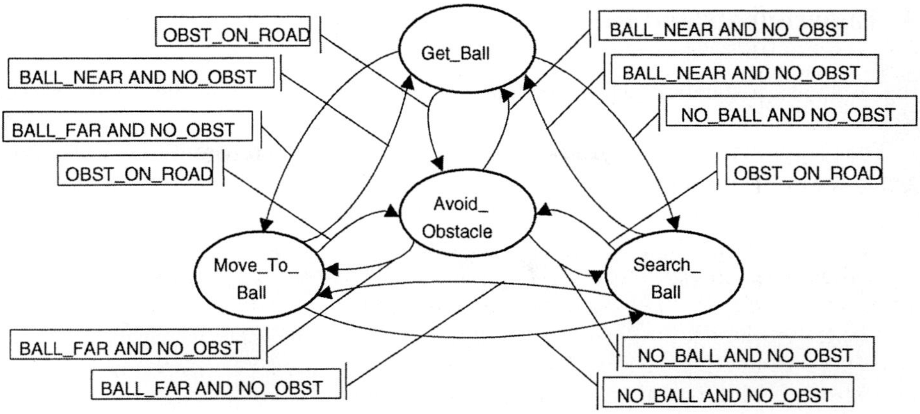

Fig. 4. State transition

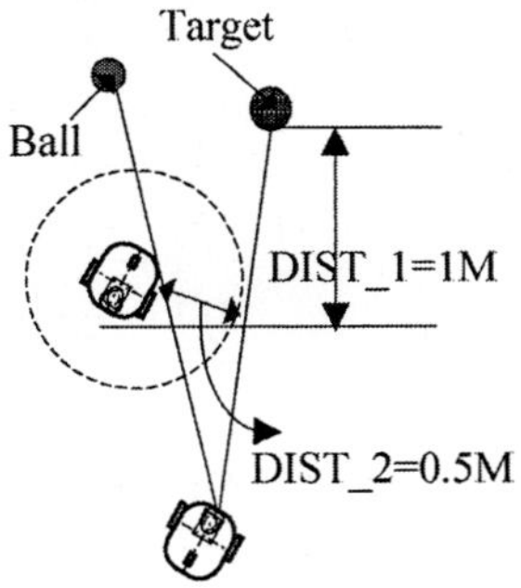

Fig. 5. Local path planning

Formula 1,2,3 can calculate the translational velocity (V) and rotational velocity (ω). In formula 1, DIST means the distance between robot and the goal calculated by visual information. The units of V and ω are mm/sec, degree/sec. P_1, P_2, P_3 are parameters. Ang means the angle of the goal position in robot's local coordinate as shown in Fig. 3.

When the obstacle is very near the robot, the line speed is calculated by formula 3. OBST_DIST is distance between robot and obstacle calculated according to LRF.

$$V = \begin{cases} DIST^2 / P_1 & DIST < 1m \\ 1000 & DIST >= 1m \end{cases} \tag{1}$$

$$\omega = \begin{cases} -40 & Ang >= 100^\circ \\ (90 - Ang) / P_2 & 80^\circ < Ang < 100^\circ \\ 40 & Ang <= 80^\circ \end{cases} \tag{2}$$

$$V = -P_3 / OBST\_DIST \tag{3}$$

4 Experiments and Results

Fig. 6 shows a standard action sequence of our robots played in the "CRC Soccer Robot in China".

The playground is set according to RoboCup rules. As there is only 1vs1 and 2vs2 game, the ground is a 6×5m place. The two teams are both autonomous robots. They can be started from outside but cannot be interfered during the competition.

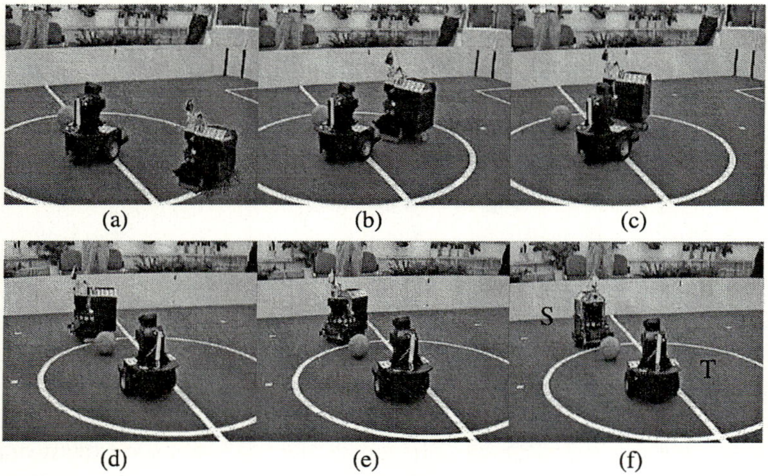

Fig. 6. Action sequence

"S" represents our robot, and "T" is the opponent. In Fig. (a), robot saw the ball, but could not manipulate it directly because of the obstacle in the way. The path-planning module generated a target behind the obstacle. Through actions in Fig. (b)-(c)-(d), robot moved to the target. At this time, robot could not see the ball, so he turned to left according to his short-time memory. Thus, after action sequence in Fig. (e)-(f), robot returned to the state of Get_Ball. Through such task decomposition, robot could finish task rapidly.

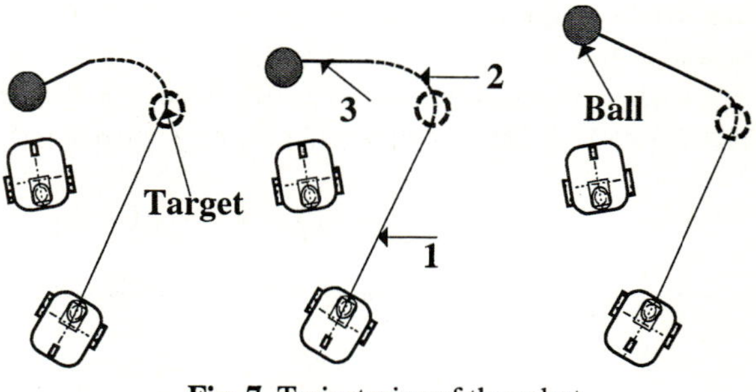

Fig. 7. Trajectories of the robot

Fig.7 shows trajectories of the robot in different situations. In the figure, 2 represents dashed line. 1 and 3 represent straight line. We can divide the whole process into three periods corresponding to three lines. In the first period, the trajectory is generally similar during many cases. The robot moves to the target rapidly and directly. In the second period, the robot moves at a certain translational speed and rotational speed. That is, the trajectory is an arc, and this ensures the smoothness of the behavior transition. After searching for some time, robot will see the ball and move to ball directly. This is the third period.

Fig.8 shows the trajectory of a real robot's motion. The environment is set like figure 7. The dark circle means an obstacle. The unit of X-axis and Y-axis is centimetre.

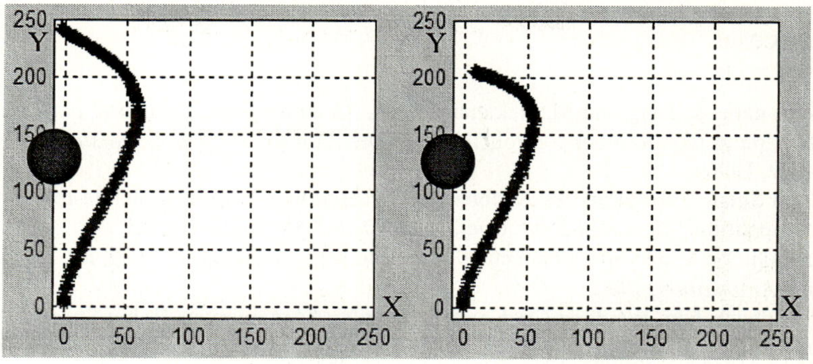

(a) (b)

Fig. 8. The robot's motion trajectory

Tab. 2 shows the time of state transition from Avoid_Obstacle state to Search_Ball state to Move_To_Ball state. These states are corresponding to the three periods of the trajectory as shown in Fig.7. The highest speed of robot is 8000mm/sec and the robot is stationary at the beginning. For each case (a and b), we made experiment for 10 times. The value in the table is an average one. From Fig.8 and Tab.2, we can see that the robot reacts rapidly and the trajectory is smooth.

Table 2. The time of motion

	Period 1	Period 2	Period 3	Total time (second)
Fig.8 (a)	3.25	1.18	1.02	5.45
Fig.8 (b)	3.23	1.15	0.82	5.2

In a word, the proposed method for multi-behavior coordination has two advantages. From the high level, there is only one behavior generated in one state and this is just the final behavior. This behavior is easily understood as this is only related to one goal or obstacle but not a coordinated one. This means the control process is simple and easy to test. From the low level, in the Avoid_Obstacle state, the robot always makes a straight-line motion and can keep a high speed in avoiding obstacles. But in some fusion method, the robot's speed is a low one when the robot is near the obstacle.

5 Conclusion

This paper introduces an AMR system applicable under dynamic environment, and adopts an FSM-based behavior selection method to solve multi-behavior coordination problem. With this method, the robot can coordinate multiple behaviors easily and react rapidly under dynamic environment. At the same time, this method simplifies the control architecture. The results of experiments and games show the validity of the method. In the future, this work should be extended to study more complex tasks such as multi-robot learning and coordination under dynamic environment.

References

1. K. Watanabe, J. Tang and M. Nakamura et al., "A fuzzy-Gaussian neural network and its application to mobile robot control," *IEEE Trans. Control Systems Technology*, Vol. 4, pp. 193-199, 1996
2. Weimin Shen, J. Adibi and R. Adobbati et al., "Building integrated mobile robots for soccer competition," *Proc. IEEE Int. Conf. Robotics and Automation*, 1998
3. J. Connell, "SSS: a hybrid architecture applied to robot navigation," *IEEE Int. Conf. Robotics and Automation*, 1992
4. 4. Changhong Fan, Weidong Chen and Yugeng Xi, "Automonous Robot Soccer System Based on Hybrid Architecture," *4th World Congress on Intelligent Control and Automation*, Shanghai, P.R.China, June 10-14, 2002
5. R. A. Brooks, "A robust layered control system for a mobile robot," *IEEE Journal of Robotics and Automation*, Vol. 2, pp. 14-23, March 1986
6. R. C. Arkin, "Towards the unification of navigational planning and reactive control," *AAAI Spring Symposium on Robot Navigation*, March 1989
7. J. Kosecka and R. Bajczy, "Discrete event systems for autonomous mobile agents," *Proc. Intelligent Robotic Systems*, pp. 21-23, July 1993
8. M. Egerstedt and Xiaoming Hu, "A hybrid control approach to action coordination for mobile robots," *Automatica*, Vol. 38, pp. 125-130, 2002
9. J. Riekki and J. Roning, "Reactive task execution by combining action maps," *Int. Conf. on Integrated Robots and Systems (IROS)*, pages 224-230, Grenoble, France, 1997
10. R. C. Arkin, "Motor schema-based mobile robot navigation," *International Journal of Robotics research*, Vol. 8, pp. 92-112, 1987
11. G. Schoner and M. Dose, "A dynamics systems approach to task level systems integration used to plan and control autonomous vehicle motion," *Robotics and Autonomous System*, Vol. 10, pp. 253-67, October 1992
12. M. Asada, H. Kitano, I. Noda, M. Veloso, "Robocop: Today and Tomorrow What we have and learned," *Artificial Intelligence*, 1999
13. M. Asada and H. Kitano, "The Robocop Challenge," *Robotics and Autonomous Systems*, Vol. 29, pp. 3-12, 1999

Trot Gait Design Details for Quadrupeds

Vincent Hugel, Pierre Blazevic, Olivier Stasse, and Patrick Bonnin

Laboratoire de Robotique de Versailles (LRV)
10/12 avenue de l'Europe, 78140 Vélizy, France
hugel@robot.uvsq.fr

Abstract. This paper presents a full description of the design of a trot locomotion that has been implemented on AIBO quadrupeds in the Sony legged league. This work is inspired by the UNSW achievements in RoboCup 2000 and 2001 in Melbourne and Seattle. The French team rebuilt a complete trot locomotion from scratch, and introduced special features that differ from the Australian original design. Many papers have already been dedicated to the work on quadruped locomotion [2–4]. However they do not detail all the parts of the design.

1 Introduction

This paper is dedicated to the design from scratch of a trot gait that allows combining linear and angular body velocities, which can make the robot walk along a sideways direction or make turns round any point. From 1999 to 2001 the French team developments in locomotion [1] gave good results but motion turns were only possible round points located on the transverse axis of the body. BY the end of 2001 it was decided to design a new trot gait algorithm to allow better maneuverability of the robot and to incorporate the option of performing a crawl-like trot introduced with success by the New South Wales University team in RoboCup 2000. The algorithms developed by the French team are inspired by the details given by the UNSW team in its 2000 RoboCup report [2] but have been rewritten and adapted to include specific features. These features include exact support phase leg trajectories that are not approximated to straight line segments, leg home positions that vary with linear body velocity to improve dynamic balance, and times of changing velocity at leg landing.
Thanks to this locomotion design, the robot should be able to adopt a high-legged trot to cover large distances at high speed, and to adopt a crawl-like gait that is very efficient for approaching and covering the ball. This paper describes the parameters used for locomotion and the details of the design.

2 Parameters Used to Tune Locomotion

2.1 Leg Trajectory Placement Parameters

Leg trajectory placement parameters permit to place the leg trajectory with respect to the body reference frame. We distinguish the horizontal parameters, the α parameter for shifting home position, and the vertical parameters.

D. Polani et al. (Eds.): RoboCup 2003, LNAI 3020, pp. 495–502, 2004.

Horizontal Distance Parameters. Figure 1 shows longitudinal and lateral spacings, and right and left shifts from geometric body centre (GBC) called ΔC_{left} and ΔC_{right}. Longitudinal spacings are taken from the limit defined by the GBC shifts.

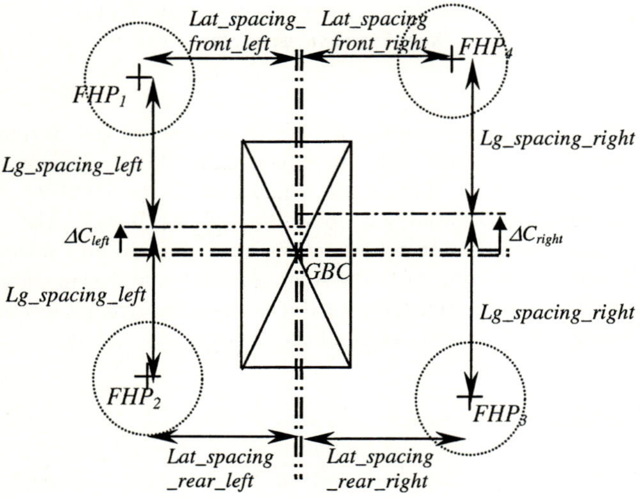

Fig. 1. Horizontal distance parameters that define fixed home positions for legs in stance phase.

Fixed home positions that serve as references are called $FHP1$, $FHP2$, $FHP3$, and $FHP4$. They are defined by the horizontal distance parameters.

Home Position Shifting Parameter. To implement a variable home position called CHP (see fig. 2) the parameter called α is introduced.

The home position shift from the fixed home position (FHP) is expressed as:

$$\overrightarrow{(FHP, CHP)} = \alpha \int \mathbf{v}.dt = \alpha.(\mathbf{v_i} + \mathbf{v_f})/2.T/2 \qquad (1)$$

where $\mathbf{v}$ is the linear velocity, T the cycle period, and $\mathbf{v_i}$ and $\mathbf{v_f}$ the linear velocities at the beginning and the end of the step.
This parameter allows to better distribute the load of the body between the two thrusting legs whatever the motion direction. Hence we get a CHP point for every leg. The CHP point is the middle point of the leg straight-line trajectory in support phase. This straight line joins landing and take-off points of the leg.

Vertical Parameters. They are the height of the front legs and the height of the rear legs with respect to the related shoulder joint.

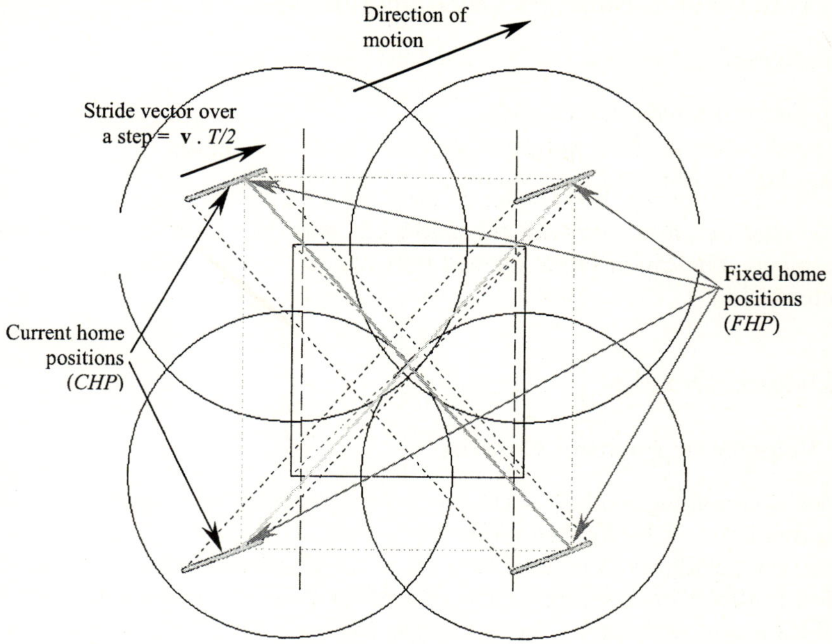

Fig. 2. Top view of the robot achieving a linear sideways motion. Velocity of the robot has two components v_x and v_y. The dotted rectangle joins the fixed home positions (FHP) as they are defined by the trajectory parameters (section 2.1). When a linear motion is required the home position varies as a function of the stride length (that is the velocity magnitude times the period of the stance phase ($T/2$)) and the direction of motion. The variable home positions are called CHP (current home positions). Here they are the middles of the linear stride segments.

2.2 Kinematic Parameters

They describe the shape of the leg trajectory over a cycle period. The kinematic parameters are the following:

- step period $= T/2$, where T is the total time of the leg cycle,
- upward phase height,
- downward phase height,
- time ratio of upward phase with respect to total air phase,
- time ratio of downward phase with respect to total air phase,
- maximal accelerations for v_x,v_y and w.

The upward phase starts when the leg leaves the ground. It terminates when the leg starts its swing phase. The downward phase starts when the leg ends the swing phase. It terminates when the leg touches ground. By varying maximal accelerations, we can vary the time of acceleration/deceleration of the robot.

2.3 Additional Parameters for Oscillation Moves

The oscillation moves that can be introduced are the following:

- up and down moves of the body,
- lateral moves of the body (right and left),
- angular moves round the transverse axis of the body.

The relating parameters are the amplitude and the phase lag of each oscillation move. The phase lag is defined with respect to the time cycle of a leg trajectory.

3 Design Details

3.1 Velocity or Position Control

The decision making module of the robot can run velocity or position control, and switch between both when needed.
Velocity control allows to request for a set of combined velocities, v_y, v_x, and w. Position control allows to request for a set of combined displacements (Dx, Dy, $D\theta$). This means that after a series of steps, the body centre should have moved by the displacement vector (Dx, Dy), and the body should have turned an angle equal to $D\theta$. In position control mode, it is possible to specify maximal velocities during the required displacement.

3.2 Adjusting Velocities to Robot's Leg Workspaces

In velocity control the velocities to be adjusted are the velocities the robot must reach. In position control the velocities to be adjusted are:

- the maximal velocities the robot should not exceed along the displacement required. They are given by the decision module.
- the successive velocities the robot must use to achieve the required displacements. These velocities must converge towards zero.

• **Use of a "workspace" calibration table.** For adjusting velocities a calibration table taking leg workspace boundaries into account is used. This calibration table can be illustrated by a 3D space where point coordinates represent body displacements $(\Delta x, \Delta y, \Delta\theta)$ over a step. A triplet of constant velocities (v_x, v_y, w) is linked to the displacements $(\Delta x, \Delta y, \Delta\theta)$ by

$$(v_x, v_y, w) = \Delta T_{step}.(\Delta x, \Delta y, \Delta\theta), \tag{2}$$

where $\Delta T_{step} = T/2$.

In the $(\Delta x, \Delta y, \Delta\theta)$ 3D space, we can calculate a closed 3D surface whose points represent the optimum set of triplet $(\Delta x, \Delta y, \Delta\theta)_{opt}$ the leg can achieve while remaining inside its workspace.

In other terms, components of points located inside the closed surface are displacements $(\Delta x, \Delta y, \Delta\theta)$ that can be achieved by the robot over one step.

Hence if the set of velocities required is located out of the 3D surface, the robot cannot achieve them. They must be reduced. Hence the decision module commands the robot to reach the optimal triplet of velocities given by

$$(\Delta x, \Delta y, \Delta\theta)_{opt}/\Delta T_{step}, \tag{3}$$

where $(\Delta x, \Delta y, \Delta\theta)_{opt}$ is defined by

$$(\Delta x, \Delta y, \Delta\theta)_{opt} = k.\Delta T_{step}.(v_x, v_y, w). \tag{4}$$

where k is a reduction factor.

Directions of vectors $(\Delta x, \Delta y, \Delta\theta)$ and (v_x, v_y, w) remain the same.

Of course the triplet of velocities remains unchanged if inside the 3D surface. The 3D surface can be computed at boot time using the trot parameters set.

• **How the workspace calibration table is built.** The calibration table contains 3D points of the 3D surface. The table is built using 3D discrete spherical coordinates. Starting with $(0, 0, 0)$ the number of points explored grows radially until the surface is reached. Beyond this surface the set of displacements cannot be achieved over a step.

Given a triplet of displacements $(\Delta x, \Delta y, \Delta\theta)$ over a step, it is marked as being inside the 3D surface if landing and take-off points belong to the workspace of the leg in support. Using this technique $(\Delta x, \Delta y, \Delta\theta)$ is adjusted automatically, and the decision module does not have to set a stride parameter for the gait.

In summary, the 3D surface ensures that leg trajectories will not go beyond the leg working space. And adjusting ΔT_{step} to a minimal value permits to set maximal forward velocity.

3.3 Update Velocity at Next Step or After Next Step

Once a pair of legs is on the ground, the velocity profile that has been planned cannot be changed all along this leg pair support phase.

When the request comes in the first half of the current step, new velocities are taken into account at starting the next support phase (i.e. the next step). When the request comes in the second half of the current step, new velocities are taken into account at starting the after next support phase (i.e. after the next step). This is because we consider that we do not have enough time to re-plan the stance phase trajectory of the current legs in the air.

• **Defining velocities at the end of a step.** A step is defined by an initial and a final triplet of velocities (see section 3.5 for generating support stance phase), and by the landing point. Knowing the initial velocity of the next step and before starting it, the problem is to get the final velocity to define the step completely. This final velocity must be known in advance ($T/4$ before starting

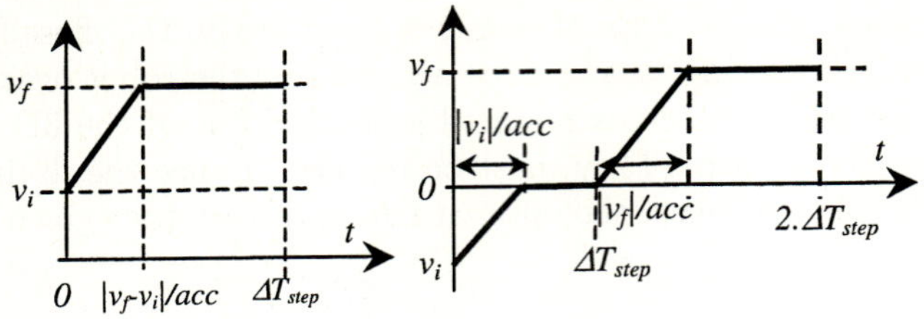

Fig. 3. Linear velocity profiles used to deal with velocity control. v_i is the velocity at the beginning of the step. v_f is the velocity at the end of the step.

the next step at least). All three velocities (v_x, v_y, w) follow a linear profile (see figure 3).

The rate of increase in velocity is defined by a maximal acceleration that is fixed. There is a different acceleration for every velocity v_x, v_y, and w. They have been set so that the robot can reach maximal velocity from still motion in a step time period. This implies: $\Delta T_{step} > |v|/acc$, where v represents whatever velocity, and acc the related acceleration (positive value).

• **Defining velocities in position control.** In position control, the goal is different. Here we do not want the robot to reach a certain set of velocities but we want it to halt motion after trotting the requested distance and turning the requested angle. And to allow re-entrance of velocity control with position control and conversely, position control is designed by updating velocities for every leg switch, like in velocity control. Velocities must converge towards 0 since the robot must halt motion once distances and angle have been achieved.

Let be $(Dx, Dy, D\theta)$ the displacements to achieve in position control. Velocities are updated 2 times a cycle period in the middle of two consecutive steps. At these times of update, the remaining displacement $(Dx, Dy, D\theta)_{remain}$ to perform is calculated The displacements already achieved are simply subtracted from the previous displacements.

Velocities are initialized to $(Dx, Dy, D\theta)_{remain}/\Delta T_{step}$ and filtered using workspace 3D surface (see section 3.2). Then velocities are optimized in order to finish the displacements as soon as possible. In case the initial velocity is too high and distance requested to be covered too small, it can appear that the robot oversteps the point to reach. But even so, the position control algorithm makes the robot joint the point by making it trot in the reverse direction.

3.4 Updating the Landing Point for the Next Step or the after Next Step

Knowing velocities at the beginning and the end of the step considered (i.e. velocities at leg landing and take-off) and by intergrating the velocity profile described in the previous section, we compute the displacements $(\Delta x, \Delta y, \Delta\theta)$ over this step. Then landing point coordinates are computed the following way: if $(\Delta x, \Delta y)$ stands for the linear displacement over a step (legs numbered 1 and 3 for instance) and $\Delta\theta$ for the angle to rotate, we have A and A' respectively

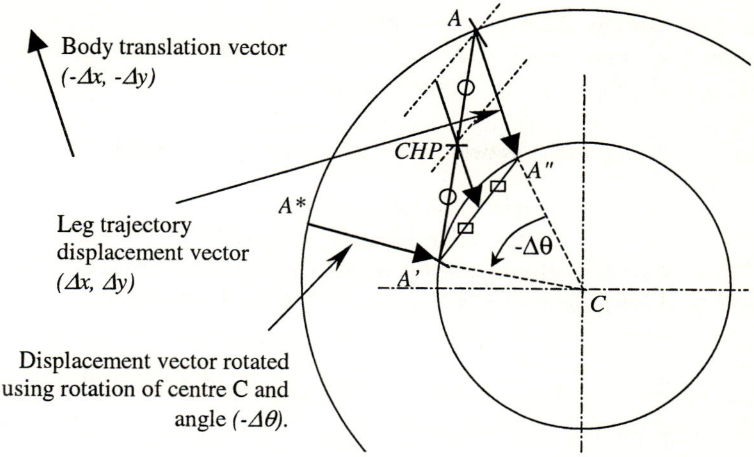

Fig. 4. A are $A^{'}$ are the landing and take-off points of the step. The body turns round C (GBC shifted) with an angle $\Delta\theta$. Hence the leg tip turns (-$\Delta\theta$). If we start with the translation, we get to $A^{'}$ by passing through $A^{"}$. If we start with the rotation, $A^{"}$ is the image point obtained by rotation of leg landing point A. Then the translation of vector $A^{*}A^{'}$ must be performed.).

landing and take-off points of the first leg (1) of the diagonal pair (figure 4), and B and B' referring to the second leg (3), expressed as:

$$CHP_1 = FHP_1 + \alpha(\Delta x, \Delta y)$$
$$CHP_3 = FHP_3 + \alpha(\Delta x, \Delta y)$$

$$\begin{aligned}
A_x &= \Delta x/2 + CHP_{1x} - (-\Delta y/2 + CHP_{1y}).tan(\Delta\theta/2)\\
A_y &= \Delta y/2 + CHP_{1y} + (-\Delta x/2 + CHP_{1x}).tan(\Delta\theta/2)\\
A_x^{'} &= -\Delta x/2 + CHP_{1x} - (-\Delta y/2 + CHP_{1y}).tan(\Delta\theta/2)\\
A_y^{'} &= -\Delta y/2 + CHP_{1y} + (-\Delta x/2 + CHP_{1x}).tan(\Delta\theta/2)\\
\\
B_x &= \Delta x/2 + CHP_{3x} - (-\Delta y/2 + CHP_{3y}).tan(\Delta\theta/2)\\
B_y &= \Delta y/2 + CHP_{3y} + (-\Delta x/2 + CHP_{3x}).tan(\Delta\theta/2)\\
B_x^{'} &= -\Delta x/2 + CHP_{3x} - (-\Delta y/2 + CHP_{3y}).tan(\Delta\theta/2)\\
B_y^{'} &= -\Delta y/2 + CHP_{3y} + (-\Delta x/2 + CHP_{3x}).tan(\Delta\theta/2)
\end{aligned} \tag{5}$$

These equations refer to figure 4 that describes how landing and take-off points of the first leg of the diagonal support are defined knowing translation and turn-in-place motions of the body ($\Delta x, \Delta y, \Delta\theta$) over the step.

3.5 Computing Current Cartesian Point of Leg Trajectory

The reference frame considered is the body reference frame whose plane xy is horizontal and whose centre is C (C is the point shifted from the geometric body centre GBC by ΔC along the longitudinal axis).

• **Stance phase.** At every time interval of the stance phase displacements $(dx, dy, d\theta)$ performed from time of landing until present time are calculated by integrating the velocity profile, knowing velocities at the beginning and the end of the step.

Then the stance point is calculated using these displacements by:

$$
\begin{aligned}
cpx &= LandingPoint.x - dx \\
cpy &= LandingPoint.y - dy \\
current\_stance\_point.x &= cos(d\theta).cpx + sin(d\theta).cpy \\
current\_stance\_point.y &= -sin(d\theta).cpx + cos(d\theta).cpy \\
current\_stance\_point.z &= front\_height\_or\_rear\_height
\end{aligned}
\tag{6}
$$

• **Air phases.** They are implemented using the heights to move the leg up and down, and the next landing point.

4 Conclusion and Perspectives

Trot parameters can be adjusted easily by updating the calibration table. Hence it is possible to implement high-legged trot or the trot introduced by the UNSW team where the fore part of the body is lowered. However the whole locomotion is based on the trot gait. And it would be great to be able to switch between trot gait and other gaits such as crawl gaits that feature always three legs on the ground. Crawl gaits can be useful for approching maneuvers when slow and precise moves are required. In addition maybe gaits faster than trot can be introduced. These gaits involve the control of ballistic phases. New studies will focus on how to switch between trot, crawl and faster gaits continuously.

References

1. Hugel, V.: On the Control of Legged Locomotion Applied to Hexapod and Quadruped Prototypes, PhD Thesis (in French), Université de Paris VI, Nov. 1999.
2. Hensgt, B., Ibbotson, D., Bao Pham, S., Sammut, C.:The UNSW United 2000 Sony Legged Robot Software System, School of Computer Science and Engineering, University of New South Wales, 2000 RoboCup SONY legged league report.
3. Veloso M,Lenser S.,Vail D.,Roth M.,Stroupe A. and Chernova S.: CMPack-02: CMU's Legged Robot Soccer Team, Carnegie Mellon University, 2002 RoboCup SONY legged league report.
4. Upenn, IROS 2001 Hawai. Chitta, S., and OStrowski, J.P.: New Insights into Quasi-static and Dynamic Omnidirectional Quadrupedal Walking, IROS 2001.

The High-Level Communication Model for Multi-agent Coordination in the RoboCupRescue Simulator

Stef B.M. Post, Maurits L. Fassaert, and Arnoud Visser

Computer Science Department, University of Amsterdam, Kruislaan 403
1098 SJ Amsterdam, The Netherlands
Tel: ++ 31 20 525 7532, Fax: ++31 20 525 7490
{sbmpost,mfassaer,arnoud}@science.uva.nl
http://www.science.uva.nl/~arnoud/research/roboresc/

Abstract. In this article we will concentrate on the communication problems in a multi-agent system, operating within the 'RoboCupRescue' Simulator system. To cope with the limited communication between the center and the agents in the field, we separate the communication in two layers that focus on synchronizing world models with different levels of detail, responsiveness and range. In this article we will explain the requirements and methods used in the high-level communication that distributes summaries of the current situation in different sectors of the map.

1 Introduction

The world is often shocked by catastrophic events. These may range from natural occurrences, such as earthquakes or floods, to urban riots or terrorist attacks. Crisis management is essential under such circumstances. Unfortunately, real large catastrophes do not only mean thousands of deaths or injured people, but also hit the communication and civil protection infrastructure, jeopardizing all but the most carefully prepared rescue plans. There is always more to be done in order to minimize such damage in the future. It is possible that the new solutions may be aided by the use of advanced technologies such as robotic rescue operators and artificially intelligent expert systems.

In this article we will investigate multi-agent communication within the 'Robo CupRescue Simulator System' [1]. The RCRSS simulates a small piece of a real world environment, in which a disaster takes place. The simulation starts with an earthquake, after which several buildings in the disaster map collapse. This causes buildings to catch fire, roads to be blocked and civilians to get buried under the debris. Simulated rescue agents need to respond in the most appropriate way, in order to minimize damage.

In section 2 we will determine what kind of information system is needed for the agents to accomplish this goal. In section 3 we will then explain the details of how we implemented this system. Finally in section 4 we will show that our design does indeed meet our requirements.

D. Polani et al. (Eds.): RoboCup 2003, LNAI 3020, pp. 503–509, 2004.

2 Requirements

2.1 Agent Objectives

There are three different types of intelligent entities in the simulator system: civilians, agents, and centers. Civilians represent ordinary people who can not be controlled by our multi-agent system, they are programmed to flee from danger and run to refuge buildings. The main objective of agents and centers is to save as many civilians as possible and to minimize damage due to fires. The agents and centers are separated in three operational types: the fire brigade to put out fires, ambulance units to rescue civilians from under the debris and to transport them to refuges, and police units to clear the roads.

The agents perform the actual rescue tasks. This requires a fair amount of localized intelligence. There are many practical problems to overcome like getting to the location of the next task, making sure a fire brigade has enough water to put out the fire and making sure there is room in an ambulance to pick up more wounded. Simple operational subprograms take care of the most basic logistics. We call these subprograms 'behaviors'. Selecting which behavior to execute is a decision making process that can be handled in many different ways. In our approach all the decision making is done by the localized intelligence of the agent. The suggested hierarchy of the centers standing above the agents is used only for spreading information and not for coordinating task assignment. The agents decide which other agents to cooperate with and what tasks to perform. To do this they need an overview of the global situation. They can not base this overview on their limited visual observations alone. This is why communication is required to update this overview.

2.2 Situation Awareness

The coordinating part of each agent relies heavily on information. Because of the limited visual range of agents, information gathering will be an active task. Some of the available agents will be patrolling the disaster space looking for problems that are not yet known by the agents. After a risk area is identified, a risk assessment is made. This consists of the measuring of the extends of a fire, the number of collapsed buildings, the number of buried civilians and their survival chance and finally the number of blocked roads. Based on this information a distribution of agents to tasks can be made.

The information that is observed by an agent is combined with the information that is communicated to it into a world model [2], which is used in the decision making process. This world model is based on summaries of what agents encounter in the field so it will not be a perfectly detailed model, nor should it be. Cutting down on the number of variables in the world will make the job of the decision making process easier. It is also a more realistic simulation of what goes on in a real disaster control center.

The information in the summaries can be limited by aggregating the details for certain areas. By grouping the elements on the map (roads, building) of the simulated city by a algorithm described in [?], the information that has to be exchanged can reduced considerably. We call the grouped elements on the map sectors. A sector can be seen as a possible problem area. Observations of properties like burning buildings, blocked roads, collapsed buildings can be summed up to give a factor of these problems for all sectors. The positions of all agents are reported and send along with the sector values. Together these values form an information package, that is sufficient for an agent to decide which problem area to go to. By making this decision for other agents as well, an agent can find out what other agents are going to be aiding it there.

The information in the summaries does not need to be detailed in the sense that agents can plan their actions with it. For this the agents use their observations supplement by a more direct method of communication. The execution of the tasks and the communication required for this is not the focus of this paper. The coordination of what problem area to focus on is separated from the cooperation that is required to fulfill this task.

2.3 Information Usage

The right number of agents should cooperate to accomplish a task. Too many agents dealing with a single fire is wasteful and hampers secondary goals like patrolling or containment of another fire. A decent estimation of the required team composition has to be made. To do this the information in the world model needs to be accurate.

The precision of the summaries is best secured by making enough observations. Some agents are reserved to patrol the area and make these observations. The combination of these observations into one state vector has to be done by weighing new data based on trustworthiness. In this process visual observations are trusted over communicated ones. The time of observations is remembered and compared with the time of arrived communication. Only communicated information that is newer than the last observation should be integrated in the overview. Only the observations that were made since the last communication, should be transmitted to other agents.

Agents use the world model to choose their next task. Because of the limited time and number of agents during a simulation, this has to be done in a very efficient way. It is important that fire agents travel towards a new fire as soon as it has been detected. Because of this the information in the world model has to be recent. On the other hand big decisions like switching teams do not have to be reconsidered as often as decisions on what action to take next. A reaction to every small change in the overview would make it harder for agents to finish their tasks. This is why agents choose new tasks and teams only after new global information is available and therefore this information does not have to be updated every cycle. As an extra bonus combining the data of sectors over a period of time, further reduces the amount of information transmitted concerning overviews by discretizing it in the time dimension. Still the information needs to reflect sudden developments accurately, so the period we chose for these updates is the minimum number of cycles required to synchronize the summaries between agents.

Table 1. Agents discard messages they do not want to hear, using only sender and receiver identities contained within the message; Tell messages can only be heard by agents of the same type, i.e an ambulance-center that controls ambulance agents, where n is the number of platoon agents that are in control of a center; Say messages can only be heard by agents that are within a $30m$ radius while the visual range is $10m$.

AGENT CATEGORY	AK_TELL/AK_SAY	KA_HEAR
Platoon agent	4	4
Center agent	$2 * n$	$2 * n$

Table 2. Message format for agents; sending relates to AK_TELL and AK_SAY messages, whereas receiving relates to KA_HEAR messages. Note that the message size is limited to 80 bytes.

sending:	SENDER_ID	MESSAGE	
receiving:	RECEIVER_ID	SENDER_ID	MESSAGE

To prevent misunderstandings in the cooperation of agents the information in the agents needs to match. The centers not only fulfill this requirement by distributing the information, but also an active role in maintaining a common view. They accumulate observations from the agents they have a direct communication channel with, which are agents of the same type. Next they combine these into one summary by negotiating with each other. The same common summary is then send back to all agents at the same time. This guarantees that all agents base their task and team decisions on the same information.

3 Solution

This section will explain the working of the communication protocol which is an attempt to make the world model as accurate as possible. The protocol consists of two different communication layers. In the low level communication layer, 'low level information' is transferred between agents that are near to each other. The high level communication layer distributes 'high level information' amongst all of the platoon and center agents. Low level information refers to basic observations that are directly relayed to other agents in the field, whereas high level information is first extracted from the low level information before it is distributed. The two communication layers match nicely on the two different kinds of message primitives provided by the kernel.

The message primitives are implemented by means of SAY and TELL messages. SAY messages can be used to talk to another agent and it only reaches agents in a limited range. However it does not matter what kind of agent is on the receiving side. The opposite is true for TELL messages where agents must be of the same type in order to communicate, but it arrives at all agents and centers of the same type regardless of distance. For both message primitives certain restrictions are imposed on the amount of messages that may be sent or received during each simulation cycle. Moreover each message should adhere to a special format. The details have been summarized in tables 1 and 2.

3.1 The Communication Protocol

Within the communication protocol TELL messages are used for the high level communication layer and SAY messages for the low level communication layer. The presence of communication restrictions during each cycle leads to the definition of 'communication phases'. An agent can only enter a new phase after at least one cycle has passed since the previous phase was entered. For high level communication, each phase has the property not to violate the restrictions that were mentioned in previous section. For example platoon agents will not produce more than four messages during a phase. It is also ensured that the receiving party is ready to accept all incoming messages without the need to throw them away. In particular centers indicate that they are ready for reception by sending a single TELL message. Thus within the high level layer, agents wait until all participating agents have reached the appropriate phase before they exchange information. The following phases have been defined:

- Platoon agents: AC_EXCHANGE, AA_SAY
- Centers: AC_EXCHANGE, CC_READY, CC_EXCHANGE, CA_READY

These phases are entered in this specific order (reading from left to right) and when the centers enter the last phase, one high level communication cycle has finished. Then a new communication cycle starts and each agent enters the first phase again. Experiments showed that 5 kernel cycles were needed to finish 1 high level communication cycle. So during the simulation there are $300/5 = 60$ high level cycles.

The first two letters of each phase describe what kind of agents participate in the communication. For example the prefix AC means that there is communication between a platoon agent and a center agent. The suffix gives an indication of what happens during a particular phase. The following list clarifies this:

- AC_EXCHANGE
  Platoon agents provide their centers with new world information, after which they obtain summaries that have been produced by the centers. Soon after, the platoon agents enter the AA_SAY phase, enabling them to do low level communication.
- CC_READY
  The centers prepare to enter the CC_EXCHANGE phase by first entering the CC_READY phase. In this phase they notify other centers that they are ready. This is when all AC_EXCHANGE messages have been processed.
- CC_EXCHANGE
  The centers distribute the new world information among each other, and merge the into summaries. Then they enter the CA_READY phase.
- CA_READY
  Platoon agents prepare to enter the AC_EXCHANGE phase, when their centers notify that they are ready.

So what actually happens is that agents communicate with their centers, exchanging information. Then the centers communicate with each other while the platoon agents can do low level communication with agents in their vicinity. Once the centers are ready the process starts over again. We want to emphasize that this high level communication scheme is synchronous; all agents know about the same information at approximately the same time. This is very useful because by using high level information (summaries) one can predict what other agents are doing. Another feature is that if for some reason the high level communication layer breaks, then the platoon agents remain in the AA_SAY state so the low level layer is still operational. Table 3 summarizes the differences between the high level and the low level communication layer.

Table 3. High level communication versus low level communication. A summary contains high level information, whereas object data equals low level information.

property	HIGH LEVEL	LOW LEVEL
datastructure used :	summaries	object data
communication pattern :	platoon agents to centers and back	only between platoon agents
agent types :	homogeneous	heterogeneous
message primitive :	TELL messages	SAY messages
reception :	synchronous, nothing lost	uncertain

4 Research Methods

4.1 Proving the Design Goals

More important than our ranking in the RoboCup Rescue competition is proving that we reached our design goal presented in this document. Our goal was making sure that the information provided by the proposed communication system is accurate, and recent enough to base decisions on and lastly that it matches whenever it is used to make decisions. We have measured the total percentages of buildings that are known to be burning, during a simulation that is typical for a competition. The map we used is known is Kobe and it is measured over 300 cycles with a couple of stationary agents on different sides of the map.

4.2 Results

To measure the precision we have to compare the world models of an agent in the field with the actual situation. It is expected that the error will be large in the first few iterations. As agents scout out the map their approximation of the real situation should improve fast and keep up with new developments like spreading fire. The amount of spreading fire as known to the kernel is shown in figure 1. This amount is compared to the knowledge of agents. We have measured the amount of fire known by two agents and show that it exceeds the amount of fire known by a single agent.

The amount of known fire of an agent in communication with another agent exceeds the amount of known fire of a single agent. What is more important is that the error between the amount of actual fire and the amount of known fire is a lot smaller. Because large, hard to extinguish fires are easier to spot from far away the difference mostly consists of young fires that are easier to extinguish and therefore more important to take action against.

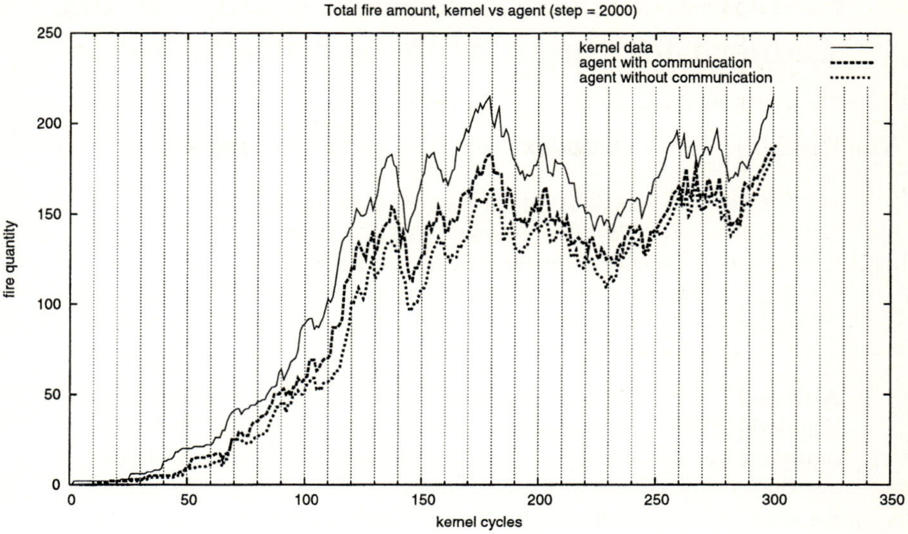

Fig. 1. Two agents patrolling together increase their knowledge of burning buildings over that of a single patrolling agent. The vertical lines mark high level communication cycles.

5 Conclusion

With an early version of this approach we have competed in the 2003 version of RoboCup Rescue Simulation League [3]. Unfortunately technical problems and flaws in our situation negotiation algorithm prevented us from seeing the effects of our approach reflected in our competition scores. We have shown that it is now working as expected and hope to see improvements in our competition results because of it.

Acknowledgments

This paper is based on preliminary research on RoboRescue Simulation done by: S. Post, K. Luikx, J. Uijlings, R. Geertzema and V. Mahabir

References

1. T. Morimoto
`http://ne.cs.uec.ac.jp/~morimoto/rescue/manual/manual-0_00alpha1.ps.gz`
2. M.L. Fassaert, S.B.M. Post, A. Visser "The common knowledge model of a team of rescue agents", 1th International Workshop on Synthetic Simulation and Robotics to Mitigate Earthquake Disaster, Padova, Italy, 6 July 2003
3. A. Visser, S.B.M. Post, M.L. Fassaert, "The communication approach of the 'UvA Rescue C2003'-team", in D. Polani, B. Browning, A. Bonarini, K. Yoshida (Eds.), RoboCup 2003, Lecture Notes on Artificial Intelligence, Springer Verlag, Berlin.

Pseudo-local Vision System Using Ceiling Camera for Small Multi-robot Platforms

Yasuhiro Masutani, Yukihisa Tanaka, Tomoya Shigeta, and Fumio Miyazaki

Graduate School of Engineering Science, Osaka University
Toyonaka, Osaka 560-8531, Japan
soccer@robotics.me.es.osaka-u.ac.jp
http://robotics.me.es.osaka-u.ac.jp/OMNI/

Abstract. *Pseudo-local vision system*, which simulates visual information derived from an on-board camera of mobile robot based on a ceiling camera image, is proposed. It consists of a vision server and a client module which communicate with each other in the SoccerServer-like protocol. An image processing module for the on-board camera in the control program is replaced with this system. The simulated visual information is not a two-dimensional image data but a one-dimensional array which represents the nearest edge in each direction around the robot. However, it contains some of essential information of the on-board camera image. This concept was implemented for our robot system for the RoboCup Small-Size League. The server can transmit the edge data to 10 clients 30 times per 1 second. The time lag between grabbing image on the server and extracting visual information on the client is about 10[ms].

1 Introduction

Coordination of multiple autonomous robots with vision is an interesting research theme. However, autonomous robots equipped with an on-board camera tend to be large and require large experimental environment like the RoboCup Middle-Size League. On the other hand, we developed a small multi-robot platform equipped with an on-board camera and participated in the RoboCup Small-Size League. Since it is not easy to load all of the system on the robot for reasons of space, power, CPU capability, and so on, the video signal of the on-board camera is transmitted by radio link to PC outside the field and is processed on it (Fig. 1 [1]). Accordingly trouble in the video radio link (e.g. interference) is fatal for our system and we could not demonstrate the fruit of our research in the RoboCup2001 competition.

To avoid such a trouble, we propose a *pseudo-local vision system*, which generates the on-board camera information approximately from a ceiling camera image and has compatible function with an image processing module for the on-board camera. It is a client/server system. The server periodically sends visual information computed from the ceiling camera image to the clients which control the robots. Using the earlier version of this system, we participated in the RoboCup2002 competition (Fig. 2[3]).

D. Polani et al. (Eds.): RoboCup 2003, LNAI 3020, pp. 510–517, 2004.

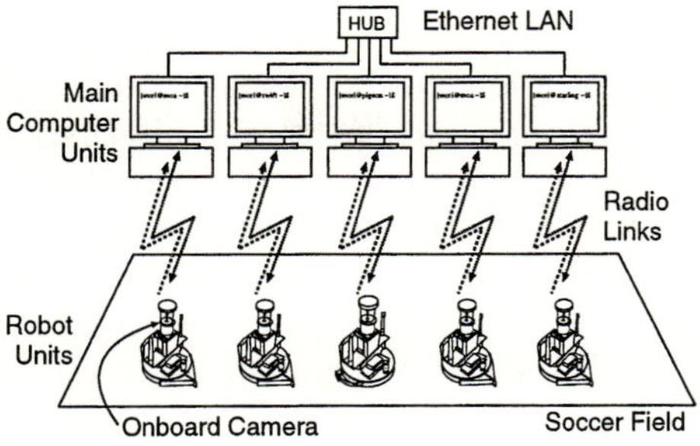

Fig. 1. Real local vision system

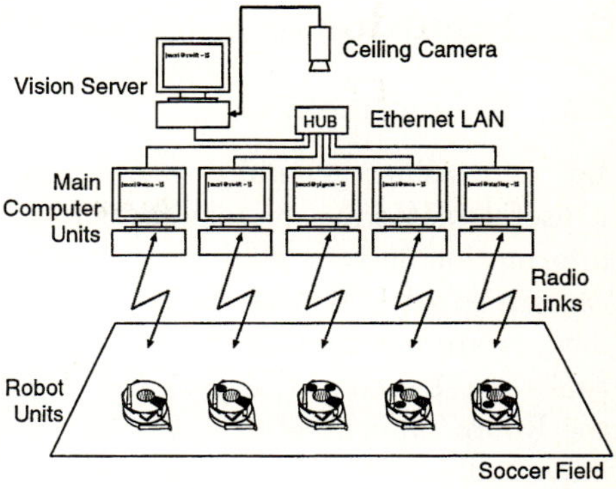

Fig. 2. Pseudo-local vision system

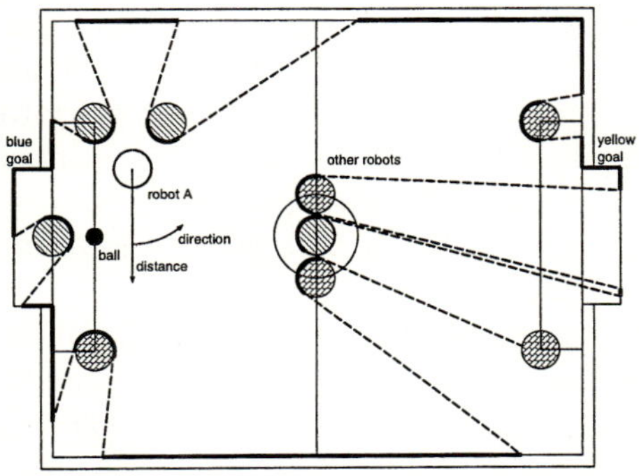

Fig. 3. Nearest edge

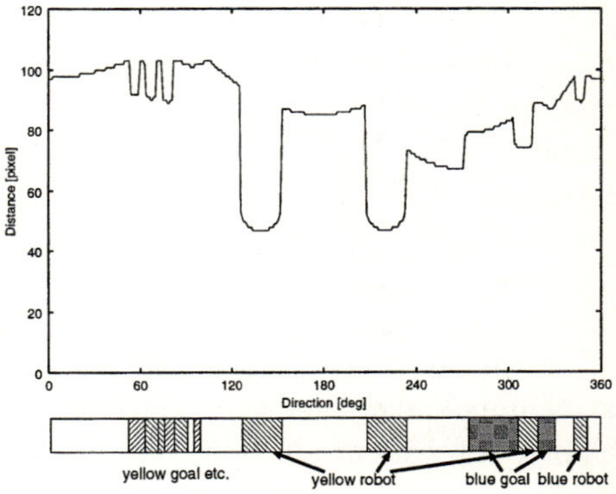

Fig. 4. Pseudo-local vision information

In such a system, there are some methods of generating the on-board camera information from the ceiling camera image. It is hard to make two-dimensional image data and to send it for 10 clients 30 times for 1 second. Therefore, in this paper, we propose a method of sending one-dimensional array data which represents the nearest edge on the floor in each direction around the robot. Fig. 3 and Fig. 4 show examples of the nearest edge of robot A and the one-dimensional array data respectively. We call this edge data *pseudo-local vision information.* It is not a real image (two-dimensional data), but contains some of essential information of the on-board camera image. For example, it represents a hidden object and overlapping objects. This system is useful as a research platform of autonomous mobile robot not limiting to the RoboCup.

This paper is organized as follows. In section 2 we discuss functions required for the pseudo-local vision system. Next the developed system is explained in section 3. In section 4 we evaluate the system. Finally conclusion and future works are described in section 5.

2 Discussion about Required Functions

2.1 Requirement

Let us consider generally requirements for the pseudo-local vision system. As a useful platform, we require that (P1) It can simulate image or alternative information derived from the on-board camera, (P2) Program code for on-board camera robot is almost available without change, (P3) Special equipments and devices are not necessary, and (P4) It can provide data for multiple robots 30 times per 1 second (same as the standard video rate). Moreover, as a tool for the RoboCup competition, we require that (R1) All teams can share the server impartially, (R2) It can serve 10 robots maximum with small time lag, and (R3) The protocol is easy to use.

2.2 Solutions

Since what is derived from the ceiling camera image is two-dimensional information on the floor plane, it is impossible to directly convert it into the on-board camera image of three-dimensional object on the floor. Thus, it is necessary to re-construct an image from the positional data derived from the ceiling camera and three-dimensional geometric models. Using recent high-performance processor, one can make simulated images with high fidelity at high speed. However, it is hard to make 10 images 30 times per 1 second.

If the server can output the simulated image as analog video signal, whole system including the frame grabber can be tested and our team does not need to change software at all. However, other teams which do not have on-board camera system need additional hardware to use the pseudo-local vision system. In case that the server transmits the simulated image as digital data, under condition, 8bit/pixel, 640×480pixel/frame, and 30frame/s without data compression for 10 clients, the necessary communication capacity is about 700[Mbps], which is too much for the standard Ethernet.

2.3 Proposed Method

According to our experience of participating in the RoboCup Small-Size League with robots equipped with omni-directional on-board camera, the most significant in the on-board camera information is the free space around the robot and the type of object beyond it in each direction. In such situations, a hidden object and overlapping objects are important problems. Therefore, we propose a method of using a one-dimensional array which represents the nearest edge on the floor around the robot in each direction as a simulated local vision information instead of the two-dimensional real image data. We call the array *pseudo-local vision information*. For example, in Fig. 3 bold lines show the nearest edge of robot A in each direction. Fig. 4 shows the contents of one-dimensional array of the nearest edge of robot A. The horizontal axis means the direction of $0 \sim 360$[deg]. The upper part shows the distribution of distance to the nearest edge. The lower part shows the distribution of type of the nearest edge.

3 Developed System

3.1 Structure

The developed pseudo-local vision system is a client/server system. It consists of the vision server and the client module which is a part of the robot control program (Fig. 5). Two parts are connected with network. Maximum of 10 clients can connect with the server considering that two teams use the server together in the RoboCup competition. The vision server derives position and orientation of all robots and ball from the ceiling camera image, computes the nearest edge data for each robot, and send it to each robot periodically. The client module communicates with server and converts the edge data into a form which can be used by the upper modules. The module replaces the image processing module for the on-board camera.

3.2 Protocol

The communication protocol is similar to that of the SoccerServer[5] on UDP/IP. The method of initializing is same as the SoccerServer. At first the client sends `(init)` message to the server and then receives visual information from the port specified by the server. The vision server accepts a team color (blue or yellow) and a robot number ($1 \sim 5$) in `(init)` message unlike the SoccerServer.

The visual information sent by the vision server consists of the time when the image is grabbed, the nearest edge array, and the ball data. The reason why the ball data is separated is that it is smaller than other objects. The format of the visual information is shown as follows.

`(see` *Time* `((edge)` *EdgeSstring*`) ((ball)` *Distance Direction Size*`))`

Time Time when the image is grabbed.

EdgeString Character string which represents the nearest edge data. The edge data in one direction is represented by three characters. The first character means the type of edge. The second and third characters means the distance of the edge on the image plane in hexadecimal notation [pixel]. The method of representing the type of edge for robot has three options as follows.
- The edge types of all robots are same.
- The team color is distinguishable by the edge type.
- The robot number is distinguishable by the edge type.

Distance Distance to the ball on the image plane [pixel]. In case that the ball is hidden, `-1` is set.

Direction Direction of the ball [deg].

Size Size of the ball on the image plane in angle [deg].

An example of the visual information is shown as follows.

`(see 20306.148 ((edge) r59r5ar5av3e··· ) ((ball) 61 121 3))`

EdgeString contains 1080 characters in our current version.

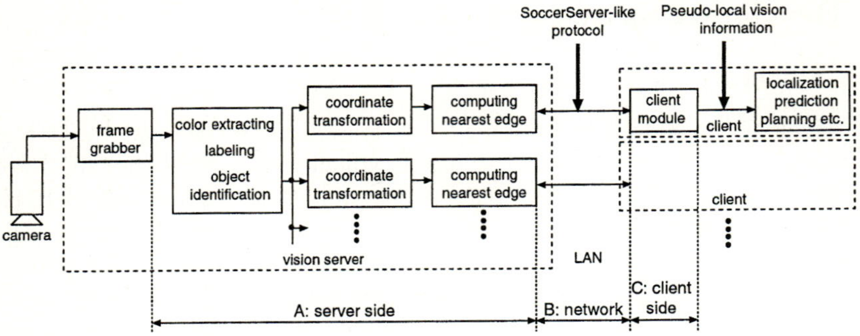

Fig. 5. Structure of the pseudo-local vision system

3.3 Vision Server

The vision server extracts color markers on the robots from the ceiling camera image, distinguishes them, and derives their positions and orientations. It also measures the position of ball. A inexpensive frame grabber is employed to digitize video signal. Modified CMVision [4] is embedded in our program to obtain color region in the image. Moreover, Kalman filter is introduced to improve the quality of information.

After that the processes are carried out for each client separately. At first the coordinates of all objects are transformed into the processed robot's coordinate system. In this system the nearest edge is found from intersection points between the object boundary and line through the origin. The boundary of robot and ball is modeled as a circle. Next the distance of the nearest edge is converted into the distance on the image plane in integer.

In the current implementation, the view angle is 360[deg] (omni-directional), the direction is quantized every 1[deg], the conversion of the distance on the floor into the distance on the image plane is based on the model of omni-directional hyperboloidal mirror used for our on-board camera. The conversion model is the following form.

$$R = \mathrm{int}(f^{-1}(r)), \qquad f(R) = \frac{R}{A - B\sqrt{C + R^2}} \tag{1}$$

where r is the distance on the floor plane from the robot center, R is the distance on the image plane from the image center, A, B, and, C are constants, $\mathrm{int}(x)$ is function to make x integer. In the current version all they are fixed. However, in the future version, they will be variable.

The above mentioned processes are carried out for maximum of 10 clients. After computing of the nearest edge for all clients, they are sent to the clients in random order in the format specified by the protocol.

3.4 Processes on Client Side

The client module replaces the on-board camera information with the information made from the edge data sent by the server. In our program for robot control, the multi-thread library is employed, the lowest image processing (grabbing image frame, extracting color, labeling, and AND operation with template) is assigned to one thread. Thus this thread is replaced with a new thread with the client module. The other threads need no change. Based on output of the substituted thread, the upper threads can execute detecting free space around the robot, self-localization, estimation of the ball position, and so on in the same manner as on-board camera case [2].

3.5 Clock Synchronization between Server and Client

In order to use image of a moving object, it is indispensable to precisely detect the time when the image is grabbed. Since the vision server which grabs image and the client module which uses visual information are running on different machines generally, clock synchronization between them is necessary. We found that a method of using the system clock of Linux and NTP (Network Time Protocol) does not give enough precision for this purpose. Therefore, we developed a method of time measurement and clock synchronization using RDTSC (ReaD Time Stamp Counter) of Pentium processor referring to Bernstein's clockspeed[6]. As a result, we can measure time among multiple machines to a precision of under one millisecond. Using this, time lag in the system is evaluated, control method of considering time lag is carried out.

4 Evaluation

The above mentioned system is implemented on Linux. In this experiment for evaluation, the server PC equipped with Pentium IV 2[GHz] grabs the ceiling camera image with frame grabber using BTB878 in 640×480 [pixel]. 10 client programs are connected with the server. Some clients are executed on the same machine. The performance of PC for clients is various. The client whose performance is measured is executed alone on a machine with Pentium III 500[MHz]. The server PC and client PCs are connected with 100Base-TX.

4.1 Example of Vision Processing

Fig. 6 is a simulated on-board camera image of robot A in Fig. 3, which is made from the nearest edge array shown in Fig. 4 by drawing it on omni-directional image plane. This robot is also equipped with an omni-directional camera. An image taken with the camera at the same point is shown in Fig. 7. Comparing the two images, there is no information of the part beyond the nearest edge in Fig. 6. However, except for this point, the nearest edge data well simulates the real image. Note that Fig. 6 is not used in the control program but is help to human to understand the situation.

4.2 Processing Period and Time Lag

The server can accept 10 clients and has capability of processing image and sending information simultaneously every 33[ms] under the above mentioned

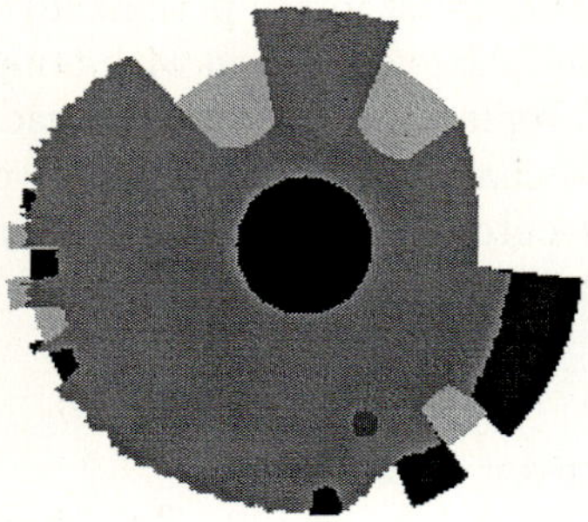

Fig. 6. Pseudo-local vision image

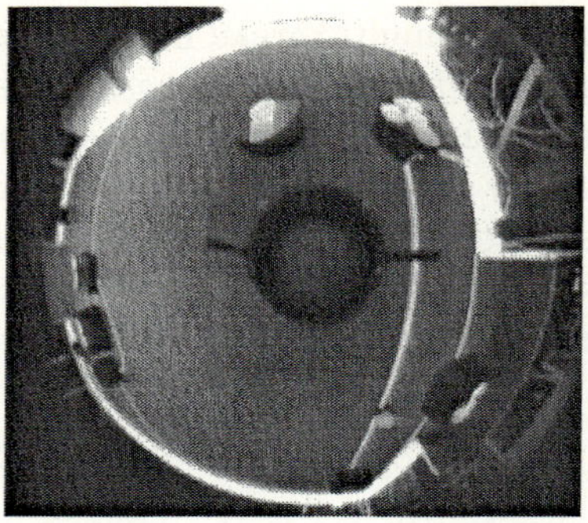

Fig. 7. On-board camera image

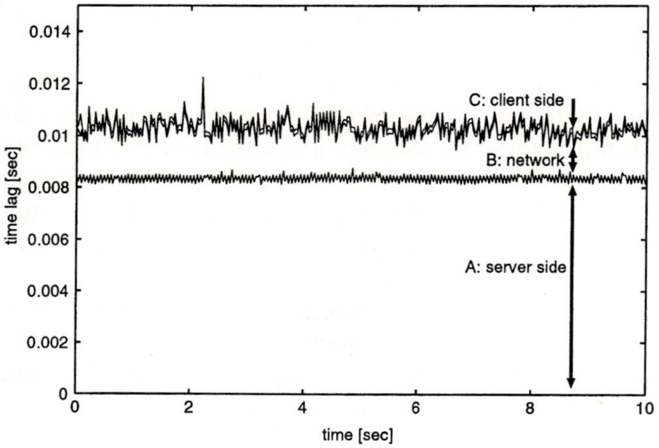

Fig. 8. Time lag in vision processing

condition. Fig. 8 shows the trend of time lag between grabbing image on the server and extracting visual information on the client. A, B, and C in the figure mean the time lags on server side, network, and client side respectively (see also Fig. 5). The time lag is almost constant, the total is about 10[ms]. Time lag from time when the shutter is released is added 33[ms] to it, which is spent in grabbing image.

In case that our robot control program for the on-board camera is executed on PC with Pentium III 500[MHz], the time lag between grabbing the image in 320 × 240[pixel] and extracting information is about 23[ms]. Introducing the high-performance PC for the vision server, the time lag of the pseudo-local vision is smaller than that of the real local vision, even if the server works for 10 clients.

4.3 Resolution and Hidden Ball

In case that the ceiling camera image is grabbed in 640×480 [pixel], the resolution is almost uniform all over the field, one pixel corresponds to 5×5[mm] on the field. On the other hand, in case that the image of our omni-directional on-board camera is grabbed in 320×240[pixel], one pixel corresponds to 4.4×4.4[mm] on the field at the nearest and available point, almost same as the ceiling camera case at point 220[mm] far from the robot, worse than it at farther point. However, it is only an example. If the view angle of the on-board camera is narrower than 360[deg] and/or the resolution is higher, the current version cannot simulate in enough precision. Moreover, there is a problem that the robots sometimes hide the ball from the the ceiling camera. It means that the system cannot completely simulate a property of the on-board camera that a nearer object is easier to observe. To overcome these problems, multiple ceiling cameras will be introduced in the future work.

5 Conclusion

In this paper we propose the practical method of simulating the on-board camera information using the ceiling camera for multiple mobile robot platform. Although the simulated visual information is simple, it represents some of essential problem of the on-board camera information. Evaluating the developed system, we found that the processing period and the time lag are small enough for the system to be used practically. Actually we participated in the RoboCup2002 using the earlier version of proposed system.

In the near future, we will open the source code of the vision server to the public on the Internet. By using this framework we may be able to realize "the Pseudo-Local Vision League". Moreover it will be a bridge between the simulation league and the real robot leagues in the RoboCup.

References

1. D. Sekimori et al.: The Team Description of the Team OMNI, RoboCup 2001: Robot Soccer World Cup V, Springer (2002)
2. D. Sekimori et al: High-Speed Obstacle Avoidance and Self-Localization for Mobile Robots Based on Omni-Directional Imaging of Floor Region, RoboCup 2001: Robot Soccer World Cup V, Springer (2002)
3. Y. Masutani et al.: Team OMNI in 2002 — Pseudo Local Vision Approach —, RoboCup 2002: Robot Soccer World Cup VI, Springer (2003)
4. J. Bruce et al.: Fast and Inexpensive Color Image Segmentation for Interactive Robots, Proceedings of 2000 IEEE/RSJ Int. Conf. on Intelligent Robots and Systems (2000)
5. SoccerServer, `http://sserver.sourceforge.net/`
6. D.J.Bernstein: clockspeed, `http://cr.yp.to/clockspeed.html`

Using Model-Based Diagnosis to Build Hypotheses about Spatial Environments

A Response to a Technical Challenge

Oliver Obst*

Universität Koblenz-Landau, AI Research Group
Universitätsstr. 1, D–56070 Koblenz, Germany
fruit@uni-koblenz.de

Abstract. We present a method to build a hypothesis on the condition of the environment in which a robotic multi-agent team moves. Initially the robots have a default assumption about the conditions of the floor and on how moving under these condition works. For certain parts of the environment however, the default assumption may be wrong and moving around does not work in the expected way. Now the robotic team builds a hypothesis on the conditions of the yet unvisited part of the environment in a way similar to computing a diagnosis for electrical circuits. Resources can be saved by avoiding areas that possibly also contain obstacles.

1 Introduction

In RoboCup Simulation League, the simulated robots are situated in a two-dimensional world where objects move according to some fixed rules, which are known to the players. Usually, during a regular tournament, these rules are not going to change. Additionally to the regular tournament during RoboCup 2001 there was an evaluation round where parts of the physics of the simulation was changed before the matches started, without that programmers knew of this in advance. The change in the physics had the effect that dashing on the upper half of the field resulted in only half of normal speed for all the players. Some of the teams did manage these matches better than others, but to the best of our knowledge none of the teams could come up with a diagnosis of what happened on the field, though this was immediately visible for human spectators.

In this paper, we present a solution to the problem how a team of robotic agents can come up with a hypothesis of what might globally be wrong with the environment in similar situations. Hypotheses should entail the currently observed behavior and provide some kind of "forecast" for those areas that no other member of the team visited so far. As time proceeds the hypothesis will be refined to match the actual situation more closely. With a hypothesis about the condition of the environment, single robots can try to avoid areas that potentially contain obstacles, take the shortest way out of these areas, or instead enter these areas to (dis-)prove the hypothesis dependent on the strategy of the team.

* This research is supported by the grant *Fu 263/6-1* from the German research foundation *DFG*.

D. Polani et al. (Eds.): RoboCup 2003, LNAI 3020, pp. 518–525, 2004.

2 From Problems to Hypotheses

What we essentially want the robots to do is related to the approach of model-based diagnosis: abnormalities have to be identified and a kind of explanation for these abnormalities should be generated (see for instance [4]). Though we identify some differences between doing model-based diagnosis for devices and building hypotheses for robots in the first part of this section, we can apply the same procedure to compute the result in our approach.

2.1 Differences between Model-Based Diagnosis for Devices and Building Hypotheses for Robots

Though at first sight both problems seem to be very similar, there are differences with respect to some general points:

Devices vs. Space. Model-based diagnosis is usually used for physical devices consisting of several components. The "device" we are interested in in our approach is space, i.e. the floor on which the robots move. To use a procedure like model-based diagnosis, we artificially have to introduce components like smaller areas which make up space.

Diagnosis vs. Hypothesis. Even if we introduce kinds of devices in both approaches there is a difference with respect to result that should be computed: By taking a description of the system and a description of the behavior of the system, model-based diagnosis is usually used to explain the observed behavior by providing a diagnosis which states possible faulty components. For our robots, we observe the – possibly faulty – behavior of some of the components (parts of an area) and by this observation, we want to build a hypothesis on the behavior of the whole system (the complete area).

2.2 Identifying Abnormalities

The only way our robots can deduce the presence of an obstacle at their current location is by moving forward and using position estimation or odometric information, that is robots can execute a kind of `move` operation and recognize if it succeeded or if it failed. A description of the (usual) behavior of the `move` operation, an observation of the behavior of the `move` operator and some knowledge concerning the accuracy of these both are necessary to decide if a robot actually has a problem or not.

3 Representation of the Environment

The assumptions made in the previous section enable us to do two things: Firstly, robots can clearly identify a part of the field as being defective if `move` does not work in a particular situation. Secondly, robots can exchange information on abnormal tiles by sending logical facts to each other. These informations need only to be sent in unexpected situations. By doing so, robots exchange and collect information about tiles on the field

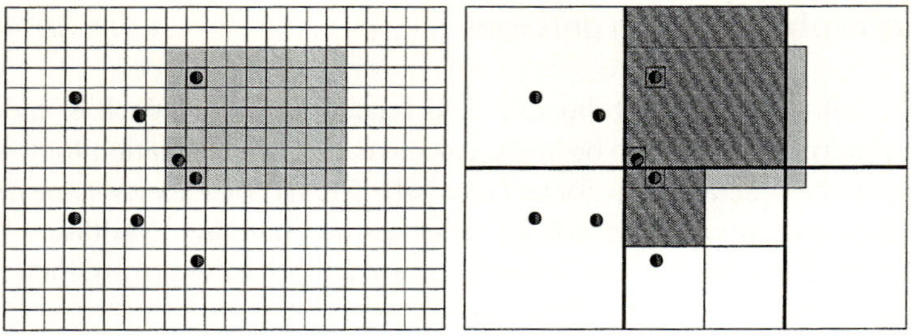

Fig. 1. Left: Area partitioned into atomic tiles with robotic team. On the gray tiles, robots cannot move as fast. **Right:** Created hypothesis from this situation.

they already visited. To get an idea of how the yet unvisited tiles could be like, we set up rules that relate tiles of different levels to each other. The basic assumption here is that if there is an unexpected change in the environment, this change usually concerns areas larger than the size of our atomic tiles. If there is a tile containing an obstacle, its unvisited neighbors will probably also contain obstacles. In the next subsection we are going to explain how we want to relate tiles of different levels to each other. We will assume that the area is partitioned into squares, but other ways of partitioning are possible. In the second part of this section we are going to describe the logical rules we are using for the relation between tiles.

3.1 Hierarchical Layering of Tiles

To build a hypothesis on defective tiles, we use a hierarchical layering of tiles inspired by quad trees [6]. In the case of squared tiles, we use 2×2 atomic squares (level 0) to make up a higher-level square (level 1), 2×2 squares of level 1 to make up a square of level 2 and so forth. The last layer in this hierarchy covers the whole area accessible for the robots. Like the atomic tiles, each of the higher-level tiles gets a symbolic label so that it can be identified uniquely.

The idea behind building a hierarchical representation using tiles is the following: Initially, all robots have the default assumption that `move` succeeds in the whole area. If, for a tile of a certain level, it is known that it covers only unknown (smaller) tiles and at least one tile where `move` succeeds, our hypothesis should be that `move` will work for the complete tile. If the tile covers only unknown smaller tiles and at least one tile where `move` fails, our hypothesis should be that `move` will fail for the complete tile. Tiles containing both kind of smaller tiles (and possibly unknown tiles) have to be split up into tiles of the next lower level. For an example hypothesis see the graphics on the right in Fig. 1.

3.2 Logical Representation

Each tile in our representation can be identified by its level and its coordinates, so the tiles are the "components" of the field. Logically, a tile is an atom $a_{l,x,y}$ with level l, where x and y denote the coordinate of the respective tile in that level. The non-atomic tiles contain lower level tiles. These connections including the resulting assumptions on the state of respective tiles can be described using logical rules.

In model-based diagnosis, a model of a device is used to compute the expected output given a particular input. Discrepancies between expected and the actual output are used to detect faults in the system. In our setting, the attempt to move on a certain position can be regarded as "input" to an atomic tile. Thus, we denote the fact that the robot tries to move on location $a_{0,i,j}$ by $\text{move}(a_{0,i,j}, i)$.

The result of this attempt can be regarded as "output" of an atomic tile, and so we denote the success of the operation `move` by $\text{move}(a_{0,i,j}, o)$, while a failure of this operation is denoted by $\neg\text{move}(a_{0,i,j}, o)$.

There can be different kinds of faults in a system like this: in our approach, we are aiming at faulty tiles, the components in the system. Possible are also faults in the actuators or in the sensors measuring the output of the actuators; however for now we ignore them.

Definition 1. *An atomic tile $a_{0,i,j}$ is called defective (or abnormal), if it is known that the `move` operation fails on $a_{0,i,j}$ (written: $\text{ab}(a_{0,i,j})$). An atomic tile $a_{0,i,j}$ is called normal if it is known that the `move` operation succeeds on $a_{0,i,j}$ (written: $\neg ab(a_{0,i,j})$).*

In our approach we use two definitions concerning the connections between atomic and higher level tiles: Definition 2 deals with implications for a higher level tile from knowing that the `move` operation succeeds on an atomic tile, while Definition 3 describes implications for higher level tiles from knowing that the `move` operation failed. A third possibility for an atomic tile is that no robot visited the tile so far, so that there is no knowledge on the behavior of the `move` operation on the respective tile.

Definition 2. *A tile of level l ($l > 0$) containing a normal atomic tile is not abnormal.*

This relation between atomic tiles and higher level tiles can be expressed used logical formulæ. The number of formulæ for each tile is dependent on the level of the tile and the used topology. In the case of square tiles the branching factor is 4, i.e. a tile of level one contains four level 0 tiles, while a tile of level two contains sixteen level 0 tiles.

Example 1. In the case of square tiles consisting of four lower level tiles the rules for Definition 2 look like this:

$$
\begin{aligned}
\neg ab(a_{1,0,0}) &\leftarrow \text{move}(a_{0,0,0}, o) \wedge \text{move}(a_{0,0,0}, i).\\
\neg ab(a_{1,0,0}) &\leftarrow \text{move}(a_{0,0,1}, o) \wedge \text{move}(a_{0,0,1}, i).\\
\dots\dots\dots &\leftarrow \dots\dots\dots\dots\dots\dots\dots\dots\\
\neg ab(a_{2,0,0}) &\leftarrow \text{move}(a_{0,0,0}, o) \wedge \text{move}(a_{0,0,0}, i).\\
\dots\dots\dots &\leftarrow \dots\dots\dots\dots\dots\dots\dots\dots
\end{aligned}
$$

Definition 3. *(At least) one of the tiles containing an abnormal atomic tile is abnormal.*

Example 2. In the case of square tiles consisting of four lower level tiles the rules for Definition 3 look like this:

$$
\begin{array}{lcl}
ab(a_{0,0,0}) \vee ab(a_{1,0,0}) \vee ab(a_{2,0,0}) & \leftarrow & \neg \mathrm{move}(a_{0,0,0}, o) \wedge \mathrm{move}(a_{0,0,0}, i). \\
ab(a_{0,0,1}) \vee ab(a_{1,0,0}) \vee ab(a_{2,0,0}) & \leftarrow & \neg \mathrm{move}(a_{0,0,1}, o) \wedge \mathrm{move}(a_{0,0,1}, i). \\
\cdots\cdots & \leftarrow & \cdots\cdots \\
ab(a_{0,3,3}) \vee ab(a_{1,1,1}) \vee ab(a_{2,0,0}) & \leftarrow & \neg \mathrm{move}(a_{0,3,3}, o) \wedge \mathrm{move}(a_{0,3,3}, i). \\
\cdots\cdots & \leftarrow & \cdots\cdots
\end{array}
$$

The number of formulæ of this kind necessary is equal to the number of atomic tiles for the complete area. The right hand side of each formula denotes that `move` failed on an atomic tile, while on the left hand side we have a disjunction of all tiles that contain the respective atomic tile – including the atomic tile itself. The length of the right hand side of these formulæ is logarithmic with respect to the number of atomic tiles.

4 Building a Hypothesis

Once the set of clauses as described in the last chapter is available, collected knowledge from moving around and from communication and a theorem prover can be used to solve the problem of building the actual hypothesis. We need to compute models of the given set of clauses, where the extension of the *ab*-literal is minimal. That means the theorem prover used should find solutions for the given clauses so that there exists no solution containing a subset of true *ab*-literals. A procedure to compute minimal diagnosis with a theorem prover can be found in [1], the theorem prover `NIHIL`[1] we have been using to compute hypotheses is based on this procedure. We briefly explain some basics for this procedure in the following subsection; in principle any other procedure for computing minimal models could be used.

4.1 Model-Based Diagnosis with Hyper Tableaux

Usually, for a diagnosis we need a system description (*SD*), a set of components of the system (*COMP*), and an observation (*OBS*). As stated earlier, a component c can be abnormal ($ab(c)$), or it can behave normal ($\neg ab(c)$). According to Reiter [9], a diagnosis is defined as follows:

Definition 4 (Reiter 87). *A Diagnosis of* $(SD, COMP, OBS)$ *is a set* $\Delta \subseteq COMP$*, such that* $SD \cup OBS \cup \{ab(c) | c \in \Delta\} \cup \{\neg ab(c) | c \in COMP - \Delta\}$ *is consistent.* Δ *is called a* Minimal Diagnosis, *iff it is the minimal set (wrt.* $\subseteq$*) with this property.*

As we shall see later, the set of observations is simply a set of logical facts. A basic ingredient to the diagnosis method mentioned above are hyper tableaux [2]. Basically, a hyper tableau is a finite ordered tree T where all nodes, besides the root node, contain literals from a finite set S of ground clauses. A branch b in T is called regular if each literal in b occurs at most once, otherwise it is called irregular. A tree T is regular iff all its branches are regular. Branches containing containing contradicting literals are labeled as closed, and as open otherwise. A tableau is closed if each of its branches is closed, otherwise it is open. Computing minimal diagnoses is possible by finishing all open branches and collecting the *ab*-literals from open branches.

[1] NIHIL = New Implementation of Hyper In Lisp. Thanks to Peter Baumgartner for providing helpful information on the calculus and on using his system.

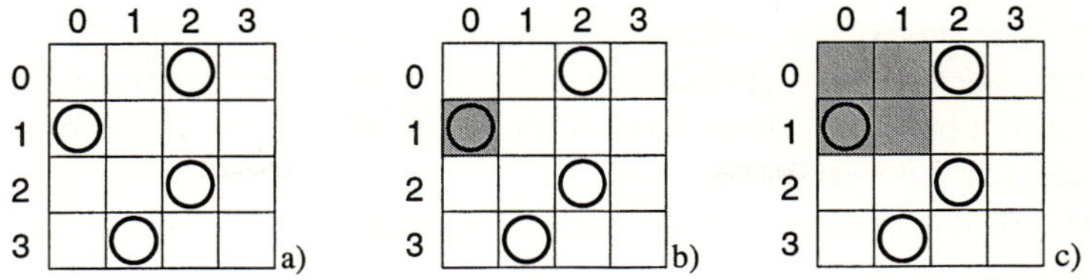

Fig. 2. Example with one robot on a defective tile (0,1) and two possible hypotheses.

4.2 Selecting a Hypothesis

Even if the computed diagnosis are minimal, for a given situation there could be several minimal ones. Consider a robot put on the center of a square finding the atomic tile on its place defective. In each level of the hierarchy of tiles there is one tile surrounding the atomic tile that could be a hypothesis for being a defective tile. The least conservative hypothesis covers only atomic tiles which are known to be defective. The most conservative hypothesis expands to the largest tiles that contain at least one abnormal, but no normal atomic tile.

Example 3. Consider the four robots from Fig. 2. The facts about their current locations are like this:

$$\begin{array}{ll} \mathrm{move}(a_{0,0,1}, i) \leftarrow \top. & \bot \leftarrow \mathrm{move}(a_{0,0,1}, o). \\ \mathrm{move}(a_{0,2,0}, i) \leftarrow \top. & \mathrm{move}(a_{0,2,0}, o) \leftarrow \top. \\ \mathrm{move}(a_{0,1,3}, i) \leftarrow \top. & \mathrm{move}(a_{0,1,3}, o) \leftarrow \top. \\ \mathrm{move}(a_{0,2,2}, i) \leftarrow \top. & \mathrm{move}(a_{0,2,2}, o) \leftarrow \top. \end{array}$$

So it is known that tile $a_{0,0,1}$ is defective, whereas tiles $a_{0,2,0}$, $a_{0,1,3}$ and $a_{0,2,2}$ are known to be working. There is nothing known about the other tiles. The rules describing the relation of tiles between the different levels are selected as in Example 1 and Example 2. One diagnosis as shown in Fig. 2 b) ($ab(a_{0,0,1})$) will be computed, and the other one as shown in Fig. 2 c) ($ab(a_{1,0,0})$). To select the most conservative hypothesis, one has to choose the diagnosis with the least number of faults. If there is more than one diagnosis with this property, the one containing tiles $a_{l,i,j}$ with the largest levels l is the most conservative one.

4.3 Why Logic?

Looking at the hypothesis and the way it gets computed the question arises why one should do it this way instead of using the quad trees directly. There are different answers to this question.

Firstly, we have chosen to use a theorem prover because it was available. Creating the rules to represent the environment was done automatically by a simple program, so the programming effort to solve the problem was very small. But there are other advantages of this approach:

Flexibility. The approach using a theorem prover is flexible with respect to the selected topology, because the representation is independent of the computation. A robot moving in a building consisting of several different rooms can select a topology dependent on the room it is in.

Robustness. The use of logic for communicating facts about the environment can add robustness to the negotiation process among different team members. We assume that robots communicate an observation only if it contradicts their current hypothesis about the environment. By receiving known information robots can possibly still improve their current hypothesis, because it can be deduced that a previous message must have been lost.

Use of Additional Knowledge. By using additional knowledge about potential obstacles, the selection of hypotheses can be controlled. This is possible because the theorem prover computes all minimal diagnosis, from which we can select an appropriate one. By choosing the hypothesis containing the tiles with the largest levels we select the most conservative one. Dependent on knowledge or assumptions about the environment we can also prefer hypothesis with obstacles of a certain size, if applicable.

However, a disadvantage of our approach is that with larger areas and increasing number of tiles calculation of hypothesis becomes slow. This is due to the fact that the number of formulæ increases quadratic with the number of tiles.

5 Related Work

As mentioned earlier, model-based diagnosis [9, 4] is usually concerned with computing an explanation for a certain behavior of technical devices. Logical descriptions of the device and of the components of the device are used to predict the expected behavior of the device, given a particular input. The diagnostic task is to identify possible faulty components by comparing the actual with the predicted output. The *ab*-literal is used to denote faulty components. Approaches as described in [1] take the system description and the observation of the behavior as a set of clauses to compute models so that the extension of the *ab*-literal is minimal. Besides the reasoning strategies the assumptions made for the reasoning process are a matter of concern, see for example [5]. For an overview about symbolic diagnosis in general see [7].

As in our approach to specify and implement a team of agents [8], the authors in [3] use a logical approach to specify the knowledge of an agent. The agent architecture ALIAS is introduced, where agents are equipped with hypothetical reasoning capabilities. Hypothesis are raised by an agent and it is checked if the hypothesis complies with the other agents knowledge, contrary to our approach where the knowledge is collected prior to raising hypotheses.

6 Conclusions and Future Work

In this paper we showed a way to build hypotheses about the environment of robotic teams with a theorem prover. A hypothesis about unknown areas can be useful to guide the selection of actions, so that resources can possibly be saved. The logical description of the environment can be generated automatically for different topologies. Because the method to actually compute the hypothesis is separated from the description of the environment, topologies could be switched at run time. Should the time of computation take too long due to a large number of atomic tiles the approach can be used to compare different kinds of topologies and to be reimplemented in a more efficient fashion later on. In our approach we looked at changes in the environment only, incorporating sensor and actuator faults into our model is left to future work. Other points for future work are methods to recover from message loss, and using methods to acquire knowledge from teammates without communication. For large areas with a low density of facts a logical description with variables could be tried.

References

1. Peter Baumgartner, Peter Fröhlich, Ulrich Furbach, and Wolfgang Nejdl. Semantically Guided Theorem Proving for Diagnosis Applications. In M. E. Pollack, editor, *15th International Joint Conference on Artificial Intelligence (IJCAI 97)*, pages 460–465, Nagoya, 1997. Morgan Kaufmann.
2. Peter Baumgartner, Ulrich Furbach, and Ilkka Niemelä. Hyper Tableaux. In *Proc. JELIA 96*, number 1126 in Lecture Notes in Artificial Intelligence. European Workshop on Logic in AI, Springer, 1996.
3. Anna Ciampolini, Evelina Lamma, Paola Mello, Cesare Stefanelli, and Paolo Torroni. An implementation for abductive logic agents. In Evelina Lamma and Paola Mello, editors, *AI*IA 99: Advances in Artificial Intelligence*, LNAI 1792, pages 61–71. Springer, Berlin, 2000.
4. Luca Console and Pietro Torasso. A spectrum of logical definitions of model-based diagnosis. *Computational Intelligence*, 7(3):133–141, 1991.
5. Dieter Fensel and Richard Benjamins. Assumptions in model-based diagnosis. In B.R. Gaines and M.A. Musen, editors, *Proceedings of the 10th Banff Knowledge Acquisition for Knowledge-Based Systems Workshop, KAW'96*, pages 5/1–5/18, Calgary, 1996. Department of Computer Science, University of Calgary, SRDG Publications.
6. Raphael A. Finkel and Jon Louis Bentley. Quad trees: A data structure for retrieval on composite keys. *Acta Informatica*, 4:1–9, 1974.
7. Peter J.F. Lucas. Symbolic diagnosis and its formalisation. *The Knowledge Engineering Review*, 12(2):109–146, 1997.
8. Jan Murray, Oliver Obst, and Frieder Stolzenburg. Towards a logical approach for soccer agents engineering. In Peter Stone, Tucker Balch, and Gerhard Kraetzschmar, editors, *RoboCup-2000: Robot Soccer World Cup IV*, number 2019 in Lecture Notes in Artificial Intelligence, pages 199–208. Springer, Berlin, Heidelberg, New York, 2001.
9. Raymond Reiter. A theory of diagnosis from first principles. *Artificial Intelligence*, 32(1):57–95, April 1987.

Self-localization Method Using Two Landmarks and Dead Reckoning for Autonomous Mobile Soccer Robots

Akira Motomura[1], Takeshi Matsuoka[2], and Tsutomu Hasegawa[1]

[1] Department of Intelligent Systems, Graduate School of Information Science and Electrical Engineering, Kyushu University
6-1-10 Hakozaki, Higashi-ku, Fukuoka 812-0053, Japan
{motomura,hasegawa}@irvs.is.kyushu-u.ac.jp
http://fortune.is.kyushu-u.ac.jp/

[2] Department of Electrical Engineering, Faculty of Engineering, Fukuoka University
8-19-1 Nanakuma, Jonan-ku, Fukuoka 814-0180, Japan
matsuoka@ctrl.tec.fukuoka-u.ac.jp
http://ctrl.tec.fukuoka-u.ac.jp/

Abstract. We propose a new method of self-localization using two landmarks and dead reckoning for a soccer robot equipped with an omni-directional camera as a local vision sensor. This method requires low computational cost. Thanks to rapid process, the system can afford to run multiple localization process in parallel resulting robust and accurate localization. An experimental result in the field of Robocup Middle-size league indicates that the approach is reliable.

1 Introduction

Many problems are left unsolved to develop an autonomous mobile robot working in a dynamically changing environment. Especially, self-localization is the most fundamental problem. Many solutions developed so far are classified into two categories, i.e., relative method and absolute method. Relative method estimates pose of the robot by the integration of incremental motion using information obtained by intrinsic on-board sensors. A typical technique of relative localization is dead reckoning[1]. This method cannot be used for long-distance navigation because it suffers from errors caused by various disturbances. The localization error grows over time. Absolute method calculates the absolute pose of the robot using various landmarks. The localization error is suppressed when measurements is available. One problem in this method is that movement of small distances cannot be calculated because of rather lower resolution of sensed data as compared with resolution of motion. Another problem is that the robot cannot always observe sufficient number of landmarks due to occlusion.

Both approaches have drawbacks. To cope with the drawbacks, absolute method and relative method are often combined[4][10][12][13][14]: Kalman filter[2][3][7][8] and Monte Carlo Localization (MCL)[6][9][11][15] are often intro-

D. Polani et al. (Eds.): RoboCup 2003, LNAI 3020, pp. 526–533, 2004.

duced. Kalman filtering, MCL and other localization methods are comparatively reviewed in [16].

In self-localization in the soccer field of RoboCup Middle-size league, the goals and the corner posts can be used as landmarks. The mutual geometrical relationship among these landmarks is known. These field features are mainly distinguishable by their color. If three landmarks are recognized by vision sensor, the pose of the robot can be obtained by triangulation based on the measured direction of them with respect to the robot frame[5]. However, robots often fail to simultaneously recognize three different landmarks during the soccer game: landmarks are easily occluded by opponent robots and teammate robots which are always moving around and are rushing to the ball.

We developed a new self-localization method for a mobile robot and applied it to the soccer robot in RoboCup Middle-size league. Initial samples of the pose of the robot are distributed on a circumference of a circle determined by measured direction of two landmarks. Then the correct pose is determined through iterative evaluation steps. This method requires low computational cost. This enables multi-process of localization running simultaneously. In addition to this, reinitialization of samples is performed when samples become highly improbable. Thus robustness and accuracy of localization is guaranteed.

This paper is structured as follows. Principle of our localization method is explained in Sect. 2. In Sect. 3, we explain techniques for satisfying constraints of real-time response. Sect. 4 describes our soccer robot and an experimental result is included in Sect. 5.

2 Principle of Localization

Throughout this section, we assume that observed landmarks are recognized correctly and that their position in the soccer field is precisely known. When a robot detects two landmarks and measures angles of their directions with respect to a robot centered coordinate system, the robot is positioned at somewhere on a circle which passes these two landmarks while satisfying that the angle at the circumference is the measured angle (Fig. 1). Orientation of the robot is determined as a function of its position on the arc and measured direction of the landmarks. To uniquely determine position and orientation of the robot, dead reckoning of the successive motion of the robot is used as followings. Discretizing position on the arc, we obtain a set of candidates of pose of the robot $P(t) = \{p_i(t)|i = 1 \dots N\}$. N is at most 360 in our current implementation. Successively the robot moves for a short period Δt, and then detects plural landmarks and measures angles of direction. Pose displacement of the robot for this short period is also obtained using odometry with measured rotation of wheels. Adding this displacement to each candidate of the pose, we obtain a set of renewed candidates $P(t + \Delta t)$.

Localization error of each renewed pose is evaluated by difference of direction angle to the landmarks: comparison is made between computed direction to each landmark from the renewed pose and actually measured direction by the robot (Fig. 2). Error value is accumulated over time by equations:

$$AngleError(t+\Delta t) = \sum_{k=1}^{m} a_k(t+\Delta t) + \sum_{k=1}^{n} d_k(t+\Delta t) \quad (1)$$

$$m = {}_nC_2, \quad n \mid number\ of\ observed\ landmarks.$$

$$Error(t+\Delta t) = AngleError(t+\Delta t) + Error(t) \quad (2)$$

In these equations, $a_k(t)$ is an absolute value of difference between measured observation angle of two landmarks from the robot and computed one, $d_k(t)$ is an absolute value of difference between measured direction angle and computed one for each landmark respectively.

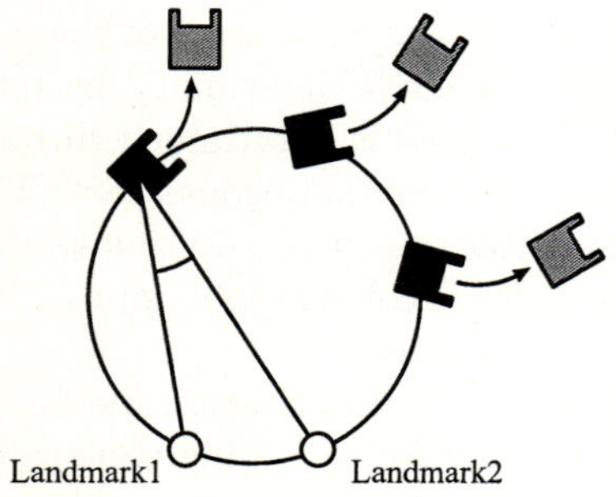

Fig. 1. Initial and renewed candidates of robot pose

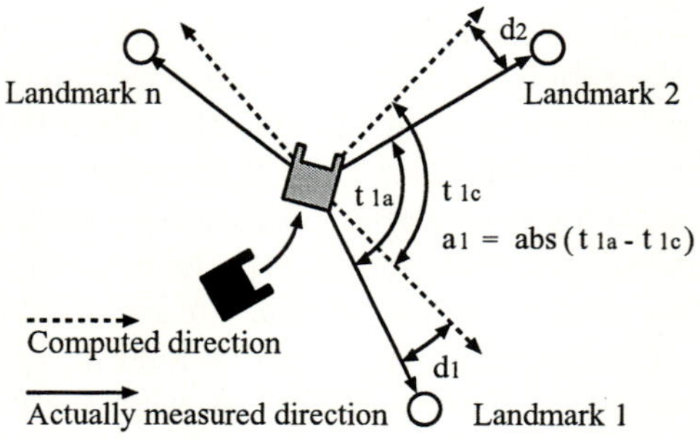

Fig. 2. Localization error of each renewed pose

$Error(0)$ is cleared to be 0 at the beginning of the localization process. Then evaluation of all the candidate for the pose of the robot is repeated based on visual measurement of the landmarks and odometry in every short period Δt. Error value obtained by Eq. 2 increases gradually. Increase of accumulated error value is slow for those candidates which are closely located near the correct pose, while increase is rapid for those which are far from the correct pose. Finally correct pose is determined uniquely as the one having the smallest error accumulation.

In the initial step, the robot needs to detect two landmarks to make a circle corresponding to the candidates of the pose of the robot. On the other hand, in subsequent evaluation step, a robot should just discover one landmark at least. If plural landmarks are observed by the robot, observation of landmarks effectively functions to evaluate errors, resulting earlier determination of the unique pose of the robot.

3 Localization Process

Localization is formalized as a selection process of the best one in a set of candidates based on the error evaluation over time. Error evaluation is performed for every candidates in every short time period Δt. Generally number of candidates in an initial set is large, outliers should be eliminated as soon as possible to reduce the number of candidates to be evaluated. We have developed several methods to satisfy constraints of real-time response of the soccer robot.

3.1 Early Elimination of Inadequate Candidates for Pose

As the robot moves, accumulated error increases rapidly for these candidates located far from the correct pose. So we introduce threshold values for the accumulated error, the error in observation angle of two landmarks, and the error in observation direction to each landmark. As soon as an error exceeds the corresponding threshold value, the candidate is immediately removed from the set of pose thus reducing computational cost (Fig. 3).

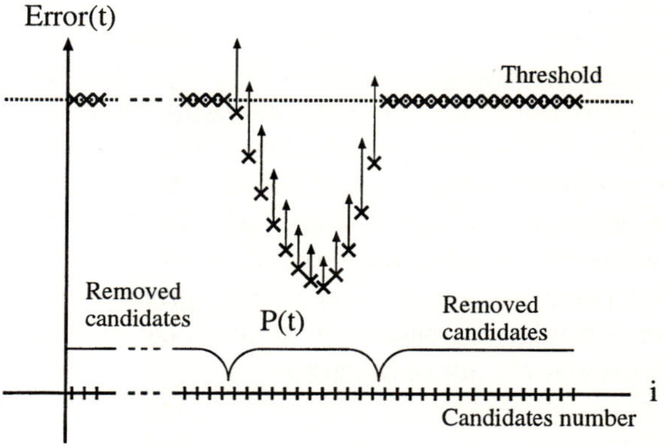

Fig. 3. Accumulation of errors

3.2 Selection of Two Landmarks among Available Landmarks

There are eight landmarks available in the soccer field: four corner poles and four goal posts. Two adequate landmarks must be selected whenever more than three are observed by an omni-directional camera of the robot to make the initial set of candidates of the pose of the robot $P(0)$. This selection is made in real time based on a table prepared in off-line as followings. An initial set of possible pose $P(0)$ is composed of points obtained by discretization of a circle: the circle is defined by two landmarks and observation angle of them by a robot. Maximum number of elements is 360 in our current implementation. Those located out of the field are eliminated at the beginning. Therefore, those two landmarks which generates least number of initial pose in the field are selected. Considering the symmetry of the field and location of landmarks, there are eleven possible pair of two landmarks. Given an observation angle of landmarks from the robot, number of discretized points on the circle within the field is enumerated for each pair. Using 180 discretized observation angles from 1 degree to 180 degrees, a table is structured containing number of possible candidates for eleven pairs of landmarks. Using this table, selection of two landmarks is made in real-time.

3.3 Control of Localization Processes

As described so far, the localization process is composed of generation of pose candidate and successive selection of unique pose by traveling with dead reckoning. Due to various errors, the pose of the robot may not be determined uniquely even if the robot continues traveling. Further extension of travel distance causes increase of accumulated error in dead reckoning. This would result erroneous pose even if the pose should be determined uniquely. In such case, the localization process should be aborted and a new process should be initiated. In the current implementation, the running process is aborted if the number of eliminated candidate exceeds a threshold value and the minimal error exceeds a threshold value.

On the contrary, once the pose is uniquely determined in the course of traveling, a new localization process is initiated at the point while the former process is continued. In this way the robot maintains multi-process of localization running simultaneously (Fig. 4). We compute a weighted sum of poses obtained in each process as a best estimate of the pose of the robot. Reciprocal of accumulated error obtained in Eq. 2 for each pose is used as a weight value. Number of processes cannot grow limitlessly due to the real-time constraint of the soccer robot. In addition, a lifetime is defined for a localization process, since an accumulated error of the dead reckoning increases over time. In the current implementation, number of the processes is limited to three.

4 Robot System

Our soccer playing robot is shown in Fig. 5. The robot is driven by two powered wheels and is equipped with an omni-directional vision system. Two additional free wheels with rotary encoders are installed to the robot for odometry with least slippage.

An omni-directional camera is mounted on the top of the robot enabling panoramic view around it (Fig. 6). The camera is composed of a CCD camera with a hyperboloidal omni-directional mirror. Obtained image frame is processed by an electronics board (IP5005, Hitachi).

5 Experiment

An experiment has been made in a half of Robocup soccer field composed of a goal and a corner pole on the green carpet (Fig. 7). Two goal posts (A and B) on both side of a goal and a corner pole (C) are used as landmarks. Performance of the proposed method is compared with conventional two methods.

A field player robot is initially placed at point a. Then it is manually pushed to follow a given reference test trajectory $a-b-c-d-e-f-g-h-a$ composed

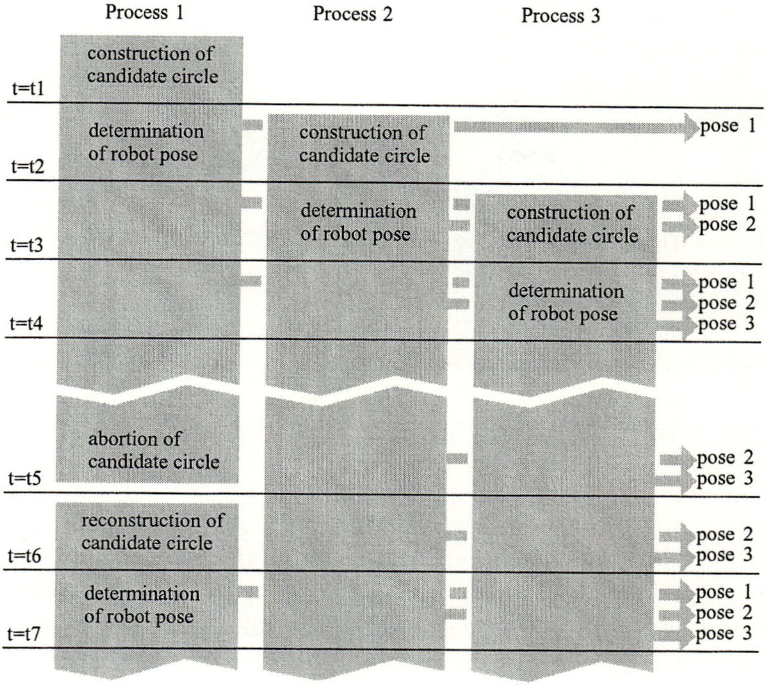

Fig. 4. Multi-process of localization

Fig. 5. Our soccer robot for RoboCup Middle-size league

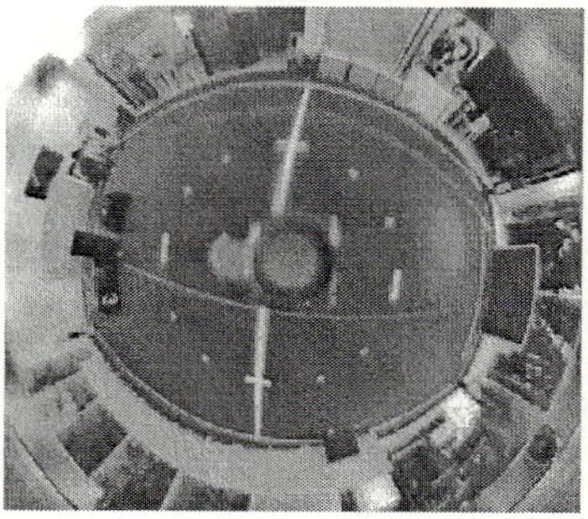

Fig. 6. Panoramic view of omni-directional camera

of straight line segments and rectangular corners. In Fig. 8, we show performance of localization for three different cases: 1) localization by the proposed method with two landmarks(A, B), 2) localization by the conventional triangulation with three landmarks(A, B, C), and 3) localization by conventional dead reckoning.

In the first case, the robot succeeds in localization after traveling for distance of 75 cm, and keep going with good accuracy till the end of the trajectory. In the second case, when all three landmarks are correctly recognized (a–b–c–d, h–a),

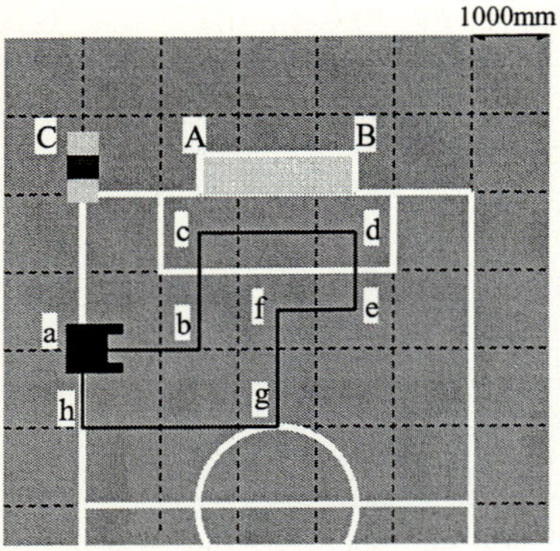

Fig. 7. Experimental setup

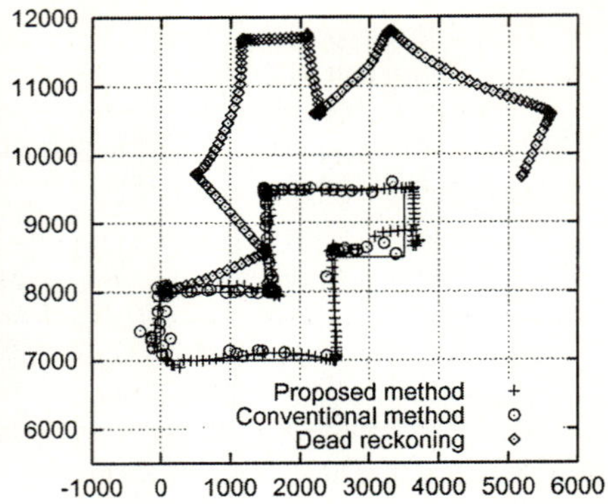

Fig. 8. Experimental result

the robot succeeds in self-localization with good accuracy. On the trajectory $d-e-f-g-h$, the robot often fails to recognize the landmark A and then fails in localization. In the third case, localization by dead reckoning suffers large error. In our proposed method, the amount of movement required to decide the pose of the robot becomes small when three landmarks can be observed, compared with the case of observation of two landmarks.

6 Conclusions

We have presented an self-localization method based on two landmarks and dead reckoning for autonomous mobile soccer robot. The effectiveness of the proposed method was proved by an experiment. The presented method becomes reliable in robot soccer compared with the other technique such as triangulation with three landmarks. Especially the method has a high adaptability for localization of a keeper robot: although the keeper can always observe two goal posts behind, it cannot almost observe other landmarks.

References

1. James L. Crowley: *Control of Displacements and Rotation in a Robot Vehicle.* In IEEE Proc. of The 1989 Int. Conf. on Robotics and Automation (1989)
2. Lindsay Kleeman: *Optimal Estimation of Position and Heading for Mobile Robots Using Ultrasonic Beacons and Dead-reckoning.* In IEEE Proc. of The 1992 Int. Conf. on Robotics and Automation (1992) 2582–2587
3. Frédélic Chenavier and James L. Crowley: *Position Estimation for a Mobile Robot Using Vision and Odometry.* In IEEE Proc. of The 1992 Int. Conf. on Robotics and Automation (1992) 2588–2593
4. Yasushi Yagi, Yoshimitsu Nishizawa and Masahiko Yachida: *Map-Based Navigation for a Mobile Robot with Omnidirectional Image Sensor COPIS.* In IEEE Trans. on Robotics and Automation, Vol.11, No.5 (1995) 634–648

5. Margrit Betke and Leonid Gurvits: *Mobile Robot Localization Using Landmarks.* In IEEE Trans. on Robotics and Automation, Vol.13, No.2 (1997) 251–263
6. Wolfram Burgard, Andreas Deer, Dieter Fox and Armin B. Cremers: *Integrating Global Position Estimation and Position Trackig for Mobile Robots: The Dynamic Markov Localization Approach.* In Proc. of the 1998 IEEE/RSJ Int. Conf. on Intelligent Robots and Systems (1998) 730–736
7. Puneet Goel, Stergios I. Roumeliotis and Gaurav S. Sukhatme: *Robust Localization Using Relative and Absolute Position Estimation.* In Proc. of the 1999 IEEE/RSJ Int. Conf. on Intelligent Robots and Systems (1999) 1134–1140
8. Jens-Steffen Gutmann, Thilo Weigel and Bernhard Nebel: *Fast, Accurate, and Robust Self-Localization in Polygonal Environments.* In Proc. of the 1999 IEEE/RSJ Int. Conf. on Intelligent Robots and Systems (1999) 1412–1419
9. Frank Dellaert, Dieter Fox, Wolfram Burgard. and Sebastian Thrun: *Monte Carlo Localization for Mobile Robots.* In IEEE Proc. of The 1999 Int. Conf. on Robotics and Automation (1999) 1322–1328
10. Cyril DROCOURT, Laurent DELAHOCHE, Claude PEGARD and Arnaud CLERENTIN: *Mobile Robot Localization Based on an Omnidirectional Stereoscopic Vision Perception System.* In IEEE Proc. of The 1999 Int. Conf. on Robotics and Automation (1999) 1329–1334
11. Scott Lenser and Manuela Veloso: *Sensor Resetting Localization for Poorly Modeled Mobile Robots.* In IEEE Proc. of The 2000 Int. Conf. on Robotics and Automation (2000) 1225–1232
12. Iwan Ulrich and Illah Nourbakhsh: *Appearance-Based Place Recognition for Topological Localization.* In IEEE Proc. of The 2000 Int. Conf. on Robotics and Automation (2000) 1023–1029
13. Arnaud Cerentin, Laurent Delahoche, Claude Pegard and Eric B. Gracsy: *A Localization Method Based on Two Omnidirectional Perception System Cooperation.* In IEEE Proc. of The 2000 Int. Conf. on Robotics and Automation (2000) 1219–1224
14. Ashley W. Stroupe, Kevin Sikorski and Tucker Balch: *Constraint-Based Landmark Localization.* In Pre-Proc. of The 2002 Int. Robocup Symposium (2002) 1–16
15. Hans Utz, Alexander Neubeck, Gerd Mayer and Gerhard Kraetzschmar: *Improving Vision-Based Self-Localization.* In Pre-Proc. of The 2002 Int. Robocup Symposium (2002) 17–32
16. Jens-Steffen Gutmann and Dietei Fox: *An Experimental Comparison of Localization Methods Continued.* In Proc. of the 2002 IEEE/RSJ Int. Conf. on Intelligent Robots and Systems (2002)

Speed-Dependent Obstacle Avoidance by Dynamic Active Regions

Hans-Ulrich Kobialka and Vlatko Becanovic

Fraunhofer Institute for Autonomous Intelligent Systems, Sankt Augustin, Germany

Abstract. An obstacle avoidance approach is introduced that has dynamic active regions. The dynamic regions are adapted to the current speed of the robot and there are different active regions used, one for speed reduction and one of turning away from obstacles. The overall strategy of this approach is that the robot can drive with high speed which will be reduced in front of an obstacle in order to do a sharper turn.

1 Introduction

Successful obstacle avoidance has to consider the abilities of the vehicle to slow down and turn. When driving at high speed, a robot needs more time to stop and making sharp curves becomes impossible. So if the robot slows down, the amount of possible trajectories increases. In presence of an obstacle, it has to be decided whether the robot should slow down and turn more sharply closer to the obstacle, or keep its speed but deviate more from the desired path.

The potential field method [8, 6] is commonly used for autonomous mobile robots in recent years. Basically it builds up an artificial potential field consisting of repulsive forces which pull the robot away from obstacles, and an attractive force directed towards a target direction. If the position of obstacles and target(s) is well-known and stable, the potential field can be used for planning a trajectory towards the target, e.g. [15]. In case of unknown environments where the robot can sense obstacles only in its near vicinity, the potential field method can only be used for local obstacle avoidance, i.e. compute the next direction to go for. The same holds true for dynamic environments, like RoboCup [1], where obstacles (e.g. robots) change their positions continuously.

The potential field method regards the robot to be holonomic, i.e. it can drive at any moment in any direction. Obviously this is not true for robots depending on their speed and mass. Non-holonomic, e.g. differentially steered, robots have to turn first before they can move in the desired direction. Obstacle avoidance methods, like VFH+ [14] and the Curvature Velocity method [11], consider this by cost functions. Using these cost functions the costs for each direction is computed. The dynamics and kinematics of a robot are considered as only feasible directions are evaluated using the cost function. A differentially steered robot can drive only small curvatures at high speed, so only these directions are considered. Finally, the direction with minimal cost is selected.

D. Polani et al. (Eds.): RoboCup 2003, LNAI 3020, pp. 534–542, 2004.

In VFH+, cost increases if a direction leads towards obstacles, deviates from the goal direction, or deviates from the current steering direction. The robot only slows down if every direction is blocked. So if a robot should drive along a corridor and turn into a narrow door, it may miss it because the speed isn't reduced. The same problem has been experienced for the Curvature Velocity method [9], and in [12] for the Dynamic Window approach [5].

The Lane Curvature method [9] chooses a free lane by minimizing an objective function which has to implement the trade-off between avoiding obstacles, heading for a target position, and maximizing speed. Among these three aspects, speed has the lowest priority, so the direction with minimum costs may demand to slow down the robot. This is similar to the effects of our approach.

In our approach, collisions are avoided by reducing the linear speed which again enables the robot to turn away from obstacles and towards the target direction. Both, speed reduction and turning, influence each other:

- speed reduction enables the robot to turn sharper, and
- turning to free space enables the robot to increase its speed again.

Speed dependency is mainly introduced by enlarging/shrinking the active region. The active region is an area that moves with the robot, usually a circle around the robot, see e.g. [6, 13]. In [12], a channel around a path (produced by a A* planner) is used. Only obstacles within the active region are considered for obstacle avoidance. An obstacle outside the active region is ignored as it is not regarded as a possible hazard.

Speed reduction and turning use different active regions. This is another outstanding characteristic of our approach. This brings the advantage that they can be tuned independently.

Speed reduction and turning are implemented as distinct behavior modules. In compliance with the Dual Dynamics architecture [7, 10, 3, 2], each module has a so called activation dynamics and a target dynamics. The target dynamics computes the motor commands by which the module wants to pursue its goals. The activation dynamics computes the activation value of this behavior, ranging in from 0 to 1. The activation value is used as a weight in order to superimpose concurrently active behavior modules. An activation value of 0 means that the module is not active, i.e. it should not influence the motors in any way. The activation of an obstacle avoidance module expresses the danger of having a collision; when approaching an obstacle it rises continuously from 0 to 1 and, after turning away, back to 0 again.

We will describe the speed reduction module *SlowDown* (section2) and the module *TurnAway* which makes the robot turning away from an obstacle (section 3). Section 4 tells how the outcome of these two modules is superimposed with the motor values of the goal-oriented behaviors. Our experience with speed-dependant obstacle avoidance is sketched in section 5.

2 Speed Reduction

2.1 Active Region

The active region of the module *SlowDown* is in front of the robot regarding its current direction of movement. In case of a differentially steered robot this active region is either at its front, or, if it moves backwards, at its back. The active region is shaped like a rectangle because obstacles within this region would collide with the robot as it moves forward. The length and the width of the active region is computed as follows:

$$length_{AR} = speed_{Robot} \cdot \Delta t + mDist2Obstacle$$
$$width_{AR} = width_{Robot} + mDist2Obstacle \tag{1}$$

where $width_{Robot}$ is the robot's width, $speed_{Robot}$ is the robot's current speed, Δt is some time span (e.g. 1 sec) and *mDist2Obstacle* is the minimum distance within which the robot should react on obstacles.

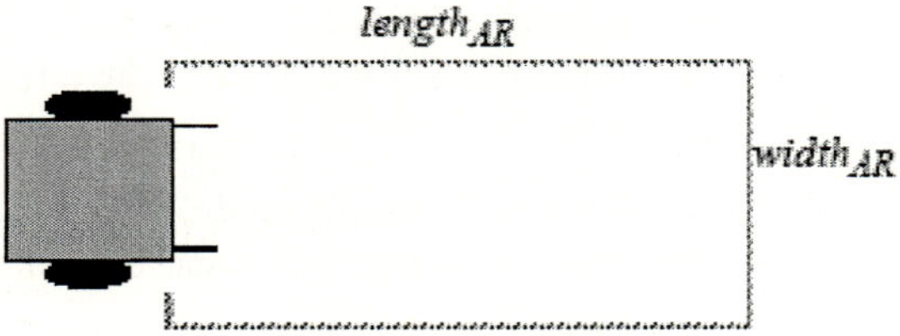

Fig. 1. The active region of *SlowDown*

2.2 Collision Danger

While the robot senses its environment (e.g. using ultrasonic distance measurement), it may detect a number of obstacle points. The possible danger of colliding with an obstacle point which is measured inside the active region, decreases exponentially depending on its distance $Dist_i$ to the robot.

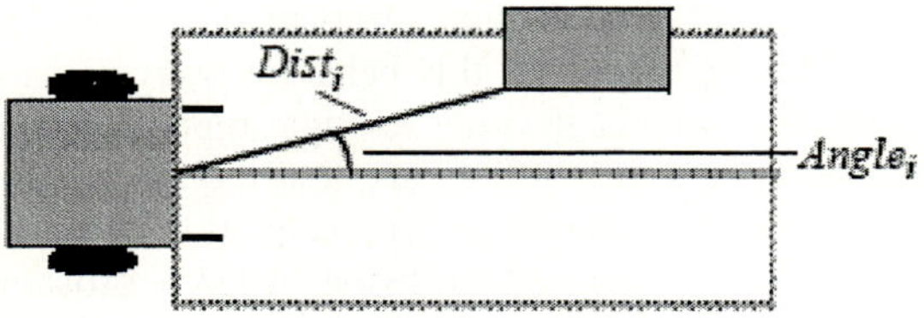

Fig. 2. An obstacle point is viewed with $Angle_i$ and $Dist_i$

$$F_{repulsive_i} = \begin{cases} e^{(-K_1 \cdot Dist_i)} \cdot weight_i & \text{if } Obstacle_i \text{ is in active region} \\ 0 & \text{else} \end{cases} \tag{2}$$

The weight of the obstacles is computed during several stages. First, it is proportional to the cosine of the obstacle's angle $Angle_i$.

$$weight_i^{**} = K_P \cdot \cos(Angle_i) + w_{\min} \tag{3}$$

where $w_{\min}$ denotes another safety margin.

If the robot drives at high speeds, the obstacles at the side (having low values for $\cos(Angle_i)$) should gain more importance. This is achieved by the following step

$$weight_i^* = \begin{cases} \max(weight_{critical}, weight_i^{**}) & \text{if } Dist_i \leq RobotSpeedK \\ weight_i^{**} & \text{else} \end{cases} \quad (4)$$

Finally, the weight is multiplied with a constant depending on how the obstacle has been classified. For instance, in the RoboCup domain there used to be a wall which has to be touched in order to get a ball away from it. So this kind of obstacle should not repel the robot as much as, for instance, other robots. If the ball lies between the robot and its own goal, it has to be avoided in order not to shoot self goals; so in this situation K_{ball} is very high, otherwise zero.

$$weight_i = \begin{cases} K_{ball} \cdot weight_i^* & \text{if } Obstacle_i \text{ belongs to the ball} \\ K_{wall} \cdot weight_i^* & \text{if } Obstacle_i \text{ belongs to a wall} \\ K_{robot} \cdot weight_i^* & \text{else} \end{cases} \quad (5)$$

2.3 Activation Dynamics

The activation dynamics of *SlowDown* computes its activation value which expresses the need for speed reduction for the current situation. Basically, it is the maximum of all repulsive "forces" clipped to 1.

$$a_{target} = \min(1, \max_i(F_{repulsive_i})) \quad (6)$$

This target value is low-passed by an ordinary differential equation

$$\dot{a}_{SlowDown} = T \cdot (a_{target} - a_{SlowDown}) \quad (7)$$

$$T = \begin{cases} T_{reactNow} & \text{if } \quad RobotSpeed \geq 70 \\ T_{reactFast} & \text{if } 70 > RobotSpeed \geq 40 \\ T_{reactModerate} & \text{else} \end{cases}$$

where the time constant T (which expresses how fast $a_{Slowdown}$ should follow a_{target}) depends on the current speed of the robot.

2.4 Target Dynamics

The target dynamics of *SlowDown* computes the maximum speed which is admissible for a certain situation. Basically it reduces the maximum Speed $Speed_{max}$ depending on $a_{Slowdown}$; especially if $a_{Slowdown}$ is zero, $Velocity_{SlowDown}$ equals $Speed_{max}$. This is done by a weighted sum where $w_{SlowDown} >> w_{Goal}$ ensures speed reduction in case of high values of $a_{Slowdown}$. In situations when turning is dangerous, i.e. will lead to collisions, the robot has to drive backwards for a short while.

$$Velocity_{SlowDown} = \begin{cases} \dfrac{w_{Goal} \cdot Speed_{max} + w_{SlowDown} \cdot a_{SlowDown} \cdot 0}{w_{Goal} + w_{SlowDown} \cdot a_{SlowDown}} & \text{if } a_{SlowDown} < 1 \\ 0 & \text{if } a_{SlowDown} = 1 \\ -b & \text{if turning is dangerous} \end{cases} \tag{8}$$

3 Turning away from an Obstacle

3.1 Active Region

The active region of the module *TurnAway* is oriented towards the current target direction.This region differs from the active region of *SlowDown*; especially it is independent of the current orientation of the robot.

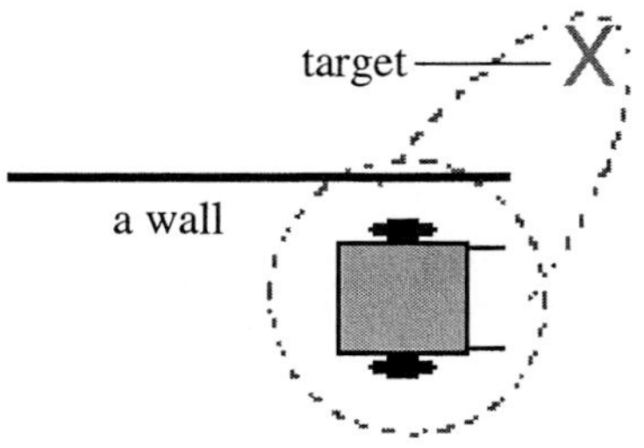

Fig. 3. The active reiogn of TurnAway

Also the shape of the active region differs from the one of *SlowDown*: first, there is a circle around the robot; if there are no obstacle point within, the robot can turn, otherwise not. Second, there is an active region oriented towards the target. It is shaped like a parabola of a width (near the robot) and a length defined like

$$length_{AR} = \min(speed_{Robot} \cdot \Delta t, dist2destination) \tag{9}$$
$$width_{AR} = width_{Robot} + mDist2Obstacle$$

Similar to the active region of *SlowDown*, the length of the parabolic active region depends on the robot's speed. In addition, the length is clipped to *dist2destination*, the distance between the robot and its target. This is because the robot should not care about obstacles being behind its target. Otherwise such obstacles would cause the robot to turn away before reaching its target. For instance, in RoboCup, the target may be the ball and there may be other robots behind the ball. These should not be avoided if a robot approaches the ball.

For the same reason the parabolic shape was chosen for the active region: the robot should not turn away (but only slow down), if there are other robots in the vicinity of the ball. This example illustrates that the active region should be shaped in an application specific way. For other applications than RoboCup, other shapes could be advantageous.

3.2 Collision Danger and Activation Dynamics

The collision danger and activation dynamics of *TurnAway* are computed in the same way as for *SlowDown* (see equations (2) to (7)). However it should be noted that

- a different active region is used, and
- the used constants are different.

This way, the modules *TurnAway* and *SlowDown* can be tuned independently (e.g. scope, reaction time, the strength of reaction) which helped us a lot while tuning the system.

3.3 Target Dynamics

The target dynamics of TurnAway computes the direction which is recommended to the robot by the obstacle avoidance module. Here we use a potential field approach: Each obstacle point has, beside $Dist_i$ and $Angle_i$, a $direction_i$ which is a unit vector pointing from the obstacle to the front of the robot.

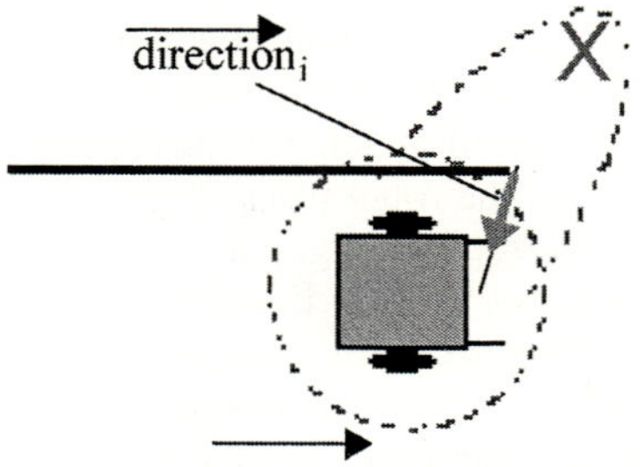

Fig. 4. Direction_i the repulsive unit vector

When computing the weighted vector sum, using $Frepulsive_i$ as weights, we get again a unit vector $direction_{TurnAway}$, the direction where *TurnAway* recommends to turn.

$$\overrightarrow{direction_{TurnAway}} = \frac{\sum F_{repulsive_i} \cdot \overrightarrow{direction_i}}{\sum F_{repulsive_i}} \tag{10}$$

4 Superposition of Goal-Oriented Behaviors and Obstacle Avoidance

After the obstacle avoidance modules have computed their output (activation values of *SlowDown* and *TurnAway* together with the recommended direction and the maximal admissible linear speed), we have to tell how the final motor commands result from this.

In the computation of the maximal admissible linear speed, the activation value of *SlowDown* is already used. So, the only thing left to be done is to clip the speed of the goal-oriented behavior(s) by this speed.

$$Velocity = \min(Velocity_{Goal}, Velocity_{SlowDown}) \tag{11}$$

Compromising between $direction_{TurnAway}$ and the direction wanted by the goal-oriented behaviors ($direction_{Goal}$) is done again by a weighted sum of vectors. Giving $direction_{Goal}$ the weight 1, $Weight_{TurnAway}$ has to be >> 1 in order to let $direction_{TurnAway}$ dominate if $a_{TurnAway}$ increases. This gives *TurnAway* the power to suppress $d_{desired}$ if obstacles are very close (i.e. if $a_{TurnAway}$ gets close to 1).

$$\overrightarrow{direction} = \frac{1 \cdot \overrightarrow{direction}_{Goal} + Weight_{TurnAway} \cdot a_{TurnAway} \cdot \overrightarrow{direction}_{TurnAway}}{1 + Weight_{TurnAway} \cdot a_{TurnAway}} \tag{12}$$

5 Results

The dynamic active region approach has been used since April 2001 by our RoboCup robots participating in the middle-size league. Each robot is equipped with 5 infrared distance sensors, two of them can sense obstacles up to 80 cm in front of the robot while the other three have a range of 50 cm. Obstacle points are remembered up to 1.5 seconds in an occupacy grid; as the robot turns, it gets a scan of obstacle points in its vicinity. Thus, the robot is quite short-sighted and the sensory input is sometimes incomplete, i.e. obstacles are not sensed. Even with this limited sensory input, our approach worked even in very crowded and dynamic situations. We limited the maximum speed of our robots to 160 cm per second because of the limited scope of our sensors and due to the fact that other robots also move fast. Under other conditions we expect our approach to work at higher speeds.

Figure 5 shows a scenario where a robot wants to drive behind the ball but finds its way obstructed by obstacles. The resulting trajectory is shown at the right side of figure 5: the points are drawn in equidistant time steps; it can be seen that the robot slows down in the vicinity of obstacles.

6 Conclusions

A novel obstacle avoidance approach is introduced that has dynamic active regions. The dynamic regions are adapted to the current speed of the robot. There are different active regions used, one for speed reduction and one of turning away from obstacles. The overall strategy of this approach is that the robot can drive with high speed which will be reduced in front of an obstacle in order to do a sharper turn.

The obstacle avoidance module is placed in the system architecture in a way that it can not be overruled by the behavior system. Also, the behavior system may choose among several possible paths by using the obstacle avoidance module as an oracle. This creates a highly flexible system that has shown good results in experiments, even with low dimensional sensor data of limited range.

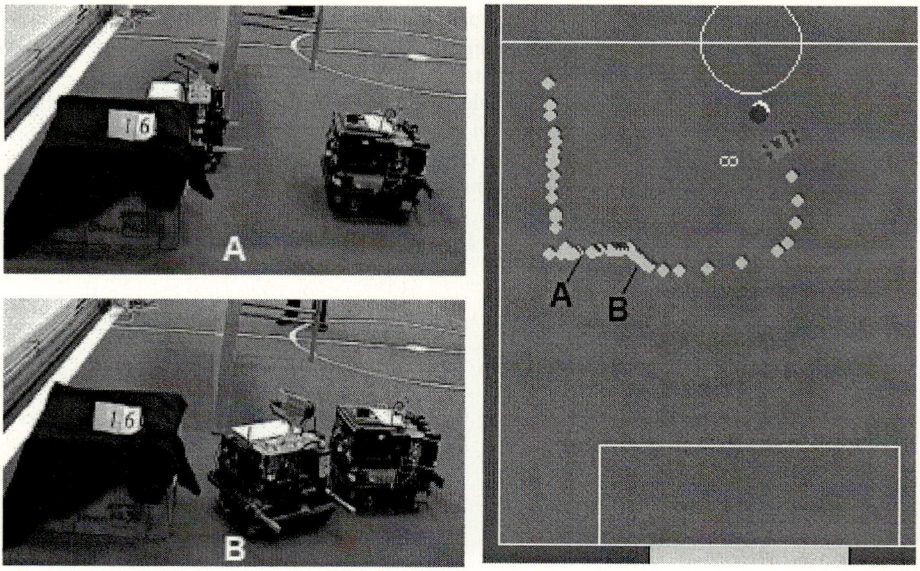

Fig. 5. A robot going for a ball while avoiding obstacles. The two snapshots at the left show the robot at positions (A and B) marked in the robot's trajectory shown at the right hand side

References

1. http://www.robocup.org
2. S. Behnke, R. Rojas, G. Wagner, 'A Hierarchy of Reactive Behaviors handles Complexity', *Workshop 'Balancing Reactivity and Social Deliberation in Multi-Agent Systems'* at the 14th European Conference on Artificial Intelligence (ECAI), 2000.
3. A. Bredenfeld, H.-U. Kobialka, 'Team Cooperation Using Dual Dynamics' *Balancing reactivity and social deliberation in multi-agent systems,* Lecture notes in computer science 2103, pp. 111 - 124, Springer, 2001.
4. A. Bredenfeld, G. Indiveri, 'Robot Behavior Engineering using DD-Designer', *Proc IEEE International Conference on Robotics and Automation (ICRA 2001),* pp. 205-210, 2001.
5. D. Fox ,W. Burgard, S. Thurn, 'The Dynamic Window Approach to Collision Avoidance' *IEEE Transactions on Robotics and Automation*, 4:1, 1997.
6. S.S. Ge, Y.J. Cui, 'Dynamics Motion Planing for Mobile Robots Using Potential Field Method', *Autonomous Robots* 13, 207-222, 2002.
7. H. Jaeger, T. Christaller, 'Dual Dynamics: Designing behavior systems for autonomous robots', *Artificial Life and Robotics*, 2:108-112, 1998.
8. O. Khatib, 'Real-time obstacle avoidance for manipulators and mobile robots', *The International Journal of Robotics Research* 5(1):90-98, 1986.
9. N.Y. Ko, R. Simmons, 'The Lane-Curvature Method for Local Obstacle Avoidance', *Proc. Intelligent Robots and Systems (IROS)*, 1998.
10. H.-U. Kobialka, H. Jaeger, 'Experiences Using the Dynamical System Paradigm for Programming RoboCup Robots', Proc. *AMiRE 2003* (2nd International Symposium on Autonomous Minirobots for Research and Edutainment), pp 193-202, 2003.
11. R. Simmons, 'The Curvature-Velocity Method for Local Obstacle Avoidance', *Proc. IEEE Int. Conf. on Robotics and Automation (ICRA 1996)*, pp. 3375-3382, 1996.

12. C. Stachniss, W. Burgard, 'An Integrated Approach to Goal-directed Obstacle Avoidance under Dynamic Constraints for Dynamic Environments', *Proc. IEEE/RSJ International Conference on Intelligent Robots and Systems (IROS 2002)*, 2002.
13. I. Ulrich, J. Borenstein, 'VFH+: Reliable Obstacle Avoidance for Fast Mobile Robots', *IEEE Int. Conf. Robotics and Automation (ICRA 1998)*, pp. 1572-1577, 1998.
14. I. Ulrich, J. Borenstein, 'VFH*: Local Obstacle Avoidance with Look-Ahead Verification', *IEEE Int. Conf. Robotics and Automation (ICRA 2000)*, pp. 2505-2511, 2000.
15. T. Weigel, J.-S. Gutmann, M. Dietl, A. Kleiner, B. Nebel, 'CS-Freiburg: Coordinating Robots for Successful Soccer Playing', *IEEE Transactions on Robotics and Automation* 18(5):685-699, October 2002.

Using the Opponent Pass Modeling Method to Improve Defending Ability of a (Robo)Soccer Simulation Team

Jafar Habibi, Hamid Younesy, and Abbas Heydarnoori

habibi@sharif.edu, hyounesy@cs.sfu.ca, aheydarnoori@cs.uwaterloo.ca

Abstract. Modeling agents' behavior has always been a challenge in multiagent systems. In a competitive environment, predicting future behaviors of opponents helps to make plans to confront their actions properly. We have used the RoboCup soccer server environment [1] to design a coach, capable of analyzing simulated soccer games and making decisions to improve teammate players' behavior during the games. We will introduce our "Opponent Pass Modeling" method which makes a model of opponent team's passing behavior to guard opponent players and as a result, to improve the defending behavior of our team. We will also describe a new approach to evaluate coach algorithms using soccer server log-files and LogCoach tool.

1 Introduction

The Simulation League of RoboCup provides a standard competition platform where teams of software agents play against each other in a simulated soccer game [2]. In robotic soccer tournaments, a team of agents plays against another team of agents for a single, short (typically 10-minute) period. The opponents' behaviors are usually not observable prior to this game and so they must be analyzed and modeled during the game. A common challenge for agents in multiagent systems is trying to predict what other agents are going to do in the future. Such knowledge can help an agent determine which of its current action options are most likely to help it achieve its goals.

So far, there has been some good works done on opponent behavior modeling in robotic soccer environment. Ideal-model-based behavior outcome prediction (IMBBOP)[3] is a technique which predicts an agent's future actions in relation to the optimal behavior in its given situation. An online opponent model recognition method is introduced in [4,5], in which the coach uses a number of pre-defined opponent models to provide the agents with proper team plans. More recently, some methods focus on unsupervised autonomous learning of the sequential behaviors of agents, from observations of their behavior [6].

In this paper, we will introduce the "Opponent Pass Modeling" method which uses statistical information about opponent players' successful passes to mark them during a game. We will also describe a new approach to test this method using soccer server log-files, by the means of some useful software tools designed

D. Polani et al. (Eds.): RoboCup 2003, LNAI 3020, pp. 543–550, 2004.

in *Sharif-Arvand* soccer simulation team [7] including LogCoach. As a platform for our coach algorithms, we have designed and implemented a coach, which is capable of gathering a rich collection of statistical information during soccer simulation games. This statistical information then can be used for doing analysis in a game to make decisions to improve players behavior [7, 8].

This article is organized as follows. Section 2 will describe the opponent pass modeling method from coach-side view which is more about gathering and analyzing statistical information and from players-side view which focuses on utilizing the results provided by coach. Section 3 introduces our testing approach using soccer server log-files with the tools implemented in Sharif-Arvand team. In section 4 we suggest some minor changes to adapt this method with the standard coach language. Section 5 presents conclusions and directions for future work.

2 Modeling Opponents' Passing Behavior

In a typical soccer simulation game, single player attacks will rarely result in scoring a goal; It is simply because of the fact that opponent team already benefits from players defending the area. So, players use passing techniques during an attack to keep the ball from opponents and make the situation ready for scoring. Figure 1 shows two sample effective methods applied by good teams during previous RoboCup competitions. The models have been recognized by observing different log-files of the games played by these teams. In each case, attacking players have used passes to open teammates to change the region of attention (where the defenders have crowded) making them free to position better for next passes. Numbers on the arrows indicate steps of scenario, and for simplicity, only the important movements have been shown in each step. It can be seen that if the defenders had marked the attackers properly they could have avoided surrendering a score. Marking, generally means guarding a player to prevent it from advancing the ball towards the goal, making an easy pass or getting the ball from a teammate.

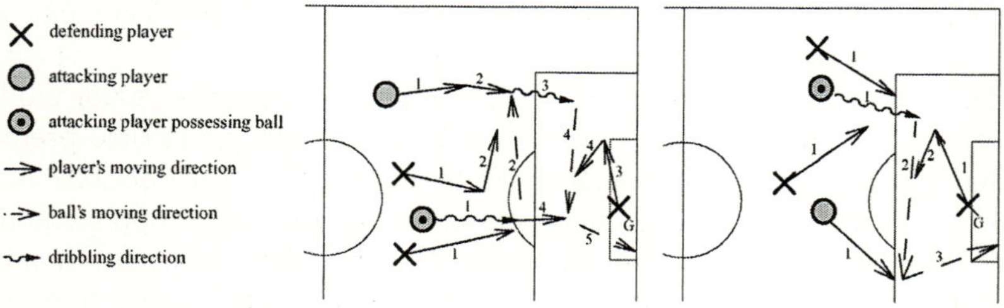

Fig. 1. Two samples of successful group attacks

To determine which players should be marked, we take advantage of the fact that teams are likely to show similar behaviors in similar cases. So, the players may use a similar sequence of actions when they are in similar situations. During a game, coach considers the successful passes of opponent players to model their passing behavior. A successful pass can be defined as a ball transfer between two teammate players. Players will use the results later, to mark opponents during a game.

2.1 Coach-Side Analysis

A coach client has the advantage of receiving noise free data about the field from soccer server. So it is the best agent to centrally collect and analyze information in a game. In a typical layered design for coach, the lower layers are responsible for communicating with the soccer server, storing the received data in proper data structures and recognizing the skills done by players. For example, skills like possessing the ball or shooting the ball can be detected by comparing the position and the speed of the ball and players during a sequence of cycles; or for detecting passes which are a more complex skill, we use the lower detected skills like the current possessor of the ball and the last player who kicked the ball; if they both belong to the same team the ball has been passed, otherwise it has been intercepted. Then, based on these primary information in the lower layers, more intelligent algorithms for the coach have been designed.

In our method, the coach uses information about successful passes of opponent players to build a simple model for their passing behavior. As it can be seen in figure 2(a), the coach stores the passes that opponents have done in the defense area as well as initial coordinates of the receivers of the passes. The defense area is defined as a portion of the field near our own goal and is about one third of the field length. This information is then sent to the players in proper cycles. Here is a simple pseudo code showing this part of the method:

```
for each successful pass of opponent {
  if source [or destination] of the pass is inside defense area {
    T1 = PassStartTime
    T2 = PassEndTime
    P = PassReceiverPlayer
    SuccessfulPassList.Store(BallPos(T1), BallPos(T2), PlayerPos(P, T1))
  }
}
```

2.2 Players-Side Utilization

Now it is players' turn to use the information recently received from coach. The basic idea is to mark the opponent who is the most probable to receive a pass from the player possessing the ball. So, when an opponent player possesses the ball in our defense area we check whether the current position of that player is near to one of the BallPos_1s in the list received from coach; i.e. the starting point of the passes. If so, the player may send a similar pass and we should mark the player who will probably receive the pass. As shown in figure 2(b) we choose the nearest player to the PlayerPos corresponding to that pass; i.e. the nearest

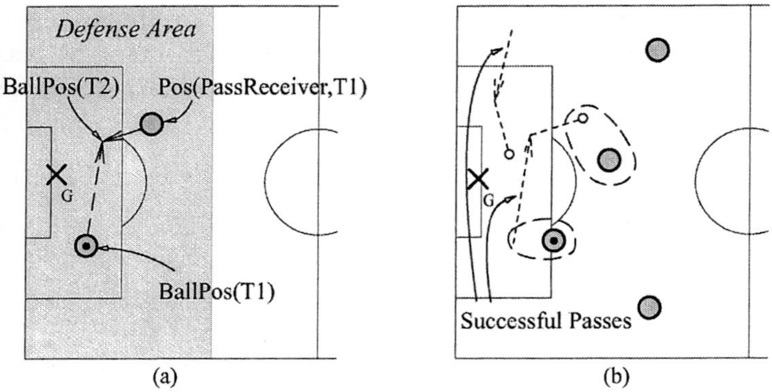

Fig. 2. (a) Coach collects information about successful passes (b) Players decide about the opponent to be marked

plater to the guessed pass destination. Now, one of our players (e.g. the nearest player) should mark the selected opponent player. Pseudo code below shows the described algorithm in more details:

```
if BallOwner is an opponent and BallPos is inside defense area {
  POpp = BallOwner
  for each item in the successful-pass list (i)
    if IsNear(Pos(POpp),BallPos_1[i])) {
      for all opponent players (P[j])
        if IsNearest(Pos(P[j]), PlayerPos[i])
          MarkOpponentPlayer(j)
    }
}
```

Lots of optimizations can be done to improve the collected information and the decision of the players when implementing this algorithm. In case more than one BallPos_1 is nominated, the opponent player which has less defense players near will be chosen. Also, in addition to original coordinates of successful passes, we store the symmetric coordinates related to the fields' x-axis; because of the fact that, teams usually have symmetric formations which causes the players to play symmetrically. We can also compute values for passes according to their regional importance or consequential effect in the game (e.g. surrendering a goal). In players-side, we only consider opponents who can receive the predicted pass in a point near the previous successful pass (near BallPos(T2)). Note that this method does not rely on the individual players (their uniform number). So, players' substitutions will not affect the result.

3 Using LogCoach for Evaluation

Using results of games is not necessarily a good way to evaluate most coach algorithms. It is because of the fact that decisions made by a coach, are usually high level commands which need to be executed by players. So, even if coach makes decisions properly, they won't be effective unless players can utilize them during a game, which will require well programmed players in skills, tactics, etc. For example, considering our pass-behavior modeling algorithm, even if players correctly predict opponents who will receive passes, they need a good marking skill to mark opponents. If for any reason, a player fails to mark an opponent properly, not only it will not be able to stop the attack, but also one of the defenders uselessly will be hanging around.

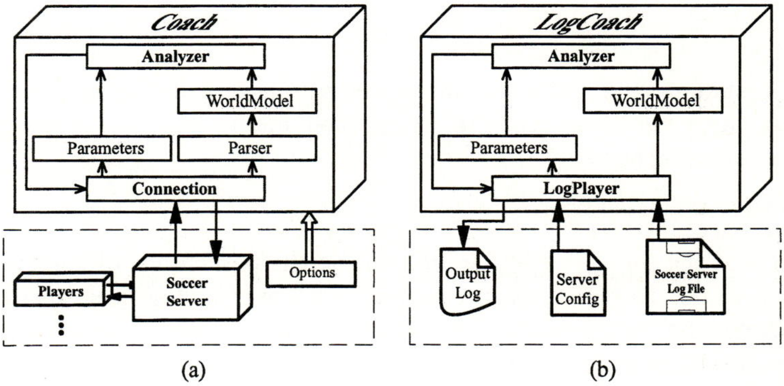

Fig. 3. (a) Layered structured of Sharif-Arvand coach (b) LogCoach structure

So, we have designed and implemented LogCoach tool. Figure 3 presents the structure of Sharif-Arvand online coach client and LogCoach. In the online coach, connection layer is responsible for gathering data by communicating with soccer server. While, in the LogCoach, LogPlayer module uses soccer server output log-files to provide the the game data and stores the output as graphical information which can be viewed using a visualizer tool called CoachDebugger later. So, rather than running real soccer simulation games, we can use stored log-files as the input data, without any changes required in other layers.

Using the LogCoach instead of coaching actual games, offers some benefits and some limitations. Instead of running a complete soccer simulation game a log-file can be processed in a few seconds; so it increases speed of development and testing. Also, combined with the CoachDebugger it is a great help for testing algorithms while enabling us to analyze the coach decisions by observation. An other main advantage of using logfiles is that exactly the same input data is available between the changes of the algorithms allowing it to see if new code segments really improve the algorithm. However, without a real game, it is impossible to have active behavior like interacting with the players or adapting the parameters according to the result of the previous decisions in a game.

To test and evaluate the correctness and effectiveness of our opponent pass modeling algorithm, we developed a simple testing module using the LogCoach. The idea is basically to find out how many opponent passes could be avoided during a game, and more precisely, how many goals could have been avoided if players had used the opponent pass modeling algorithm to mark opponent players. Actually, we put our players' predicting algorithm in the test module and use the coach's statistical information to predict players who will receive passes:

```
When an opponent possesses the ball in our defense area {
  Predict the player who will receive the pass
  Store the predicted player and other game status information
  GuessesCount++
}

For each successful pass of an opponent in our defense area {
  if the pass receiver is the predicted one
    CorrectGuesses++
}
```

In the end the $\frac{CorrectGuesses}{GuessesCount} \times 100$ will be the percentage of correct guesses during the game. To test the algorithm we used over 40 log files most of which were from the last two RoboCup official games. We chose the games whose total scored goals were more than 10 and we considered the defeated team as our team for which the coach was used. The result is presented in Table 1. As you can see the results are satisfactory; even in the worst case, about half of the opponents' passes in our defense area could be predicted and probably avoided. Although it does not mean that avoiding the passes could completely stop the attacks, but they sure could deviate the opponents' main plan.

Table 1. Evaluation results based on correct pass predictions

	Game Result	GuessCount	CorrectGuess	Correctness
Worst Case	1-11	23	10	43%
Best Case	0-23	34	31	91%
Average (rounded)	0-14	26	19	74%

Doing some changes in the evaluation function, the results were even more interesting. This time we just considered the passes which consequently resulted in a goal.

```
When opponent scores a goal {
  P1 = Player who scored the goal
  P2 = The last player who sent a pass to P1
  GoalGuessesCount ++
  if we had already predicted P1 to receive a pass from P2
    GoalCorrectGuesses++
}
```

Table 2 shows results of the new evaluation approach. Correctness column shows the percentage of the goals which could be avoided if the scorer was marked before receiving the last pass. Again we insist that this does not mean that the defending team could survive from all the successfully predicted attacks, specially because of their weakness in skills and tactics comparing with the opponents players.

Table 2. Evaluation results based on prediction of passes resulted in a goal

	Game Result	GoalGuessesCount	GoalCorrectGuesses	Correctness
Worst Case	1-11	11	5	45%
Best Case	0-23	23	19	83%
Average	0-14	14	9	67%

4 Adapting for the Standard Coach Language

Standard Coach Language is a general language for communication between a coach and players during a soccer simulation game. The idea is that a coach uses a unified standard language to communicate with teammate players, so it can be used for coaching any team that can understand this language.

The syntax of the standard coach language allows the coach to define rules, regions, constraints and instructions [1]. Our opponent pass modeling method can be easily adapted to be used with a standard language based coach. For each successful pass, it defines regions around the starting point of the pass and starting point of the pass receiver. Then it defines a rule for the teammate players to mark the opponent inside a region, related to where the ball exists. Here is an example showing usage of the standard coach language for this method (based on soccer server version 8.0).

```
(definer "Reg1" (arc (pt X_FROM Y_FROM) 0 R 0 360))
(definer "Reg2" (arc (pt X_PASS_REC Y_PASS_REC) 0 R 0 360))

(definerule "rule1" model
((and (time >= INITIAL_TIME) (bpos "Reg1") (bowner opp {X}))
(do our {X} (markl "reg2"))) )
```

where X_FROM & Y_FROM will be coordinates of starting point of the pass, X_PASS_REC & Y_PASS_REC will be initial position of the pass receiver. R and INITIAL_TIME are constant values (e.g. 5 and 1000) indicating the radius of the regions and the initial time to be waited before executing the rule (a minimum time, coach needs to collect more data). "markl" is an instruction defined in the standard language, tells the players to mark the line from the ball to a specific region.

5 Conclusions and Further Works

In this paper, we described a new method to model the passing behavior of opponent teams, based on statistical information about previous successful passes.

We saw that the model then can be used by players to guard opponents properly and as a result to improve their defense ability. We presented a new approach to evaluate coach algorithms using soccer server log-files instead of running actual games. We also showed how the opponent pass modeling method can be adapted to be used with standard coach language.

We are currently working on using this method to decide about team formation. For example if coach observes situations in which there are more opponent players to be marked than the number of defenders it may decide to improve the defense line by changing number or placement of defenders in the formation.

References

1. Chen, M., et al.: RoboCup Soccer Server User Manual for Soccer Server Version 7.07 and later. *http://sserver.sourceforge.net/docs/*, last visited: Jan 2003
2. Kitano, H., et al.: The RoboCup Synthetic Agent Challenge. In Proceedings of the 15th International Joint Conference on Artificial Intelligence (1997)
3. Stone, P., Riley, P., Veloso, M.: Defining and Using Ideal Teammate and Opponent Models. In Proceedings of the Twelfth Innovative Applications of Artificial Intelligence Conference (2000) 1040-1045
4. Riley, P., Veloso, M.: Coaching a Simulated Soccer Team by Opponent Model Recognition. In Proceedings of the Fifth International Conference on Autonomous Agents (2001) 155-156
5. Riley, P., Veloso, M.: Recognizing Probabilistic Opponent Movement Models. RoboCup 2001: Robot Soccer World Cup V, Springer-Verlag New York (2002) 453-458
6. Kaminka, K., et al.: Learning The Sequential Coordinated Behavior of Teams from Observations. In Proceedings of the RoboCup-2002 Symposium (2002)
7. Habibi, J., et al.: Sharif-Arvand Simulation Team. RoboCup 2000: Robot Soccer WorldCup IV, Springer-Verlag New York (2001) 433-436
8. Habibi, J., et al.: Coaching a Soccer Simulation Team in RoboCup Environment. In The 1st EurAsian Conference on Advances in Information and Communication Technology (EurAsia-ICT 2002), Springer-Verlag (2002) 117-226
9. Drucker, C., et al: As Time Goes by: Using Time Series Based Decision Tree Induction to Analyze The Behavior of Opponent Players. RoboCup 2001: Robot Soccer World Cup V, Springer-Verlag New York (2002) 325-330

Topological Navigation in Configuration Space Applied to Soccer Robots

Gonçalo Neto, Hugo Costelha, and Pedro Lima

Instituto de Sistemas e Robótica
Instituto Superior Técnico
Av. Rovisco Pais, 1 – 1049-001 Lisboa, Portugal
{gneto,hcostelha,pal}@isr.ist.utl.pt
http://socrob.isr.ist.utl.pt

Abstract. This paper describes a topological navigation system, based on the description of key-places by a reduced number of parameters that represent images associated to specific locations in configuration space, and the application of the developed system to robotic soccer, through the implementation of the developed algorithms to RoboCup Middle-Size League (MSL) robots, under the scope of the SocRob project (*Soccer Robots* or *Society of Robots*). A topological map is associated with a graph, where each node corresponds to a key-place. Using this approach, navigation is reduced to a graph path search. Principal Components Analysis was used to represent key-places from pre-acquired images and to recognize them at navigation time. The method revealed a promising performance navigating between key-places and proved to adapt to different graphs. Furthermore, it leads to a robot programming language based on qualitative descriptions of the target locations in configuration space (e.g., Near Blue Goal with the Goal on its Left). Simulation results of the method application are presented, using a realistic simulator.

1 Introduction

The problem of robot navigation is, perhaps, one of the key issues in mobile robotics. Roughly, it consists in driving a robot through a given environment, using the information from his sensors.

The most common form of solving this problem is to construct a world model from the sensorial information. Based on this model, it is relatively easy to apply control algorithms and drive the robot to its target locations. One of the problems with this approach is the amount of computational effort required to store the above-mentioned world model. Even when that is not an issue, most of the times the sensorial information is not as exact as it would be desirable.

In this line of thought, approaching the problem in a more qualitative manner seems to be quite a promising alternative, in which the relevant places of the world are determined from their appearance. Vision sensors are normally used to extract the relevant information, as opposing to non-vision sensors which are widely used in the construction of world models. The problem of appearance

D. Polani et al. (Eds.): RoboCup 2003, LNAI 3020, pp. 551–558, 2004.

based methods for navigation is surely not a closed one and has been addressed in various manners, in a recent past.

In [1] the localization of the robot in a topological map is obtained by analyzing images in the frequency domain. On the other hand, [2] uses the original input images and [3] uses the color histograms of such images. An interesting idea is to use Principal Component Analysis methods to extract the most important characteristics from an image or group of images, as proposed in [4].

This paper introduces the application of topological navigation methods to robotic soccer. We have developed a flexible method that allows the robot to pursue its navigation objectives, during a match, using a qualitative approach, making the interaction with higher knowledge levels natural. The methodology used in solving the overall problem is summarized in Fig. 1.

The outline of this paper is as follows: Section 2 describes the PCA methods used to construct the topological map. Section 3 presents the techniques used for localization and navigation. In Section 4 the application of the previous methods to robotic soccer is explained and Section 5 presents the main results. Finally in Section 6 some conclusions are drawn and future research perspectives explained.

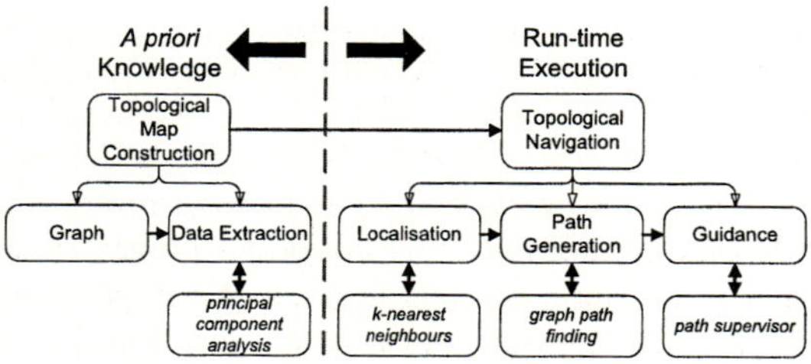

Fig. 1. Summary of methods used in this work

2 Map Construction

The first step of the topological navigation approach is to build a representation for the topological map, based on which the robot will navigate. Several representations of this map can be used; in our case we have chosen a directed graph, where the nodes represent the key-places in the map and the transitions represent the functions used to navigate between key-places.

This kind of representation is general enough to be used in a large number of applications. In our case, the nodes will be identified with groups of postures in the configuration space (x, y, θ) and the transitions correspond to basic functions like *move back* or *move forward*.

After defining the topological map, the information required to represent each of the nodes must be gathered. As this is an appearance based method, we first start to acquire a set of images P that represents the space where the robot will navigate. This set of images has to be general enough to represent all the areas of the configuration space where we want the robot to navigate.

The following step is to use Principal Component Analysis (PCA), also known as Karhunen-Loeve (KL) expansion [5, 6], to compress the information in P. At the end of the expansion, a basis for the η-dimensional that best approximates the acquired images, in the least squares sense, is obtained. This method is equivalent to retain the directions with the larger variance (i.e., the largest amount of information) of the data set.

Furthermore, it's possible to define the error we are making in the approximation, by the following expression:

$$E[\varepsilon^2] = \sum_{i=\eta+1}^{M} \lambda_i \tag{1}$$

or, in percentage terms:

$$\xi = \frac{\sum_{i=\eta+1}^{M} \lambda_i}{\sum_{i=1}^{M} \lambda_i} \tag{2}$$

where η corresponds to the number of eigenvalues chosen to represent the eigenspace. These expressions provide a criterion to choose the number of eigenvectors to be used.

Having explained the idea behind the approximation, an iterative procedure can be used to compute the principal images, as explained in [5]. We call *principal space* to the space obtained after all these calculations.

The next step in the construction of the topological map is to associate the previously gathered information with the nodes of the graph. In this part of the method, we start by projecting each image in the principal space, associating the projection with the node of the graph that the corresponding image represents. Of course, it is essential that the input images are previously grouped according to the node they best characterize.

We can think of the obtained groups of projections as classes of patterns, which allows the localization problem to be formulated as a pattern classification one.

3 Topological Navigation

With the tools to represent the topological map available, it is possible to define the methods to use in the navigation itself. So, we start by presenting the localization problem and then move to the discussion of the path planning and path execution problems.

It must be pointed out that navigation procedures, as opposed to map construction, must be executed in real-time.

As previously stated, the localization of a certain image in a node of the graph can be formulated as a classification problem. To solve this problem, a known classifier can be used. We chose to classify the images in the *k-nearest neighbor* [7] sense. In this classifier, the current input image projection is compared to the projections of all the images used in the map construction (also known as learning) stage of the method. We then retain the k closest images and classify the image in the most represented class among the selected k images.

A path to the objective node of the robot was obtained using widespread search algorithms [8]. Of course, an heuristic could be used to improve the search efficiency.

The other step in the navigation process concerns guidance, in which we must ensure that the robot accomplishes the goals set by the path planner. In this context, it is straightforward to achieve such goals, since the path is defined as a sequence of primitive transition functions. The guidance loop for this approach is described through the flowchart in Fig. 2. The problem with this solution is

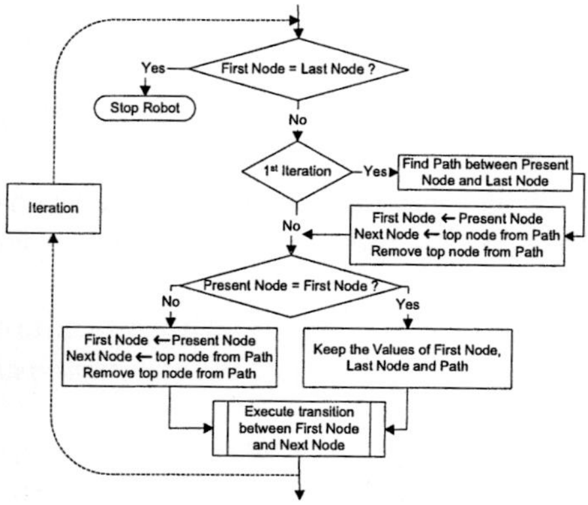

Fig. 2. Open loop guidance supervisor

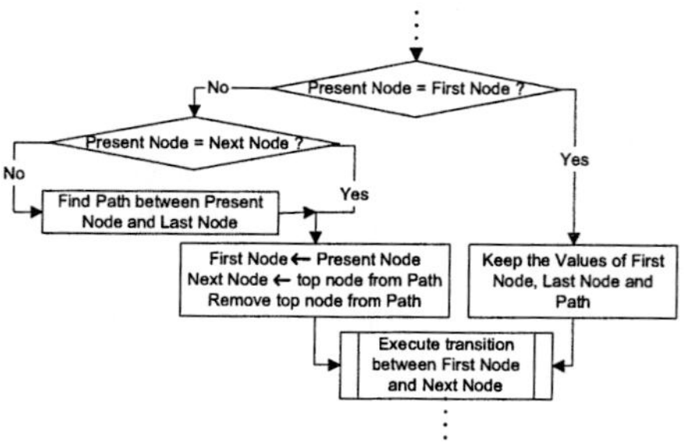

Fig. 3. Modified open loop guidance supervisor, to include path replanning

that, if a transition fails and the robot finds itself, e.g., in a node which does not belong to the path, it simply cannot fulfill its objectives. A straightforward solution for this problem is to generate a new path each time the robot detects a failure in a transition, as explained in the flowchart of Fig. 3.

4 Application to Robotic Soccer

Our team of robots, the RoboCup MSL *ISocRob* team, consists of four *Nomadic Super-Scout II* robots, which have, among other sensors, a front camera and moves based on a differential drive kinematic structure.

In order to better test the previously described navigation method, we developed a simulator, which could generate images similar to the front camera of our robots. This simulator was implemented in *Virtual Reality Modeling Language* (VRML), and field textures were used to improve the realism. The images obtained were RGB images with 320×240 pixels taken over a field with $10 \times 5[m]$.

Figure 4 compares a simulator image and the image seen by the robot's front camera, in a real situation.

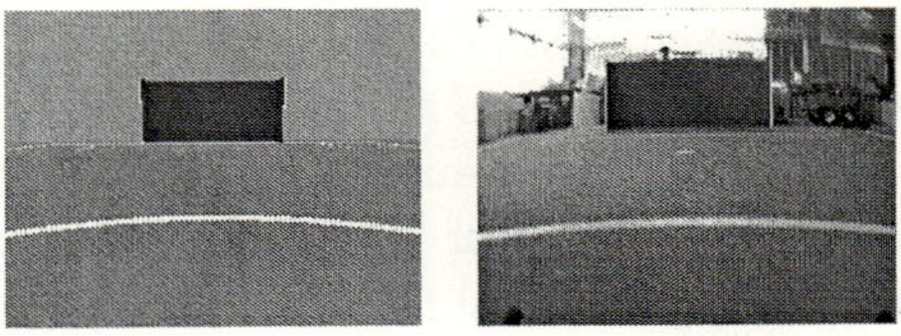

Fig. 4. Simulated vs Real images

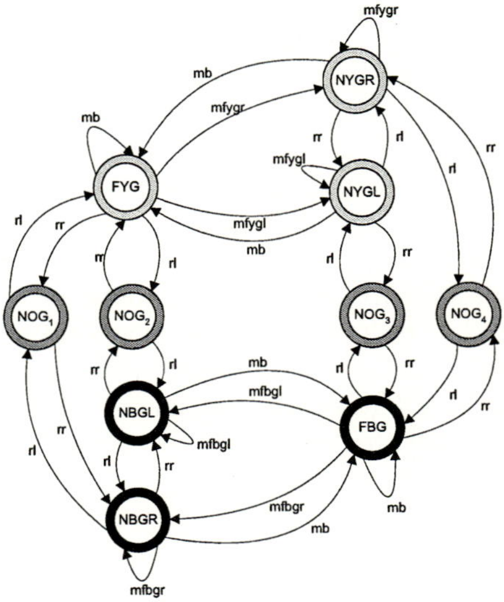

Fig. 5. Graph representation of the Topological Map

4.1 Topological Map

To construct the map, we considered 3 nodes where the BLUE goal was visible, 3 nodes where the YELLOW goal was visible and 4 nodes where no goal was visible. We used 7 different transitions to travel between these nodes.

The nodes where defined considering the relative position of the goal w.r.t the robot. We used four nodes for the situation where there is no goal on the images to avoid conflicts and ambiguities in the definition of the graph. However, the difference between these nodes can only be determined based on the history of the robot path accomplished so far.

The transitions where defined specifying the angular and/or linear speed, and the relative position which we want to maintain of a specific goal (if any) w.r.t. the robot (e.g., move forward keeping the blue goal in the right of the robot).

The graph that was used in the robots navigation is shown in Fig. 5. After defining the graph, we extracted a set of 528 training images and applied the *KL* transform. To determine how many principal components to use, we computed all the 528 eigenvalues and used equation (2) to compute the number of eigenvalues required to achieve $\xi < 10\%$, leading to 46 principal components.

In Fig. 6, a plot comparing the mean square error for the training set and for a testing set is presented. The latter was obtained in random field positions and the error was calculated for several principal spaces with different numbers of eigenvalues, between 0 and 60.

4.2 Navigation

For localization, we used, as previously stated, the *k-nearest neighbor classifier*, with $k = 5$ an associated Euclidean metric, in the principal space. We obtained several classification results using a large number of postures from most regions of the configuration space. In Section 5 we explain some of those results.

The overall algorithm was tested by running it on the simulator, generating a new goal node randomly each time the last one was accomplished. Furthermore, we created another graph having the same nodes and transitions topology but where the images associated with each of the nodes were different — we only considered an image to be "near a goal" when it was taken at a distance of $2\,m$ from that goal, as opposing to the first situation where this distance was extended to $5\,m$ (mid-field situation). This setup also allowed us to test the percentage of times a given transition would fail for a given representation. Some of the results from those tests are presented in Section 5.

5 Experimental Results

The principal space was obtained from a set of training images. Obviously, not all images with the same size live on this space and, in general, they are far from it. In fact, by neglecting the smaller eigenvalues we are doing an acceptable approximation of the eigenspace obtained from the training images but, on the other hand, the principal space might not be so good so as to approximate conveniently other images of the same size.

In Fig. 6 we compare the mean square error for training and test sets. As expected, the error for the training set converges to a constant value. This value corresponds to the mean square distance between the test set and the eigenspace. For the training set the mean square error converges towards zero. The implemented classifier was also tested, as mentioned in Section 4, by using a large number of postures on the soccer field, with small distances between them, which ensures a good representation of the whole configuration space. We present an example of a 2 dimensional cut in the configuration space, with the coordinate y fixed in the zero value. Refer to Fig. 8 for information on the frame used.

The localization tests led to the conclusion that the classifier is not immune to the presence of outliers, which can make the path execution a bit more troublesome, justifying the need for a supervisor. However, the results are much better for 5 neighbors than the use of a single *nearest neighbor* classifier.

As for the execution of the complete navigation method, we stated that it was mostly successful in travelling between key-places of the topological map, due to

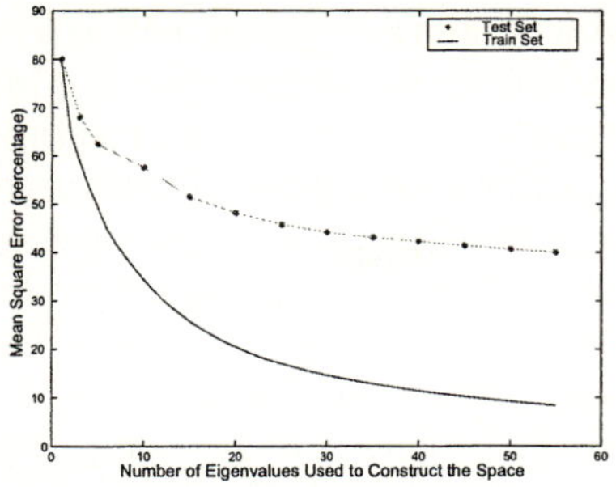

Fig. 6. Mean square error in the train and test sets

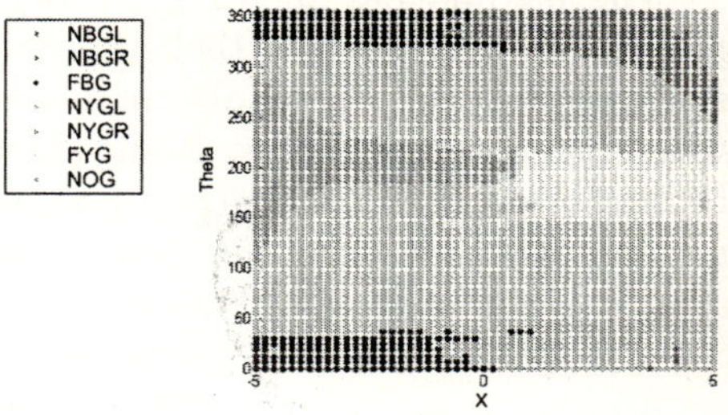

Fig. 7. Example of the classification/localization

the usage of an execution supervisor. In fact, some of the transitions proved to have a failure rate higher than 50%, which suggests that there are some implicit transitions on the graph which were not defined *a priori*. This was, in fact, expected, since the nodes did not quite correspond to the configuration space regions we intended them to represent.

A more serious situation that occurred, due to the existence of these implicit transitions, were *live-locks* — situations where the robot stays in a loop from where it cannot get out, unless a new goal node is given. Other than that, the method was always successful in achieving the proposed key-places.

In Fig. 8 we can see a trajectory of the robot in a multi-objective situation, and where the border between *Far Goal* nodes and *Near Goal* nodes was set at the distance of $2\,m$ from the goal. It is interesting to underline angular velocity control to ensure that the robot keeps the goal on its right or left, leads the robot to align itself with the goal-posts.

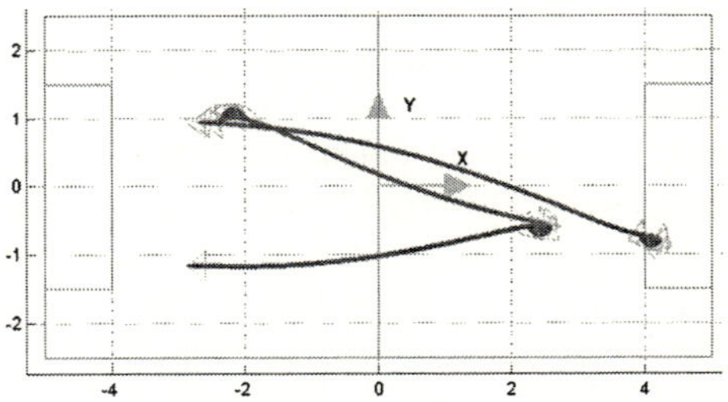

Fig. 8. Example of a trajectory

6 Conclusions and Future Work

This paper addressed the application of topological navigation methods to robotic soccer. We introduced an appearance based method which can navigate between different regions of the configuration space, represented by key-places of a topological map.

The method has shown promising results, when navigating between key-places. Although some of the transitions displayed high failure rates, the inclusion of the supervisor was able to deal with most of those situations, making the robot reach its goal nodes. However, some parts of the algorithm still need to be improved. The main one is the occurrence of *live-locks*, not in the topological map but as the result of the method application, due to specific failure cycles in the transition execution. We will investigate the application of Discrete Event Supervision techniques [9] to the detection and prevention of such cycles.

Another part of the method with open research topics is image compression. Actually, we would like to show it is possible, for this application, to use a smaller number o eigenvectors, compressing the needed information much more. This goal is most likely to be accomplished by reducing the size of the acquired images and/or using omnidirectional cameras instead of the front camera.

We are currently committed to use the developed methodology to solve the *RoboCup Challenge 4 - Play with an arbitrary FIFA ball* of RoboCup 2003 MSL. The idea is to use PCA to extract the most important features describing the ball, Topological Navigation to move the robot close to the ball and, finally, metric-based navigation already implemented in the ISocRob team to take the ball to the desired goal.

References

1. Ishiguro, H., Tsuji, S.: Image-based memory of environment. IEEE/RSJ Int. Conf. Intelligent Robots and Systems (1996) 634–639
2. Horswill, I.: Polly: A vision-based artificial agent. Proc. Nat. Conf. Artificial Intelligence (1993) 824–829
3. Blaer, P., Allen, P.: Topological mobile robot localization using fast vision techniques. Proc. of the 2002 IEEE Int. Conference on Robotics & Automation (2002)
4. Gaspar, J., Winters, N., Santos-Victor, J.: Vision-based navigation and environmental representations with an omnidirectional camera. IEEE Trans. on Robotics and Automation **16** (2000)
5. Murakami, H., Kumar, V.: Efficient calculation of primary images from a set of images. IEEE Transactions on Pattern Analysis and Machine Intelligence **4** (1982) 511–515
6. Murase, H., Nayar, S.K.: Visual learning and recognition of 3-d objects from appearence. International Journal of Computer Vision **14** (1995) 5–24
7. Mitchell, T.: Machine Learning. 1st edn. Computer Science Series. McGraw-Hill, Singapore (1997)
8. Russel, S., Norvig, P.: Artificial Intelligence: a Modern Approach. 1st edn. Prentice Hall Series on Artificial Intelligence. Prentice-Hall, New Jersey (1995)
9. Cassandras, C., Lafortune, S.: Introduction to Discrete Event Systems. Discrete Event Dynamic Systems. Kluwer Academic Publishers (1999)

A Fuzzy Reinforcement Learning for a Ball Interception Problem

Tomoharu Nakashima, Masayo Udo, and Hisao Ishibuchi

Department of Industrial Engineering, Osaka Prefecture University
Gakuen-cho 1-1, Sakai, Osaka, 599-8531
{nakashi,udo,hisaoi}@ie.osakafu-u.ac.jp

Abstract. In this paper, we propose a reinforcement learning method called a fuzzy Q-learning where an agent determines its action based on the inference result by a fuzzy rule-based system. We apply the proposed method to a soccer agent that intercepts a passed ball by another agent. In the proposed method, the state space is represented by internal information the learning agent maintains such as the relative velocity and the relative position of the ball to the learning agent. We divide the state space into several fuzzy subspaces. A fuzzy if-then rule in the proposed method represents a fuzzy subspace in the state space. The consequent part of the fuzzy if-then rules is a motion vector that suggests the moving direction and velocity of the learning agent. A reward is given to the learning agent if the distance between the ball and the agent becomes smaller or if the agent catches up with the ball. It is expected that the learning agent finally obtains the efficient positioning skill.

1 Introduction

Fuzzy rule-based systems have been applied mainly to control problems [1]. Recently, fuzzy rule-based systems have also been applied to pattern classification problems. There are many approaches to the automatic generation of fuzzy if-then rules from numerical data for pattern classification problems.

Reinforcement learning [2] is becoming a more and more important research field for acquiring the optimal behavior of autonomous agents. One of the most well-known reinforcement learning methods is Q-learning [3]. In the original Q-learning, it is assumed that both a state space and an action space is discretely defined. The optimal discrete action is obtained for each discrete state in a learning environment through updating the mapping from a state-action pair to a real value called Q-value.

In order to deal with continuous state space and action space, various methods have been proposed such as tile coding, neural networks, linear basis functions and so on (see [2] for detail). Fuzzy theory has been also successfully applied in order to extend the Q-learning to fuzzy Q-learning. For example, Glorennec [4] proposed a fuzzy Q-learning algorithm for obtaining the optimal rule base for a fuzzy controller. Horiuchi et al. [5] proposed a fuzzy interpolation-based Q-learning where a fuzzy rule base is used to approximate the distribution of

D. Polani et al. (Eds.): RoboCup 2003, LNAI 3020, pp. 559–567, 2004.

Q-values over a continuous action space. In [5], action selection was performed by calculating Q-values for several discrete actions and then selecting one action through the roulette wheel selection scheme.

In this paper, we propose a fuzzy Q-learning that can deal with a continuous state space and a continuous action space. Q-values are calculated using fuzzy inference from the sensory information of the learning agent. We apply the proposed method to a soccer agent that tries to learn to intercept a passed ball. That is, it tries to catch up with a passed ball by another agent. In the proposed method, the state space is represented by internal information that the learning agent maintains such as the relative velocity and the relative position of the ball. We divide the state space into several fuzzy subspaces. We define each fuzzy subspace by specifying the fuzzy partition of each axis in the state space. A reward is given to the learning agent if the distance between the ball and the agent becomes smaller or if the agent catches up with the ball. Simulation results show that the learning agent can successfully intercept the ball over time.

2 Ball Interception Problem

The problem we solve in this paper is called a ball interception problem, where the task of the learning agent is to follow the passed ball by another agent. This problem is illustrated in Fig. 1. First, the passer approaches the ball to kick it. Then learning agent tries to catch up with the passed ball. Let (x_a, y_a) and (x_b, y_b) be the absolute position of the learning agent and the ball, respectively. We also denote the velocity of the learning agent and the ball as (v_{ax}, v_{ay}) and (v_{bx}, v_{by}), respectively. Suppose that the learning agent moves at the speed of (v_{ax}, v_{ay}) from the position (x_a, y_a) at the time step 0, and the learning agent can intercept the ball at the time step t. The position of the ball at the time step 0 is (x_b, y_b). Here we do assume that there is no noise nor friction in the movement of the objects. The positions of the learning agent and the ball at the time step t are $(x_a + v_{ax}t, y_a + v_{ax}t)$ and $(x_b + v_{bx}t, y_b + v_{bx}t)$, respectively. In order for the learning agent to successfully intercept the ball, it is necessary that the following two conditions hold:

$$x_a + v_{ax}t = x_b + v_{bx}t, \tag{1}$$

and

$$y_a + v_{ay}t = y_b + v_{by}t. \tag{2}$$

Thus, we have

$$t = \frac{x_a - x_b}{v_{bx} - v_{ax}} = \frac{y_a - y_b}{v_{by} - v_{ay}}. \tag{3}$$

The objective of the ball interception problem can be viewed as the minimization of the time to intercept the ball. There is a constraint that the learning agent can not move at more than some pre-specified maximal speed. Let us denote the maximal speed of the learning agent as $V_{\max}$. Then the ball interception problem can be rewritten as follows:

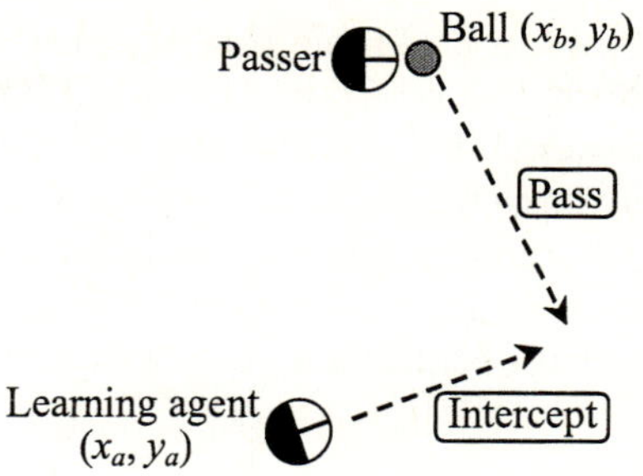

Fig. 1. Ball interception problem.

[Ball interception problem]
Minimize t,
subject to $V_{\max} \leq \sqrt{v_{ax}^2 + v_{ay}^2}$.

3 Q-Learning

In the conventional Q-learning, a Q-value is assigned to each state-action pair. The Q-value reflects the expected long-term reward by taking the corresponding action from the action set A in the corresponding state in the state space S. Let us denote $Q_t(s, a)$ as the Q-value for taking action a in state s at time step t. When a reward r is obtained immediately after taking action a, the Q-value $Q_t(s, a)$ is updated as follows:

$$Q_{t+1}(s, a) = (1 - \alpha) \cdot Q_t(s, a) + \alpha \cdot (r + \gamma \hat{V}_t), \quad (4)$$

where α is a learning rate, γ is a discount factor, and $\hat{V}$ is the maximum Q-value in the state s' after taking the action a in the state s at the time step t, which is defined as

$$\hat{V} = \max_{b \in A} Q_t(s', b), \quad (5)$$

where s' is the next state assuming that the agent took action a in the state s.

The action selection in the Q-learning is done considering a trade-off between exploration and exploitation of the state-action pairs. The roulette wheel selection scheme is often used for selecting an action where the following Boltzmann distribution is used with the selection probability $P(s, a)$ of the action a:

$$P(s, a) = \frac{\exp(Q(s, a)/T)}{\sum_{b \in A} \exp(Q(s, b)/T)}, \quad \text{for } a \in A. \quad (6)$$

The main procedure of the Q-learning is as follows. First, the learning agent observes the state of the environment. Next, action selection is performed according to the Q-values for the observed state. Then, Q-values corresponding to the selected action is updated using the reward from the environment. This procedure is iterated until some pre-specified stopping criterion is satisfied.

The Q-values are stored in a Q-table that is referred by the learning agent for retrieving Q-values for action selection. One difficulty in the Q-learning is so-called *curse of dimensionality*. That is, the number of Q-values to be stored in the Q-table is intractably large when the state space is large. Another problem is that the conventional Q-learning can not be applied when the state space and/or the action set is continuous. In order to overcome this problem, we propose fuzzy Q-learning that can deal with continuous state and continuous action.

4 Fuzzy Q-Learning

4.1 Fuzzy If-Then Rule

Let us assume that we would like an agent to learn the optimal behavior in a continuous state space with continuous actions. We also assume that the state space in the fuzzy Q-learning is described by n-dimensional real vector $\mathbf{x} = (x_1, \ldots, x_n)$ and there are m representative values W_k, $k = 1, \ldots, m$, for determining a single continuous action. In the fuzzy Q-learning, we use fuzzy if-then rules of the following type:

$$R_j : \text{If } x_1 \text{ is } A_{j1} \text{ and } \ldots \text{ and } x_n \text{ is } A_{jn} \text{ then } \mathbf{w}_j = (w_{j1}, \ldots, w_{jm}), \quad j = 1, \ldots, N, \tag{7}$$

where R_j is the label of the fuzzy if-then rule, A_{ji}, $i = 1, \ldots, n$, is a fuzzy set for a state variable, $\mathbf{w}_j$ is a consequent real vector of the fuzzy if-then rule R_j, and N is the number of fuzzy if-then rules. As fuzzy sets we can use any type of membership functions such as triangular, trapezoidal, and Gaussian type. Each element of the consequent vector $\mathbf{w}_j$ corresponds to the weight for a typical vector $\mathbf{a} = (a_1, \ldots, a_m)$ in the continuous action space.

4.2 Action Selection

From a state vector $\mathbf{x} = (x_1, \ldots, x_n)$, the overall weights of each typical point in the continuous output space is calculated through fuzzy inference as follows:

$$W_k = \frac{\sum_{j=1}^{N} w_{jk} \cdot \mu_j(\mathbf{x})}{\sum_{j=1}^{N} \mu_j(\mathbf{x})}, \qquad k = 1, \ldots, m, \tag{8}$$

where $\mu_j(\mathbf{x})$ is the compatibility of a state vector $\mathbf{x}$ with the fuzzy if-then rule R_j that is defined by a multiplication operator as follows:

$$\mu_j(\mathbf{x}) = \mu_{j1}(x_1) \cdot \mu_{j2}(x_2) \cdot \ldots \cdot \mu_{jn}(x_n), \tag{9}$$

where $\mu_{jk}(x_k)$ is the membership function of the fuzzy set A_{jk}, $k = 1, 2, \ldots, n$ (see (7)). While various schemes for action selection such as Boltzmann selection and ϵ-greedy selection can be used as in the conventional Q-learning, we use a fuzzy inference scheme for selecting the action of the learning agent. That is, we do not use the explorative action selection method but the exploitative action selection method that is strictly determined by the fuzzy inference of the fuzzy rule base in (7). The similar approach was used in [6] where his proposed TPOT-RL method was used for optimizing the packet routing problem without any explorative action selection but only with greedy action selection.

The final output o is calculated as

$$o = \frac{\sum_{k=1}^{m} a_k \cdot W_k}{\sum_{k=1}^{m} W_k}. \tag{10}$$

4.3 Updating Fuzzy If-Then Rules

After the selected action was performed by the learning agent, the environment provides it with either a reward or a punishment according to the resultant state of the environment. Assume that the reward r is given to the learning agent after performing the selected action which is determined by (10). Weight values of each fuzzy if-then rule is updated by

$$w_{jk}^{\text{new}} = (1 - \alpha'_{jk}) \cdot w_{jk}^{\text{old}} + \alpha'_{jk} \cdot (r + \gamma \cdot W_{\max}), \tag{11}$$

where r is a reward, $W_{\max}$ is the maximum value among W_k, $k = 1, \ldots, m$ before the update, γ is a positive constant, and α' is an adaptive learning rate which is defined by

$$\alpha'_{jk} = \alpha \cdot \frac{\mu_j(\mathbf{x})}{\sum_{s=1}^{N} \mu_s(\mathbf{x})} \cdot \frac{W_k}{\sum_{t=1}^{m} W_t}, \tag{12}$$

where α is a positive constant.

5 Computer Simulations

In order to apply the fuzzy Q-learning to the ball interception problem, we use the fuzzy if-then rules of the following type:

$$\begin{aligned} R_j : \text{If } x_r \text{ is } A_{j1} \text{ and } y_r \text{ is } A_{j2} \text{ and } v_{rx} \text{ is } A_{j3} \text{ and } v_{ry} \text{ is } A_{j4} \\ \text{then } \mathbf{v} = (\mathbf{v}_1, \mathbf{v}_2, \mathbf{v}_3, \mathbf{v}_4) \text{ with } \mathbf{w}_j = (w_{j1}, w_{j2}, w_{j3}, w_{j4}), \\ j = 1, \ldots, N, \end{aligned} \tag{13}$$

where (x_r, y_r) and (v_{rx}, v_{ry}) is the relative position and the relative velocity of the ball to the learning agent, $\mathbf{v}_j = (v_{jx}, v_{jy})$ is the velocity of the learning speed by the fuzzy if-then rule R_j, and $\mathbf{w}_j = (w_{j1}, w_{j2}, w_{j3}, w_{j4})$ is the vector of the recommendation degree for each velocity. Thus the following relations hold:

$$x_r = x_b - x_a, \tag{14}$$

$$y_r = y_b - y_a, \tag{15}$$

$$v_{rx} = v_{bx} - v_{ax}, \tag{16}$$

$$v_{ry} = v_{by} - v_{ay}. \tag{17}$$

As fuzzy sets for each state variable, we use triangular type of fuzzy partitions in Fig. 2. Since we partition each state variable into three fuzzy sets, the total number of combinations of the antecedent fuzzy sets is $N = 3^4 = 81$.

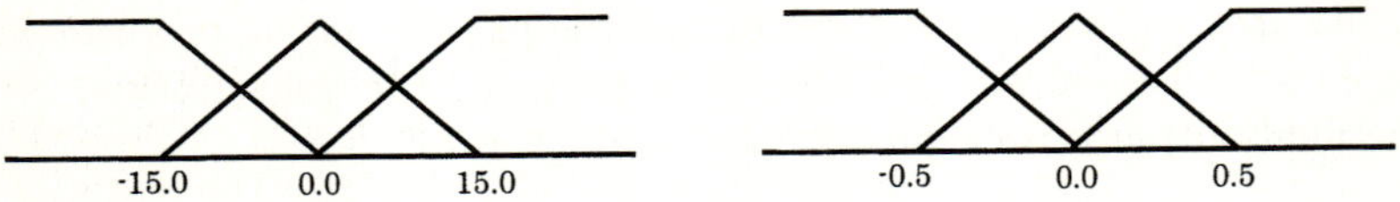

(a) Fuzzy partition for x_r and y_r. (b) Fuzzy partition for v_{rx} and v_{ry}.

Fig. 2. Fuzzy partitions.

Each of the consequent part $\mathbf{v} = (\mathbf{v}_1, \mathbf{v}_2, \mathbf{v}_3, \mathbf{v}_4)$ represents the recommended velocity of the learning agent by R_j. In this paper, we use four combinations of v_{ax} and v_{ay} to calculate the motion vector of the learning agent. They are determined according to the maximum velocity of the learning agent in the constraint of the ball interception problem (see Section II). We use the following four recommended velocities (also see Fig. 3):

$$\mathbf{v}_1 = (v_x^{\max}, 0), \tag{18}$$
$$\mathbf{v}_2 = (0, v_y^{\max}), \tag{19}$$
$$\mathbf{v}_3 = (-v_x^{\max}, 0), \tag{20}$$
$$\mathbf{v}_4 = (0, -v_y^{\max}). \tag{21}$$

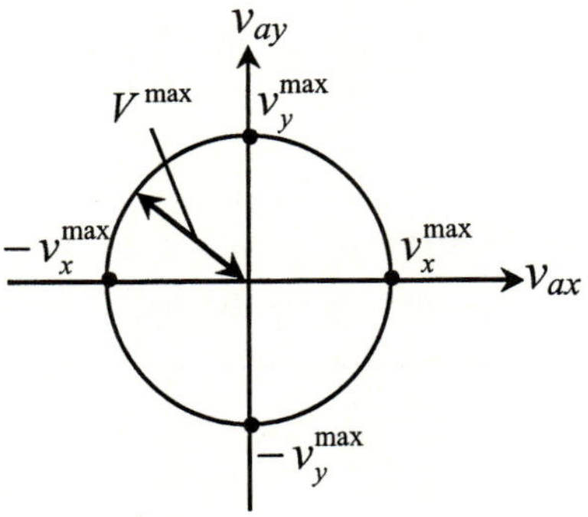

Fig. 3. Four typical velocities in the continuous action space.

After receiving the state vector $\mathbf{s} = (x_r, y_r, v_{rx}, v_{ry})$, the motion vector of the learning agent is determined using the weights for the recommended velocities as follows:

$$W_k = \frac{\sum_{j=1}^{81} w_{jk} \cdot \mu_j(\mathbf{s})}{\sum_{j=1}^{81} \mu_j^t}, \qquad k = 1, 2, 3, 4, \tag{22}$$

where $\mu_j(\mathbf{s})$ is the compatibility of the fuzzy if-then rule R_j to the state vector $\mathbf{s}$ and defined as the product operator as follows:

$$\mu_j(\mathbf{s}) = \mu_{j1}(x_r) \cdot \mu_{j2}(y_r) \cdot \mu_{j3}(v_{rx}) \cdot \mu_{j4}(v_{ry}), \tag{23}$$

where $\mu_{jk}(\cdot)$, $k = 1, 2, 3, 4$ is the membership of the antecedent fuzzy set A_{ji}.

The action is determined by the interpolation among the recommendation degrees of the four typical velocities. That is, the velocity of the learning agent (v_{ax}, v_{ay}) is determined by the following equation:

$$v_{ax} = \frac{W_1 \cdot v_x^{\max} - W_3 \cdot v_x^{\max}}{W_1 + W_2 + W_3 + W_4}, \tag{24}$$

$$v_{ay} = \frac{W_2 \cdot v_y^{\max} - W_4 \cdot v_y^{\max}}{W_1 + W_2 + W_3 + W_4}. \tag{25}$$

In this paper, we consider two types of reward to the learning agent. One is task reward r_t that is given when the learning agent can successfully intercept the passed ball. Since the task reward is sparsely given to the agent, we use an intermediate reward r_i that is given to the learning agent when the learning agent can reduce the distance between the passed ball and the learning agent. In our computer simulations in this paper, we specified those rewards as $r_t = 5$ and $r_i = 1$. Note that when the learning agent goes away from the passed ball, the negative value is given (i.e., $r_i = -1$) to the learning agent as the punishment.

The consequent weight vector $\mathbf{w}_j = (w_{j1}, w_{j2}, w_{j3}, w_{j4})$, $j = 1, \ldots, 81$, is updated by the following equation:

$$w_{jk}^{\text{new}} = (1 - \alpha'_{jk}) \cdot w_{jk}^{\text{old}} + \alpha' \cdot (r + \gamma \cdot W_k), \quad k = 1, 2, 3, 4, \tag{26}$$

where γ is the discount rate, and α'_{jk} is the learning rate, and r is the total reward to the learning agent. α'_{jk} and r are determined by the following equations:

$$\alpha'_{jk} = \alpha \cdot \frac{\mu_j(\mathbf{s})}{\sum_{l=1}^{81} \mu_l(\mathbf{s})} \cdot \frac{W_k}{\sum_{m=1}^{4} W_m}, \tag{27}$$

$$r = r_i + r_t. \tag{28}$$

We applied the proposed fuzzy Q-learning to RoboCup server 7.09 which is available from the URL *http://sserver.sourceforge.net/*. In our computer simulations, one trial ends when the learning agent can successfully intercept the passed ball or the maximum time step is reached. We specified the maximum time step as $t_{\max} = 300$ simulator cycles. Before the computer simulations, we set the initial values for the recommendation degree $\mathbf{w}_j = (w_{j1}, w_{j2}, w_{j3}, w_{j4})$ in the fuzzy if-then rule R_j randomly from the unit interval $[0, 1]$.

We performed the computer simulation for 300 trials. Every 25 trials we examined the performance of the fuzzy Q-learning by making the learning agent intercept the passed ball with the fixed learned fuzzy if-then rules for 50 trials. We show simulation results of the fuzzy Q-learning in Fig. 4. Fig. 4 shows the success rates over the 50 runs of performance evaluation with the fuzzy if-then rules fixed. From Fig. 4, we can see that the number of the successful intercept increases as the number of trials increases. The learning curve in Fig. 4 did not monotonically increase to the number of trials. This is because the RoboCup server employs some degree of noise in the movement of objects such as the ball and agents and in the perception such as the information on the location and velocity of the objects. Also we observed the effect of the result of learning at the previous trials on the performance of succeeding learning. For example, when the learning agent successfully intercept the passed ball and received the positive reward, the next trial is likely to succeed as well. On the other hand, the learning agent is not likely to succeed when the negative reward was given to the agent in the previous trial.

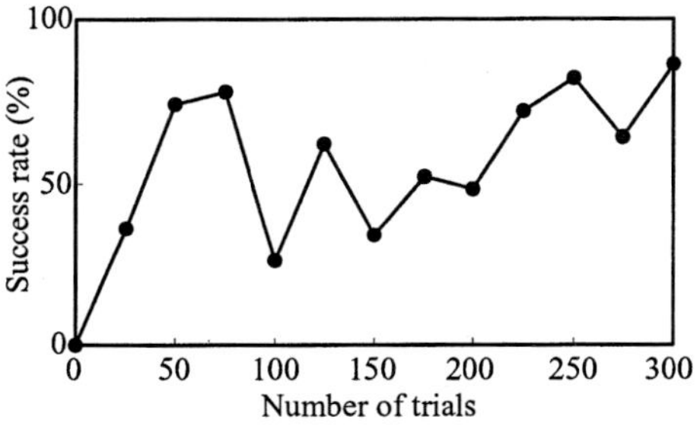

Fig. 4. Simulation results.

6 Conclusion

In this paper, we proposed a fuzzy Q-learning that can deal with the continuous state space and the continuous action space. In the fuzzy Q-learning, a fuzzy rule base is maintained that is used to calculate the recommendation degree by the fuzzy inference scheme. The action was determined by the exploitation-only scheme that includes no exploration procedure of continuous actions. The reward was used for updating the recommendation degrees of fuzzy if-then rules. In the computer simulation, we applied the fuzzy Q-learning to the ball intercept problem. Simulation results showed that the learning agent can gradually learn to successfully intercept the passed ball by another agent.

References

1. M. Sugeno, "An Introductory Survey of Fuzzy Control", *Information Science*, Vol. 30, No. 1/2 pp. 59-83, 1985.
2. R. S. Sutton and A. G. Barto, *Reinforcement Learning: An Introduction*, MIT Press, 1998.

3. C. J. C. H. Watkins and P. Dayan, "Q-Learning", *Machine Learning*, Vol. 8, pp. 279–292, 1992.
4. P. Y. Glorennec, "Fuzzy Q-Leaning and Dynamical Fuzzy Q-Learning", *Proc. of FUZZ-IEEE'94*, pp. 474–479, 1994.
5. T. Horiuchi, A. Fujino, O. Katai, and T. Sawaragi, "Fuzzy Interporation-Based Q-Learning with Continuous States and Actions", *Proc. of FUZZ-IEEE'96*, pp. 594–600, 1996.
6. P. Stone, *Layered Learning in Multiagent Systems – A winning Approach to Robotic Soccer*, MIT Press, 2000.

Intelligent Control of Autonomous Mobile Soccer Robot Adapting to Dynamical Environment

Nobuyuki Kurihara, Ryotaku Hayashi, Hikari Fujii,
Daiki Sakai, and Kazuo Yoshida

Yoshida Lab, System Design Engineering, Keio University
3-14-1, Hiyoshi, Kohoku-ku, Yokohama, Kanagawa 223-8522, Japan

Abstract. This paper deals with an intelligent control of an autonomous mobile robot, which can adapt to dynamical environments. RoboCup soccer robots are chosen as demonstration targets. An important issue in robotic soccer is how to respond to dynamical environment. This study presents an intelligent control method based on System-Life concept for the autonomous mobile robot system. In other words, it is presented how to design the activating controller, sensing element, processing element and so on. Furthermore, Kalman filter and Euler's method are applied to estimate the motion of the ball. First, the soccer robot system is developed based on System-Life concept. Second, the intelligent control method is applied to them. Finally the experiment is carried out on RoboCup Soccer field. It is demonstrated the field player succeeds in approaching moving ball and goalkeeper prevents moving ball from netting. The validity of the proposed method is verified.

1 Introduction

Recently, specialized robots have been installed in factories. It is expected that autonomous mobile robots will be used in dynamical environment. Although there have been considerably many researches on the design methods and adaptation to dynamical environment for autonomous mobile robots, the results of these researches have not necessarily been put into practice. In such classical planning methods as sense-model-plan-action framework, the environment was supposed to be static [1]. Working in dynamical environment is very hard for autonomous robots. For achieving intellectual action in dynamical environment, it is necessary to use such decision making based on sensory information as Subsumption Architecture [2] and [3]. However, these methods adapt an artificial system to the environment unilaterally and there is not a concept of symbiosis with natural systems. Natural systems include some information in gene. The information is influenced through the evolutionary process of adaptation to the environment. And in natural systems the information on their existences and evaluation is embedded. In the design of artifacts, the information of a system is not necessarily embedded, but it is possessed by the designer. It will be effective to embed this information into artificial systems. The above-mentioned concept was proposed as System-Life Concept (SLC) [4] and [5] which is applied to the design concept of the soccer robots system. In this concept, the System-Life Information (SLI) is provided for designing well-balanced artificial systems, so that an effective behavior control can be achieved.

D. Polani et al. (Eds.): RoboCup 2003, LNAI 3020, pp. 568–575, 2004.

2 Design of Autonomous Soccer Robot System

2.1 Application of SLC

The soccer robot system is designed to play soccer game on the field of RoboCup MSL. The system consists of Sensing, Processing, Activating and Expression mechanisms. The sensing mechanism recognizes environment in real-time. The processing mechanism selects the most appropriate behavior with environmental information. The activating mechanism drives actuators. The expression mechanism is a communication system to realize cooperative behavior. SLI comprises environmental model, memory, estimated state, purpose, and self-evaluation. Figure 1 shows the scheme of the System-Life architecture, for the robot.

Section 2.2, 2.3, 2.4 and 2.5 describe the sensing mechanism, the processing mechanism, the activating mechanism and the expression mechanism, respectively.

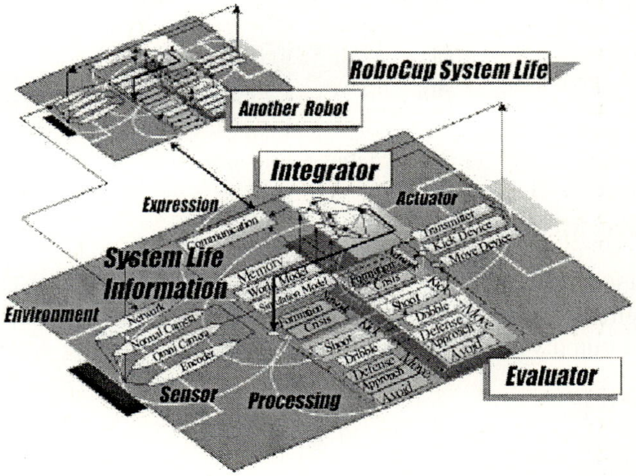

Fig. 1. System life concept on RoboCup soccer

2.2 Sensing Mechanism

The sensing mechanism is constructed by a normal color CCD camera, an Omni-directional vision system, encoders of DC motors and receiver of wireless LAN. The sensors for external environment are only cameras. Two image frames (256*220 pixels) from these normal and omni cameras are combined by field-mix circuit. The combined image (256*440 pixels) is shown in Fig.2. This combined image is the object of image processing. Robots recognize the distance (r) and the direction (θ) to the ball and two goals on polar coordinate shown in bottom half of Fig.2. In this paper, for decreasing image-processing time, only the image from the Omni-directional camera is used to recognize environment. Additionally, image-processing area $\mathbf{A}_{np}$ is narrowed by G_{n-1} and A_{n-1} calculated by S_{n-1} when robots recognize S_n where

S_n : the state in some situation such as direction and distance to the ball and two goals
S_{n-1} : the state recognized 1 sampling time before S_n
$\mathbf{A}_{\mathbf{np}}$: the processing area when robot recognizes S_n
G_{n-1} : the gravity point of each recognized object: ball and two goals
A_{n-1} : the area of each recognized object

This situation is shown in Fig.2. Each image-processing time is decreased to 40ms.

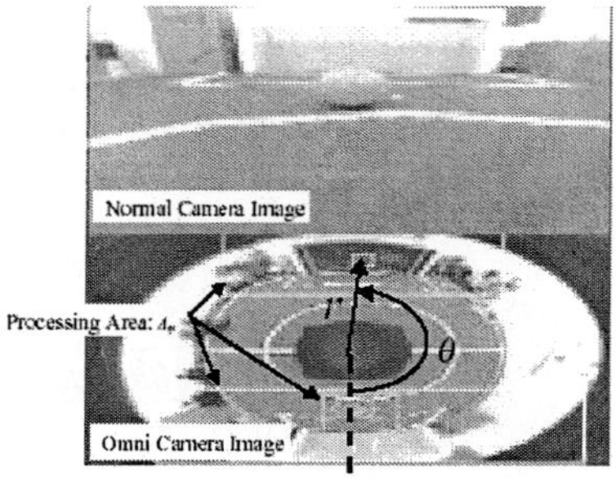

Fig. 2. Processing image

2.3 Processing Mechanism

Pentium II 450MHz CPU is used for the processing mechanism. This mechanism receives the environmental information obtained from the sensing mechanism, selects an action and outputs control signal. In this paper the following action modules are provided.

For field player
  Ball approach, Dribble, Shoot and so on
For goalkeeper
  Defense based on S_n, Defense based on the predicted state S_{n+j} and so on

These modules are allocated in the same level as shown in Fig.1, where the dribble module is selected. Robots can adapt to dynamical environment by selecting these modules in real-time.

2.4 Activating Mechanism

Activating mechanism is constructed by a kicking device and two DC motors to drive robot. The robot can run about 1.5m/s and kick the ball about 1.6m/s. The activating mechanism works according to the control signal from the processing mechanism.

2.5 Expressing Mechanism

The expressing mechanism is constructed by a wireless LAN. In this paper the *Info*(n) is sent and received to realize a cooperative behavior. The *Info*(n) is Ball-position, Self-position, Self-evaluating value and Degree of risk for loss.

3 Extended Kalman Filter for Predicting Ball-Position

Accurate information of the ball-position (X_B, Y_B) is necessary to estimate ball-position on the absolute coordinate (Σ_A). The ball-position on Σ_A is calculated by the self-position (X_R, Y_R) and the ball-position (r,θ) on the relative coordinate (Σ_R). However, the self-position and the ball-position on Σ_R obtained from camera image include errors. Therefore, a method for predicting ball-position with only the ball-position on Σ_R is considered.

3.1 Modeling of Ball Dynamics

The dynamics of a ball is formulated as Eq.(1) on the polar-coordinate shown in Fig.3.

$$\begin{aligned}
&\begin{pmatrix} r \\ \dot{r} \\ \theta \\ \dot{\theta} \end{pmatrix} = \begin{pmatrix} 0 & 1 & 0 & 0 \\ 0 & -\dfrac{c}{m} & 0 & r\dot{\theta} \\ 0 & 0 & 0 & 1 \\ 0 & 0 & 0 & -\left(\dfrac{2\dot{r}}{r}+\dfrac{c}{m}\right) \end{pmatrix} \begin{pmatrix} r \\ \dot{r} \\ \theta \\ \dot{\theta} \end{pmatrix} + \begin{pmatrix} 0 \\ -c/m \\ 0 \\ -c/m \end{pmatrix} w \\
&\begin{pmatrix} r \\ \theta \end{pmatrix} = \begin{pmatrix} 1 & 0 & 0 & 0 \\ 0 & 0 & 1 & 0 \end{pmatrix} \begin{pmatrix} r \\ \dot{r} \\ \theta \\ \dot{\theta} \end{pmatrix} + v
\end{aligned} \tag{1}$$

where

m: mass of a ball c: damping coefficient w,v: uncorrelated noise each other

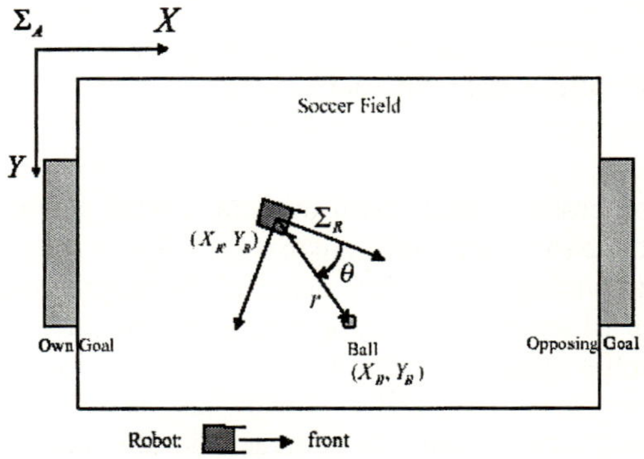

Fig. 3. Prediction coordinate

3.2 Predicting Ball-Position

In order to estimate and predict the ball-position an extended Kalman filter and Euler's method are used. Equation (3) indicates the algorithm of the extended Kalman filter for Eq. (2). By applying Euler's method to the estimated state, the ball-position in the future is predicted. To predict ball-position at intervals of Δt, Eq.(4) is used.

$$
\begin{aligned}
x_{t+1} &= f_t(x_t) + G_t w_t \\
y_t &= h_t(x_t) + v_t
\end{aligned}
\tag{2}
$$

$$
\begin{aligned}
&\hat{x}_{t+1/t} = f_t(\hat{x}_{t/t}) \\
&\hat{x}_{t/t} = \hat{x}_{t/t-1} + K_t\left[y_t - h_t(\hat{x}_{t/t-1})\right], \quad \hat{x}_{0/-1} = \overline{x}_0 \\
&K_t = P_{t/t-1}\hat{H}_t^T\left[\hat{H}_t P_{t/t-1}\hat{H}_t^T + R_t\right]^{-1} \\
&P_{t+1/t} = \hat{F}_t P_{t/t}\hat{F}_t^T + Q_t, \quad P_{0/-1} = \Sigma_0 \\
&P_{t/t} = P_{t/t-1} - P_{t/t-1}\hat{H}_t^T\left[\hat{H}_t P_{t/t-1}\hat{H}_t^T + R_t\right]^{-1}\hat{H}_t P_{t/t-1} \qquad \hat{F}_t = \left(\frac{\partial f_t}{\partial x_t}\right)_{x=\hat{x}_{t/t}} \quad \hat{H}_t = \left(\frac{\partial h_t}{\partial x_t}\right)_{x=\hat{x}_{t/t-1}}
\end{aligned}
\tag{3}
$$

$$
\begin{pmatrix} r \\ \dot{r} \\ \theta \\ \dot{\theta} \end{pmatrix}_{t+\Delta t} =
\begin{pmatrix}
1 & \Delta t & 0 & 0 \\
0 & 1-\frac{c}{m}\cdot\Delta t & 0 & r\dot{\theta}\cdot\Delta t \\
0 & 0 & 1 & \Delta t \\
0 & 0 & 0 & 1-\left(\frac{2\dot{r}}{r}+\frac{c}{m}\right)\cdot\Delta t
\end{pmatrix}
\begin{pmatrix} r \\ \dot{r} \\ \theta \\ \dot{\theta} \end{pmatrix}_t
\tag{4}
$$

4 Control Method

This section describes the method for action selection as shown in Fig.1. The method is divided into 6 steps. Figure 4 shows the action selection concept.

Step1 : *Info*(n) obtained from the sensing mechanism is sent to the processing mechanism.

Step2 : S_n is recognized and S_{n+j} is predicted with the method mentioned in section 3.2. S_{n+j}, S_n, S_{n-i} (i=1,2,…,10), the satisfaction values of system objective and E_n: self-evaluation value are elements of SLI. In this paper the objective is to carry the ball to the opposing goal and to defend the own goal.

Step3 : the integrator shown in Fig.4 selects an action module by S_{n-i}, S_n and S_{n+j} .

Step4 : the processing mechanism outputs control signal to the activating and the expression mechanisms.

Step5 : the activating mechanism executes the action based on the control signal which the processing mechanism outputs.

Step6 : the self-evaluating value is fed back to the integrator. Each action module has the self-evaluator as shown in Fig.1. If the self-evaluation value is satisfactory, the action is carried over without the decision by the integrator. Otherwise, the action is stopped and the integrator selects another action module.

By repeating Steps 1 to 6, it is possible to select an action module adapting to the environment.

The feature of this control method is to use the integrator and the self-evaluator. Additionally, it also has characteristics of the robustness against lacking image information from the sensing mechanism, which is achieved by using S_{n-i}, S_{n+j} and so on.

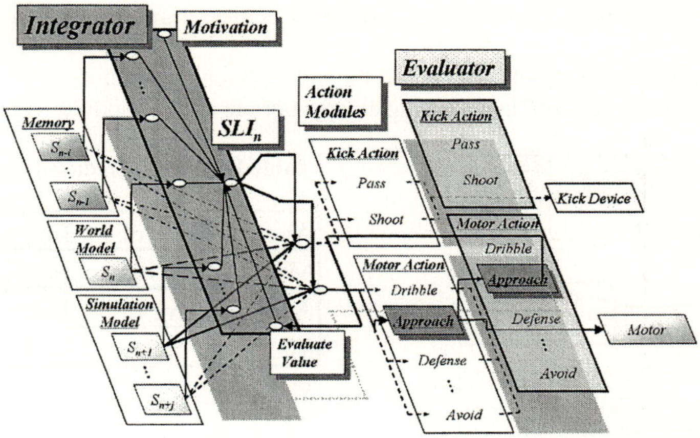

Fig. 4. Action selection scheme

5 Experimental Results

5.1 Experiment of Prediction

The precision of the prediction method is examined. The experiment is performed under the condition that a ball is rolled at the speed about 500 mm/s, and the ball-positions after 1 second and 3 seconds are predicted. In the experiment, the position of the robot is not changed. The prediction result is shown in Fig.5. The average and the maximum errors of r and θ after 1 second are shown in Table.1. These errors are mainly caused by the errors of the ball-position estimated from camera image. In Table.1, the error of r is large, while the error of θ is small. Therefore, the prediction result is accurate enough for the robots, since they are designed to behave using mainly θ information.

5.2 Experiment of Behavior Control

Experiment is carried out to compare two kinds of the behavior control methods using only S_n the present environmental information and using predictive information. The initial positions of the ball and the robot are shown in Figs. 6 and 7. Where the unit is [mm]. The initial position of the ball is (6000, 3500) in the experiment using field player and the initial position of ball is (3000, 500) in the experiment using goalkeeper, and the ball is rolled toward the left side of the own goal at initial velocity, 1.6 m/s and 2.0 m/s, respectively. The ball and robot are started at the same time. The experimental results are shown in Figs. 8 to 11. The goals in these figures are own goal.

In the experiment using a field player, the ball goes through the front of the robot and then the robot follows the ball, in the case of only using S_n. On the contrary, in the case of using SLI, the robot can shoot directly by predicting the path of the ball.

In the experiment using a goalkeeper, the robot moves right to defend own goal, and then moves left after goal is scored, in the case of using S_n. On the other hand, in the case of using SLI, the robot moves left immediately and a defensive action is achieved. Furthermore, a ball clearing action is achieved and the action selection is also smooth.

These experimental results show that the behavior control adapting to the dynamical environment is realized by selecting an appropriate action module with SLI.

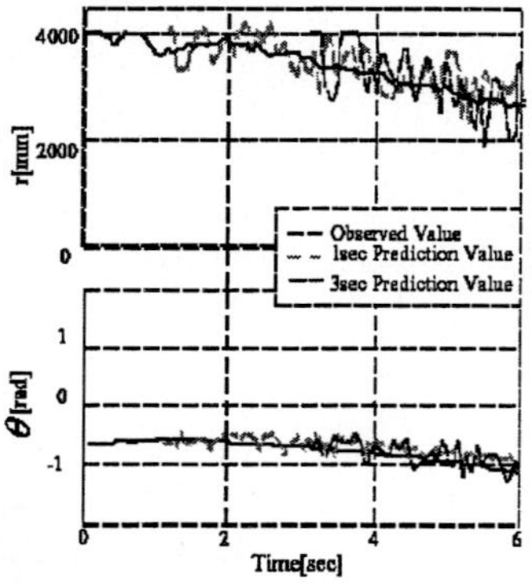

Fig. 5. Result of predicted ball-position

Table 1. Errors of predicted Ball-position

Average Error of r	[mm]	280
Max Error of r	[mm]	752
Average Error of θ	[rad]	0.12
Max Error of θ	[rad]	0.23

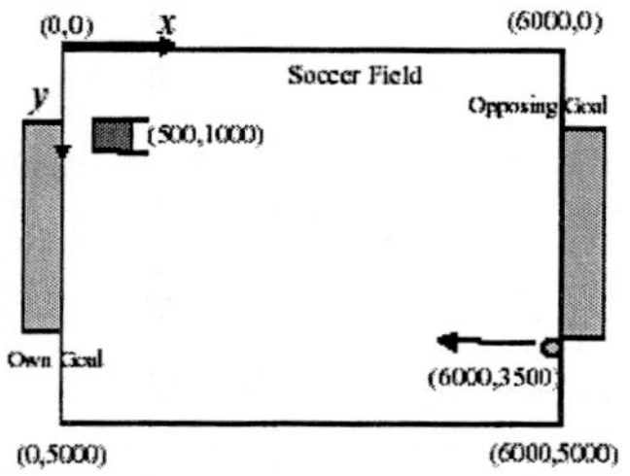

Fig. 6. Field player starting position

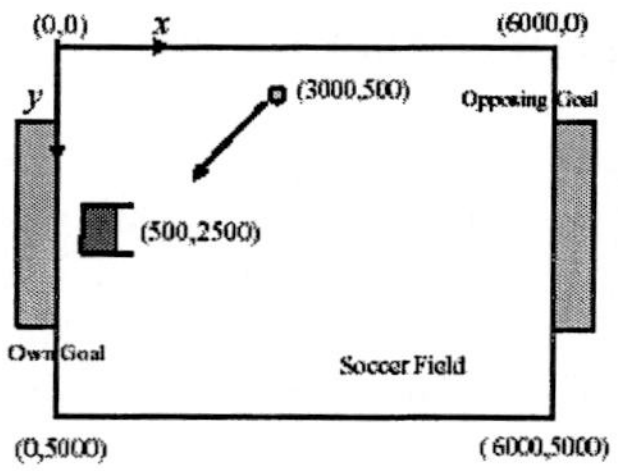

Fig. 7. Goalkeeper starting position

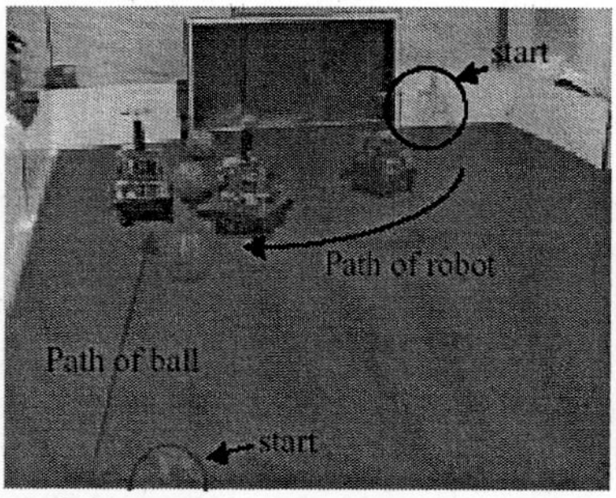

Fig. 8. Result of experiment without SLI (Field player)

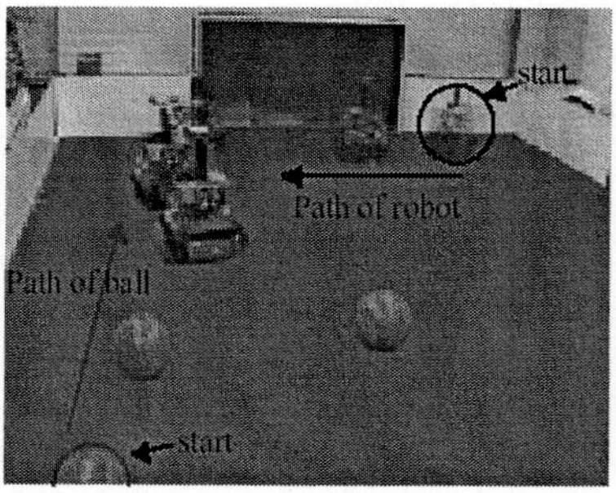

Fig. 9. Result of experiment with SLI (Field player)

6 Conclusion

In this paper, the purpose was the realization of intelligent control for autonomous mobile robots in the dynamical environment. The object of this research was the soc-

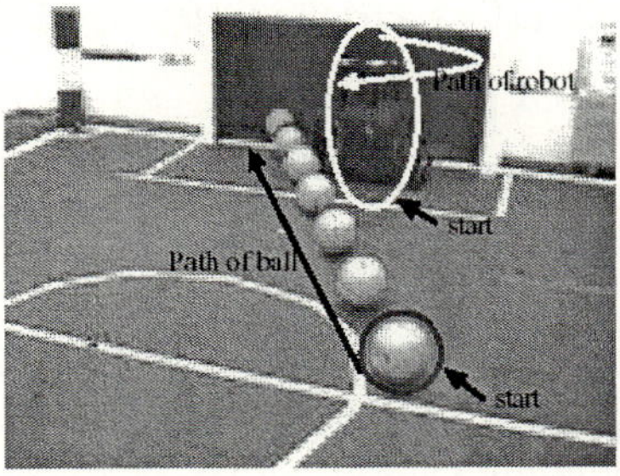

Fig. 10. Result of experiment without SLI (Goalkeeper)

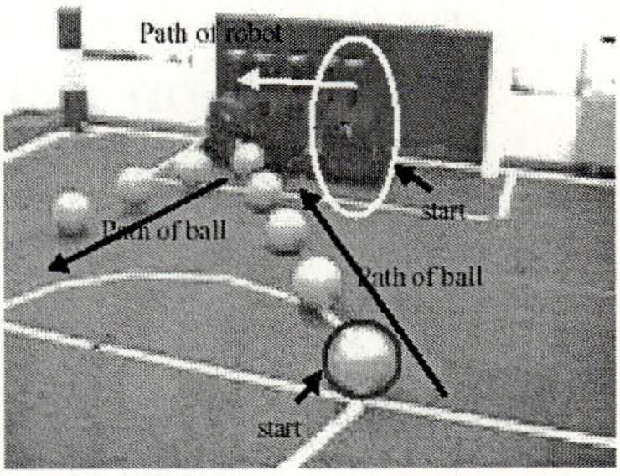

Fig. 11. Result of experiment with SLI (Goalkeeper)

cer robot of RoboCup middle size robot league. The robots were designed based on the System-Life Concept where there are sensing, activating, processing, expression mechanisms and system life information. The behavior control method was applied to action module selection using an integrator. The experiments demonstrate that field players and goalkeeper can cope with the dynamical environment and the proposed method is effective. The proposed method was applied to soccer robots at RoboCup 2002 in Fukuoka and the validity of this method in dynamical environment was demonstrated.

This concept leads to the design methodology such that the system itself possesses explicitly the information on its objective and evaluation with respect to action control and decision-making. The design methodology is expected to cope with unpredicted situation and will be applied to various fields as an intelligent control methodology.

Future works are decrease of designing difficulty, emersion of new action adapted to environment and realization of self-repairing and self-organization.

References

1. Fikes, R.E. and Nilsson, N.J., "STRIPS: A New Approach to the Application of Theorem Proving to Problem Solving" Artificial Intelligence, Vol.2, pp,189-208, 1971
2. Brooks, R.A, "A Robust Layered Control System for a Mobile Robot" IEEE Robotics and Automation, Vol2, No.1, pp.14-23, 1986
3. Rosenshein, S., "Formal Theories of Knowledge in AI and Robotics" New Generation Computing, Vol3, No.4, pp.345-357, 1985
4. Yoshida, K., "System-Life", Proc. General Conf. JSME, No.98-1, pp.113-114, 1998 (in Japanese)
5. Yoshida, K., "Intelligent Control Using Cubic Neural Network", Systems Control and Information, Vol.43, No.7, pp.328-336, 1999 (in Japanese)

A Hierarchical Multi-module Learning System Based on Self-interpretation of Instructions by Coach

Yasutake Takahashi[1,2], Koichi Hikita[1], and Minoru Asada[1,2]

[1] Dept. of Adaptive Machine Systems, Graduate School of Engineering
[2] Handai Frontier Research Center
Osaka University
Yamadagaoka 2-1, Suita, Osaka, Japan
{yasutake,khikita,asada}@er.ams.eng.osaka-u.ac.jp
http://www.er.ams.eng.osaka-u.ac.jp

Abstract. We propose a hierarchical multi-module leaning system based on self-interpretation of instructions by coach. The proposed method enables a robot to decompose (i) a long term task which needs various kinds of information into a sequence of short term subtasks which need much less information through its self-interpretation process for the instructions given by coach, (ii) to select sensory information needed to each subtask, and (iii) to integrate the learned behaviors to accomplish the given long term task. We show a preliminary result of a simple soccer situation in the context of RoboCup.

1 Introduction

Reinforcement learning (hereafter, RL) is an attractive method for robot behavior acquisition with little or no *a priori* knowledge and higher capability of reactive and adaptive behaviors [2]. However, single and straightforward application of RL methods to real robot tasks is considerably difficult due to its almost endless exploration which is easily scaled up exponentially with the size of the state/action spaces, that seems almost impossible from a practical viewpoint.

Fortunately, a long time-scale behavior might be often decomposed into a sequence of simple behaviors in general, and therefore, the search space is expected to be able to be divided into some smaller ones. Connell and Mahadevan [3] decomposed the whole behavior into sub-behaviors each of which can be independently learned. However, task decomposition and behavior switching procedure are given by the designers. Takahashi and Asada [4, 5] proposed a multi-layered RL system. The modules in the lower networks are organized as experts to move into different categories of sensor output regions and learn lower level behaviors using motor commands. In the meantime, the modules in the higher networks are organized as experts which learn higher level behaviors using lower modules. However, this system tends to produce not only purposive behavior learning modules but also many non-purposive ones, and as a result, to require large computational resources.

D. Polani et al. (Eds.): RoboCup 2003, LNAI 3020, pp. 576–583, 2004.

When we develop a real robot which learns various behaviors in its life, it seems reasonable that a human instructs or shows some example behaviors to the robot in order to accelerate the learning before it starts to learn. Whitehead [6] showed that instructions given by coach significantly encourages the learning and reduces the learning time. This method, called LBW (Learning By Watching), reduces the exploration space and makes learner have experiences to reach the goal frequently. Asada et al. [1] proposed a method, called LEM (Learning from Easy Mission). The basic idea is that a learning scheduling such as a robot starts to learn in easy situations to accomplish a given task at the early stage and learns in more difficult situations at the later stage accelerates learning the purposive behaviors. They applied this idea to a monolithic learning module. In order to cope with more complicated tasks, this idea can be extended to a multi-module learning system. That is, the robot learns basic short term behaviors at the early stage and learns complicated long term behaviors at the later stage based on instructions given by coach.

In this paper, we propose a behavior acquisition method based on hierarchical multi-module leaning system with self-interpretation of coach instructions. The proposed method enables a robot to

1. decompose a long term task into a set of short term subtasks,
2. select sensory information needed to accomplish the current subtask,
3. acquire a basic behavior to each subtask, and
4. integrate the learned behaviors to a sequence of the behaviors to accomplish the given long term task.

We show a preliminary result applied to a simple soccer situation in the context of RoboCup.

2 A Basic Idea

There are a learner and a coach in a simple soccer situation (Fig. 1). The coach has *a priori* knowledge of tasks to be played by the learner. The learner does not have any knowledge on tasks but just follows the instructions. After some instructions, the learner segments the whole task into a sequence of subtasks, acquire a behavior for each subtask, find the purpose of the instructed task, and acquire a sequence of the behaviors to accomplish the task by itself. It is reasonable to assume that the coach will give instructions for easier tasks at the early stage and give ones for complicated tasks at the later stage although it does not have any *a priori* knowledge about the learning system on the agent.

Fig. 2 shows a perspective of development of the learning system through instructions given by coach at three stages. When the coach gives new instructions, the learner reuses the learning modules for familiar subtasks, generates new learning modules for unfamiliar subtasks at lower level and a new module for a sequence of behaviors of the whole instructed task at the upper level. After the learning at one stage, the learner adds newly acquired learning modules to the learning module database. The learning system iterates this procedure from easy tasks to more complicated ones.

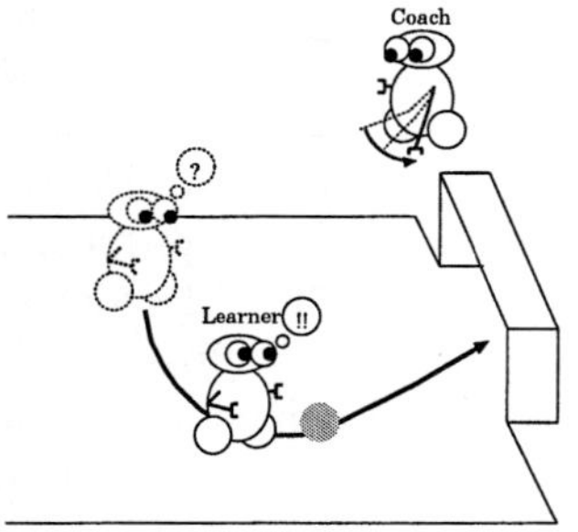

Fig. 1. Basic concept: A coach gives instructions to a learner. The learner follows the instruction and find basics behaviors by itself

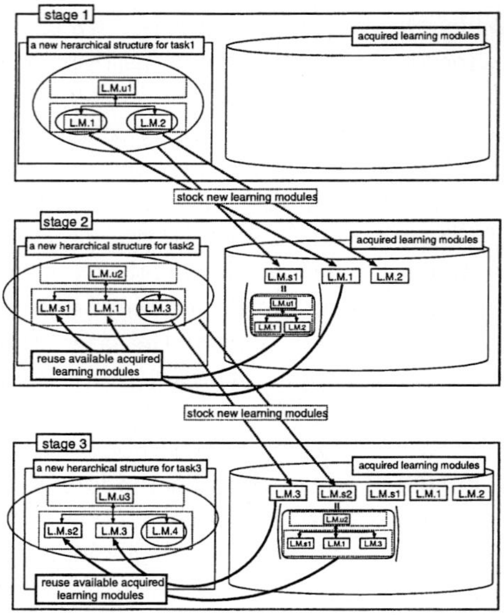

Fig. 2. A perspective of of development of the learning system with staged instructions

3 Hierarchical Multi-module Learning System

3.1 An Architecture

The basic idea of multi-layered learning system is similar to [4, 5]. The details of the architecture has been extended. The robot prepares learning modules of one kind, makes a layer with these modules, and constructs a hierarchy between the layers. The hierarchy of the learning module's layers can be regarded as a role of task decomposition. Each module has a forward model (predictor) which represents the state transition and reward models ($\hat{\mathcal{P}}^a_{ss'}$, $\hat{\mathcal{R}}^a_{ss'}$), and a behavior learner (policy planner) which estimates the state-action value function ($Q(s, a)$) based on the forward model in an RL manner (Fig. 3(b)). The state and the action are constructed using sensory information and motor command, respectively at the bottom level.

The input and output to/from the higher level are the goal state activation and the behavior command, respectively, as shown in Fig. 3. The goal state activation g is a normalized state value, and $g = 1$ when the situation is the goal state. When a module receives the behavior command b from the higher modules, it calculates the optimal policy for its own goal, and sends an action command to the lower module. The action command at the bottom level is translated to an actual motor command, then the robot takes an action in the environment.

3.2 A Learning Procedure

The steps of the learning procedure are as follows:

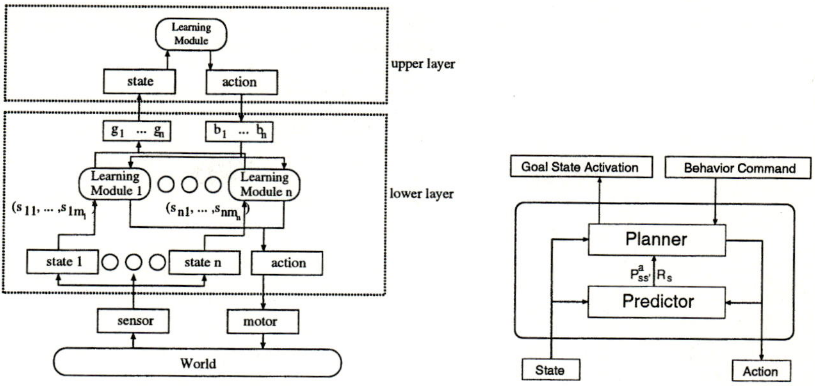

(a) whole system (b) module

Fig. 3. A multi-layered learning system

1. Coach instructs some example behaviors to accomplish a task.
2. Learner evaluates the availability of learned behaviors to accomplish the task by watching the examples.
3. The learner segments the task into subtasks, and produces new learning modules at the lower layer if needed, and learns the behavior for each.
4. The learner produces a learning module at the higher layer and learns the whole behavior to accomplish the task.
5. Go to step 1.

3.3 Availability Evaluation

The learner needs to evaluate the availability of learned behaviors which help to accomplish the task by itself because the coach neither knows what kind of behavior the learner has already acquired directly nor shows perfect example behavior from the learner's viewpoint. The learner should evaluate a module valid if it accomplishes the subtask even if the greedy policy seems different from the example behavior. Now, we introduce $\overline{Q}$ in order to evaluate how suitable the module's policy is to the subtask as follows:

$$\overline{Q}(s, a_e) = \frac{Q(s, a_e) - min_{a'} Q(s, a')}{max_{a'} Q(s, a') - min_{a'} Q(s, a')}, \quad (1)$$

where a_e indicates the action taken in the instructed example behavior. $\overline{Q}$ becomes larger if a_e leads to the goal state of the module while it becomes smaller if a_e leaves the goal state. Then, we prepare a threshold $\overline{Q}_{th}$, and the learner evaluates the module valid for a period if $\overline{Q} > \overline{Q}_{th}$. If there are modules whose $\overline{Q}$ exceeds the threshold $\overline{Q}_{th}$ simultaneously, the learner selects the module which keeps $\overline{Q} > \overline{Q}_{th}$ for longest period among the modules (see Fig. 4).

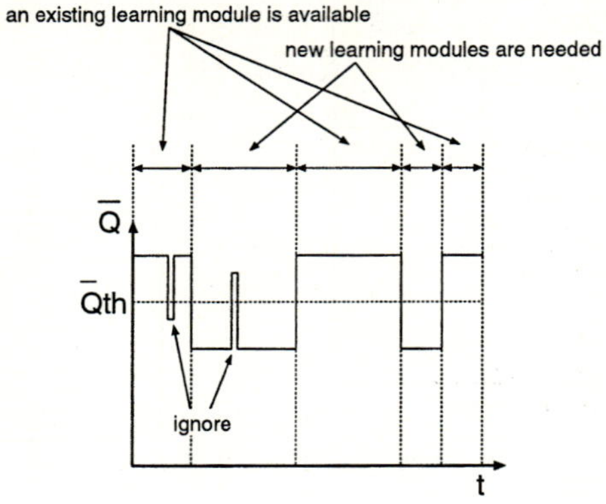

Fig. 4. Availability identification during the given sample behavior

3.4 Producing New Learning Modules

If there is no module which has $\overline{Q} > \overline{Q}_{th}$ for a period, the learner creates a new module which will be assigned to the not-learned-yet subtask for the period. In order to assign a new module to such a subtask, the learner identifies the state space and the goal state. The following shows the steps briefly.

1. Prepare a set of state spaces $\boldsymbol{S}$ and, set their priorities as $\boldsymbol{S}_i : i = 1, 2, \cdots$.
2. For each state space $\boldsymbol{S}_i$,
   (a) Estimate a goal state space $\boldsymbol{G}$ in the state space $\boldsymbol{S}_i$ based on the instructed example behaviors.
   (b) If the estimated goal state space $\boldsymbol{G}$ covers all of the state space $\boldsymbol{S}_i$, increment i and goto step (a).
   (c) Construct a learning module and calculate Q values.
   (d) Check the performance of the learned behavior for the subtask. If the success rate is low, increment i and go to step (a).
3. Add a new module based on the state space $\boldsymbol{S}_i$ and the goal state space $\boldsymbol{G}$.
4. Check the availability of modules over the given task. If there is a period where there is no available module, go to step 1.
5. Exit.

State Variables Selection. We introduce heuristics and set priorities to the set of state spaces as follows:

1. Only a few state variables are needed for all subtasks even if large number of state variables are necessary for the whole task: We limits the number of variables to only three in this study.
2. Higher priority is assigned to the state variable which changes largely from the start to the end during the example behaviors because it can be regarded as an important variable to accomplish the subtask.

3. Higher priority is assigned to the state space which has smaller average of entropy $H(s,a)$ (see equation 2) of the state transition probability $P^a_{ss'}$ for the experienced transition. The reason is that the learning module acquires a more purposive behavior with more stable state transition probability which has lower entropy.

$$H(s,a) = -\sum_{s' \in S} P^a_{ss'}(s,a,s') log_2 P^a_{ss'}(s,a,s') \quad (2)$$

Goal State Space Selection. It is hard to specify the goal state of the subtask with limited number of experiences of example behaviors. We need other heuristics here.

– A state variable of the goal state tends to be the maximum, the minimum, or the medium.
– If the value of a variable has no consistent one at the terminate state of the example behavior, the variable is independent of the goal state.

The system produce a reward model based on these heuristics.

3.5 Learning Behavior Coordination

After the procedures mentioned above, there should be necessary and sufficient modules at the lower layer, then the learning system puts a new learning module at the upper layer, and the module learns to coordinate the lower modules. The upper module has a state space constructed with the goal state activations of the lower modules. A set of actions consists of the commands to the lower modules.

4 Experiments

4.1 Setting

Fig. 5 (a) shows a mobile robot we have designed and built. The robot has an omni-directional camera system. A simple color image processing is applied to detect the ball area and an opponent one in the image in real-time (every 33ms). Fig. 5 (b) shows a situation with which the learning agent can encounter and Fig. 5 (c) shows the simulated image of the camera with the omni-directional mirror mounted on the robot. The larger and smaller boxes indicate the opponent and the ball, respectively. The robot has a driving mechanism, a PWS (Power Wheeled Steering) system.

4.2 Learning Scheduling and Experiments

The robot receives instructions for the tasks in the order as follows:

Task 1: ball chasing
Task 2: shoot a ball into a goal without obstacles
Task 3: shoot a ball into a goal with an obstacle

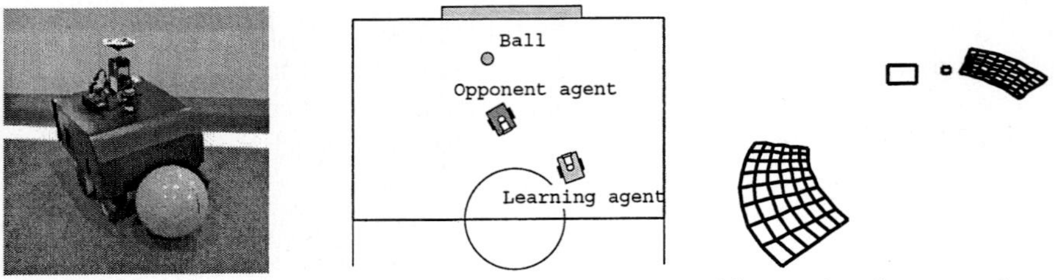

(a) A real robot and a ball (b) Top view of the field (c) Simulated camera image with a omni-directional mirror

Fig. 5. Real robot and simulation environment

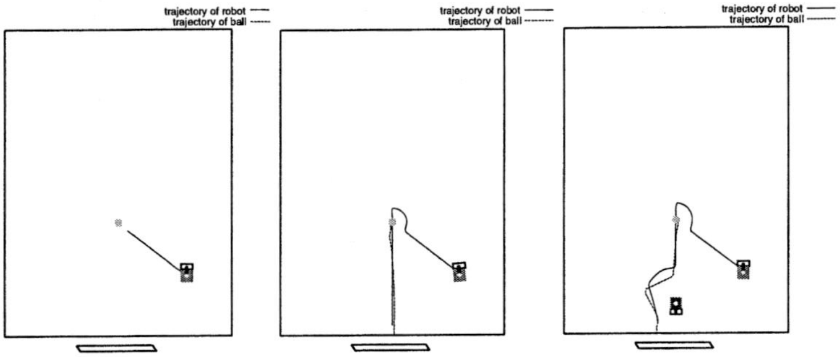

(a) Task 1 (b) Task 2 (c) Task 3

Fig. 6. Example behaviors for tasks

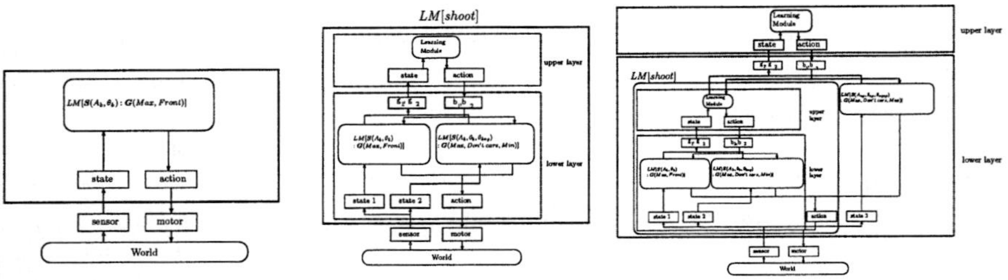

(a) Task 1 (b) Task 2 (c) Task 3

Fig. 7. The acquired hierarchical structure

Figs. 6 (a), (b), and (c) show the one of the example behaviors for each task. Figs. 7 show the constructed systems after the learning of each task. First of all, the coach gives some instructions for the ball chasing task. According to the learning procedure mentioned in **3**, the system produce one module which acquired the behavior of ball chasing. At the second stage, the coach gives some instructions for the shooting task. The learner produces another module which has a policy of going around the ball until the directions to the ball and the goal become same. At the last stage, the coach gives some instructions for the

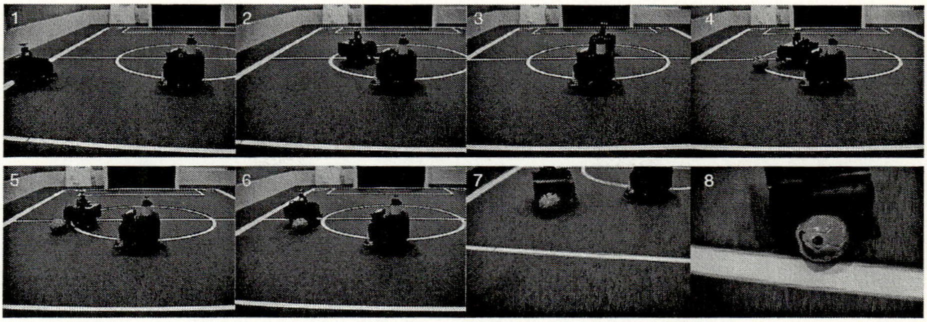

Fig. 8. A sequence of an experiment of real robots (task3)

shooting task with obstacle avoidance. The learner produces another module which acquired the behavior of going to the intersection between the opponent and the goal avoiding the collision. Fig.8 shows a sequence of an experiment of real robots for the task.

References

1. M. Asada, S. Noda, S. Tawaratsumida, and K. Hosoda. Vision-based reinforcement learning for purposive behavior acquisition. In *Proc. of IEEE Int. Conf. on Robotics and Automation*, pages 146–153, 1995.
2. Jonalthan H. Connell and Sridhar Mahadevan. *ROBOT LEARNING*. Kluwer Academic Publishers, 1993.
3. Jonalthan H. Connell and Sridhar Mahadevan. *ROBOT LEARNING*, chapter RAPID TASK LEARNING FOR REAL ROBOTS. Kluwer Academic Publishers, 1993.
4. Y. Takahashi and M. Asada. Vision-guided behavior acquisition of a mobile robot by multi-layered reinforcement learning. In *IEEE/RSJ International Conference on Intelligent Robots and Systems*, volume 1, pages 395–402, 2000.
5. Y. Takahashi and M. Asada. Multi-controller fusion in multi-layered reinforcement learning. In *International Conference on Multisensor Fusion and Integration for Intelligent Systems (MFI2001)*, pages 7–12, 2001.
6. Steven D. Whitehead. Complexity and cooperation in q-learning. In *Proceedings Eighth International Workshop on Machine Learning (ML91)*, pages 363–367, 1991.

Building Aunt Hillary: Creating Artificial Minds with 'Neural Nests'

Mike Reddy and Stuart Lewis

School of Computing, University of Glamorgan
Trefforest, Pontypridd,, RCT. CF37 1DL Wales, UK
{mreddy,sflewis}@glam.ac.uk
http://www.comp.glam.ac.uk/staff/mreddy/

Abstract. This paper describes work in progress to enable a real robot to recreate trail following of ants engaged in pheromone-reinforced recruitment to food gathering. Specifically, it is proposed that development of a set of macro-behaviours for creating and following a trail can be achieved by use of micro-behaviours in a simulated environment to develop a novel neural architecture – the Neural Nest - for learning without explicit representation. A simulated 'neural nest' has been tested to determine the feasibility for ant colonies to encode higher-level behaviours for controlling a physical robot. In our experiments, the emergent behaviour from reinforcement of interactions between unsupervised simple agents, can allow a robot to sense and react to external stimuli in an appropriate way, under the control of a non-deterministic pheromone trail following program. Future work will be to implement the architecture entirely on the physical robot in real time.

1 Introduction

Swarm Intelligence (SI) research has been useful for applying models of adaptive behaviours of complex biological structures, consisting of simple entities (e.g. termites, ant nests, and beehives), to computational problems that require prohibitively expensive search algorithms [1]. One of the key strengths of SI is that elements have flexibility to interact in unpredictable ways, to facilitate optimisation. Specifically, Ant Colony Optimisation (ACO) [2] and Ant Colony Routing (ACR) [3] have been proposed, which use the analogy of pheromones in stylised environments to govern search and routing strategies. However, these techniques have, so far, been mostly applied to specific topological/geographical problems (e.g. the Travelling Salesman Problem [2], Communications Networks [3-5], etc). It should be noted that other approaches, such as novel neural networks [6] or evolutionary algorithms [7] may be more successful, where potential solutions are numerous, or uniformly distributed [1].

SI has also been applied to the problem of coordinating groups of autonomous robots to collaborate on a task [8]; the concept of simulated ants being mapped directly onto sets of robots situated in a simulated or real environment. The power of SI has, therefore, been in coordinating these situated agents at the macro-level without the

D. Polani et al. (Eds.): RoboCup 2003, LNAI 3020, pp. 584–591, 2004.

need for planning or explicit communications. However, most existing ant colony simulations do not attempt to address the complex emergent control of individual physical ants (robotic or biological). We believe that the strength of SI – its use of micro-components to optimise a complex macro-system – may also contribute to the control of individual micro-components, such as a single physical robot, rather than just simulating them as part of a larger macro environment. Many adaptive robot control architectures have been proposed, to control agents at the micro-level, that is, within the individual robot; these include traditional and recurrent neural networks and evolutionary programming, etc. (A good summary can be found in [9, 10]). Ethology, in the form of Tinbergen's instinct centres [11] has inspired a learning classifier approach using Genetic Algorithms (GAs) [12]. However, there has been no attempt to make direct use of SI to control robots at the micro-level; namely to implement an autonomous learning robot using an Ant Colony System (ACS). Simulation of ant behaviour may yet prove useful for understanding the emergent properties of complex biological structures within the human brain. The neural nest, proposed in this paper, is an ant colony inspired implementation of a real-time neural network, without a pre-defined symbolic representation or the need for supervised learning. To allow emergent intelligence, it is, vital that a physical platform be available to reproduce the conditions for evolutionary development, and to enable evaluation of the ant-powered brain. Finally, we describe integrating the symbolic ant brain with the physical robot.

2 Previous Work

2.1 Aunt Hillary – A Reprise

In his seminal work "Gödel, Escher, Bach – an Eternal Golden Braid" [13], Douglas R. Hofstadter describes Aunt Hillary, a conscious ant colony, who consists of signal propagating and symbol manipulation activities using ants as base units interacting with the colony environment to shape the activities of ants into higher level concepts. This is not at odds with more symbolic representations. However, it is difficult to perceive how behaviours could evolve, when we only consider large grained views of intelligent systems. While Aunt Hillary serves principally as a clever analogy to explain how neurons can collaborate to produce symbol manipulation and, ultimately, human consciousness, the concept is an inspiring one, raising many interesting questions, especially when the brain can be considered a network of real-time processes (or neurons), with a flow of control from sensors to motors. Symbolic and subsymbolic techniques make no attempt to describe how higher-level concepts could be implemented in the structures of the brain, and draw a sharp distinction with the physical nervous system. It must also be remembered that it is not just brains, but also bodies (sensors, actuators, etc.) that have co-evolved. Biological systems have not only learned to adapt, but also adapted to learn.

2.2 AntNest – A Reprise

Real ants have limited senses and communication, but manage to evolve optimal solutions to search and routing problems, using pheromones trails, which result in a positive feedback effect. Several simulations of this activity exist, but most have a simple rule-based approach for controlling ant behaviour. Furthermore, these simulations are limited, especially where the strength, degradation and effectiveness of scents can be critical for success. Our previous work, AntNest, has been particularly successful due to using two scents: one for food-bearing ants (food scent), and one for foraging ants (nest scent) [14]. The revised rules in AntNest are as follows:

1) Ants leave the nest in a random direction, leaving a nest scent in the process.
2) Random search influenced by the food scent level, if any, leads the ant to food.
3) Some of the food is acquired and the ant begins to lay food scent towards home.
4) Random movements, influenced by the nest scent level, lead the ant homeward.
5) The ant then returns to step 1.

AntNest was originally conceived as a means for visualising Economics problems, by modeling a biological simulation of an ant colony; ants had several roles in addition to food gathering: nursing babies, clearing the dead from the nest, attacking enemies and sleeping. However, we were inspired by the similarity of the simulations to the architectures of neurons: axons and dendrites. Several features of current neural network architectures have been abstracted from the biological reality: real neurons react to rates of fire, and do so in a continuous time-frame; they are also excitatory or inhibitory in nature, whereas artificial neurones can generate positive and negative values, mediated by weightings (again positive or negative) at the receiving neuron. The research question was whether we could use competing ant colonies to implement a neural architecture that was more closely inspired by the qualities of real neurons: ant trails being analogous to neural links, while the rate of flow of ants from source to nest representing the rate of fire of a neuron.

3 Neural Nests – A Proposal

In this section, a connectionist reinforcement learning architecture is described, which may enable an autonomous robot to acquire motor-sensory navigation strategies without explicit representation. The use of an embedded ACS to implement a novel neural architecture has several advantages: Firstly, it allows us to experiment with combining the strengths of both SI and neural networks; Secondly, it allows the possibility of applying SI techniques in a direct manner, within a single entity, namely a physical robot ant in a real setting; Thirdly, it satisfies an aesthetic need to explore the ability for complex systems, such as the human brain, to evolve and manipulate meaningful symbols using simple components, without explicit prior representation as is the case for many other solutions to robot control.

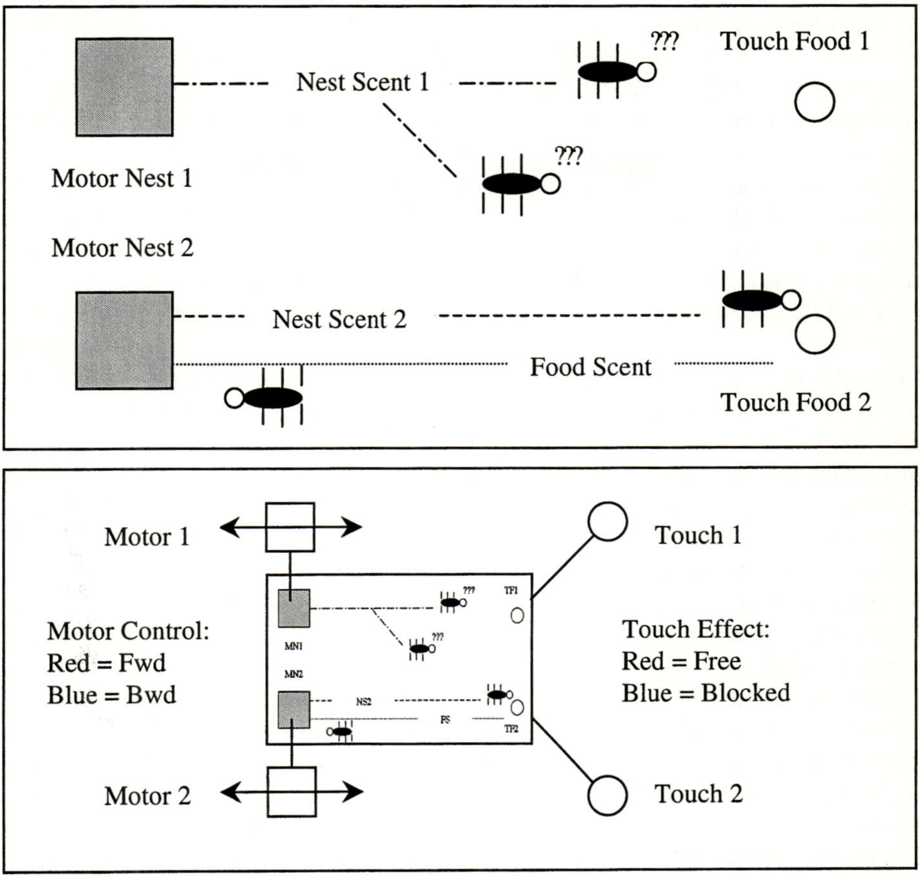

Fig. 1. Neural Nest Internals: Food Sources and Motor Nests are passive entities inside the environment, which represent sensor inputs and motor outputs respectively. Ants collecting food and returning to the nest will cause the external robot to act accordingly depending on their behavioural properties

Furthermore, a tentative cognitive model of unsupervised neural learning is described – the Neural Nest – which uses a set of simple micro-agents to recreate the linkage and firing behaviours of biological neurons to demonstrate the feasibility of our approach to real autonomous robot control. It was recognised that a physical implementation in a real robot was needed as early as possible, rather than working purely with simulation, and then attempting to port to a robot afterwards. Simulation does not throw up the real-world errors and inconsistencies that are present with even the most simplest physical environments; this is especially important as we consider the body evolution to be key to brain evolution. However, the neural nest architecture is a slow one at present – certainly not capable of working in real-time – so a simulated robot was modelled in a closed environment. Initially, this robot had just two touch sensors and two motors.

The basic currency of the Neural Nest architecture is the collection of food from food sources linked to external sensors on a real or simulated robot. Nests are associated with external actuators (in this case, motors on the robot) and the collection of food and its return to the nest provides a signal processing capability that will control the robot's behaviour. The first experiment in the Neural Nest simulator was to see whether the robot would more sensibly with two motor nests and two sensor food sources (See Figure 1). This first experiment showed the gradual grafting of the two touch sensors onto their respective motors. The early effect of this was to drive the robot into obstacles rather than away from them. However, over time a cross-link occurred that allowed the robot to be a little more effective at avoiding walls, when approaching them with one of the sensors. The second experiment involved adding light sensors to the robot, in order for it to engage in line following. We added food sources related to position of the sensor array on the front of the vehicle to the previous simulation arrangement, placing the new food sources in comparable positions to their placement on the physical robot described in Section 4. Again in simulation the architecture generated by the simulation did not show the bipolar symmetry that most animals display. This may be a fault in the model, such as scent evaporation, or faults in the simulated environment[1] but it may also be due to a very limited time available for the system to evolve a solution before testing. The model generated is shown in Figure 2. From observation of the simulated robot, what appeared to be happening was as follows:

1) The right motor stayed on continuously, so directions and steering came from the activities of the left motor, unless the right touch sensor was activated;
2) The left motor stayed on continuously, but reacted to both the left touch sensor and the left and centre light sensors.
3) When a line was detected by either of these light sensors the left motor would stall. If both detected a line the left motor would reverse, turning the robot left.

This gave a limited ability to sense and follow lines, while also maintaining the ability to react partially to the touch sensors. The effect of this architecture on the physical environment is detailed in Figure 4.

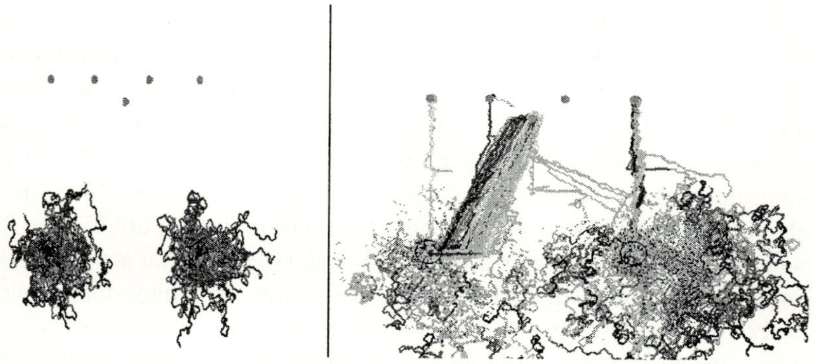

Fig. 2. Touch and Light Simulator: (a) The two motor nests search for food; (b) Neural architecture is complete with the left-hand nest also using the left light sensor

[1] This could be the random number generation skewing movement.

4 Robot AUNT[2] – The Macro-level

Robot AUNT is a tracked robot with two touch sensors on the front corners and three light sensors mounted facing down at the front of the canopy. The experiments ran on a white board placed on the ground and fenced off with walls, which the touch sensors would detect. A retractable white-board marker allowed the robot to lay a trail for the light sensors to follow. The initial problem to be faces was whether such a pen-based trail on a whiteboard could act as a suitable environment for testing pheromone following with a robot. Initial experiments were performed using a hand-coded program to test the feasibility of a real physical implementation of pheromone following. Note: This was done for various reasons, but the most significant was to test whether I could get the tracks to perform a rudimentary form of evaporation – the pen marks will not fade the way that a scent trail would, and there is only one robot, rather than thousands. In fact, the robot tracks provided an excellent mechanism for reproducing the effect of scent evaporation, even if not in the way it would in reality[3].

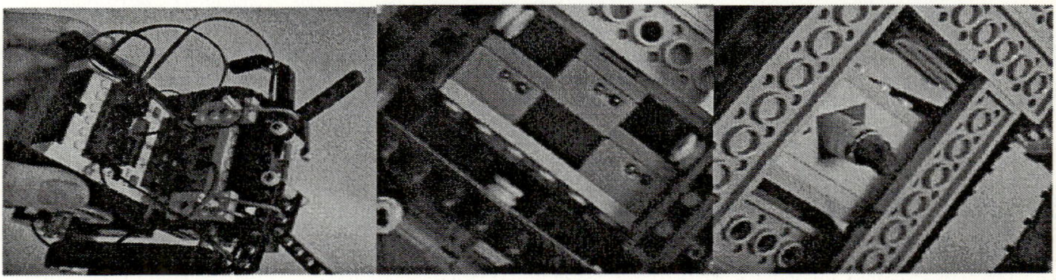

Fig. 3. Robot AUNT: (a) Touch sensors mounted on the corner; (b) Three light sensors facing down; (c) A retractable pen mounted at the pivot point to mark trails

The key issue was then to determine whether the neural nest architecture could be used to control a physical robot. In order to do this effectively, an abstraction of the simulated network was implemented as a set of probabilistic choices in the control of robot motors, dependent upon sensor values. We simulated the arrival of ants as decisions to go forwards and backwards, with the delay between sensing and acting proportional to the average time for an ant to travel between the food source and the nest. Clearly, this is not a perfect reproduction of the actual nest architecture, but allowed a simple implementation on the robot. As was true for the simulation, the robot tended to veer to one side and occasionally trapped itself in corners. The next stage of the experiment was to implement the architecture described in Figure 2, using light and touch sensors. In order to do this, Robot AUNT would need to be able to lay trails autonomously. Modifications were made to the environment to allow the robot to detect whether it was at a food source (shiny surfaces) or the nest (black surface). The

---

2 AUNT stands for Autonomous Unsupervised Neural Turtle – in honour of Aunt Hillary. The robot is a turtle rather than an ant because of its previous incarnation as a drawing robot.

3 It is a little known fact that white-board marker is easier to wipe off once it is completely dry. The efficacy of the tracks to wipe the board clean was proved beyond doubt when we forgot to change the pen for one experiment and returned to find the board eerily clean.

program was modified so that the pen would be lowered once the robot had found food, and raised when it had returned successfully to the nest. Given that there was no help from a number of ants leaving trails, Robot AUNT did mark out trails to and from the nest (see Figure 4). It also performed slightly better at following lines than the simulation did[4].

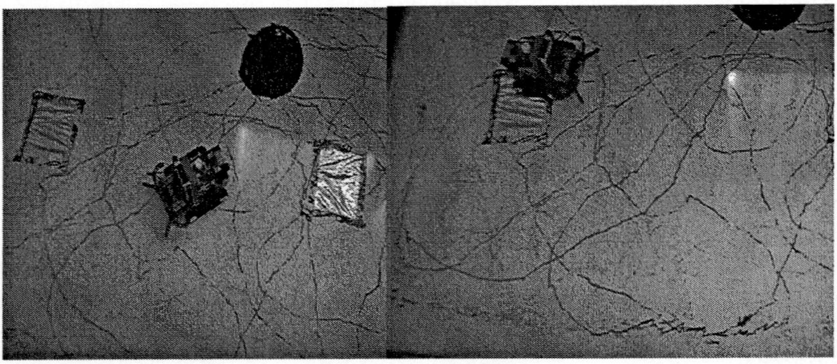

Fig. 4. Robot AUNT II: (a) Searching for food by following lines from previous journeys; (b) Carrying food back to the nest, leaving a scent trail in white-board pen

5 Conclusions

A cognitive model of unsupervised neural learning has been described – the Neural Nest – which uses a set of simple micro-agents to control the macro-behaviours of a physical robot. We have seen a neural nest link touch and light sensors to motors in a way that is effective in both a simulated and real robot, with a degree of sensor fusion. Ant trails and food foraging have been used as the basis for a computational model of how neurons forge links and activate by rate of firing and excitatory and inhibitory behaviours. Results are encouraging, but tentative, with more quantifiable analysis needed to ensure that this is a valid technique for evolving robotic control systems. However, the current scent model needs some revision, as nests switch between different sensors, with only one good link at a time. Sensitivity analysis of the range of producible behaviours is also needed. Furthermore, the resulting robot search pattern is prone to the lack of positive reinforcement, where real ants would be surrounded by recruited colleagues in scent trail generation; our studies only used one macro-robot. Further research into the use of multiple robots may show more clearly the effectiveness of this macro and micro approach to robot control. While this may not produce the most efficient solution to the problem, it is our hope that such an evolutionary approach may also eventually lead to a better abstract representation of the link between low-level neural functioning and higher order behaviours of a robot.

---

[4] This is most probably due to the need to use eight fixed directions in the simulation, where a physical robot would be changing direction more uniformly using real motors.

References

1. Bonabeau, E., Dorigo, M., Theraulaz, G.: Swarm Intelligence: From Natural to Artificial Systems. Oxford Univ. Press, New York, (1999).
2. Dorigo, M., Gambardella, L. M.: Ant colonies for the traveling salesman problem. BioSystems 43, 73-81 (1997).
3. Schoonderwoerd, R., Holland, O., Bruten, J., Rothkrantz, L.: Ant-based load balancing in telecommunications networks. Adapt. Behav. 5, 169-207 (1997).
4. Heusse, M., Guérin, S., Snyers, D., Kuntz, P.: Adaptive agent-driven routing and load balancing in communication networks. Adv. Compl. Syst. 1, 237-254 (1998).
5. Di Caro, G., Dorigo, M.: AntNet: Distributed stigmergetic control for communications networks. J. Artif. Intell. Res. 9, 317-365 (1998).
6. Chen, K.: A simple learning algorithm for the traveling salesman problem. Phys. Rev. E 55, 7809-7812 (1997).
7. Baluja, S., Caruana, R. In Prieditis, A., Russell, S.: Proc. 12th Intl Conf. Machine Learning. Morgan Kaufmann, Palo Alto. (1995). 38-46.
8. Kube, C. R., Zhang, H.: Collective robotics: from social insects to robots. Adaptive Behavior 2, (1994). 189-218.
9. Dorigo, M. (Guest ed): Special Issue on Learning Autonomous Robots. IEEE Transactions on Systems, Man, and Cybernetics-Part B Vol.26, No.3, (1996).
10. Dorigo, M., Colombetti, M.: Robot Shaping: An Experiment in Behavior Engineering. MIT Press. (1997).
11. Tinbergen, N.: The Study of Instincts. Oxford University Press. (1966).
12. Dorigo, M., Schnepf, U.: Genetics-based Machine Learning and Behiour Based Robotics: A New Synthesis. IEEE Transactions on Systems, Man, and Cybernetics. Vol. 23, No. 1, (1993) 141-154.
13. Hofstadter, D.: Gödel, Escher, Bach: An Eternal Golden Braid.
14. Reddy, M., Kedward, P.: Ant Economics: What simple agents can teach us about supply, demand and eating your Dead. Procs. of ANTS'2000 - From Ant Colonies to Artificial Ants: 2nd International Workshop on Ant Colony Optimization. Brussels. Belgium. 8-9th September 2000.

Autonomous Robot Controllers Capable of Acquiring Repertoires of Complex Skills

Michael Beetz, Freek Stulp, Alexandra Kirsch, Armin Müller, and Sebastian Buck

Munich University of Technology, Department of Computer Science IX
{beetzm,stulp}@in.tum.de
http://www9.in.tum.de/agilo

Abstract. Due to the complexity and sophistication of the skills needed in real world tasks, the development of autonomous robot controllers requires an ever increasing application of learning techniques. To date, however, learning steps are mainly executed in isolation and only the learned code pieces become part of the controller. This approach has several drawbacks: the learning steps themselves are undocumented and not executable.

In this paper, we extend an existing control language with constructs for specifying control tasks, process models, learning problems, exploration strategies, etc. Using these constructs, the learning problems can be represented explicitly and transparently and, as they are part of the overall program implementation, become executable. With the extended language we rationally reconstruct large parts of the action selection module of the AGILO2001 autonomous soccer robots.

1 Introduction

Programming sophisticated low-level control skills as well as action selection strategies for autonomous robots acting in dynamic and partially observable environments is both tedious and error prone. Autonomous robot soccer, which has become a standard "real-world" test-bed for multi robot control, provides a good case in point. In robot soccer (mid-size league) two teams of four autonomous robots – one goal keeper and three field players – play soccer against each other. The key characteristics of mid-size robot soccer is that all sensing and action selection is done on-board.

Competent soccer play entails, besides other capabilities, the skillful execution of various navigation tasks such as defending, dribbling, moving past opponents, and intercepting the ball. To realize them, robots could use computational means to infer which control signal should be issued to arrive at a desired state, how long it will take to get there, and how promising it is in the current situation to try to arrive at this state.

Because of the dynamical and adversarial nature of a soccer play and physical subtleties such as friction on different surfaces and the weight of the robot, programming procedures for these reasoning tasks is very difficult. An attractive alternative is the development of control systems that can acquire and adapt such procedures automatically. Obviously, such a control system must learn at various levels of abstraction. It must learn process models from sensory data such as the effects of control signals on the dynamical state, and optimize control laws, e.g. for approaching the ball. It must acquire models of control routines including their success rates and time requirements to decide whether or not to go for a ball or defend the goal. Finally, it must learn the situation-specific selection of appropriate actions.

D. Polani et al. (Eds.): RoboCup 2003, LNAI 3020, pp. 592–599, 2004.

In current practice, such learning tasks are typically considered and solved in isolation. A programmer gathers a set of examples for learning or provides an exploration strategy. Then a learning system, a feature language in which the acquired data are to be encoded, and a suitable parameterization of the learning system are selected. Subsequently, the learning step is executed, the learned routine transformed into an executable piece of code, and provided as part of a software library. The controller can then use the learned routine by calling it as a subroutine. This approach has several disadvantages: the learning steps themselves are undocumented and not automatically executable, and therefore difficult to reproduce.

In this paper, we extend an existing control language to automate this process. Using the constructs of this language, the system model can be specified, and the learning problems can be represented explicitly and transparently, both becoming an integrated part of the overall program. Therefore, the learning problems become executable, documentation of models is integrated in the program, modularity is higher, and analysis of the system is simplified.

Several programming languages have been proposed and extended to provide learning capabilities. Thrun [8] has proposed CES, a C++ software library that provides probabilistic inference mechanisms and function approximators. Unlike our approach a main objective of CES is the compact implementation of robot controllers. Programmable Reinforcement Learning Agents [2] is a language that combines reinforcement learning with constructs from programming language such as loops, parameterization, aborts, interrupts, and memory variables. A difference with our method is that learning tasks can also be specified in our language, and that the learning methods are not confined to reinforcement learning only.

In the remainder of this paper we proceed as follows. The next section describes how process models can be specified. In section 3 it is shown how learning problems can be specified, and in section 4 we give a case example: the AGILO2001 controller. We conclude with a detailed outlook on our future research investigations.

2 Modeling the System

In this section we specify models of different components of the robot and its environment, and show how they can be made explicit in our language. The first class of models we present are process models, such as the robot's dynamics and sensing processes. The second class models the robot's control routines.

2.1 Specifying the Environment Process

define controlled process robot field process
  static environment field model
  environment process ball proc., robot proc., team-mate proc. $_{1,2,3}$, opponent proc. $_{1,2,3,4}$
  sensing process $state \rightarrow$ camera-image $\times$ odometry-reading $\times$ $message$

The construct above specifies that the "robot field process" is a controlled process (we use the dynamic system model, described in [6]), consisting of a static field model and environment and sensing process. These two processes can themselves be decomposed into subprocesses. In the model of the AGILO RoboCup control system the environment process consists of the dynamics of the robot, which specifies how the control inputs change the physical state of the robot. The other subprocesses of the environment process are those that change the physical states of the team mates, the ball, and the opponent robots. All these processes are largely independent of each other. The only interactions between them are collisions and shooting the ball.

By using this construct the model of the environment is an explicit part of the overall program. This ensures that documentation about the model is integrated, and that modularity of different processes is enforced. Not only does this construct model the real environment, it can also be used as an environment simulator specification. To realize this simulator, the process models need to be implemented. A straight-forward method of realizing this is by manually implementing procedures for each process. If these process models suffice for predicting the effects of control signals on the game situation then they constitute an ideal simulator for the system. In section 3 it will be shown that by using learning task constructs, the manual implementation can be replaced by a learning procedure, which can be integrated into the overall program as well.

2.2 Specifying the Control Process

In this section we provide the control system with models of its control routines. These models enable the robot to do more sophisticated action selection and thereby improve its behavior with respect to the given performance measure. We provide this model in two pieces. First, the control task that specifies *what* has to be done and second, the control routine that specifies *how* it has to be done. The rationale behind separating these two aspects is that a task can be accomplished by different routines that have different performance characteristics.

Control Tasks specify *what* the robot should be capable of doing. For example, the robot should be capable of going to the ball, intercepting a ball, and defending. A control task can be specified in two ways: first, we can specify it using a start state and a description of goal states. An example of such a control task is the following one: reach the position of the ball facing the opponent goal, which we represent as $\langle x = x_{ball}, y = y_{ball}, \theta = \theta_{toball}\rangle$. The set of possible control tasks can be constrained as $\{\langle is, gs\rangle | is \in \mathcal{IS} \wedge gs \in \mathcal{GS}\}$. For example, we might constrain this control task to situations where the ball is not in the own penalty area. In addition, we can specify a probability distribution over the possible set of tasks. This probability distribution affects the expected performance of the control routines.

An alternative way of stating a control task is to specify it using the start state and an objective function, a performance measure that the robot should try to maximize. For example, for attacking, an appropriate performance measure might be the expected time needed to shoot a goal, which should be minimized.

control task achieve-dynamic-state
  task specification $\langle x = x_{ball}, y = y_{ball}, \theta = \theta_{toball}\rangle$
  process signature $\langle x, y, \theta\rangle \times \langle x_g, y_g, \theta_g\rangle \rightarrow \langle v_{tra}, v_{rot}\rangle$
  control process specification
    goal $(\langle x_g, y_g, \theta_g\rangle) \wedge$ achievable$(\langle x_g, y_g, \theta_g\rangle)$ $\Rightarrow$ *active* (navigate $(\langle x_g, y_g, \theta_g\rangle)$)
    goal $(\langle x_g, y_g, \theta_g\rangle) \wedge \neg$achievable$(\langle x_g, y_g, \theta_g\rangle)$ $\Rightarrow$ *fails* (navigate $(\langle x_g, y_g, \theta_g\rangle)$)
    active (navigate $(\langle x_g, y_g, \theta_g\rangle)$) $\wedge$ timed out $\Rightarrow$ *fails* (navigate $(\langle x_g, y_g, \theta_g\rangle)$)
    active (navigate $(\langle x_g, y_g, \theta_g\rangle)$) $\wedge$ $\langle x, y, \theta\rangle = \langle x_g, y_g, \theta_g\rangle$
      $\Rightarrow$ *succeeds* (navigate $(\langle x_g, y_g, \theta_g\rangle)$)

Let us consider, as an example, the control task specification shown above. The task specification states the desired values for the state variables x, y, and θ. The process signature indicates that the current dynamic state and the desired dynamic state are mapped into control inputs for the translational and rotational velocity of the robot. The control process specification then states when the control task should be activated and the different possible outcomes of performing the control task.

Control Routines specify *how* the robot is to respond to sensory input or changes in the estimated state in order to accomplish a given control task. We might have different routines for a given control task, which have different performance characteristics. Control routines first of all consist of an implementation, a procedure that maps the estimated state into the appropriate control signals. Besides the implementation we have the possibility to specify models for the control routine, encapsulating the procedural routine within a declarative construct which can be used for reasoning and manipulating. Suppose we had a model consisting of decision rules that can identify situations in which the control routine is very likely to succeed and very likely to fail. We could then apply these rules in order to decide whether to activate the routine in the current situation or terminate it if the routine is likely to fail. This is not possible with only a procedural representation of the control routine.

3 Adding Learning Capabilities

In the last section we have specified process models and implemented them as procedures. However, it is often difficult and tedious to program them manually. In this section we will describe how the corresponding learning problems and the mechanisms for solving them can be specified. There are a number of program pieces that can be learned rather than being specified by programmers. In our robot soccer application, the robot learns, for example, the process model for its dynamics. Also, there are many opportunities for learning the implementation as well as the models of control routines.

We will restrict ourselves to learning problems that can be solved by function approximation. In particular, we will look at problems that can be learned by artificial neural networks and by decision tree learning algorithms. Our primary example will be learning how to achieve a given dynamic state, which is representative for learning in the context of control routines.

To state learning problems we must provide several pieces of information. The type of function that is to be learned, whether it is a process model or a control task. We must also specify the robot to be controlled and the environment it is to act in. In addition, the learning problem specification contains a specification of a problem specific learning system, and the data gathering process. This information is provided using the macro *define-learning-problem*.

define learning problem achieve-dynamic-state
    control-routine *achieve-dynamic-state-3*
    control-task achieve ($\langle x = x_{ball}, y = y_{ball}, \theta = \theta_{toball} \rangle$)
    dynamic system agilo-robot-controller
    data-collector achieve-state-data-collector
    learning-system achieve-state-learner

The learning problem specification shown above specifies these information pieces for the task *achieve dynamic state*. In the remainder of this section, we introduce the representational structures for specifying the different components of learning problems and then discuss issues in the implementation of these constructs.

3.1 Specifying Data Collection

The first step in learning is the data collection. To specify data collection we must state a control task or sample distribution and we have to provide monitoring and exception handling methods for controlling data collection. CLIP [1] is a macro extension of LISP, which supports the collection of experimental data. The user can specify where data should be collected, but not how.

In order to acquire samples we have to define a sample distribution from which samples are generated. During data collection the perception, control signals, and control tasks of each cycle are written to a log file. The data records in the log files can then be used for the learning steps.

define task distribution achieve-dynamic-state
  with global vars
    translation-tolerance ← 0.03
    max-time-steps ← 20
  for initial state $\langle x, y, \theta, v_{tra}, v_{rot}\rangle \leftarrow$ ⟨ -4.5, 1.5, 0.3, [0-0.6], [0-0.7] ⟩
    for goal state $\langle v_{tra:g}, v_{rot:g}\rangle \leftarrow$ ⟨ [0-0.6], [0-0.7] ⟩
      with setup *setstate* $\langle x, y, \theta, v_{tra}, v_{rot}\rangle$, *setgoal* $\langle v_{tra:g}, v_{rot:g}\rangle$

The task distribution specification above defines a distribution for the initial state consisting of the robot's pose and translational and rotational velocity, as well as the goal state of an episode, which consists of the desired translational and rotational velocity. The body of the episode consists of setting the robot in its initial state and then setting the desired dynamic state as a goal for the controlled process.

The data collection cycle starts with an initialization phase in which a task is sampled from the task distribution. In the distribution above this is the *setstate* command in the *with setup* statement. The data collector monitors the initialization to detect whether or not it has been successful. Unsuccessful initialization is particularly likely when the learning task is performed using the real robots. Upon successful initialization the data collection phase is activated. In the data collection phase the perception and control signals are collected. The data collector then detects the end of the episode and decides whether or not the episode is informative and should be recorded.

Thus, in order to collect data reliably and efficiently it is necessary to specify special purpose code for the initialization phase, for failure handling, and resetting the robot. These code pieces can be provided using the *define data collection functions*.

3.2 Specifying Learning Systems

Learning problems can be solved with very different learning systems such as artificial neural networks or decision tree learners. In order to specify a task-specific learning systems three pieces of information must be provided. First, we have to specify how the collected data are to be transformed into patterns that are used as input for the learning system. An example is the transformation of global coordinates to an robot-centric coordinate system. Second, a function for parameterizing the learning system. For an artificial neural network for instance we have to define the topological structure of the network. Finally, we must provide a function that transforms the output of the learning system into an executable function that can be employed within the robot control program. Thus, the input patterns for the *achieve-dynamic-state* control task can be defined as follows:

define state space achieve-dynamic-state
    input-features (($v_{tra:0}$ (percept:v_{tra} :time t))
                ($v_{tra:g}$ goal-state:v_{tra}))
    output-values (($v_{tra:1}$ (percept:v_{tra} :time $(t+1)$)))
    output-function (v_1 float)

The macro declares the training pattern to consist of the feature $v_{tra:0}$, which is taken to be the v_{tra} of the percept component of a log file entry and $v_{tra:g}$ that is the v_{tra} of the goal state of the log file entry. The output value of the pattern, called $v_{tra:1}$, is the translation velocity of the subsequent entry in the log file. The output of the learned function will be $v_{tra:1}$.

3.3 From Partial Specification to Executable Program

We have implemented our representational structures using LISP's macro facilities and provided them in the context of *Structured Reactive Controllers (SRCs)* [3]. The LISP-based parts are equipped with an abstract robot interface consisting of C libraries loaded as shared objects. The system uses shared memory communication as a coupling to the low-level control processes. To date, we have provided the functionality of two learning systems: the decision tree learning algorithm C4.5 and the Stuttgart Neural Network Simulator (SNNS).

An important implementational aspect is how we get from declarative problem specifications to executable and effective learning processes. For this and other purposes control routines, process models, and control routine models are represented as first class objects that computer programs can reason about and manipulate. One property of these objects is that they can have an implementational status, such as *to-be-learned*. The distinct characteristic of this program is that in the beginning it only partially specifies the behaviour, leaving unspecified behaviour to be learned at a later stage.

Transforming this partial specification to an executable control program proceeds in three steps. In the first phase a problem collector traverses the top-level control routine and collects all references to procedures with the status *to-be-learned* and stores them in the set of learning problems. In the second phase, the routines are learned. We consider a routine as learnable if it is *to-be-learned* and does not call any code *to-be-learned*. Therefore, the procedures are learned in an order that respects the dependencies between them, until all routines are learned. Finally, the complete program can be run to produce the behaviour, for instance playing robot soccer.

4 A Rational Reconstruction of the AGILO2001 Controller

With the extended language we have rationally reconstructed large parts of the action selection module of the AGILO autonomous soccer robots, with processes, data collection, learning system generation specified in constructs. The control program resulting from this reconstruction can automatically learn a repertoire of routines for playing robot soccer competently. In this section the learning process is demonstrated.

In the bootstrap learning phase the first learning task that is tackled is the learning of the robot dynamics. This learning task acquires a process model for the environment and does not depend on the solution of any other learning problems. To learn the dynamics we use the log files obtained from earlier RoboCup games and transform them into a pattern file for the learning system. For learning, we train an artificial neural network using the SNNS system in order to approximate the mapping: $state_i \times action_i \mapsto state_{i+1}$. In [4] this procedure is described in more detail.

The next learning task that the robot can address is how navigation tasks are to be achieved. We have already seen in the specification of this task in section 2.2. This is learned in a simulated environment, which only requires the learned robot dynamics. To solve the learning task as a function approximation problem we provide the robot with streams of control signals that generate good navigation trajectories. From this data the robot learns the mapping $state_{current} \times state_{goal} \mapsto action_{current}$, using a multi layer neural network. For a more detailed description see [5].

Once the controller can execute the navigation task, the robot learns a model of the task. Given the simulated dynamics and the controller learned in the two previous steps, the robot is given many automatically generated navigation tasks to execute. The duration of the task is recorded, and this training data is used to learn the mapping $state_{current} \times state_{goal} \mapsto time\_to\_complete$, again using a multi layer neural network. This mapping is necessary to coordinate the behaviour of our robots, the next step in creating the complete AGILO controller. How this is done is discussed in [5].

Since all these learning tasks are specified with constructs in the program, adaptation of the program is simplified. Learning tasks can be exchanged, without requiring a redesign of the program structure. The program just needs to check if any constructs are *to-be-learned*, and relearn them and any depending constructs. Another case example that shows the strength of making learning tasks explicit is when hardware is upgraded. Recently our robots have received new control-boards. Instead of having to repeat all learning steps to model these new boards manually (involving undocumented scripts, manual data copying and transformation, and so forth), we only have to provide the system with a log-file, and set all relevant learning tasks to *to-be-learned*. The dependencies between the learning tasks ensure that first the robot dynamics will be learned. Given these dynamics, which can be used in the simulator, the navigation tasks can be relearned. Given the navigation routines, models of the navigation routines can be learned, and so on, until the complete executable AGILO2001 controller is acquired. We are currently testing these procedures. Furthermore, the scope of learning tasks is being extended beyond the AGILO2001 controller, to include additional control tasks such as dribbling with fake movements, defending, shooting, and others. In addition, we are starting to tackle reinforcement learning problems.

5 Conclusions

In this paper, we have extended a control language with constructs for explicitly representing (1) the physical system that is to be controlled and (2) the learning problems to be solved. In the extended language entities such as control tasks, process models, learning problems, and data collection strategies can be represented explicitly and transparently, and become executable. In the learning and execution phase, the entities are first class objects that control programs cannot only execute but also reason about and manipulate. These capabilities enable robot learning systems to dynamically redesign learning problems. We have also sketched the implementation of the learning phase and showed how a bootstrap learning method can automatically complete a partially defined control program by solving the stated learning problems. The extensions that we have presented are expressive enough to rationally reconstruct most of the AGILO2001 action selector. Other complex control systems need to be implemented using our approach and the conciseness and expressivity of our constructs need to be assessed and analyzed. We are just starting to incorporate optimizing learning techniques such as reinforcement learning into our approach.

We see the main impact of our framework along two important dimensions. From a software engineering perspective, the language extensions allow for transparent implementation of learning steps and abstract representation of complex physical systems. These aspects are typically not adequately addressed in current control systems, which makes them hard to understand and adapt to new requirements and conditions. The second dimension, which we find much more exciting, is the use of the framework as a tool for investigating more general and powerful computational models of autonomous robot learning. The programmability of learning systems, the modifiability of state representations, the possibility of reparameterizing learning systems, and the executability of learning specifications within the framework enables us to solve complex robot learning tasks without human interaction. The framework thereby enables us to investigate adaptive robot control systems that can autonomously acquire sophisticated skills and competent task control mechanisms for a variety of performance tasks.

References

1. S. Anderson, D. Hart, J. Westbrook, and P. Cohen. A toolbox for analyzing programs. *International Journal of Artificial Intelligence Tools*, 4(1):257–279, 1995.
2. D. Andre and S. Russell. Programmable reinforcement learning agents. In *Proceedings of the 13th Conference on Neural Information Processing Systems*, pages 1019–1025, Cambridge, MA, 2001. MIT Press.
3. M. Beetz. Structured Reactive Controllers – a computational model of everyday activity. In O. Etzioni, J. Müller, and J. Bradshaw, editors, *Proceedings of the Third International Conference on Autonomous Agents*, pages 228–235, 1999.
4. S. Buck, M. Beetz, and T. Schmitt. M-ROSE: A Multi Robot Simulation Environment for Learning Cooperative Behavior. In *Distributed Autonomous Robotic Systems 5, LNAI*. Springer-Verlag, 2002.
5. S. Buck, M. Beetz, and T. Schmitt. Planning and Executing Joint Navigation Tasks in Autonomous Robot Soccer. In *5th International Workshop on RoboCup, Lecture Notes in Artificial Intelligence (LNAI)*. Springer-Verlag, 2001.
6. T. Dean and M. Wellmann. *Planning and Control*. Morgan Kaufmann Publishers, San Mateo, CA, 1991.
7. D. McDermott. A Reactive Plan Language. Research Report YALEU/DCS/RR-864, Yale University, 1991.
8. S. Thrun. Towards programming tools for robots that integrate probabilistic computation and learning. In *Proceedings of the IEEE International Conference on Robotics and Automation (ICRA)*, San Francisco, CA, 2000. IEEE.

A New Odometry System to Reduce Asymmetric Errors for Omnidirectional Mobile Robots

Alireza Fadaei Tehrani[1,2], Ali Mohammad Doosthosseini[3],
Hamid Reza Moballegh[1,3], Peiman Amini[1,3], and Mohammad Mehdi DaneshPanah[1,3]

[1] Robotic Center of Isfahan University of Technology
[2] Mechanical Engineering Department, Isfahan University of Technology (IUT)
[3] Electrical Engineering Department, Isfahan University of Technology (IUT)

Abstract. This work describes an investigation to reduce positioning error of 3 wheel middle size robot by using a modified odometry system. In this technique the positioning sensor (shaft encoders) are mounted on 3 free-running wheels so the slippage of the driving wheels does not affect the measurements of the sensors. This will result in decreasing the cumulative error of the system. This mechanism accompanying by omnidirectional vision system presents reliable and accurate self-localization method for any 3 wheel driving robot. Experimental results have shown performance improvement up to 86% in orientation error and 80% in position error.

1 Introduction

Self-localization is one of the most important issues in mobile robots. Different methods have been suggested for self-localization, naming a few, vision based self-localization, laser range finders, odometry [1], ultrasonic approach, gyroscope [5] and global positioning system (GPS). These methods mainly differ in their robustness, measuring accuracy, speed, cost and ease of implementation.

In odometry, robot displacement and change of direction are measured comparing to previous position. This method has acceptable results on platforms where robot moves on a smooth surface. Ease of implementation, relatively cheap and light weight instrument are some advantages of this method.

One of the platforms in which robots need self-localization is RoboCup middle size league (MSL). In this environment, because of predefined colors for different objects and smooth surface, vision based self-localization and odometry are assumed to be effective [6]. Laser range finders were also a good solution until the field walls were not removed [9]. Using vision based self-localization alone has some drawbacks:

Field landmarks are limited to flags, goals and border lines. Because of moving objects in the field, a few land marks may not be seen by robot. In addition, errors in object detection and the nonlinear map (from image units (pixels) to physical units (centimeters)) cause unreliable output. Robust image processing algorithms can enhance the precision, but the algorithms are complex and time consuming.

Adding another sensor and using sensor data fusion is assumed to be a suitable method for precision enhancement, odometry sensors are an example [1]. In case where sufficient landmarks can not be detected, odometry results would be helpful.

D. Polani et al. (Eds.): RoboCup 2003, LNAI 3020, pp. 600–610, 2004.

Variation in odometry results is smooth, this can help to detect and avoid sudden faulty reports of vision based self-localization. In the other hand, odometry errors are cumulative and regarding slippage, sensor outputs could be of low confidence.

In this article, three improvements are introduced for decreasing slippage error and enhancing sensor precision. The designed mechanical structure is also presented. Then two methods for position estimation are introduced and discussed in detail. Finally the implementation of a linear controller is described. The aim of this work is to implement a reliable navigation system, using modified odometry accompanying omnidirectional vision based self-localization.

2 Modified Odometry Sensors

Asymmetric errors are those errors which are caused by slippage and field unevenness. In MSL RoboCup, due to the flat playground, the highest portion of asymmetric error in odometry is caused by slippage when the robot accelerates or deaccelerates.

The designed structure includes an omni-directional system in which three omni-wheels are placed under the robot $120°$ apart.

Fig. 1.

Fig. 2.

These free wheels are then coupled with shaft encoders and are assembled on a flexible structure, which ensures their firm contact with ground. The symmetry point of this structure coincides with the robot's center of mass and the wheels have an angular distance of $60°$ from the active (driving) wheels (Fig. 1,2). Most slippage occurs on the driving wheels during acceleration and changing direction.

Separating the sensors from driving wheels, decreases the slippage of the sensor wheels to a great extent, which results in decreasing the cumulative error of the system. In addition, special designed multi-roller wheels have vibration amplitude of less than 0.2 mm, also slippage decreases notably on carpet playground.

3 Robot Kinematics Model [3]

P_o is defined as the vector representing robot center of mass in (Fig.3). The position and tangential vector of each odometry wheel in the frame fixed to the robot are:

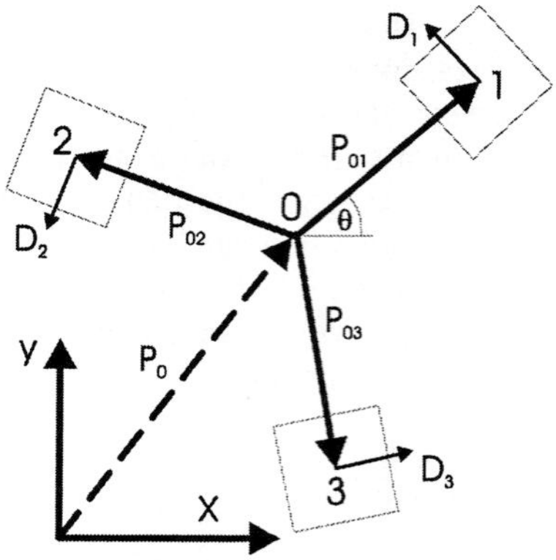

$$P_{o1} = L\begin{bmatrix}1\\0\end{bmatrix} \quad , \quad P_{o2} = R(\frac{2\pi}{3})P_{o1} \quad , \quad P_{o3} = R(\frac{4\pi}{3})P_{o1} \tag{1}$$

$$D_1 = \begin{bmatrix}0\\1\end{bmatrix} \quad , \quad D_2 = -\frac{1}{2}\begin{bmatrix}\sqrt{3}\\1\end{bmatrix} \quad , \quad D_3 = \frac{1}{2}\begin{bmatrix}\sqrt{3}\\-1\end{bmatrix} \tag{2}$$

Fig. 3.

where L is the distance of wheels from the center of mass.

Using the above notations, the wheel position and velocity vectors can be expressed with the rotation matrix $R(\theta)$ by the following equations:

$$R_i = P_o + R(\theta)P_{oi} \tag{3}$$

$$V_i = P_o + \dot{R}(\theta)\dot{P}_{oi} \tag{4}$$

The angular velocity of each wheel can be expressed as:

$$\dot{\varphi}_i = \frac{1}{r}V_i^T(R(\theta)D_i) \tag{5}$$

where $\dot{\varphi}_i$ is the angular velocity and r is the radius of odometry wheels.

Substituting V_i from (4) into (5) yields:

$$\dot{\varphi}_i = \frac{1}{r}[\mathrm{P}_o^T R(\theta)D_i + \mathrm{P}_{oi}^T \dot{R}^T(\theta)R(\theta)D_i] \tag{6}$$

where the second term in the right hand side is the tangential velocity of the wheel. This tangential velocity could be also written as $L\dot{\theta}$, so from (6), we have:

$$L\dot{\theta} = \mathrm{p}_{Oi}^T \dot{R}^T(\theta)R(\theta)D_i \tag{7}$$

From the kinematics model of the robot, it's clear that the wheel velocities are linear functions of the robot linear and angular velocities.

$$\begin{bmatrix} \dot{\varphi}_1 \\ \dot{\varphi}_2 \\ \dot{\varphi}_3 \end{bmatrix} = \frac{1}{r} \begin{bmatrix} -\sin\theta & \cos\theta & L \\ -\sin(\frac{\pi}{3}-\theta) & -\cos(\frac{\pi}{3}-\theta) & L \\ \sin(\frac{\pi}{3}+\theta) & -\cos(\frac{\pi}{3}+\theta) & L \end{bmatrix} \begin{bmatrix} \dot{x} \\ \dot{y} \\ \dot{\theta} \end{bmatrix} \qquad \dot{\Phi} = \frac{1}{r} W.\dot{S} \tag{8}$$

4 Self-localization Using Odometry

In this section two self-localization methods based on shaft encoder outputs are introduced, which are called differential method and direct method.

4.1 Differential Method

In this method, the outputs of shaft encoders are differentiated temporally resulting the individual wheel velocities. Using the kinematics equations, the robot linear and angular velocities are computed. Integrating these velocities, one can extract the robot position. The algorithm for this method can be described as fallows:

Using kinematics equation of the robot (8), x , y and θ are the parameters to be computed by having $\dot{\varphi}_i$ from shaft encoders. Solving equation (8) using the inverse of W results:

$$\begin{bmatrix} \dot{x} \\ \dot{y} \\ \dot{\theta} \end{bmatrix} = r \begin{bmatrix} \frac{\cos(\pi/3+\theta)-\cos(\pi/3+\theta)}{3\sin\pi/3} & \frac{-\cos\theta-\cos(\pi/3+\theta)}{3\sin\pi/3} & \frac{\cos\theta+\cos(\pi/3-\theta)}{3\sin\pi/3} \\ \frac{\sin(\pi/3+\theta)+\sin(\pi/3+\theta)}{3\sin\pi/3} & \frac{-\sin\theta-\sin(\pi/3+\theta)}{3\sin\pi/3} & \frac{\sin\theta+\sin(\pi/3-\theta)}{3\sin\pi/3} \\ \frac{1}{3L} & \frac{1}{3L} & \frac{1}{3L} \end{bmatrix} \begin{bmatrix} \dot{\varphi}_1 \\ \dot{\varphi}_2 \\ \dot{\varphi}_3 \end{bmatrix}$$

$$\dot{S} = rW^{-1}.\dot{\Phi} \tag{9}$$

where $\dot{\varphi}_i$ can be calculated from the current encoder samples, compared with previous samples. Using numerical integration methods on $[\dot{x} \quad \dot{y} \quad \dot{\theta}]^T$ (knowing the initial position of the robot), the new position can be obtained.

Although this method is very simple, but due to inherent differentiation operations, it is not reliable in long term. The main drawback of differential method is the error accumulation. Some of important errors are: error in measuring time intervals, errors caused by the quantized nature of shaft encoders and output errors due to floating point computations.

Practically, time is divided into equal intervals, ΔT , in which the robot wheels' velocities are assumed to be constant. Considering these constant speeds, both robot linear and angular velocities are determined, and the head angle and robot position would be grown with the extracted differential values.

The assumption of constant velocities in each time interval ΔT , will cause some error, which in turn will decrease the reliability of the computation results as time passes. In addition, limited precision of shaft encoders will result in some kind of quantization error which therefore leads to inaccuracy, the last point is that this error might increase due to digital computations. In order to decrease these errors, one possible solution is to define ΔT as small as possible and to increase shaft encoder resolutions. Needless to remark that this solution will increase the calculation's overhead.

4.2 Direct Method

The differential method was described as digitally differentiating the output of shaft encoders, applying the kinematics equations of the robot to the result of differentiation and integrating over the results.

Because of quantized nature of the values in time, integration and differentiation can not be implemented without approximation. If these operations could be eliminated and results can be driven directly from the measured angle by shaft encoders, more accuracy in calculation results are then achieved.

Using the equations of motion (9), in order to calculate x, y, θ directly, the following integral should be computed:

$$\dot{S} = r\int W^{-1}.\dot{\Phi}\ dt \tag{10}$$

where

$$S = \begin{bmatrix} x \\ y \\ \theta \end{bmatrix} = \begin{bmatrix} x(\varphi_1,\varphi_2,\varphi_3) \\ y(\varphi_1,\varphi_2,\varphi_3) \\ \theta(\varphi_1,\varphi_2,\varphi_3) \end{bmatrix} \tag{11}$$

system of equations in (10), results:

$$\mathrm{x} = r\int \frac{1}{3\sin(\frac{\pi}{3})}\left[(\cos(\frac{\pi}{3}+\theta)-\cos(\frac{\pi}{3}-\theta))\,\dot{\varphi}_1 + (-\cos(\theta o - \cos(\frac{\pi}{3}+\theta))\,\dot{\varphi}_2 + (\cos(\theta c - \cos(\frac{\pi}{3}-\theta))\,\dot{\varphi}_3\right] dt \tag{12}$$

$$\mathrm{y} = r\int \frac{1}{3\sin(\frac{\pi}{3})}\left[(\sin(\frac{\pi}{3}-\theta)-\sin(\frac{\pi}{3}+\theta))\,\dot{\varphi}_1 + (-\sin(\theta i - \sin(\frac{\pi}{3}+\theta))\,\dot{\varphi}_2 + (\sin(\theta s - \sin(\frac{\pi}{3}-\theta))\,\dot{\varphi}_3\right] \mathrm{dt} \tag{13}$$

$$\theta = r\int \frac{\dot{\varphi}_1+\dot{\varphi}_2+\dot{\varphi}_3}{3\mathrm{L}}\,\mathrm{dt} \tag{14}$$

Further simplification of (14) results:

$$\theta = \frac{r}{3\mathrm{L}}\left[\int \dot{\varphi}_1\,\mathrm{dt} + \int \dot{\varphi}_2\,\mathrm{dt} + \int \dot{\varphi}_3\,\mathrm{dt}\right] = \frac{r}{3\mathrm{L}}(\varphi_1+\varphi_2+\varphi_3) \tag{15}$$

It's apparent from (15), that θ could be easily calculated from shaft encoder outputs, but it is impractical to extract x or y as direct functions of φ_1, φ_2, φ_3, from equations (12), (13). This could be demonstrated with the following example:

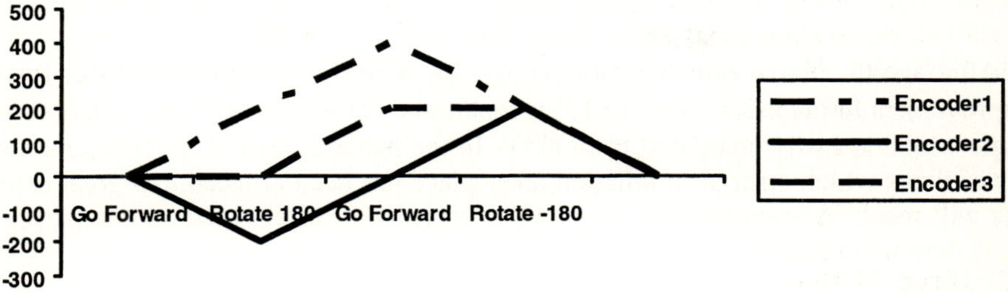

Fig. 4. The robot has the displacement of 2L however the shaft encoder outputs are the same as initial position.

In the above example, robot has started moving from an initial position. After moving a straight path with length L, it rotates 180° and then moves straight the same length along the previous direction and then rotates -180°. It is clear from (Fig.4) that for both initial and end points, $\varphi_1 = \varphi_2 = \varphi_3 = 0$, while the robot has a displacement of 2L. Therefore direct computation of x and y from $\varphi_1, \varphi_2, \varphi_3$ is not practical, which necessitates some restrictions on x and y to be extractable directly.

As it is understood from equations (12), (13) the coefficients of $\dot{\varphi}_i$ are functions of θ which itself is a function of time. Assuming θ to be independent of time, equations (12), (13) could be rewritten as follows:

$$x = \frac{r}{3\sin\frac{\pi}{3}}\Big[\int\Big(\cos(\frac{\pi}{3}+\theta) - \cos(\frac{\pi}{3}-\theta)\Big)\dot{\varphi}_1\,dt + \int\Big(-\cos\theta - \cos\Big(\frac{\pi}{3}+\theta\Big)\Big)\dot{\varphi}_2\,dt + \int\Big(\cos\theta + \cos\Big(\frac{\pi}{3}-\theta\Big)\Big)\dot{\varphi}_3\,dt\Big]$$
$$= \frac{r}{3\sin\frac{\pi}{3}}\Big[\Big(\cos\Big(\frac{\pi}{3}+\theta\Big) - \cos\Big(\frac{\pi}{3}-\theta\Big)\Big)\varphi_1 + \Big(-\cos\theta - \cos\Big(\frac{\pi}{3}+\theta\Big)\Big)\varphi_2 + \Big(\cos\theta + \cos\Big(\frac{\pi}{3}-\theta\Big)\Big)\varphi_3\Big] \tag{16}$$

$$y = \frac{r}{3\sin\frac{\pi}{3}}\Big[\int\Big(\sin(\frac{\pi}{3}-\theta) + \sin(\frac{\pi}{3}+\theta)\Big)\dot{\varphi}_1\,dt + \int\Big(-\sin\theta - \sin\Big(\frac{\pi}{3}+\theta\Big)\Big)\dot{\varphi}_2\,dt + \int\Big(\sin\theta - \sin\Big(\frac{\pi}{3}-\theta\Big)\Big)\dot{\varphi}_3\,dt\Big]$$
$$= \frac{r}{3\sin\frac{\pi}{3}}\Big[\Big(\sin\Big(\frac{\pi}{3}-\theta\Big) + \sin\Big(\frac{\pi}{3}+\theta\Big)\Big)\varphi_1 + \Big(-\sin\theta - \sin\Big(\frac{\pi}{3}+\theta\Big)\Big)\varphi_2 + \Big(\sin\theta - \sin\Big(\frac{\pi}{3}-\theta\Big)\Big)\varphi_3\Big] \tag{17}$$

Assuming θ to be independent of time, will result in $\dot{\theta}$ to be zero, thus:

$$\dot{\theta} = 0 \Rightarrow \frac{1}{3L}(\dot{\varphi}1 + \dot{\varphi}2 + \dot{\varphi}3) = 0 \Rightarrow \dot{\varphi}1 + \dot{\varphi}2 + \dot{\varphi}3 = 0 \tag{18}$$

Equation (18) shows that robot position can be extracted directly if the robot has had no rotation while moving to the new position. If $\dot{\theta}$ changes to a value other than zero, the origin of the coordinate system should be moved according to the calculated robot displacement vector and the computation should be started over again. In practice the condition on $\dot{\theta}$ is satisfied if the difference between the new head angle and the last stored angle exceeds some certain threshold.

Although the above equations are extracted assuming θ to be constant, for further improvement in precision, θ in (16,17) is replaced dynamically with the instantaneous angle which is easily computed from (15). In the method described above, the number of times which error accumulation takes place has been reduced to a great extent, this will result in improving the overall performance of direct method comparing to the differential method.

5 Robot Controller

In [10] it has been shown that two identical PID controllers for robot position and orientation are suitable for controlling a robot. If the residual error is negligible the integrator can be omitted. This controller has shown to be robust enough for controlling a soccer player robot. P and D coefficient are adjusted manually.

5.1 Position Controller Architecture

In order to implement the position controller, the position error vector is determined as follows:

$$\vec{e} = \begin{bmatrix} x \\ y \end{bmatrix} - \begin{bmatrix} x' \\ y' \end{bmatrix} \tag{19}$$

$$\vec{V} = k_d \frac{d\vec{e}}{dt} + k_p \vec{e} \tag{20}$$

where $\vec{V}$ expresses the output of the position controller for driving units, which it's components on each driving wheel are extracted with:

$$V_i = \vec{V}^T . D_i \tag{21}$$

5.2 Head Angle Controller

Assuming that the head angle of the robot is β and the desired head angle is θ.The angle controller is then determined as follows:

$$w = k_p e_\theta + k_d \frac{de_\theta}{dt} \tag{22}$$

where $e_\theta = \theta - \beta$ is the angle error of the robot.

5.3 Scaling

The output of the position or angle controller in Equations (21), (22) may exceed the range of the applicable voltages on the motors. In order to deal with such a situation, a scaling algorithm is believed to be efficient.

If the output vector of position controller was out of range of maximum motor voltage (v_{Max}), it is scaled down with the following scale factor:

$$\alpha = \frac{v_{Max}}{\max|V_i|} \tag{23}$$

For the angle controller, another scaling method is used, in which the output of the angle controller is clipped to $\beta v_{\max}$ ($0<\beta<1$). The reason for using such threshold is that a relatively high angle error would result in high motor voltages and this in turn affects the output of position controller.

5.4 The Final Motor Controller

The final applicable voltages on the motors are computed as:

$$u_i = v_i + w \tag{24}$$

The following scale factor is used for u_i before being applied to each motor if $\max|u_i| > u_{\max}$:

$$\gamma = \frac{u_{Max}}{\max|u_i|} \tag{25}$$

6 Experimental Results

In order to check the improvement of the odometry system introduced in this article compared to similar systems, a test condition was suggested to measure the position and head angle error.

In the designed test condition, care was taken to simulate the real situation. The test condition and the results are presented in the following two sections. At the end, the effect of the odometry sensor in reducing vision based self-localization overhead will be discussed.

6.1 Test Condition

UMBmark method is a popular benchmark for odometry systems [8], but the following test condition was used for simulating robot behavior more closely.

In this test, the robot is placed at one end of the field, and then the robot moves straight to the other end (10 meters) while rotating $360°$. The position and head angle error are then measured. This procedure is repeated several times, and the mean values are computed.

In the following figure, shaft encoder outputs of an ideal odometry system under the test condition are presented.

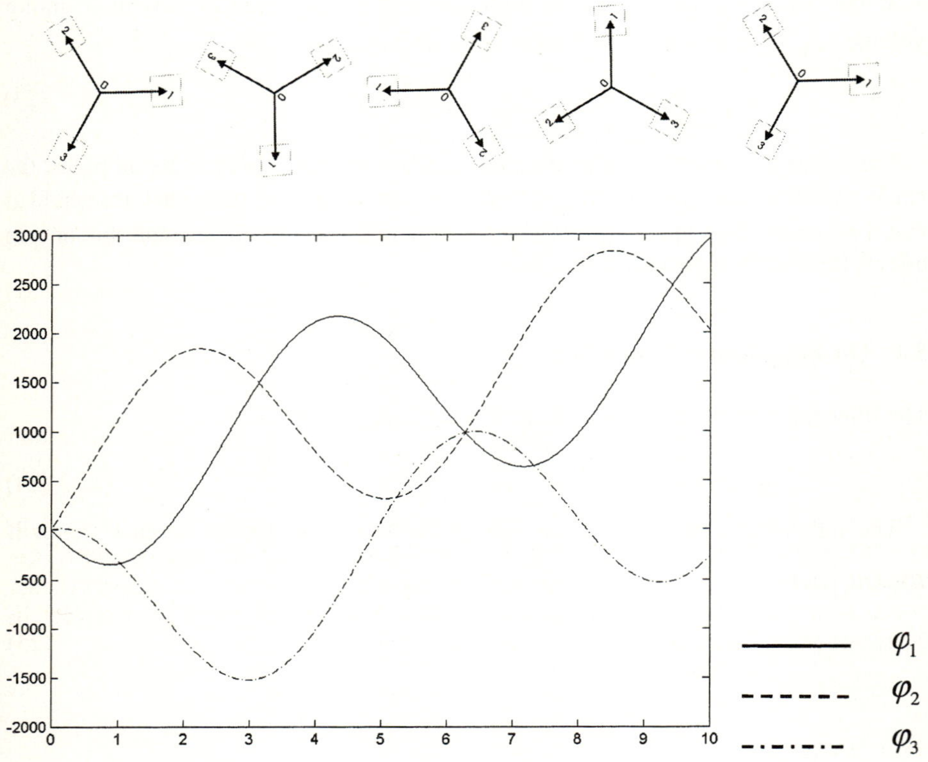

Fig. 5. Up: Position and orientation of robot while moving straight in 2.5 meters intervals. Down: The output of shaft encoders plotted over the length of 10 meters under the test condition.

6.2 Test Results

The following table demonstrates a comparison between robots with the ordinary odometry sensor, which the shaft encoders are connected to driving wheels, and the next generation robot in which the improved odometry system was installed on. The direct method was used for position calculations.

Using the improved sensors, the relative position error is 1.5%, while the relative orientation error is 1.1%. So, the new odometry sensors have improved the orientation error to 86%, and the position error to 80%. In this way we have reduced the asymmetric errors to a great extent.

Table 1. The absolute and relative position and angle error of ordinary and improved sensors.

	Position Error	Orientation Error	Relative Pos. Error	Relative Orientation Error
Ordinary sensors	75cm	30°	7.5%	8.2%
Improved sensors	15cm	4°	1.5%	1.1%

Considering the position accuracy, the vision based self-localization can be updated with longer time intervals, while the odometry results are used for self-localization. This in turn enables us to perform accurate, complicated and time consuming vision based self-localization algorithms.

Having relatively precise information about orientation is also used in searching algorithms of stationary objects (goals and flags), which causes the image processing algorithms more robust.

7 Conclusions

Joint visualization – odometry localization of moving robots has attracted the attention of many researchers. In this work we intended to improve the performance of the existing algorithms by proposing a new odometry system with lower asymmetric errors. In our approach there are free wheels accompanying the driving wheels solely to measure the robot rotation and displacement. Hence slippage is reduced considerably and error accumulation diminishes. New methods for converting measurements into desirable data are also proposed to avoid the amplification of inaccuracies in differentiations. Test results verified the improvements clearly.

References

1. Fredric Chenavier,James L. Crowley: "Position Estimation for a Mobile Robot Using Vision and Odometry", *Proceeding of IEEE International Conference on Robotics and Automation*
2. Johann Borenstein: "Experimental Results from Internal Odometry Error Correction with the OmniMate Mobile Robot ", *IEEE Transactions on Robotics and Automation, Vol. 14, NO. 6, December 1998 963*
3. Tamás Kalmár-Nagy, Rafaello D'Andrea, Pritam Ganguly: " Near-Optimal Dynamic Trajectory Generation and Control of an Omnidirectional Vehicle " ,*Cornell University April 8 2002*
4. Agostino Martinelli, " Possible Strategy to Evaluate the Odometry Error of A Mobile Robot ",*Proceeding of The 2001 IEEE/RSJ International Conference On Intelligent Robots and*
5. K. Komoriya, E. Oyama, "Position Estimation of a Mobile Robot Using Optical Fiber Gyroscope (OFG)", *Proc. Int. Conf. Intell. Robot. Syst. (IROS'94), Munich, Germany, pp.143-149, Sept. 12–16, 1994.*
6. Ashly W. Stroup,Kevin Sikorki,and Tuker Balch: "Constrained – Based Landmark Self-localization" ,*The 2002 International RoboCup Symposium*

7. Kok Seng Chong ,Lindsay Kleeman, " Accurate Odometry and Error Modeling for a Mobile Robot", Proceeding of the 1997 IEEE International Robotics and Automation
8. J. Borenstein and L. Feng, "UMBmark: A Benchmark Test for Measuring Dead-Reckoning Errors in Mobile Robots", *Proc. 1995 SPIE Conf. Mobile Robots.*
9. Hans Utz, Alexander Neubeck , Gerd Mayer, Gerhard Kraetzschmar. "Improving Vision-Based Self Localization" ,*The 2002 International RoboCup Symposium*
10. Keigo Watanabe, "Control of an Omnidirectional Mobile Robot", *Second International Conference Acknowledge-Based Intelligent Electronic Systems,21-23 April*

Texture-Based Pattern Recognition Algorithms for the RoboCup Challenge

Bo Li and Huosheng Hu

Department of Computer Science, University of Essex
Wivenhoe Park, Colchester CO4 3SQ, United Kingdom
{bli,hhu}@essex.ac.uk

Abstract. Since texture is a fundamental character of images, it plays an important role in visual perception, image understanding and scene interpretation. This paper presents a texture-based pattern recognition scheme for Sony robots in the RoboCup domain. Spatial frequency domain algorithms are adopted and tested on a PC while simple colour segmentation and blob-based recognition are implemented on real robots. The experimental results show that the algorithms can achieve good recognition results in the RoboCup Pattern Recognition challenge.

1 Introduction

In many applications, texture-based pattern recognition typically uses discrimination analysis, feature extraction, error estimation, cluster analysis (statistical pattern recognition), grammatical inference and parsing (syntactical pattern recognition). Besides colour, texture is a fundamental character of natural images, and plays an important role in visual perception. The basic methods for texture-based recognition are combinations of both spatial and frequency domains.

Texture-based segmentation algorithms divide the image into regions yielding different statistical properties. Such methods assume that the statistics of each region are stationary and that each region is extended over a significant area. However, most image regions do not present stationary features in the real world. Also, meaningless small-sized regions related to stains, noise or punctual information might appear. Consequently, methods relying on a priori knowledge of the number of textures in a given image [7] [11] often failed, because if any unexpected texture region appears, like the ones related to shadows or boundaries, a wrong fusion of two non-related regions is forced. Unsupervised segmentation [1][3] does not rely on such knowledge, but it is slow because it requires a computationally expensive additional stage to calculate the correct number of regions. Gabor filters and Gabor space are very useful tools in spatial frequency domains. There is some similarity between Gabor process and human vision [8][9][10][6]. Multi-resolution and multi-band methods, as well as a combination of statistical and learning algorithms are very useful in this area [8]..

In the area of texture recognition and segmentation, image modulation and multiband techniques are very useful tools to analyse texture-based patterns. Havlicek has done a lot in AM-FM modulation for texture segmentation [3][4][13]. Images with multi-texture are processed by a bank of Gabor filters. Using the filter responses,

D. Polani et al. (Eds.): RoboCup 2003, LNAI 3020, pp. 611–620, 2004.

Dominant Component Analysis and Channelised Component Analysis are implemented for texture decomposition. There are also other researchers using similar methods.

The rest of the paper is organized as follows. Section 2 describes briefly texture-based recognition, including the pattern recognition challenge in RoboCup and the texture analysis process. In section 3, the issues related to colour filters and spatial domain processes are investigated. Then, both multi-band techniques and Gabor filters are proposed for texture segmentation in section 4. Section 5 presents image modulation modules based on multi-dimensional energy separation. Experimental results are presented in section 6 to show the feasibility and performance of the proposed algorithms. A brief conclusion and future work are given in section 7.

2 Texture-Based Pattern Recognition

2.1 Pattern Recognition Challenge in RoboCup

In the Sony robot challenge of RoboCup 2002, each Sony robot is expected to recognize the targets with chessboard texture pattern and individual geometrical shapes, which are placed at random positions and orientations. To implement this task, we define a function F to represent the pattern recognition system that can be used in this challenge. More specifically, its inputs are images with dimension of $M \times N$, and its output for each image is the element of a name set or a code set. For example, a name set S={*triangle, square, rectangle, T, L*} may be defined here for the recognition challenge of Sony robots in the RoboCup competition. Figure 1 shows 4 targets to be identified in the challenge. The size, rotation angle and background are random.

(a) A square target (b) A T-shape target (c) A triangle target (d) A L-shape target

Fig. 1. Target patterns in the RoboCup Sony Robot Challenge

The whole process of recognition can be decomposed to four stages. There are many methods that can be applied at each stage. In general, complex methods can do well, but are expensive. Simple methods are required for real-time systems. More specifically, we have four stages in recognition as follows:

- Stage 1 -- *Image processing* stage can enhance image quality and make segmentation or edge detection much easier by removing noises and blurs images. Some images that are not focused well can be sharpened.
- Stage 2 -- *Image conversion* includes filter process, DFT (Discrete Fourier Transform) process, DWT (Discrete Wavelet Transform) process and so on. Its main purpose is to make properties of images easy to be extracted. For instance, skin

colour can be used to identify hands of humans. In the RoboCup Sony robot league, the competition field is colour-based. Under natural environment with complex backgrounds, Sony robots may be unable to find objects easily by using colour information only. Further complex methods are required [5].

- Stage 3 -- *Information extraction* to obtain useful information for recognition. In most cases, it should output data in a special format for processing at a later stage. For instance, blob-based methods should output useful information for all blobs, including position, size, and so on. In some cases this part is called coding, as it is describing objects with numeric properties.
- Stage 4 -- *Pattern recognition* converts numeric properties to most-like objects and output their name. The process depends on the previous results. If input data is easy to be discriminated, this process can be very simple.

2.2 Texture Analysis Process

Texture is a fundamental character of natural images. It plays an important role in visual perception and provides information for image understanding and scene interpretation. Figure 2 (a) shows the target texture used in the experiments.

Figure 2 (b) shows the magnitude of the DFT result from Figure 2 (a). The result is cantered, logarithmically compressed and histogram stretched. From this result, the frequency distribution of the pattern is clear. However, it is difficult to design a filter for recognition. Figure 2 (c) shows an original image captured by the robot.

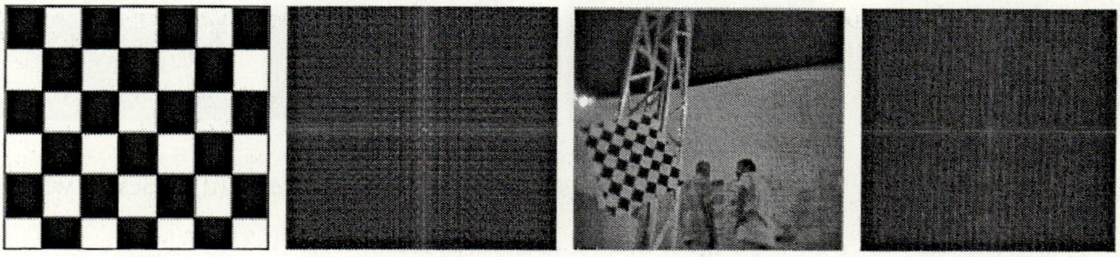

(a) Chessboard (b) DFT magnitude (c) An original image (d) Along Y-axis

Fig. 2. Texture Analysis Process

Figure 2(d) shows the DFT magnitude for luminance in Figure 4. Because of its complex background, no filtering hint for the pattern can be seen from figure 5. It is impossible to separate the pattern from complex background by using a simple filter in the frequency domain.

3 Colour Filter and Spatial Domain Process

A colour filter can be used to obtain pixels with the concerned colour. Spatial domain processes such as morphology filters cost little, but are very effective for smoothing binary images. Following processes are all based on the image shown in Figure 2(c). The colour filter here are LUT (Look Up Table) based, which was implemented on the robots for the purpose of good performance. Figure 3 shows the result by the LUT method. The concerned colour is white. Figure 4 shows the Horizontal Signature, i.e. the image projection onto the x-axis. The first peak is the dark area and the second part is mainly white part.

It is not easy to separate white and black blobs from the background on the basis of the signatures alone. Figure 5(a) shows the result for black blobs.

Fig. 3 An image

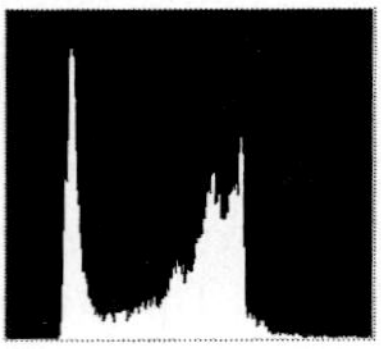

Fig. 4. Horizontal Signature

With LUT results, edges can be extracted. Figure 5(b) shows the image with edges of white blobs. The edge detection here is morphology based, i.e. binary calculation. Figure 5(c) shows an image with edges of black blobs. Figure 5(d) shows the overlay of black and white edges. The main problem with spatial processes is that they are limited by the size of masks. In fact, the target frequencies are fixed with masks. It is therefore difficult to extract useful information for recognition, especially with complex background.

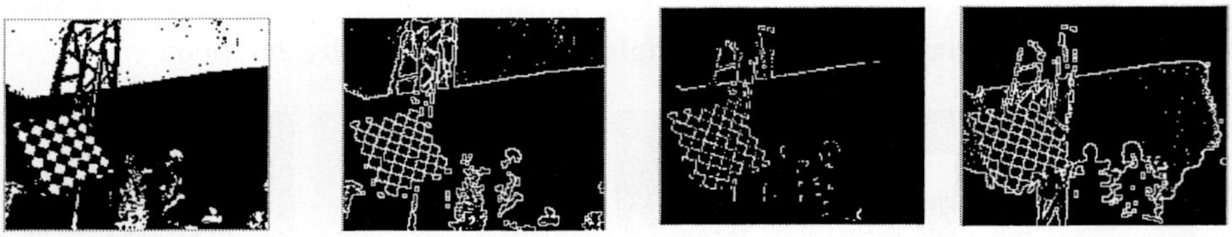

(a) Black blobs (b) White blobs edges (c) Black blobs edges (d) Black & white

Fig. 5. Spatial domain process

4 Multi-band Techniques and Gabor Filters for Texture Segmentation

In this section, the research focus is placed on multi-band techniques in order to overcome the problem existing in spatial processes described in the previous section. Gabor filters are adopted for texture segmentation in the RoboCup domain. As presented in [6], multi-band techniques are based on the following model:

$$f(m,n)=\sum_{q=1}^{Q} f_q(m,n) \tag{1}$$

It assumes that an image $f(m, n)$ can be decomposed into Q components and each component is based on a given frequency band. As we know, Gabor filters are wavelets based narrow band filters. With a bank of Gabor filters, it is possible to isolate components from one another. In some cases, they can be used to detect the direction of target's rotation. Gabor wavelets have following properties:

- Gabor functions achieve the theoretical minimum space-frequency bandwidth product; that is, spatial resolution is maximised for a given bandwidth.
- A narrow-band Gabor function closely approximates an analytic function. Signals convolved with an analytic function are also analytic, allowing separate analysis of the magnitude (envelope) and phase characteristics in the spatial domain.
- The magnitude response of a Gabor function in the frequency domain is well behaved, having no side lobes.
- Gabor functions appear to share many properties with the human visual system.

4.1 Two-Dimensional Gabor Function: Cartesian Form

The Gabor function is extended into two dimensions as follows. In the spatial frequency domain, the Cartesian form is a 2-D Gaussian formed as the product of two 1-D Gaussians:

$$G_C(\omega_x, \omega_y, \omega_{Cx'}, \omega_{Cy'}, \theta, \sigma_x, \sigma_y) = G(\omega_{x'}, \omega_{Cx'}, \sigma_{x'})G(\omega_{y'}, \omega_{Cy'} \sigma_{y'}) \quad (2)$$

where θ is the orientation angle of G_C, $x' = x\cos\theta + y\sin\theta$, and $y' = -x\sin\theta + y\cos\theta$.

In the spatial domain, G_C is separable into two orthogonal 1-D Gabor functions that are respectively aligned to the x' and y' axes:

$$g_C(\omega_x, \omega_y, \omega_{Cx'}, \omega_{Cy'}, \theta, \sigma_x, \sigma_y) = g(\omega_{x'}, \omega_{Cx'}, \sigma_{x'})g(\omega_{y'}, \omega_{Cy'} \sigma_{y'}) \quad (3)$$

An image could be represented as

$$f(x, y) = \sum_{n_x=-\infty}^{\infty} \sum_{n_y}^{\infty} \sum_{m_x=-\infty}^{\infty} \sum_{m_y=-\infty}^{\infty} \beta_{m_x, m_y, n_x, n_y} h_{m\ , m_y, n_x, n_y}(x, y) \quad (4)$$

where $h_{m\ , m_y, n_x, n_y}(x, y) = g(x - n_x X, m_x \Omega_x, \sigma_x) g(y - n_y Y, m_y \Omega_y, \sigma_y)$; σ_x, σ_y, X, Y, Ω_x, Ω_y are constants; and $X\Omega_x = Y\Omega_y = 2\pi$. Assume that the parameters are chosen appropriately, approximation to $\beta_{m_x, m_y, n_x, n_y}$ are obtained by using

$$\hat{\beta}_{m_x, m_y, n_x, n_y} = f(x, y) h_{m\ , m_y, n_x, n_y}(x, y) \approx \beta_{m_x, m_y, n_x, n_y} \quad (5)$$

4.2 Gabor Filtering Results

In this research, a bank of 72 Gabor filters are implemented on images captured by Sony robots. The filters are of 9 frequencies. For each frequency, there are 8 orientation angles. The parameters of the filters can be set for different applications. Figure 6(a) shows the filter property of its real part and the orientation is 0. Figure 6(b) shows anther filter property of its real part. It has higher frequency and the orientation is 90. Figure 6(c) shows the product of Gabor filter.

The responses with only low frequency features remain. By adjusting the frequency and orientation parameters, some results can be used as texture recognition or segmentation. Figure 6(d) shows a suitable frequency with an orientation of 0-degree.

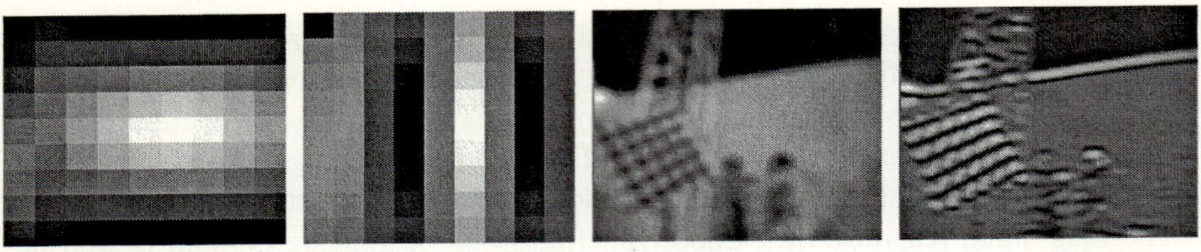

(a) Filter properties (b) Filter properties (c) Gabor Filter (d) Suitable frequencies

Fig. 6. Gabor filtering results

(a) (b) (c) (d) (e)

Fig. 7. Different responses of filters at different orientations

Figure 7 shows different responses by filters with different orientations. Orientation is very important for filtered results. The texture region is stressed greatly in several image results. The next problem is how to find a good result in object recognition without human supervision. It will be discussed in the next section.

5 Image Modulation Models and Multidimensional Energy Separation

The unsupervised image segmentation is very important for computer vision and image understanding, especially for robotic systems. Comparing with colour segmentation, texture based segmentation is much more challenging. Image modulation models are very useful in this area, which may be used to represent a complicated image with spatially varying amplitude and frequency characteristics as a sum of joint amplitude-frequency modulated AM-FM components [4].

5.1 Multidimensional Energy Separation

For any given image *f(n,m)*, there are infinitely many distinct parts of functions $\{\hat{a}(n,m), \nabla\hat{\varphi}(n,m)\}$. As an example, we could interpret the variation in *f(n,m)* exclusively as frequency modulations by setting $a(n,m)=\max|f(n,m)|$ and $\varphi(n,m)=\arccos\left(\frac{f(n,m)}{a(n,m)}\right)$. Teager-Kaiser energy operator (TKEO) could be used to estimate the AM-FM function.

For a 1-D signal *f(n)* the discrete TKEO is defined by following:

$$\Psi[f(n)]=f^2(n)-f(n+1)f(n-1) \tag{6}$$

When applied to a pure cosine signal $f(n)=A\cos(\omega_0 n+\phi)$, the TKEO yields $\Psi(f(k))=A^2\omega_0^2$, a quantity that is proportional to energy required to generate the displacement *f(n)* in a mass-spring harmonic oscillator. For many locally smooth signals such as chirps, damped sinusoids, and human speech formats, the TKEO delivers

$$\Psi(f(n))=\hat{a}^2(n)\dot{\varphi}_e(n) \tag{7}$$

where $\hat{a}(n)$ and $\dot{\varphi}_e(n)$ are good estimates of an intuitively appealing and physically meaningful part of modulations. $\{a(n),\dot{\varphi}(n)\}$ satisfies $f(n)=a(n)\cos[\varphi(n)]$. The quantity $a^2(n)\dot{\varphi}^2(n)$ is known as the Teager energy of the signal *f(n)*. A 2-D discrete TEKO is:

$$\Phi[f(n,m)]=2f^2(n,m)-f(n-1,m)f(n+1,m)-f(n,m-1)f(n,m+1) \tag{8}$$

$$|\hat{U}(n,m)|=\arcsin\sqrt{\frac{\Phi[f(n+1,m)-f(n-1,m)]}{4\Phi[f(n,m)]}} \tag{9}$$

$$|\hat{V}(n,m)|=\arcsin\sqrt{\frac{\Phi[f(n,m+1)-f(n,m-1)]}{4\Phi[f(n,m)]}} \tag{10}$$

$$\hat{a}(n,m)=|\hat{a}(n,m)|=\sqrt{\frac{\Phi[f(n,m)]}{\sin^2\left[|\hat{U}(n,m)|^2\right]\sin^2|\hat{V}(n,m)|}} \tag{11}$$

Fig. 8. An example of TEKO

For a particular pair of modulation functions $\{a(n,m),(n,m)\nabla\varphi\}$ satisfying $f(n,m)=a(n,m)\cos(\varphi(n,m))$, the operator $\Phi[f(n,m)]$ approximates the multi-dimensional Teager energy $a^2(n,m)|\nabla\varphi(n,m)|^2$. For images that are reasonably locally smooth, the modulating functions selected by the 2-D TKEO are generally consistent with intuitive expectations. With the TKEO, the magnitudes of the individual amplitude and frequency modulations can be estimated using the energy separation analysis (ESA), as shown by equations (9), (10) and (11).

Fig. 8 shows the magnitude with the TEKO. The operation is based upon the image in Figure 6(d) and the luminance level was adjusted here for the purpose of presentation.

5.2 Multi-component Demodulation

Multi-component demodulation is based on the following model:

$$f(h,m)=\sum_{q=1}^{Q}a_q(n,m)\cos[\varphi_q(n,m)]=\sum_{q=1}^{Q}f_q(n,m) \tag{12}$$

One popular approach for estimating the modulating functions of the individual components is to pass the image *f(n,m)* or its complex extension *z(n,m)* through a bank of band pass linear Gabor filters mentioned above. Each filter in the filter bank is called a channel. For a given input image, each channel produces a filtered output that is named as the channel response.

Suppose that $h_i(n,m)$ and $H_i(n,m)$ are respectively the unit pulse response and frequency response of a particular filter bank channel. Under mild and realistic assumptions, one may show that, at pixels where the channel response $y_i(n,m)$ is dominated by a particular AM-FM component $f_q(n,m)$, the output of the TKEO is well approximated by

$$\Phi[y_i(n,m)] \approx a_q^2(n,m)\left|\nabla \varphi_q(n,m)\right|^2 \left|H_i[\nabla \varphi_q(n,m)]\right|^2 = \Phi[f_q(n,m)]\left|H_i[\nabla \varphi_q(n,m)]\right|^2 \quad (13)$$

In fact, the results of the filters can be passed into this conversion for the AM magnitude separation. Figure 9 shows the AM modulation result from Figure 15 and Figure 16. No luminance level was adjusted.

(a) (b) (c) (d) (e)

Fig. 9. AM modulation result from Figure 7

For real-time requirement, to select a suitable frequency and orientation of the filter can be based on pre-known data and energy separation. The Sony robot can measure an approximate distance to a target with an infrared range sensor. This means that the frequency and orientation of the target can be estimated by a robot system, which is very helpful for the selection of a suitable filter.

6 Experimental Results

Information extraction from a well-pre-processed image is a simple task. From the image in either Fig. 7 or Fig. 9both edge-based and blob-based methods can work well. Harris algorithm [2][12]can be used to get the corner and edges. In contrast, a blob-based algorithm can obtain the exact position and the number of blobs in a texture pattern, especially for the image in Fig. 9. Then, the pattern can be recognized based on the position of corners or blobs. The method can be described as follows:

Step 1: Find one edge of the shape & rotate it to make the edge parallel with Y-axis.
Step 2: Calculate the distribution of blob or corner positions on X-axis and Y-axis.
Step 3: Recognize the pattern by using the rules as follows:

- If position numbers are equal on both X-axis and Y-axis, it is a square target.
- If position numbers are equal on X and Y respectively, it is a rectangle target.
- If position numbers in the middle of X are bigger than both side, and position numbers of one side of Y are bigger than another side, it is a T-shape target. Otherwise, it is a triangle target.
- If position numbers of one side are bigger than another side on both X and Y axes, which is a L-shape target.

Figure 10(a) shows the edges detected from the image in Figure 7(e). Figure 10(b) shows the result of linearisation. Figure 10(c) shows the result of rotation. In fact, from the result of linearisation, it is clear that the pattern is a square. Ambient methods can avoid problems caused by viewpoints. For complex shapes, complex algorithms are necessary.

It is estimated that for each operation of a Gabor filter, with an image dimension $M \times N$, filter dimension $W \times H$, the number of basic calculations can be up to $(M-W+1)(N-H+1)W^2H^2$. For the experiment above, $M=176$, $N=144$, $W=9$, $H=9$, it costs about 149905278 basic calculation, 0.5 second on a Pentium III 500 PC. If it runs on the robot with a MIPS 4000 100 CPU, it will take about 1.5 second. The cost can be reduced by 64% if reducing the filter dimension to 7x7. For the new version of Sony AIBO robot with a supercore CPU, it is possible to implement the filtering processes at a rate of 6Hz, which becomes necessary for complex environments.

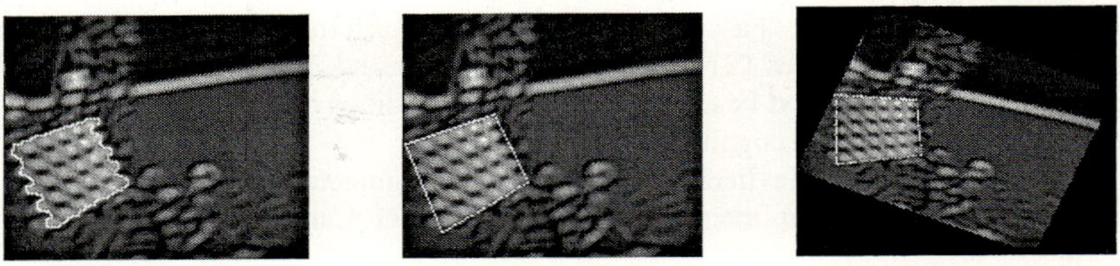

(a) Edge detection **(b)** After linearisation **(c)** After rotation

Fig. 10. Experimental results

7 Conclusions and Future Work

In this paper, we have developed spatial frequency domain algorithms that can recognize shapes in chessboard texture in a cluttered environment. Unlike simple algorithms that require clear background and less interfere; spatial frequency domain algorithms can recognize shapes from a cluttered background. These algorithms have been successfully implemented on a PC. In contrast, simple colour segmentation and blob-based recognition are implemented on real robots to satisfy the real-time requirements. When the background is less cluttered, good recognition results can be achieved reliably.

In our future work, the spatial frequency algorithms will be implemented on real robots for real-time performance, especially object recognition under a complex environment. A leaning system will be integrated in order for the Sony robot to recognize colour objects under complex and cluttered environments.

References

1. C. Bouman and B. Liu, "Multiple resolution segmentation of textured images", IEEE Transactions on Pattern Anal. Machine Intel l., 13 (2), pages 99-113, 1991
2. C.G. Harris and M.J. Stephens. "A combined corner and edge detector", *Proceedings Fourth Alvey Vision Conference, Manchester*, pages 147-151, 1988
3. J. P. Havlicek, A. C. Bovik, and D. Chen, "AM-FM image modelling and Gabor analysis", Visual Information Representation, Communication, and Image Processing, 343–385

4. J.P. Havlicek and A.C. Bovik, "Image Modulation Models", in Handbook of Image and Video Processing, A.C. Bovik, ed., Academic Press, pages 305-316, 2000
5. B. Li, H. Hu and L. Spacek, An Adaptive Colour Segmentation Algorithm for Sony Legged Robots, Proc. 21st IASTED Int. Conf. on Applied Informatics, pp. 126-131, Innsbruck, Austria, 10-13 February 2003
6. B. S. Manjunath, G. M. Haley, W.Y. Ma, "Multiband techniques for texture classification and segmentation" in Handbook of Image and Video Processing, A.C. Bovik, ed., Communications, Networking, and Multimedia Series by Academic Press, pp. 305-316, 2000
7. M. Pietikinen and A. Rosenfeld, "Image segmentation using pyramid node linking", IEEE Trans. on Systems, Man and Cybernetics, SMC-11 (12), pages 822-825, 1981
8. J. Puzicha, T. Hofmann, and J. Buhmann, "Histogram clustering for unsupervised segmentation and image retrieval", Pattern Rec. Letters, Vol. 20, No. 9, pp. 899–909, 1999
9. T. Reed and H. du Buf, "A review of recent texture segmentation and feature extraction techniques", CVGIP: Image Understanding, Vol. 57, No. 3, pages 359–372, 1993
10. Y. Rubner and C. Tomasi, "Texture-based image retrieval without segmentation", Proceedings of ICCV, pages 1018–1024, 1999
11. O. Schwartz and A. Quinn, "Fast and accurate texture-based image segmentation", Proceedings of ICIP96, Vol. 1, pages 121-124, Laussane-Switzerland, 1996
12. J. Shi and C. Tomasi. "Good Features to Track", Proceedings of IEEE Conference on Computer Vision and Pattern Recognition, 1994
13. T. Tangsukson and J.P. Havlicek, "AM-FM image segmentation", Proceedings of IEEE International Conference on Image Processing, Vancouver, Canada, pages 104-107, 2000

An Open Robot Simulator Environment

Toshiyuki Ishimura, Takeshi Kato, Kentaro Oda, and Takeshi Ohashi

Dept. of Artificial Intelligence, Kyushu Institute of Technology
isshi@mickey.ai.kyutech.ac.jp

Abstract. At present, various kinds of robots such as AIBO, ASIMO and etc, are available in public. However, the development of robots is still having some difficulties since of their complexity, continual changes of environments, limitation of resources and etc. To overcome this problem, robot developers often use the simulator that allows to program and test robots' program effectively under ideal environmental conditions where specified various conditions can easily be reproduced. It is still difficult to realize the simulator regardless of its usefulness, because the cost of simulator implementation seems the unexpected cost in the development of robots. As a result, it is need to realize the open robot simulation environment in which any kind of robots can be simulated. This paper focuses on vision-based robot simulation environment and describes a method to construct it. Finally, we implemented a simulator for Robocup Sony 4-Legged League by using this method.

1 Introduction

Nowadays various robots are available in public such as AIBO, ASIMO and etc. However, the development of robots is still having some difficulties because of their complexity, continual changes of environments, limitation of resources and etc. Considering environment changes in vision-based robot, for instance, changes of the lighting condition in a specific environment will affect robot's behavior seriously. To understand the problems of the robot's strategies in the real environment, it needs to check robots strategies in exactly the same environment conditions. Because of in each testing time, sensory values such as camera images and the effectors will change. There are two types of robot simulators in order to solve this problem. One aims to simulate the robot mechanical behavior with accurate robot model data and the other aims to simulate the vision of the robot in order to test robot strategy. The simulator allows developers to program and test robots' program effectively in the ideal environment where specified conditions can easily be reproduced. It is still difficult to realize the simulator regardless of its usefulness since the cost of simulator implementation can be an unexpected cost in robot development. Can overcome this problem by realizing the open robot simulation environment as it can simulate any type of robots.

This paper focuses on vision-based robot simulation environment and describes a method to construct it. The next section describes concept of open robot simulator. Section 3 shows architecture of this method. Section 4 presents an implementation of the environment. Section 5 shows evaluation of the simulator for Robocup Sony 4-Legged League. Finally, section 4 concludes with a discussion of our method.

D. Polani et al. (Eds.): RoboCup 2003, LNAI 3020, pp. 621–627, 2004.

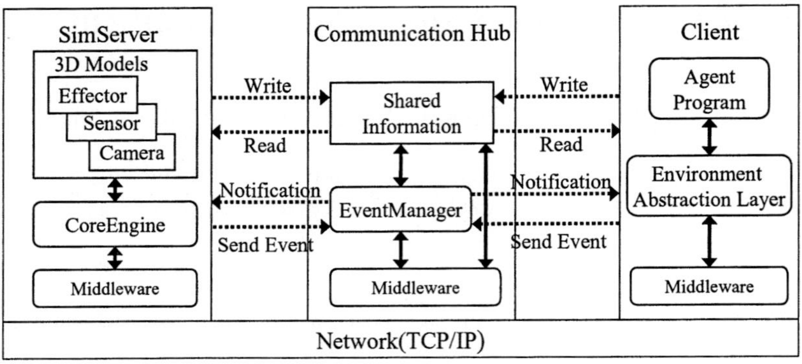

Fig. 1. The Simulation Environment Architecture

2 System Design

The purpose of the method is to improve the development efficiency of the vision-based robots' strategies. To achieve this, we implemented an open robot simulation environment so that different robot developers can introduce different robots to the environment in order to realize the functions of the simulator. The overall policy of the method is as follows.

- Openness is the main focus in the system design stage. With high openness, developers can manipulate the simulation environment easily to treat various kinds of robots. It also allows customizing simulation environment according to the real environment to introduce new robots, new planning strategy, new color modules, new collision detection module, new image processing module, etc
- Reproducibility - Real robot always changes the behavior according to the environmental condition changes such as lighting condition. Since testing of robot strategy consumes number of development cycle consists of coding and debugging, the simulator must support to reproduce a certain situation.
- Distribution in a Network - Since individual robots exist distributed in a physical space, the environment must support distribution in a network.
- Minimize the Requirements - When migrating between real robot and the simulator, it needs to reduce the cost of migration. In our method it is possible, since the robot's strategy program having less modification and its easiness to modify

3 Architecture

Fig 1 illustrates the system architecture, which adopts client/server model and introduces two servers called the simulation server (*SimServer* for short) and the communication hub (*ComHub* for short). In this model, there are multiple heterogeneous robot agents, which are implemented in different program languages and can participate to this environment because clients connect to the server via the network.

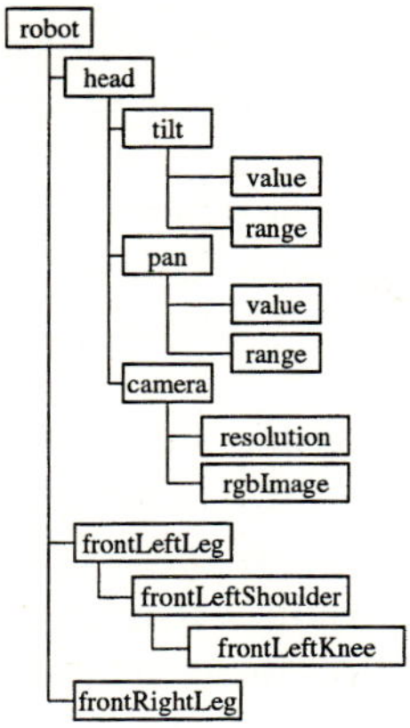

Fig. 2. A Tree Representation Example of SimServer's Objects


```xml
<?xml version="1.0" encoding="UTF-8" ?>
<Environment version="0.1" xmlns="http://www.asura.ac/xethereal">
  <Leaf name="/Robo0/robot/head/tilt/value" type="double" scalar="true">
   <i>32.5</i>
  </Leaf>
 <Leaf name="/Robo0/robot/head/pan/value" type="double" scalar="true">
  <i>11.4</i>
 </Leaf>
 <Leaf name=" /Robo0/robot/leftLeg" type="double" scalar="false" length=3>
  <i>1.5</i> <i>131.4</i> <i>231.1</i>
 </Leaf>
  ....
</Environment>
```


Fig. 3. An Example XML Output of Environment Information

3.1 Simlation Server

The *SimServer* manages all the objects in the simulated environment and synthesizes virtual camera images and simulates the robots' effector. To reduce the cost of realization of the simulator, the system provides a class library of fundamental component which are implemented with opened interface in order to operate easily. The following list show the main part of the class library.

Robot holds cameras, effectors and its position and orientation.

Camera holds parameters such as view angle, resolution and so on, synthesized images by the Core Engine.

Effector keeps current value and the range of values. It also can hold any sub-effector and camera as children.

This library allows users (i.e. robot developers) to construct the virtual robot by few steps; combination of any object according to real robot, customization of object parameter such as camera resolution and the range of effectors, and corresponding to existing 3D shape model data of the robot. The *SimServer* manages all the objects in tree structure and provides name space according to that structure so as to developer can get information user friendly. The robot's strategy program and developers can access all the objects and its all attribute information by that name. Fig 2 shows an example of tree structure of the Sony ERS-210.

Any object in the *SimServer* can be appended and removed in runtime to configure the simulation environment dynamically. For example, when a new robot participates in the simulator, the *SimServer* create new objects according to robot template on demand.

3.2 Communication Hub

The communication hub (*ComHub* for short) holds whole environmental information including from each robot's estimated position in the soccer field to camera images generated through the *SimServer* .

Snapshot. The hub also can take a snapshot of its contents. The snapshot is XML formatted so that programmers can modify and export. (Fig 3 gives an example of the XML output result.)

Dynamic Connection/Disconnection of Clients. This is one reason why we introduce the hub. In development cycle, we frequently kill/run the clients. Using this hub, the system can continue the execution regardless of disconnection because the last state of a robot is still hold even if it was crashed.

Dual Communication. We provide two communication mechanisms - one is synchronous operations: `read()` and `write()` operations to the tree. These operations are fairly simple. To read value on the path name of /robot/head/camera/rgbImage is `read("/robot/head/camera/rgbImage")`. It returns a byte array container. To gain the performance, bulky versions of these operations are provided. The other is an asynchronous event mechanism which supports send and listen event operations. All events must be sent with a string label to address the contents. The event receiver can set a filter to receive interested events and reduce bandwidth by specifying a regular expression. Only events matches the regular expression can be received.

A communication pattern may be used in common is like that 1) a client receives an update event, and 2) issues `read()` operations to get its interested nodes, 3) send an finish read event. 4) After receiving the finish read event, a client who wants to write data start to write the nodes through `write()` operation. After that, 5) it sends an update event. The tree held in the communication hub works as a shared memory. Asynchronous event mechanism can be used as a synchronization mechanism between data producer and consumer.

3.3 Functionality

The following functionalities enhance the simulation environment: plugin user module, a script language, persistence of simulation environment state, communication among the agents and visualization of view frustum.

First, developers can insert fragment of program as plug-in module to the virtual robot in order to customize its behavior. The environment provides interfaces for plug-in modules such as image transformation plug-in, effector plug-in and constraints plug-in. By using this functionality, user can realize strange camera image, reduction of effector speed, etc without time consuming task. For instance, if a simulator should provide the YUV image, the robot developer only prepare a plug-in which transform from RGB to YUV color space. In fact, our team implemented the plug-in described a few codes, since Sony ERS-210 generates YUV camera image. Since every plug-in module is as the *SimServer* 's object, they can be appended / removed and enabled / disabled by scripting at runtime.

Second, to provide interaction between developers and the *SimServer* , a script language likes the S-Expression is introduced. The following enumeration shows typical features of the scripting:

– Getting and setting values of all the objects in the environment
– Loading new objects and plugin at runtime

```
(let
  (set /Ball/loc/x
    (plus /ribo0/loc/x
      (/ (dist /robo0 /robo1) 2)
    )
  )
  (set /Ball/loc/y
    (plus /ribo0/loc/y
      (/ (dist /robo0 /robo1) 2)
    )
  )
)
```

Fig. 4. An Example of script language

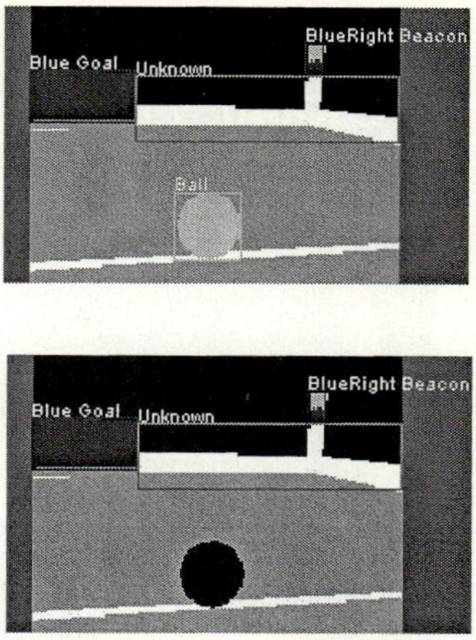

Fig. 5. An Example of image transformation plug-in

- Removing all the objects and plugins at runtime
- Configuration of system setting such lighting condition, frame rate of the simulation
- Easy to introduce new commands

This allows us to test the robot's strategy in the exactly the same environment because the simulator can reproduce it repeatedly. Fig 4 shows an example code to place a ball between two robots (named `robo0` and `robo1`).

Third, the *ComHub* can export its snapshot to a file as described the above. By using this feature, the system can reproduce a certain situation according to the snapshot. This functionality increases reproducibility.

Forth, by using the *ComHub* 's feature that holds whole environmental information as shared information, the system provides communication among the each client. The *ComHub* allows the *SimServer* and every client to put different information to it. When putting information, the *ComHub* notifies this update to the *SimServer* and each client.

Finally, when using the active camera, enhancement of camera motions is important to recognize the virtual environment. The simulator provides functionality to visualize the view frustum that each of the cameras is now seeing. This visualization is useful in order to adjust and tune-up camera motion.

4 Implementation

Initially, we implemented the *SimServer* and the *ComHub* by using Java and Java3D [2]. The *SimServer* is shown in Fig 6 and it consists of four components; global view, command line panel, the tree view, local camera views.

On the global view, user can change his own view by mouse operations. The command line panel allows us to interact with the simulator by the script language described in the above. The tree view shows information about all the objects in the simulation environment.

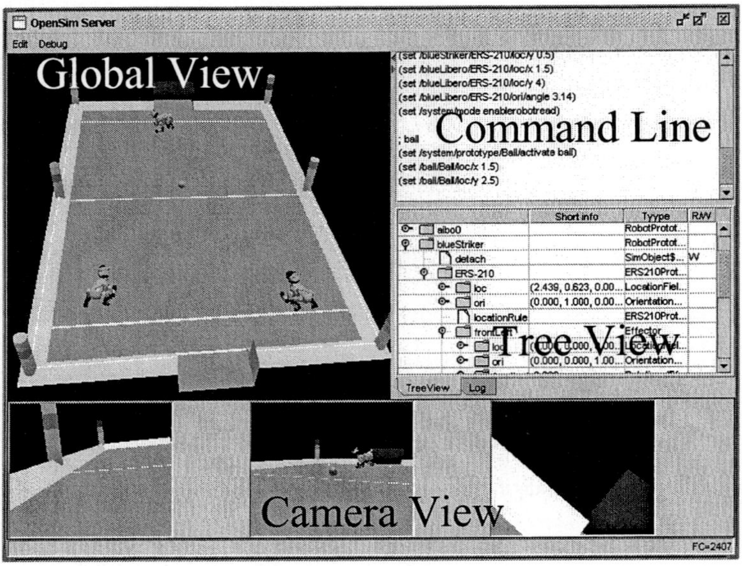

Fig. 6. The overview of the *SimServer*

Second, to evaluate the method, we implemented a simulator for the Robocup Sony 4-Leggued League on the environment. Then we succeed in migration from real robot's strategy program [1] to the simulator with minor modification.

5 Evaluation

We measured the frame rate of the simulator on the PC Intel Pentium 4, 2.8GHz, 1024MB Ram with ATI RADEON 9700 Pro video card and three robot agents connected to the simulator via 100MB LAN. As a result, each client worked at almost 6.2 fps (frame/seconds).

At this time, the main differences arise in the simulator environment against the real field one are that lack of a robot physical model, accurate sensory values, ideal effectors (no slippery walking), no collision to the ball, robots and the field boundaries, noiseless synthesized local vision images, relatively rich computation resources rather than ERS-210, lesser inter-robot communication latency. Most of the factors would be dealt with the introduction of new plug-in modules (like a noisy camera filter). But depending on applications, these differences could break precision of a target simulation and make the results unusable. However, as far as we have tested the simulator from 1 years ago, in our application practice shows its strong effectiveness. The following is the some practical examples.

1. Multi-agents coordination programming: The simulator can accommodate multiple-agents with its inter-agent communication facility. A programmer checked an inter-agent information sharing mechanism, ball occlusion test (Fig 5), a lot of team formation strategy. Through this process, the programmer successfully created a coordination algorithm which applied in a real game.

2. Vision module debugging: A programmer discovered a serious vision code bug which appears very occasional. With the simulator, debugging can be done effectively because of rich development environment.
3. Robustness test: A lot of the audience wears colorful clothes which can easily confuse the robots vision system. To minimize the influence, we introduced a filter which eliminates too high markers. The simulator can create an ideal bad environment in a fraction of time.
4. Education: For aibo programming beginners, we provided the simulation environment as a primary test bed. This accelerates their learning curve. The simulator never hurts real robots so the learner can try new things freely.

6 Conclusion

On the vision-based robot simulator, it is important to reduce the cost of robot strategy programming. To achieve this, we proposed the open robot simulation environment to accommodate any kinds of robots by using standard Java, distribution in a network with TCP/IP, minimized the requirements through simple communication operations, providing rich information through the tree structured external representation, openness with communication hub and plug-in facility in the simulator and rich debugging facility.

In the four-legged league in RoboCup 2002, our system had not been implemented. Now we have a good fundamental to experiment new planning and coordination strategy and so on. At this time, the primary implementation has been done. As a future work, we have to evaluate the system in practical and to introduce another sensor such as omni-directional camera and laser range sensor, etc. On the other hand, the system has been yet considered physical effects in the simulation, nevertheless, we plan to introduce simple method by using collision detection.

The authors would like to express their gratitude to the Information-technology Promotion Agency (IPA), Japan for sponsorship of this project and Professor Takaichi Yoshida for supporting.

References

[1] Kentaro Oda, Takeshi Ohashi, Takeshi Kato, Toshiyuki Ishimura, Yuki Katsumi, The Kyushu United Team in the Four Legged Robot League, in Robocup-2002: Robot Soccer World Cup VI, page.452, 2002

[2] Java 3D(TM) API Home Page. http://java.sun.com/products/java-media/3D/

Application of Parallel Scenario Description for RoboCupRescue Civilian Agent

Kousuke Shinoda[1,2], Itsuki Noda[1,2,3], Masayuki Ohta[1], and Susumu Kunifuji[2]

[1] Cyber Assist Research Center, AIST, 135-0064 Aomi, Koto-ku Tokyo, Japan
[2] Japan Advanced Institute of Science and Technology
923-1292 asahidai, Tatunokuti Nomi Ishikawa, Japan
[3] PRESTO, Japan Science and Technology Corporation

Abstract. We propose a novel agent framework to describe behaviors of the general public in rescue simulations and implement an application for "Risk-Communication for disaster rescur". Conventional agent description languages are designed to model intellectual behaviors of human that solve a task to achieve a single goal. In a disaster situation, however, it is difficult to model civilians' behaviors such as goal-oriented problem-solving. Instead of such a formalization, we introduce the "Parallel Scenario Description" approach that models agents' behavior as an action pattern or plan of situations. We call these "Scenarios". In the proposed framework, behaviors are divided into multiple scenarios for each goal by *Posit* and Posit operator, in which behavior rules are grouped based on situations where the rules are active. The problem solver PS^2 constructs a rule-set of behavior dynamically according to the situation of the environment and the agent's state. The framework is implemented as civilian agents for RoboCupRescue Simulation to adapt to a general civilian simulation. Moreover, we implemented refuge simulation for disaster rescue simulations to realize "Risk-Communication".

1 Introduction

RoboCupRescue Simulator[1, 2] is designed to maximize contribution to the society and attain high throughput in research. We aim to utilize this simulator for "Risk-Communication" of a disaster rescue. That is, it is very important to simulate civilian agents suitably for simulating realistic phenomena in disaster simulation. This is true because furthers that civilian agents comprise the majority of agents in the simulated world and behave for multiple and ambiguous purposes. Moreover, the general public are themselves, our target people.

In this paper, we model the general public as agent suitably for disaster rescue simulator (RoboCupRescue Simulator) and implement some applications. In disaster rescue simulation, civilian are numerous elements: calculation cost becomes an obstacle to scalability. Moreover, human actions are very complicated. One important attribute of autonomous agents is the ability to behave for multiple and ambiguous goals in rescue simulation, including disaster rescue simulation. Therefore, the behavior design of civilian agents is a key issue from the viewpoint of disaster rescue simulation.

D. Polani et al. (Eds.): RoboCup 2003, LNAI 3020, pp. 628–636, 2004.

The "Risk-Communication" is promotion for education about some risk related with various phenomena. We aim to perform "Risk-Communication" using RoboCupRescue simulator in that disaster rescue is important. It is important for applications about disaster rescue to design many civilian agents. So, we implement a civilian agent for RoboCupRescue Simulator and simulate various situations used by these agents.

2 Behavior Description: Parallel Scenario Description

We motivate "Parallel Scenario Description" by describing the difficulty of agent design of a civilian agent in RoboCupRescue Simulation. We introduce an agent framework, that we call "Parallel Scenario Description". This framework has the following features: 'Scenario-based Behavior Description', 'Light-weight Processing', and 'Prototyping Programming'.

Agents' behaviors have various styles from reflective action to deep-thinking decision making accompanied with long-term plans. The agent behavior descriptions are expressed as reactive planning [3] or deliberative planning [4, 5]. However, these planning approaches present some problems. The former has difficult describing long-term planning; the latter presents difficulty in maintaining true expression of an environment and wastes large amount of calculation time in identifying each condition of action rules.

In rescue simulation, on the other hand, human tend to behave as a custom action patterns and plans of a heuristic knowledge to achieve a specific task. It is not easy to decide own action with deep-thinking. In this research, we model human behavior as a collection of such action patterns or plans, and call it a "scenario"; We then behavior description by **scenario-based behavior description**. Scenario-based behavior description is based on scenarios. A "scenario" is an action pattern or plan to achieve a specific task, e.g, "move to a safe location" and "look for refuge". With individual scenarios, designers of agent behavior(scenario-writer) write action rules that are utilized for achieving a specific goal.

We propose *Posit* and PS^2 for an agent framework based on "Parallel Scenarios Description". *Posit* is rule description syntax and PS^2 is problem solver, which accomplishes multiple purposes(scenarios) in parallel, suitable for simulation of civilian. Civilian agent is implemented with *Posit*, Posit Operator and PS^2.

2.1 Posit: Part Of SITuation

Posit is a framework of behavior rule description for PS^2, and a behavior rule set under a specific situation. *Posit* means a piece of situation of agent. "Scenario" is a set of *Posit* and edges, which represent a temporal transition between each *Posit*. Figure 1 shows syntax of *Posit*.

Action rules, in this *Posit*, consist of *Condition*, *Activation* and *Action*. These rules are applicable in the specific situation, and are candidates for firing when

```
Posit ::= (defposit PositName Rule* )
 Rule ::= (defrule RuleName  :condition CondiForm
                             :activition   ActivForm
                             :action       ActioForm )


CondiForm ::= ([LogicOP] CondiForm*)|(InSensor ...)
ActivForm ::= CalcForm
ActioForm ::= ([LogicOP] ActioForm*)|(OutSensor ...)
  LogicOP ::= and | or | not | progn
 InSensor ::= ⟨input sensor name with prefix "?"⟩
OutSensor ::= ⟨output sensor name with prefix "!"⟩
 CalcForm ::= ⟨number⟩|Var|(CalcOP CalcForm CalcForm)
   CalcOP ::=  + | − | ∗ | / |%
      Var ::= ⟨variable name with prefix "*"⟩
```

Fig. 1. Syntax of Posit

the *Condition* is satisfied. *Activation* shows a value function, which is expressed as a numerical value or formula of *Action.* This *Activation* value is utilized as a value for conflict resolution in decision-making. Finally, the agent process system executes the *Action*'s rule that is selected by some selection rule[1]. The *Action* part has two types of operations: one is primitive action, e.g., move, search, say, and so on; the other is *Posit* Control operator. The primitive action is used to request an agent to output to external environment mainly. The latter will be able to manage an internal environment. In *Posit*,a word starting with "?" indicates a passive action; the same one with "!" is a positive action, and "*" is variable. These are shown Fig. 1.

2.2 Posit Operator

Posit is a collection of rules, which are usable in a spacific situation. This means that *Posit* is a piece of situation, which agent faces. That is, the current situation of an agent is expressed as a set of *Posit* is active. In this study, we call this set the "current-situation". Posit operator the manages "current-situation" by replacing *Posit* in accordance with the situation and means temporal-transition of the situation of an agent. Scenario-writer writes these operations in *Posit* clearly. An Agent can control its own rule candidates used by this *Posit* operator.

[1] e.g., max selection, roulette selection

Posit controller has two basic operators: one is (add_posit ...), the other is (remove_posit ...). In addition, there is reserved operator; e.g. (transit_posit A B), (start_scenario S), or some other.

2.3 Parallel Scenario and PS^2: Parallel Scenario Problem Solver

An agent's simple behavior with *Posit* can be shown using a state-transition diagram. This state-transition diagram shows the basic routine-work or behavior pattern which accomplishes a single goal or a simple task in the short term. We call its state-transition a "Scenario". An agent behavior represents a collection of *Posit* and transitions among them by Posit operator. It is changed by changing the component of its collection.

PS^2 is an inference engine that infers the manipulate *Posits* and determine behaviors according to Scenarios. The key concept of PS^2 is *situated-ness* evaluation, that is, not to say that rules are evaluated in a cycle; only rules in active Scenarios are evaluated, where activeness of *Posits* are handled explicitly as situational manipulations. In addition, activation of a *Posit* is operated by another *Posit* action slot specifically.

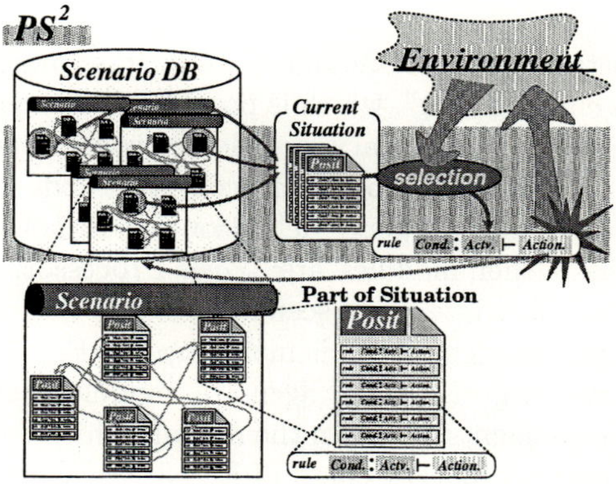

Fig. 2. The processing flow of Decision Making of an Agent's Behavior with PS^2

In PS^2, an entire scenario is stored in "Scenario DB", a form by which a set of active *Posit* is selected into "Current Situation"(CS). Only rules in CS are candidates to be applied in the current situation. Figure 2 shows a flow of information and control in PS^2. A basic cycle to determine one's own action is as follows: Condition parts of all rules in CS are tested for the current environment. Then, for each rule that passes this test, the activation value of these rules is calculated under the current situation. Finally, a rule is selected used by a value of *Activation*, and fire. This basic cycle is repeated until no rule can be applied in the current environment.

3 Application for Social Simulation

3.1 Civilian Agent for RoboCupRescue

Here, we would like to providing explanation of application with our proposed agent architecture: Civilian Agent for RoboCupRescue. Figure 3 shows construction of modules for civilian agent.

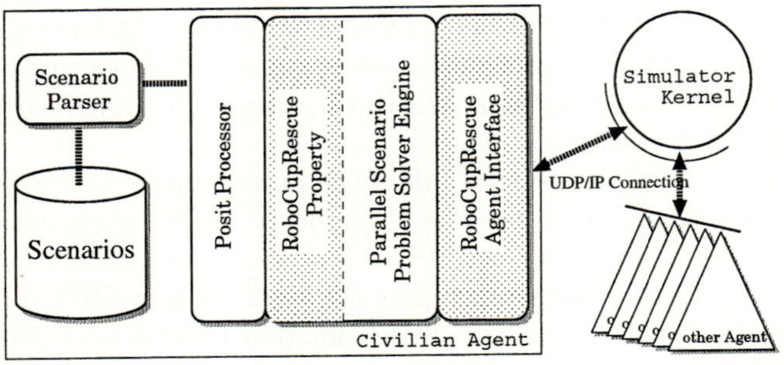

Fig. 3. CivilianAgent Architecture of RoboCupRescue Simulator

The agent consists of several modules. First, the basic level module is "RoboCupRescue Agent Interface", which is provided by RoboCupRescue Simulator as a sample program. It has library modules for connecting with a simulator kernel and a parser of communications protocol, etc. Next, level one is a PS^2. This module needs a specific property definition for RoboCupRescue Simulator to implement this. Other modules are: *Posit* Processor, which manages a current-situation, and which interprets scenarios.

We extended the definition of *Posit*, is mentioned by Fig. 1, to RoboCupRescue Simulator implementing of civilian agent. A module with the necessity of implementing for civilian agents newly, in the module which constitutes an agent is the area, which is covered by gray color, of Fig. 3. We show syntax of *Posit* is the extended definition for RoboCupRescue Simulator, as follows:

3.2 Refuge Action Simulation

We implement a sample application to simulate a refuge action. It is an agent who takes refuge during a disaster. This sample aims at simulating a refuge the general public when an urban disaster is assumed to have occurred; it utilizes RoboCupRescue Simulator as a field of simulation. Concretely, we assume implementation of three kinds of refuge methods: a safe refuge, a guided point refuge, and an instruction absorption refuge. We intend to find from minimum experimentation that refuge time was different by each refuge method. An example of civilian behavior and its *Posit* is as follows:

```
   InSensor ::= ?see | ?hear | ?know
  OutSensor ::= !search | !move | !say
  LocalData ::= (Object :cond CompForm)
     Object ::= building :x :y :id :dist :fieryness :broken
              | refuge :id :dist
              | road :x :y :id :dist :width :brocked
              | humanoid :x :y :id :dist :stamina :hp :damage :buriedness
              | world :time
              | self :x :y :id :position :stamina :hp :damage :buriedness
   CompForm ::= (CompOP CalcForm CalcForm)
     CompOP ::= < | <= | == | >= | >
```

Fig. 4. Definition of Local Data for RoboCupRescue

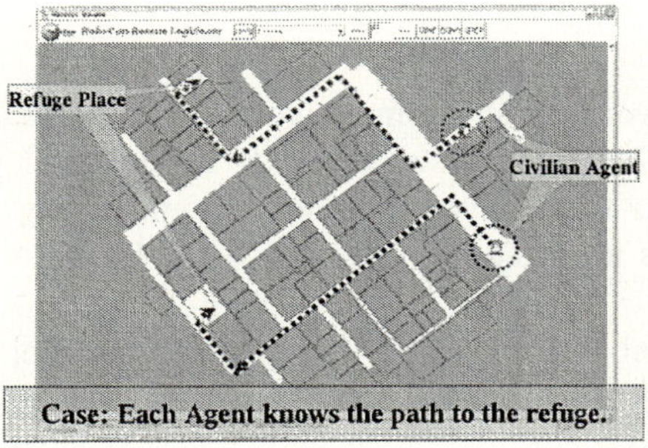

Fig. 5. 'Agent A' and 'Agent B': both civilian knew the surrounding map

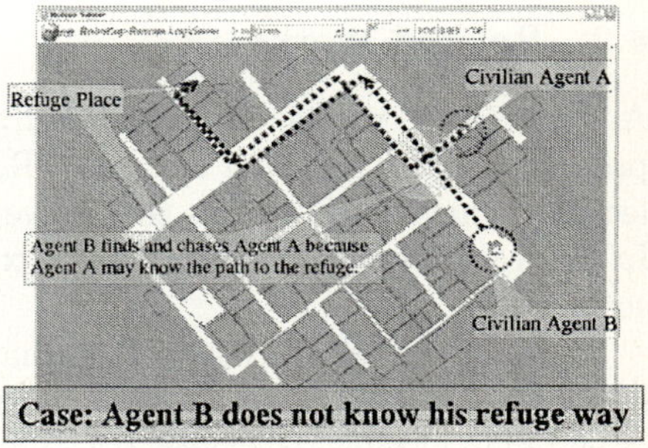

Fig. 6. Agent B: this agent did not know surrounding map, but 'Agent A' knew surrounding map detail, so that 'Agent B' chases 'Agent A' to find the refuge

```
(defposit search_chase_target
  (defrule select_chage_target
    :condition (?see ``people A'')
    :activity 10    :action (!move to: ``position of its people''))
  (defrule s_chase_target
    :condition (and (?know ``people A'')(?see ``his position''))
    :activity 20    :action (!move to: ``his position''))
  (defrule clear_chase_target
    :condition (and (?know ``people A'')(not(?see ``his position'')))
    :activity 20    :action ``lost a target''))
```

Figure 5 & 6 shows the refuge action of an agent. Their difference is their own detailed knowledge of the surrounding map.

4 Discussion

4.1 Contribution of Parallel Scenario Description for a Disaster Rescue Simulation Agent

This paper presents a novel agent architecture for large-scale agent-based simulation. Key contributions include: (i) Situated-ness description, (ii) Scenario-based description, (iii) Parallelism and Multiplicity of ordinary human behavior.

These contributions provide several merits for large-scale agent-based simulation. First, a scenario writer is able to design agent behavior in each situation independently according to the scenario and to define dependence among scenarios later. Second, scenarios can be defined individually, and join agent behavior later. Thus, the scenario writer can complicate agent behavior by degrees as a prototyping programming. Third, our framework selects appropriate rules according to circumstances, so that this provides light-weight processing for agents' processes. Fourth, *Posit* is a set of rules under a specific situations. Scenario-writer reuse a *Posit* in other scenarios.

4.2 Related Works

Among related works, our proposed framework differs from conventional system: production systems, Soar and ACT-R, in that these systems aim to model intellectual behavior of humans that solve tasks to achieve a simple goal effectively. In contrast, our approach aims to express knowledge of human behavior as a scenario.

A reactive system: RAPs[3, 6], may enable application to represent parallel scenarios. In RAPs, a task is deconstructed into several sub-tasks that can be executed in parallel. However, RAPs does not offer flexible mechanisms to resolve conflicts of application of multiple rules like activation in *Posit*. For example, it is hard to implement behaviors for an emergency situation that override all other tasks and behaviors. Compared with this, we can realize such behaviors by adding comprehensive activation values to rules for the emergency.

The scenario-type description of agents' behavior in *Posit* was inspired from the work of Interaction Design Language Q[7, 8]. Language Q can handle multiple scenarios. However, its behavior description is complicated because Q has only a "goto" operator. Figure 7 shows, for example, the difference of creating a behavior combining scenario A and scenario B. In the case of *Posit*, a scenario-writer need only add a new edge between A and B. In the case of Q, however, a scenario-writer sets new scenario is combined A and B, and creates complex state-transition. Thus, when scenario-writer wants to add new scenario, writer needs to consider an existent scenario and add new condition in one. Compared with this, scenario writer can create new additional scenario independently, and add several state-transition edges among existent scenarios and new ones in our framework.

Posit shares the mechanism to handle "situated-ness" with GAEA[9, 10]. Both works treat a complex situation as a collection of atomic situations pieces;

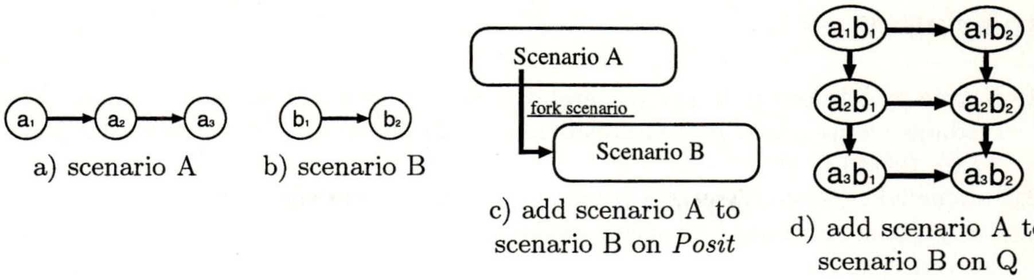

Fig. 7. The difference of rule description flexibility between Q and *Posit*: Agent behavior becomes complex

final behaviors then emerge as composites of atomic behaviors from these situations. However, because GAEA includes full implementation of Prolog, it is too heavy to apply to numerous agents' simulation.

5 Conclusion

This paper proposed a novel agent behavior description and framework using "Parallel Scenarios Description". Agent description languages have been studied intensively, but are not yet proposed as a convenient language used for RoboCupRescue simulation. To utilize knowledge of social scientists, we developed *Posit* (Part Of SITuation) and PS^2 (Parallel Scenarios Problem Solver), a parallel scenario description language, to capture behavior patterns in a specific situation, and to connect edge each small scenario.

We implemented some disaster rescue agent using its behavior model, because our proposed aims to model general public, who has multiple and ambiguous goals, as civilian agent. Concretely, we implemented a civilian agent for RoboCupRescue and a disaster refuge simulation as an application. We intended to use civilian agnet as agent simualtion tools for Risk-Communication, etc.

Moreover, our approach has some demerits. It is not easy to write behavior rules on a text-based interface. Moreover, it is difficult to decide activation values for each rule considering the relation among different *Posit*, and so on. That is, we focus on several point as Feature Work. as following:

- Communicate model base on RoboCupRescue Civilian ACL[11]: To do cooperation work with other rescue agent.
- Another application: for example, Rescue Agent etc.
- Development Environment:To promote of civilian agent development for social scientist, Visual Programming Environment etc..
- Learning Mechanism: To change *Activation* function to acquire appropriate action of civilian agent, and management balance of each scenario.

References

1. Kitano, H.: Robocup rescue : A grand challenge for multi-agent system. In: Proceedings of the Third International Conference on Multi-Agent Systems(ICMAS-2000). (2000) 5–11
2. Rescue HP: (`http://www.r.cs.kobe-u.ac.jp/robocup-rescue/`)
3. Firby, R.J.: Adaptive Execution in Complex Dynamic Worlds. PhD thesis, Yale University (1989)
4. Newell, A.: Unified Theories of Cognition. Harvard University Press (1990)
5. Anderson, J.R., Matessa, M., Lebiere, C.: Act-r: A theory of higher level cognition and its relation to visual attention. HUMAN-COMPUTER INTERACTION **12** (1997) 430–462
6. Firby, R.J.: Task networks for controlling continuous processes. In: the Second International Conference on AI Planning Systems. (1994)
7. Ishida, T., Fukumoto, M.: Interaction design language q : The proposal(in japanese). In: Proceeding of JSAI'01. (2001)
8. Ishida, T.: Q: A scenario description language for interactive(to appear). IEEE Computer (2002)
9. Nakashima, H., Noda, I., Handa, K.: Organic programming language gaea for multi-agents. In: Proceedings of International Conference on Multi-Agent Systems 96, AAAI Press (1996) 236–243
10. Nakashima, H., Noda, I.: Dynamic subsumption architecture for programming intelligent agents. In: Proceedings of International Conference on Multi-Agent Systems 98, AAAI Press (1998) 190–197
11. Noda, I., Takahashi, T., Morita, S., Koto, T., Tadokoro, S.: Language design for rescue agent. In: Proceedings of RoboCup 2001 Symposium. (2001)

RoboCup Advanced 3D Monitor

Carla Penedo, João Pavão, Pedro Nunes, and Luis Custódio

Instituto de Sistemas e Robótica
Instituto Superior Técnico
Av. Rovisco Pais 1, 1049-001 Lisboa, Portugal
{ccfp,jpp,pmgn}@rnl.ist.utl.pt, lmmc@isr.ist.utl.pt

Abstract. RoboCup Advanced 3D Monitor is a three-dimensional application for visualizing and debugging games of the RoboCup Soccer Simulation League. This paper discusses the issues pertaining the implementation of this monitor using OpenGL, a standard API for rendering high-performance 3D graphics. Our application provides a true 3D soccer game experience maintaining a healthy balance of realistic animation features and high-speed rendering achieved by the implementation of specific computer graphics techniques. Besides its main usefulness as a visualization tool, this monitor may be used as a supporting tool for the development of other robotics techniques. To illustrate this, two of such techniques are discussed here: sensor fusion and Markov localization methods.

Keywords: three-dimensional monitor, simulation league, levels of detail, markov localization, sensor fusion.

1 Introduction

The RoboCup *soccer server* simulates a 2D virtual field in which two agent teams play a soccer match. Although the environment is two-dimensional, the *soccer server* has proved to be an adequate platform for the development of realistic 3D visualization tools.

In this paper, the RoboCup Advanced 3D Monitor (RA3DM), a three-dimensional monitor for the RoboCup Simulation League is introduced. RA3DM aims to turn the visualization of simulated soccer games more realistic and entertaining. As we were developing RA3DM we realized that the RoboCup *soccer server* provides an interesting testbed for prototyping algorithms which could be used afterwards on real robots. A Markov localization algorithm and a sensor fusion method were successfully implemented in this environment and the latter has already been adapted to real robots of the middle-size league. Our monitor was extremely useful for developing/testing/debugging the implemented algorithms.

RA3DM is implemented on OpenGL, a low-level graphics library which is designed to be used with C and C++ programming languages. One of the main advantages that OpenGL presents is its independence from operating and windowing systems. This feature enabled the development of versions for the Linux, MacOS X and Windows platforms with only a little extra effort.

D. Polani et al. (Eds.): RoboCup 2003, LNAI 3020, pp. 637–644, 2004.

The remainder of this paper is organized as follows. In the next section the global architecture of the distributed system is briefly described. In Section 3 we present the main features of the implementation of RA3DM. Section 4 explains how RA3DM was used for the application of sensor fusion and Markov localization algorithms in the RoboCup Simulation League. Finally, in Section 5 conclusions are drawn and future work is discussed.

2 System Overview

The RoboCup simulation system is controlled by a central process — the *soccer server* — that runs the whole simulation, sending and receiving all the relevant information to and from all the client programs connected to it [1]. An extra connection was added to enable us to visualize the internal state of a soccer player. The RA3DM connects to a special port on the *soccer player* program to be monitored via a dedicated UDP socket. The global architecture employed in our implementation is illustrated in Fig. 1.

The *soccer server* program sends information to all its connected monitors every 0.1 s. However, the majority of hardware available today (both graphic cards and main processors) is capable of delivering full-screen image update rates (or frame rates) well above 10 Hz (i.e., 10 frames per second). This fact led us to an architecture that maximizes drawing performance by allocating the maximum possible CPU time updating the scene. To accomplish this we use two permanently running threads [2]. One of the threads is responsible for all transactions with the *soccer server* such as sending commands issued by the user and receiving data concerning the state of the simulation. The other thread will update the displayed 3D scene as fast as the underlying hardware allows. This thread is also responsible for managing all user interaction through the use of menus, mouse and keyboard.

3 Implementation

3.1 Player

RA3DM uses a skeletal animation system which allows for three main animations to be performed by the player: walk, run and kick (see Fig. 2). A skeleton is defined by a hierarchy of bones connected by joints with a position and an orientation, arranged in a tree structure. The model, usually a single mesh of

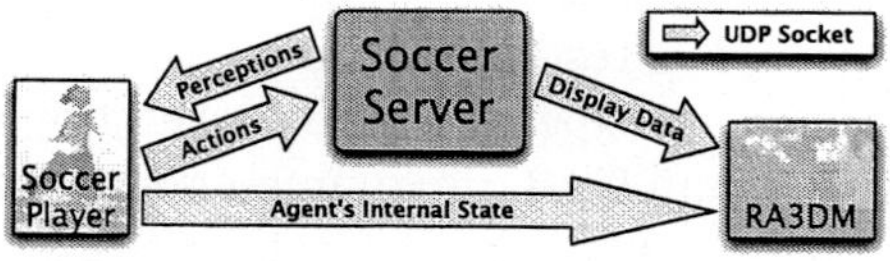

Fig. 1. Global architecture of the distributed system.

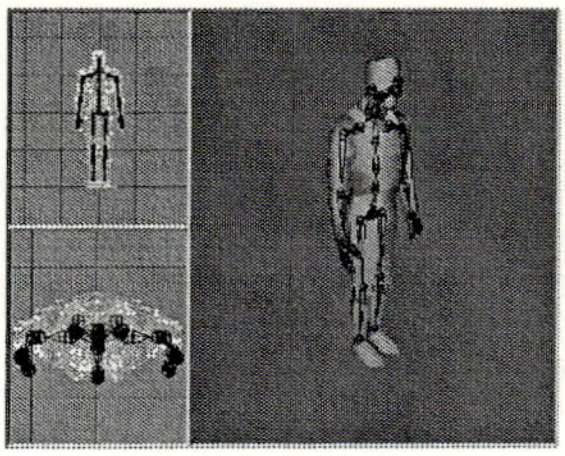

Fig. 2. Player model and its bone structure.

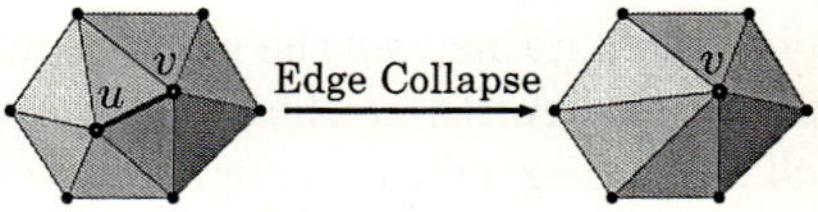

Fig. 3. Collapsing vertex u onto v.

vertices or polygons, is attached to the skeleton through a process known as *skinning*. When the latter is animated, the mesh is deformed accordingly, and therefore, the model is continually morphed to match the movements of the skeleton.

The animation process is simplified by keyframe interpolation techniques. A keyframe animation is created by interpolating frames (snapshots in time) between several successive keyframes that define specific pivot points in the action. Each keyframe contains the values of the rotations of each joint of the skeleton. The mesh is stored only once, along with the skeleton and the keyframes of the animation, resulting in a very compact representation. Each animation is created according to an uniform and constant speed that can be modulated directly in RA3DM to obtain different running or walking velocities.

3.2 Levels of Detail

The computation and storage requirements for complex scenes, typical in some computer graphics applications, far exceeds the capacity of modern hardware. In order to accelerate the rendering of such scenes, approximations of decreasing complexity (levels of detail) are produced similarly to the initial model. The simpler versions contain fewer details that can not be noticed when the object is farther away [3]. Most of the best techniques of polygonal simplification are based on Hughes Hoppe's progressive mesh work [4]. One possible simplification is achieved by continuously applying an *edge collapse* operator that merges two edge vertices into one, thus removing that edge.

Vertex Remove. This operation takes two vertices u and v (the edge $\overline{uv}$) and "moves" or "collapses" one of them onto the other [5]. Figure 3 illustrates a polygon before and after the application of the edge collapse operator. The following steps explain how this operation is implemented:

1. Remove any triangles that have both u and v as vertices (i.e., remove triangles on the edge $\overline{uv}$).
2. Update the remaining triangles that use u as a vertex to use v instead.
3. Remove vertex u.

The removal process is repeated until the target polygon count is reached.

Selection of the Next Vertex to Collapse. When selecting a vertex to collapse, the basic idea is to preserve (as far as possible) the global appearance of an object, trying to cause the smallest visual change to it. Despite the considerable number of algorithms that determine the "minimal cost" vertex to collapse at each step, they are, in general, too elaborate to implement. A simple approach for this selection process may be to just consider the cost of collapsing an edge defined as its length multiplied by a curvature term (the latter having half the weight of the former in this calculation). The curvature term for collapsing an edge $\overline{uv}$ is therefore determined by comparing dot products of face normals in order to find the triangle adjacent to u that faces furthest away from the other triangles that are along $\overline{uv}$. Equation (1) expresses this idea, where T_u is the set of triangles that contain u, $T_{\overline{uv}}$ refers the set of triangles that contain both u and v, and f_N and n_N are the face normals of triangles f and n.

$$\text{cost}(u, v) = \|u - v\| \times \max_{f \in T_u} \left\{ \min_{n \in T_{\overline{uv}}} \left\{ \frac{1 - f_{\mathrm{N}} \cdot n_{\mathrm{N}}}{2} \right\} \right\} \tag{1}$$

The algorithm described throughout this section can be summed up as follows: while the current polygon count is greater than the desired target number, select a candidate edge to collapse (according to its associated cost) and apply the edge collapse operator to it.

Player Simplification. In practice, the polygon simplification algorithm produces very reasonable results. Our soccer player originally contained 347 vertices and 639 triangles which, after the application of this algorithm, were reduced to 68 vertices and 130 triangles. These two versions correspond to the highest and the lowest level of detail used in RA3DM. The former is used when the camera is closer to the player while the other is selected if it is far away. In between we consider intermediate levels obtained by the continuous variation of the total number of vertices used to draw the player. In this way, it is possible to avoid an undesired effect, named *popping*, that can be seen as an abrupt change in the detail present in the object's shape caused by the considerable difference in the number of vertices.

3.3 Cameras

There are three distinct types of views into the field available to the end user: static, user controlled and automatic cameras. A view similar to the display of the official 2D soccer monitor (i.e., aerial view, pointing towards the center of the soccer field) is available, as well as a completely custom camera that the user controls using the mouse and the keyboard. Next we describe two of the most interesting automatic cameras:

TV camera has a fixed position slightly raised on one side of the field and constantly follows the ball with a small delay delivering image sequences similar to those captured by a human camera operator. This camera zooms in and out as the subject being shot approaches and moves away from the camera;

Event-driven cinematic camera consists of a series of cameras placed in some strategic points of the 3D scene. Some of these cameras have a fixed position whereas others are capable of some dynamic motion. At any given point in time during the simulation of a soccer match, each one of these predefined cameras will be assigned a subjective value representing the quality and usefulness of the image sequence provided by it. The measurement of this value and its assignment to the cameras is taken care of by a *director* agent through the use of some heuristics. This agent also decides how to point the cameras to the scene and how much zoom to use. After this evaluation, an *editor* agent in charge of producing the sequence of images seen on screen will decide (based on the values produced by the *director*) which camera to choose and for how long. Different *director* and *editor* agents can be defined. By changing the way the *director* rates the footage and the way the *editor* chooses among them will effectively change the resulting cinematic style, thus giving a totally different experience to the user [6]. For example, if two players are fighting for the ball in the midfield, the director may choose to zoom in one of the closest cameras to make a close-up and give that camera a high priority value. The editor will then be able to use that footage, always taking into account the minimum and maximum time intervals allowed between cuts.

4 Applications

In this section we describe the implementation of sensor fusion and Markov localization algorithms in the framework of the RoboCup Simulation League. Our 3D monitor was a valuable resource allowing the visualization of the player's internal state and the gathering of experimental results. RA3DM includes a low quality viewing mode that was extensively used in order to facilitate the data observations.

4.1 Sensor Fusion

Sharing information among robots increases the effective instantaneous perception of the environment, allowing accurate modeling. Two known sensor fusion approaches were tested in the RoboCup Simulation League: the Stroupe and the Durrant-Whyte methods [7].

To perform sensor fusion the soccer server's noise model is approximated by a two-dimensional Gaussian distribution $N(\mu, C_L)$, where μ is a vector representing the calculated position of the object and C_L is a diagonal matrix that denotes the variance along both axis (see Fig. 4). The variance along the axis that points from the robot towards the observed object (σ_{maj}) is calculated based on the quantization made by the *soccer server*. Thus, σ_{min} represents the variance along the perpendicular axis and is based on the maximum error in angle that an observation can have.

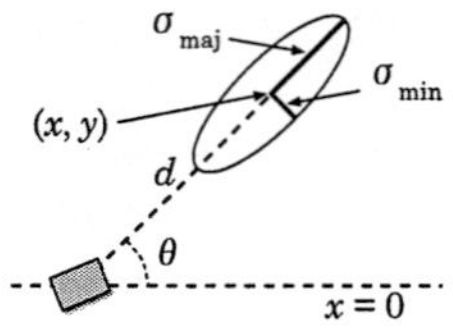

Fig. 4. Distribution parameter definitions: mean (x, y), angle of major axis (θ), major and minor standard deviations $(\sigma_{maj}, \sigma_{min})$ and distance to mean (d).

In order to use sensor fusion we must exchange sensor information between team members. This exchange provides a basis through which individual sensors can cooperate, resolve conflicts or disagreements, or complement each other's view of the environment. In this case, an agent communicates information (μ, C_L and θ) about seen objects to other agents.

Our goal was to compare the efficiency of the two mentioned sensor fusion methods. The first approach simply merges the Gaussian distributions of the observations made by the robot with the ones made by the others. The second method takes into account the last known position of the object and tests if the readings obtained from several sensors are close enough to make the fusion. When this test fails, no fusion is made and the sensor reading with less variance is chosen. The conditions in which this test fails and succeeds are presented in Fig. 5.

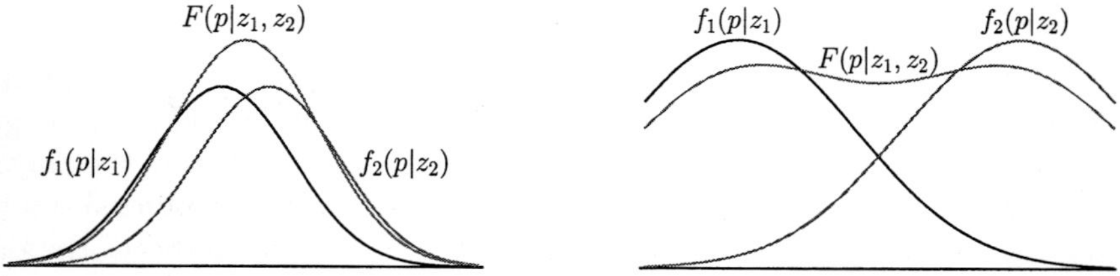

Fig. 5. Two Bayesian observers with joint posterior likelihood indicating agreement and disagreement.

The ball in Fig. 7 (a) with the black and white texture represents the real ball position communicated by the *soccer server*. The farthest ball from the real position represents the position where the goalkeeper, on the other side of the field, sees the ball. The position communicated by a team member to the goalkeeper is represented by the other dark ball. Finally, the white ball represents the fusion between heard and seen ball obtained with the Stroupe method. As can be noticed, the newly calculated position represents a better estimate than the observed value.

4.2 Markov Localization

For mobile robots, localization is the process of updating the pose of a robot, given information about its environment and the history of its sensor readings.

An implementation of the Markov localization method was applied to self-localize a player in the RoboCup Simulation League [8].

At any point in time, Markov localization maintains a position probability density (belief) over the entire configuration space of the robot based on an incoming stream of sensor data (observations) and an outcome of actions. This probability framework employs multi-modal distributions for the robot belief enabling the representation of ambiguous situations by considering multiple hypotheses in parallel.

This particular implementation of Markov localization uses a fine-grained geometric discretization to represent the position of the robot. A position probability grid is three-dimensional, as each possible location l is defined by a tuple $\langle x, y, \theta \rangle$ representing the robot position and orientation. The principle of the position probability grid approach ascribes to each cell of the grid the probability of the robot being located in that cell, denoted by $Bel(L_t = l)$. Figure 6 illustrates the structure of a position probability grid and the correspondence between grid and field positions in the RoboCup simulation system. Each layer of the grid assigns all possible poses of the robot with the same orientation.

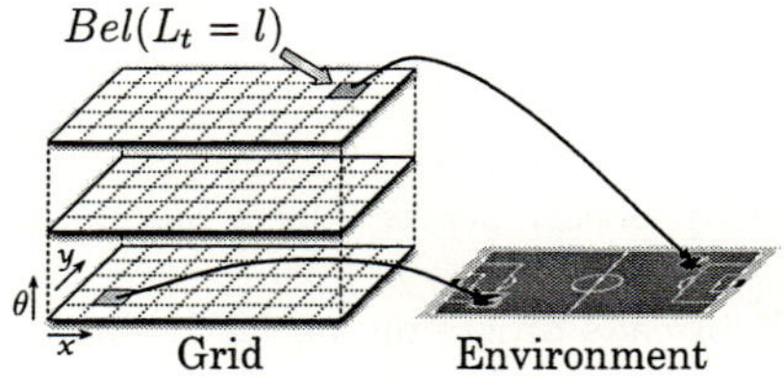

Fig. 6. Transformation of grid coordinates into field coordinates.

When used as a debugger, RA3DM enables us to inspect the internal state of a player (more precisely, its position probability grid) in real time. Figure 7 (b) shows a screen capture of RA3DM while it was monitoring the goalkeeper player. The cells with a belief value above a certain threshold are drawn in a shade of blue and the one with the highest belief at each time step is distinguished with a red color.

Experimental results show that, generally, the Markov localization method keeps the robot position error bound within very reasonable limits. For example, we obtained a medium error of 2.44 m for a grid of cells of 2.63 × 1.70 meters.

5 Conclusions and Future Work

Watching simulated soccer games on a realistic 3D monitor such as RA3DM is far more motivating than on the traditional 2D monitor. We showed that the RA3DM is a powerful tool that can be used as an aid for testing and debugging of prototype applications that may later be employed on real robots. Another possible application that we are considering consists of using RA3DM to simulate

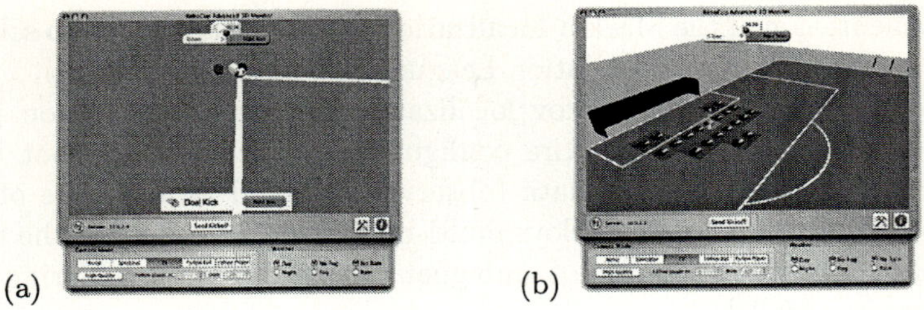

Fig. 7. (a) Ball localization using Stroupe's sensor fusion method. (b) Goalkeeper self localization using Markov estimation.

humanoid stereo vision. This way, the soccer playing agents could receive images supplied by RA3DM instead of the visual perceptions in the form of strings provided by the *soccer server*.

There are some features we would like to add in the future, such as sound effects, the recording of parts of a game to instant replay them later on and the addition of a commentary system.

References

1. Noda, I., *et al.*: RoboCup Soccer Server: Users Manual for Soccer Server Version 7.07 and later. (2001)
2. Barney, B.M.: POSIX threads programming (2001) Lawrence Livermore National Laboratory.
3. Krus, M., Bourdot, P., Guisnel, F., Thibault, G.: Levels of detail & polygonal simplification. ACM Crossroads **3.4** (1997) 13–19
4. Hoppe, H., DeRose, T., Duchamp, T., McDonald, J., Stuetzle, W.: Mesh optimization. Computer Graphics **27** (1993) 19–26
5. Melax, S.: A simple, fast, and effective polygon reduction algorithm. Game Developer Magazine (1998) 44–49
6. Hawkins, B.: Creating an event-driven cinematic camera. Game Developer (2002)
7. Nunes, P., Marcelino, P., Lima, P., Ribeiro, M.I.: Improving object localization through sensor fusion applied to soccer robots. In: Proc. of the Robótica 2003 Portuguese Scientific Meeting, Lisbon (2003) 51–58
8. Penedo, C., Pavão, J., Lima, P., Ribeiro, M.I.: Markov localization in the RoboCup Simulation League. In: Proc. of the Robótica 2003 Portuguese Scientific Meeting, Lisbon (2003) 13–20

RoboCup Rescue Simulation: Methodologies Tools and Evaluation for Practical Applications

Alessandro Farinelli, Giorgio Grisetti, Luca Iocchi,
Sergio Lo Cascio, and Daniele Nardi

Dipartimento di Informatica e Sistemistica
Università di Roma "La Sapienza"
Via Salaria 113, 00198 Roma, Italy
{farinelli,grisetti,iocchi,lo.cascio,nardi}@dis.uniroma1.it

Abstract. The activities of search and rescue of victims in large-scale disasters are not only highly relevant social problems, but pose several challenges from a scientific standpoint. In this context, the RoboCup-Rescue project focused on the problems of bringing aids immediately after a large disaster, and aims at creating system based on AI and Robotics technologies, where heterogeneous agents (software, robots, human beings) interact in a cooperative manner.

In this paper we present the achievements of a research project, based on the RoboCup Rescue simulator, carried out in Italy in collaboration with the Italian Fire Department. The overall project goal is to devise tools to allow monitoring and supporting decisions which are needed in a real-time rescue operation in a large scale disaster, and to provide a methodology for evaluation of multi-agent system which considers not only the efficiency of a system, but also its robustness when conditions in the environment change, as well as other features, such as the ability to acquire a precise and coherent representation of the disaster scenario.

1 Introduction

Search and rescue of victims in large-scale disasters is a very relevant social problem, and pose several challenges from a scientific standpoint. When earthquakes, eruptions or floods happen, a considerable organizational capability to aid the disaster victims as fast as possible is required. This task is rather difficult since often different secondary disasters (e.g. fires, damages in the transportation and communication systems) connected with the main one, occur, which make the correct execution of a rescue plan a priori decided impossible. Moreover, one of the critical issues for a proper response to emergencies is to have timely and reliable information on the disaster scenario, thus enabling for a more effective use of the resources available for the rescue operations [3].

In the recent past significant research initiatives have been undertaken in Japan [5], in the USA [3], and also in Italy [4], that specifically focus on the problem of developing software tools to support the management of this kind of emergency and, more specifically, to design a support system for search and

D. Polani et al. (Eds.): RoboCup 2003, LNAI 3020, pp. 645–653, 2004.

rescue operations in large-scale disasters. These tools are intended both for on-line support during the actual operations as well as for previsional analysis and training.

In this paper we present the achievements of a research project [4], based on the RoboCup Rescue simulator, carried out in Italy in collaboration with the Italian Fire Department and the University of Geneva and Politecnico di Milano. The overall goal of the project was to devise tools to allow monitoring and supporting decisions which are needed in a real-time rescue operation in a large scale disaster.

The activities needed to reach the project's goals were: (i) the creation of a simulation environment, (ii) the design of a framework for Cognitive Agent Development and (iii) a definition of a evaluation methodology. Since in this paper we want to focus more on the evaluation aspects for a MAS in the rescue domain than on the agent development, the agent development tools are just sketched, in order to give a basic for understanding the experimental settings. The first two issues are described in the next section, the evaluation methodology is outlined in section 3.

2 Project Description

In this section we provide a general overview of the project "Real-time planning and monitoring for search and rescue operations in large-scale disasters"[1], that has been developed in collaboration with other Italian Universities (University of Genova and Politecnico di Milano) and the VVF (Italian Fire Department) [4].

The goal of this project has been to develop a prototype tool, based on the RoboCup Rescue simulator, to allow monitoring and supporting decisions which are needed in a real-time rescue operation in a large scale disaster, by integrating competences and tools already available to the VVF and using as a case-study event the Marche and Umbria earthquake in Fall 1997.

The development of a system with the desired characteristics requires to integrate in an effective way, the three following main components: (1) modeling of events related, in a direct way or not, to the disaster; (2) acquiring and integrating data coming from different heterogeneous sources; (3) modeling/monitoring/planning the resources used in the intervention. The simulator developed in the framework of RoboCup-Rescue considers simultaneously all of the three elements above-mentioned, offering an environment for experimentation, which provides completely innovative characteristics compared to those of the existing applications related to this field. For these reasons the RoboCup Rescue simulator has been chosen as a basic tool in the project development for monitoring and planning rescue operations.

In order to ground our project in a real scenario we have chosen to access the data of the Umbria and Marche earthquake (1997), and in particular of the city of Foligno, that is one of the most important cities in that region. Foligno

[1] Funded under the program Agenzia 2000 of the Italian Consiglio Nazionale delle Ricerche.

is located in a flat region of eastern Umbria. Its urban structure is characterized by a medieval center surrounded by more recent suburbs; in particular we focused our attention on an area of about 1 km^2 in the city center. In the area under consideration there are no high-rise buildings; most recent structures are mid-rise, in the four- to nine-story range. All large, multi-story buildings were constructed of reinforced concrete. Oldest buildings were mainly constructed of rubble-work, whereas only few structures are steel frame buildings or wood buildings. There are no industrial structures; most buildings are housing estates having variously-shaped plants. The road network is quite irregular, with not very large roads and narrow alleys (see Figure 1).

Moreover, by considering and analyzing the structures and the strategies currently adopted by the Fire Department (VVF), we have acquired not only a significant body of expertise on the rescue operations, but also the opportunity to set up a prototype experimental setting to show the results of the project.

In order to reach the project goal, two main operations have been performed:

- the set up of an initial environment situation, simulating a disaster in Foligno,
- the definition of multi-agent systems acting in the simulated world performing rescue operations.

In the following we shortly describe two tools that address the above issues, while in Section 3 we focus on the experiments that has been preformed with the systems created on those tools. Such tools have to be considered in this context, mainly for the support given in the rapid development of the MAS systems on which the experiments have been run, and for a better understanding of the settings.

GIS Editor. For running the RoboCup Rescue simulator on a new scenario, a map of the chosen simulation environment in a format suitable for the simulator is needed, moreover some other parameters, like the fire ignition points, or the earthquake magnitudo level, has to be specified. In order to match the above requirements we realized a graphical tool, for generating RoboCup Rescue configuration files: GIS Editor, a screenshot of which is shown in Figure 1.

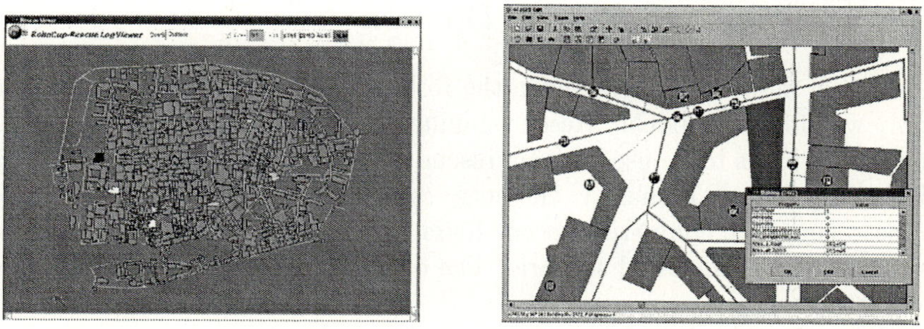

Fig. 1. The map of Foligno(right). A GIS editor screenshot(left)

Cognitive Agent Development Kit. In order to reach good performances in the post-earthquake disaster situation agents needs to exhibit both planning and cooperation capabilities, since the abilities of a single individual agent are often not enough for fighting an expanding disaster. Another item to be considered while developing a team of rescue agents is the need of integrating partial and noisy information coming from the agents, in order to assess a global situation, on which to perform the resource allocation. The Cognitive Agent Development Kit (CADK) is a tool that allows for the rapid development of a team of agents that perform rescue operations in the simulated world. Moreover the CADK provides some debugging facilities for run-time inspecting the internal agent state and the communication state. We used the CADK for experimenting different combination of methods for approaching the above problems in a structured way.

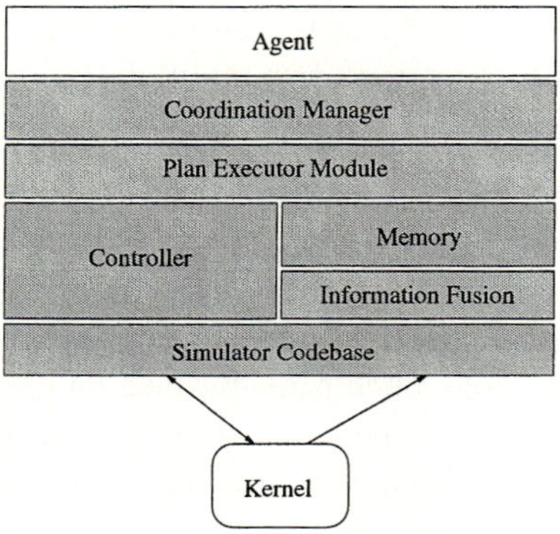

Fig. 2. The Cognitive Agent Development Kit architecture

The architecture of an agent built on the CADK is sketched in Figure 2. The CADK itself is built on the RoboCup Rescue Agent Development Kit [1], by adding coordination, planning and information integration layers.

3 Experimental Evaluation of Multi-agent Systems in the RoboCup Rescue Domain

Evaluation of multi-agent systems in the RoboCup Rescue domain is important not only within the RoboCup Rescue simulation competitions, but also for evaluating actual plans to be used during rescue emergencies. Evaluation of MAS in the RoboCup Rescue domain is currently carried out within international contests, by rating each competing rescue team with a score representing the result of its activity in a simulated scenario. The one with the highest measured score is the contest winner.

In the real world, however, events not always develop in a known and predictable way, since unexpected charges in the operative conditions and failures can occur at every time. The evaluation rule used in RoboCup Rescue simulation competitions is applied in a standard fixed situation and does not take into account the ability of a multi-agent system to work under troublesome situations and its ability to adapt to non-standard operative conditions.

The evaluation method proposed here is based on [2] and allows the analysis of a rescue team in a more realistic way, by analyzing the performance of a multi-agent system in terms of efficiency under normal conditions, as well as in terms of reliability and robustness under changing working conditions.

3.1 Experimental Setting

To acquire a measure of the reliability and the robustness of a MAS, a series of simulations have been executed under changing operative conditions. These tests give a measure of the system adaptability to unexpected situations. The changing parameters that we have considered are: (i)perception radius, (ii)number of agents and (iii) errors in the communication system. Each parameter characterizes a particular series of simulations, referred respectively as the visibility test, the disabled agents test and the noisy communications test.

The Visibility Test. In outdoor environments, visibility conditions are extremely variable. Rescue operation can be needed every time of the day, also in the night. Thus, it is necessary to probe the activity of a system also in these situations. The visibility test is performed by executing five simulations, each with decreasing perception radius, modeling activities under different visibility conditions (i.e. twilight, night time, fog). In this test the changing conditions are on the perception range of each agent, that is 30 meters under normal conditions, and we have performed experiments for the same multi-agent system also using 20, 10, 5 and 3 meters of perception range.

The Disabled Agents Test. In a real emergency situation, it can happen that an agent suddenly become not operational for some reason (for example a mechanical failure of its vehicle or its equipment); this test analyzes the reactions of a system against new operative conditions in which some of the operative agents are disabled.

The disabled agents test is composed of five simulations: in the normal conditions all the agents are active, for the other simulations one, two, three, and four of the *best* agents for each force are disabled. The choice of the best agents to disable is based on the number of tasks performed: for each force, the agent that has completed more tasks in shorter time will be disabled.

The Noisy Communication Test. Agent cooperation in the rescue domain is mainly attained by radio communications among coordination centers and between a coordination center and the operative agents. In real conditions communication transmissions are not free from network failures, or human misunderstandings. This test verifies the robustness of an analyzed multi-agent system by introducing errors in the communication channel, thus preventing messages to reach their destination. The noisy communication test is composed of five phases: under normal conditions there are no errors in the communication channel, while in the other simulations 1/10, 1/3, 1/2, and 9/10 of the sent messages are lost.

3.2 Performance Measures

The performance of a rescue multi-agent system is measured in terms of efficiency and reliability. The **efficiency** is directly evaluated by the formula used in RoboCup-Rescue tournaments, which is:

$$V = (P + S/S_0) * \sqrt{B/B_0}$$

where P is the number of living agents, S is the remaining hit points (health level) of all agents, S_0 is the total hit points of all agents at initial, B is the area of houses that are not burnt and B_0 is the total area at initial; the higher the value of V for a rescue system, the better the results of the rescue operation.

The **reliability** describes how much system efficiency is affected by the worsening of operative conditions, and how much it depends on the values V assumed in the simulation sequence of a single test. It is evaluated with the linear regression slope formula:

$$LRS = \frac{\sum_{i=0}^{N-1}(x_i - x_m) * (y_i - y_m)}{\sum_{i=0}^{N-1}(x_i - x_m)^2}$$

where (x_i, y_i) are the coordinates of a point in a Cartesian system, (x_m, y_m) the average values of these coordinates, N the number of points considered.

Since each point of the graph represents the value of V obtained in the i-th phase, this formula is applied for evaluating the reliability of a system, by using $x_i = i$ and $y_i = V(i)$. For example, if in the visibility test V assumes values described in the following table:

Simulation	0	1	2	3	4
Visibility	30m	20m	10m	5m	3m
V	23.3	12.6	17.1	10.7	11.0

the efficiency is 23.3, while the reliability, calculated by the LRS formula, is -2.65 . It is reasonable to have a negative value of the LRS, since usually the effectiveness of the agents decreases with the worsening of the operative conditions. A little absolute value of the LRS means a good degree of reliability of the system to adverse situations.

3.3 Performance Comparison

The values presented in the previous section are of little significance if not compared to the measures obtained from simulations of other rescue systems. The performance comparison allows to establish the effectiveness of a new technique over the previous ones or over the state-of-the-art.

In the following section it is shown as an example the performance evaluation executed on four different rescue-systems created with the ADK tool. The analyzed MAS are distinguished for the information integration and resource allocation techniques employed, as shown in the following table:

	Simple fusion integration	No fusion integration
Static resource allocation	MAS1	MAS2
Dynamic resource allocation	MAS3	MAS4

For these four multi-agent systems, the controlled experimentations are executed, providing the results shown in the following tables:

Visibility test	Effic.	Rel.
MAS1	23.3	-2.65
MAS2	18.8	-2.06
MAS3	18.8	-1.99
MAS4	11.6	-1.67

Disabled agents	Effic.	Rel.
MAS1	23.3	-6.18
MAS2	18.8	-5.23
MAS3	18.8	-5.09
MAS4	11.6	-4.88

Noisy communic.	Effic.	Rel.
MAS1	23.3	-3.18
MAS2	18.8	-3.95
MAS3	18.8	-4.04
MAS4	11.6	-2.79

In each test there is a rescue system which gets the best value about efficiency and another one which is better in reliability. Rarely in these tests the same rescue system is the best for the two measures, since usually more sophisticated techniques means less robustness to nonstandard operative conditions.

To get some more meanings from the previous results, a graphical representation is presented, sorting the results of each test both by Efficiency and by Reliability, as shown in the following picture.

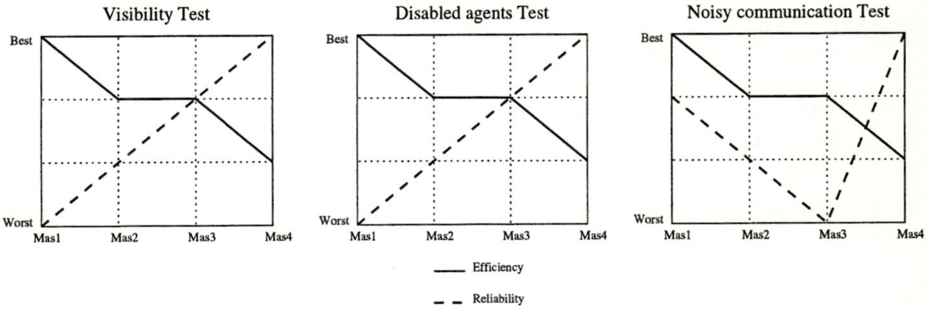

Fig. 3. Performance comparison

At last, it is not easy to identify which system has the best overall performance. In the first test, the visibility one, MAS1 is the best one in terms of efficiency, but it gets the worst rating about reliability. MAS2 and MAS3 have the same efficiency value, and are jointly ranked in the second place. MAS4, which is the worst system in terms of efficiency, it's the best one with respect to reliability. Looking at the graph, it can also be seen that MAS3 offers the best compromise between efficiency and reliability, since it is second in both of the two measures. In the noisy communication test, MAS1, which has the best efficiency value, is also a good system in terms of reliability, ranking in the second place; in this case, it seems to be better than the other ones.

This example shows that the choice of the best system is hard to cast in absolute terms. Depending on the application of the evaluated systems, it should be selected the system which offers the best score with respect to efficiency, reliability, or to a (weighted) combination of the two, but this is still an open problem.

Currently, in RoboCup Rescue tournaments, competing teams are evaluated for their activities in standard situations, thus measuring only their efficiency. It should be pointed out that rescue teams in real life barely operate in normal conditions, and official tournaments should also evaluate the ability of a rescue system to face unattended situations.

4 Conclusions

We have presented the current development of the project "Real-time planning and monitoring for search and rescue operations in large-scale disasters". The aim of this research has been to develop a tool to support search and rescue operations in large scale disasters. In particular, we built our system upon the RoboCup Rescue Simulator, that provides an environment for experimentation of multi-agent technology in the framework of the RoboCup initiative. We have addressed a specific application domain: the disaster scenario recorded after the earthquake of Umbria and Marche. The availability of the RoboCup simulator has been extremely valuable for the development of the present project, providing an experimental setting that can be effectively used for developing a prototype implementation.

The main results of the project has been:

1. a set of tools for improving the development of rescue systems based on the RoboCup Rescue simulator, namely: the GIS Editor and the Cognitive Agent Development Kit.
2. the definition of an evaluation system for MAS in order to analyze different rescue strategies

The research developed in this project has provided a significant use of agent technology in the design of tools supporting the acquisition of information as well as the planning of activities when there is the need to act immediately with partial information about the situation, as in a typical emergency scenario.

Evaluating the activity of a MAS is an hard task since it involves the analysis of several parameters, and the system evolves in unpredictable ways; moreover the rescue operations has to be performed in situations in which the agent adaptability to different environment settings is required. The proposed evaluation system allows for an easy comparison of different MAS, providing an easy understandable graphical representation of the performances, and considers the robustness a key factor as well as the agent efficiency under normal conditions for the overall performance measure.

While the prototype developed has been designed for demonstration and not intended for actual operation, it shows the potential benefits of an integrated approach to the simulation and monitoring of a real search and rescue scenario. While it is premature to consider the effectiveness of the tool in the management of operation, both the analysis of past scenarios as well as the training of personnel seem to be already suitable for application.

References

1. Micheal Bowling. Robocup rescue: Agent developement kit. *available at official RoboCup Rescue Web Site,*, 2000.
2. G. A. Kaminka, I. Frank, K. Arai, and K. Tanaka-Ishii. Performance competitions as research infrastructure: Large scale comparative studies of multi-agent teams. *Journal of Autonomous Agents and Multi-Agent Systems*, 2002. To appear.
3. J. Llinas. Information fusion for natural and man-made disasters. In *Proc. of 5th International Conference on Information Fusion*, 2002.
4. D. Nardi, A. Biagetti, G. Colombo, L. Iocchi, and R. Zaccaria. Real-time planning and monitoring for search and rescue operations in large-scale disasters. In *Technical Report University "La Sapienza" Rome*, 2002. `http://www.dis.uniroma1.it/~rescue/`
5. S. Tadokoro and et al. The robocup rescue project: a multiagent approach to the disaster mitigation problem. *IEEE International Conference on Robotics and Automation (ICRA00), San Francisco*, 2000.

An Efficient Need-Based Vision System in Variable Illumination Environment of Middle Size RoboCup

Mansour Jamzad and Abolfazal Keighobadi Lamjiri

Sharif University of Technology
Department of Computer Engineering
Azadi Ave. Tehran, Iran
jamzad@sharif.edu, keighoba@ce.sharif.edu

Abstract. One of the main challenges in RoboCup is to maintain a high level of speed and accuracy in decision making and performing actions by the robot players. Although we might be able to use complicated hardware and software on the robots to achieve the desired accuracy, but such systems might not be applicable in real-time RoboCup environment due to their high processing time. This is quite serious for the robots equipped with more than one vision systems.
To reduce the processing time we developed some basic ideas that are inspired by a number of features in the human vision system. These ideas included *efficient need-based vision*, that reduces the number of objects to be detected to a few objects of interest with the minimum needed accuracy, *introduction of static and dynamic regions of interest*, which proposes the most probable areas to search for an object of interest, *an experimentally reliable method for color segmentation in variable illumination situation*, and finally, the *usage of some domain specific knowledge* that is used in detecting and tracking a unique safe point on the ball.
We have implemented these methods on RoboCup environment and satisfactory results were obtained.

Keywords: RoboCup, Object Detection, Need-based Vision, Variable Illumination.

1 Introduction

Human vision system, including its ability to recognize objects is based on a combination of image processing, volitional interpretation of colors and shapes, according to a prior knowledge and beliefs. Some of the main capabilities of the human vision system can be listed as follows:

1. The attention of human vision can go towards a particular object or area in the scene, extracting detailed and precise information about it. For example, in real soccer, during the game, a player has *close-up looks* at the opponent goal when he is about to shoot, but in many other situations he has just rough estimates for objects' locations.

D. Polani et al. (Eds.): RoboCup 2003, LNAI 3020, pp. 654–661, 2004.

However, except for a few systems that consider the problem of visual attention, computer vision systems have to process the entire image (scene) to locate the objects of interest.
2. Usually, human vision is a *Need-based Vision* mechanism; i.e., in order to perform a certain action, we receive the necessary information from our environment.
3. Human vision system can perfectly adjust itself to correctly determining colors in different illumination levels. Designing a machine vision system with the same capability of humans is not trivial. Extensive research on this issue has been done, some of which are listed in the area of color constancy (e.g. [6,7]).
   However, since in RoboCup 2003, the rules allow variable illumination, therefore, in this paper we describe a method for estimating field illumination and color segmentation accordingly.
4. Our vision system uses the information and natural/logical relation of events in the environment and continuously adds it to our knowledge. This process, plays a key role in our fast and efficient interpretation of the scene and understanding of the environment. If we could model and implement such a fast and extendable knowledge-base for the robot, then a robot can correctly recognize objects in similar situations. However, current knowledge-based robot vision systems are far beyond that of humans.

The mobile robots used in RoboCup usually have only one front view CCD camera that has a field of view of about 45 to 90 degrees, or an omni-directional viewing system that can view 360 degree around the robot. Our robots has both viewing systems ([1,3]). Real-time processing of 30 frames per second is a challenging problem for the robots. In order to get a near real-time speed for the vision system, we introduce intelligent methods inspiring a number of the above mentioned features in human vision system. In our method a mobile robot can use its processing power efficiently by extracting only the required information with minimum needed accuracy in different situations.

2 How Can a Robot See Intelligently

In present stage of RoboCup research, a robot vision system may be considered intelligent if it can perform some degree of human intelligence used by a human player during a real soccer game. In the following sections we address a number of these features and propose methods towards exploiting them.

2.1 Static and Dynamic Regions of Interest

We have introduced a fast search mechanism based on an idea of *Perspective Jump Points* [1] for finding objects in the RoboCup field. In our previous work, we used only one pattern of jump points with a perspective distribution as shown in figure 1(a) for finding all objects in the field. These jump points are distributed in such a way that no matter where the ball (i.e. the smallest object) might be located in the field, at least 5 such jump points will be located on the ball. Although it is possible to find a minimum number of such jump points and their distribution, this optimal model would not be applicable because of existing shadows under, labels on, and brightness on top of the ball. So, we experimentally determined the distribution of all maps of jump points (figure 1), which are discussed later on.

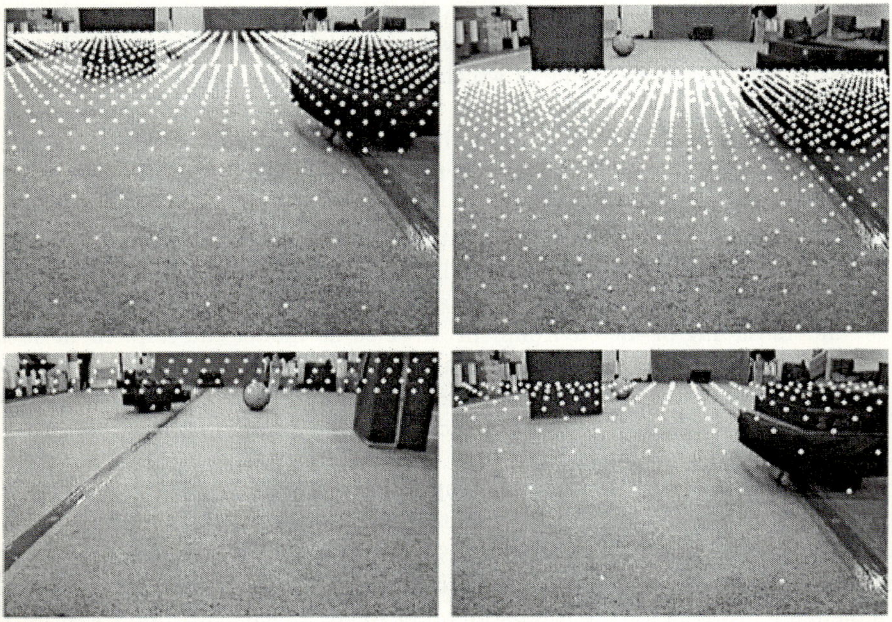

a b
c d

Fig. 1. (a) Map of jump points for the front view CCD camera used for detecting the ball. (b) - (d) Map of jump points used for detecting goal objects, field lines and player robots, respectively.

Since in RoboCup each object has a unique color, by introducing the idea of jump points, we have reduced the problem of color segmentation of the whole image to the problem of examining the color of pixels at jump points followed by a novel idea of region growing [1].

However, in our way towards modeling the efficient need-based features of human vision for a robot, the robot vision system should be able to detect any object of interest in a least possible time. The areas on which we search to find a class of objects is called *Predefined* or *Static Regions of Interest (ROI)*.

Table 1 and 2 show the frequency of time that the omni-vision camera and front view camera needed to use the static and dynamic ROI. For example, for searching the ball by omni-directional view camera, only in 40 % of times we need to use its static ROI that contained 3400 jump points (i.e. figure 1 (a)) but, in the rest (60 %) of time the Dynamic ROI was used to find the ball. Similar results are given for robots and also the field lines for front vision camera as well. The right most column of these tables show the processing time needed for object detection using jump points in static ROI.

We used a SONY VAIO laptop with Intel 800 MHZ CPU as the processor on each robot.

Table 1. Processing time needed for detecting objects using Static ROI in omni-directional viewing system.

Class of JPs	% of Usage	No of JPs	Detection Time(ms)
Ball	40%	3400	16
Robot	30%	500	5
Field Lines	Not Used	2600	11

Table 2. Processing time needed for detecting objects using Static ROI in front viewing system.

Class of JPs	% of Usage	No of JPs	Detection Time(ms)
Ball	20%	1100	160
Robot	10%	220	145
Field Lines	Not Used	1600	172

The idea of ROI, leads us to consider a separate map of jump points for each class of objects such as ball, goals, robots, field lines, etc. For example, if a goal is seen by a robot's front view CCD camera, it always appears at the topmost area of the image plane. Therefore in order to find the goal in this image we only need to check the color of jump points shown in figure 1 (c). Similarly the distribution of jump points for field lines and robots are shown in figure 1 (b) and (d), respectively.

However, we extended the idea of predefined search areas for objects (static ROI), and introduced *Dynamic ROI*. Suppose a robot is moving near the borders of the field or near the corner posts. In these situations, processing of a large sector of the omni-directional image that shows spectators and the background area would be a great waste of processing time. In these situations, Dynamic ROI is determined by using the *localization information* (that enables us to discard areas outside the field) or by *tracking of objects* in the field. As an example of the later case, we restricted search areas in the omni-directional image to just one half of the captured round image for the ball, when ball has been seen far in previous frames and it is most probable to appear in nearby locations. Tables 3 and 4 show a few Dynamic ROIs and the reduction percent in the number of search points obtained in omni-directional and front viewing systems, respectively.

2.2 Need-Based Vision

When we decide to do something and that job requires information from the environment, we start looking at specific objects to satisfy our needs. In addition, the information retrieved from the objects of interest may have low precision at first, and if higher resolution is required, we get that in close-up looks. In most current computer vision systems, usually there are a number of processing layers, each of which provides input data for another layer. The results are available when the topmost layer finishes processing its input data [4].

Table 3. Average decrease in number of search points in omni-directional vision system for Dynamic ROIs.

ROI Type	Usage	Decrease in No of JPs
Only Obstacles on the Moving Direction of Robot	60%	50%
Obstacle Tracking	60%	80%
Ball Within the Angle of its Previous Location	50%	70%
Ball Tracking	40%	95%

Table 4. Average decrease in number of search points in front vision system for Dynamic ROIs.

ROI Type	Usage	Decrease in No of JPs
Obstacle Tracking	50%	70%
Ball Tracking	50%	95%
Goal Tracking	80%	90%

However, inspiring need-based vision in human beings, we have developed a dynamically configurable vision system that is capable of providing just the needed information about the environment, i.e. detecting specific objects of interest, and specific precision of information. These requirements are provided by the playing algorithms of our robot. For example, while dribbling, the robot may need to get only the information about the nearest opponent player(s) in the direction of its movement with the highest possible accuracy.

Determining these sensing needs by playing algorithms of the robot can potentially be quite complex, as long as the robot can have a good degree of inter-agent communication and cooperation mechanisms. In this regard, suppose a teammate gets the possession of the ball; the sensory needs of other players will then change, that is, they do not need to look for the ball.

3 Using Domain Knowledge in Tracking the Ball

In RoboCup, the accuracy of the playing algorithms highly depend on the accuracy of ball detection, i.e. its distance and angle with respect to the robot. In this way, we need to know the (x, y) coordinate of ball's contact point with the field. This contact point is estimated as the middle point of the lower edge of a surrounding rectangle or bounding box around the ball [1].

3.1 Defining a Safe Border Point on the Ball

In practice, due to shadow under the ball and also the bright spots on top of the ball caused by projector illumination sources, in most cases, it is very difficult to find the bottom most and topmost points on the ball with relative accuracy. But in such situations, there is not much noise on the rightmost and leftmost parts of the ball. This means that these two parts can be detected more accurately.

Therefore, we introduced an algorithm for estimating the bounding box around the ball by detecting its leftmost and rightmost positions [2].

3.2 Tracking the Ball

Having determined a unique point on the border of the ball (i.e. the leftmost point on its horizontal diameter) we can compute the speed vector of the ball by detecting the same unique point in consecutive frames. Using this speed vector, we can track the ball by predicting the position of the bounding box around the ball in consecutive frames. As we discussed in section 2, this box is considered to be *dynamic ROI* for the ball. Therefore, to find the ball, we can first examine the area inside dynamic ROI and an area around it in the direction of speed vector. However, if the ball was not found, we can check the jump points of the static ROI of the ball (i.e. figure 1(a)).

4 Color Constancy in Variable Illumination

According to the rule changes of RoboCup 2003, variable illumination in a limited range is allowed[1]. In this situation, we have to consider the luminance of the field according which color segmentation shall be carried on.

4.1 Estimating the Field Illumination Level

In order to measure the field luminance, we mounted a piece of the same green carpet used for the field on top of robot body such that the omni-vision system of our robot can always see that. Figure 2(d) shows one of our robots with a green piece of carpet mounted on the right side of its body.

Since the green color of this piece of carpet will always be seen in fixed position in the image plane, we selected a small rectangle of size 10×30 pixels on the image plane corresponding to this piece of carpet. Field average illumination is estimated to be the average Y (i.e. Y component of YIQ color model [5]) value of all pixels inside the above mentioned rectangle.

However, the reason for using the green color as a reference for measuring the field illumination, is because the whole soccer field is covered with green carpet, and thus green is the dominating color in RoboCup environment.

4.2 Real Time Updating of Scene Illumination

In order to determine the scene illumination in real time, in each loop of our vision algorithm, we calculate the average intensity (i.e. $Y_{average}$) for a set of

[1] From 800 to 1200 Lux.

pixels inside the green color marker. If this intensity is much different for its previously calculated value, a new average value is determined. In this way, the estimation for field illumination is updated upon changes in the environment illumination.

4.3 Using Average Illumination Estimate for Color Segmentation

For color segmentation we used the HSY color model (i.e. components $HandS$ are taken from HSI and Y is taken from YIQ color model [5].) Since the Y axis is perpendicular to HS planes, therefore, if the value of Y is known in an environment, then we can find the range for H, S and Y that best segment the standard colors in RoboCup. However, to give a solution for variable estimation situation, we used two set of lamps on the roof of our RoboCup field. One set of lights were always on, but we controlled the brightness of the other set in seven different levels. At each level, we manually selected a few areas on each standard color in RoboCup (i.e. red, green, yellow, etc) and determined the average values for H, S and Y for that colors. Figures 2 (a),(b) and (c) show these averages in each illumination level for H, S and Y.

Now, during the real games, since our vision system is dynamically updating the filed illumination (i.e. $Y_{average}$), by locating this $Y_{average}$ on the horizontal axis of figure 2, we can approximate the corresponding H, S and Y for each color and segment the image accordingly.

a b
c d

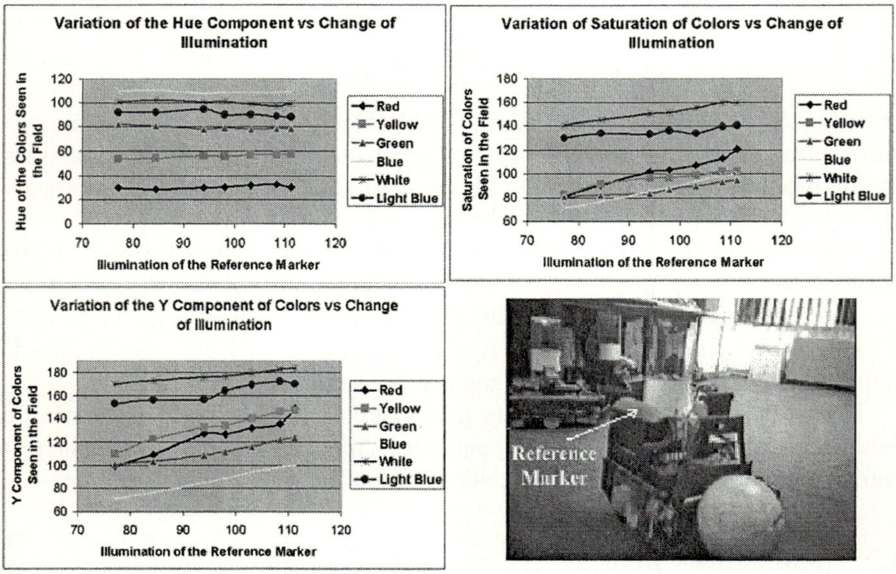

Fig. 2. (a)-(c) The average value of H, S and Y components in different levels of filed illumination for standard colors of RoboCup. (d) One of our robots having a piece of green carpet installed on it as reference marker.

5 Conclusion

Achieving the final goal of RoboCup that is "by year 2050 a team of robots will play against human world soccer champion" needs enormous amount of investment in related research fields, some of which can be named as mechanical design, accurate control, sensor fusion, multi agent cooperative behavior, etc., and especially a fast vision system.

In this paper, we introduced some basic concepts on how to develop an efficient vision system for robots in RoboCup environment. We introduced and implemented ideas of need-based vision, reduction of search areas in images acquired by robot's CCD cameras by means of static (predefined) and dynamic ROI, a color segmentation method in variable illumination. These methods were designed with the goal of reducing the processing time while maintaining a reliable level of accuracy.

Although our approach is in its early stages of making an intelligent robot vision system, but we believe that continuing research in this domain can lead to a level of intelligence that can be compared with that of human vision in a real soccer field.

Acknowledgments

We would like to thank a lot of the students in mechanics, hardware and software, who have worked hard in the past few years to construct the robots of Sharif CE team. In particular we would like to name Reza Zamninasab, Arash Farzan, Mansour Nurani, Mohamad Mahmoodi, Vahid Mirjalili, Reza Ghorbani and Moslem Kazemi. We are very grateful to all our sponsors and also we should give our sincere gratitude to all top ranking managers of Sharif University of Technology whose serious support made this work possible.

References

1. Jamzad M., et al., *A Fast Vision System for Middle Size RoboCup*, RoboCup-2001: Robot Soccer World Cup V, Peter Stone, et al (Eds), Springer, pp.76-85, 2001,
2. Jamzad M., Keighobadi Lamjiri A., *Towards an Intelligent Vision System for Mobile robots in RoboCup Environment*, To appear in Proc. of 2003 IEEE International symposium on Intelligent Control, Houston, TX, USA, Oct. 2003.
3. Jamzad M., et al., *An Omni-directional Vision System for Localization and Object Detection in Middle Size RoboCup*, Proc. of the IASTED International Conference, Applied Modeling and Simulation, Cambridge, MA, USA, pp.401-407, 2003.
4. Draper B., Hanson A., Riseman E., *Knowledge-Directed Vision: Control, Learning, and Integration*, Proc. IEEE Signals and Symbols, 84(11), pp.1625-1637, 1996.
5. Gonzalez R.C., and Woods R.E., *Digital Image Processing*, Addison-Wesley, 1993.
6. Finlayson G., Hordley S., Hubel P., *Color by Correlations: A Simple, Unifying Framework for Color Constancy*, IEEE Trans on Pattern Analysis and Machine Intelligence, Vol.23, pp. 1209-1211, 2001.
7. Funt B.V., Drew M.S., Ho J., *Color constancy from mutual reflection*, Int. Journal of Comp. Vision, Vol.6, pp.5-24, 1991.

Filling the Gap among Coordination, Planning, and Reaction Using a Fuzzy Cognitive Model

Andrea Bonarini, Matteo Matteucci, and Marcello Restelli

Politecnico di Milano Artificial Intelligence and Robotics Lab
Department of Electronics and Information
Politecnico di Milano, Milan, Italy
{bonarini,matteucc,restelli}@elet.polimi.it

Abstract. Coordination, planning, and reactivity are important for successful teams of autonomous robots, in dynamic adversarial domains. In this paper, we propose a fuzzy cognitive model to integrate coordination, planning and reactive behaviors in a team of cooperating robots. In our architecture, behavioral modules are used as high-level macro-actions that compose structured plans defined by a flexible multi-agent coordination system. The use of an unifying cognitive model provides an effective tool for seamless integration of reactive and deliberative components in the robots, gaining as much as possible from the presence of different skills and competencies in the team. The control model is designed to be tuned and adapted on-line so that the team strategies and the role of robots in the control schemata can be automatically modified to face different opponent teams, and changes in robot capabilities.

1 Introduction

Since some years, the architecture of autonomous robots integrates the traditional planning activity, which provides goals for the robot, with behavior-based architecture that implements simple and fast control modules. In designing this kind of hybrid architectures, most of the issues arise from the connection between the abstract and physical level representations used respectively in the deliberative and reactive components of the system [6].

Usual solutions propose an ad-hoc integration, without any high-level cognitive model as a uniform substratum for the overall system. The architecture we are proposing in this paper is aimed at integrating coordination, planning, and reactivity using such a cognitive approach where agents have internal models, intentions, goals, and can still react to unexpected events. This cognitive model is implemented by *fuzzy predicates* that we use to represent *concepts*.

In our architecture, the perception-action loop is performed through an abstraction process that, starting from raw data gathered from sensors, produces the high-level concepts used in planning, coordination, and execution. On the first stage of this process, we have the *Sensing* activity: intelligent sensors (i.e., sensors with some computational capabilities) process the raw data stream into

D. Polani et al. (Eds.): RoboCup 2003, LNAI 3020, pp. 662–669, 2004.

features to be used on the next stage for *Data Interpretation*. For instance, omnidirectional vision systems [2] provide relative positions and distance of colored objects in the Robocup domain.

Data interpretation is obtained using *MAP* (*MAP Anchors Percepts*) [4] to realize sensor fusion, anchoring, and self-localization. MAP is suitable for enhancing the local internal model of each agent in a Multi-Agent System (MAS) also with information coming from other agents obtaining, in such a way, global distributed sensing, and map integration. The results of this data interpretation is a real-valued internal model on which are evaluated basic fuzzy predicates that describe the percepts of the robot in the higher level formalism used for knowledge processing both in deliberative and reactive components.

The use of fuzzy predicates gives the possibility to represent both interpretation of data coming from sensors, and high level abstractions coming from a deliberative process involving planning, coordination and communication. The selection of a fuzzy representation makes it possible to face, with an intrinsically robust model, uncertainty and approximation, unavoidable in applications interfaced with the real world. At the same time, fuzzy predicates can be quite effectively automatically adapted to changing situations. Our robots represent by fuzzy predicates also their intentions and goals, which are in this way integrated with data in a unique, simple, adaptive model.

Coordination among robots is defined using the deliberative component described in the next section. In this activity, we consider also aggregated information about the skills of each robot and their appropriateness in a given situation. These are parameters that can be easily defined and tuned on-line in a relatively short time, and can affect the behavior of the single robot and of the whole team. This results in a team able to adapt on-line to different types of opponents, and to possible performance degradation of its members. On the reactive side of the control loop, we have a behaviors management system that maps fuzzy predicates into actions to be executed by the robot actuators. Actions are weighted considering their applicability, intentions and goals, so that the behavior of a robot depends both on actions proposed by reactive module, desires, and intentions.

In the next section we give a brief description of the deliberative component in charge of coordination and long-term planning in our architecture. Section 3 introduces the behavior management system used in our robotics applications, and in the section that follows we describe how deliberative and reactive component can be easily integrated by mean of the underlying cognitive model that we have previously introduced. Section 5 presents a robotic application in which we have adopted our cognitive approach as a case study.

2 SCARE: The Coordination System

Cooperation holds a very important role in multi-agent system applications. To face the typical issues of these applications, we have implemented *SCARE* (*Scare Coordinates Agents in Robotic Environments*) [5] a general architecture for coordination in multi-robot domains. SCARE is able to deal with:

heterogeneity when a MAS is made up of agents with different skills, our architecture exploits these differences in order to improve the overall performance

communication coordination policy may change according to the amount of information that can be exchanged among agents and according to the network connectivity

adaptation in order to grant the autonomy of the system, the coordination mechanism is able to dynamically modify its parameters in reaction to environment changes

Using SCARE, the MAS application developer has to identify the macro-activities that the agents can carry out: *jobs* and *schemata*. A schema is a complex activity consisting of sequences of jobs that require the collaboration of two or more agents. A job is defined by the goals that should be achieved, and each agent is free to select the procedures to achieve these goals according to its own capabilities.

The job assignment process is carried out by a special kind of agent called *meta-agent* ($\hat{A}$), through two phases: *decision making* and *coordination*. In the *decision making phase*, $\hat{A}$ computes the suitability of an agent for each activity, through the composition of several parameters, all but the second implemented by fuzzy predicates:

cando define when the activity can take part in the assignment process;

attitude define how much the skills of an agent are useful for the activity;

chance define the situation where the agent has good possibilities to succeed in the activity;

utility define the situation where the activity is useful for the agent team;

success define the goal achievement situation for the activity;

failure define the situation where the activity should be stopped because of unrecoverable failure.

An activity terminates when the success or failure condition is verified. If an agent is idle, $\hat{A}$ tries to assign it to some activity. $\hat{A}$ firstly evaluates, for each activity, the cando predicates in order to reject those activities that cannot take place. For each remaining activity, $\hat{A}$ evaluates utility and chance predicates, and the agent's attitude, thus obtaining indicators to take the decision. $\hat{A}$ obtains an ordered list of activities (*agenda*) by solving the optimization problem, and can start the coordination phase.

The *coordination phase* considers that the agents are not working individually and that they must cooperate to achieve the MAS goals. If we simply assign each agent to the most suitable activity according to the decision making phase, it may happen that the assigning process does not satisfy some *coordination constraint*, such as: *cardinality* – for each job the designer sets the minimum and maximum cardinality (i.e., the minimum and maximum number of agents that can be assigned to the job at the same time) – and *schema coverage*, a schema can be assigned only if there is a suitable agent for each functionality.

In this phase, $\hat{A}$ searches for the best job allocation to agents by observing coordination constraints. How this can be achieved depends on the structure of the MAS communication system and is better discussed in [5]; in that paper we also show some configurations which match different qualities and topologies of the communication network.

At the end of this process, each agent is assigned to a job, and new fuzzy predicates are produced by SCARE to be used in the behavior management system.

3 BRIAN: The Behavior Management System

Especially in dynamic environments, the main approach to robot control design is the so called behavior-based architecture, where the robot is controlled by the implicit cooperative activity of behavioral modules. Each module operates on a small subset of the input space implementing a relatively simple mapping from sensorial input to actions.

In our behavior management system *BRIAN* (*Brian Reactively Implements AgeNts*) [3], we face the issue of controlling the interactions among modules by decoupling them with context conditions described in terms of internal state, environmental situation, goals, and communications with other agents. All these concepts are defined using the cognitive model introduced in Section 1, based on fuzzy predicates. Integration and coordination among behavior modules is achieved using two sets of fuzzy predicates associated to each of them: *CANDO* and *WANT* conditions. CANDO conditions are used to decide whether a behavior module is appropriate to the specific situation. For instance, in order to consider to kick a ball into the opponent goal, the agent should have the ball control, and it should be oriented towards the goal.

WANT conditions represent the motivation for an agent to execute a behavior in the present situation. They may come either from the environmental context, or from strategic goals.

The use of these two different sets of conditions allows the designer to design a dynamic network of behavior modules defined by means of context predicates. This is our main improvement with respect to usual behavior-based architectures; we do not have any complex, predefined interaction schema that has to take into account all possible execution contexts.

Since tasks have typically different priorities, it may happen that, in particular situations, some behavioral modules would like to filter the actions proposed by less critical modules. Let us consider, for example, the case of the *AvoidObstacle* behavior; it is implemented by rules which become active when an obstacle is detected on the path followed by the robot, and they produce actions which modify the current trajectory in order to avoid collisions. The problem that we must face is that the actions proposed by the *AvoidObstacle* module are composed with the output of the other behavioral modules, thus possibly producing unexpected and undesirable results. One possible solution consists in disabling any other behavior module while the *AvoidObstacle* is active. This approach achieves the aim of avoiding collisions with other objects, but it has some drawbacks. If the *AvoidObstacle* module is the only active behavior module, it avoids the obstacles by taking a trajectory that is independent from the one that would have been produced by the other behavior modules, thus degrading the performances. Moreover, if we have several levels of priority, with several behavioral modules for each level, this approach turns out to be very awkward.

To overcome the previously described problem, we have adopted a hierarchical organization of the behavior modules Any module at level l receives as input, besides the sensorial data, also the actions proposed by the modules which belong to the level $l-1$. Behaviors (except those at the first level) can predicate on the actions proposed by modules at lower levels, inhibit some actions, and propose other actions. For the i-th behavior of level l we have:

$$B_i^l : I_i \times A^{l-1} \mapsto A_i^l \cup \overline{A}_i^l, \qquad (1)$$

where A^{l-1} is the set of the actions proposed by all the modules at level $l-1$, A_i^l is the set of the action proposed by the behavior, while $\overline{A}_i^l$ is the set of actions that should be inhibited.

With this organization, behavioral modules with higher priority must be placed in the higher levels of the hierarchy, so that they can modify the proposed actions in order to manage urgencies. For example, if we place the *AvoidObstacle* module at the second level, we could write rules of this kind: "*if* I am facing any obstacle that is near *and* I would like to move forward *then* cancel all the actions that propose a forward movement". Whenever the robot has no obstacle in its proximity, the *AvoidObstacle* module passes all the proposed actions unchanged to the following level.

Apparently this design approach may seem in contrast with the behavior-independency principle, but it is actually the opposite. Consider the *KeepBall* behavior, taken again from the RoboCup domain. The goal of this behavior is to hold the ball inside the kicker while the robot is moving. If we think that, when the robot takes the ball, it is of primary importance to maintain the ball possession, we must put the *KeepBall* module at the highest level, so that it can, if necessary, modify the proposed actions as much as it is needed for keeping the ball inside the kicker. In a single level architecture, the most effective way of implementing this *KeepBall* behavior is to embed it in each behavioral module, thus complicating the rules and really breaking the principle of modularity. In our architecture, dependencies among modules are explicit in the hierarchical structure, and the interface is limited to the action proposed by the modules.

4 Embedding Plans in Reactive Control

Several robotic control architectures present both planning and reactive modules. The main difference between these approaches is the interaction between the two modules. In many approaches [1] [7], the planning module simply activates and deactivates behaviors in order to achieve more abstract goals. In our architecture, the execution of plans is completely integrated in the reactive behavior engine.

SCARE interacts with BRIAN sending to it two kinds of predicates: *Coordination predicates*, which influence the enabling conditions, and *Perceptual predicates*, which are used by the parametric behavioral modules. Coordination predicates refer simply to the job that is assigned to the agent. This information allows to activate only those behavior modules that are necessary to the execution of the assigned job. Perceptual predicates are all the information that BRIAN needs to execute the job, and it is not directly available through the world model. At each iteration, SCARE sends both the Coordination and Perceptual predicates, and it monitors the state of execution of the assigned job. When a termination condition (either success or failure) occurs, SCARE starts a new job assignment process, in order to identify the best activity, considering the multi-agent context.

5 Experimental Results

We have successfully employed our architecture in several robotic applications, where we have verified the versatility of the proposed approach. Both in mono and multi-agent systems, we exploited the benefits deriving from the interaction between the coordination, planning, and reactive modules. The use of a conceptual model allows an easy and quick development of the high-level control structures. In the following, in order to show the effectiveness of our method, we describe a case study in the RoboCup context.

RoboCup is a multi-robot domain with both cooperative and adversarial aspects, suited to test on the effectiveness of our approach. In the coordination layer we have defined several jobs (*RecoverWanderingBall*, *BringBall*, *Defense*, etc.) and schemata (*DefensiveDoubling*, *PassageToMate*, *Blocking*, etc.). These activities are executed through the interactions of several behavioral modules (see Figure 1). We have organized our modules in a five levels hierarchy. At the lower level we have put the behavioral modules whose activation is determined also by the information coming from the planning layer. At this level we find both purely reactive modules and parametric modules (the latter can be distinguished by a couple of round brackets after their name). The higher levels contain only purely reactive behavioral modules, whose aim is to manage critical situations (such as avoiding collisions, or leaving the area after ten seconds). The activation of these modules does not depend on the high level activity. In order to give an idea of how the whole system works, we present a couple of examples.

Let us consider the behavior *MarkOpponent*: the goal of this behavior is to obstruct the way between the ball and a certain opponent robot. Once SCARE decides that our robot must mark an opponent, there are several ways in which this can be achieved by BRIAN. We could define the parametric behavioral module *GoBetween*, that takes as input the polar coordinates of two points and has the goal to put the robot in a position along the joining line. SCARE activates this behavior which receives as input the polar coordinates of the ball and of the opponent. Using another approach, SCARE could compute the polar coordinates of a point on the joining line and then activate the *GoToTarget* module.

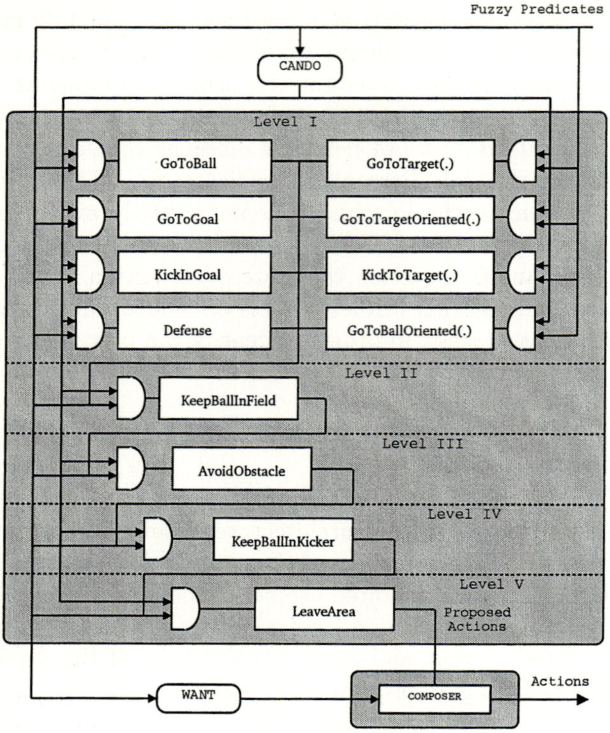

Fig. 1. The behavior configuration used in RoboCup

A further approach, consists in employing a planning algorithm in order to find a free trajectory that leads on a point belonging to the line that joins the ball with the opponent. The trajectory will be followed by the robot through the activation of the *GoToTarget* module with different values of its parameter. These three modalities are ordered by increasing complexity from the SCARE point of view, that implies a simplification for what concerns the behavior modules. In fact, the implementation of the *GoBetween* module is more difficult than the implementation of the *GoToTarget* module. In the third approach it is not even necessary the interaction with the *AvoidObstacle* module, since it is all managed at the planning level. On the other hand, in environments where situations change very quickly, it would be necessary to always replan everything. We have adopted the second approach because, by exploiting the potentiality of both SCARE and BRIAN, it keeps low the complexity of the whole system.

6 Conclusion

Robot control architectures have deeply evolved in the last twenty years, but there is still considerable debate over the optimal role of internal representation.

In this paper we have presented the fuzzy cognitive model we use to integrate in a uniform framework the deliberative and reactive components of multi-agents systems. The cognitive model we propose, integrates coordination, planning and reactive behaviors providing a common cognitive substratum for a team of robots where behavioral modules are used as high-level macro-actions that compose structured plans defined by a flexible multi-agent coordination system.

From the multi-agent system perspective, the use of such unifying cognitive model provides an effective tool for seamless integration of the heterogeneous members in the team, gaining as much as possible from the presence of different skills and competencies in the robots. Moreover, all the elements in the knowledge processing level are based on simple fuzzy predicates that can easily be managed, designed and adapted. Doing it this way, the control model can be easily designed to be tuned and adapted on-line so that the team strategies and the role of robots in the control schemata can be automatically modified to face different opponent teams, and changes in robot performances.

References

1. R.C. Arkin. Towards the unification of navigational planning and reactive control. In *Proc. of the AAAI Spring Symposium on Robot Navigation*, pages 1–5, 1989.
2. A. Bonarini. The body, the mind or the eye, first? In M. Asada M. Veloso, E. Pagello, editor, *RoboCup 98–Robot Soccer World Cup III*, pages 695–698, Berlin, D, 2000. Springer Verlag.
3. A. Bonarini, G. Invernizzi, T.H. Labella, and M. Matteucci. An architecture to co-ordinate fuzzy behaviors to control an autonomous robot. *Fuzzy Sets and Systems*, 134(1):101–115, 2002.
4. A. Bonarini, M. Matteucci, and M. Restelli. A framework for robust sensing in multi-agent systems. In *RoboCup 2001 - Robot Soccer World Cup V*, Berlin, D, 2001. Springer Verlag.
5. A. Bonarini and M. Restelli. An architecture to implement agents co-operating in dynamic environments. In *Proceedings of AAMAS 2002 - Autonomous Agents and Multi-Agent Systems*, pages 1143–1144, New York, NY, 2002. ACM Press.
6. S. Coradeschi and A. Saffiotti. Anchoring symbols to sensor data: preliminary report. In *Proceedings of the 17th AAAI Conf*, pages 129–135, Austin, Texas, 2000.
7. K. Konolige, K.L. Myers, E.H. Ruspini, and A. Saffiotti. The Saphira architecture: A design for autonomy. *Journal of experimental & theoretical artificial intelligence: JETAI*, 9(1):215–235, 1997.

Toward an Undergraduate League for RoboCup

John Anderson[2], Jacky Baltes[2], David Livingston[3],
Elizabeth Sklar[1], and Jonah Tower[1]

[1] Department of Computer Science, Columbia University
New York, NY, 10027, USA
{sklar,jpt2002}@cs.columbia.edu

[2] Department of Computer Science, University of Manitoba
Winnipeg, Manitoba, R3T 2N2, Canada
{andersj,jacky}@cs.umanitoba.ca

[3] Department of Electrical and Computer Engineering, The Virginia Military Institute
Lexington, VA, 24450, USA
livingstondl@mail.vmi.edu

Abstract. This paper outlines ideas for establishing within RoboCup a league geared toward, and limited to, undergraduate students. Veterans of RoboCupJunior are outgrowing the league as they enter college and this has motivated us to develop a league especially for undergraduate students – the *ULeague*. The design of the league, presented here, is based on a simplied setup of the Small-size league by providing standard Vision and Communication packages.

1 Introduction

With the rise in popularity of RoboCupJunior (RCJ) [4], a growing base of participants are graduating from high school and wanting to continue with RoboCup but are unable to because they do not have the resources required to enter the senior leagues. Some of these students may be attending a university that has an existing RoboCup team, so that they will perhaps have an opportunity to participate on a senior league team as advanced undergraduates. Other students attend universities where there is no RoboCup senior team and/or no robotics lab; and for these students, participation as undergraduates is not an option. We are thus motivated to create a RoboCup league especially for undergraduates, and we have spent recent months designing and prototyping this league – the *ULeague*.

The goal of the ULeague is to provide a stepping stone from RoboCupJunior to participation in the Small-size (F180) or Mid-size leagues. There is a significant leap in both expertise and resources necessary to be a competitive entrant in either of these leagues compared to the Junior league. The sophistication and costs of the robots used in the F180 league may be prohibitive. A typical Small-size team has robots that have omni-directional drives, dribble bars and powerful kicking mechanisms. Such a robot has four to six high quality motors and a powerful on-board processor to control them. Each one of these robots costs around US$3,000. Added to this is the cost of high quality video cameras. These and other expenses typically drive the cost for a team to around US$20,000–US$30,000.

D. Polani et al. (Eds.): RoboCup 2003, LNAI 3020, pp. 670–677, 2004.

The ULeague is intended to provide a scaled-down and cheaper version of the Small-size league problem for undergraduate students. To achieve this goal, the ULeague factors out the most complex aspects of the Small-size league, namely vision processing and communication, by providing common platforms for these tasks, so teams may focus on robot development.

In addition to practical rationale, the ULeague also provides unique research challenges and opportunities. In the F180 league, it is possible for multiagent aspects of the game to have little impact on the performance of a team. A powerful dribble bar and kicker or other combination of physical features makes it possible for a single robot to be dominant. The physical constraints placed on robots in the ULeague are intended to force teams to rely more on coordination and cooperation to make progress in the game. Moreover, since a common architecture is employed, the ULeague can be thought of as a physical version of a simulation league. It is hoped that the ULeague will encourage collaboration with teams from the RoboCup Simulator League. This could mean faster dissemination of the research and progress made by teams in the Simulator league to physical robot competitions.

We have chosen the Small-size league as our primary model, since this league requires the least amount of space and the least expense in terms of equipment[1]. Given this, we have identified two major stumbling blocks for teams entering the Small league: *vision* and *communication*. So for the ULeague our idea is to provide a standard solution for these two aspects, provided by league organizers, and have teams build the rest (see figure 1).

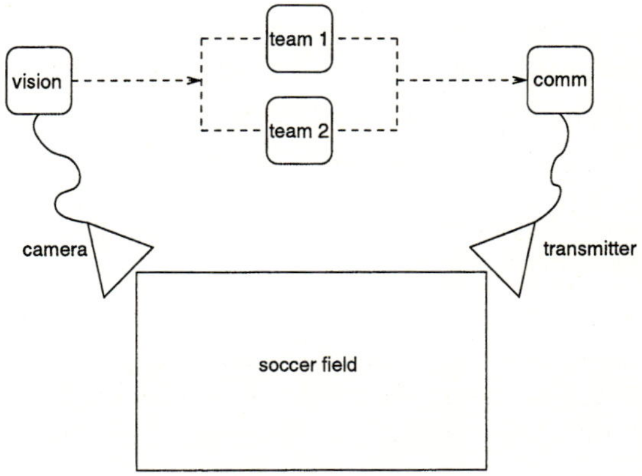

Fig. 1. High-level architecture of the ULeague. The dashed lines represent an Ethernet connection

2 Vision

The ULeague will use a standard vision software package to make it easier for teams to enter the ULeague and to speed up setup time at a competition. The current video server is the *Doraemon* package developed by Baltes [2], which is open source software

[1] As opposed to the Mid-size, Four-Legged or Humanoid leagues.

released under the Gnu Public Licence[2]. Doraemon has been in development for over three years and has been used by several robotic teams in international competitions, including RoboCup. Doraemon includes real-time camera calibration, color calibration and object tracking components.

Several features separate Doraemon from similar software written for other global vision robotic soccer teams. It supports full 24-bit color. It supports maximum field capture rates of 50 (PAL) or 60 (NTSC) fields per second. It includes sophisticated Tsai camera calibration [5], allowing the system to calibrate the geometry of the scene from any view, which means that it is not necessary to have the camera mounted directly overhead relative to the playing field, nor is it necessary to add a wide-angle lens to the camera. It tracks objects in two or three dimensions (the latter requires using multiple cameras or stereoscopic vision); and it employs a clean object-oriented design that makes it easy to define different types of robots. Currently robots using two colored markers or bar codes are available, but there are also more sophisticated object recognizers that use only a single colored spot on the robot [2]. The developers are currently working on a pattern-recognition process using neural networks that does not require any markers [3].

Installation and setup of Doraemon consists of four phases: (1) setting up the camera and the viewing angle, (2) calibrating the camera geometry, (3) calibrating the rotation-translation matrix, and (4) calibrating colors. Each phase is detailed below.

2.1 Setup of the Camera

In the F180 league, each team provides their own video camera and mounts it in their desired position. Most teams choose to place the camera directly overhead. This means that the local organizing committee must provide a physical structure above the playing field onto which teams can mount their cameras. However, the limited angle of view requires that a wide-angle lens be mounted on the camera in order to have a view of the whole playing field. Doraemon has more complex camera calibration routines and is not limited to overhead views (see figure 2).

Fig. 2. A sample view of the playing field

[2] `http://sourceforge.net/projects/robocup-video`

2.2 Camera Calibration

Doraemon's camera calibration uses the well-established Tsai camera calibration [5] which is popular among computer vision researchers. It is also suitable for global vision in robotic soccer since it can compute the calibration from a single image. The Tsai camera calibration computes six external parameters (x, y and z of the camera position as well as angles of roll, pitch and yaw) and six internal parameters (focal length, center of the camera lens, uncertainty factor S_x, and κ_1, κ_2 radial lens distortion parameters) of a camera using a set of calibration points. Calibration points are points in the image with known world coordinates. In practice, Tsai calibration requires at least 15 calibration points.

Doraemon uses a fast, robust and flexible method for extracting calibration points from the environment. A simple colored calibration carpet is used. The user selects a colored rectangle and specifies the distance in the x and y direction between the centers of the rectangle. Doraemon's calibration is iterative, so it can compensate for missing or misclassified calibration points.

2.3 Rotation Matrix

The Tsai calibration results in the overlay of a world coordinate system over the scene. In principle this is sufficient to play soccer. However, Doraemon uses an additional rotation and translation matrix to establish the world coordinate system. Instead of being forced to compute the twelve parameters, the rotation and translation matrix allows one to rotate, scale, shear and translate the playing field. This results in a set of linear equations with three unknowns and allows a rotation and translation matrix to be established with only three points. The four corner points of the playing field are used (resulting in one more point than required), and a least-squared error approximation of the matrix is produced.

2.4 Colors

Doraemon uses a sophisticated 12-parameter color model that is based on red (R), green (G) and blue (B) channels as well as the difference channels red-green (R-G), red-blue (R-B) and green-blue (G-B). Simple thresholding of the R, G and B channels is sensitive to the brightness of the color. Even on a robotic soccer playing field with its controlled lighting, it is difficult to detect more than four colors robustly using only these channels. Other color models that have been proposed in the literature are less sensitive to brightness (e.g. the HSI, YUV, or SCT models). However, these models are computationally expensive, which greatly impacts the performance of the video server. The color model used in Doraemon attempts to balance the best of both worlds. The difference channels in the color model are similar to the hue values in the HSI or similar models, but have the advantage that they can be computed faster than the HSI model.

2.5 Output of the Video Server

Doraemon transmits the position, orientation and velocity of all objects being tracked to the team clients listening on the Ethernet. The messages are transmitted in ASCII via

UDP broadcast in a single package. Each line is terminated by an end of line character (`\n`). For the ULeague, each message contains eleven lines: two lines of header information, one line for ball information and eight lines for robot information. An example is shown in figure 3.

LineNum	content
0	9 1073821804 0.00139797
1	-76.3836 -1820.48 2356.39
2	1 ball NoFnd 971.056 840.711 35 0 -2.31625 58.1464
3	0 b0 Found 1185.59 309.187 100 0.0596532 436.282 -43.083
4	0 b1 Found 1158.59 508.198 100 0.0596532 499.282 285.083
5	0 b2 Found 1086.95 1009.187 100 0.0596532 499.282 285.083
6	0 b3 Found 2185.59 309.187 100 0.0596532 499.282 285.083
7	0 y0 Found 989.95 304.588 100 -0.10185 413.528 -1.08564
8	0 y1 NoFnd 1689.95 1004.588 100 -0.10185 413.528 -1.08564
9	0 y2 Found 189.95 704.588 100 -0.10185 413.528 -1.08564
10	0 y3 Found 1789.95 1304.588 100 -0.10185 413.528 -1.08564

Fig. 3. Sample output message from Doraemon

The first line of each message contains (a) the number of objects that the video server is currently tracking; (b) the absolute frame number; and (c) the time difference in seconds between this message and the previous message. A client can use the absolute frame number and time difference value to determine if any frames were dropped by the video server.

The second line of each message contains the coordinates (C_x, C_y, C_z), in millimeters, of the camera with respect to the real-world coordinate system. This is used in distributed vision or stereoscopic vision applications and will not be used in the ULeague. The example shows that in this case, the camera was mounted 2.3m above the playing field (line 1 in figure 3).

Following this package header, there is one line for each object that the video server is tracking[3]. Each object line contains the following information:

- the type of object (`0`=robot, `1`=ball);
- the name of the object;
- whether the object was found in the image or if the video server did not find the object and predicted the positions based on previous motion (`Found` or `NoFnd`);
- the x, y, and z coordinates of the object, in millimeters;
- the orientation of the object in radians; and
- the velocity of the object in the x and y directions.

The names used for each of the nine ULeague objects are `ball` for the ball, `b0` through `b3` for the four robots on the blue team and `y0` through `y3` for the four robots on the yellow team (lines 2-10 in figure 3). The example shows that ball (`ball`) was not found and that the video server's best estimate of where the ball is is the position $(x = 971, y = 840, z = 35)$. A ball has no orientation and the video server always gives it an orientation of 0 radians. Note that the height (z-coordinate) of all objects is fixed if only a single camera view is used. For example, the height of the ball is 35mm and the

[3] In the case of the ULeague, this will always be 9.

height of the robot in the next line (below) is 100mm. The best guess for the velocity of the ball is a vector $(dx = -2, dy = 58)$. The example also shows that the robot `b0` was found at position $(x = 1185, y = 309, z = 100)$. The orientation of the robot is approximately 0 degrees and its motion is given by the vector $(dx = 436, dy = -43)$. Note that since this particular robot is a differential-drive type robot, it is not surprising that the direction of the velocity and the orientation of the robot coincide. The actual velocity of the robot is approximately 43 cm/s. The information for robots `b1` through `y3` is in the same format as described above for robot `b0`.

3 Communication

The ULeage Communications component consists of two paths. One path goes from each team's client program to the Communication Server. We refer to this as the input, or *read*, path. The second path goes from the Communication Server to the robots, and we refer to this as the output, or *write*, path. We have defined protocols for both paths. The Communication Server contains two threads, one for reading messages from clients and one for writing messages to robots.

Input messages are passed along an Ethernet link, connecting each team's computer to the computer where the Communication Server is running (see figure 1). The Comm Server listens for messages from clients on a socket. The clients send ASCII messages of the form: `[name] : [msg] \n` where `[name]` is the name of the robot (as above, `b0` through `b3` and `y0` through `y3`) and `[msg]` is an 8-bit (one byte) message to be sent to the specified robot, i.e., a number between 0 and 255 (however value 0 is reserved as a NULL command, described below). Thus an example of a complete message would be: `y0:123\n`. The Comm Server maintains an 8-byte command buffer, one byte per robot. Each time an input message is received, the Comm Server updates the byte corresponding to the robot specified in the message.

Output messages are transmitted to robots using a standard Infra-Red (IR) transmitter connected to the computer via a serial or USB port. The output thread writes continuously to the output port, sending messages of the form: `[START] [command-buffer] [CHKSUM]` where `[START]` is a one-byte start value (255) and `[CHKSUM]` is a stop byte with an included 3-bit checksum (so the values of the checksum will range from 0 to 7). The checksum computation is taken from the Lego networking protocol as implemented in BrickOS[4].

4 Platform

The ULeague will not use a standard platform like the RoboCup Four-Legged league. However, we have agreed that we need to choose standard platform specifications in order to keep teams on an equal plane regarding costs of the robots. Thus we provide maximum specifications for processor capability, in terms of size (RAM) and speed. This allows the option either to purchase a pre-designed robot kit or to build one from

[4] BrickOS is a free OS for the Lego Mindstorms RCX. For more details, see `http://brickos.sourceforge.net`

basic components. Several popular robotic kits that are within the range we are currently testing include: *Basic Stamp Board of Education (BOE Bot)*[5], *Handyboard*[6] and *LEGO Mindstorms 2.0*[7].

5 Rules of Play

The field set-up and rules of play are based on those of the Small-size league [1], with adjustments as outlined below.

Field. The field of play is rectangular, 274cm by 152cm in size. The playing surface is heavy green felt or carpet. There are goals at the long ends of the field: 70cm wide by 22cm high by 18cm deep. The field of play is marked with lines and does not have walls (see figure 4). The organizers will place a digital video camera above the field, which will send images to a central vision computer as described in section 2. The organizers will also place IR transmitter(s) around the field, to send messages to all the robots on the field (see section 3). The organizers will also supply a means to transmit the referee's signals to the robots; this will be a keyboard/mouse interface to the communications computer.

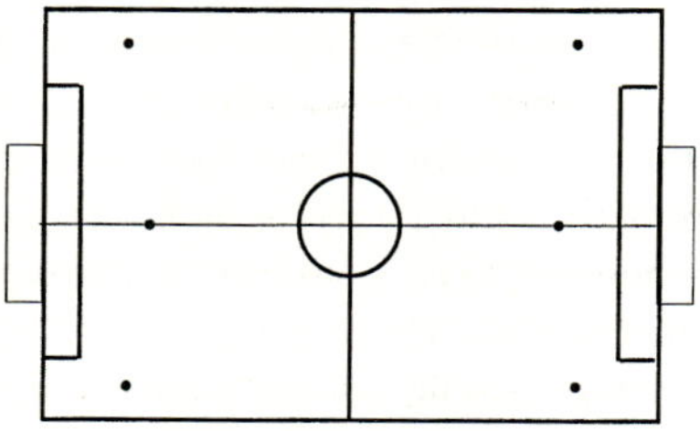

Fig. 4. The ULeague field of play

Robots and Ball. The ball is a standard orange-colored golf ball. Each team consists of four robots. Teams may designate one of these robots to be a *goalie*, but this is not required. Each robot must fit inside a cylinder 22cm in diameter and 22cm in height. Robots are intended to use wheels or rubber tracks for locomotion. Metal spikes and velcro are specifically prohibited for the purpose of locomotion. Robots do not communicate with each other directly; however, their client programs may communicate with each other through sockets on the team computer. Before a game, each of the two teams has a color assigned, either yellow or blue. The teams change colors and ends of the field for the second half of a match. The organizers will supply circular markers of each color, to be placed on the top of each robot so that they are visible by the overhead camera.

Play. A human referee controls the game. The match lasts two equal periods of 10 minutes, with a half-time interval of 10 minutes. The clock will run at all times, with

[5] `http://www.parallaxinc.com`
[6] `http://www.handyboard.com`
[7] `http://www.legomindstorms.com`

allowances only made for major stoppages as per the FIFA laws. The game begins with a kick-off by the team that wins the coin toss. The other team kicks off to begin the second half of the game. The ball is out of play when it has completely crossed any of the field boundaries. The referee may stop all the robots using the referee interface, retrieve the ball and resume play. A goal is scored when it fully crosses the goal line, between the goal walls. The winning team is the one that scores the greater number of goals during a match.

6 Summary

We have presented our design for a new undergraduate league within RoboCup, as a means for students who grow too old for RoboCupJunior to stay involved in the initiative and as an entry level for new undergraduates and universities to gain experience with RoboCup soccer. As well, the ULeague gives undergraduates who have tried the RoboCup Simulator League a chance to experiment with physical robots without needing to build a sophisticated and expensive hardware setup. The ULeague architecture consists of a common platform for Vision and Communication. Both will be provided by the league organizers at any competition venues.

There are still several open questions relating to the design of the ULeague. We propose a field without walls, but how will this work in practice? Our proposed communication mechanism has not been tested in a RoboCup competition yet; again, how will this work in practice? The exact restrictions on robot platforms have not been defined yet; what should these be?

The first full exhibition of this league will be held at RoboCupJunior-2003, in Padua, Italy (July 2003). Presently, we have teams from three universities developing the league (University of Manitoba, Canada; Columbia University, USA; Virginia Military Institute, USA). Teams from Australia, Iran and Singapore are also following along with the development. We hope to open the league to any interested parties in 2004. Regularly updated information and a discussion board can be found on our web page: `http://www.robocupjunior.org/uleague`.

Acknowledgements

We gratefully acknowledge the work of the many students involved at the universities of each of the co-authors.

References

1. http://www.itee.uq.edu.au/~wyeth/f180rules.htm
2. Jacky Baltes. Yuefei: Object orientation and id without additional markers. In *RoboCup-01: Robot Soccer World Cup V*, New York, 2002. Springer.
3. Jacky Baltes and John Anderson. Learning orientation information using neural nets. In *submitted to Robocup-03 Symposium*, 2003.
4. E. Sklar, A. Eguchi, and J. Johnson. RoboCupJunior: learning with educational robotics. In *Proceedings of RoboCup-2002: Robot Soccer World Cup VI*, 2002.
5. Roger Y. Tsai. An efficient and accurate camera calibration technique for 3d machine vision. In *Proceedings of IEEE Conference on Computer Vision and Pattern Recognition*, 1986.

A Probabilistic Framework for Weighting Different Sensor Data in MUREA

Marcello Restelli[1], Domenico G. Sorrenti[2], and Fabio M. Marchese[2]

[1] Politecnico di Milano, Dept. Elettronica e Informazione
piazza Leonardo da Vinci 32, 20135, Milano, Italy
restelli@elet.polimi.it
[2] Universitá degli Studi di Milano - Bicocca, Dept. Informatica
Sistemistica e Comunicazione, via Bicocca degli Arcimboldi 8, 20126, Milano, Italy
{sorrenti,marchese}@disco.unimib.it

Abstract. We shortly review a mobile robot localization method for known 2D environments, which we proposed in previous works; it is an evidence accumulation method where the complexity of working on a large grid is reduced by means of a multi-resolution scheme. We then elaborate a framework to define a set of weights which takes into account the different amount of information provided by each perception, i.e. sensor datum. The experimental activity presented, although the approach is independent on the sensory system, is currently based on perceptions coming from omnidirectional vision in an indoor environment.

1 Introduction

In previous works [1] [2] we introduced a method for robot localization, named MUREA (MUlti-Resolution Evidence Accumulation). It is an evidence accumulation method where the complexity of working on a large grid, i.e. at a useful accuracy of the localization estimate, is reduced by means of a multi-resolution scheme. This work has some correlations with other known works. In the following we very shortly review the ones which are more related to our, outlining the limitations which our proposal tries to overcome. *Grid-based Markov Localization* [3] operates on raw sensor measurements and, when applied to fine-grained grids, could turn into a very expensive computation. Our method can be classified as a grid-based method too; we propose to circumvent the problem with the use of a multi-resolution scheme. *Scan matching* techniques [4] [5], which make use of *Kalman Filtering*, are sensible to small errors in data association. Our proposal solves jointly the data association problem as well as the generation of a pose estimate. *Monte Carlo Localization* [6] have problems in recovering from unexpected events with respect to the motion model. Our approach reacts immediately, i.e. in the same activation, whenever the sensor data does not match the a priori pose estimate; in such case it turns into a global localization. The works mentioned so far, including ours, share the capability to handle so-called dense data sets.

D. Polani et al. (Eds.): RoboCup 2003, LNAI 3020, pp. 678–685, 2004.

2 Localization Algorithm

The method we proposed has three main components: the map of the environment, the perceptions and the localization engine. The environment is represented by a 2D geometrical map that can be inserted in the system through a configuration file or can be built by a SLAM system, e.g. as in [5]. The map is made up of simple geometrical primitives, like points, lines, circles, etc. On the other hand, we have sensor data. The localization engine takes in input the map of the environment as well as the perceptions, and outputs the estimated pose(s) of the robot (if the environment and/or the perceptions have inherent ambiguities).

In evidence accumulation the difficulty is in working at high resolution in order to get an accurate estimate. We divide the search-space in subregions (hereafter cells), to which we associate a counter. Each cell represents a localization hypothesis. Each perception increases the counter associated to a cell if some point, inside that cell, can be found, which is *compatible*, see [1] for details, with both the perception and the model. Then, on the basis of the votes collected by each cell, the method selects the ones which are more likely to contain the correct robot pose. Those cells undergo a refinement phase, which is where the multi-resolution comes into play. This process is further iterated on the refined hypotheses until a termination condition is matched, see [1] for details. The *compatibility verification* is in charge of checking whether a perception, e.g. a point perception, can be an observation of a certain map element, e.g. a line, for the pose associated to the cell. The more the votes of a cell, the more trustable is the match between the perceived world and its model, and the higher are the chances that the robot pose falls inside that cell. Therefore we search for the maximum in the vote accumulator; this is done at each level of resolution, which in turn increases at each iteration of the process.

3 The Weights of the Perceptions

According to what has been presented so far, the compatibility verification of a perception concludes with a vote or a non-vote to a cell, which in turn implies to increase or not, by a constant value, the cell counter (here we have the main difference with respect to other Bayes-based approaches). This is unsatisfactory since the amount of information carried by each perception is not the same. We think that the problem of detecting the perceptual saliency, i.e. the informativeness, of each sensor datum (i.e. perception) in order to give it more credit is quite general and applies to other approaches (including the method proposed in our previous work) as well. Our proposal is to analyze how much each perception is useful for the localization process and determine a criterion to define the weight that each perception should have in the voting phase. In this paper, we propose a voting approach which estimates the probability that each cell contains the actual robot pose. In this way, the selection phase will choose those cells that have an high probability of being the correct one.

There are two main factors that determine the perceptual saliency of a perception: the *informativeness* and the *reliability*. The information carried by a perception is directly proportional to the number of localization hypothesis that it excludes. This implies that perceptions which vote for few cells are more informative; on the other hand, a perception that votes all the cells provides no information, whilst a perception that votes only one cell provides the maximum information. The reliability of a perception is linked with the probability that this perception carries erroneous information, for localization purposes. We try to capture this concept with the probability of the correctness of a perception.

Definition 1. *A perception is* correct *if it votes, i.e. is compatible with, the cell containing the actual robot pose.*

3.1 Probabilistic Analysis

We want to identify which weight to associate to each perception, in order to select the cells that are most likely to contain the actual robot pose.

Definition 2. *Let Λ be the set of cells into which the C-Space is divided, and N the cardinality of Λ.*

Definition 3. *Let P be the set of perceptions acquired at time t.*

Definition 4. *Let Φ_c be the subset of all the perceptions acquired at time t which vote for cell c: $\Phi_c = \{p \in P | p \text{ votes } c\}$.*

Definition 5. *Let Ψ_c be the subset of all the perceptions acquired at time t which do not vote for cell c: $\Psi_c = P \setminus \Phi_c$.*

Definition 6. *Let r be the cell containing the actual robot pose (a priori unknown).*

Definition 7. *Let $(c = r)$ be the event where the generic cell c is actually corresponding to cell r. The complement of the event $(c = r)$ is represented by $(c \neq r)$.*

Definition 8. *Let C_p be the event where perception p votes cell c, i.e. $p \in \Phi_c$.*

Definition 9. *Let V_p be the number of cells voted by p.*

Definition 10. *Let E_π be the event where all the perceptions belonging to the set π are not correct, i.e. they do not vote for r. π can contain a single perception; in this case we write E_p. With $\overline{E_\pi}$ we mean the event where the perceptions in π are correct.*

We want to evaluate the probability of event $(c = r)$. A priori, i.e. without having any information from the perception system, this value is the same for every cell.

$$Pr[(c = r)] = 1/N \tag{1}$$

Let us compute the same probability, given that some perceptions are compatible with cell c while others are not.

$$Pr[(c = r)|\Phi_c, \Psi_c] = \frac{Pr[(c = r), \Phi_c, \Psi_c]}{Pr[\Phi_c, \Psi_c]} \tag{2}$$

$$= \frac{Pr[(c = r), \Phi_c, \Psi_c|(c = r)] \cdot Pr[(c = r)] + Pr[(c = r), \Phi_c, \Psi_c|(c \neq r)] \cdot Pr[(c \neq r)]}{Pr[\Phi_c, \Psi_c|(c = r)] \cdot Pr[(c = r)] + Pr[\Phi_c, \Psi_c|(c \neq r)] \cdot Pr[(c \neq r)]}$$

$$= \frac{Pr[\Phi_c, \Psi_c|(c = r)] \cdot Pr[(c = r)]}{Pr[\Phi_c, \Psi_c|(c = r)] \cdot Pr[(c = r)] + Pr[\Phi_c, \Psi_c|(c \neq r)] \cdot Pr[(c \neq r)]} \tag{3}$$

First we applied Bayes'formula (eq. (2)), then we develop both the numerator and the denominator through the total probability lemma, and finally we deleted the second addendum (since $Pr[(c = r)|(c \neq r)] =_{def} 0$) and obtained the expression of eq. (3). If we substitute eq. (1) in eq. (3) and divide both the numerator and the denominator by $Pr[\Phi_c, \Psi_c|(c \neq r)]$, we obtain:

$$Pr[(c = r)|\Phi_c, \Psi_c] = \frac{\frac{Pr[\Phi_c, \Psi_c|(c=r)]}{Pr[\Phi_c, \Psi_c|(c\neq r)]} \cdot \frac{1}{N}}{\frac{Pr[\Phi_c, \Psi_c|(c=r)]}{Pr[\Phi_c, \Psi_c|(c\neq r)]} \cdot \frac{1}{N} + \frac{N-1}{N}} = \frac{\Sigma_c}{\Sigma_c + (N-1)}, \tag{4}$$

where $\Sigma_c = \frac{Pr[\Phi_c, \Psi_c|(c=r)]}{Pr[\Phi_c, \Psi_c|(c\neq r)]}$.

The Numerator of $\boldsymbol{\Sigma_c}$. The numerator of Σ_c is the probability that the perceptions which belong to Φ_c vote c and those that belong to Ψ_c do not, given that c is the cell containing the actual robot pose. This probability equals the probability that the Φ_c perceptions are correct, and that the other Ψ_c perceptions are not correct (see def. 1).

$$Pr[\Phi_c, \Psi_c|(c = r)] = Pr[\overline{E_{\Phi_c}}, E_{\Psi_c}] = \prod_{p \in \Phi_c} (1 - Pr[E_p]) \cdot \prod_{p \in \Psi_c} Pr[E_p], \tag{5}$$

Eq. (5)exploits the fact that these events, i.e. whether the perceptions are correct or not, are statistically independent (the noise, which makes them vote or not a cell c, acts independently on each perception).

The Denominator of $\boldsymbol{\Sigma_c}$. As far as the denominator of Σ_c is concerned, if we assume that the perceptions vote independently a cell c that does not contain the actual robot pose, we can introduce the following factorization:

$$Pr[\Phi_c, \Psi_c|(c \neq r)] = \prod_{p \in \Phi_c} Pr[C_p|(c \neq r)] \cdot \prod_{p \in \Psi_c} Pr[\overline{C_p}|(c \neq r)] \tag{6}$$

The probability that a perception p votes a cell c, given that c does not contain the actual robot pose, is:

$$Pr[C_p|(c \neq r)] = Pr[C_p|(c \neq r), E_p] \cdot Pr[E_p] + Pr[C_p|(c \neq r), \overline{E_p}] \cdot Pr[\overline{E_p}] =$$

$$= \left(\frac{V_p}{N-1}\right) \cdot Pr[E_p] + \left(\frac{V_p - 1}{N-1}\right) \cdot (1 - Pr[E_p]) = \frac{V_p - (1 - Pr[E_p])}{N-1} \tag{7}$$

The probability that a perception p votes a cell c is equal to the ratio of the number of cells voted by p (V_p) and the total number of cells (N). Given that $c \neq r$ the number of possible cells becomes $N-1$. If we also know that the perception p is not correct, c could be any of the Vp cells voted by p. On the other hand, if we know that the perception p is correct, there are only $V_p - 1$ cells that are voted by p and are different from r.

Let us now consider the probability that p does not vote c, given that c does not contain the actual robot pose; its value is:

$$Pr[\overline{C_p}|(c \neq r)] = Pr[\overline{C_p}|(c \neq r), E_p] \cdot Pr[E_p] + Pr[\overline{C_p}|(c \neq r), \overline{E_p}] \cdot Pr[\overline{E_p}] =$$
$$= \left(\frac{N - V_p - 1}{N-1}\right) \cdot Pr[E_p] + \left(\frac{N - V_p}{N-1}\right) \cdot (1 - Pr[E_p]) = \frac{N - V_p - Pr[E_p]}{N-1}, \quad (8)$$

which comes from considerations similar to those used for eq. (7).

Σ_c Expression. Finally, using the results of eq. (5), (6), (7) and (8) we can obtain the expression for Σ_c and, from it, the expression of $Pr[(c = r)|\Phi_c, \Psi_c]$ (see eq. (4)):

$$\Sigma_c = \prod_{p \in \Phi_c} \frac{(N-1) \cdot (1 - Pr[E_p])}{V_p - (1 - Pr[E_p])} \cdot \prod_{p \in \Psi_c} \frac{(N-1) \cdot Pr[E_p]}{(N - V_p) - Pr[E_p]}. \quad (9)$$

It is worthwhile observing that the introduction of Σ_c allows each perception to provide a contribution, i.e. a factor of Σ_c, which is independent on the contribution of the other perceptions. Moreover, from eq. (4) we can notice that $Pr[(c = r)|\Phi_c, \Psi_c]$ is a function which is monotonically larger than Σ_c. This implies that cells which have higher values of Σ_c will also have higher probabilities of being the cell that contains the actual robot pose.

In the following section we will show how this result leads to the implementation of a weighting method that can be used in our localization approach.

3.2 Implementation

We would like, for each perception, to just add a quantity to the voted cells. To this aim we apply the logarithmic operator to Σ_c and, since it maintains the ordering, we will be able to discriminate among the localization hypotheses.

$$A_c = \ln(\Sigma_c) = \sum_{p \in \Phi_c} ln \frac{(N-1) \cdot (1 - Pr[E_p])}{V_p - (1 - Pr[E_p])} + \sum_{p \in \Psi_c} ln \frac{(N-1) \cdot Pr[E_p]}{(N - V_p) - Pr[E_p]}. \quad (10)$$

A_c is the accumulator value, after the voting phase. From eq. (10) we can see that each perception must be associated to a couple of weights:

$$w_p = \begin{cases} ln \frac{(N-1) \cdot (1 - Pr[E_p])}{V_p - (1 - Pr[E_p])}, if p \in \Phi_c \\ ln \frac{(N-1) \cdot Pr[E_p]}{(N - V_p) - Pr[E_p]}, if p \in \Psi_c \end{cases} \quad (11)$$

In order to compute the weights for each perception p, we have to know the values of N, V_p, and $Pr[E_p]$, for each resolution level. Moreover, we would like to know these values before starting the localization process. For this reason, the set P is partitioned in subsets g called *groups*, e.g. the group of the "blue direction perceptions", the group of the "white point perception between 50cm and 100cm" and so on. In the off-line phase, for each group and for each resolution level, the three values are estimated and the weights are computed by means of eq. (11). These values will then be used as an approximation of the actual (unknown) values. The rationale behind this is that all the perceptions in a group are quite homogeneous, so that their V_p and $Pr[E_p]$ values can be approximated by the group values, determined off-line. This process occurs at each resolution level since the weights change accordingly with it.

Definition 11. *N^l is the number of cells at level l.*

Definition 12. *V_g^l is the number of cells that a perception of the group g votes at level l.*

Definition 13. *$Pr[E_g^l]$ is the probability that a perception of the group g does not vote the cell containing the actual robot pose, at level l.*

The estimation of V_g^l is accomplished considering one perception $p \in g$ and testing its compatibility with all the cells at level l. Since the number of cells voted by a perception is independent of the sensors and time of acquisition, this estimation process can be carried out off-line. The same does not hold for $Pr[E_g^l]$. The value of this probability is a priori unknown, since it is related to the sensor that generates the perception, and, above all, it may change with time. At each level l, after the voting phase, $Pr[E_p^l]$ is estimated for each perception p, by considering the probabilities $Pr[(c = r)|\Phi_c, \Psi_c]$ for those cells that p has not voted. Since we know the value $Pr[(c = r)|\Phi_c, \Psi_c]$ only for the cells which have been analyzed (i.e. which resulted from the refinement of the cells selected al level $(l-1)$), and in order to normalize the sum of the $Pr[(c = r)|\Phi_c, \Psi_c]$, we propose to use the following approximation instead:

$$Pr[E_p^l] \approx \frac{\sum_{c \in \Xi_{\overline{p}}^l} Pr[(c = r)|\Phi_c, \Psi_c]}{\sum_{c \in \Xi^l} Pr[(c = r)|\Phi_c, \Psi_c]} = \frac{\sum_{c \in \Xi_{\overline{p}}^l} \frac{e^{A_c}}{e^{A_c} + N^l - 1}}{\sum_{c \in \Xi^l} \frac{e^{A_c}}{e^{A_c} + N^l - 1}} \tag{12}$$

where Ξ^l is the set of the cells analyzed at the current level l, and $\Xi_{\overline{p}}^l$ is the subset of Ξ^l whose elements are the cells that have not been voted by p. This approximation is justified because we do not analyze the cells with a low probability of being the correct cell. The estimate of $Pr[E_g^l]$ is computed as the average $\widehat{Pr[E_g^l]}$ of the $Pr[E_p^l], \forall p \in g$, which, by means of eq. (11), allows to determine the weights for the next iteration of the localization process.

4 Localization Experiments

We made some experiments with a robot equipped with an omnidirectional vision system, based on a multi-part mirror [7]. This mirror allows the vision system to have a resolution of about $40mm$, for distances under $6m$. We put the robot into a RoboCup middle-size league playground. We defined three map elements of type *line* and attribute *blue* to describe the blue goal and four lines to describe the borders of the half-field used. The vision sensor produces two types of perceptions: white point perception and blue direction perception.

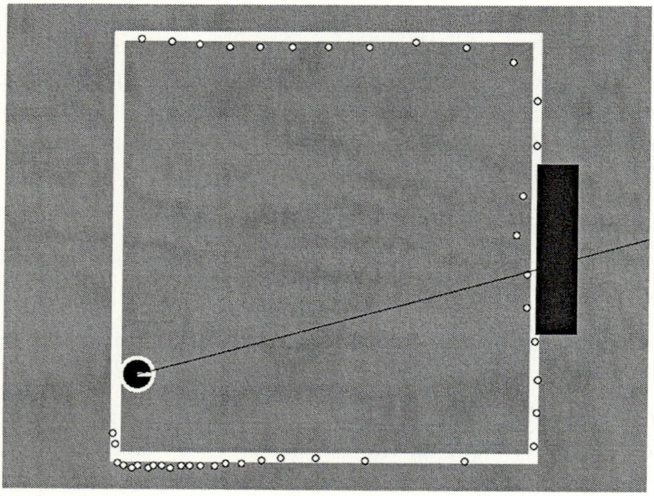

Fig. 1. White point and blue direction perceptions re-drawn on the map, altogether with the map elements; the black and white circle represents the actual robot pose

Table 1. Values of the positive weights for some group (columns) and for some level (rows)

Lev	Blue	White 0-100	White 100-200	White 200-300
1	0.00	0.00	0.00	0.00
5	0.38	0.00	0.00	0.01
10	1.37	0.34	0.49	0.74
15	1.93	1.63	5.49	5.75

Table 2. Values of the negative weights for some group (columns) and for some level (rows)

Lev	Blue	White 0-100	White 100-200	White 200-300
1	0.00	0.00	0.00	0.00
5	-1.35	0.00	0.00	-0.12
10	-2.05	-1.28	-1.50	-1.74
15	-2.16	-2.11	-2.30	-2.30

We devised a preliminary experiment for validating the definition of the weights. In table 1 and table 2, we have reported, respectively, the positive and negative weights associated with some group of perceptions at some levels of resolution. We can notice that perceptions begin (i.e. when considering large cells) with a zero weight for both positive and negative weights. The reason is that the localization hypotheses are too wide and the perceptions vote all the cells, bringing no information about the robot pose. On the other hand, as the resolution increases, we have that the positive weights grow, while the negative weights decrease their value. The values of table 1 and table 2 were computed following the procedure described in section 3.2, with the values of $Pr[E_g^l$ initialized to 0.1. In the experiment reported in Figure 1 we had several white point perceptions and only one blue direction perception (which coulb be outweighted by the white ones). If we had considered the white point perceptions only, we would have had ambiguities, because of the environment symmetries. In such case (all the perceptions have the same weight) the localization ends up with four distinct poses, at the same level of likelyhood: $(465cm, 143cm, 180^o)$, $(101cm, 220cm, 270^o)$, $(34cm, -143cm, 0^o)$, $(382cm, -229cm, 90^o)$. With the weighting mechanism here presented, the correct solution could be reached: actual value $(25cm, -150cm, 0^o)$, estimated value $(26cm, -138cm, -1^o)$.

References

1. Restelli, M., Sorrenti, D.G., Marchese, F.M.: Murea: a multi-resolution evidence accumulation method for robot localization in known environments. In: proceedings of the International Workshop on RoboCup - Robot Soccer World Cup VI, Springer-Verlag, LNCS series (to appear)
2. Restelli, M., Sorrenti, D.G., Marchese, F.M.: A robot localization method based on evidance accumulation and multi-resolution. In: proceedings of 2002(IEEE/RSJ) Int. Conf. on Intelligent Robots and Systems (IROS2002). (2002)
3. Burgard, W., Fox, D., Hennig, D., Schmidt, T.: Estimating the absolute position of a mobile robot using position probability grids. In: AAAI/IAAI, Vol. 2. (1996) 896–901
4. Gutmann, J., Burgard, W., Fox, D., Konolige, K.: An experimental comparison of localization methods. In: proceedings of the IEEE/RSJ International Conference on Intelligent Robots and Systems (IROS98). (1998)
5. Lu, F., Milios, E.: Globally consistent range scan alignment for environment mapping. Autonomous Robots **4** (1997) 333–349
6. Dellaert, F., Fox, D., Burgard, W., Thrun, S.: Monte carlo localization for mobile robots. In: proceedings of the IEEE International Conference on Robotics and Automation (ICRA99). (1999)
7. Marchese, F.M., Sorrenti, D.G.: Omni-directional vision with a multi-part mirror. In: proceedings of the International Workshop on RoboCup. Volume 2019 of LNAI., Springer-Verlag (2001) 179–188 P. Stone, T. Balch, G. Kraetzschmar Eds.

Plays as Team Plans for Coordination and Adaptation

Michael Bowling, Brett Browning, Allen Chang, and Manuela Veloso

Computer Science Department, Carnegie Mellon University, Pittsburgh PA, 15213-3891, USA
{mhb,brettb,mmv}@cs.cmu.edu, allenc@andrew.cmu.edu

Abstract. Coordinated action for a team of robots is a challenging problem, especially in dynamic, unpredictable environments. In the context of robot soccer, a complex domain with teams of robots in an adversarial setting, there is a great deal of uncertainty in the opponent's behavior and capabilities. We introduce the concept of a *play* as a team plan, which combines both reactive principles, which are the focus of traditional approaches for coordinating actions, and deliberative principles. We introduce the concept of a *playbook* as a method for seamlessly combining multiple team plans. The playbook provides a set of alternative team behaviors which form the basis for our third contribution of *play adaptation*. We describe how these concepts were concretely implemented in the CMDragons robot soccer team. We also show empirical results indicating the importance of adaptation in adversarial or other unpredictable environments.

1 Introduction

Coordination and adaptation are two of the most critical challenges for deploying teams of robots to perform useful tasks. These challenges become especially difficult in environments involving other agents, particularly adversarial ones, not under the team's control. In this paper, we examine these challenges within the context of robot soccer [6], a multi-robot goal-driven task in an adversarial dynamic environment. The presence of adversaries creates significant uncertainty for predicting the outcome of interactions particularly if the opponent's behavior and capabilities are unknown a priori, as is the case in a robot soccer competition. As such, this task encapsulates many of the issues found in realistic multi-robot settings.

Despite this unpredictability, most robot soccer approaches involve single, static, monolithic team strategies (e.g., see robot team descriptions in [1].) Although these strategies entail complex combinations of reactive and deliberative approaches, they can still perform poorly against unknown opponents or in unexpected situations. With the uncertainty present in the task, such situations are common. An alternative approach uses models of opponent behavior, constructed either before or during the competition [5], which are used then to determine the best team response. A model may be used in a reactive fashion to trigger a pre-coded static strategy, or in a deliberative fashion through the use of a planner [7]. Although these techniques have had success, they have limitations such as the requirement for an adequate representation of opponent behavior. For a completely unknown opponent team, constructing an a prior model of their strategy is impractical.

Here, we take a novel approach based on observing our own team's effectiveness rather than observing the opponent's behavior. We replace a single monolithic team

D. Polani et al. (Eds.): RoboCup 2003, LNAI 3020, pp. 686–693, 2004.

strategy, with multiple team plans that are appropriate for different opponents and situations, which we call *plays*. Each play defines a coordinated sequence of team behavior, and is explicit enough to facilitate evaluation of that play's execution. A *playbook* encapsulates the plays that a team can use. Each execution of a play from the playbook can then be evaluated and this information collected for future play selection. Successful plays, whose successes may be attributed to weaknesses in the opponent or particular strengths of our team, are selected more often, while unsuccessful plays are ignored.

2 Overview

The work described in this paper was fully implemented on our CMDragons'02 small size league (SSL) robot team. We competed with our robots at the RoboCup 2002 International competition in Fukuoka, Japan. As with other SSL teams, our small robots utilize perceptual information from an overhead color camera, and an off-field computer for centralized team processing. Hence, our approach does not yet address issues of distributed coordination per se. Due to space limitations, we do not go into the details of the larger architecture. Instead, we refer the reader to [3] and [4].

From the perspective of strategy, each robot can perform a range of individual skills. Each individual skill is encapsulated as a tactic, and all tactics are heavily parameterized to provide a wide range of behavior. The role of strategy is to assign tactics, with suitable parameters, to each robot. The robots then execute the tactic actions each and every frame. Hence, the strategy layer provides the coordination mechanism and executes one instance for the entire team and must meld individual robot skills into powerful and adaptable team behavior. Tactics can be classified as either active or non-active. An active tactic is one that attempt to manipulate the ball in some manner. Active tactics include `shoot`, `steal`, and `clear`, while example non-active tactics include `position_for_loose_ball`, `defend_line`, and `block`. Parameters are tactic specific, but example parameters often include target points, regions, or information affective behavior such as whether to aim or include deflections etc. Each is itself a complex interaction between the robot control layer that maintains robot-specific information, navigation, and motion control.

3 Play-Based Strategy

The main question addressed in this work is: "Given a set of effective and parameterized individual robot behaviors, how do we select each robot's behavior to achieve the team's goals?" This is the problem addressed by our strategy component.

3.1 Goals

The main criterion for team strategy is performance. However, a single, static, monolithic team strategy that maximizes performance is impractical. Indeed, in adversarial domains with unknown opponents, optimal static strategies are unlikely to exist. Therefore we break down the performance criteria into more achievable subgoals. The subgoals are to *(i)* Coordinates team behavior, *(ii)* Executes temporally extended sequences of action, *(iii)* Allow for special behavior for certain circumstances, *(iv)* Allow ease of human design and augmentation, *(v)* Enable exploitation of short-lived opportunities, and *(vi)* Allow on-line adaptation to the specific opponent.

The first four goals require plays to be able to express complex, coordinated, and sequenced behavior among teammates. In addition, plays must be human readable to make strategy design and modification simple (a must at competitions!). These goals also requires a capable of executing the complex behaviors the play describes. The fifth goal requires the execution system to recognize and exploit fortuitous opportunities not explicitly described by the play. Finally, the last goal requires the system to improve its behavior over time. These goals, although critical to robot soccer, are also of general importance for coordinated agent teams in other unpredictable or adversarial environments. We have developed a play-based team strategy, using a specialized play language, to meet these goals. We describe the major components of this system, specifically play specification, execution, and adaptation, in the following sections.

3.2 Play Specification

Plays are specified using the play language, which is in an easy-to-read text format (e.g., Table 1, Plays use keywords, denoted by all capital letters, to mark different pieces of information. Each play has two components: *basic information* and *role information*. The basic information describes when a play can be executed (“`APPLICABLE`”), when execution of the play should stop (“`DONE`”), and some execution details (e.g., “`FIXEDROLES`”, “`TIMEOUT`”, and “`OROLE`”). The role information (“`ROLE`”) describes how the play is executed, making use of the tactics described above (see Section 2). We describe these keywords below.

The `APPLICABLE` keyword denotes the state of the world under which a play can be executed. It defines the state of the world through a conjunction of high-level predicates following the keyword. Multiple keywords, on separate lines, define a logical DNF where the result of each line forms a disjunction. Examples of common predicates include `offense`, `defense`, `their_ball`, where the meaning of the predicate should be apparent. The ability to form logical DNF's means that we can choose exactly which conditions a play can be operate under.

Unlike classical planning, the level of uncertainty when running real robots makes it difficult to predict the outcome of a particular plan. Although, a play does not have effects, it does have termination conditions. Termination conditions are specified by the keyword `DONE` followed by a result (e.g., `aborted`) and a conjunctive list of high-level predicates similar to the applicability conditions. Plays may have multiple `DONE` conditions, each with a different result, and a different conjunction of predicates. Whenever *any* `DONE` condition is satisfied, the play terminates. In the example play in Table 1, the only terminating condition is if the team is no longer on offense. In this case the play's result is considered to have been `aborted`. In addition to the termination conditions, a play may be terminated by a timeout or by completing the sequence of tactics for each role. Timeouts, the length of time which can be overridden with the `TIMEOUT` keyword, are necessary to prevent the team becoming irretrievably stuck attempting an action that is not succeeding. Completions, defined by the keyword `completed`, means the play terminated correctly but did not lead to a goal score. Finally, a play is considered to have `succeeded` or `failed` whenever a goal is scored during the play for, or against, the team. These play results form the basis for evaluating the success of the play for the purposes of adaptation.

Table 1. Two example plays involving sequencing of behaviors. The left play is a special purpose play that only executes when the ball is in an offensive corner of the field.

```
PLAY Two Attackers, Corner Dribble 1
APPLICABLE offense in_their_corner
DONE aborted !offense
TIMEOUT 15

ROLE 1
  dribble_to_shoot {R {B 1100 800} ...
  shoot A
  none
ROLE 2
  block 320 900 -1
  none
ROLE 3
  position_for_pass { R {B 1000 0}...
  none
ROLE 4
  defend_line {-1400 1150} ...
  none
```

```
PLAY Two Attackers, Pass
APPLICABLE offense
DONE aborted !offense
OROLE 0 closest_to_ball

ROLE 1
  pass 3
  mark 0 from_shot
  none
ROLE 2
  block 320 900 -1
  none
ROLE 3
  position_for_pass {R {B 1000 0}...
  receive_pass
  shoot A
  none
ROLE 4
  defend_line {-1400 1150}...
  none
```

Roles are the active component of each play, and each play has four roles corresponding to each non-goalie robot on the field. Each role contains a list of tactics with associated parameters for the robot to perform in sequence. As tactics are heavily parameterized, the range of tactics can be combined into nearly an infinite number of play possibilities. Table 1 shows an example play where the first role executes two sequenced tactics. First the robot dribbles the ball out of the corner and then switches to the shooting behavior. Meanwhile the other roles execute a single behavior for the play's duration. Sequencing implies an enforced synchronization, or coordination between roles. Once a tactic completes, all roles move to their next behavior in their sequence (if one is defined). Thus, in the example in Table 1, when the player assigned to pass the ball completes the pass, then it will switch to the mark behavior. The receiver of the pass will simultaneously switch to receive the pass, after which it will try to execute the shooting tactic.

3.3 Play Execution

The play execution module is responsible for instantiating the active play into actual robot behavior. Instantiation consists of many key decisions: role assignment, role switching, sequencing tactics, opportunistic behavior, and termination. Role assignment is dynamic, rather than being fixed, and is determined by uses tactic-specific methods. To prevent conflicts, assignment is prioritized by the order in which roles appear. Thus, the first role, which usually involves ball manipulation, is assigned first and considers all four field robots. The next role is assigned to one of the remaining robots, and so on. The prioritization provides the execution system the knowledge to select the best robots to perform each role and also provides the basis for role switching. Role switching is a very effective technique for exploiting changes in the environment that alter the effectiveness of robots fulfilling roles. The executor continuously reexamines the role assignment for possible opportunities to improve it as the environment changes. Although, it has a strong bias toward maintaining the current assignment to avoid oscillation.

Sequencing is needed to move the entire team through the list of tactics in sequence. When the tactic executed by the *active player*, the robot whose role specifies a tactic related to the ball, succeeds then the play transitions *each* role to the next tactic in their relative sequence. Finally, opportunistic behavior accounts for unexpected fortuitous situations where a very basic action would have a valuable outcome ie. when an opportunity to shoot directly on goal presents itself. Thus, opportunistic behavior enables plays to have behavior beyond that specified explicitly. As a result, a play can encode a long sequence of complex behavior without encumbering its ability to respond to unexpected short-lived opportunities. Finally, the play executor checks the play's termination criteria, the completion status of the tactics, and the incoming information from the referee to determine if the play has completed, and with what result.

3.4 Play Selection

The final facet of the playbook strategy system is the mechanism for play selection and adaptation of play selections given experience. Our basic selection scheme uses the applicability conditions for each play to form a candidate list from which one play is selected at random. To adapt play selection, we modify the probability of selecting a play using a weighting scheme. We describe this mechanism, along with experimental results, in more detail below.

4 Playbook Adaptation

Playbook adaptation is the problem of adapting play selection based on past execution to find the dominant play, or plays, for the given opponent and the history of execution. In order to facilitate the compiling of past outcomes into the selection process, we associate with each play a weight, $w_{p_i} \in [0, \infty)$. For a given set of applicable plays, A, the weights are normalized to define a probability mass distribution for selecting each play as,

$$Pr(\text{selecting } p_i) = \frac{w_{p_i}}{\sum_{p_j \in A} w_{p_j}}.$$

Playbook adaptation involves adjusting the selection weights given the outcome of a play's execution. An adaptation rule is a mapping, $W(\mathbf{w}, p_i, o) \rightarrow [0, \infty)$, from a weight vector, a selected play, and its outcome, to a new weight for that play. These new weights are then used to select the next play.

There are a number of obvious, desirable properties for an adaptation rule. All things being equal, more successes or completions should increase the play's weight. Similarly, aborts and failures should decrease the weight. In order for adaptation to have any effect, it also must change weights drastically to make an impact within the short time-span of a single game. This leads us to the basic rule that we implemented for the RoboCup 2002 competition uses a weight multiplication rule, where each outcome multiplies the play's previous weight by a constant. Specifically, the rule is,

$$W(\mathbf{w}, p_i, o) = C_o w_{p_i},$$

With C_o fixed to $C_{\texttt{succeeded}} = 4$, $C_{\texttt{completed}} = 4/3$, $C_{\texttt{aborted}} = 3/4$, $C_{\texttt{failed}} = 1/4$.

4.1 Evaluation

Our strategy system was used effectively during RoboCup 2002 against a wide range of opponents with vastly different strategies and capabilities. Throughout, we observed that our team quickly honed in on the plays that worked within a few minutes. We predominantly began each game with a uniform prior. That is, with $w_i = 1$ for each play i. As the competition progressed, we developed more capable plays, in a reasonably efficient manner helped by the readability of the play language (our final playbook contained around 20 plays, including specialized plays for penalties, free kicks etc). Although we have no specific results, anecdotally, adaptation appears to make the team work as good as the performance of the best play given the underlying tactics. Thus, in situations where the tactics cannot perform their assigned objective, say when playing a very good team, play adaptation does not improve the performance of the team. However, it does not hinder performance either.

In order to more scientifically understand the capabilities and limitations of the play approach, we constructed a number of simplified scenarios to evaluate adaptation performance. These scenarios compare whether multiple plays are actually necessary, and also examine the usefulness of playbook adaptation. We compared four simple offensive plays paired against three defensive tactics. Only two offensive robots were used against one defensive robot, where the offensive plays consist of the various combinations of `shoot` with and without aiming for the active role, and `position_for_rebound` or `screen` for the supporting role. The defensive robot executed a single tactic, which was one of `block`, `active_def`, or `brick` where the robot did not move. In all cases, the robots start in the usual "kick off" position in the center of the field. For each scenario 750 trials were performed in our UberSim SSL simulator [2]. A trial was considered a success if the offense scored a goal within a 20s time limit.

Table 2 shows the play comparison results. Each trial is independent, and so the maximum likelihood estimate of each play's success probability is the ratio of successes to trials. Note that there is no static strategy that is optimal against every possible opponent even in this simplified scenario. Our results support the notion that play-based strategies are capable of exhibiting many different behaviors with varying degrees of effectiveness. For instance, the screen plays, one of which was shown in the example trace, are effective against an "active" defender which tries to steal the ball from the offense, because the non-shooting attacker is capable of setting screens for the non-shooting attacker. On the other hand, the screen plays are less effective against a "blocking" defender which guards the goal.

To explore playbook adaptation we use a playbook containing all four offensive plays against a fixed defender running either `block` or `active_def`. We initially used

Table 2. Play comparison results. For each scenario, the percentage of successes for the 750 trials is shown. The boldfaced number corresponds to the play with the highest percentage of success for each defensive behavior.

Play tactic 1	tactic 2	`block`	`active_def`	`brick`
shoot(aim)	position_for_rebound	**72.3%**	49.7%	**99.5%**
shoot($no\ aim$)	position_for_rebound	66.7%	57.3%	43.1%
shoot(aim)	screen	40.8%	59.0%	92.4%
shoot($no\ aim$)	screen	49.2%	**66.0%**	72.0%

the algorithm described above, but discovered an imperfection in the approach. Due to the strength of the reinforcement for a completed play, it is possible for a moderately successful but non-dominant play to quickly gain reward and dominate selection. This phenomenon did not occur in competition due to the larger disparity in plays against a given opponent and lower success probabilities. The issue is a lack of normalization in the weight adjustment to account for play selection probabilities. Therefore, we included a normalization factor in the weight updates. Specifically, we used the following rule,

$$W(\mathbf{w}, p_i, o) = \begin{cases} w_{p_i} 2/\Pr(p_i) \text{ if } o = \texttt{Succeeded} \\ w_{p_i} \Pr(p_i)/2 \text{ if } o = \texttt{Failed} \end{cases},$$

where $\Pr(p_i)$ is the probability assigned to p_i according to $\mathbf{w}$.

To evaluate the performance of the algorithm, we compare the expected success rate (ESR) of using this adaptation rule against a fixed defensive behavior. We used the results in Table 2 to simulate the outcomes of the various play combinations. All the weights are initialized to 1. Figure 1(a) and (b) show the ESR for play adaptation over 100 trials, which is comparable to the length of a competition (approximately 20 minutes). The lower bound on the y-axis corresponds to the ESR of randomly selecting plays and the upper bound corresponds to the ESR of the playbook's best play for the particular defense. Figure 1(c) shows the probabilities of selecting each play over time when running the adaptation algorithm.

As graphs (a) and (b) indicate in Figure 1, against each defense the overall success rate of the offense quickly grows towards the optimal success rate within a small number of trials. Likewise graph (c) shows that against the `block` defense, the probability of selecting either of two plays with comparatively high individual success rates quickly dominates the probability of selecting the two less successful plays. Clearly, the algorithm very quickly favors the more successful plays.

These results, combined with the RoboCup performances, demonstrate that adaptation can be a powerful tool for identifying successful plays against unknown opponents. Note the contrast here between the use of adaptation to more common machine learning approaches. We are not interested in convergence to an optimal control policy. Rather, given the short time limit of a game, we desire adaptation that achieves good results quickly enough to impact the game. Hence a fast, but non-optimal response is desired over a more optimal but longer acting approach.

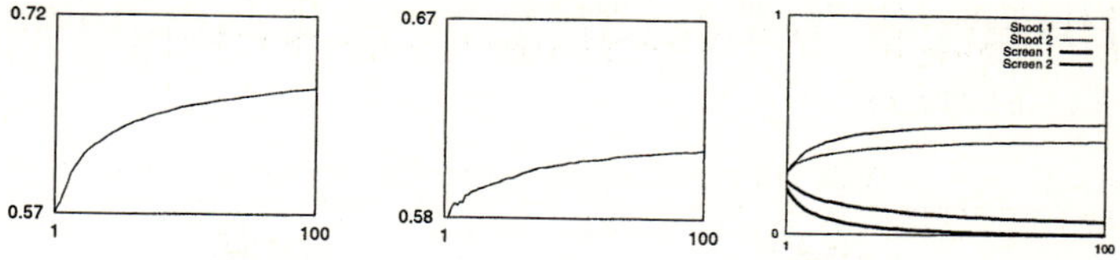

(a) (b) (c)

Fig. 1. (a), (b) show ESR against block and active_def, (c) shows expected play success probabilities against block. These results have all been averaged over 50000 runs of 100 trials.

5 Conclusion

We have introduced a novel team strategy engine based on the concept of a play as a team plan, which can be easily defined by a play language. Multiple, distinct plays can be collected into a playbook where mechanisms for adapting play selection can enable the system to improve the team response to an opponent without prior knowledge of the opponent. The system was fully implemented for our CMDragons robot soccer system and tested at RoboCup 2002, and in the controlled experiments reported here. Possible future directions of research include extending the presented play language, enhancing the play adaptation algorithm.

Acknowledgements

This research was sponsored by Grants Nos. DABT63-99-1-0013, F30602-98-2-013 and F30602-97-2-020. The information in this publication does not necessarily reflect the position of the funding agencies and no official endorsement should be inferred.

References

1. Andreas Birk, Silvia Coradeschi, and Satoshi Tadokoro, editors. *RoboCup 2001: Robot Soccer World Cup V*. Springer Verlag, Berlin, 2002.
2. Brett Browning and Erick Tryzelaar. Ubersim: A multi-robot simulator for robot soccer. In *Proceedings of AAMAS*, 2003.
3. James Bruce, Michael Bolwing, Brett Browning, and Manuela Veloso. Multi-robot team response to a multi-robot opponent team. In *ICRA Workshop on Multi-Robot Systems*, 2002.
4. James Bruce and Manuela Veloso. Real-time randomized path planning for robot navigation. In *Proceedings of IROS-2002*, pages 2383–2388, Switzerland, October 2002.
5. S.S. Intille and A.F. Bobick. A framework for recognizing multi-agent action from visual evidence. In *AAAI-99*, pages 518–525. AAAI Press, 1999.
6. Itsuki Noda, Shóji Suzuki, Hitoshi Matsubara, Minoru Asada, and Hiroaki Kitano. RoboCup-97: The first robot world cup soccer games and conferences. *AI Magazine*, 19(3):49–59, Fall 1998.
7. Patrick Riley and Manuela Veloso. Planning for distributed execution through use of probabilistic opponent models. In *ICAPS-02, Best Paper Award*, Toulouse, France, April 2002.

Progress in Learning 3 vs. 2 Keepaway

Gregory Kuhlmann and Peter Stone

Department of Computer Sciences
The University of Texas at Austin
Austin, Texas 78712-1188
{kuhlmann,pstone}@cs.utexas.edu
http://www.cs.utexas.edu/~{kuhlmann,pstone}

Abstract. Reinforcement learning has been successfully applied to several subtasks in the RoboCup simulated soccer domain. *Keepaway* is one such task. One notable success in the keepaway domain has been the application of SMDP Sarsa(λ) with tile-coding function approximation [9]. However, this success was achieved with the help of some significant task simplifications, including the delivery of complete, noise-free world-state information to the agents. Here we demonstrate that this task simplification was unnecessary and further extend the previous empirical results on this task.

1 Introduction

RoboCup simulated soccer has been a popular test-bed for studying reinforcement learning algorithms over the years. In principle, modern reinforcement learning methods are reasonably well suited to meeting the challenges of RoboCup simulated soccer; and RoboCup soccer is a large and difficult instance of many of the issues which have been addressed in small, isolated cases in previous reinforcement learning research. Despite substantial previous work (e.g., [10, 7]), the extent to which modern reinforcement learning methods can meet these challenges remains an open question.

This article builds upon the work of Stone & Sutton [9] who began scaling reinforcement learning up to RoboCup simulated soccer by considering a subtask of soccer involving fewer than the full 22 players. In particular, they consider the task of *keepaway*, a subproblem of RoboCup soccer in which one team, the *keepers*, tries to maintain possession of the ball within a limited region, while the opposing team, the *takers*, attempts to gain possession. Parameters of the task include the size of the region, the number of keepers, and the number of takers. We have recently incorporated the framework for this domain into the standard, open-source RoboCup soccer simulation software [4].

In their previous work, Stone & Sutton [9] apply episodic SMDP Sarsa(λ) with linear tile-coding function approximation (CMACs [1]) to the keepaway task. The learners choose not from the simulator's primitive actions (e.g. kick, dash, and turn) but from higher level actions constructed from a set of basic skills (implemented by the CMUnited-99 team [8]).

D. Polani et al. (Eds.): RoboCup 2003, LNAI 3020, pp. 694–702, 2004.

Keepers have the freedom to decide which action to take only when in possession of the ball. A keeper in possession may either hold the ball or pass to one of its teammates. Keepers not in possession of the ball are required to select the **Receive** option in which the fastest player to the ball goes to the ball and the remaining players try to get open for a pass.

The keepers' set of state variables are computed based on the positions of: the keepers K_1–K_n and takers T_1–T_m, ordered by increasing distance from K_1; and C, the center of the playing region. Let $dist(a,b)$ be the distance between a and b and $ang(a,b,c)$ be the angle between a and c with vertex at b. For 3 keepers and 2 takers, we used the following 13 state variables: $dist(K_1,C)$; $dist(K_2,C)$; $dist(K_3,C)$; $dist(T_1,C)$; $dist(T_2,C)$; $dist(K_1,K_2)$; $dist(K_1,K_3)$; $dist(K_1,T_1)$; $dist(K_1,T_2)$; $\mathrm{Min}(dist(K_2,T_1), dist(K_2,T_2))$; $\mathrm{Min}(dist(K_3,T_1), dist(K_3,T_2))$; $\mathrm{Min}(ang(K_2,K_1,T_1), ang(K_2,K_1,T_2))$; $\mathrm{Min}(ang(K_3,K_1,T_1), ang(K_3,K_1,T_2))$.

The behavior of the takers is relatively simple. The two fastest takers to the ball go to the ball while the remaining takers try to block open passing lanes.

Using this setup, Stone & Sutton [9] were able to show an increase in average episode duration over time when keepers learned against hand-coded takers. They compared their results with a **Random** policy that chooses among its options with uniform probability, an **Always Hold** policy, and a hand-coded policy that uses a decision tree for pass evaluation. Experiments were conducted on several different field sizes. In each case, the keepers were able to learn policies that outperformed all of the benchmarks. Most of their experiments matched 3 keepers against 2 takers. However, they also showed that their results extend to the 4 vs. 3 scenario.

In the RoboCup soccer simulator, agents typically have limited and noisy sensors: each player can see objects within a 90° view cone, and the precision of an object's sensed location degrades with distance. However, to simplify the task, Stone & Sutton [9] removed these restrictions. The learners were given 360° of noiseless vision. Here, we demonstrate that these simplifications are unnecessary: the agents are able to learn successful policies despite having sensor noise and limited vision. We also extend the results to larger teams and provide further insights into the previous results based on additional controlled experiments. One of our key observations is that a large source of the problem difficulty is the fact that multiple agents learn simultaneously: when a single agent learns in the presence of pre-trained teammates, it is able to do so significantly more quickly.

2 Experimental Setup and Results

This section addresses each of the following questions with focused experiments in the keepaway domain:

1. Does the learning approach described above continue to work if the agents are limited to noisy, narrowed vision?
2. How does a learned policy perform in comparison to a hand-coded policy that has been manually tuned?
3. How robust are these policies to differing field sizes?

4. How dependent are the results on the state representation?
5. How well do the results scale to larger problems?
6. Is the source of the difficulty the learning task itself, or the fact that multiple agents are learning simultaneously?

2.1 Limited Vision

Limited vision introduces two challenges with respect to the complete vision setup of Stone & Sutton. First, the agents must model their uncertainty in the world state and make the appropriate decisions based on those uncertainties. Second, the agents must occasionally take explicit information-gathering actions to increase their confidence in the world state.

To model state uncertainty, a player stores a certainty factor along with each state variable that decays over time. When the player receives sensory information, it updates its world state and resets the certainty factor. If the keeper with the ball does not have reliable information about the position of its teammates, then we force the player to hold the ball and turn its neck until it has enough information to make a pass.

Using this method, we attempted to reproduce the results of Stone & Sutton for 3 vs. 2 keepaway on a $20m \times 20m$ field but without the simplification of unrestricted vision. In their work, keepers were able to learn policies with average episode durations of around 15 seconds. However, learning with noisy, narrowed vision is a more difficult problem than learning with complete knowledge of the state. For this reason, we expected our learners to hold the ball for less time. However, since these same difficulties impact the benchmark policies, the salient question is whether or not learning is still able to outperform the benchmarks.

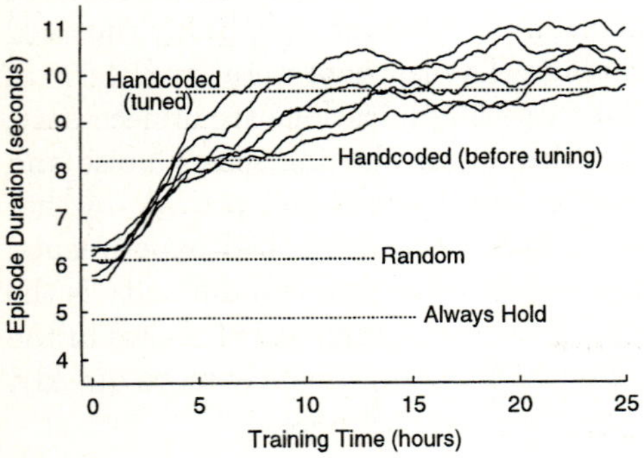

Fig. 1. Learning curves and benchmarks for limited vision: 3 keepers vs. 2 takers.

We ran a series of 6 independent learning trials in which the keepers learned while playing against the hand-coded takers. In each run, the keepers gradually improved their performance before leveling off after about 25 hours of simulator time. The learning curves are shown in Figure 1. We plotted all 6 trials to give a sense of the variance.

All of the learning runs were able to outperform the **Always Hold** and **Random** benchmark policies. The learned policies also outperformed our **Hand-coded** policy which we describe in the next section.

2.2 Comparison to Hand-Coded

In addition to the **Always Hold** and **Random** benchmark policies described previously, we compared our learners to a new **Hand-coded** policy. In this policy, the keeper in possession, K_1 assigns a score to each of its teammates based on how "open" they are. The degree to which the player is open is calculated as a linear combination of the teammate's distance to its nearest opponent, and the angle between the teammate, K_1, and the opponent closest to the passing lane. The relative importance of these two features are weighted by the coefficient α. If the most open teammate has a score above the threshold, β, then K_1 will pass to this player. Otherwise, K_1 will hold the ball for one cycle.

This **Hand-coded** policy was designed to use only state variables and calculations that are available to the learner. We chose initial values for α and β based on educated guesses. We tuned these values by experimenting with values near our initial guesses. Altogether, we tried about 30 combinations, before settling on our final tuned values.

We ran a few thousand episodes of our tuned **Hand-coded** policy and found that it was able to keep the ball for an average of 9.6 seconds per episode. Also, for comparison, we tested our **Hand-coded** policy before manual tuning. This policy was able to hold the ball for an average of 8.2 seconds. From Figure 1 we can see that the keepers are able to learn policies that outperform our initial **Hand-coded** policy and exhibit performance roughly as good as (perhaps slightly better than) the tuned version. We examined the **Hand-coded** policy further to find out to what degree its performance is dependent on tuning.

2.3 Robustness to Differing Field Sizes

Stone & Sutton already demonstrated that learning is robust to changes in field sizes [9]. Here we verify that learning is still robust to such changes even with the addition of significant state uncertainty. We also benchmark these results against the robustness of the **Hand-coded** policy to the same changes. Overall, we expect that as the size of the play region gets smaller, the keepers will have a harder time maintaining possession of the ball regardless of policy. Here we compare the **Hand-coded** policy to learned policies on five different field sizes. The average episode durations for both solutions are shown in Figure 2. Each value for the learned runs was calculated as an average of six separately learned policies.

	Keeper Policy	
Field Size	Hand-coded	Learned ($\pm 1\sigma$)
30×30	**19.8**	18.2 ± 1.1
25×25	**15.4**	14.8 ± 0.3
20×20	9.6	$\mathbf{10.4 \pm 0.4}$
15×15	6.1	$\mathbf{7.4 \pm 0.9}$
10×10	2.7	$\mathbf{3.7 \pm 0.4}$

Fig. 2. Average possession times (in simulator seconds) for hand-coded and learned policies on various field sizes.

As can be seen from the table, the hand-coded policy does better on the easier problems ($30m \times 30m$ and $25m \times 25m$), but the learned policies do better on the more difficult problems.

A possible explanation for this result is that the easier cases of keepaway have more intuitive solutions. Hence, these problems lend themselves to a hand-coded approach. However, without any impetus to choose a simpler approach, learned policies tend to be more asymmetric and irregular. This lack of rhythm seems to lead to suboptimal performance on easier tasks.

In contrast, when the keepers are forced to play in a smaller area, the "intuitive" solution breaks down. The hand-coded keepers tend to pass too frequently, leading to missed passes. In these more difficult tasks, the trained keepers appear to find "safer" solutions in which the ball is held for longer periods of time. This approach leads to fewer missed passes and better overall performance than the hand-coded solution.

2.4 Changing the State Representation

A frequent challenge in machine learning is finding the correct state representation. In all of the experiments reported so far, we have used the same state variables as in Stone & Sutton's work, which were chosen without any detailed exploration [9]. Here we explore how sensitive the learning is to the set of state variables used.

As a starting point, notice that our **Hand-coded** policy uses only a small subset of the 13 state variables mentioned previously. Because the **Hand-coded** policy did quite well without using the remaining variables, we wondered if perhaps the unused state variables were not essential for the keepaway task.

To test this theory, we performed a series of learning runs in which the keepers used only the five variables from the hand-coded policy. Figure 3 shows the learning curves for six runs. As is apparent from the graph, the results are very similar to those in Figure 1. Although we found that the keepers were able to achieve better than random performance with as little as one state variable, the five variables used in the hand-coded policy seem to be minimal for peak performance. Notice by comparing Figures 1 and 3 that the keepers are able to learn at approximately the same rate whether the nonessential state variables are present or not.

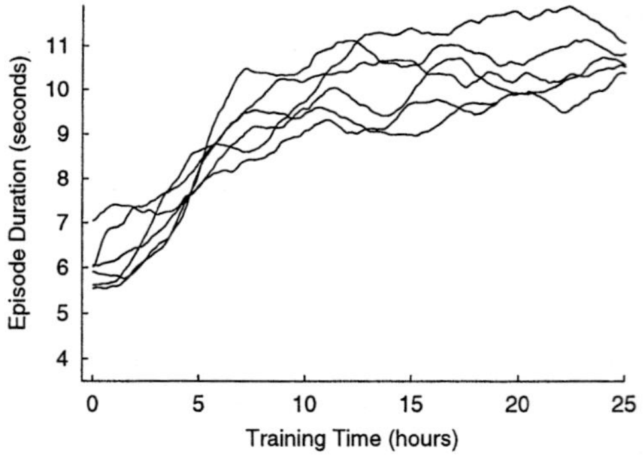

Fig. 3. Learning with only the 5 state variables from the **Hand-coded** policy.

To explore this notion further, we tried adding additional state variables to the original 13. We ran two separate experiments. In the first experiment, we added 2 new angles that appeared relevant but perhaps redundant. $ang(K_1, C, K_2)$; $ang(K_1, C, K_3)$. In the second experiment, we added 2 completely irrelevant variables: each time step, new values were randomly chosen from $[-90,90]$ with uniform probability.

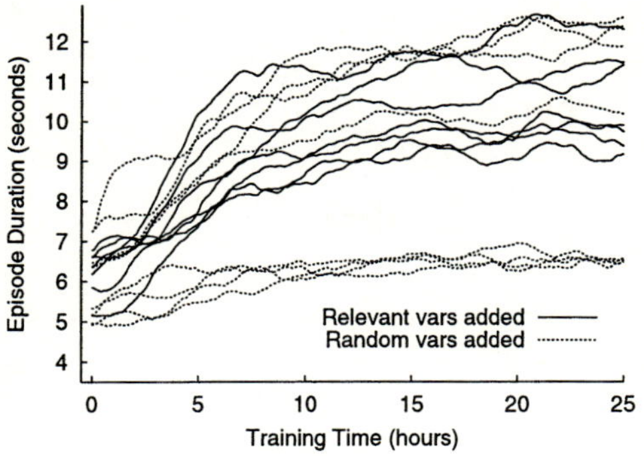

Fig. 4. Learning with the original 13 state variables plus an additional two.

From Figure 4, we can see that the learners are not greatly affected by the addition of relevant variables. The learning curves look roughly the same as the ones that used the original 13 state variables (Figure 1). However, the curves corresponding to the additional random variables look somewhat different. The curves can clearly be divided into two groups. In the first group, teams are able to perform about as well as the ones that used the original 13 variables. In the second group, the agents perform very poorly. It appears that agents in the second group are confused by the irrelevant variables while the agents in the first group are not. This distinction seems to be made in the early stages of learning (before the 1000th episode corresponding to the first data point on the graph). The learning curves that start off low stay low. The ones that start off high continue to ascend.

From these results, we conclude that it is important to choose relevant variables for the state representation. However, it is unnecessary to carefully choose the minimum set of these variables.

2.5 Scaling to Larger Problems

In addition to our experiments with 3 vs. 2 keepaway, we ran a series of trials with larger team sizes to determine how well our techniques scale. First we performed several learning runs with 4 keepers playing against 3 hand-coded takers. We compared these to our three benchmark policies. The results are shown in Figure 5. As in the 3 vs. 2 case, the players are able to learn policies that outperform all of the benchmarks.

We also ran a series of experiments with 5 vs. 4 keepaway. The learning curves for these runs along with our three benchmarks are shown in Figure 6. Again, the learned policies outperform all benchmarks. As far as the authors are aware, these experiments represent the largest scale keepaway problems that have been successfully learned to date.

From these graphs, we see that the learning time approximately doubles every time we move up in size. In 3 vs. 2, the performance plateaus after roughly (by eyeballing the graphs) 15 hours of training. In 4 vs. 3, it takes about 30 hours to learn. In 5 vs. 4, it takes about 70 hours.

2.6 Difficulty of Multiagent Learning

A key outstanding question about keepaway is whether it is difficult as an individual learning task, or if the multiagent component of the problem is the largest

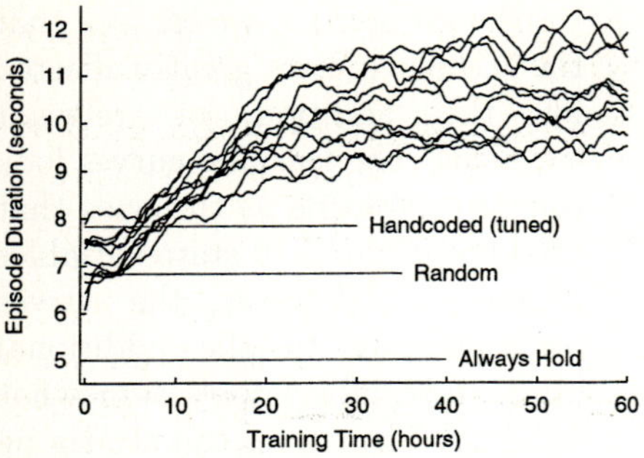

Fig. 5. Training 4 Keepers against 3 takers with benchmarks.

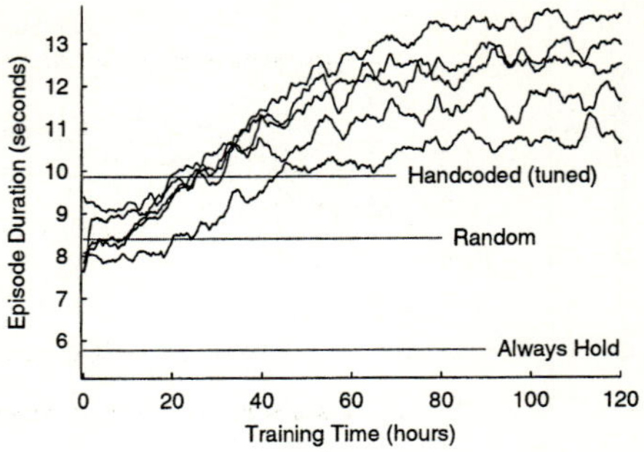

Fig. 6. Training 5 Keepers against 4 takers with benchmarks.

source of difficulty. To see how the number of agents learning simultaneously affects the overall training time, we ran a series of experiments in which a subset of the keepers learned while the remaining teammates followed a fixed policy learned previously. We ran each experiment three times. The learning curves for all nine runs are shown in Figure 7.

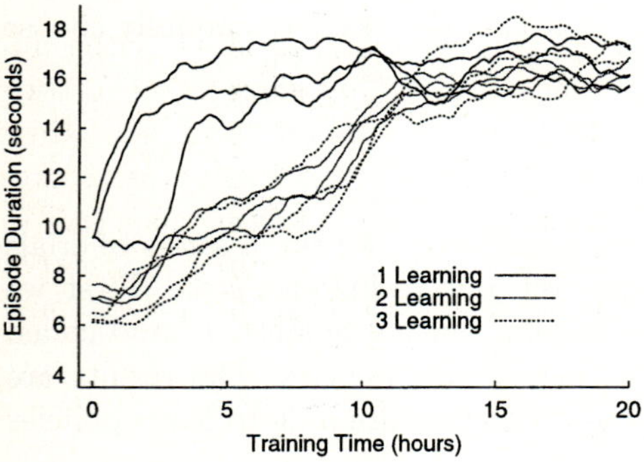

Fig. 7. Learning curves for varying number of keepers learning simultaneously.

From the graph we can see that the learning curves for 2 learning agents and 3 learning agents look roughly the same. However, the runs with only 1 player learning peak much sooner. Apparently, having pre-trained teammates allows an agent to learn much faster. However, if more than one keeper is learning, the presence of a pre-trained teammate is not helpful. This result suggests that multiagent learning is an inherently more difficult problem than single agent learning, at least for this task. In the long run, all three configurations' learned policies are roughly equivalent. The number of learning agents does not seem to affect the quality of the policy, only the rate at which the policy is learned.

3 Related Work

Several previous studies have used keepaway soccer as a machine learning testbed. Whiteson & Stone [11] used neuroevolution to train keepers in the SoccerBots domain [3]. The players were able to learn several conceptually different tasks from basic skills to higher-level reasoning using a hierarchical approach they call "concurrent layered learning." The keepers were evaluated based on the number of completed passes. Hsu & Gustafson [5] evolved keepers for 3 vs. 1 keepaway in the much simpler and more abstract TeamBots simulator [2]. Keepers were trained to minimize the number of turnovers in fixed duration games. It is difficult to compare these approaches to ours because they use different fitness functions and different game dynamics.

More comparable work to ours applied evolutionary algorithms to train 3 keepers against 2 takers in the RoboCup soccer simulator [6]. Similar to our work, they focused on learning keepers in possession of the ball. The keepers chose from the same high-level behaviors as ours. Also, they used average episode duration to evaluate keeper performance. However, because their high-level behaviors and basic skills were implemented independently from ours, it is difficult to compare the two learning approaches empirically. Additional related work is discussed in [9].

4 Conclusion and Future Work

Taken together, the results reported in this paper show that SMDP Sarsa(λ) with tile-coding scales further and is more robust than has been previously shown. Even in the face of significant sensor noise and hidden state, it achieves results at least as good as those of a tuned hand-coded policy.

The main contribution of this paper is a deeper understanding of the difficulties of scaling up reinforcement learning to RoboCup soccer. We focused on the keepaway task and demonstrated that players are able to improve their performance despite having noisy, narrowed vision. We also introduced a new hand-coded policy and compared it for robustness to our learned policies. We demonstrated the difficulty of scaling up current methods and provided evidence that this difficulty arises mainly out of the fact that several agents are learning simultaneously.

References

1. J. S. Albus. *Brains, Behavior, and Robotics.* Byte Books, Peterborough, NH, 1981.
2. Tucker Balch. Teambots, 2000. `http://www.teambots.org`.
3. Tucker Balch. Teambots domain: Soccerbots, 2000. `http://www-2.cs.cmu.edu/~trb/TeamBots/Domains/SoccerBots`.
4. RoboCup Soccer Simulator Maintenance Group. The robocup soccer simulator, 2003. `http://sserver.sourceforge.net`.
5. W. H. Hsu and S. M. Gustafson. Genetic programming and multi-agent layered learning by reinforcements. In *Genetic and Evolutionary Computation Conference*, New York,NY, July 2002.
6. Anthony Di Pietro, Lyndon While, and Luigi Barone. Learning in RoboCup keepaway using evolutionary algorithms. In W. B. Langdon, E. Cantú-Paz, K. Mathias, R. Roy, D. Davis, R. Poli, K. Balakrishnan, V. Honavar, G. Rudolph, J. Wegener, L. Bull, M. A. Potter, A. C. Schultz, J. F. Miller, E. Burke, and N. Jonoska, editors, *GECCO 2002: Proceedings of the Genetic and Evolutionary Computation Conference*, pages 1065–1072, New York, 9-13 July 2002. Morgan Kaufmann Publishers.

7. M. Riedmiller, A. Merke, D. Meier, A. Hoffman, A. Sinner, O. Thate, and R. Ehrmann. Karlsruhe brainstormers – a reinforcement learning approach to robotic soccer. In Peter Stone, Tucker Balch, and Gerhard Kraetszchmar, editors, *RoboCup-2000: Robot Soccer World Cup IV*. Springer Verlag, Berlin, 2001.
8. Peter Stone, Patrick Riley, and Manuela Veloso. The CMUnited-99 champion simulator team. In M. Veloso, E. Pagello, and H. Kitano, editors, *RoboCup-99: Robot Soccer World Cup III*, pages 35–48. Springer Verlag, Berlin, 2000.
9. Peter Stone and Richard S. Sutton. Scaling reinforcement learning toward RoboCup soccer. In *Proceedings of the Eighteenth International Conference on Machine Learning*, pages 537–544. Morgan Kaufmann, San Francisco, CA, 2001.
10. Peter Stone and Manuela Veloso. Team-partitioned, opaque-transition reinforcement learning. In Minoru Asada and Hiroaki Kitano, editors, *RoboCup-98: Robot Soccer World Cup II*. Springer Verlag, Berlin, 1999. Also in *Proceedings of the Third International Conference on Autonomous Agents*, 1999.
11. Shimon Whiteson and Peter Stone. Concurrent layered learning. In *Second International Joint Conference on Autonomous Agents and Multiagent Systems*, July 2003.

Distributed Control of Gait for a Humanoid Robot

Gordon Wyeth and Damien Kee

School of Information Technology and Electrical Engineering
University of Queensland, St. Lucia, Queensland, 4072, Australia
{wyeth,damien}@itee.uq.edu.au

Abstract. This paper describes a walking gait for a humanoid robot with a distributed control system. The motion for the robot is calculated in real time on a central controller, and sent over CAN bus to the distributed control system. The distributed control system loosely follows the motion patterns from the central controller, while also acting to maintain stability and balance. There is no global feedback control system; the system maintains its balance by the interaction between central gait and "soft" control of the actuators. The paper illustrates a straight line walking gait and shows the interaction between gait generation and the control system. The analysis of the data shows that successful walking can be achieved without maintaining strict local joint control, and without explicit global balance coordination.

1 Introduction

Humanoid robots typically require coordinated control of a large number of joints. In most existing implementations of humanoid robots, coordination is achieved by the use of a central control computer that interfaces to all sensors and actuators providing local control of joint positions and torques as well as global control of balance and posture. This paper describes a distributed approach to control and coordination that provides local control of position and torque at each joint in a fashion that maintains global balance and posture.

1.1 Paper Overview

After a brief description of related work, the paper describes the GuRoo robot that forms the basis for the later experiments. Details of the architecture and design of the robot are followed by a description of the computing system that supports the distributed control system. The paper then describes the approach to distributed control and provides details of gait generation, including results gathered from a straight line walk.

2 Related Work

The OpenPino project [Yamasaki, 2000], utilizes a very centralized approach to humanoid control. An onboard SH2 micro-controller is responsible for taking high level commands from a PC, such as walk forward, stop etc, and converts them into position

D. Polani et al. (Eds.): RoboCup 2003, LNAI 3020, pp. 703–711, 2004.

information. These positions are converted into PWM signals by a single CLPD, responsible for controlling all 26 degrees of freedom.

The University of Waseda's humanoid, WABIAN, employs a similar control system, with high level commands generated by the onboard Pentium 166Mhz computer[Yamaguchi, 1998]. The resulting velocity profiles are fed into one of two 16 channel D/A boards via an ISA bus, which supply the motor drivers the required signals to actuate the motors. This system is physically centralized with all computational equipment and motor drivers located in the torso.

Similarly, H6 from the University of Tokyo makes use of an onboard PIII 700Mhz computer to generate high level commands [Nishiwaki, 2000]. A pair of Fujitsu I/O controller, similar to WABIAN's, generates the control signals necessary for the motors. Individual motor drivers supplying the necessary power are located physically close to each motor.

2.1 The GuRoo Project

GuRoo is a 1.2 m tall, fully autonomous humanoid robot designed and built in the University of Queensland Robotics Laboratory [Wyeth, 2001]. The robot has a total mass of 34 kg, including on-board power and computation. *GuRoo* is currently capable of a number of demonstration tasks including balancing, walking, turning, crouching, shaking hands and waving. The robot has performed live demonstrations of combinations of these tasks at various robot displays.

The intended challenge task for the robot is to play a game of soccer with or against human players or other humanoid robots. To complete this challenge, the robot must be able to move freely on its two legs. Clearly, the robot must operate in a completely autonomous fashion without support harnesses or wiring tethers. The current GuRoo robot cannot withstand the impacts associated with playing soccer, but serves as an excellent platform for research into the design of balance behaviors and dynamic gait control. The location and axis of actuation of each joint can be seen in Figure 1.

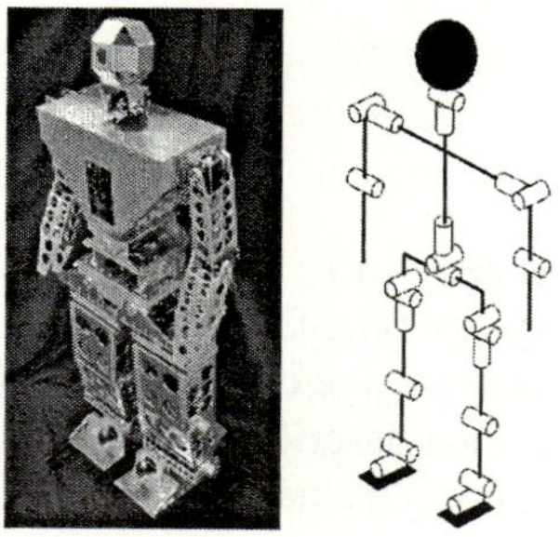

Fig. 1. The GuRoo Humanoid robot and the degrees of freedom of each joint.

2.2 Electro-Mechanical Design

The key element in driving the mechanical design has been the choice of actuator. The robot has 23 joints in total. The legs and abdomen contain 15 joints that are required to produce significant mechanical power, most generally with large torques and relatively low speeds. The other 8 joints drive the head and neck assembly, and the arms with significantly less torque and speed requirements.

The 15 high power joints all use the same motor-gearbox combination consisting of a Maxon RE 36 motor with a gearbox reduction of 156:1. The maximum continuous generated output torque is 10 Nm. Each motor is fitted with an optical encoder for position and velocity feedback. The 8 low power joints are Hi-Tec RC servo motors model HS705-MG. with rated output torque to 1.4 Nm.

The motors that drive the roll axis of the hip joints are supplemented by springs with a spring constant of 1 Nm/degree. These springs serve to counteract the natural tendency of the legs to collide, and help to generate the swaying motion that is critical to the success of the walking gait.

Power is provided by 2 × 1.5Ah 42V NiCd packs for the high power motors, and 2 x 3Ah 7.2 V NiCd battery packs for computing and servo operation. The packs are chosen to give 20 minutes of continuous operation.

2.3 Sensing

The position feedback from the encoders on the high power joints provides 867 encoder counts per degree of joint motion. In addition, each DSP can measure the current to each motor. Provision has also been made for inertial and balance sensors, as well as contact switches in the feet and in the joints.

3 Distributed Control Network

A distributed control network controls the robot, with a central computing hub that sets the goals for the robot, processes the sensor information, and provides coordination targets for the joints. The joints have their own control processors that act in groups to maintain global stability, while also operating individually to provide local motor control. The distributed system is connected by a CAN network.

3.1 Central Control

The central control of the robot derives the joint velocities required to perform the walking gait. A PIII 1.1GHz laptop currently calculates velocities for all 23 degrees of freedom in real time. The velocities are passed along a serial link to a distribution board which serves as a bridge between the serial bus and Control Area Network (CAN) Provision has been made to port this control to a Compaq IPAQ mounted on the robot to enable true autonomy, free of any tethers.

3.2 Joint Controllers

All joint controllers are implemented using a TMS320F243 Digital Signal Processor from Texas Instruments, a 16 bit DSP designed for motor control. The availability of the CAN module in this series, along with bootloader programmable internal Flash memory makes the device particularly attractive for this application. Five controller boards control the 15 high power motors, each board controlling three motors. A sixth controller board controls the eight RC servo motors. Figure 2 outlines the interaction between the various node on the control network.

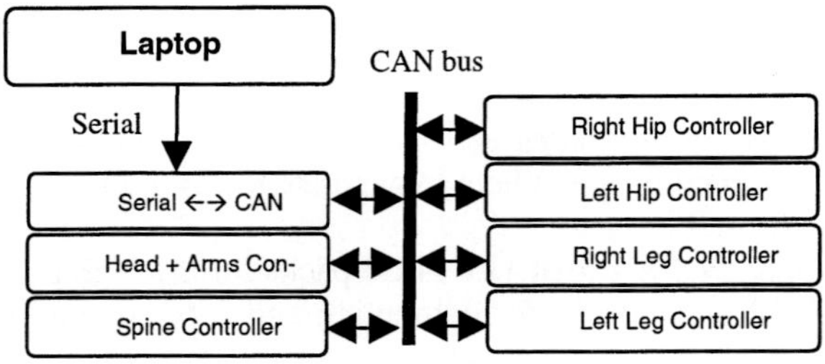

Fig. 2. Block diagram of the distributed control system.

4 Software

The software consists of four main entities: the global movement generation code, the local motor control, the low-level code of the robot, and the simulator. The software is organized to provide a standard interface to both the low-level code on the robot and the simulator. This means that the software developed in simulation can be simply re-compiled to operate on the real robot. Consequently, the robot employs a number of standard interface calls that are used for both the robot and the simulator including reading the encoders and setting PWM values as well as the transfer of CAN packets.

4.1 Gait Generation

The gait generation module is responsible for producing realizable trajectories for the 23 joints so that the robot can perform basic behaviors, such as standing on one leg, crouching and walking. The most important properties of the trajectories are that they are smooth and that they can be linked together in a smooth fashion. Smoothness of motion implies less disturbance to the control of other joints. Based on these criteria, a normalized joint movement as shown in Figure 3 is applied to all motor trajectories.

The trajectories are generated from a parameterized sinusoidal curve where ω is the desired joint velocity, θ is the total joint angle to move and T is the period of that movement. The trajectory contrasts with typical trajectories generated for robotic manipulators, which typically focus on smoothness of the end effector motion rather than smoothness at the joint.

Trajectories for each of the motors may be coordinated by using the same beginning time for the motion and specifying the same period for the trajectory. Trajectories may be naturally linked as the velocities all reach zero at the beginning and the end of a motion. Section 5 will illustrate how trajectories may be coordinated and linked to perform a walking operation.

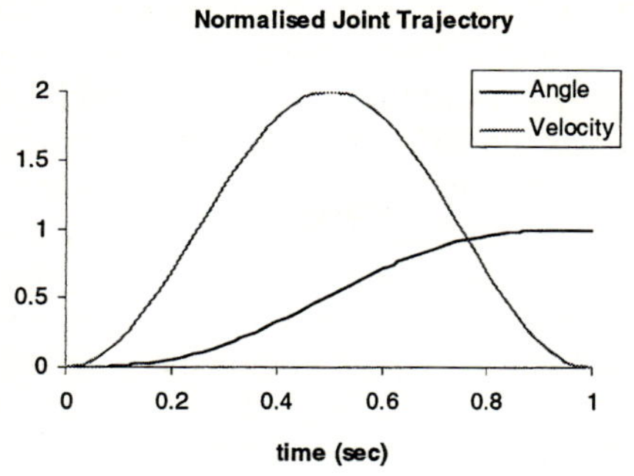

$$\omega = \frac{\theta}{T}\left(1 - \cos\left(\frac{2\pi t}{T}\right)\right)$$

Fig. 3. The trajectory used for a total joint movement of 1 radian over a period of 1 second.

4.2 Joint Controller Software

There are two types of joint controller boards used in the robot – five controller boards control the fifteen high power motors and one controller controls the eight low power motors. The controller software for the low power motors is a single interrupt routine that is triggered by the arrival of a CAN packet addressed to the controller's mailbox. The routine reads the CAN mailbox for the change in position sent by the gait generation routine. The PWM duty cycle that controls the position of the RC servos is varied accordingly.

The control loop for the high power controllers has two interrupt routines. As for the low power controller, an interrupt is executed upon receipt of trajectory data in the CAN mailbox. The data is used to set the velocity setpoints for the motor control routine. There is also a periodic interrupt every 500 μs to run the motor control software. The motor control routine compares the error between velocity setpoint and the encoder reading and generates a PWM value for the motor based on a Proportional-Integral control law. The routine also checks the motor current against the current limits, and adjusts the PWM value to prevent over-current situations.

The PI control law on each joint has been hand tuned to provide both good trajectory following for typical velocity input profiles, and spring-damper model impedance to torque disturbances from gravity and the cross-coupling torques from other joints. The "soft" response of this control law to disturbance prevents torques being transmitted throughout the robot and helps to maintain global stability. The disadvantage is that the controller suffers from position error that must be accounted in the gait generation software. Section 5 illustrates how this potential liability is turned to an asset in the generation of a dynamically stable walk.

4.3 Low-Level Code

The lowest level of code on the robot provides direct access to the sensors and communication system. The level of abstraction provided by function calls at this level aids in the cross development of code between the simulator and the real robot.

4.4 Simulator

The simulator is based on the *DynaMechs* project [McMillan, 1995], with additions to simulate specific features of the robot such as the DC motors and motor drives, the RC servos, the sensors, the heterogeneous processing environment and the CAN network. These additions provide the same interface for the dynamic graphical simulation as for the joint controller and gait generation code. The parameters for the simulator are derived from the CAD models and the data sheets from known components. These parameters include the modified Denavit-Hartenberg parameters that describe the robot topology, the tensor matrices of the links and the various motor and gearbox characteristics associated with each joint. The surface data from the CAD model is also imported to the simulator for the graphical display.

For the high power DC motor joints, the simulator provides the programmer with readings from the encoders and the current sensors, based on the velocities and torques from the dynamic equations. In the case of the RC servos, the simulator updates the position of the joints based on a PD model with a limited slew rate. The programmer must supply the simulator with PWM values for the motors to provide the control. The simulator provides fake interrupts to simulate the real events that are the basis of the control software.

The simulator uses an integration step size of 500μs and updates the graphical display every 5ms of simulated time. When running on 1.5 GHz Pentium 4 under Windows 2000, the simulation updates all 23 joints at a very useable 40% of real time speed.

5 Walking

The robot can walk with a step rate of 1 Hz using a step length of 100 mm. The walk is open-loop; there is no feedback from the joint controllers to the gait generation software. The lack of global feedback, combined with the absence of a global balance sensor presents a substantial challenge in walking algorithm design.

5.1 Walking Algorithm

The robot uses a simplified version of a typical human gait. In particular, it limits the swing of the legs to prevent balance disturbance as this cannot be corrected without global balance control. In order to minimize the accelerations of the torso, head and arms (which make up 1/3 of the mass of the robot), the robot maintains a constant relative position of the torso, such that the face of the torso is always normal to the direction of travel. The allowable roll of the torso is also limited. The stabilization of the torso also reduces disturbances from gravity to the control of the leg joints.

Before walking, the robot loads each motor against gravity by performing a slight squat that introduces a 6 degree ankle pitch, with the knee and hip pitch joints set to keep the torso upright. The initial loading of the joints reduces the likelihood of backlash in the gearheads.

The walking gait commences with a side-to-side sway generated from the roll axes of the ankles and hips. The sway frequency of 0.5 Hz is sympathetic with the spring mass system formed by the ankle controllers with the mass of robot. The sway sets up the pattern of weight transfer from one foot to the other necessary to swing the legs alternately to achieve walking. At each extreme of the sway, the inertia of the upper body ensures the ZMP (Zero Moment Point) of the robot lies within the support polygon formed by the support foot, even thought the centre of mass may not. This action places requires less torque from the hip and ankle roll actuators, as the motion due to gravity brings the robot away from the extreme of each sway.

Fig. 4. Frontal view of the walking process.

Once the sway is sufficient to leave no ground reaction force on the non-supporting foot, the non-supporting leg is lifted using the pitch axes of the hip, knee and ankle. Between the lifting and lowering of the non-supporting leg, each yaw axis motor twists, so that the non-supporting leg swings forward to create the step. When the swing leg contacts the ground, the robot is dynamically stable, with the centre of mass over the supporting foot in the frontal plane, but in front of the toes in the sagittal plane. The robot then swings across to the other foot, repeating the sequence and progressing with the walk.

5.2 Analysis of Results

The motion of the robot is best analyzed by comparing the desired velocity from the gait generation module to the actual velocity at each joint. Figure 5 shows this comparison for the motion of the hip, knee and ankle in the roll, pitch and yaw axis. The graphs are initialized midway through the double support phase, with both legs in contact with the ground. The graphs comprise one second of data, describing the right leg as it moves from the double support phase, through the swing phase back to the double support phase.

At the point $t = 0.25$ s, the swing leg starts to lift and loses contact with the ground. With the hip roll axis of the swing leg no longer contributing to the support of the robot, the spring located in this axis briefly dominates the actual velocity causing the overdamped oscillation seen at hip joint at this time.

Once the foot leaves the ground the ankle roll motor switches from driving the leg from the foot, to driving the foot from the leg. This large decrease in relative inertia results in a brief increase in the magnitude of the ankle roll velocity. The foot has a relatively low inertia compared with the rest of the robot, and as such the PI controller has little trouble following the desired velocity until the foot again makes contact with the ground. The robot reaches the extreme of each sway at $t = 0.5$ s, where all motion in the roll plane ceases. The swing leg is now theoretically fully lifted, although the knee and hip pitch do not reach their desired positions until T=0.6s.

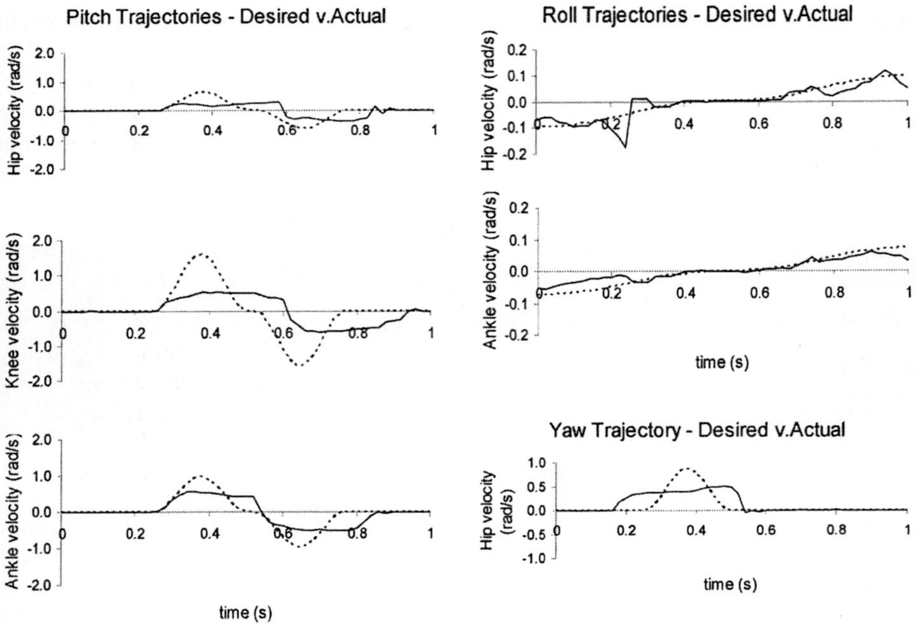

Fig. 5. Desired vs Actual joint angles for straight line walking over a 1 second period. Graphs start during the double support phase and follow the right leg through the swing phase.

The actual joint velocity profile for each pitch motor in the swing leg shows the increasing effect of gravity and leg inertia through the swing stage. The integral term in the PI controller seeks to eliminate steady state error, and as such, dominates the actual velocity, driving each pitch motor to its desired position. As the motor does not reach the maximum desired velocity, it is forced to lengthen the movement time to ensure the areas under each graph are equal. The low proportional term results in the poor tracking of the desired velocity, but enables the joint to better deal with external disturbances.

The comparison of the actual velocity with the desired velocity for each pitch motor in each leg degrades from the ankle to the knee to the hip. The ankle need only accelerate the foot, whereas the knee must accelerate the foot and lower leg. The hip pitch must accelerate the entire leg during the swing phase.

The motion of the yaw axis as the swing leg is lifted, propels the robot forward. When the yaw motion occurs on the support leg, the momentum of the robot causes the joint to overshoot its position. The swing leg is then lowered placing the robot into a double support phase. The friction of two feet against the ground and the weight of the robot on the support leg prevents the yaw axis positional error from being resolved. This provides a pre-loading of the joint that supports the motion of the next of yaw swing phase.

As the robot sways back to the other side, weight is gradually released from the supporting foot, until the torque acting on the joint overcomes the co-efficient of friction between the foot and the floor. This is not necessarily the point at which the swing leg loses contact with the ground. By time the leg has resolved this error, the joint is experiencing the yaw motion associated with the swing leg twist. As a result, the area under the curve for the hip yaw during the swing phase is greater than the desired area.

Contact with the ground is achieved at T=0.85s and once made, the robot returns to the double support phase of its gait. Both the hip and ankle of the swing leg now assist the support leg roll motors to sway the robot across to the other foot, and in the process gradually switch the roles of the support and swing leg

6 Conclusions

This paper has illustrated that a humanoid robot can walk without the need for explicit global feedback, or tightly controlled joint trajectories. By combining a group of loosely coordinated control systems that use "soft" control laws with smooth trajectory generation, the robot can use the natural dynamics of its mechanical structure to move through a gait pattern. The work in this paper shows sound walking performance that can only improve with the augmentation of global inertial sensors and feedback paths.

References

[McMillan, 1995] S. McMillan, Computational Dynamics for Robotic Systems on Land and Underwater, PhD Thesis, Ohio State University, 1995.

[Nishiwaki, 2000] K. Nishiwaki, T. Sugihara, S. Kagami, F. Kanehiro, M. Inaba and H. Inoue, Design and development of research platform for perception-action integration in humanoid robot: H6, International Conference on Intelligent Robots and Systems, IROS 2000

[Wyeth, 2001] G. Wyeth, D. Kee, M. Wagstaff, N. Brewer, J. Stirzaker, T. Cartwright, B. Bebel. Design of an Autonomous Humanoid Robot, Proceedings of the Australian Conference on Robotics and Automation (ACRA 2001), 14-15 November 2001,Sydney

[Yamaguchi, 1998] J. Yamaguchi, S. Inoue, D. Nishino and A. Takanishi, Development of a bipedal humanoid robot having antagonistic driven joints and three DOF trunk, Proceedings International Conference on Intelligent Robots and Systems, IROS 1998

[Yamasaki, 2000] F. Yamasaki, T. Matsui, T. Miyashita, and H. Kitano. PINO the Humanoid that Walk, Proceedings of First IEEE-RAS International Conf on Humanoid Robots, CDROM 2000

Predicting Away Robot Control Latency

Sven Behnke[2], Anna Egorova[1], Alexander Gloye[1],
Raúl Rojas[1], and Mark Simon[1]

[1] Freie Universität Berlin, Institute for Computer Science
Takustraße 9, 14195 Berlin, Germany
[2] International Computer Science Institute
1947 Center St., Berkeley, CA, 94704, USA
`http://www.fu-fighters.de`

Abstract. This paper describes a method to reduce the effects of the system immanent control delay for the RoboCup small size league. It explains how we solved the task by predicting the movement of our robots using a neural network. Recently sensed robot positions and orientations as well as the most recent motion commands sent to the robot are used as input for the prediction. The neural network is trained with data recorded from real robots.

We have successfully field-tested the system at several RoboCup competitions with our *FU-Fighters* team. The predictions improve speed and accuracy of play.

1 Introduction

The time elapsed between making an action decision and perceiving the consequences of that action in the environment is called the control delay. All physical feedback control loops have a certain delay, depending on the system itself, on the input and output speed and, of course, on the speed at which the system processes information.

In the RoboCup small size league a global vision module is used to determine the positions of all robots on the field. In order to control the behavior of the robots in a way that is appropriate for the situation on the field we need the exact positions of them at every moment. Because of the delay inherent in the control loop, however, the behavior control system actually reacts to the environment four frames ago (about 132 ms). When moving faster – our robots drive at a speed of up to 2 m/s – the delay becomes more significant as the error between the real position and the position used for control grows up to 20 cm.

In order to correct this immanent error we have developed a neural network which processes the positions, orientations, and the action commands sent to the robots during the last six frames. It predicts the actual positions of the robots. These predictions are used as a basis for control. We use real recorded preprocessed data of moving robots to train the network.

The concept of motor prediction was first introduced by Helmholtz when trying to understand how humans localize visual objects (see [6]). His suggestion

D. Polani et al. (Eds.): RoboCup 2003, LNAI 3020, pp. 712–719, 2004.

was that the brain predicts the gaze position of the eye, rather than sensing it. In his model the predictions are based on a copy of the motor commands acting on the eye muscles. In effect, the gaze position of the eye is made available before sensory signals become available.

The paper is organized as follows. The next section gives a brief description to our system architecture. Then we explain how the delay is measured and we present some other approaches to eliminate the dead time. Section 4 describes architecture and training of the neural network used as a predictor. Finally, we present some experimental results and some plans for future research.

2 System Architecture

The small size league is the fastest physical robot league in the RoboCup competition relative to the field size. Top robot speeds exceed 2m/s and acceleration is limited by the traction of the wheels only, hence a robot can cross the entire field in about 1.5 sec. This is possible, since the sensing is done mainly by a camera overlooking the field and behavior control is done mainly by an off-the-field computer. Action commands are sent via a wireless link to the robots that contain only minimal local intelligence. Thus, the robot designer can focus in this league on speed, maneuverability, and ball handling.

Our control system is illustrated in Fig. 1. The only physical sensor we use for behavior control is one S-Video camera[1]. It looks at the field from above and produces an output video stream, which is forwarded to the central PC. Images are captured by a frame grabber and given to the vision module.

The global computer vision module analyzes the images, finds and tracks the robots and the ball and produces as output the positions and orientations of the robots, as well as the position of the ball. It is described in detail in [3].

Based on the gathered information, the behavior control module then produces the commands for the robots: desired rotational velocity, driving speed and direction, as well as the activation of the kicking device. The central PC then sends these commands via a wireless communication link to the robots. The hierarchical reactive behavior control system of the FU-Fighters team is described in [2].

Each robot contains a microcontroller for omnidirectional motion control. It receives the commands and controls the movement of the robot using PID controllers (see [4]). Feedback about the speed of the wheels is provided by the pulse generators which are integrated in each motor.

3 Delay: Measurement, Consequences, and Approaches

As with all control systems, there is some delay between making an action decision and perceiving the consequences of that action in the environment. All

[1] There are various other sensors on the robots and in the system, but they aren't used for behavior control. For example, the encoders on the robots are only used for their motion control.

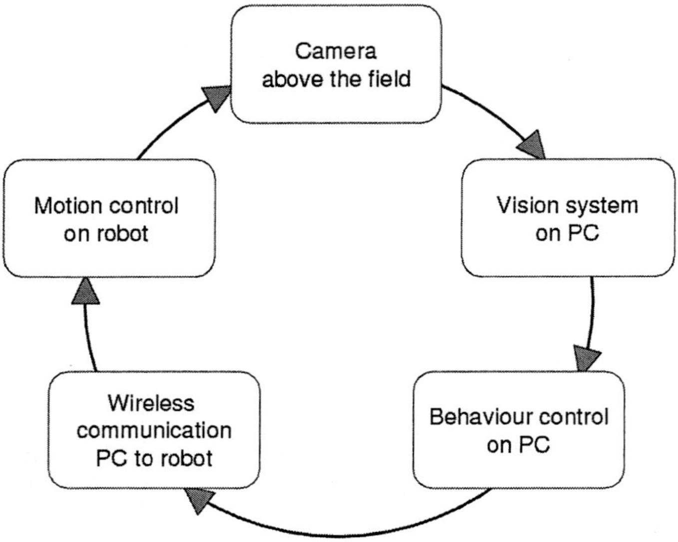

Fig. 1. The feedback control system. All stages have a different delay. But only the overall delay is essential for the prediction.

stages of the control loop contribute to the control delay that is also known as dead time. Although image capture, computer vision, and behavior control run at a rate of 30 Hz, motion control runs at 80 Hz, and the entire system is optimized to minimize delay, all the small delays caused by the individual stages add up to about 100 to 150 ms. For the purposes of control, all delays can be aggregated to a single dead time. This is illustrated in Fig. 2(a). If a robot moves at 2 m/s and the dead time is 100 ms, the robot will travel a distance of 20 cm before we perceive the consequences of an action decision. This causes problems when robots move fast, producing overshooting or oscillations in movement control.

In order to measure the system delay, we use the following simple technique. We let the robot drive along the field with a speed that is determined by a sinusoidal function. This means, the robot moves back and forth with maximum speed in the middle of the path, slowing down and changing direction at both turning points. We then measure the time between sending the command to change the direction of motion and perceiving a direction change of the robot's movement.

There are many possibilities to counter the adverse effects of the control delay. The easiest way would be to move slower, but this is frequently not desirable, since fast movement is a decisive advantage in play. Another possibility would be to reduce precision requirements, but this would lead to unnecessary collisions with other players and uncontrolled ball handling.

Control researchers have made many attempts to overcome the effects of delays. One classical approach is known as Smith Predictor [8]. It uses a forward-model of the plant, the controlled object, without delays to predict the con-

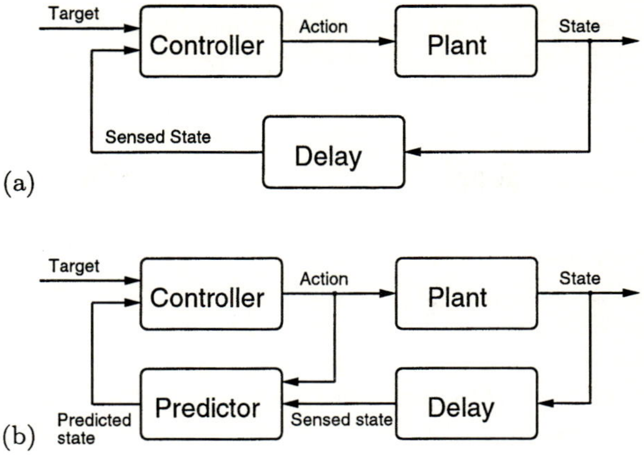

Fig. 2. Control with dead time: (a) control is difficult if the consequences of the controller actions are sensed with significant delay; (b) a predictor that is trained to output the state of the plant can reduce the delay of the sensed signal and simplify control.

sequences of actions. These predictions are used in an inner control loop to generate sequences of actions that steer the plant towards a target state. Since this cannot account for disturbances, the plant predictions are delayed by the estimated dead time and compared to the sensed plant state. The deviations reflect disturbances that are fed back into the controller via an outer loop. The fast internal loop is functionally equivalent to an inverse-dynamic model that controls a plant without feedback. The Smith Predictor can greatly improve control performance if the plant model is correct and the delay time matches the dead time. It has been suggested that the cerebellum operates as a Smith Predictor to cancel the significant feedback delays in the human sensory system [7]. However, if the delay exceeds the dead time or the process model is inaccurate, the Smith Predictor can become unstable.

Ideally, one could cancel the effects of the dead time by inserting a negative delay of matching size into the feedback loop. This situation is illustrated in Fig. 2(b), where a predictor module approximates a negative delay. It has access to the delayed plant state as well as to the undelayed controller actions and is trained to output the undelayed plant state. The predictor contains a forward model of the plant and provides instantaneous feedback about the consequences of action commands to the controller. If the behavior of the plant is predictable, this strategy can simplify controller design and improve control performance.

A simple approach to implement the predictor would be to use a Kalman filter. This method is very effective to handle linear effects, for instance the motion of a free rolling ball [5]. It is however inappropriate for plants that contain significant non-linear effects, e.g. caused by the slippage of the robot wheels or by the behavior of its motion controller. For this reason, some teams use a Extended Kalman-Bucy Filter [9] to predict non-linear systems. But this approach requires a good model of the plant. We propose to use a neural network as a predictor for the robot motion, because this approach doesn't require an explicit model and can easily use robot commands as additional input for the prediction. This allows predicting future movement changes before any of them could be detected from the visual feedback.

4 Neural Network Design

Since we have no precise physical model of the robot, we train a three layer feed-forward network to predict the robot motion. The network has 42 input units, 10 hidden units, and 4 output units. The hidden units have a sigmoidal transfer function while the transfer function of the output units is linear.

We train the network with recorded data using the standard backpropagation algorithm [1]. A great advantage of the neural network is that it can be easily trained again if something in the system changes, for example if a PID controller on board the robot is modified. In this case, new data must be recorded. However, if the delay itself changes we only have to adjust the selection of the target data (see below) before retraining.

4.1 Input Vector

The input presented to the neural network is based on position and orientation of the robot as perceived by the vision module during the last six frames, as well as the last few motion commands sent to the robot. Some preprocessing is needed to simplify the prediction task. The preprocessing assumes translational and rotational invariance. This means that the robot's reaction to motion commands does not depend on its position or orientation on the field. Hence, we can encode its perceived state history in a robot-centered coordinate system.

The position data consists of six vectors – the difference vectors between the current frame and the other six frames in the past, given as (x, y)-coordinates. The orientation data consists of six angles, given as difference of the robot's orientation between the current frame and the other six ones in the past. They are specified as their sine and cosine. This is important because of the required continuity and smoothness of the data. If we would encode the angle with a single number, a discontinuity between $-\pi$ and π would complicate the training. The action commands are also given in a robot-centered coordinate system. They consist of the driving direction and speed as well as the rotational velocity. The driving direction and velocity are given as one vector with (x, y)-coordinates, normalized by the velocity.

Preprocessing produces seven float values per frame, which leads to a total of $7 * 6 = 42$ input values for the neural network.

4.2 Target Vector

The target vector we are using for training the network consists of two components: the difference vector between the current position and the position four frames forward in the future and the difference between the current orientation and the orientation four frames ahead. They are encoded in the same format as the input data.

4.3 Data Collection and Constraints

Data for training the network is generated by moving a robot along the field. This can be done by manual control using a joystick or a mouse pointer, or by specialized behaviors developed for this purpose.

To cover all regions of the input space, the robot must encounter all situations that could happen during game play. They include changing speed over a wide range, rotating and stopping rapidly, and standing still. We also must make sure that the robot drives without collisions, e.g. by avoiding walls. This is necessary because the neural network has no information about obstacles and hence can not be trained to handle them. If we would include such cases in the training set, the network would be confused by conflicting targets for the same input. For example driving freely along the field or driving against a wall produce the same input data with completely different target data.

We could solve this problem by including additional input features for the neural network, e.g. a sensor for obstacles, and thus handle also this situation, but this would complicate network design and would require more training data to estimate additional parameters.

5 Results

We have extensively tested the neural network for the prediction of the position and orientation of the robots since its first integration into the FU-Fighters' system. It performs very well and we have nearly eliminated the influence of the delay on the system.

To demonstrate the effect of the prediction on robot behavior, we have tested one particular behavior of the robot: drive in a loop around the free kick points. We compare the performance of the neural predictor to a linear predictor that has been trained on the same data.

The histograms in Fig. 3 show that the neural network prediction has much more samples with small errors than the linear prediction. The average position error for the linear prediction is 5.0471 cm and 2.8030 cm for the neural network prediction. The average orientation error for the linear prediction is 0.1704 rad (9.76°) and 0.0969 rad (5.55°) for the neural network prediction.

6 Conclusion and Future Work

We have successfully developed, implemented, and tested a small neural network for predicting the motion of our robots. The prediction compensates for the system delay and thus allows more precise motion control, ball handling, and obstacle avoidance. To make the perception of the world consistent, predictions

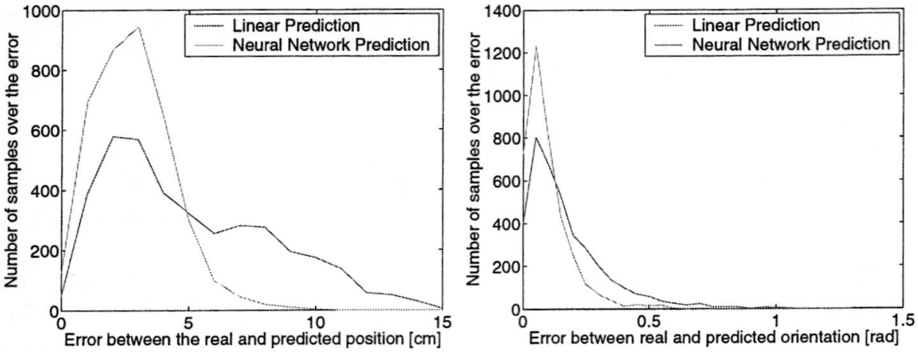

Fig. 3. Comparison between the histograms of linear and neural network predicted robot position (left) and orientation (right) error. Both histograms have about 3000 samples.

are not only used for own robots, but also for robots of the opponent team and the ball. However, here the action commands are not known and hence simpler predictors are used. We employ Kalman filters to model linear effects.

For advanced play, it would be beneficial to anticipate the actions of opponent robots, but this would require learning during a game. Such online learning is dangerous though, because it is hard to automatically filter out artifacts from the training data, caused e.g., by collisions or dead robots.

Another possible line of research would be to apply predictions not only to basic robot motion, but also to higher levels of our control hierarchy, where delays are even longer.

Finally, one could also integrate the neural predictor into a simulator as a replacement for a physical robot model. A simulator allows quick assessment of the consequences of actions without interacting with the external world. If there are multiple action options during a game, this 'mental simulation' could be used to decide which action to take.

References

1. Rojas, Raúl: Neural Networks – A Systematic Introduction. Springer Verlag, Heidelberg, 1996.
2. Behnke, Sven; Frötschl, Bernhard; Rojas, Raúl; Ackers, Peter; Lindstrot, Wolf; de Melo, Manuel; Schebesch, Andreas; Simon, Mark; Spengel, Martin; Tenchio, Oliver: Using Hierarchical Dynamical Systems to Control Reactive Behavior. Lecture Notes in Artificial Intelligence **1856** (2000) 186–195.
3. Simon, Mark; Behnke, Sven; Rojas, Raúl: Robust Real Time Color Tracking. Lecture Notes in Artificial Intelligence **2019** (2001) 239–248.
4. Rojas, Raúl; Behnke, Sven; Liers, Achim; Knipping, Lars: FU-Fighters 2001 (Global Vision). Lecture Notes in Artificial Intelligence **2377** (2002) 571–574.

5. Veloso, Manuela; Bowling, Michael; Achim, Sorin; Han, Kwun; Stone, Peter: CMUnited98: RoboCup98 SmallRobot World Champion Team. *RoboCup-98: Robot Soccer World Cup II*, pp. 61–76, Springer, 1999.
6. Wolpert, Daniel M.; Flanagan, J. Randall: Motor Prediction. *Current Biology Magazine*, vol. 11, no. 18.
7. Miall, R.C.; Weir, D.J.; Wolpert, D.M.; Stein, J.F.: Is the Cerebellum a Smith Predictor? *Journal of Motor Behavior*, vol. 25, no. 3, pp. 203–216, 1993.
8. Smith, O.J.M.: A controller to overcome dead-time. *Instrument Society of America Journal*, vol. 6, no. 2, pp. 28–33, 1959.
9. Browning, B.; Bowling, M.; Veloso, M.M.: Improbability Filtering for Rejecting False Positives. *Proceedings of ICRA-02, the 2002 IEEE International Conference on Robotics and Automation*, 2002.

Towards a Probabilistic Asynchronous Linear Control Theory

Daniel Polani

Adaptive Systems Research Group
Department of Computer Science
University of Hertfordshire
d.polani@herts.ac.uk

Abstract. A framework for asynchronous stochastic linear control theory is introduced using a simple example motivated by the early RoboCup soccer server dynamics. Worst and average case scenarios are studied and it is demonstrated that they fit smoothly into the framework of standard synchronous control theory.

1 Introduction

In the early versions of the RoboCup Soccer Simulator, a central problem was the synchronization of the soccer clients with the soccer simulator. The games were run in real-time, but the UDP/IP protocol used does not guarantee that the communication packets will reach the other side such as to guarantee a well-defined temporal relation between the agent state and the soccer server state [11]. An important question was how a consistent world view could be maintained [12]. Commands sent by the agents to the server, as well as the agent world model might not be synchronized with the server cycle. In tournaments from 1999 onwards, faster machines, networks and better synchronization algorithms, like that developed by CMU [14] increasingly made this view obsolete and allowed agents to become synchronized with the simulator.

In the present paper, we will revisit the original problem. But why return to a problem which seems to be no longer relevant? There are several reasons for that: First, the existence of precisely timed world states is a model which is a coarse approximation of physical reality. In newer versions of the soccer server simulator it is considered to introduce a continuous time, where actions will be incorporated into the physical dynamics *as they arrive*; this means that they will not be buffered to be processed only at the fixed time steps of the simulation update. Second, although current hardware robots are often organized in synchronized time steps due to technical reasons, it is far from clear that, say, their sensorics and processing systems should obey this principle of equal and consistent timing. Different concepts of timing and of time processing exist in biological systems and may, among other aspects, be responsible for the phenomenon of consciousness [6]. Third, the concept of what constitutes a worldly "reality" is a generally intriguing question. In the extreme cases of relativity theory, the

D. Polani et al. (Eds.): RoboCup 2003, LNAI 3020, pp. 720–728, 2004.

concept of "now" loses all its meaning without additional assumptions. The unsynchronized relation between world and agent time in the early soccer server can be interpreted as a kind of "unorganized relativistic system" For this reason, we believe that the question of studying systems with a stochastic asynchrony bears an interest in its own right. In the present paper, we will study a linear control model for a strongly specialized class of stochastic asynchronous control systems as a starting point for a generalized theory of stochastic asynchronous control.

Asynchronous systems are of interest in the development of highly integrated systems where tight timing schedules make asynchrony an integral part of the dynamics [2]; such systems are usually of digital nature and analyzed using Petri networks [9]. In continuous systems, the typical approach is linear [13] or nonlinear [10] control theory, typically assumed to be synchronous. Another line of research studies the dynamics in spiking neural networks [8] exhibiting phenomena of asynchronicity not unlike technical systems. Hassibi et al. [4] introduce a method for the control of asynchronous dynamical systems. The method considers asynchronous events in the control loop whose behaviour is only restricted by an event rate. In their work, few assumptions are made and stochastical aspects of time delays are not modelled. These, however, will be part of the present work. The relation between models studied in this paper and switched systems [7] as well as Markovian jumping systems [1] will be studied in a later paper.

2 A Scenario

2.1 The Synchronization

We concentrate on a simple scenario in the framework of the early soccer server to illustrate the approach. Assume that at integer times $t = 1, 2, 3\ldots$ the server world state is updated according to its dynamics [3], taking into account all the agent action commands that arrived before t. After time t, the server sends out its updated world state to the agents. Upon receiving this update, the agents decide on an action to take and send it to the server which will incorporate it into the simulation step at time $t+1$. We now make several assumptions: 1. The server sends out world state information after each simulation step. This does not conform to the standard setting of the soccer server, but serves to simplify our present considerations. 2. The world state information sent out at time t is processed by the agent which then issues an action command. 3. The action command from 2., called a_t, has a probability of p of reaching the server in time for the next simulation step. In that case, it will be included in calculation of the next world state. a_t has a probability of $q = 1-p$ of not reaching the server in time, and instead after time $t+1$. What now happens depends on whether the following action command a_{t+1} reaches the server in time or not. If a_{t+1} is late again, then at time $t+2$ the server performs action a_t already in its buffer. If a_{t+1} is in time, though, it will overwrite a_t in the server buffer and will be performed at time $t+2$ (this is the case shown in Fig. 1).

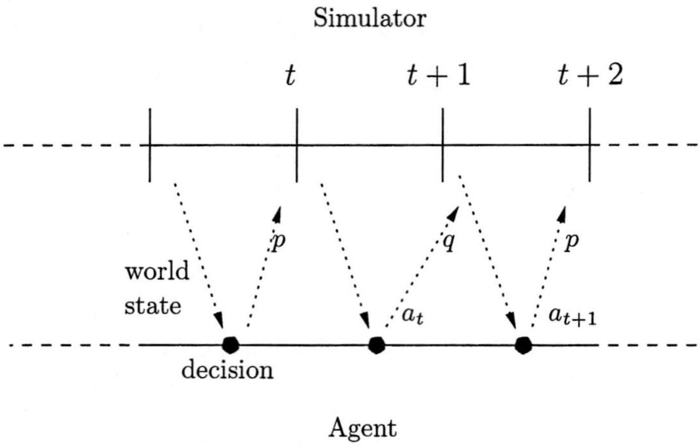

Fig. 1. The simulation-control cycle. The arrows from the server time-line to the agent denote sensor world information sent from server to agent, the arrows from the agent time-line to the server denote the action commands. With probability p, action a_t reaches the server in time for the update at $t+1$, with probability $q = 1 - p$, it is too late and stays in the buffer until after $t+1$. If the following command a_{t+1} is late again, a_t in the buffer will be processed at time $t+2$. The present figure, however, shows the case where a_{t+1} reaches the server in time and overwrites a_t in the buffer; thus, at time $t+2$ a_{t+1} will be carried out.

2.2 An Example for a Control Variable

As example, let us assume that we wish to control the orientation ϕ of the agent, in approximation for small angles so that we can assume that $\phi \in \mathbb{R}$. The goal is now a linear control rule that will stabilize ϕ to 0. Consider the synchronous control case. If the orientation after time t is ϕ_t, then sending $a_t = \Delta\phi_t$ results in a new orientation $\phi_{t+1} = \phi_t + \Delta\phi_t$ after the next update. The control rule is linear in ϕ: $a_t := -c\phi_t$, the update rule $\phi_{t+1} = \phi_t - c\phi_t$. The best possible control is achieved setting $c := 1$, immediately stabilizing the angle at 0.

Consider now the asynchronous case with a synchronization pattern as in Sec. 2.1. Represent the state of the system by the vector $(\phi_t, \phi_t^B)^T$. Here, ϕ_t^B is the possible action from before time t stored in the server buffer. If no action is stored set ϕ_t^B to 0. ϕ_t^B is not known to the agent. As in the synchronous case, set $a_t := -c\phi_t$. Now there are two possibilities: 1. With probability p, the action will reach the server in time to be processed at $t+1$. The new system state will now be $(\phi_{t+1}, \phi_{t+1}^B)^T = (\phi_t - c\phi_t, 0)^T$. The 0 in the second component indicates the empty action buffer. 2. With probability $q = 1-p$, the action will be late for the update at $t+1$. Thus, the new orientation will be obtained using the action in the server action buffer, and the present command will end up in the action buffer: $(\phi_{t+1}, \phi_{t+1}^B)^T = (\phi_t + \phi_{t+1}^B, -c\phi_t)^T$ In case 1., the transition matrix for the states is given by $T_p = \left(\begin{smallmatrix} 1-c & 0 \\ 0 & 0 \end{smallmatrix}\right)$ In case 2., the transition matrix is given by $T_q = \left(\begin{smallmatrix} 1 & 1 \\ -c & 0 \end{smallmatrix}\right)$.

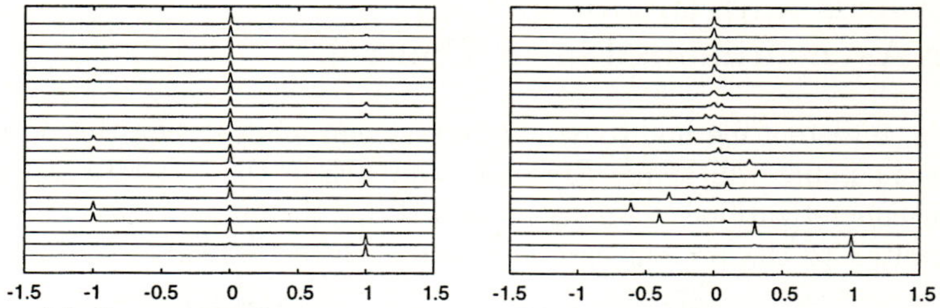

(a) Control coefficient $c = 1$ (b) Control coefficient $c = 0.7$

Fig. 2. Example for asynchronous control of the orientation if the probability p that the action command reaches the server in time is $p = 0.1$. The time axis is $t = 0$ to $t = 20$ from bottom to top; the diagram shows the probability that the agent has a certain orientation if it issues a change orientation command $-c\phi$ in each step. For more details, see text.

2.3 Solving the Control Problem: Example

The linear control of a single variable is the simplest possible control scenario in the synchronous case and a special instance of a control problem. In our example $c := 1$ solves the problem. In the asynchronous case, this is not anymore the case. Assume an extreme delay probability, e.g., that the probability that the action command is received in time $p = 0.1$.

Figure 2(a) shows the dynamics of the system. From bottom to top the state information for times $t = 0, \ldots, t = 20$ are shown. The graphs show the probability distribution that the agent has a given orientation at a certain time. The bottommost line shows the probability distribution at time $t = 0$. The agent starts with the orientation $\phi_0 := 0$ and the server with an empty action buffer. The command to correct the orientation by $-c\phi$ reaches the server in time only with probability $p = 0.1$. We see that in the second graph from the bottom. With probability 0.1, the agent has assumed the desired orientation of 0. In the rest of the cases, the agent has remained in its original orientation since the action command has not reached the server.

Now, if still in state $\phi = 1$, the agent will reissue a reorientation command while there is already such a command in the queue. In the following step $t = 2$, the agent will then have assumed orientation 0, because now the original $t = 0$ reorientation action has finally reached the server (or, either, it had already reached $\phi = 0$ and does not need reorientation). However, since in 90% of the cases the agent had reissued a reorientation command in time step $t = 1$, the system overshoots; this process repeats itself in the next cycles. It takes considerable time until the probability for $\phi = 1$ or $\phi = -1$ becomes negligible. While therefore in the undelayed and synchronous control problem it makes sense to set $c = 1$, this setting is too "aggressive" in the asynchronous case. For comparison, the same control problem is shown in Fig. 2(b) for $c = 0.7$.

3 Towards a Methodology for Asynchronous Linear Control

Is there a consistent method to handle the asynchronous stochastic control case? The present paper will demonstrate an approach in the concrete example, because the concretization is more useful in bringing the point across. The principle is not limited to the present case and can easily be extended to different and more general systems. Since we do not have a single-valued system variable like in deterministic control theory, but a probability distribution of the system variable, we have a variety of choices for the criterium according to which we wish to control. Here, we will discuss the worst case scenario and the average case scenario.

3.1 The Worst Case Scenario

The worst case scenario is not concerned with the probability that a certain state is assumed, as long as it is assumed; the worst possible state (i.e. the state that is the farthest away from the desired one) determines the quality of the control. To explain this, consider the case of Fig. 2(a). Here, even for large times t there is still a nonzero probability for the orientation still to be at $\phi \in \{-1, 1\}$, i.e. as far away from the target state as in the beginning. Though this probability decays exponentially, for the worst case scenario this is not relevant. It is the slowest decaying branch of the probability dynamics that counts. In the worst case scenario, one is interested to control the system in such a way as to maximize the decay of the slowest decaying branch. The present section demonstrates the approach, using the matrix lub-norm ("lub" = lowest upper bound).

Decay Calculation via LUB-Norm in a Simple Case. Assume for a minute that p were 1, i.e. we had a synchronized system with the transition matrix as above $T_p = \left(\begin{smallmatrix} 1-c & 0 \\ 0 & 0 \end{smallmatrix}\right)$. Then, for a general initial state x_0, the state at time t is given by $T_p^t x_0 = \underbrace{T_p \ldots T_p}_{t \text{ times}} x_0$. An upper bound for the worst case is obtained via the lub-norm $\| \,.\, \|$ of matrices[1]. Obviously $\|T_p^t\| = |1-c|^t$. The strongest decay in this case is obtained by $c := 1$ as done above.

The General Case. In our scenario, neither p nor q vanish. Thus the set of states that may have a nonzero probability at a time step t is given by $\{x \mid x = T_{k_t} T_{k_{t-1}} \ldots T_{k_1} x_0, \text{with } k_i \in \{p, q\} \text{ for } i = 1 \ldots t\}$. This is the set of all possible states x arising via all possible combinations of t delayed and undelayed simulation updates. Call such a sequence of updates a *simulation run.* We now show how one can obtain increasingly accurate decay boundaries which can be

[1] For a matrix A, the lub-norm is defined as $\|A\| = \sup_x \frac{\|Ax\|_2}{\|x\|_2}$ with $\| \,.\, \|_2$ the Euclidean metric. One obtains $\|A\| = \sqrt{\lambda_{\max}}$ with $\lambda_{\max}$ the largest eigenvalue of $A^T A$. The lub-norm fulfils the criterium of *submultiplicativity,* $\|AB\| \leq \|A\| \|B\|$ [5].

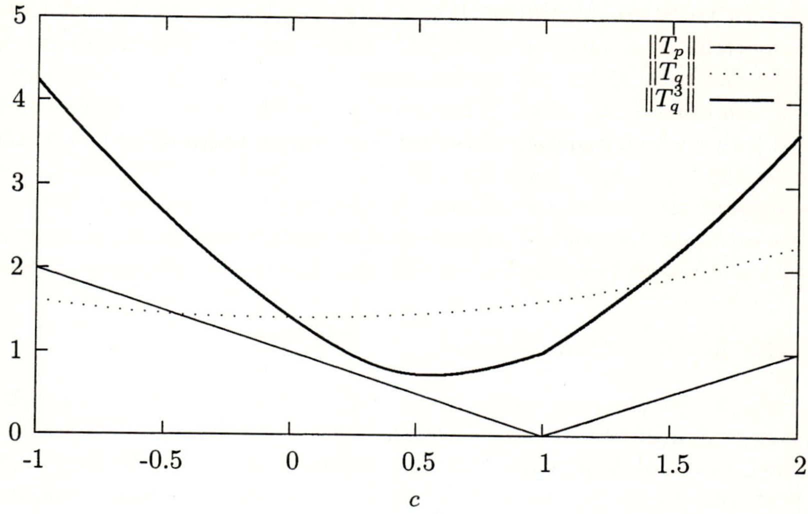

Fig. 3. The values of $\|T_p\|$, $\|T_q\|$ and $\|T_q^3\|$for different values of c.

used to determine the optimal control variable c. The coarsest possible boundary can be calculated via the property of submultiplicativity of the lub norm (Footnote 1). We obtain for the state after any possible t-step simulation runs $\|x_t\|_2 = \|T_{k_t}T_{k_{t-1}}\dots T_{k_1}x_0\| \leq \|T_{k_t}\|\|T_{k_{t-1}}\|\dots\|T_1\|\|x_0\|_2 \leq \left(\max_{k\in\{p,q\}}\|T_k\|\right)^t\|x_0\|_2$.

In this approximation, the decay rate is bounded by $\max_{k\in\{p,q\}}\|T_k\|$. We know $\|T_p\| = |1-c|$ and have to calculate the lub-norm of $T_q = \left(\begin{smallmatrix}1 & 1\\ -c & 0\end{smallmatrix}\right)$. One obtains $T_q^T T_q = \left(\begin{smallmatrix}1+c^2 & 1\\ 1 & 1\end{smallmatrix}\right)$. This gives the set of eigenvalues $\{\frac{c^2}{2} + \frac{\sqrt{c^4+4}}{2} + 1, \frac{c^2}{2} - \frac{\sqrt{c^4+4}}{2} + 1\}$, the square root of the larger of the which is the lub-norm of T_q. The results are shown in Fig. 3. The larger of the two values $\|T_p\|$ and $\|T_q\|$ is the estimate for the worst-case decay factor achievable by selecting a given value c. The figure shows that for no choice of c, $\|T_q\|$ becomes smaller than 1. I.e., it is not possible to construct a decay rate below 1 considering only a one-step lookahead. However, by iterating T_q three times, there are, in fact, values of c for which the lub norm of T_q^3 drops below 1 (Fig. 3[2]).

We therefore see that the lub-norm provides us only with an upper bound for the worst case scenario; if we wish more accurate boundaries, one has to construct the possible simulation runs (sequences of ps and qs) and calculate the pertinent lub-norms. Calculating the lub-norms for $T_pT_pT_p$, $T_pT_pT_q$, $T_pT_qT_p$, $T_pT_qT_q$, $T_qT_pT_p$, $T_qT_pT_q$, $T_qT_qT_p$, $T_qT_qT_q$ gives the plot shown in Fig. 4. This procedure can be extended to longer sequences of p and q and used to determine the pertinent control coefficient c.

[2] An alternative way of showing that would have been to consider the eigenvalues of T_q and seeing that their modulus is smaller than 1; however, this does not allow us to identify the best value of c.

Reduction to State Variable. The above calculation has still not exploited further options to reduce the boundary. Since we are not actually interested in having the total norm $\|x\|$ of the state vector decaying, but actually only the first component, x_1 which is the orientation (the second component is the buffered action which we are not interested in), what we are actually seeking is to maximize the worst case decay for $PT_{k_t}T_{k_{t-1}} \dots T_{k_1}$ where P is the projection of the current total state vector onto the orientation component. For a more accurate estimate, the control variable c should minimize the lub-norm of above term. We will not proceed to do that here, but the idea is analogous to Sec. 3.1.

3.2 The Average Case Scenario

The average case scenario differs from the worst case one that probabilities are incorporated in calculating the state vector whose behavior is going to be controlled. A simulation with t updates results in a vector $T_{k_t}T_{k_{t-1}} \dots T_{k_1}x_0$ with probability $k_t k_{t-1} \dots k_1$ with $k_i \in \{p, q\}$ (note that we abuse notation by using p and q as index names for the T, but as numerical values for the k).

For the expectation value for the state vector we have to sum up over all the possible probabilities and combination of p and q, obtaining $\mathbf{E}x = \sum_{(k_t,\dots,k_1)\in\{p,q\}^t} k_t \dots k_1 T_{k_t} \dots T_{k_1} x_0 = (pT_p + qT_q)^t x_0$, where the first equation is the probability-weighted sum over all possible mutually disjoint t-step histories and the second equation arises from the binomial theorem. This quantity is a matrix operator obtained by averaging T_p and T_q according to their weight and iterating it t times. To this operator, the methods from Sec. 3.1 can be applied again in a straightforward manner. Thus, we can smoothly incorporate the average case scenario in the methodology developed above. The same holds for the projection P of the average state on the orientation component.

4 Conclusion

The methodology developed here to describe and solve the linear stochastic asynchronous control problem has been described given a specific example, to make it more accessible, but is not restricted to it. It smoothly encompasses both description of worst case and average case. Standard analysis methods from linear control control theory [13] can be easily applied to the formalism, thus making available the toolbox of standard control theory to the stochastic asynchronous case, thereby complementing the formalism of standard control theory.

5 Summary and Outlook

A formalism to solve a linear stochastic asynchronous control problem was presented in the framework of a simple, but paradigmatic case. It was discussed in view of the worst and average case scenario. Future work will study the relation of the present framework to existing work on switched systems and Markovian jumping models [7, 1]. In addition, it will focus on the relation of information

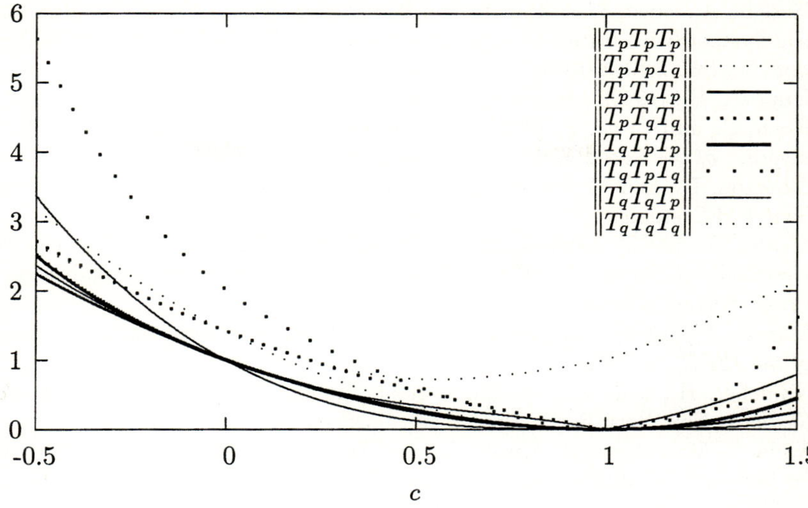

Fig. 4. Plot of the lub-norms for all three-step simulation runs. Since it is difficult to distinguish the varieties of different graphs, it should be mentioned that the maximum of the lub-norms is dominated by $T_qT_pT_q$ for smaller values of c and by $T_qT_qT_q$ for larger values of c.

theory, control and time and combine it to extend the information-theoretic perspective of control theory from [16, 15] towards a better understanding of how information and time are related in a control loop. We believe that these questions are not just of mainly academic interest, but will instead lead to a better fundamental understanding of the role of time in intelligent information processing.

Acknowledgement

The author thanks the anonymous reviewers for their helpful comments.

References

[1] Boukas, E. K., and Yang, H., [1995]. Stability of discrete time linear systems with Markovian jumping parameters. *MCSS*, 8:390–402.

[2] Cortadella, J., Kishinevsky, M., Burns, S. M., and Stevens, K., [1999]. Synthesis of asynchronous control circuits with automatically generated relative timing assumptions. *ICCAD*, 324–331.

[3] Corten, E., Heintz, K. D. F., Kostiadis, K., Kummeneje, J., Myritz, H., Noda, I., Riekki, J., Riley, P., Stone, P., and Yeap, T., [1999]. *Soccerserver Manual Ver. 5 Rev. 00 beta (for Soccerserver Ver.5.00 and later).* `http://www.dsv.su.se/~johank/RoboCup/manual/`, March 19, 2001

[4] Hassibi, A., Boyd, S. P., and How, J. P., [1999]. Control of Asynchronous Dynamical Systems with Rate Constraints on Events. *38th IEEE Conference on Decision and Control*, 2:1345–1351.

[5] Huppert, B., [1990]. *Angewandte lineare Algebra.* De Gruyter.

[6] Koch, C., and Crick, F., [2001]. The neural basis of consciousness. In *Intl. Encyclopedia of the Social and Behavioral Sciences*, 2600–2604. Elsevier.

[7] Liberzon, D., and Morse, A., [1999]. Basic Problems in Stability and Design of Switched Systems.

[8] Maass, W., [2003]. Computation with spiking neurons. In Arbib, M. A., editor, *Handbook of Brain Theory and Neural Networks*, 1080–1083. Cambridge: MIT Press. Second edition.

[9] Murata, T., [1989]. Petri Nets: Properties, analysis and applications. *Proceedings of the IEEE*, 541–580.

[10] Nijmeijer, H., and van der Schaft, A. J., [1990]. *Nonlinear Dynamical Control Systems.* New York Berlin Heidelberg: Springer-Verlag.

[11] Noda, I., Matsubara, H., Hiraki, K., and Frank, I., [1998]. Soccer Server: A Tool for research on Multi-Agent Systems. *Journal of Applied Artificial Intelligence*, 12(2–3).

[12] Polani, D., Weber, S., and Uthmann, T., [1998]. The Mainz Rolling Brains in RoboCup '98: A Direct Approach to Robot Soccer. In *Workshop "RoboCup" at the KI-98.*

[13] Sontag, E. D., [1990]. *Mathematical control theory; Determistic finite dimensional systems (Texts in Applied Mathematics)*, vol. 6. New York: Springer-Verlag.

[14] Stone, P., Veloso, M., and Riley, P., [1999]. The CMUnited-98 Champion Simulator Team. In Asada, M., and Kitano, H., editors, *RoboCup-98: Robot Soccer World Cup II.* Berlin: Springer Verlag. To Appear.

[15] Touchette, H., and Lloyd, S., [2000]. Information-Theoretic Limits of Control. *Phys. Rev. Lett.*, 84:1156.

[16] Touchette, H., and Lloyd, S., [2001]. Information-theoretic approach to the study of control systems. *IEEE Transactions on Information Theory.* Submitted. `http://xxx.lanl.gov/abs/physics/0104007`, Feb 2003

Recognizing and Predicting Agent Behavior with Case Based Reasoning

Jan Wendler[1] and Joscha Bach[2]

[1] Zuse Institute Berlin, Takustraße 7
14195 Berlin, Germany
wendler@zib.de

[2] Institut für Informatik, Humboldt-Universität zu Berlin
Unter den Linden 6, 10099 Berlin, Germany
bach@informatik.hu-berlin.de

Abstract. Case Based Reasoning is a feasible approach for recognizing and predicting behavior of agents within the RoboCup domain. Using the method described here, on average 98.4 percent of all situations within a game of virtual robotic soccer have been successfully classified as part of a behavior pattern. Based on the assumption that similar triggering situations lead to similar behavior patterns, a prediction accuracy of up to 0.54 was possible, compared to 0.17 corresponding to random guessing. Significant differences are visible between different teams, which is dependent on the strategic approaches of these teams.

1 Introduction

Our work is concerned with the analysis of the behavior of agents in a highly dynamic and heterogeneous domain– virtual robotic soccer (RoboCup [4]). Here, two teams of 11 agents each connect to a server that simulates their environment and their 'bodies' in discrete time-steps of 100 ms [11]. The agents have a limited field of view and partially incomplete information about their surroundings. Agents within a team may communicate, but this is limited.

This paper addresses a way of automatically classifying and an attempt at predicting the behavior of a team of agents, based on external observation only. A set of conditions is used to distinguish behaviors and to partition the resulting behavior space. From observed behavior, team specific behavior models are then generated using Case Based Reasoning (CBR) [5, 6]. These models, which are derived from a number of virtual soccer games, are used to predict the behavior of a team during a new game.

This work is based on a diploma thesis by Uwe Müller [9] and a doctoral thesis by one of the authors [15].

2 Approach

There have been previous attempts at recognizing behavior within RoboCup, especially for automatically commenting on soccer games (see, for instance, André

D. Polani et al. (Eds.): RoboCup 2003, LNAI 3020, pp. 729–738, 2004.

et al. [1], Voelz et al. [14], Matsubara et al. [7], Frank et al. [2] and Raines et al. [12]). Wünstel et al. (see, for instance, [18]) use self organizing feature maps to highlight typical behavior elements of teams of agents on a global scale. In Miene and Visser [8] a universal approach for specifying and recognizing behavior is introduced based on spatial and temporal relations.

Relatively little has been published in the RoboCup domain concerning the prediction of individual moves of soccer agents, however, Riley and Veloso [13] use behavior models to predict movements of opponents in standard soccer situations by maintaining probability distributions of the positions of individual agents.

In [17], one of the authors has used a CBR based behavior model to imitate the successful behavior of agents (specifically, the passing behavior).

2.1 Terms

A behavior model is a structure $\mathfrak{M} = [\mathfrak{C}; \mathfrak{B}; f]$, where $\mathfrak{C}$ is a nonempty set of causes, $\mathfrak{B}$ is a nonempty set of behaviors, and f is a function $f : \mathfrak{C} \rightarrow \mathfrak{B}$. Behavior models may be based on actions or situations. Actions are not directly observable in our approach and can not always be determined because they can not always be distinguished by their outcomes. Therefore, we base our approach on situations and assume that similar situations correspond to similar behavior.

Behaviors may be modeled implicitly or explicitly; for instance, the agents of our team AT Humboldt 98 [3, 10] have used an implicit model based on their own interception algorithm to estimate the intercept behavior of opponents and teammates.

A situation S_t^i is what the agent i knows of a world situation W_t. This situation is generated by combining several observations. An observation O_t^i is a part of a world situation W_t, interpreted by an agent i, and may be erroneous and incomplete. A world situation consists of the features that determine the state of the agents' environment at a given time-step t, such as the positions of agents and the ball, their speed vectors, the game-state, the score and the stamina values of the agents.

An (observed) behavior is a process that extends over several adjacent situations and may be described by events. A behavior is caused by the actions of multiple agents and what follows from these actions. The part of the data that is considered to lead to a behavior is what we call a *trigger*. As trigger we apply an extract (relevant data from an observers point of view) of the situation at the start time of the according behavior.

Similar behaviors are grouped into behavior patterns, and likewise, sets of triggers that lead to the same behavior pattern are grouped into trigger patterns.

To describe behaviors, we resort to a sequence of values that we call a behavior-essence, with $\texttt{behaviors}_E$ being the set of all behavior essences. Thus, we can describe the behavior model as $\mathfrak{M} = [\texttt{situations}; \texttt{behaviors}_E; f]$, with $f : \texttt{situations} \rightarrow \texttt{behaviors}_E$.

2.2 Using CBR

f is implemented by a CBR system, which consists of a case base, a similarity measure defined between the cases, a method for updating the case base, and a method for predicting behavior.

To collect cases, behaviors needs to be specified and recognized (section 3). The behavior essences and its triggers are specified (sections 4.1, 4.2) and cases which consist of a trigger and a behavior essence are generated (section 4.3).

Because triggers and behavior essences are determined by a sequence of values, combined similarity measures are used (section 4.4). Each combined measure consists of local similarities between two individual values and a weighted sum of these local similarities. For the behavior similarities, weights and values are specified by the designer. For triggers, only local similarities are given by the designer, and weights are determined automatically.

Based on the case base and the similarity measure the methods to predict behaviors and to update the case base are introduced (section 4.5).

3 Behavior Recognition

The basic recognition process is based on behavior patterns. Behaviors are constructed from events, and they can be recognized by performing a pattern matching of these events against the observations. The general behavior recognition algorithm can be found in [15].

Whenever behaviors contain each other, the longest recognizable behavior is used (for instance, within a dribbling, further instances of dribbling may occur. Currently, we only examine behaviors involving the ball (i.e. positioning behavior such as the building of a pass chain are not considered).

3.1 Specification of Behaviors

Behaviors are specified by defining and arranging the events they are constructed of. We have specified behaviors for passing, dribbling, goal-kicking, clearing and other. We will give a short overview how the pass behavior has been specified. The exact specification can be found in [15] and [16].

For the pass behavior it is necessary that one player controls the ball exclusively during a small time interval. After that the ball mustn't be controlled by any player for some time. Finally the ball needs to be exclusively controlled by another player. Further conditions are used to distinguish a pass from a mere ball transfer, like that the ball is departing from the kicker with enough speed and other.

By adding further conditions regarding ball speed and player movement further differentiation between direct passes, indirect passes and passes with approaching movement is possible. Diagonal passes, back passes and transversal passes can be identified by taking the angle of the ball movement into account.

A behavior fails if an opponent gets into possession of the ball, or if the game-state changes in such a way that an opponent will receive ball control, for instance, when the ball has been shot out of the field. We have added conditions to classify both cases for the four behavior patterns pass, dribbling, goal-kick and clearing.

3.2 Abstraction Levels for the Description of Behaviors

A behavior instance is (from the point of view of an observer) completely determined by a sequence of world states. We discern four levels of abstraction:

- General description: on this level we use names like pass, dribbling or other.
- Manifestation description: this introduces sub-categories like direct pass etc.
- Source-target description: the source of all ball transfers is given by the kicking agent. The target may be a receiving agent, the goal, a position on the field etc.
- Detailed description: this consists of all relevant features of the behavior, that is, of the values of all important attributes. For passing, these are the kicking and the receiving players, the start and end time of the pass, the ball speed and the movement of the receiver.

Of these levels, the source-target description and the detailed description are used for the behavior prediction, while the general and manifestation description maybe important to describe and predict higher level strategies.

4 Prediction of Behaviors

4.1 The Trigger Patterns

The prediction of behaviors is based on the recognition of associated triggers, which are assumed to cause the agents to start the corresponding behavior. Unfortunately, the comparison between triggers that account for all relevant attributes is computationally expensive, let alone the recognition of similarities. To make the comparison of triggers feasible in real-time, only the most important attributes are considered. The primary attributes for a *pass* are *the position of the initiating player* p_1, *the vector from the initiating to the receiving player* $\overrightarrow{p_1p_2}$, *the angle* $\measuredangle(\overrightarrow{p_1p_3};\overrightarrow{p_1p_2})$ *to the first opponent* p_3 *to the right, the distance* $|\overrightarrow{p_1p_3}|$ *to* p_3, *the angle* $\measuredangle(\overrightarrow{p_1p_4};\overrightarrow{p_1p_2})$ *to the first opponent* p_4 *to the left* and *the distance* $|\overrightarrow{p_1p_4}|$ *to* p_4.

For dribbling, clearing and goal-kicking, similar attributes have been defined. For additional tests, player number, game state, name of opponent team and time-step are included as *secondary attributes.*

4.2 Behavior Essences

Just like triggers, behavior essences consist of a sequence of primary and secondary attributes. Only the primary attributes are used for the similarity calculations. For passing, the primary attributes are *the ball speed*, *the direction of ball movement, the movement of the receiver, the duration* and *success or failure.*

The reason for failure is examined as secondary attribute.

For other behaviors, similar behavior essences have been specified.

4.3 Generation of Cases

A case assigns a behavior essence to a trigger:

$$case = (T; BE) : T \in \texttt{triggers}; BE \in \texttt{behaviors}_E$$

Let us examine the generating process of a case. As mentioned in section 2.1, the agents each perceive a partial extract of a world-situation W_t as an observation O_t and react with actions A_t that contribute to the world-situation W_{t+1}. This results in a sequence of world situations within the environment. In every time-step, the modeling agent observes the multi-agent system and receives a sequence of observations, which is subsequently evaluated until a complete behavior B_k^{k+l} is recognized. For this behavior, the situation at its starting point is examined to determine the respective T_k. Eventually, the pair of recognized trigger and behavior essence BE_k^{k+l} is added to the case base.

4.4 Similarity Measures

To determine the similarity between triggers of the same trigger pattern and between behavior essences of the same behavior essence pattern two similarity measures are required:

$$\text{similarity}_\text{T} : \texttt{triggers} \times \texttt{triggers} \rightarrow \mathbb{R}_{[0,1]}$$

$$\text{similarity}_\text{B} : \texttt{behaviors}_E \times \texttt{behaviors}_E \rightarrow \mathbb{R}_{[0,1]}$$

These are defined using a weighted sum of the local similarities that correspond to the individual primary attributes of the triggers and behaviors. The weights are specific to each pattern, that is, they are the same for all triggers or behaviors of the same pattern, respectively. Thus, for each pattern a sequence of weights is required.

The local similarities are functions that return values $\in \mathbb{R}_{[0;1]}$; as an example, we give the local similarity function for the positions v_k, v_l of initiating players for the pass trigger pattern:

$$\text{sim}_{\text{T},1}^{\text{pass}}(v_k, v_l) = \begin{cases} 1 & \text{if } |(v_l - v_k)| < 2 \\ 0 & \text{if } |(v_l - v_k)| > 15 \\ 1 - \frac{|(v_l - v_k)| - 2}{15 - 2} & \text{else} \end{cases}$$

Similar functions exist for the other primary attributes of the pass trigger pattern.

While the local similarities and the weights for behavior patterns are defined by the designer, the weights for trigger patterns are determined automatically using the following method. The goal of this method is, given a set of models $\mathbb{M}$ and a finite set of data $\mathbb{D} \subset \mathbb{X} \times \mathbb{Y}$, to find the model $M^* \in \mathbb{M}$ that best describes the functional relationship between the input values $X \in \mathbb{X}$ to the output values $Y \in \mathbb{Y}$. $\mathbb{D}$ is separated into a set of base data and a set of test data. Each model is then evaluated in combination with the base data for every test item. The best model M is determined by calculating the prediction error of all models and choosing the one with the minimal error.

The course of a game is taken as test data $\mathbb{TD} \subset \texttt{situations} \times \texttt{behaviors}_E$ and the best weights model W^* is then determined using

$$W^* = \underset{W_i \in \texttt{weights}_\text{T}}{\operatorname{argmax}} \frac{1}{|\mathbb{TD}|} \sum_{(S_j, BE_j) \in \mathbb{TD}} \text{similarity}_\text{B}\,(BE_j, \text{predict}(W_i, caseBase, S_j))$$

where predict is the behavior prediction function that returns an estimate for the behavior $BE \in \texttt{behaviors}_E$, based on a sequence of weights for the trigger $W_i \in \texttt{weights}_\text{T}$, a case base $caseBase \subset \texttt{cases}$ and a situation $S_j \in \texttt{situations}$.

$$\text{predict} : \texttt{weights}_\text{T} \times 2^{\texttt{cases}} \times \texttt{situations} \rightarrow \texttt{behaviors}_E$$

Thus, this method determines the sequence of weights that maximizes the similarity between predicted and recognized behavior for a set of test data $\mathbb{TD}$. The sequence of weights depends on the basis data and test data and therefore on the modeled team.

The complexity of this calculation depends linearly on the cardinality of the test set $\mathbb{TD}$, on the cardinality of the set of weight sequences $\texttt{weights}_\text{T}$ and the complexity of predict. By assuming that identical attributes have identical weights, we may reduce the space of $\texttt{weights}_\text{T}$ from $\mathbb{R}_+^{20}$ (because there are 20 primary attributes) to $\mathbb{R}_+^5$ (there are only 5 different primary attributes). If the weights for each attribute are limited to natural numbers $\leq n$, only $(n+1)^5$ sequences of weights have to be tested. Still, this is too computationally expensive for the given predict function.

Instead of varying the weights between 0 and n, a distribution of n weight units on k primary attributes is considered, i.e. n weight points are completely distributed to the k attributes, $\texttt{weights}_\text{T} = \{(w_1, \ldots, w_k) : w_i \in \mathbb{N} \wedge \sum_{i=1}^{k} w_i = n\}$. Thus, the set of weight sequences has $\binom{n+(k-1)}{k-1}$ elements, and for 10 weights and 5 primary attributes, there are 1001 weight sequences.

4.5 Selecting Cases from the Case Base

The function f of the behavior model $\mathfrak{M} = [\texttt{situations}; \texttt{behaviors}_E; f]$ is primarily determined by the function predict:

$$f(S) = \begin{cases} \text{predict}(W, caseBase, S) & \text{if } S \text{ is a decision situation} \\ \perp & \text{else} \end{cases}$$

that is, f gives results only in decision situations - this is the case when an agent of the modeled team controls the ball. Furthermore, decision situations have to be at least 4 time-steps apart (this is the minimal dribble duration).

In a decision situation, first potential triggers are identified. For passing this are up to 10 triggers, for goal-kick up to one trigger, and for dribbling and clearing exactly 6 triggers.

For every identified trigger, the case base is searched for cases with similar triggers. The similarity leads to an assessment of the cases. The best cases are selected.

From these cases, the best case is chosen using preferences, which are partly derived from the secondary attributes. Cases are preferred, if e.g.

- they reflect the same game state as the trigger
- they have the same opponent team as the trigger
- they have the same initiating agent as the trigger

The degree of preference depends on a bonus value for each criterion. The bonus values are determined automatically and in the same manner as the trigger weights.

The case that maximizes the sum of similarity and bonus value is chosen. Finally, the case is adapted according to the situation at hand: by comparing the observed trigger and the trigger stored in the case, the speed and angle of the ball movement, as well as the behavior duration are adjusted.

The function predict returns the adapted behavior essence of the case that maximizes the sum of similarity and bonus value.

The case base is extended by the current case (i.e. the case derived from the trigger and the actually observed behavior) if the similarity between the predicted behavior and the observed behavior is smaller than a fixed value σ_{limit} (i.e. the behavior essences differ significantly). Deletion or modification of cases does not take place. The similarity measure is updated by recalculating the sequence of weights for all trigger patterns.

5 Evaluation

For testing our approach, we made use of the data set derived from the GermanOpen RoboCup competition of 2001, which consisted of 44 games by 12 teams.

5.1 Behavior Recognition

We have found that using our approach on the full set of 44 games, we managed to classify between 96.4 and 99.7 percent of all ball controlling behaviors. Of these, passing amounted on average to 43.4 percent, dribbling to 24.8 percent, clearing to 20.2 percent and goal kick to 2 percent. Ball combat was observed in 8.2 percent of the time, one-two passes in 1.1 percent. Only in 1.4 percent of the time, non-classifiable ball transfers were observed, with 0.1 percent of completely unrecognized behavior.

The results of the algorithm have been verified against an independent, detailed manual classification of the same data.

5.2 Prediction of Behavior

Prediction of behaviors can be done at different abstraction levels (see section 3.4). We use here the "detailed description"-level and concentrate our experiments on pass behaviors. For every pass the predicted pass instance is compared with the actual performed pass instance. For the comparison the similarity measure for behaviors are used.

Using the described approach, a number of experiments for the prediction of ball-handling behaviors were conducted, especially to determine the relationship between prediction accuracy and the number of cases and the number of weight units. In detail we will only present the experiment regarding the dependency of the behavior model on the modeled team. For more experimental results, especially on the prediction of passing partners, see [15].

When using different numbers of weight units, it becomes clear that the distribution of weights on the individual attributes differs with the modeled team, i.e. teams differ in the importance they apply to the individual attributes.

During the experiments we discovered that a small number of weight units is sufficient to find a good team specific weight sequence. With more than 10 weight units, no further substantial improvements in prediction accuracy could be made using our approach.

The accuracy of prediction can be improved by increasing the number of cases. When using between 900 and 1000 cases (8 games), a plateau in prediction accuracy was reached, and the inclusion of more cases did not result in visible improvements any more. The maximum average prediction accuracy for the team "Brainstormers01" amounted to 0.54 (using 928 cases), however, for a game of the "Brainstormers01" against a different team, the accuracy was 0.46, using the same case base.

Dependency of Behavior Model on Modeled Team. To establish that the models obtained using cross validation are indeed team specific, we have used the "Brainstormers01" model to predict the behavior of other teams in the tournament. We have found that the behavior models are indeed specific for the modeled team.

The experiment was done for all teams who reached the final round. The case bases were generated based on the first five games per team. The weight sequences are determined for a randomly chosen game (among the 5). The prediction is evaluated for all final games of the team a) using the case base of the modeled team, and b) using the case base of the team "Brainstormers01".

The prediction using the team specific model is in all cases more accurate than the one with the 'alien' "Brainstormers01" model. The relatively high similarity between the prediction results of "Brainstormers01" to "DrWeb", "MRB" suggest that these teams use strategies for behavior selection that bear similarity to that of "Brainstormers01".

"Aras", "FCPortugal" and "RoboLog2k1", on the other hand, seem to use strategies that are very different from those of "Brainstormers01".

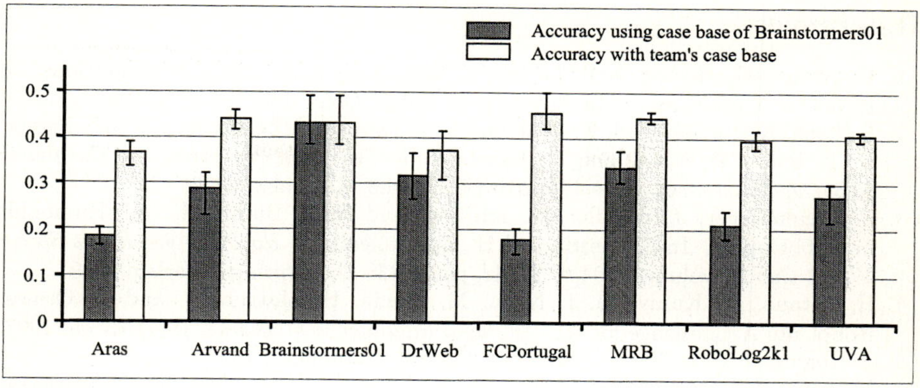

Fig. 1. Prediction accuracy for eight teams using a team specific model and the model of the "Brainstormers01" team

5.3 Conclusion and Outlook

We have found that, using our approach, ball controlling behaviors can be successfully classified, but predicted only with an accuracy between 0.39 and 0.54. However, this compares favorably to random guessing, which returns an accuracy of only 0.17. To obtain these results, a case base with about 1000 entries and a similarity measure using 10 weight units are sufficient.

It can be shown that the prediction model is specific for each team, whereby the use of the model of another team points to the use of similar strategies.

The limited accuracy of the results is due to two aspects: First, there are limitations that stem from the assumptions of the approach, which is ignorant of internal states of the agents, such as an incomplete world model, previous communication between agents or the execution of long-term plans. On the other hand, the description of the cases has been severely restricted to few parameters, which was necessary to limit the complexity of the modeling process.

Our approach only attempts to determine the behavior with the maximum probability. A more general description can be achieved by learning the probability distribution of the behaviors that are triggered by each trigger pattern.

Still, we consider the approach as being successful. We expect that some improvements over our current results can be made with alternative case descriptions. Substantially better results will probably be obtained with the simulation of internal states of the agents, thus performing a transition from a reactive to a context dependent model.

Acknowledgments

This research has been supported by the German Research Society, Berlin-Brandenburg Graduate School in Distributed Information Systems (DFG grant no. GRK 316). Most of this work was done at the Humboldt-Universität zu Berlin.

We thank Prof. H.-D. Burkhard and Kay Schröter for their contributions, fruitful discussions and support.

References

1. E. André, G. Herzog, and T. Rist. Generating Multimedia Presentations for RoboCup SoccerGames. In *RoboCup-97:Robot Soccer World Cup I*, 1997.
2. I. Frank, K. Tanaka-Ishii, K. Arai, and H. Matsubara. The Statistics Proxy Server. In T. Balch, P. Stone, and G. Kraetzschmar, editors, *Proceedings of the fourth RoboCup Workshop*, pages 199–204, Melbourne, Australia, 2000.
3. P. Gugenberger, J. Wendler, K. Schröter, and H.-D. Burkhard. AT Humboldt in RoboCup-98. In M. Asada and H. Kitano, editors, *RoboCup-98: Robot Soccer World Cup II*, volume 1604 of *LNAI*, pages 358–363. Springer-Verlag, 1999.
4. H. Kitano, Y. Kuniyoshi, I. Noda, M. Asada, H. Matsubara, and E. Osawa. RoboCup: A Challenge for AI. *Artificial Intelligence Magazine*, 18(1):73–85, 1997.
5. J. Kolodner. *Case-Based Reasoning.* Morgan Kaufmann, 1993.
6. M. Lenz, B. Bartsch-Spörl, H. D. Burkhard, and S. Wess. *Case Based Reasoning Technology. From Foundations to Applications.* LNAI. Springer, 1998.
7. H. Matsubara, I. Frank, K. Tanaka-Ishii, I. Noda, H. Nakashima, and K. Hasida. Automatic Soccer Commentary and RoboCup. In M. Asada and H. Kitano, editors, *RoboCup-98: Robot Soccer World Cup II*, LNAI. Springer, 1999.
8. A. Miene and U. Visser. Interpretation of spatio-temporal relations in real-time and dynamic environments. In *Proceedings of the 5th International Symposium RoboCup*, Seattle, WA, USA, 2001. The RoboCup Federation.
9. U. T. Müller. Beschreiben und Erkennen von Verhaltensmustern beim simulierten Fußballspiel. Diploma thesis, Humboldt Universität zu Berlin, Germany, 2002.
10. P. Müller-Gugenberger and J. Wendler. AT Humboldt 98 – Design, Implementierung und Evaluierung eines Multiagentensystems für den RoboCup-98 mittels einer BDI-Architektur. Diploma thesis, Humboldt Universität zu Berlin, 1998.
11. I. Noda, H. Matsubara, K. Hiraki, and I. Frank. Soccer Server: A tool for research on multiagent systems. *Applied Artificial Intelligence*, 12:233–250, 1998.
12. T. Raines, M. Tambe, and S. Marsella. Automated Assistants to Aid Humans in Understanding Team Behaviors. In M. Veloso, E. Pagello, and H. Kitano, editors, *RoboCup-99: Robot Soccer World Cup III*, LNAI. Springer, 2000.
13. P. Riley and M. Veloso. Planning for Distributed Execution Through Use of Probabilistic Opponent Models. In *IJCAI-2001 Workshop PRO-2: Planning under Uncertainty and Incomplete Information*, 2001.
14. D. Voelz, E. Andre, G. Herzog, and T. Rist. Rocco: A RoboCup Soccer Commentator System. In M. Asada and H. Kitano, editors, *RoboCup-98: Robot Soccer World Cup II*, LNAI, pages 50–60. Springer, 1999.
15. J. Wendler. *Automatisches Modellieren von Agenten-Verhalten – Erkennen, Verstehen und Vorhersagen von Verhalten in komplexen Multi-Agenten-Systemen.* PhD thesis, Humboldt Universität zu Berlin, Germany, Feb. 2003.
16. J. Wendler and J. Bach. Case based behavior assessment in simulated robotic soccer. In G. L. H. D. Burkhard, T. Uthmann, editor, *Modellierung und Simulation menschlichen Verhaltens*, volume 163 of *Informatik-Berichte*, pages 134–147. Humboldt-Universität zu Berlin, 2003.
17. J. Wendler, G. A. Kaminka, and M. Veloso. Automatically Improving Team Cooperation by Applying Coordination Models. In B. Bell and E. Santos, editors, *2001 AAAI Fall Symposium Series, Symposium: Intent Inference for Collaborative Tasks*, pages 84–90, Menlo Park CA, 2001. AAAI Press / The MIT Press.
18. M. Wünstel, D. Polani, T. Uthmann, and J. Perl. Behavior Classification with Self-Organizing Maps. In P. Stone, T. Balch, and G. Kraetzschmar, editors, *RoboCup-2000: Robot Soccer World Cup IV*, LNAI, Germany, 2001. Springer.

Case Based Game Play in the RoboCup Four-Legged League Part I The Theoretical Model

Alankar Karol[1], Bernhard Nebel[2], Christopher Stanton[1], and Mary-Anne Williams[1]

[1] Faculty of Information Technology, University of Technology, Sydney, NSW 2007, Australia
[2] Institut für Informatik, Albert-Ludwigs-Universität Freiburg
Georges-Köhler-Allee, Geb. 52, D-79110 Freiburg, Germany

Abstract. Robot Soccer involves planning at many levels, and in this paper we develop high level planning strategies for robots playing in the RoboCup Four-Legged League using case based reasoning. We develop a framework for developing and choosing game plays. Game plays are widely used in many team sports e.g. soccer, hockey, polo, and rugby. One of the current challenges for robots playing in the RoboCup Four-Legged League is choosing the right behaviour in any game situation. We argue that a flexible theoretical model for using case based reasoning for game plays will prove useful in robot soccer. Our model supports game play selection in key game situations which should in turn significantly advantage the team.

1 Introduction

Robot Soccer involves planning at many levels. In this paper we are concerned with developing high level planning strategies for robots playing in the RoboCup Four-Legged League. We develop a framework for developing and choosing game plays. Game plays are widely used in many team sports e.g. soccer, hockey, polo, basketball and rugby. There are several important differences between robot soccer in the Four-Legged League and human soccer, e.g. all the players on a team have the same physical capability so specialisation cannot exploit individuals physical talents. This is in contrast to the simulated league where it is possible for the players to have different attributes and physical capabilities. In addition because of their poor sensors, relative to humans, the AIBO robots possess a limited capability to predict detailed actions of others and hence there is little advantage to be gained from certain moves, e.g. *disguising a kick*.

Robot Soccer is a relatively new research initiative and in terms of its development it is in its infancy. One of the current challenges for robots playing in the RoboCup Four-Legged League is choosing the right behaviour, e.g. *the best kick*, in any game situation. Soccer is all about positioning; being in the right place at the right time. If a robot implements a kick then it needs to be in the best position to obtain maximum power and control.

We argue that a flexible theoretical model for using case based reasoning [8] for game plays will prove useful in the Four-Legged League. Our model will support game

D. Polani et al. (Eds.): RoboCup 2003, LNAI 3020, pp. 739–747, 2004.

play selection in key game situations which should, in turn, significantly advantage the team. Case based reasoning has been used in other robot soccer leagues for various purposes. Most of the focus in strategy development has been in the *Simulated League*. The Simulated League lends itself to cased based techniques and to machine learning approaches because of the speed and flexibility of developing virtual robots as well as the ease and practicality of data collection during actual games.

In 1999 Wendler *et al* [10] developed a case based reasoning approach to action selection in the Simulation League, whilst Wendler, Kaminka and Veloso [11] provided a general theoretical case based reasoning model for robot coordination between team members. More recently, Gabel and Veloso [3] described a highly sophisticated case based approach to enhance the performance of the online coach in the Simulated League. Their system allows knowledge about previous match performances to be incorporated into the online coach's decision making process.

In this paper we propose a simple and robust cased based reasoning model for the RoboCup Four-Legged League that can be customised and enhanced. In a companion paper we will describe some experimental results that evaluate the model's performance using the UTS Unleashed Robot Soccer System (http://magic.it.uts.edu.au/unleashed).

2 Robot Soccer Game Play

A major benefit of developing a case base for robot soccer game plays is that it will result in a powerful knowledge base containing important knowledge about the game. A case base of game plays can capture creative genius and enduring principles of how to play the game for the purpose of teaching robots to play soccer more effectively.

Cases can describe set situations, like *kickoffs*, as well as running and passing game plays, attacking moves, and defensive formations. For any game situation, in our model, game plays are chosen based on the similarity between the current state of play and a collection of cases in the knowledge base. As our robot soccer multiagent system evolves, more ambitious new game play cases can be added, for example, as the robots become better at passing we can develop cases for *pass blocking*.

The game play strategies embedded in the cases can blend the lessons of the past with best guesses for future matches, and as such they incorporate some key elements for a winning game. The game play cases in the knowledge base can be selected depending on the type of game required. For example, if it is known that a particular opposition plays a certain style of game then the case base used could reflect specific tactics and strategies to counter that style.

Game play cases allow teams to string several plays together that take advantage of a teams' strengths. In other words, game play cases can form the building blocks of larger plans. Set game plays could prove critical for the success of team. In human games the difference between winning and losing is often the successful execution of set game plays in both offense and defense. In human soccer it has been calculated that as many as 40% of all goals scored are from set game play situations.

3 Case Based Reasoning in Conceptual Spaces

Categorisation of information helps robots reduce the complexity of the information they need to acquire and manage during their lifetime. In addition, the ability to categorise gives rise to broad powers of explanation. For example, without the ability to categorise, robots would not be capable of representing visual information beyond the pixel level, and as a result would not develop a world model that could support even simple forms of object recognition and reasoning. The ability to form and manipulate categories enhances robots capacity for problem solving, communication, collaboration etc. We expect robots to respond appropriately to information acquired through their sensory systems. The ability to categorise new sensory information and to anchor it to objects in the real world allows a robot to behave *sensibly* in previously unencountered situations[6].

For the purpose of this paper we believe that categorizing game situations will assist robots play better soccer. We use the conceptual spaces approach [4, 6, 5] to categorization driven by similarity measures.

Few concepts or categories of objects can be specified using necessary and sufficient conditions: Mathematical entities like *triangles* can, but almost all everyday objects, like *chairs* for example, defy explicit definition.

A similarity based approach to categorization is more widely applicable to robot soccer than explicit rules, because soccer playing robots need to make useful generalizations about previously unencountered situations. To play soccer well robots cannot be *hardwired* they must be able to respond appropriately to situations that were not foreseen at design time.

Conceptual spaces are multidimensional spaces that can be used to describe both physical and abstract concepts and objects. In contradistinction to the use of explicit (causality) rules to describe the relationship between objects, conceptual spaces adopt a similarity-based approach to categorization.

The main idea is that objects are categorized according to how similar they are to a prototype or (cluster of) exemplar(s). For instance, the colour *yellow* is more similar to *green* than it is to *blue*. For the purpose of robot soccer strategies we are interested in identifying prototypical or important game states and measuring the similarity across different game states.

Conceptual spaces are geometrical structures based on quality dimensions. Quality dimensions correspond to the ways in which stimuli/features are judged to be similar or different. Judgments of similarity and difference typically generate an ordering relation of stimuli/features, e.g. judgments of *level of control of the ball* generate a natural ordering from "weak" to "strong" [4]. There have been extensive studies conducted over the years that have explored psychological similarity judgments by exposing human subjects to various physical stimuli.

Objects are characterized by a set of qualities or features $\{\mathbf{q}_1, \mathbf{q}_2, ..., \mathbf{q}_n\}$. Each feature $\mathbf{q}_i$ takes values in a domain $\mathbf{Q}_i$. For example, the distance from a robot to the goal can take values in the domain of positive real numbers. Objects are identified with points in the conceptual space $\mathbf{C} = \mathbf{Q}_1 \times \mathbf{Q}_2 \times \ldots \mathbf{Q}_n$, and concepts/categories are regions in conceptual space.

For the purpose of problem solving, learning and communication, robots can adopt a range of conceptualizations using different conceptual spaces depending on the cognitive task at hand. For this reason we develop various meta-level strategies that determine the cases to consider and a number of pertinent similarity measures for our application in robot soccer.

For our current purpose, and without loss of generality, we often identify a conceptual space $\mathbf{C}$ with $\mathbf{R^n}$, but hasten to note that conceptual spaces do not require the full richness of $\mathbf{R^n}$. For example, in two of our similarity measure given in Section 4 we measure the distance between two objects on the soccer field using Euclidean distance, however we also develop a third qualitative similarity measure based on a partitioning of the field into strategic regions where each region can be given a weighting that represents its strategic importance (see Figure 1).

Similarity relations are fundamental to conceptual spaces [7]. They capture information about similarity judgments. In order to model some similarity relations we can endow a conceptual space with a distance measure; A distance measure d is a function from $\mathbf{C}$ x $\mathbf{C}$ into $\mathbf{T}$ where $\mathbf{C}$ is a conceptual space and $\mathbf{T}$ is a totally ordered set. Distance measures lead to a natural model of similarity; the smaller the distance between two objects in conceptual space, the more similar they are. The relationship between distance and similarity need not be linear, e.g. similarity may decay exponentially with distance.

A categorization results in a partitioning of a conceptual space into (meaningful) subregions. The geometrical nature of conceptual spaces coupled with representations for prototypes, and the ability to manipulate dimensions independently of one another ensures that they provide a highly flexible and practical representation of context-sensitive case-based reasoning.

Our cases consist of *prototypical situations* and *important situations* that are encountered during a soccer match. For example: kick off, a single attacking player in the goal penalty area, a player with the ball in a goal-end corner, or a player with the ball on the field border.

The cases have been developed over the last year through observation of the NUbot [1] team in practice matches and in competition matches at RoboCup 2002, and more recently during the practice matches of UTS Unleashed!. In Part II, the sequel to this paper, the cases will be refined and tested using experimentation. During the experiments robots will be placed in preselected positions and their behaviour monitored. Successful sequences of actions that lead to positive results will be adopted and incorporated into the cases.

Our aim is to develop a collection of cases to create a conceptual space for the purpose of providing strategic decision making assistance to robots. To that end we must define the appropriate quality dimensions, i.e. features, which will prove crucial for the similarity measure, and then the similarity measure itself. In addition, we identify some meta-level features which can be used to determine the set of cases that should be considered during a game.

Each of our cases consists of a set:

$$\text{Case} = \{\text{Field}, \text{Possession}\}$$

The State of the Field is described in terms of absolute coordinates with the center of the field prescribed as the origin of the coordinate system. The positive y-axis is directed along the opponents goal. In such a system we can then denote the position of the players by an ordered pair (x, y).

The state of the field is defined by the position of the players. We denote the set of our players by P and the opponents by the set P', where

$$P = \{p_1, p_2, p_3, p_4\} \text{ and } P' = \{p'_1, p'_2, p'_3, p'_4\}$$

Here $p_i = (x_i, y_i)$ and $p'_i = (x'_i, y'_i)$ are the absolute x and y coordinates of the ith player.

The Degree of Possession is another important dimension in our case. The degree of possession is a measure of which team possesses the ball and what the nature of that possession is.

Table 1. Degree of Possession

Possession	degree
no possession	0
no possession but in a scrum	1
possession but in a scrum	2
possession and clear	3
possession by the opponent team in the scrum	-2
possession by the opponent team and clear	-3

This numerical degree given to an otherwise qualitative characteristic will allow us to use it effectively in calculating a similarity between a current situation and our case-base.

In addition to the object level features we also use several meta-level features (or global parameters) that can aid in the selection of the appropriate strategy. For example, these meta-level features can help us identify a subset of the possible cases that should be considered in a given situation. Furthermore, they can also be used to resolve conflicts when two or more cases are "equally" similar.

The Situation involves a numerical evaluation that represents the context of the player with a high degree of possession. If the player with control of the ball has no obstacles between herself and the goal, then the team's situation is said to be "wide open" and is given a degree of 2. If the player with control of the ball has obstacles between herself and the goal, then the team's context is *clear* and assigned a degree of 1 and if the player is in a scrum then the situation is given a degree of 0. Negative numerical values are attributed to the above situations if the ball is in possession of the opponent team. A player's *situation* differs to a player's *degree of possession* in that the situation is determined by the global state of the field, whilst degree of possession only concerns the immediate vicinity of the robot.

The Score and the Time to Game Completion are two important meta-level parameters that can be used to determine the set of cases that should be considered. In this way the score and the time left in the game can influence the strategy. For example, an unfavourable score and a short time remaining might induce the robots to take more adventurous actions. In contrast a favourable score and a short time remaining could induce a more defensive behaviour.

Countering Opponent Team Strategies. This parameter can be advantageous in the case where reliable information about the opponent teams strategies can be obtained. For example if the other team is known to play a strong attacking game, then it would be in the interests of a team to ensure that they maintained possession of the ball at the cost of pushing the ball forward.

4 Measure of Similarity and Action

As mentioned earlier an appropriate measure of similarity is essential for developing strategies based upon past experiences. The performance of case based reasoning is strictly dependent on the quality of the similarity measure adopted. We intend to verify our similarity measures' effectiveness via experiment and for that reason we have developed several measures of similarity; two quantitative measures and one qualitative measure.

The quantitative similarity measures that we chose to make a correlation with a prototypical case and the current situation on the field are calculated by minimising the Euclidean norm.

Let N be the number of cases. Then the field in the j_{th} case can be represented as:

$$C_j = \{P_{ij}, P'_{ij}\}$$

We let the current field situation be represented by $C = \{Q_i, Q'_i\}$. If we now want to determine the similarity between a case C_j and the current situation, we have to come up with a pairing between the players in the case and the current situation. We will use a permutation π for this purpose.

For any given case C_j and given permutation π, we find the distance between the players

$$P_{ij}Q_{\pi(i)} = \sqrt{(x_{P_{ij}} - x_{Q_{\pi(i)}})^2 + (y_{P_{ij}} - y_{Q_{\pi}(i)})^2}$$

We construct a 4×4 matrix as follows

$$\begin{bmatrix} P_{1j}Q_{\pi(1)} & P_{1j}Q_{\pi(2)} & P_{1j}Q_{\pi(3)} & P_{1j}Q_{\pi(4)} \\ P_{2j}Q_{\pi(1)} & P_{2j}Q_{\pi(2)} & P_{2j}Q_{\pi(3)} & P_{2j}Q_{\pi(4)} \\ P_{3j}Q_{\pi(1)} & P_{3j}Q_{\pi(2)} & P_{3j}Q_{\pi(3)} & P_{3j}Q_{\pi(4)} \\ P_{4j}Q_{\pi(1)} & P_{4j}Q_{\pi(2)} & P_{4j}Q_{\pi(3)} & P_{4j}Q_{\pi(4)} \end{bmatrix}$$

A corresponding matrix can be constructed for the opponent team as well. We now look at two different methods of defining a quantitative similarity measure. Both of these methods can be easily extended to include a weighting given to the position of the players.

Method 1. The similarity measure is achieved by calculating the following:

$$\min_{k,\pi}\left(\sum_i\sum_k P_{i1}Q_{\pi(k)},\cdots,\sum_i\sum_k P_{ij}Q_{\pi(k)}),\cdots,\sum_i\sum_k P_{iN}Q_{\pi(k)}\right)$$

If the minimum is found for C_l and π, then we say that in terms of the field configuration the lth case with player pairing π is the most similar to the current situation.

Method 2. In this method we first find the maximum distance between any two players:

$$d_j \max \ = \ \max\left(P_{ij}Q_{\pi(k)}\right) \qquad i,k = 1,\cdots 4.$$

The rationale behind this is that the similarity between two cases should be based on the time required to move all robots into the positions from one case to the other. Since each robot can move independently it can be done in parallel, hence it is enough if we look for the maximum distance.

The similarity measure is then found by: $\min\limits_{k,\pi}(d_1,\cdots,d_j,\cdots,d_N)$

Since we have to consider all pairings between our own players and between the opponent players, we have to compare all in all $N\times(4!+4!)$ (sub-) cases with our current situation, i.e., we have to look at $48\times N$ (sub-)cases. If we want to apply this method to larger teams, e.g., with 11 players, one clearly needs to employ different methods because we would have to consider approx. $(11!+11!)\times N$ cases which is approximately equal to $8.0\times10^7\times N$ cases. However, fortunately, one could use *minimal-weight perfect matching* [2] techniques at this point, which are polynomial in the number of players.

However it is important to note that our case actions are based solely upon the position of our team's players since we have no control over the movement of the opponent team's players. So in order to have a more comprehensive strategy we include in our analysis a subset of similar cases. Suppose that the similarity measure for the state of the field gives us a minimum distance D from a particular case. Instead of simply considering the best matched case, we could also consider cases which satisfy $D+\epsilon$ where ϵ is small. The best case can then be selected from this subset by taking a meta-feature such as the *situation* into account.

The qualitative measure of similarity is based upon the minimum number of moves that the players have to make to go from a particular situation to one of the situations in our case-base. To achieve this numerical value we divide the field into 30 strategic regions. For simplicity we assume that the various regions are rectangular regions. The sectors are created by dividing the field into three vertical regions and then each region is subdivided into four smaller regions (along the horizontal direction). The four corners thus generated are further subdivided into four regions. Each region or sector of the field is given a number to uniquely identify the region. The diagram of the field is shown in Figure (1).

We can then classify the regions by assigning them a number. Our *goal region* is given the number 0 and then starting from the field on our side we number the regions left to right. Each player is then given a corresponding number which is equal to the number of the region it is found in. We define our similarity measure by calculating the

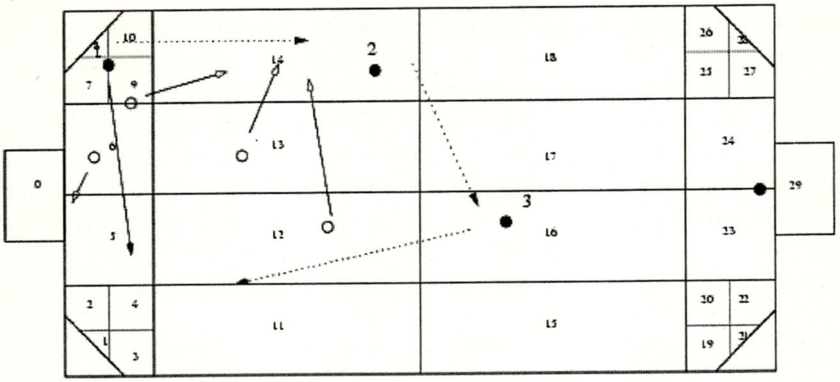

Fig. 1. The soccer field divided into 30 strategic regions and an example of game play scenario

minimum number of moves our players have to make to be in the same position as the case being compared with. We weight the moves as: a move towards the opponents goal is given a weight of $+2$, a move backward towards our goal is given a weight of -2 and a move to the left or the right is given a weight of $+1$.

5 Example

Let us consider a prototypical case in a soccer match; the player and the ball are stuck in a corner with the player facing away from the field and obstructed by an opponent player. The situation is represented in Figure (1) which illustrates the motion of the ball and the movement of the players. Our team players are represented by fully shaded circles while the opponent team players are represented by hollow circles. Dotted arrows indicate the motion of the ball and the arrows with a shaded head show the movement of our team players. The unshaded arrows indicate the motion of the opponent team players.

In this particular example, the strategy is to kick the ball backwards. As a result of the motion of the ball the opponent players move along the direction of the moving ball. The receiving player (Player 2) then kicks the ball back to Player 3. This motion leaves the situation wide open and without any obstructions and allows Player 1 to position itself in front of the opponents goal.

6 Discussion

Robot Soccer involves planning at many levels, and in this paper we developed a theoretical case based reasoning model for robot soccer strategies for the RoboCup Four-Legged League.

We argued that a flexible theoretical model for using case based reasoning for game plays will prove useful in robot soccer. Our model will support game play selection in common and key game situations which should in turn significantly advantage the team.

One of the current challenges for robots playing in the RoboCup Four-Legged League is choosing the right behaviour in any game situation. Our model allows robots to develop and choose game plays for any game situation.

We adopted the conceptual spaces framework which relies on the determination of prototypical situations and a measure of similarity across all situations. We developed three similarity measures for our model; two quantitative and a strategically oriented qualitative measure.

Having developed a theoretical model we intend putting it to the test using experimental evaluations, and have begun to develop our experimental framework using the UTS Unleashed! Robot Soccer Multiagent System.

One of the challenges for our future work is to extend our model to handle incompleteness and uncertainty. Throughout our discussion we have assumed that the robots world model is reasonably, but not perfectly, accurate, however in reality much of the information required to choose the best matching case may be simply *unknown* even in the case where robots can communicate. We expect our experimentation to reveal the best way to design for high level incompleteness and uncertainty and plan to address it using techniques given in Liu and Williams[9].

References

1. Chalup, S., Creek, N., Freeston, L., Lovell, N., Marshall, J., Middleton, R., Murch, C., Quinlan, M., Shanks, G., Stanton, C., Williams, M-A.: When NUbots Attack! The 2002 NUbots Team Report, Newcastle Robotics Laboratory Report 2002, http://robots.newcastle.edu.au/NUbotFinalReport.pdf.
2. Cook, W., Rohe, A.: Computing minimum-weight perfect matchings. INFORMS Journal on Computing.**11, 2**, 138–148, 1999
3. Gabel, T., Veloso, M.: Selecting Heterogenous Team Players by Case-Based Reasoning: A Case Study in Robotic Soccer simulation. CMU-CS-01-165. December, 2001.
4. Gardenfors, P.: Conceptual Spaces: The Geometry of Thought. A Bradford Book, MIT Press, Cambridge Massachusetts, 2000.
5. Gärdenfors, P., Williams, M-A.: Reasoning about Categories in Conceptual Spaces. Proceedings of the Joint Conference of Artificial Intelligence. Morgan Kaufman. San Francisco, 385–392, 2001.
6. Gärdenfors, P. and Williams, M-A, Building Rich and Grounded Robot World Models from Sensors and Knowledge Resources: A Conceptual Spaces Approach, in the Proceedings of the International Symposium on Autonomous Mini-robots for Research and Edutainment, 123 – 133, 2003.
7. Hahn and Ramscar, Similarity and Categorization, Oxford University Press, 2001.
8. Case-Based Reasoning Kolodner, J., San Francisco, California: Morgan Kaufmann, 1993.
9. Liu, W., Williams, M-A.: Trustworthiness of Information Sources and Information Pedigrees. Intelligent Agents VIII. Series: Lecture Notes in Computer Science. Volume. 2333, Springer Verlag. Berlin, 290 - 307, 2002.
10. Wendler, J., Gugenberger, P., Lenz, M.: CBR for Dynamic Situation Assessment in an Agent-Oriented Setting. ECAI, 1998.
11. Wendler, J., Kaminka, Gal A., Veloso, M.: Automatically Improving Team Cooperation by Applying Coordination Models, 2001.

How Contests Can Foster the Research Activities on Robotics in Developing Countries: Chile – A Case Study

Javier Ruiz-del-Solar and Juan Cristóbal Zagal

Department of Electrical Engineering, Universidad de Chile
{jruizd,jzagal}@ing.uchile.cl

Abstract. The aim of this article is to describe our experience in the participation and organization of robot contests, and to show how these actions have increased the activities on robotics in Chile. We describe the annual Latin American IEEE Robotics Competition, we present the IEEE Latin American Robotics Council, we explain our participation in RoboCup, and we present our activities concerning robotic courses for children.

1 Introduction

The purpose of RoboCup and in general of robot contests is to foster the activities on robotics and to push the field and inspire future research. The aim of this article is to describe our experience in the participation and organization of robot contests, to show how these actions have increased the activities on robotics in Chile, and finally to give our view about the effect of this kind of activities. We believe that our experience can be useful for other developing countries.

In this article we first show the current state of the robotics field in Chile and we contrast it against the current state of other disciplines (section 2). For doing that we carried out an extensive research study, which considered investigation on Internet, on public paper databases, on national science databases, as well as direct contact with the researchers. We show that the robotics field is underdeveloped in our country, and that its development degree is low compared with other research fields.

Afterwards we show how robot contests in general, and RoboCup in particular, can help in reverting this situation. In section 3 we give concrete examples of that. We describe the annual Latin American IEEE Robotics Competition, we present the IEEE Latin American Robotics Council, we explain our participation in RoboCup, we present our activities concerning robotic courses for children, and finally we show how all these activities have largely increased not only our own activities in the field, but also the activities on robotics in Chile and Latin America.

We consider that the development of robotics is important not only by its own, but also because it serves as a motivation to work in related technologies as sensoric, electronics, power systems, signal processing, automatic control, intelligent systems, software agents, just to mention a few.

Finally, in section 4 some conclusions of this work are given.

D. Polani et al. (Eds.): RoboCup 2003, LNAI 3020, pp. 748–756, 2004.

2 Problem Definition

2.1 Research in Chile

The state of research in science and technology in Chile will be analyzed on hand of the number of papers produced by researchers working in Chile, the impact of these papers, as well as indicators of the local technological development (e.g. patents indices). The main information source will be CONICYT, the National Chilean Council of Science and Technology [1].

Chile is a country of 15 millions inhabitants with a per capita income of US$ 4346, an alphabetization rate of 95.2% and an expenditure in R&D of US$377.2 millions (table 4.4 in [2]). There are 7.2 thousand researchers in Chile, and 22575 ISI publications were generated during period 1981-2000, the number of citations was 152070 (table 4.4 in [2]). According to CONICYT the impact index of the publications generated in Chile is 6.74, which is well compared to similar indices of Spain (6.78), Argentina (5.63), Hungary (6.57), Portugal (5.99), Mexico (5.48) and Brazil (4.99), just to show some examples (USA has the largest index: 16.34). Chile has the highest research indices in Latin America, with 11.94 ISI papers per 100 thousand inhabitants and 0.978 ISI papers per researcher (tables 4.8 and 4.9 in [2]).

However, if we look carefully on the Chilean research indices, we will notice that science is much more developed than technology. While in the period 1981-2000 only 658 ISI paper where published in engineering, 2120 were published in astrophysics and astronomy, 4079 in biology, 3138 in chemistry and 1397 in physics (table 4.1 in [2]). Moreover, if we observe the number of requested patents in Chile, an indicator of technological development, we will see that in year 2000, 407 patents were requested by Chilean residents, while 3276 were requested by non-Chilean residents (table 4.16 in [2]). The number of patents requested per every 10 thousand inhabitant is only 0.32 (table 4.20 in [2]). Just as a last fact, Chile was classified number 20 in the IMD 2002 World Competitiveness Yearbook [3], which is a very good position. However, the IMD report indicates that the main drawback of Chile in order to raise his relative position is on its low R&D indices.

As a final conclusion we can say that although sciences in Chile shows good development level, engineering and in general technology has a low development level.

2.2 Study about the State of Robotics in Chile: Methodology

The aim of our study was to determine the state of the robotics field in Chile. For doing that we required information about: (i) researchers working in robotics in Chile and their institutions, (ii) the international and national publications on the field, (iii) the research projects, and (iv) courses on robotics given to undergraduate and graduate students. For carrying out our study we have made the following sequential actions:

1. All main Chilean universities were asked for their activities in robotics and for the researchers working on robotics and related fields. These researchers were directly contacted by e-mail.
2. An exhaustive Internet investigation using the search engines google and todocl [4] was carried out. All Chilean sites containing the words "robot" and "robót", which are the seeds for robotics related words in English (robot, robotic, etc.) and in Spanish (robótica, robótico, etc.), were analyzed.

3. All government-funded projects in robotics, which are listed in the databases of CONICYT, were analyzed.
4. Using the information obtained with these three described actions, a first database of researchers working on robotics in Chile was built.
5. The personal web pages of these researchers were analyzed.
6. Finally, we looked for the international papers published by these researchers, using the ISI databases [5] and "Citeseer" [6].

2.3 Study about the State of Robotics in Chile: Results

This first study about the state of robotics in Chile gives the following results:

- **International Publications:** The number of international publications on robotics is very low. Only 3 papers were published in international journals (2 of them correspond to joint collaborations with foreign institutions) [7, 8, 9] and 14 papers in international conferences. The publications were done between years 1985 and 2002. The situation is shocking. These first results clearly indicate that the field is underdeveloped in Chile and that their development degree is low in comparison with other research fields in Chile (see section 2.1).
- **National and Latin American publications:** 13 papers were published in national journals (non ISI indexed) and 53 in conferences. The publications were done between 1983 and 2002. We believe that these results show that either the level of the publications is below international standards or the researchers have no interest or no means to publish outside Chile. Even if the later is the case, the number of publications is still rather low.
- **Research Projects:** We found just 15 projects on the field between years 1987 and 2002. 7 correspond to projects of the National Fund for Science and Technology Development, FONDECYT [10], 3 correspond to projects of the Fund for Fostering Scientific and Technological Development, FONDEF [11], 2 correspond to FONDECYT-DOCTORAL projects and the last 3, to projects funded by universities. FONDECYT and FONDEF are the two main national research program from CONICYT, and according to the Chilean standards, these projects have normally a better level than projects funded by universities. However, we find a contradiction on these numbers. We have 10 (7+3) good level projects, but just 3 international journal papers and 14 international conference papers generated by them. This fact could indicate either that the projects were not successful in terms of the obtained results or that the researchers had no interest on publishing.
- **Courses:** At this moment we can find just 5 courses on robotics given by 3 Chilean universities. At the end of this year this offer will increase to 7. In two more years this number will reach to 9 courses given by 4 different universities.

By analyzing the presented results we can conclude the following:

- By all different possible measures we can affirm that the robotics field is underdeveloped in Chile. Its development degree is low compared with other research fields. Even if we think that the results of this study are not accurate and that they have a deviation of 100%, the obtained numbers are extremely poor.
- These first results contrast with our experience on which robotics generates high interest in Chilean engineering students, especially from fields like electrical and

mechanical engineering, as well as computer science. Just 2 examples: (i) from 70 students entering our department this year, 31 have interest on robotics and computer vision (13 as first choice, 9 as second choice and 9 as third choice), and (ii) our robotics course, given by the first time this semester, has 25 students.

- Taking into account the number of publications, research projects, as well as courses on robotics, we can establish that four universities concentrate most of the activities performed on the field of robotics in Chile. They are (in alphabetic order) Universidad de Chile (Dept. of E.E.), Universidad de Santiago de Chile (Depts. of Computer Science, E.E. and Industrial Eng.), Universidad Técnica Federico Santa María (Dept. of Electronics) and Pontificia Universidad Católica de Chile (Depts. of Mech. Eng. and E.E.). In this four universities no more than 10 researchers are active in the field.

2.4 Understanding This Situation

Chile has a low level of industrial development and its economy is based either on the service industry (basic services, banking industry, tourism industry, etc.) or in the exploitation of natural resources (mining industry, forestry, fishing, wine industry, farming industry, etc.). Moreover, given the current per capita income, the cost of the manual labor is still low, although not as low as in south Asia or other Latin American countries. For these reasons robotics related technologies have not attracted much attention, either in industrial or educational sectors. Another factors to mention are: (i) lack of knowledge on the importance of the field and (ii) research in robotics is usually considered as extremely expensive.

However, things are changing. Chilean economy was growing at a 7% rate during the last decade, and experts say that a condition to keep this growing rate is to introduce technology in the Chilean traditional economy areas for increasing productivity and also in the manufacturing industry for improving their development. In both cases robotics related technologies can play an important role.

3 Robot Contests Can Foster the Research Activities on Robotics

In this section we will show how robotic contest can revert the situation described in the previous section.

3.1 Latin American IEEE Student Robotics Competitions

The first Latin American IEEE Student Robotics Competition was held last year in Chile (http://ewh.ieee.org/reg/9/robotica/1stRobotContest/). The event was organized by the Department of Electrical Engineering of the Universidad de Chile and by the IEEE Region 9. The main objective of this competition was to increase the interest of engineering students in robotics.

A second objective was to stimulate student involvement in advanced technologies and their application to practical use. The competition looked for bringing the exhilaration of scientific discovery to young students, fostering technological innovation in the graduating engineers, while providing more technical activities to local IEEE students. A robotics competition was considered an activity that could help in that direction, particularly given the success of similar events in other countries. The competition was held during period 29-30 November 2002, at the Department of Electrical Engineering, Universidad de Chile, Santiago, Chile. Two separate robot competitions took place, details are provided in Spanish in [20].

The first competition ("beginners") was aimed for students starting to work in robotics and was based on the use of Lego MindStorms building blocks. In this category the proposed challenge was the design and programming of a robot, or equipment of robots, that accede and cross a simulated minefield. The robot or equipment had to detect simulated, explosive charges and to avoid stepping on them in an inadvertent form. The success criterion was the number of detected mines, the time from the departure point to the term of the mission, and the number of inadvertently detonated mines (those that damage the robot).

The second competition ("advanced") was for more experienced student groups. The competition consisted in crossing a soccer field with obstacles using any kind of legged robots (biped, hexapods, etc.). Robots could be designed and constructed by the own participants, or could be bought and then adapted (mechanical modifications, new sensors, etc.). It was understood that robots should be autonomous and in no case they could be controlled, although partially, in remote form. There were no restrictions concerning the size of robots, the power source used, neither the sensors employed (infrared, ultrasonic, laser, cameras, etc.). The success criterion was the crossing time.

The first competition day was for reception of the participants and for setup of the robots. A plenary talk on robotics technology, given by a Distinguished Lecturer from the IEEE Robotics and Automation Society was also scheduled for the first day. The second day was dedicated to the contests. Both competitions were developed in parallel, although the finals of each one were scheduled at different time. Before competitions a 5 minutes presentation of each robot to the jury and the general public by the participants was considered. After the competitions there was an exhibition of all participants robots to general public, especially to school' children.

We believe that this first Latin American competition was a success. 29 groups of 4 different countries (Argentina, Chile, Mexico and Peru) participated in the contest, totaling more than 100 engineering students and 4 high school students. Considering the short preparation time (6 months) and the long distances in Latin America, this was a good starting that for sure will be improved in the next versions of the contest. During the second competition day we had more than one thousand persons attending the event. The event was also covered by almost all Chilean newspapers (see some articles in [21]) and by the most important TV station in its main news program. Media coverage was considered very important, because it promotes public awareness of the importance of technology in society.

In general terms we believe that the main result of this event is that it will foster the activities in robotics in Chile and also in Latin America. Here we mention some concrete examples:

- The institution of an annual Latin American IEEE Student Robotics Competition as a regular activity of the IEEE Region 9. The next competition will be held on Sept. 2003 in Brazil, while in 2004 it will be held in Mexico.

- The foundation of the IEEE Latin American Robotics Council in January 2003, with the long-term mission of developing IEEE activities on robotics in Latin America. A detailed explanation can be found in section 3.2.
- The institution of an annual Chilean IEEE Student Robotics Competition, which next version will be held on August 2003 at the Department of Electrical Engineering, Universidad de Chile.
- At a national level the activities on robotics, such as the creation of robotics courses and the conforming of robot research groups, have increased in the universities where students participated in the robotic contest. As an example, based on the demands of our students we have created an introductory course on robotics, which is being given for the first time during this fall semester (March-June). During our spring semester (August-November) we will give, also for the first time, a course on mobile robotics. These two courses will be added to our related courses on mechatronics and in automatic control. Since October 2002 our "Robotics Lab" is full from students, not only from our department, but also from Computer Science and Mechanical Engineering.
- Also at national level we can see an increasing interest in similar kind of robot contests, especially on robot soccer. Recently the Pontificia Universidad Católica de Chile (Dept. of Mech. Eng.) has conformed a FIRA robot team and our department has conformed two RoboCup teams, F-180 and "Four-Legged". This activity is described in section 3.3.

3.2 IEEE Latin American Robotics Council

For giving continuity to the IEEE robot contest in Latin America in January 2003 was founded the IEEE Latin American Robotics Council [22], whose organization and operation is in charge of us. The main purpose of this council is to organize, with the support of local groups, an annual IEEE Student Robotics Competition in Latin America. The Council will promote the programming of these events, define the basis for the competitions, and interact with the local volunteers who finally will develop the activities themselves. The council will grow from this event into helping at developing more IEEE activities in robotics in Latin America.

As a first activity, the council is organizing the 2nd Latin American IEEE Student Robotics Competition (http://www.ewh.ieee.org/reg/9/robotica/2ndRobotContest/), which will be held in September 2003, in Bauru, Brazil. This competition will include beginners and advanced contests, as well as robot soccer competitions.

3.3 RoboCup Teams of the University of Chile

As a way of canalizing the increasing interest of our students on robotics activities, we decided to create two robotic soccer teams, which will participate in RoboCup, from this year on. The teams are from the small-size F180 league and from the AIBO four-legged league. In December 2002 we took that decision, in spite of we had only 7 months before RoboCup 2003 and just 4 months before the classification stage. We started from zero. We had no experience with RoboCup (no Chilean team was in RoboCup before) and we neither had all the robots.

To achieve our difficult objective we started to work in parallel, in several fronts, with a group of 35 undergraduate and graduate students. In a first stage of 'setup', one student group was in charge of the compilation of all necessary RoboCup information (official site, leagues information, rules, source codes, simulators, etc.) mainly through Internet. Another group was in charge of building the two official soccer fields, while a last group was in charge of buying the AIBO robots, and buying and adapting the robots for the F-180 league. This stage took about 25 days.

Afterward we started the second stage that consisted in the research work for writing the algorithms and designing the playing strategies for both teams. We divided our 35 students in two teams, 20 for the F-180 league and 15 for the four-legged league. This decision was based on the fact that the F-180 league requires work in hardware and software aspects, while the four-legged not. For each robot team we defined the block diagram of the final system we wanted to implement. A group of 3 to 4 students was in charge of each one of these blocks. In each category we have a special group in charge of making simulators. Given that we have 9 groups working in parallel, organizational aspects are very important. We have a hierarchical organization of the work and meetings every two days for each team, whose main objective is to coordinate the work of the teams. A graduate student from our doctoral program on automation was in charge of each team. After months of very hard work, including holidays, we finished this second stage last March.

Since March 15 we are working on the third stage, which can be characterized as "playing, playing and playing". The aim of this stage is to increase the performance of our teams by means of intensive experimentation. In this context, we participated in the RoboCup American Open with our two teams. This stage will be finished on July 2, with the participation of our four-legged team in the RoboCup 2003.

Participating in RoboCup has been an interesting challenge for us. We built a multidisciplinary team of students, which includes students with different majors like computer science, automatic control, telecommunications, electronics, power systems, signal processing and even physics. These students have been working for free and also during their vacations time. Student holidays are from December to February in Chile and universities are closed during February. We believe that this group of students will be the basis for the consolidation of our research group on robotics. It should be stressed this is the first time that in our department 35 students are working together in a single research project.

3.4 Current Projects

Things are changing very fast in our "Robotics Lab". The lab started on May 2002 and at the moment we have two doctoral students and four master degree students doing their thesis in mobile robotics. Four more master degree students are performing their final thesis in robotic vision. At the moment we have several projects in robotics. In addition to robot soccer, we are involved in the development of robotic vision systems using robotic-heads, and in the development of a friendly robot interface that interacts with humans by detecting and recognition their faces. We are building an electronic glove for the haptic control of our robot arms, and we are developing a 1.5-meter-high mobile robot, with the final goal of installing it in an important Chilean museum.

Last but not least, thinking on the future we have been working with school' children by doing practical robotics courses. Our motivation comes from the need, especially in developing countries like ours, of motivating and foments the interest of children and young people in technology as early as possible. The participation in practical courses of robotics is a highly motivating activity, because it allows the students to approach the technology in an entertained and intuitive form. Based on that we have achieved that more than 700 children and 80 school' teachers assisted to one of our one-week robot courses. As a side effect of our courses, some Chilean schools are planning the installation of their own robotics laboratories. For more information about our activities with children, in Spanish, see http://www.robot.cl/.

4 Conclusions and Projections

The main purpose of robot contests is to foster the activities on robotics and to push the field and inspire future research. The aim of this article was to describe our experience in the participation and organization of robot contests, to show how these actions have increased the activities on robotics in Chile, and finally to give our view about the effect of this kind of activities. We showed the current state of the robotics field in Chile and we contrasted it against the current state of other disciplines. Afterwards, we described some actions taken to revert the current situation: the organization of annual Latin American IEEE Robotics Competition, the creation of the IEEE Latin American Robotics Council, our participation in RoboCup, and our activities on robotics with children. After all of this we can conclude the following:

- The robotics field is underdeveloped in Chile and its development degree is very low compared with other research fields.
- Robot contests in general, and RoboCup in particular, can help in reverting this situation. The organization and the participation in robot contests have largely increase not only our own activities in the field, but also the activities on robotics in Chile and Latin America.
- Robotics related technologies could play an important role in the technological development of countries. In this context, the development of the robotics field is important not only by its own, but also as a motivation to work in related technologies.

Acknowledgements

Dr. Hugh Rudnick, Director of the IEEE Region 9, who co-organized the First Latin American IEEE Student Robotics Competition and was the driving force behind the creation of the IEEE Latin American Robotics Council.

References

1. http://www.conicyt.cl/
2. http://www.conicyt.cl/bases/indicadores/2002/index2002.html
3. http://www01.imd.ch/wcy/

4. http://www.todocl.cl/
5. http://www.isinet.com/
6. http://citeseer.nj.nec.com/cs
7. Kelly, R., Favela, J., Ibarra, J, Bassi, D., "Asymptotically Stable Visual Servoing of Manipulators via Neural Networks", Journal of Robotic Systems 17(12), 659-669, 2000.
8. Esquivel W., Chiang L., "Non-holonomic path planning among obstacles subject to curvature restrictions", Robotica. 20, 1, 48-58, 2002.
9. Mery, D., Filbert, D., "Automated Flaw Detection in Aluminum Castings Based on The Tracking of Potential Defects in a Radioscopic Image Sequence", *IEEE Trans. Robotics and Automation*, 18(6): 890-901, 2002.
10. http://www.fondecyt.cl/
11. http://www.fondef.cl/ingles/about.html
12. http://ewh.ieee.org/reg/9/robotica/1stRobotContest/bases.htm
13. http://ewh.ieee.org/reg/9/robotica/1stRobotContest/prensa.htm
14. http://ewh.ieee.org/reg/9/robotica

Grounding Robot Sensory and Symbolic Information Using the Semantic Web

Christopher Stanton and Mary-Anne Williams

Innovation and Technology Research Laboratory, Faculty of Information Technology
University of Technology, Sydney, Australia
{cstanton,Mary-Anne}@it.uts.edu.au

Abstract. Robots interacting with other agents in dynamic environments require robust knowledge management capabilities if they are to communicate, learn and exhibit intelligent behaviour. Symbol grounding involves creating, and maintaining, the linkages between internal symbols used for decision making with the real world phenomena to which those symbols refer. We implement grounding using ontologies designed for the Semantic Web. We use SONY AIBO robots and the robot soccer domain to illustrate our approach. Ontologies can provide an important bridge between the perceptual level and the symbolic level and in so doing they can be used to ground sensory information. A major advantage of using ontologies to ground sensory and symbolic information is that they enhance interoperability, knowledge sharing, knowledge reuse and communication between agents. Once objects are grounded in ontologies, Semantic Web technologies can be used to access, build, derive, and manage robot knowledge.

1 Introduction

Robots interacting with other agents in complex dynamic physical environments require sophisticated and robust concept and knowledge management capabilities if they are to solve problems, communicate, learn and exhibit intelligent behaviour [4]. The Semantic Web is touted as the next stage in the World Wide Web's evolution, and involves the use of ontologies - collections of information that formally define the relations among terms. The structured nature of ontologies allows agents, such as autonomous robots, to make sense of the knowledge available on this global network.

For autonomous agents, ontologies can provide an important bridge between the perceptual level and the symbolic level, and in so doing they can be used to ground sensory information. Ontologies can be used to help link sensory data with symbolic representations by defining the associated sensory phenomena of real-world objects. Symbolic information can then be derived for the purpose of planning or communication. A major advantage of using ontologies to ground sensory and symbolic information is that they can be used to enhance interoperability, knowledge sharing, knowledge reuse and communication between agents. Once objects are grounded in ontologies, Semantic Web technologies can be used to access and manage robot knowledge.

D. Polani et al. (Eds.): RoboCup 2003, LNAI 3020, pp. 757–764, 2004.

Knowledge management requires the formation and evolution of concepts and categories. Categorization forms the foundation of intelligent behaviour. Robots, for example, need to categorize both physical and abstract entities of interest in their environment to achieve their goals. Ontologies are a practical tool for developing a conceptual network that robots can use to classify and identify objects with a high degree of exception tolerance. The ability to classify new information is essential for reasoning, problem solving, communication and learning [4].

OWL is a semantic markup language for Web resources that can be used to define ontologies (http://www.w3.org/). Using OWL, robots can build rich and grounded world models from a wide variety of internal and external knowledge (re)sources, such as sensors, ontologies, databases, knowledge bases, the Semantic Web, web services, and other agents.

In this paper we describe a Robot Soccer Ontology (RSO) in OWL. This ontology enhances the knowledge management capability of a robot and allows it to integrate information from multiple sensory and symbolic sources as well as understand its environment to the extent of communicating with humans and other agents. We use AIBO robots that play robot soccer to illustrate our framework and approach. Our area of concern is modeling how robots utilize concepts and knowledge as a result of their interaction with heterogeneous information landscapes. One of our motivations for this work is to build a robot soccer multiagent system in which each robot *knows* it is playing soccer and understands all the important elements of the game such as it is a ballgame with a set of rules that govern how it is played.

2 Our Application – Robot Soccer

The AIBO ERS-210 is an entertainment robot produced by SONY. The AIBO has a MIPS CPU with a clock cycle of 192 MHz and a 32 megabyte main memory. It is equipped with an array of touch sensors, on positions such as the head, back, chin and paws. The AIBO's vision system is driven by a single CMOS camera, while there is also an infrared range finder. All four legs, as well as the head, have three degrees of freedom allowing for a large range of motion. Sony's freely available OPEN-R software development kit provides a set of application programming interface library files and related tools that allow both querying and control of the AIBO's range of hardware through C++ code. The ERS-210 has wireless LAN capabilities that allow inspection of the robot's internal state and access to the vast resources of the Internet.

The SONY Four-Legged RoboCup Competition is held annually. Each robot soccer team is comprised of four AIBOs; of which one is designated as the goalkeeper. Most of the robot functions and soccer behaviours are driven by visual sensory information. The soccer field and other objects of importance in the game are uniquely colour-coded. The field is *green* in colour and defined by six bi-coloured beacons; one in each corner, and two on either side of the half-way line. One goal is *blue* and the other is *yellow*. The ball is *orange*. The beacons, goals and ball are all of known size. The beacons are used by the robots to determine their location on the field. Since

AIBOs do not possess stereo vision they cannot make independent and reliable judgments about distances to objects using vision alone, so algorithms based on vision data determine distance using known dimensions of objects. The vision processing system of the robot converts raw camera images in YUV format into conceptual information about physical objects. Pixels in the initial YUV image are classified using a set of colour concepts which are of value to the robots' task of playing soccer e.g. recognising the ball, the beacons, and the goals. Classified pixels are accumulated into larger "blobs" of colour. The blobs are then attributed to specific objects, e.g. beacons.

3 Grounding Sensory and Symbolic Information

The symbol grounding problem refers to the task of linking the internal concepts used by an agent in its world model for reasoning and decision making with the real world phenomena to which they refer [7]. An agent using ungrounded information could be considered delusional because it can perceive things that do not exist while failing to perceive things that do exist. A special case of the symbol grounding problem is anchoring [3]. Anchoring involves maintaining the link between an agent's internal representation, and the physical objects represented by these symbols. For example, in the RoboCup domain, grounding (or anchoring) could involve linking an orange blob in a robot's camera image with a soccer ball.

Simple systems that are purely reactive, such as an elevator, do not require a symbolic model of their environment, and thus avoid the grounding/anchoring problem entirely. However, for more complex systems a symbolic system can allow an agent to construct an internal model of the world, which importantly allows the agent to reason, plan and predict future states - activities integral to intelligent behaviour. Any agent functioning in a physical environment that has a symbolic reasoning component must have a solution for grounding or anchoring symbols. However, such solutions tend to be restricted to the particular domain for which the agent operates. There is a need for more general, less domain-specific solutions.

Harnard [7] argues that symbolic representations must be directly grounded bottom-up using iconic and categorical representations, where both iconic and categorical representations are non-symbolic. Iconic representations can be thought of as copies of sensory projections that preserve relevant features of the projection, where as categorical representations are reduced to the features of the projection that can determine category membership. By building categories that filter features of iconic representations, within category similarity differences compress, while between-category similarity distances expand, allowing for reliable classification of category membership.

The use of categories restricts the symbol grounding problem to a reduced set of elementary symbols, where each of these elementary symbols has a corresponding real world grounded object or event [1]. High-order symbols can be built through combinations of elementary symbols, making it possible to learn new concepts from pure linguistic definition. For example, the concept of zebra could be defined as "horse + stripes", with the terms "horse" and "stripes" grounded in the real world. By grounding bottom-up high order symbolic representations can be developed from elementary symbols.

Ontologies can provide an important bridge between the low-level, sub-symbolic Perception Layer, and high-level symbolic representations in the Deliberative Layer, by defining the properties of physical objects in terms of sensory data and events. In providing libraries of structured knowledge universally accessible to machines (including autonomous robots), we can move towards a generalised solution to the grounding problem. For instance, a robot confronted with a banana for the first time could use the semantic web to discover that this yellow, curved shaped object is a non-toxic fruit, and can not move on its own,. As such, ontologies represent a general domain independent mechanism for grounding sensory and symbolic information in autonomous robot systems.

4 Concepts and Knowledge Management on the Semantic Web

Categorization involves partitioning objects into cognitively useful groups, with these groups referred to as categories or concepts. Resultant concepts can be used to build knowledge; concepts can become predicates which are used to describe the world and its behaviour in a knowledge base, e.g. Situation Calculus [8].

Concepts help robots reduce the complexity of the information they need to acquire and manage. Furthermore, concepts give rise to broad powers of explanation. For example, without concepts robots would not be capable of representing visual information beyond the pixel level, and as a result they would not develop a world model that could support even simple forms of object recognition and reasoning. The ability to form and manipulate concepts explicitly enhances robots capacity for problem solving, communication, collaboration etc, as they forage around information rich heterogeneous environments.

We expect robots to respond appropriately to information acquired through their sensory systems. The ability to categorize new sensory information and to anchor it to physical objects in the real world allows a robot to behave sensibly in previously unencountered situations. Ontologies provide a powerful and practical tool for developing feature based categorizations of physical and abstract concepts.

According to Guarino [6] an ontology is a philosophical discipline, a branch of philosophy that deals with the nature and the organisation of reality, and for Aristotle it is the science of being. In the enterprise of building artificial agents, an ontology is an explicit specification of a conceptualization which involves terms and relations in a domain [5], or a shared understanding of some domain [9].

Ontologies are collections of information that formally define the relations among terms. Communication relies on the ability to share meaning – this can be achieved via ontologies. For example, two robot soccer knowledge bases may use different identifiers for what is the same concept e.g. one might use *beacon* and whilst the other uses *marker*. An agent that wants to compare or combine information across these databases must know that the two terms mean the same thing. Ideally, an agent should have the capability to discover common/shared meanings for whatever database it encounters.

The most typical kind of Ontology for the web has a taxonomy and a set of inference rules. The taxonomy defines the class of objects and relations among them (see Appendix A for our Robot Soccer object model). Ontologies can be used for multiple purposes, from improving the accuracy of web searches to more advanced applications, such as Robot Soccer which can use ontologies to relate the information from a robot's sensors or knowledge base to the associated knowledge structures and inference rules. A markup language like OWL helps to develop programs that can tackle complicated problems whose answers require the fusion of information from multiple sources.

We describe our Robot Soccer Ontology in Appendix A. It is an explicit and formal description of the concepts (and categories of objects) in the domain of robot soccer and is composed of an object model and axioms. The concepts of interest are classes, where the properties of each concept describe important features and attributes of the concept, and restrictions on properties constrain roles of the properties. The ontology, a set of individual instances of classes, together with a set of axioms, constitutes our Robot Soccer Knowledge Base.

Classes describe concepts in the domain, and are the base objects of an ontology. In OWL each class is represented by a URI , and can be defined as a sub-class of an existing class. A description of the class can be added, as can restrictions to that class. These restrictions are based on properties. For example, the super-class `Thing` represents all objects of interest in Robot Soccer for the purpose of our application. Specific objects are instances of `Thing`, e.g., a goalie would be an instance of `Goal Keeper`, a subclass of `Player` which, in turn, is a subclass of `Thing`, while a cardinality restriction states that there is only one `Goal Keeper` per `Team`.

Axioms are used to state facts and rules about classes and their relationships. Axioms are provided in addition to class definitions which are essentially restricted forms of subsumption/equivalence axioms. OWL allows class definitions to assert the disjointness or equivalence of classes. An example of an axiom is that for all teams, there exists only one defensive player who may enter the penalty area, and that player is the designated goal keeper. Other ontologies can be incorporated into OWL using namespaces which are URIs that have been imported from other ontologies.

5 The Semantic Web and Web Services for Robots

The aim of the Semantic Web is to bring meaningful content to the World Wide Web, creating an environment where agents interact and accomplish sophisticated tasks. The Semantic Web is designed to offer access to ontologies, knowledge resources, and knowledge management tools and services. It allows people and agents, like robots, to embed meaning into web pages to enhance "understanding" of the content using the Resource Description Framework (RDF).

RDF represents data as triples; subject, predicate and its value. RDF Schema extend this binary propositional language's expressivity to that of a semantic net. RDF

Schema allow a wide variety of knowledge resources to be described. OWL is an example of a restricted semantic net that can be used to build complex ontologies. An OWL ontology is essentially a web page containing: (i) an optional owl:Ontology instance, (ii) a set of classes , (iii) a set of properties, and (iv) a set of restrictions relating the classes and properties. OWL extends RDF(S) by: (a) supporting XML Schema Datatypes rather than just string literals, (b) local restrictions, (c) enumerations, (d) class expressions, and (e) ontology and instance mapping.

Web services are self-contained, modular applications that have an explicit description of a service offerred. They can be published, accessed and used on the web. Web services are currently based on a small set of XML based standards for message transfer (i.e. SOAP), Web Service description (WSDL), and Web Service discovery (e.g. UDDI framework). Web Services not in registries can be published using Web Service Inspection Language which assists search engines to find them. Logic for the Semantic Web can be added to web pages using a combination of OWL, Resource Description Framework (RDF), RDFS (RDF Schema) and XML.

We have developed a Robot Soccer Ontology which describes the domain of Robot Soccer for the RoboCup SONY 4-Legged League. The purpose of the RSO is to assist robots to play soccer by providing a mechanism for grounding their sensory and symbolic information to physical and abstract objects and concepts. In particular, the RSO is used to allow the robots to answer the following questions:

- Am I seeing a beacon? Which beacon am I seeing?
- Am I seeing a robot? What team is the robot I am seeing on?
- Am I seeing the goal? Which goal am I seeing?
- What is the score? Is my team winning?
- Which goal should I aim for? Where is the goal?
- Where are my team-mates?
- Which kick should I use?

The RSO can be implemented in OWL and placed on the internet for the robots to access. Once assessable it can be used for the robots to acquire and share information. Some examples of "Things" in the RSO represented in OWL are given below:

Robot Class Example.

```
<owl:Class rdf:ID="Robot">
   <rdfs:label> Robot</rdfs:label>
    <rdfs:comment>
     4-Legged League robots are on the red or blue team.
    </rdfs:comment>
 <rdfs:subClassOf rdf:resource="#ConcreteObject">
    <rdfs:subClassOf>
        <owl:Restriction owl:cardinality="1">
            <owl:onProperty rdf:resource="#hasTeam">
        </owl:Restriction>
    </rdfs:subClassOf>
</owl:Class>
```

Example Instance of Robot Class.

```
<Robot>
     <team> RedTeam </team>
    <position>Attacker</position>
</Robot>
```

6 Summary

Grounding involves the creation and maintenance of linkages between objects in a robots internal world model and objects in the environment. In this paper we described methods that can be used to implement the grounding of sensory and symbolic information embedded in a robots world model using ontologies designed for the Semantic Web. We use SONY AIBO robots and the robot soccer domain to illustrate our approach.

We developed a Robot Soccer Ontology and Knowledge Base that can be used to identify and reason about real-world objects and events. For example, the RSO defines the individual colours of the colour-coded RoboCup world in terms of sensory data, such as YUV pixel values.

A major advantage of using ontologies to ground sensory and symbolic information is that they enhance interoperability, knowledge sharing, knowledge reuse and communication between agents/robots. Once objects are grounded in ontologies, Semantic Web technologies can be used to access, build, infer from, and manage robot knowledge.

References

1. Cagenolsi, A., Greco, A., and Harnard, S., 2002, "Symbol grounding and the symbolic theft hypothesis", in A. Cangelosi and D. Parisi, (eds) Simulating the Evolution of Language, London: Springer.
2. Chalup, S., Creek, N., Freeston, L., Lovell, N., Marshall, J., Middleton, R., Murch, C., Quinlan, M., Shanks, G., Stanton, C., and Williams, M-A., "When NUbots Attack! The 2002 NUbots Team Report", Newcastle Robotics Laboratory Report, http://robots.newcastle.edu.au/NUbotFinalReport.pdf
3. Coradeschi, S., and Saffioti, A., 2000, "Anchoring symbols to sensor data: preliminary report", in Proc. of the 17th National Conference on Artificial Intelligence (AAAI-2001), pp.129-135.
4. Gärdenfors, P. and Williams, M-A, Building Rich and Grounded Robot World Models from Sensors and Knowledge Resources: A Conceptual Spaces Approach, in the Proceedings of the International Symposium on Autonomous Mini-robots for Research and Edutainment, 123 – 133, 2003.
5. Gruber, T. R., 1993, "A translation approach to portable ontology specifications", Knowledge Acquisition, 5(2).
6. Guarino, N. Formal ontology, conceptual analysis and knowledge representation. International Journal of Human-Computer Studies, 43(5–6), 1995, pp. 625–640.
7. Harnard, S., 1990, "The symbol grounding problem", Physica D, Vol.42, pp.335-346.
8. McCarthy, J. and Hayes, P. J., 1969, "Some Philosophical Problems from the Standpoint of Artificial Intelligence", in D. Michie (ed), Machine Intelligence 4, New York: American Elsevier.
9. Uschold, M., and Grüninger, M., 1996. "Ontologies: Principles, methods and applications", Knowledge Engineering Review, 11(2).

Appendix A: Robot Soccer Ontology

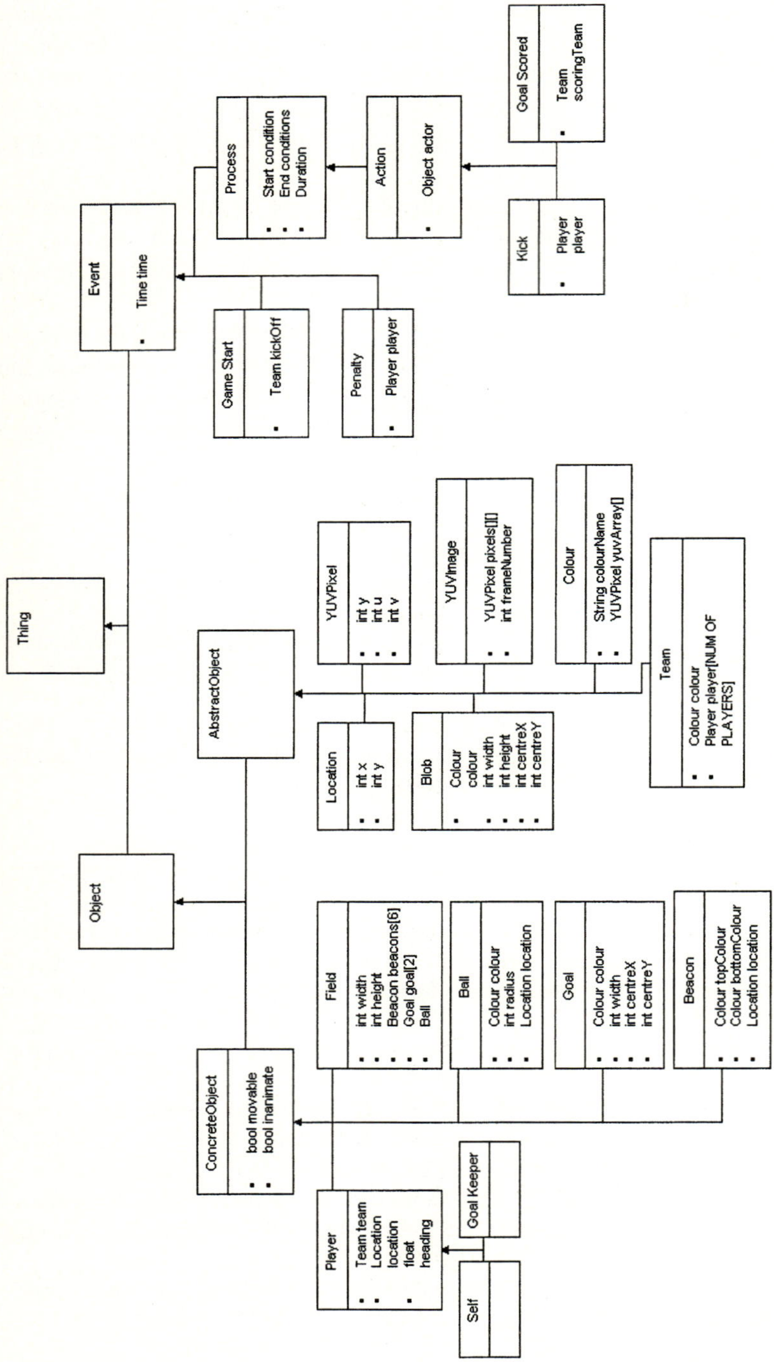

Author Index

Lecture Notes in Artificial Intelligence (LNAI)

Vol. 3123: A. Belz, R. Evans, P. Piwek (Eds.), Natural Language Generation. X, 219 pages. 2004.

Vol. 3120: J. Shawe-Taylor, Y. Singer (Eds.), Learning Theory. X, 648 pages. 2004.

Vol. 3097: D. Basin, M. Rusinowitch (Eds.), Automated Reasoning. XII, 493 pages. 2004.

Vol. 3071: A. Omicini, P. Petta, J. Pitt (Eds.), Engineering Societies in the Agents World. XIII, 409 pages. 2004.

Vol. 3070: L. Rutkowski, J. Siekmann, R. Tadeusiewicz, L.A. Zadeh (Eds.), Artificial Intelligence and Soft Computing - ICAISC 2004. XXV, 1208 pages. 2004.

Vol. 3068: E. André, L. Dybkj{\}ae r, W. Minker, P. Heisterkamp (Eds.), Affective Dialogue Systems. XII, 324 pages. 2004.

Vol. 3067: M. Dastani, J. Dix, A. El Fallah-Seghrouchni (Eds.), Programming Multi-Agent Systems. X, 221 pages. 2004.

Vol. 3066: S. Tsumoto, R. Słowiński, J. Komorowski, J.W. Grzymała-Busse (Eds.), Rough Sets and Current Trends in Computing. XX, 853 pages. 2004.

Vol. 3065: A. Lomuscio, D. Nute (Eds.), Deontic Logic in Computer Science. X, 275 pages. 2004.

Vol. 3060: A.Y. Tawfik, S.D. Goodwin (Eds.), Advances in Artificial Intelligence. XIII, 582 pages. 2004.

Vol. 3056: H. Dai, R. Srikant, C. Zhang (Eds.), Advances in Knowledge Discovery and Data Mining. XIX, 713 pages. 2004.

Vol. 3055: H. Christiansen, M.-S. Hacid, T. Andreasen, H.L. Larsen (Eds.), Flexible Query Answering Systems. X, 500 pages. 2004.

Vol. 3040: R. Conejo, M. Urretavizcaya, J.-L. Pérez-de-la-Cruz (Eds.), Current Topics in Artificial Intelligence. XIV, 689 pages. 2004.

Vol. 3035: M.A. Wimmer (Ed.), Knowledge Management in Electronic Government. XII, 326 pages. 2004.

Vol. 3034: J. Favela, E. Menasalvas, E. Chávez (Eds.), Advances in Web Intelligence. XIII, 227 pages. 2004.

Vol. 3030: P. Giorgini, B. Henderson-Sellers, M. Winikoff (Eds.), Agent-Oriented Information Systems. XIV, 207 pages. 2004.

Vol. 3029: B. Orchard, C. Yang, M. Ali (Eds.), Innovations in Applied Artificial Intelligence. XXI, 1272 pages. 2004.

Vol. 3025: G.A. Vouros, T. Panayiotopoulos (Eds.), Methods and Applications of Artificial Intelligence. XV, 546 pages. 2004.

Vol. 3020: D. Polani, B. Browning, A. Bonarini, K. Yoshida (Eds.), RoboCup 2003: Robot Soccer World Cup VII. XVI, 767 pages. 2004.

Vol. 3012: K. Kurumatani, S.-H. Chen, A. Ohuchi (Eds.), Multi-Agnets for Mass User Support. X, 217 pages. 2004.

Vol. 3010: K.R. Apt, F. Fages, F. Rossi, P. Szeredi, J. Váncza (Eds.), Recent Advances in Constraints. VIII, 285 pages. 2004.

Vol. 2990: J. Leite, A. Omicini, L. Sterling, P. Torroni (Eds.), Declarative Agent Languages and Technologies. XII, 281 pages. 2004.

Vol. 2980: A. Blackwell, K. Marriott, A. Shimojima (Eds.), Diagrammatic Representation and Inference. XV, 448 pages. 2004.

Vol. 2977: G. Di Marzo Serugendo, A. Karageorgos, O.F. Rana, F. Zambonelli (Eds.), Engineering Self-Organising Systems. X, 299 pages. 2004.

Vol. 2972: R. Monroy, G. Arroyo-Figueroa, L.E. Sucar, H. Sossa (Eds.), MICAI 2004: Advances in Artificial Intelligence. XVII, 923 pages. 2004.

Vol. 2961: P. Eklund (Ed.), Concept Lattices. IX, 411 pages. 2004.

Vol. 2953: K. Konrad, Model Generation for Natural Language Interpretation and Analysis. XIII, 166 pages. 2004.

Vol. 2934: G. Lindemann, D. Moldt, M. Paolucci (Eds.), Regulated Agent-Based Social Systems. X, 301 pages. 2004.

Vol. 2930: F. Winkler (Ed.), Automated Deduction in Geometry. VII, 231 pages. 2004.

Vol. 2926: L. van Elst, V. Dignum, A. Abecker (Eds.), Agent-Mediated Knowledge Management. XI, 428 pages. 2004.

Vol. 2923: V. Lifschitz, I. Niemelä (Eds.), Logic Programming and Nonmonotonic Reasoning. IX, 365 pages. 2004.

Vol. 2915: A. Camurri, G. Volpe (Eds.), Gesture-Based Communication in Human-Computer Interaction. XIII, 558 pages. 2004.

Vol. 2913: T.M. Pinkston, V.K. Prasanna (Eds.), High Performance Computing - HiPC 2003. XX, 512 pages. 2003.

Vol. 2903: T.D. Gedeon, L.C.C. Fung (Eds.), AI 2003: Advances in Artificial Intelligence. XVI, 1075 pages. 2003.

Vol. 2902: F.M. Pires, S.P. Abreu (Eds.), Progress in Artificial Intelligence. XV, 504 pages. 2003.

Vol. 2892: F. Dau, The Logic System of Concept Graphs with Negation. XI, 213 pages. 2003.

Vol. 2891: J. Lee, M. Barley (Eds.), Intelligent Agents and Multi-Agent Systems. X, 215 pages. 2003.

Vol. 2882: D. Veit, Matchmaking in Electronic Markets. XV, 180 pages. 2003.

Vol. 2871: N. Zhong, Z.W. Raś, S. Tsumoto, E. Suzuki (Eds.), Foundations of Intelligent Systems. XV, 697 pages. 2003.

Vol. 2854: J. Hoffmann, Utilizing Problem Structure in Planing. XIII, 251 pages. 2003.

Vol. 2843: G. Grieser, Y. Tanaka, A. Yamamoto (Eds.), Discovery Science. XII, 504 pages. 2003.

Vol. 2842: R. Gavaldá, K.P. Jantke, E. Takimoto (Eds.), Algorithmic Learning Theory. XI, 313 pages. 2003.

Vol. 2838: N. Lavrač, D. Gamberger, L. Todorovski, H. Blockeel (Eds.), Knowledge Discovery in Databases: PKDD 2003. XVI, 508 pages. 2003.

Vol. 2837: N. Lavrač, D. Gamberger, L. Todorovski, H. Blockeel (Eds.), Machine Learning: ECML 2003. XVI, 504 pages. 2003.

Vol. 2835: T. Horváth, A. Yamamoto (Eds.), Inductive Logic Programming. X, 401 pages. 2003.

Vol. 2821: A. Günter, R. Kruse, B. Neumann (Eds.), KI 2003: Advances in Artificial Intelligence. XII, 662 pages. 2003.

Vol. 2807: V. Matoušek, P. Mautner (Eds.), Text, Speech and Dialogue. XIII, 426 pages. 2003.

Vol. 2801: W. Banzhaf, J. Ziegler, T. Christaller, P. Dittrich, J.T. Kim (Eds.), Advances in Artificial Life. XVI, 905 pages. 2003.

Vol. 2797: O.R. Zaïane, S.J. Simoff, C. Djeraba (Eds.), Mining Multimedia and Complex Data. XII, 281 pages. 2003.

Vol. 2792: T. Rist, R.S. Aylett, D. Ballin, J. Rickel (Eds.), Intelligent Virtual Agents. XV, 364 pages. 2003.

Vol. 2782: M. Klusch, A. Omicini, S. Ossowski, H. Laamanen (Eds.), Cooperative Information Agents VII. XI, 345 pages. 2003.

Vol. 2780: M. Dojat, E. Keravnou, P. Barahona (Eds.), Artificial Intelligence in Medicine. XIII, 388 pages. 2003.

Vol. 2777: B. Schölkopf, M.K. Warmuth (Eds.), Learning Theory and Kernel Machines. XIV, 746 pages. 2003.

Vol. 2752: G.A. Kaminka, P.U. Lima, R. Rojas (Eds.), RoboCup 2002: Robot Soccer World Cup VI. XVI, 498 pages. 2003.

Vol. 2741: F. Baader (Ed.), Automated Deduction – CADE-19. XII, 503 pages. 2003.

Vol. 2705: S. Renals, G. Grefenstette (Eds.), Text- and Speech-Triggered Information Access. VII, 197 pages. 2003.

Vol. 2703: O.R. Zaïane, J. Srivastava, M. Spiliopoulou, B. Masand (Eds.), WEBKDD 2002 - MiningWeb Data for Discovering Usage Patterns and Profiles. IX, 181 pages. 2003.

Vol. 2700: M.T. Pazienza (Ed.), Extraction in the Web Era. XIII, 163 pages. 2003.

Vol. 2699: M.G. Hinchey, J.L. Rash, W.F. Truszkowski, C.A. Rouff, D.F. Gordon-Spears (Eds.), Formal Approaches to Agent-Based Systems. IX, 297 pages. 2002.

Vol. 2691: V. Mařík, J.P. Müller, M. Pechoucek (Eds.), Multi-Agent Systems and Applications III. XIV, 660 pages. 2003.

Vol. 2684: M.V. Butz, O. Sigaud, P. Gérard (Eds.), Anticipatory Behavior in Adaptive Learning Systems. X, 303 pages. 2003.

Vol. 2671: Y. Xiang, B. Chaib-draa (Eds.), Advances in Artificial Intelligence. XIV, 642 pages. 2003.

Vol. 2663: E. Menasalvas, J. Segovia, P.S. Szczepaniak (Eds.), Advances in Web Intelligence. XII, 350 pages. 2003.

Vol. 2661: P.L. Lanzi, W. Stolzmann, S.W. Wilson (Eds.), Learning Classifier Systems. VII, 231 pages. 2003.

Vol. 2654: U. Schmid, Inductive Synthesis of Functional Programs. XXII, 398 pages. 2003.

Vol. 2650: M.-P. Huget (Ed.), Communications in Multiagent Systems. VIII, 323 pages. 2003.

Vol. 2645: M.A. Wimmer (Ed.), Knowledge Management in Electronic Government. XI, 320 pages. 2003.

Vol. 2639: G. Wang, Q. Liu, Y. Yao, A. Skowron (Eds.), Rough Sets, Fuzzy Sets, Data Mining, and Granular Computing. XVII, 741 pages. 2003.

Vol. 2637: K.-Y. Whang, J. Jeon, K. Shim, J. Srivastava, Advances in Knowledge Discovery and Data Mining. XVIII, 610 pages. 2003.

Vol. 2636: E. Alonso, D. Kudenko, D. Kazakov (Eds.), Adaptive Agents and Multi-Agent Systems. XIV, 323 pages. 2003.

Vol. 2627: B. O'Sullivan (Ed.), Recent Advances in Constraints. X, 201 pages. 2003.

Vol. 2600: S. Mendelson, A.J. Smola (Eds.), Advanced Lectures on Machine Learning. IX, 259 pages. 2003.

Vol. 2592: R. Kowalczyk, J.P. Müller, H. Tianfield, R. Unland (Eds.), Agent Technologies, Infrastructures, Tools, and Applications for E-Services. XVII, 371 pages. 2003.

Vol. 2586: M. Klusch, S. Bergamaschi, P. Edwards, P. Petta (Eds.), Intelligent Information Agents. VI, 275 pages. 2003.

Vol. 2583: S. Matwin, C. Sammut (Eds.), Inductive Logic Programming. X, 351 pages. 2003.

Vol. 2581: J.S. Sichman, F. Bousquet, P. Davidsson (Eds.), Multi-Agent-Based Simulation. X, 195 pages. 2003.

Vol. 2577: P. Petta, R. Tolksdorf, F. Zambonelli (Eds.), Engineering Societies in the Agents World III. X, 285 pages. 2003.

Vol. 2569: D. Karagiannis, U. Reimer (Eds.), Practical Aspects of Knowledge Management. XIII, 648 pages. 2002.

Vol. 2560: S. Goronzy, Robust Adaptation to Non-Native Accents in Automatic Speech Recognition. XI, 144 pages. 2002.

Vol. 2557: B. McKay, J. Slaney (Eds.), AI 2002: Advances in Artificial Intelligence. XV, 730 pages. 2002.

Vol. 2554: M. Beetz, Plan-Based Control of Robotic Agents. XI, 191 pages. 2002.

Vol. 2543: O. Bartenstein, U. Geske, M. Hannebauer, O. Yoshie (Eds.), Web Knowledge Management and Decision Support. X, 307 pages. 2003.

Vol. 2541: T. Barkowsky, Mental Representation and Processing of Geographic Knowledge. X, 174 pages. 2002.

Vol. 2533: N. Cesa-Bianchi, M. Numao, R. Reischuk (Eds.), Algorithmic Learning Theory. XI, 415 pages. 2002.

Vol. 2531: J. Padget, O. Shehory, D. Parkes, N.M. Sadeh, W.E. Walsh (Eds.), Agent-Mediated Electronic Commerce IV. Designing Mechanisms and Systems. XVII, 341 pages. 2002.